NASA SP-4004

ASTRONAUTICS AND AERONAUTICS, 1963

Chronology on Science, Technology, and Policy

Prepared by the NASA Historical Staff,
Office of Policy Planning

Scientific and Technical Information Division   1964
NATIONAL AERONAUTICS AND SPACE ADMINISTRATION
Washington, D.C.

FOREWORD

The National Aeronautics and Space Act defined the national effort in the exploration of space—the largest, most complex research and development effort ever undertaken. It also provided for the continuation of the long tradition of research in aeronautics begun under the National Advisory Committee for Aeronautics.

During the past five years, an immense effort has been set in motion. That this effort has borne fruit is evidenced by the events of the years since the passage of the Act, some of which are recorded in the pages that follow. These events constitute at least a partial record of the raw material which is giving shape to the space effort. In some instances, they represent the growth of space technology and equipment: improvement in the percentage of successful launches; the development of large boosters capable of supporting larger payloads in more complex missions. The overall increases in reliability, sensitivity, and accuracy of equipment are all signs of the increasing breadth and depth of our knowledge of the requirements for space flight. In other instances, these events portray our increasing capacity to explore both the near and far reaches of space and to benefit from knowledge gained by exploration. Though our knowledge has increased at a dramatic rate, the demand for new information—which has become such a pressing demand for practical reasons, for defense purposes, and as part and parcel of the persistent human drive to know—threatens constantly to outstrip the supply.

The wide range of events recorded here portray the wide scope of this tremendous scientific undertaking. But perhaps even more indicative than the variety of events, is the broad social, economic, and political impact of the many projects and programs. For instance, Professor Frederic Seitz, President of the National Academy of Sciences, said recently that there was no part of university activity related to science and technology "which is not involved in a fundamental way in the space effort." Or, in another field, we note that by the end of 1963, some 65 political entities were cooperating in our international space program.

These are developments, perhaps trends, which are emerging as the Space Age becomes more and more a part of the Nation's daily life. Their parts are recorded here as past, but, more importantly, they are also prologue.

HUGH L. DRYDEN,
Deputy Administrator.

CONTENTS

	Page
Foreword by Deputy Administrator Hugh L. Dryden	III
Preface	VII
January	1
February	33
March	76
April	118
May	174
June	224
July	262
August	293
September	331
October	364
November	4 3
December	456
Appendix A: "Satellites, Space Probes, and Manned Space Flights, 1963"	501
Appendix B: "A Chronology of Major NASA Launchings, 1963"	513
Index	515

PREFACE

This chronology of the sixth year of the space age was prepared from open public sources. Like its annual ancestors, it was intended to provide a compilation of known events related to the scientific, technological, organization, and policy aspects of space exploration and exploitation. Although its index is a ready vehicle for informational re-entry it was not conceived as historical assessment. It provides a comprehensive listing of the growing welter of events in their own date and place. The pace and complexity of the challenging and sometimes dramatic endeavor as man learns and masters nature beyond planet Earth is at least chronicled in a useful form. Hopefully, it provides some of the much-needed perspective for most readers, many of whom are undoubtedly as breathless as contemporary historians.

In the preparation of *Astronautics and Aeronautics, 1963,* Mrs. Helen T. Wells carried major responsibilities for drafting and editing. The entire NASA Historical Staff read, screened, and digested available source materials and comments, including Dr. Frank W. Anderson, Jr., Miss Molly Holme, and Miss Sara Corbett. The index was drafted by Miss Nancy Ebert (ATSP). Indispensible contributions were made by the dedicated overtime work of Mrs. Dema Nappier (AFEE-I), as well as Center Historians and monitors, in particular David S. Akens (MSFC), James Grimwood (MSC), Robert Mulac (LARC), and Alfred Rosenthal (GSFC). Lloyd Robbins and Creston Whiting (ATSS-T) were unfailingly helpful in translation of Russian materials. Many busy scientific and managerial personnel were generous in their comments on preliminary drafts.

"Satellites, Space Probes, and Manned Space Flights—1963," Appendix A of this volume, was prepared by Dr. Frank W. Anderson, Jr., Assistant NASA Historian, carrying on the annual operational log compiled for previous years. Appendix B, "Major NASA Launchings, 1963," adds another year to the NASA Historical Report which now spans the first five years of NASA.

A chronology is but a preliminary tool of the historical process. Factual bookkeeping and auditing are a continuous process prefatorial to full-fledged analysis and literary presentation. Comments and criticism are welcome from any reader at any time.

EUGENE M. EMME,
NASA *Historian (ATPH)*
Office of Technology Utilization and Policy Planning.

JANUARY 1963

January 1: Effective this date, the International Code would designate all orbiting artificial satellites and space probes with Arabic numerals (1963–1, 1963–2, etc.) rather than Greek letters (1963 Alpha, 1963 Beta, etc.), National Academy of Sciences announced. New system had been agreed upon by all national members of COSPAR in May 1962. (NAS Release)
- NASA Ames Research Center was assigned direction of future Pioneer space probes, first of which would be launched in 1964. Newly created Space Sciences Div. would direct scientific aspects of the project. (San Jose *News*, 1/1/63)
- Walter C. Williams and J. C. Elms assumed duties as deputy directors of NASA Manned Spacecraft Center, under MSC Director, Dr. Robert R. Gilruth. Williams, formerly MSC Associate Director, became Deputy Director for Mission Requirements and Flight Operations; concurrently he was named Director of Flight Operations in NASA Office of Manned Space Flight. Elms was appointed MSC Deputy Director for Development and Programs, with primary management responsibility for spacecraft development. (*MSC Roundup*, 11/28/62, 1)
- Sen. Robert S. Kerr (D.-Okla.) died of heart attack in Washington. He had been Chairman of Senate Committee on Aeronautical and Space Sciences since January 1961. (*Wash. Post*, 1/2/63, A1)

January 2: Washington Post quoted informed DOD officials as saying USAF was considering converting Skybolt development program into project for exploring antisatellite problems or for use in space probes. (*Wash. Post*, 1/3/63, A9)
- Contract awarded to RCA Services Company to create the sun's intensity in two space environmental chambers. The largest chamber would measure 120 ft. high and 85 ft. in diameter; it would house the Apollo vehicle, consisting of command, service, and lunar excursion modules. The smaller chamber, measuring 85 ft. high and 65 ft. in diameter, would house the Apollo command module and provide the space environment for astronaut training. (Fact Sheet #96, MSC)

January 3: Both U.S. communications satellites, TELSTAR I and RELAY I, came to life. TELSTAR I, silent since Nov. 23, responded to signals sent by Bell Telephone Laboratories; later in the day, RELAY I, silent since first being orbited Dec. 13, responded twice to television test patterns sent from New Jersey and Maine. (*Wash. Post*, 1/4/63, A1)
- NASA was tentatively planning to extend one-day MA-9 flight of Astronaut Leroy Gordon Cooper (Maj., USAF) from 18 to 22 orbits (27 to 34 hours), John Finney of *New York Times* re-

ported. Scheduled for April 1963, MA-9 would be last flight in Project Mercury. (NYT Co., *Atlanta Constitution*, 1/3/63)

January 3: Rep. Bob Wilson (R.-Calif.) speaking for Republican Party's Advisory Committee for Space and Aeronautics, charged Kennedy Administration had failed to build a strong military space program and recommended USAF be given authority to undertake an immediate military space program with priority over NASA's manned lunar landing program: "Very little of the hardware developed by NASA can be used militarily. But the Kennedy Administration tends to lead the public to believe that the opposite is true . . . . [NASA projects such as Mercury and Apollo] could no more be converted to competitive military space systems than could a Liberty ship into an aircraft carrier or a truck into a tank . . . ." (AP, Boston *Herald*, 1/4/63)

- British sources reported U.S. would disclose details of its space launch vehicles to European Launcher Development Organization (ELDO) in exchange for cooperative work on joint space projects. (*Manchester Guardian, Wash. Post*, 1/3/63)

January 4: TELSTAR I communications satellite transmitted 10-min. television pictures across the Atlantic, the satellite's first transatlantic transmission since Nov. 23 when radiation damaged onboard circuits. Eugene S. O'Neill, director of satellite communications for Bell Telephone Laboratories, said BTL scientists "tricked TELSTAR's decoders into receiving commands" by changing the command signals slightly. Remote-control repair was performed after BTL constructed laboratory model of TELSTAR I with its faulty circuits and experimented with it to devise corrective signals. (UPI, *Wash. Post*, 1/5/63; AP, Wash. *Eve. Star*, 1/5/63, A1, A3)

- MARINER II ceased transmitting scientific data to earth. The spacecraft was 5.7 million mi. beyond planet Venus and 54.3 million mi. from earth. This was new communications record, previous record having been set by PIONEER V which in 1960 stopped transmitting scientific data at 17.7 million mi. and position signal at 22.5 million mi. (AP, Wash. *Eve. Star*, 1/5/63, A3)

- U.S.S.R. unannounced launching (1963 1-A, 1-C), with fragments re-entering from Jan. 5 through Jan. 11. Launch reported by GSFC in *Satellite Situation Report*, June 15, 1963. (GSFC *Sat. Sit. Rpt.*, June 15, 1963)

- Soviet Academy of Sciences announced that Soviet scientists had successfully bounced first radar signals off planet Mercury. Performed last summer when Mercury was 52–54 million mi. from earth, experiment was credited with confirming value of the Astronomical Unit calculated in 1961 by U.S., U.K., and U.S.S.R. using Venus as radar target. Eberhardt Rechtin of JPL called Soviet Mercury achievement "another significant step in radar exploration of the solar system." (AP, Wash. *Eve. Star*, 1/5/63, A3; *San Francisco Chronicle*, 1/5/63)

- President Kenneth S. Pitzer of Rice Univ. announced establishment of a Dept. of Space Science, first of its type. Headed by Dr. Alexander J. Dessler, new department would offer study in geomagnetism, dynamic characteristics of interplanetary space, Van

Allen radiation belts, auroras, atmospheric structure and dynamics, planetary structure, and meteoritics. (Houston *Post*, 1/5/63; AP, Newport News *Daily Press*, 1/5/63)

January 4: Astronaut M. Scott Carpenter said in Palo Alto, Calif., press conference: "People who do the spectacular things in space are few—they owe their success to many, many people behind the scenes who are not given proper recognition—scientists, technicians, industrial workers—people in all walks of life . . . ." (Palo Alto *Times*, 1/5/63)

- Soviet newspaper *Krasnaya Zvezda* (*Red Star*) quoted Edward Teller (U.S. nuclear physicist), Gen. L. Lemnitzer (USA), and other Americans regarding U.S. plans to "militarize the moon, including the delivery of an 'absolute' weapon there for the conduct of interplanetary nuclear war." (*Krasnaya Zvezda*, 1/4/63, 3)

January 5: RELAY I communications satellite made two successful intercontinental television test transmissions between Andover, Me., and Goonhilly, England, one for 23 min. and the other for an hour; teletype tests were also successfully made from Nutley, N.J., to Fucino, Italy. NASA said RELAY I's power difficulty had apparently corrected itself, but "project officials have experienced difficulties with RELAY responding properly to commands. Tests during the past three days were possible by employing special operational procedures and altering command sequences to the satellite. Experiments will continue to evaluate the satellite's communications and command systems." (AP, *Wash. Post*, 1/6/63, A8)

- A Russian Embassy official in Washington quietly collected from the U.S. State Department the Russian satellite fragment that fell in Manitowoc, Wis., on September 5, 1962. U.S.S.R. had previously ignored U.S. offer, made at U.N. last fall, to return the re-entered fragment. (*Wash. Post*, 1/8/63, 1)
- Gov. Nelson Rockefeller was preparing his proposal for establishment of New York science and technology institution, it was reported. Proposed state institution would offer both undergraduate and graduate study in science and technology. Rockefeller and state university trustees were expected to appoint committee of educators and scientists to develop detailed plans. (Orlando *Sentinel*, 1/5/63)
- AEC expenditure for military nuclear-reactor R&D was reported: $1,355,700,000 through start of current fiscal year. (*A-N-AF Journal & Reg.*, 1/5/63, 6)
- President of MIT, Julius A. Stratton, announced plans for new Center for Space Research on MIT campus. NASA would provide $3 million of the total $4 million cost. (Boston *Sun. Advertiser*, 1/6/63)

January 6: Review of Space Research, report of eight-week Space Science Summer Study sponsored by National Academy of Sciences in support of NASA, was transmitted to NASA Administrator James E. Webb. Consensus of the more than 100 U.S. scientists from Government, universities, and industry: "Of all the discoveries that have come from or can now be anticipated from man's efforts in space sciences, none more easily captures the imagination nor is more likely to focus interest and acclaim than the empirical proof that there is in this universe a biota other than our own.

"On solid scientific grounds, on the basis of popular appeal, and in the interests of our prestige as a peace-loving nation capable of great scientific enterprise . . . finding and exploring extraterrestrial life should be acclaimed as the top-priority scientific goal of our space program . . . .

"What is at stake is an opportunity to gain a new level of discussion of the meaning and nature of life." Even without "definitive evidence" of extraterrestrial life, report said there was no basis for ruling out possibility of life on Mars, Venus, or the moon; it strongly endorsed NASA policy of sterilizing interplanetary spacecraft to prevent contamination of possible extraterrestrial life.

The report also recommended that trained scientist-observers be assigned important roles in future U.S. space missions. "By his presence, man will contribute critical capacities for scientific judgment, discrimination, and analysis (especially of a total situation) which can never be accomplished by his instruments, however complex and sophisticated they become.

"Hence manned exploration of space *is* science in space, for man will go with the instruments that he has designed to supplement his capacities—to observe what is there, and to measure and describe the phenomena in terms that his scientific colleagues will clearly understand. A scientifically trained and oriented man will be essential for this purpose." Report asked NASA to take immediate steps to train scientists for space investigations so that: a "scientist-astronaut" would be member of each Project Apollo (lunar mission) crew; meteorologists could co-pilot future manned orbiting space observatories, beginning with two-man Gemini flights in 1964; biologists would be available for first manned flights to Mars; astronomers would be prepared for advent of space observatories and for maintenance and modification of these facilities.

Urging "maximum possible participation of scientists in all space missions," report outlined four specific levels of training: scientist-astronauts (men combining experience and resourcefulness of trained scientist and trained astronaut); scientist-passengers (experienced, mature scientists with adequate training in critical and emergency spacecraft operations); ground scientists (leading scientists in pertinent fields collaborating with spacecraft personnel in accomplishment of scientific mission); astronaut observers (astronauts with varying degrees of special training in making scientific observations). (NAS, *Review of Space Sciences;* NAS Releases)

January 6: AFSC announced 13-month Asset program would begin with first launch of six unmanned, non-orbiting vehicles from Cape Canaveral in mid-1963. Asset (Aerothermodynamic/elastic Structural Systems Environmental Tests) re-entry tests would provide data on environmental control, guidance and control, instrumentation, recovery techniques and equipment, structural cooling, and radar tracking under ion-sheath conditions. Recovery of the delta-wing vehicles was planned. (AFSC Release 31-R-2; UPI, *Wash. Post*, 1/6/63)

January 7: U.K. sent television signals across the Atlantic for first time via RELAY I communications satellite. Signals sent from Goonhilly Downs to Nutley, N.J., were described as "very good" and "extremely clear"; they were also clearly received at ground station of Italian space communications agency Telespazio in Fucino. (Reuters, *Chicago Trib.*, 1/8/63)

- USAF launched Thor-Agena D space vehicle from Vandenberg AFB with undisclosed payload. (UPI, *Wash. Post*, 1/8/63, A1)
- *Missiles and Rockets* reported NASA Goddard Space Flight Center was studying Mariner B and Voyager projects to determine if the center could acquire management responsibility for all or part of these planetary projects, study having been initiated at request of NASA Hq. Unnamed Hq. spokesman pointed out that JPL is now responsible for all NASA unmanned lunar and planetary programs and has heavy workload with Ranger, Surveyor, and Mariner R projects; Goddard may be asked to perform major subsystem work in Mariner B and Voyager, perhaps leading to eventual overall direction of the projects. (*M&R*, 1/7/63, 14)
- Moscow Radio said U.S. astronauts "become bourgeois exploiters [of the people] in their own right," but the Russian people had "profound respect" for U.S. spacemen, who needed "no little bravery" to let themselves be launched in rockets "whose reliability is highly dubious." (Reuters, *Balt. Sun*, 1/8/63)
- Soviet Chief Marshal of Aviation Konstantin Vershinin was quoted as saying that air-launched missiles had become basic form of weapon for Soviet aircraft, replacing bombs. (*Av. Wk.*, 1/7/63, 25)
- USAF Minuteman ICBM fell short of its planned range flight from Cape Canaveral, and USAF was conducting study to determine cause of the advanced Minuteman's malfunction. This was second flight of new, more powerful Minuteman model. (DOD Release 25-63; AP, Wash. *Eve. Star*, 1/8/63)

Early January: United Technology Center made first U.S. test-firing of cluster of large segmented solid rocket motors. Four rocket motors consisting of six segments each were fired for 14 sec., producing 140,000 lbs. of thrust. Test was to demonstrate techniques for ignition of clusters "which appear applicable to the simultaneous ignition of clusters of solid booster rockets producing millions of pounds of thrust." (Wash. *Eve. Star*, 1/14/63; *Av. Wk.*, 1/14/63, 33)

- NASA Marshall Space Flight Center awarded preliminary, 90-day contract for development of variable-thrust RL-10 rocket engine to Pratt and Whitney Aircraft. The hydrogen-powered engine would be capable of operating at as little as 12.5 per cent of its rated thrust and be capable of restarting in space. Throttleable RL-10 engines had been test-fired in feasibility studies at Pratt and Whitney plant and NASA Lewis Research Center. (Huntsville *Times*, 1/8/63)

Early January: Dr. Wernher von Braun, Director of NASA MSFC, visiting West Berlin's Technical University to receive honorary doctorate, said: "I am convinced personally that the Russians will not succeed in making a manned space flight to the moon and back before the Americans. The Russians now could send a man to the moon but that would be a one-way trip." (Newport News *Times-Herald*, 1/12/63)

January 8: NASA reported RELAY I communications satellite's low battery voltage had been result of faulty voltage regulator in one of its twin transponders. Continued tests by RCA and NASA engineers pinpointed the difficulty and also discovered that the regulator fails to function properly when it becomes too hot or too cold. Despite the difficulty engineers would attempt live television transmission via RELAY I by sending special command signals to the satellite and concentrating on the remaining good transponder. (*Wash. Post*, 1/9/63, C8)

- Task force from NASA Marshall Space Flight Center sent to Rocketdyne, Canoga Park, Calif., where F-1 engine had developed "combustion instability," William Hines reported in Washington *Evening Star*. Five-engine cluster of 1.5-million-lb. thrust F-1's would power first stage of Advanced Saturn (C-5). (Hines, Wash. *Eve. Star*, 1/9/63)
- Results of recent radarastronomy and radioastronomy studies of planet Venus were reported by Howard Simons in *Washington Post*. Studies conducted at NRL, Washington, concentrated on detecting water in atmosphere of Venus; NRL radar-astronomers and radioastronomers concluded that Venutian atmosphere contains very little, if any, water. Their research suggests surface temperature of Venus may be 600° F, as previously reported. Meanwhile, radioastronomy studies at JPL, Pasadena, indicated Venutian surface was characteristically desert; that Venus appears to rotate very slowly, perhaps only once every 250 earth days; and that Venus appears to rotate in direction opposite that of earth. These results, combined with those from MARINER II Venus probe, provided evidence that Venus is swept by winds at hundreds of mph, which in turn constantly churn up high sandstorms, JPL's Eberhardt Rechtin said. (*Wash. Post*, 1/8/63, 1)
- World's largest radiotelescope, developed by National Science Foundation, was now operational, Thomas R. Henry reported in Washington *Evening Star*. Located on Papago Indian Reservation in Arizona, radiotelescope has antenna of 300-ft. diameter. Among first objects of study were radio waves emitted from planet Jupiter; later study would be devoted to outer planets Uranus and Neptune and to distribution of neutral hydrogen in Milky Way Galaxy and nearby galaxies. (Wash. *Eve. Star*, 1/8/63)
- Charles H. Zimmerman, NASA Director of Aeronautical Research addressed Aero-Space Luncheon Club in Washington on short-range, intermediate-range, and long-range aircraft. Short-range: "I think there is a tremendous future in the helicopter and the V/STOL aircraft in the short range and feeder line application. However, in order to take full advantage of the capabilities of

these aircraft, we have to find some better way of utilizing the air space and the airport space that is available. I think we will do it. It is a matter of time and effort . . . ."

Intermediate range: "I am speaking of 2,000 miles and up . . . . We are working closely with the FAA and with the DOD in steps leading—we sincerely hope—to the development of a supersonic commercial transport . . . .

"I might point out that NASA today is opening the proposals that were submitted on feasibility studies which are a part of the overall FAA–DOD–NASA program. This is one of the steps that we are taking to help this program along . . . ."

Long range: "At this stage of history, we can send 'a man from the U.S. to Australia in about an hour in a Mercury capsule . . . .

"Now, conceivably in the future we will be able to put a number of people in some sort of hypersonic aircraft and send them to Australia in an hour. This is looking out into the blue. But I really think that mankind will not quit in this development of transportation through the air till they have made this a practical possibility . . . ." (Text)

January 8: Opening of Technical University of Berlin's Institute for Elements of Space Travel was attended by Prof. Eugen Saenger, Dr. Wernher von Braun, Prof. Hermann Oberth, and other dignitaries of West Berlin and Univ. of Berlin. Prof. Saenger, former head of Research Institute for Physics of Jet Propulsion at Stuttgart, has been nominated as director of the new institute. (*Bild* [Hamburg], 1/8/63, in MSFC *SIN*, 3/63, 5)

January 9: RELAY I communications satellite transmitted its first transatlantic television programs, sending British and French viewers clear pictures of ceremonial unveiling of Mona Lisa in Washington and 10 min. of network program "Today." (UPI, *Wash. Post*, 1/10/63, A3)

- Dr. Abe Silverstein, Director of NASA Lewis Research Center, told Chicago press conference that trouble in Centaur launch vehicle had been corrected. Centaur was proceeding at Lewis under highest national priority (DX) and would be developed on schedule to launch Surveyor lunar spacecraft and Mariner interplanetary spacecraft in 1965. Dr. Silverstein pointed out new testing philosophy for Centaur called for more extensive and qualitative ground test before flights. (Chicago *Sun-Times*, 1/10/63; Chicago *Daily News*, 1/10/63)

- Enea Bossi, international aviation pioneer, died in Dayton, Ohio. Native of Milan, Bossi built and flew his own airplane in 1908; he designed first Italian seaplane. After coming to U.S. in 1919, Bossi designed first U.S. stainless steel plane, most successful manpowered aircraft, and an early helicopter. He also invented vacuum fuel-intake system used in U.S. automobiles during 1920's. (AP, *Wash. Post*, 1/12/63)

January 10: EXPLORER XIV energetic particles satellite developed radio transmission difficulty, not repairable by remote control. Exact cause of difficulty, apparently in one of the binary counters of satellite's encoder system, was not determined. (NASA EXPLORER XIV Program. Rpt. No. 4; NASA Release 63–6)

January 10: NASA announced MARINER II Venus probe had been so successful that repeat flight scheduled for 1964 had been canceled. Interplanetary efforts would be concentrated instead on such projects as sending probe toward Mars in 1964 and later flight to Venus with advanced Mariner probe in 1965. (NASA Release 63-3; JPL Release)

- Dr. Robert C. Seamans, NASA Associate Administrator, told Washington Representatives Chapter of National Security Industrial Association: "The goals of this [accelerated national space] program and of our nation in space are to achieve a position of pre-eminence in every aspect of space science and technology for the benefit of all mankind . . . . Four main reasons underlie the tremendous effort we must make to achieve the goals stated by the President. Our position as leader of the Free World requires it. Our national security requires it. The fact that we are a practical people, dedicated to the improvement of life on earth for ourselves and our world neighbors, requires it. And man's age-old urge to explore and our national traditions of pioneering require it . . . ."

    Dr. Seamans also discussed the Pioneer space probe program: ". . . We are initiating a competition for a new probe, PIONEER, to develop a better understanding of solar flares. These probes will be used, commencing in 1964 (the International Year of the Quiet Sun), to measure the characteristics of the interplanetary medium such as magnetic field, solar plasma, solar and galactic cosmic radiation, and micrometeorites. The probe will provide information of intrinsic scientific value in addition to measurements in direct support of the manned lunar landing . . . ." (Text)

- French Scientific Research Minister Gaston Palewski told French National Assembly a satellite launching site would be established in Eastern Pyrenees Department near the Spanish border. France's first satellite was scheduled for launching in 1965; other European satellites may also be launched from the site. (Reuters, *Wash. Post*, 1/12/63)

- Finnish Foreign Ministry announced Finland had obtained permission from U.S.S.R., U.K., and other Western allies to obtain defensive missiles, forbidden since signing of Paris peace treaty in 1947. Finland would buy air-to-air missiles and squadron of MiG-21 fighters from U.S.S.R., antitank missiles from U.K. (UPI, *Wash. Post*, 1/10/63)

- White House published report by President's Science Advisory Committee, *Science, Government, and Information.* The Committee found communication of technical information to be a necessity to a healthy scientific and technical community. To this end it made a series of specific recommendations to both the technical community and to Government agencies involved in technical programs, following the general principle that technical information must be an area of major commitment on the part of the U.S. Government but not in such a way as to stifle independent efforts of the technical community. (*Science, Government, and Information*)

January 10: Titan II ICBM launched from Cape Canaveral fell short of its intended 6,300-mi. range after developing second-stage difficulties. USAF said many of the flight's objectives were achieved. (*M&R*, 1/21/63, 10)

January 11: Analysis of radar observations of planet Venus indicated Venutian surface is smoother than that of earth or the moon, it was reported. Observations were made Nov. 29–Dec. 7, 1962, by National Bureau of Standards' radartelescope at Jicamarca, Peru. (*Wash. Post*, 1/11/63, A5)

- Univ. of Pittsburgh's Chancellor Edward H. Litchfield announced plans for $30-million space research and coordination center. Center would be used for study in natural and social sciences and engineering and health areas connected with aerospace activities. (UPI, *Wash. Post*, 1/12/63, 1)
- Dr. Knox Millsaps resigned as chief scientist of USAF OAR and executive director of AFOSR. (*Av. Wk.*, 12/17/62, 25)
- Thor missiles being returned to U.S. from U.K. would be modified for use as space vehicles, USAF announced. Three of U.K.'s 60 operational Thors already had been returned to Douglas Aircraft Co. for conversion, so that they could be used in space assignments similar to those of conventional Thor space boosters. (*Wash. Post*, 1/12/63)
- USAF announced Titan II missile had been test-fired while locked in its 155-ft.-deep silo at Vandenberg AFB. Firing of "brief" duration was to test the missile's resistance to noise and vibration. (AP, Wash. *Eve. Star*, 1/11/63)
- Translation of article on Soviet VOSTOK III and IV by H. A. Vilter, East German engineer, was quoted. Of the Soviet vehicle used to place Vostok capsules in orbit, Vilter said: "The power of the six rockets of the initial stage was roughly 20 million horsepower" (or 880,000-lb. thrust)—about ⅔ power of first two Saturn vehicles, whose eight clustered engines developed 1.32-million-lb. thrust (30 million horsepower). Vilter's article appeared in East Berlin publication *Die Technik;* abstract of his report was published by Dept. of Commerce's Office of Technical Services. (Wash. *Eve. Star*, 1/11/63)
- Lockheed Aircraft Corp. and IAM broke off contract negotiations indefinitely; still unresolved was dispute over "union shop." (AP, Wash. *Eve. Star*, 1/12/63)

January 12: Houston *Chronicle* reported NASA Manned Spacecraft Center was planning 30-man space station that could stay in orbit for five years; MSC had solicited contractors' proposals for electrical power system capable of producing 40,000 watts. (Houston *Chron.* in UPI, Boston *Sunday Advertiser*, 1/13/63)

- U.S. Dept. of Commerce translation of article, "How to Fly to Mars," by Soviet scientist Prof. G. Chebotarev, was reported.

    "The 'Mars-1' interplanetary station was launched from a heavy satellite in a circular earth satellite orbit; the velocity of the satellite was 4.9 miles a second. The rocket was launched from this satellite at a velocity of 3.1 miles a second, but the rocket left the orbit at 8 miles a second.

"It headed toward Mars in a hyperbolic trajectory and left the earth's sphere of attraction at 4 miles a second. For it to assume an orbital path around Mars the station would have to be braked to a velocity of 3.2 miles a second . . . ." (Wash. *Eve. Star*, 1/12/63, A5)

January 12: David Sarnoff, writing in *Saturday Review*, said: "The technology of electronics is reaching today for summits of national, global, and space communications beyond anything conceived since the invention of movable type.

"The year 1962 brought conclusive proof of the utility of orbiting satellites to relay across ocean or wilderness the immensely high-capacity signals in the upper reaches of the radio frequency spectrum.

"The way is thus opened technically for the establishment over the next few decades of a communications system by which governments, organizations, or individuals may establish contact with anyone, anywhere, at any time, by voice, sight, or document, separately or in combination . . . ."

He predicted development of satellite communications in three phases: "*Phase I*, between 1965 and 1970, should see a global system of low-power synchronous satellites, each with a capacity of 2,000 voice channels or two television channels . . . .

"*Phase II*, between 1970 and 1980, may mark the beginning of international satellite communications between cities rather than through centralized national terminal facilities . . .

"*Phase III*, beyond 1980, envisages an all-embracing satellite communications system: direct personal transmission of voice and sight through satellites without intermediate routing . . ." (*Sat. Review*, 1/12/63, 88)

- Frederick R. Kappel, chairman of the board and chief executive officer of AT&T, was cited by *Saturday Review* as "Businessman of the Year" (1962). Mr. Kappel spearheaded AT&T's $50-million commitment to Telstar ever since it was envisioned by Bell Labs scientist, Dr. John R. Pierce, in 1954. *Saturday Review* called Telstar "more than a triumph of the modern technology of space communications. It was also a symbol wise men could approve, of a new era in human communications, a work of peace that sought only to bring men together, not to destroy them. Finally, it was the product of a promising new teamwork in space between government and industry, in this case the largest private enterprise venture in the world . . . ." (*Sat. Review*, 1/12/63, 46)
- USN launched its 16th Polaris-carrying submarine, *Nathan Hale*, equipped to fire Polaris A–3 now being developed. (AP, Wash. *Eve. Star*, 1/12/63)

January 13: In his State of the Union message to Congress, President Kennedy said:

"In these past months, we have reaffirmed the scientific and military superiority of freedom. We have doubled our efforts in space, to assure us of being first in the future. We have undertaken the most far-reaching defense improvements in the peacetime history of this country. And we have maintained the frontiers of freedom from Viet Nam to West Berlin . . . ." (Text, *Wash. Post*, 1/15/63, A10)

January 13: NASA Goddard Space Flight Center announced sodium-vapor cloud experiments by Goddard during past two years had shown wind behavior 44–50 mi. above earth becomes erratic and unpredictable. Below that altitude, winds generally follow global pattern, regularly reversing with the seasons. Region between 56- and 68-mi. altitudes is characterized by "remarkable wind sheers"—within altitude span of less than three miles, wind speed was observed to increase swiftly by more than 250 mph and even to reverse direction. Immediately above this band of maximum wind velocity, wind diminishes almost to zero. Above 70 mi., research indicated region of "strong but more uniform winds" with velocities of about 200 mph. Goddard experiments, launched on sounding rockets from NASA Wallops Station, did not extend beyond 105-mi. altitude. (Goddard Release; AP, *Wash. Post*, 1/13/63, A16)

- Moscow *Pravda* announced Yevgeny Alexandrov, head of research laboratory in Moscow Mining Institute, had discovered a new law of physics which corrects the laws of Sir Isaac Newton. Newspaper did not indicate which laws of Newton were corrected nor how they were corrected, but it said: "Up to now, mechanisms and machines of an impact nature [presumably mining equipment such as pneumatic drills] quickly went out of commission. Now their life will not only be extended immeasurably but also their power will increase many times over.

"Innumerable examples can be given," the newspaper said, but did not offer any. (AP, *Wash. Post*, 1/14/63, A3)

- Soviet Cosmonaut Pavel Popovich, boarding Soviet aircraft in Havana bound for Moscow, made the comment: "The world will soon know about the first female cosmonaut." (UPI, *Wash. Post*, 1/14/63, A3)

January 14: NASA announced signing of Memorandum of Understanding with India's Dept. of Atomic Energy, providing for cooperative U.S.-India space program. Joint scientific experiments to explore equatorial electrojet and upper-atmosphere winds from geomagnetic equator would be launched from Thumba, India, during 1963. For equatorial electrojet experiments NASA would provide nine Nike-Apache vehicles; ground launching, tracking, and telemetry equipment, ground instrumentation on loan basis; and training in U.S. for Indian personnel responsible for telemetry and launch operations. Univ. of New Hampshire would provide instrumented payloads through NASA. India would provide launching site and facilities; personnel for launch operations, telemetry, and data analysis; and supporting ground magnetic and meteorological observations. For atmospheric-wind measurements, NASA would provide four Nike-Cajun vehicles, appropriate launching device on loan basis, and training at NASA centers for Indian personnel responsible for launch operations. India would supply four sodium-vapor-release payloads, photographic equipment, launch site and facilities, personnel, and supporting meteorological data. (NASA Release 63-5)

January 14: French President Charles de Gaulle, speaking in press conference in Paris, referred to President Kennedy's offer of Polaris missiles as joint defensive weapons for NATO: "Nobody in the world, especially nobody in America, can say, if, where, when, how, and to what degree American atomic armaments will be employed to defend Europe. Therefore we will construct and if need be will employ our atomic force ourselves.

"Of course, this does not exclude the combined action of our force with an analogous allied force of the same kind, but for us integration is unthinkable in this case. . . .

"We have neither the submarines to launch them [Polaris missiles] nor the nuclear warheads to arm them. By the time we have them, what good will the Polaris be? Without doubt, by then we will have our own missiles. For us, this matter has no present application. . . ." (*Wash. Post*, 1/15/63, A1, A7)

- NASA decision to procure Atlas-Agena B vehicles directly from contractors, thus eliminating USAF as procurement agent, was predicted in *Aviation Week*. NASA would name single responsible contractor for vehicle integration and would standardize the Atlas-Agena B for space missions. NASA already had used seven of the vehicles—five for Ranger and two for Mariner—and was planning to use 20 Atlas-Agena B's over next three years—in Gemini rendezvous flights, Ogo, Oao, Ranger, and Mariner R. Prime vehicle contractors were General Dynamics/Astronautics for Atlas stage and Lockheed for Agena; USAF had vehicle integration responsibility. (*Av. Wk.*, 1/14/63, 38)
- Article in Soviet newspaper *Pravda*, reported in *Missiles and Rockets*, said the moon appears to be undergoing changes: During recent years, crater Linne has decreased to half its former size; crater Alhazen has disappeared; small craters have formed at bottom of Plato; and, at sunrise, greenish-gray spots periodically appear at bottom of some craters. *Pravda* said spots in bottom of crater Eratosthenes appear to some observers to be moving. (*M&R*, 1/14/63, 23)
- J. A. MacTaggart, managing director of Niagara Falls, Ontario, wax museum, said wax image of Astronaut John Glenn was not on display because of "astronomical" $3,000 it would cost to purchase space suit from American manufacturer. Wax figure of Soviet Cosmonaut Yuri Gagarin was on display, Soviet Government having supplied copy of his space suit after request through British Aeronautical Board. (AP, Wash. *Eve. Star*, 1/14/63)

January 15: EXPLORER XIV energetic particles satellite transmitted 38 sec. of complete data, and officials of NASA Goddard Space Flight Center were hopeful the satellite might eventually resume normal operations. EXPLORER XIV developed transmission difficulty Jan. 10, after 100 days of nearly continuous transmission. Project Manager Paul G. Marcotte of Goddard reported EXPLORER XIV received less than 10 per cent degradation from space radiation since its launch Oct. 2; project officials did not believe radiation damage to be cause of satellite's transmission malfunction. (NASA EXPLORER XIV Prog. Rpt. No. 4; NASA Release 63-6)

January 15: TELSTAR communications satellite relayed pictures of opening of East German Communist party congress to television viewers in U.S. (Chicago *Trib.*, 1/16/63)

- FAA announced Administrator Najeeb E. Halaby had told supersonic transport advisory group to restudy cost details of supersonic transport development and to prepare specific proposal on management organization of the aircraft's development. Advisory board, headed by Gen. Orval R. Cook (USAF, Ret.), had been studying supersonic commercial aircraft for 13 months. (*Wash. Post*, 1/16/63)
- Dr. Edward C. Welsh, Executive Secretary of National Aeronautics and Space Council, told National Rocket Club in Washington: "From time to time, major public attention shifts from one space project to another. . . . It would be incorrect to conclude, however, that any one of these projects or any other specific project encompasses the whole job of implementing our space policy. Rather, all of them, plus many projects not yet conceived, combined to make the national program viable, energetic and selectively expanding. . . .

    ". . . Our military space activities are just as peaceful as our non-military ones. Indeed, our space legislation labels both as 'peaceful.' Both, incidentally, are also scientific, so one cannot make the further distinction that NASA's and AEC's projects are scientific and the Defense Department's are something other than that, or vice versa. . . .

    "Whether or not enough is being done in either the military or non-military space field—and I am never satisfied—I would suggest that before one comes to a firm conclusion on such a question, he should examine in considerable detail just how much has been accomplished during the past 12 months. It is an impressive record and one need not skip over the failures or the delays in specific projects in order to develop a balance sheet which shows a remarkable degree of progress in a field in which we started late and moved slowly too long. . . .

    "In concluding . . . , I would emphasize several points: (1) It is in this nation's interest—economically, technologically, scientifically, militarily, and internationally—to carry out an accelerating space program; (2) to maintain our position of leadership in the free world we cannot afford anything less than first place in space accomplishment and there is no sound basis for complacency regarding our present position; (3) we must take risks and overcome failures in this complex endeavor and we dare not let the risks or the failures slow our momentum; and (4) in spite of all the excitement and glamor of the space program, it must be managed and operated in a tough-minded fashion so that we will get every one of its many benefits as soon and as efficiently as possible." (Text)
- USAF solicited proposals from industry for development of solid-propellant space booster capable of up to six-million-lb. thrust. USAF action stemmed from DOD-NASA agreement of December 1962 to advance the technology of large solid-propellant rocket motors. (DOD Release 53–63; AP, *Wash. Post*, 1/16/63)

January 15: R. W. Gillespie, systems engineer of Lockheed Missiles and Space Co., said in Los Angeles press conference: "America's giant program for a [manned] lunar landing should be reoriented to allow for a manned landing on Mars and a manned capture [orbital flyby] of Venus by 1973. Most of the talent now working on the lunar landing would have to be diverted for the Mars and Venus flights." He said he was "one of those who thinks there is some form of life on Mars" and that a landing there would be of more value than one on the moon; scientists already have a pretty good idea of what man would find on the moon. Gillespie was in Los Angeles for 9th annual meeting of American Astronautical Society. (L.A. *Times* to *Wash. Post*, 1/16/63, A7)

January 16: Nike-Cajun rocket launched from NASA Wallops Station to detect atomic hydrogen in the atmosphere and test theories of sodium airglow. 126-lb. payload released ozone into sodium airglow layer of the atmosphere at approximately 47-mi. altitude; faint luminosity created by ozone cloud was measured by photometers in the payload and photographed with telescopic camera at Wallops Station. (Wallops Release 63-7)

- Background briefing in Pentagon on DOD budget proposals for FY 1964 indicated a total request in new obligational authority of $53.7 billion. Of this, some $5.5 billion was earmarked for R&D for the three services: Army's main emphasis would be on the modernized Nike-Zeus antimissile system and on antiguerrilla weapons; Navy's on antisubmarine warfare; Air Force's on missile and space systems development. The $1.66 billion for RDT&E and Astronautics represented the DOD portion of the national space program. (DOD Background Briefing on the FY 1964 Defense Budget, 1/16/63)

- USAF announced the launch of an unidentified satellite with Thor-Agena D vehicle from Vandenberg AFB. (UPI, *Wash. Post*, 1/17/63, A1)

- Biweekly *Satellite Situation Report*, by NASA Goddard Space Flight Center, initiated practice of reporting only those objects launched by NASA, with no coverage in future of launchings by DOD agencies or U.S.S.R. (*Sat. Sit. Rpt.*, 1/16/63)

- 180 workers of International Brotherhood of Electrical Workers struck McDonnell Aircraft Corp., but other unions crossed the picket lines and the plant continued to operate normally. (AP, *Wash. Post*, 1/17/63, 4)

January 17: X-15 No. 3, piloted by Joseph A. Walker (NASA), reached 271,000-ft. altitude and 3,677-mph speed (mach 5.47) in flight near NASA Flight Research Center, Edwards, Calif. Flight was second highest altitude achieved with rocket-powered X-15 and fourth to reach or exceed altitude for which aircraft was designed; it was X-15's highest altitude without ventral tail fin. (NASA Release 63-8)

- President Kennedy sent FY 1964 Budget Request to Congress, recommending NASA appropriation of $5.712 billion. Of this sum, $3.19 billion was for manned space flight, with Project Apollo receiving largest single increase in NASA budget—from $435 million in FY 1963 to $1.2 billion in FY 1964. NASA Administrator

James E. Webb characterized the budget request as "austere"—it would neither speed up nor slow down the National effort to land an American on the moon in this decade.

Total FY 1964 space budget was estimated at $7.614 billion—NASA, $5.664 billion; DOD $1.668 billion; AEC, $254.3 million; Weather Bureau, $26.2 million; and National Science Foundation, $2.3 million. (NASA Budget Briefing FY 1964; *Wash. Post*, 1/18/63, A12; *M&R*, 1/21/63, 13)

January 17: INJUN I satellite, silent since Dec. 25, responded to signals of State Univ. of Iowa scientists, who did not know cause of the cosmic-ray satellite's transmission interruption. (AP, *Wash. Eve. Star*, 1/30/63)

- Nike-Cajun sounding rocket launched at Wallops Island, Va., to 46.5-mi. altitude, ejecting ozone about 10 sec. after trajectory peak. Objects of the NASA Lewis Research Center experiment were to: test excited oxygen theory of sodium airglow; detect concentration of atomic hydrogen in the atmosphere; measure altitude of sodium and hydroxyl airglow layers; and test a new technique for studying atmospheric chemical releases with rocket-borne photometers. (NASA Rpt. of S. Rkt. Launching)

- RELAY I satellite transmitted 12-min. Voice of America program as well as AP and UPI news dispatches from Nutley, N.J., to Rio de Janeiro and back. Transmissions were reported perfect, even though ordinary high-frequency radio communication with Rio was not possible because of atmosphere conditions. (*Wash. Eve. Star*, 1/17/63; AP, *Wash. Post*, 1/18/63, 4)

- Questioned at NASA Budget Briefing about NASA's substituting pickaback rendezvous target for Agena stage in Project Gemini flights, Associate Administrator Robert C. Seamans said: "That is definitely by the boards. We did study that kind of possibility. We feel it is very important to have a program where there will be two separate launches. We will put up the Agena. . ." as target object in Gemini flights training astronauts for Project Apollo lunar orbit rendezvous. (NASA Budget Briefing FY 1964 Transcript)

- At NASA FY 1964 Budget Briefing, NASA Administrator James E. Webb acknowledged reported combustion instability in F-1 engine: "It isn't just going to turn out to be a problem; it is a problem, and it has been a problem with every engine as I understand it that has ever been developed. We will solve the problem." (NASA Budget Briefing FY 1964 Transcript; *Wash. Eve. Star*, 1/23/63)

- At NASA Budget Briefing Associate Administrator Dr. Robert C. Seamans commented on rumor that NASA planned another Mercury flight after the Cooper flight scheduled for April: "You will note that the Mercury and one-day [manned orbital] effort is presumed to be completed in fiscal '64. We have a flight scheduled for this April. . . . In planning flights of that importance we always have a back-up possibility. We do have two back-up capsules as well as the two back-up Atlas boosters that could be used in the event that we don't obtain all the information that we anticipate obtaining in the Cooper flight. . . ." (NASA Budget Briefing FY 1964 Transcript, 1/17/63)

January 17: Rainer Berger, senior research scientist at Lockheed-California, Co., predicted man would probably encounter some form of extraterrestrial life within 10 years. Speaking at ninth annual meeting of American Astronautical Society, Los Angeles, Berger also suggested "greenhouse effect" of planet Jupiter might make possible the existence of warm oceans hospitable to life on that planet. (*L.A. Times, Wash. Post*, 1/18/63, 6)

- NASA Space Vehicle Review Board met at NASA Marshall Space Flight Center, board being composed of 28 members representing Marshall, Manned Spacecraft Center, and Launch Operations Center. (*Marshall Star*, 1/23/63, 3)
- NASA Langley Research Center announced selection of the Boeing Co.'s Transport Div. and the Lockheed-California Co. to negotiate two nine-month study contracts. Studies would evaluate four concepts of supersonic commercial transport investigated by NASA Langley Research Center and Ames Research Center and would provide information on feasibility of these four concepts through engineering investigations. Their evaluations would provide basis for and serve as guide to NASA's future supersonic transport research programs. (NASA Release 63-7; Langley Release)
- NASA Administrator James E. Webb announced U.S. and U.S.S.R. delegates would meet in March in Rome to work out "detailed arrangements to carry out proposals for co-operation in space." Meeting would immediately precede meeting of COSPAR, also in Rome. (NASA Budget Briefing FY 1964, Transcript; *Wash. Eve. Star*, 1/17/63)
- USAF announced series of 27 upper atmosphere chemical releases (Project Firefly III), completed in mid-December, had "ended a four-point basic research program aimed at learning more about the ionosphere." Information gathered during tests, which were made at Eglin Gulf Test Range, would eventually aid pilots of X-20 vehicle. (DOD Release 70-63)
- At the White House, President Kennedy was presented with one-eighth scale model of MARINER II Venus probe by NASA Administrator James E. Webb, JPL Director William H. Pickering, and other officials. (AP, *Wash. Post*, 1/19/63; *Wash. Eve. Star*, 1/18/63)
- USN gave name "Phoenix" to its long-range air-to-air missile to be fired from TFX fighter plane now under development by Hughes Aircraft Co. (DOD Release 71-63; AP, *Wash. Post*, 1/18/63, A5)
- NASA budget recommendation for FY 1964 indicated that NASA Electronics Center would be established in the Greater Boston (Mass.) area. (Boston *Globe*, 1/18/63)
- White House's FY 1964 budget request included $511,000 proposed appropriation to begin design of National Air Museum in the Smithsonian Institution. New building would replace temporary structure housing such historical vehicles as *Kitty Hawk*, *Spirit of St. Louis*, and Mercury spacecraft FRIENDSHIP 7. (*Wash. Eve. Star*, 1/18/63)

January 17: Dr. Soloman W. Golomb of Cal Tech, speaking before American Astronautical Society in Los Angeles, warned: "There's been a great deal of thought given to preventing contamination of the moon and planets during visits by earth space vehicles, but an even greater problem entirely overlooked is . . . microscopic organisms carried to the earth on return flight." Dr. Golomb said extraterrestrial organisms inadvertently brought to earth might be able to successfully compete with terrestrial organisms for basic materials of life. (UPI, Newport News *Times-Herald*, 1/18/63)

* Tregaron, Washington estate, was approved by D.C. Board of Zoning Adjustment as temporary headquarters for Space Communications Corp. (*Wash. Post*, 1/18/63)

January 18: James E. Webb, NASA Administrator, said in address to Charlotte (N.C.) Chamber of Commerce:

"The most important of the major reasons for undertaking a broad national program of space research and development—the one which promises the greatest rewards for mankind—is the least understood and the most difficult to explain. It is the certainty that the basic scientific knowledge gained, and the technological applications which will flow from it, will offer the greatest return on our space investment over many years ahead . . . .

"The road to pre-eminence in space may be a long one. It certainly will not end on the moon, regardless of which nation lands the first explorers there . . . .

"Much of what we need for pre-eminence in space will be acquired on the way to the moon. That is why we have chosen the lunar exploration as an immediate national goal of prime importance.

"When will we clearly achieve the pre-eminence we seek? This is not an easy question. The answer depends not only on the sustained effort and investment we make, but on what the Russians do. Their system lends itself to concentration of resources on specific tasks, but their system like any other is faced with conflicting demands for priority. At the present time we simply cannot know what level of investment they expect to make on a long-range space program. We do know we must not be *trapped* into a *narrow* one-purpose program such as one limited to a lunar landing. . . .

"Unless the Russians unveil a new rocket on the order of Advanced Saturn in the next few years, our chances for being first to explore the moon remain good . . . ." (Text)

* NASA Flight Research Center announced award of contract to Bell Aerosystems Co. for design and construction of two manned lunar-landing research vehicles to be used in conjunction with Project Apollo. Free-flight test vehicles would be capable of taking off and landing under their own power, of attaining about 4,000-ft. altitude, of hovering, and of horizontal flight. Vehicles would be used in investigation of problems that may be encountered in landing manned vehicle on the moon; results of these studies would assist in preparation of flight-crew training devices. (FRC Release)

January 18: NASA Administrator, James E. Webb, addressed North Carolina Press Association in Chapel Hill: "Our goal is to be first in every aspect of space research, development, and exploration. To achieve it we must develop the same sort of capability in the space environment that we have achieved on land, on and under the sea, and in the air, and 'be in a position second to none.'

"... I believe the challenge that faces us is to turn the scientific and technological revolution ... into more of an *evolution*—into an orderly advance into a better world of tomorrow ...." (Text)

- DOD announced new antimissile missile program, Nike X, to employ the Spring missile. (Space Log, Index supplement to *Space Bus. Daily*)

January 19: Dr. Hugh L. Dryden, NASA Deputy Administrator said, in accepting Gold Medal of the International Benjamin Franklin Society, New York: "... Space science and technology require research and development at the frontiers of almost every branch of science and technology. Any industrial nation which refuses to devote substantial effort to space exploration is incurring the hazard of future technological obsolescence of its industries, the hazard of potential loss of leadership, and the hazard of military surprise by potential enemies ...." (Text)

- Senator Margaret Chase Smith (R.-Me.) addressed the Columbian Women of George Washington Univ., Washington, D.C.:

"From ... space research and exploration will come knowledge—knowledge about the universe and its physical laws; knowledge about the earth on which we live; and knowledge about life itself.

"The space program will stretch the abilities and minds of our people for years to come. It will provide a continuing, long-term stimulant to our economy. The magnitude of the task will test the resources and cooperative will of all major elements of our society. Still, space exploration, and manned space flight in particular, offers the United States the opportunity for unparalleled progress in the future.

"Ultimately, within this century, the sum of all our efforts will give us the equipment, the knowledge and the skill to utilize space as we now utilize the seas and the air.

"This then is our motive—our objective. To develop superior competence in space which will be available for any national purpose which may be required, whether it be for the peaceful use of space for the benefit of all mankind or to keep the peace...." (*CR*, 1/24/63, A230-32)

- First S-IV stage for Saturn C-1 vehicle left Douglas Aircraft Co. for two-week trip to Cape Canaveral. This S-IV would not be launched but would be mated to S-I stage for checkout of NASA launch operation facilities at the Cape.

January 20: Team of four USAF and Lockheed experts, headed by Laurence Stoddard of Lockheed-California Co., Lunar and Planetary Sciences Laboratory, announced they would attempt to photograph moon during 37-sec. solar eclipse near Cape Town,

South Africa, Jan. 25. "Knowledge of the exact shape of the moon will give us additional information on the effects of moon gravity. This information is of great importance in planning lunar landings and computing the lifetime of manmade vehicles orbiting the moon," Stoddard said. Stoddard photographed similar eclipse last July in Senegal, West Africa. (AP, *Wash. Post*, 1/21/63, A3)

January 20: French National Center for Space Studies announced it would launch 100 sounding rockets during 1963 in program to study upper atmosphere, double the number of rockets launched by France during 1962. (*NYT* West. Ed., 1/21/63)

January 21: Saturn SA-4 vehicle left Huntsville, Ala., by barge on 10-day trip to Cape Canaveral, where it would be launched. (UPI, Birmingham *Post-Herald*, 1/22/63)

- In memorandum to FAA, DOD and Commerce, NASA, CAB, and Director of Office of Science and Technology, White House requested a report evaluating the $1 million already spent or allocated to research for supersonic transport program and "firm recommendations for possible further action." Requested report on supersonic airliner program was to be submitted before the end of this year. (UPI, *Wash. Post*, 1/22/63)

- First flight-test of X-20A's inertial guidance system was announced by AFSC. First in series of 24, test was conducted in NF-101B "Voodoo" aircraft over Eglin Gulf Test Range, Fla. (AFSC Release 31-R-5)

- Raymond L. Bisplinghoff, Director of NASA Office of Advanced Research and Technology (OART), said in speech before IAS in New York:

    "Within the NASA, our philosophy of advanced research and technology is an extension and an expansion of its predecessor, the NACA. There may be said to exist two principal responsibilities in these activities. The first is a simple extension of the NACA in which applied research, with a substantial component of basic research, is emphasized. Here, the principal effort takes place within our field research centers. . . .

    "[The second] . . . could be described as the innovation of selected advanced subsystems to permit rapid development and engineering of future systems in the shortest possible time and at minimum cost.

    "Again, the mode of operation places the principal burden of responsibility within the field centers, but now we find them closely coupled with American industry . . . ." (Text)

- Study on possibility of polluting outer space with rocket exhaust gases, made by Geophysics Corp. of America for ARPA, was reported by John W. Finney of *New York Times*. Principal conclusion of study was that there is "strong probability that the accelerated rocket program with enormously large missiles in prospect in the near future may cause modifications over large local areas or on a worldwide basis in the upper atmosphere." Modifications could include marked changes in composition, structure, and temperature of upper atmosphere and changes in electron density of the ionosphere. For example, large amounts of hydrogen could absorb ultraviolet radiation from sun and thus decrease electron density in D region of the ionosphere. (NYT Co., *Atlanta Constitution*, 1/21/63)

January 21: Reported that Russia's Aeroflot had increased its student fare discount from 25% to 50%, discount applicable to all college students, including those taking evening and correspondence courses, high school students, and persons enrolled at military or naval institutes. (*Av. Wk.*, 1/21/63, 52)

- Soviet scientist Anatoli A. Blagonravov said in *Izvestia* article that dual flights of Cosmonauts Nikolayev and Popovich in August 1962 proved that, from a medical point of view, manned flight to the moon is now possible. Neither cosmonaut has shown physiological disturbances since the prolonged space flights, he said; flights proved man could avoid dangerous radiation and could work in outer space over long periods.

  Reviewing Cosmos series of 12 scientific satellites, to date, begun by Soviets in March 1962, Blagonravov said they could be considered a preliminary stage in creating "inhabited" satellites. (*Av. Wk.*, 1/21/63, 59)

- Atomic Energy Commission awarded full-power license to operate the 60 megawatt Plum Brook research reactor to the Plum Brook Station of NASA Lewis Research Center. (LRC Release 63–3, Lewis Chronology, 1)

- $331 million, or almost two thirds, of AEC's FY 1964 budget request of $527 million for reactor development would be directed toward space and military applications. (*Av. Wk.*, 1/21/63, 31)

- National Aeronautics and Space Council's budget request for FY 1964 amounted to $525,000—decrease of $5,000 from FY 1963. (*Av. Wk.*, 1/21/63, 25)

- Eastern Air Lines survey revealed that 17.3% of its air-shuttle passengers had been converted from surface transportation on Boston-New York-Washington route because of no-reservation commuter flights. (*Av. Wk.*, 1/21/63, 52)

- A statement from the American Meteorological Society expressed the hope that the U.S. would continue to provide leadership in the challenging new field of meteorological rockets and urged support on an expanded and more enduring basis. The AMS recommended that the national program should pursue three objectives:

  (1) A U.S. Meteorological Rocket Network based on the existing network should be established and cooperative programs with other countries should be arranged.

  (2) A first-generation meteorological rocket network capable of reliably probing the atmosphere between 30 and at least 60 km. to measure winds and either pressure or temperature or density as a function of height should be produced.

  (3) A second-generation meteorological rocket network capable of reliably probing the atmosphere between 30 and 100 km. or higher, to measure winds and either pressure or temperature or density as a function of height, and electron density above 60 km. should be established. (*Bulletin of the AMS*, May 1963)

January 22: NASA Administrator James E. Webb and DOD Secretary Robert S. McNamara announced NASA-DOD agreement establishing Gemini Program Planning Board, with Dr. Robert C. Seamans, Jr. (Associate Administrator for NASA) and Dr. Brockway McMillan (Assistant Secretary of the Air Force for R&D) as co-chairmen. Planning Board would delineate NASA and DOD scientific and technological requirements and monitor the Gemini program to ensure that the requirements are met. Agreement supplemented NASA-USAF Gemini management agreement in effect since spring 1962; NASA would continue to be responsible for Project Gemini management. DOD would participate in development, pilot training, pre-flight check-out, launch operations, and flight operations "to assist NASA and to meet the DOD objectives." DOD also would assist in funding Gemini. (NASA Release 63-12; NASA-DOD Agreement; DOD Release 84-63)

- NASA-DOD agreement setting forth management responsibilities for operations in Cape Canaveral area was announced. Agreement provided that DOD would continue as the single manager of AMR, extending from the Cape to the Indian Ocean; USAF Missile Test Center, under DOD authority, would continue as host agency at existing 15,000-acre Cape Canaveral launch area; NASA Launch Operations Center would manage and serve as host agency at new 87,000-acre Merritt Island launch area, being developed primarily for use with very large launch vehicles such as NASA's Advanced Saturn (C-5). NASA and DOD would be responsible for their own logistics and administrative functions in their respective areas; regardless of location, NASA and DOD would perform most specific mission functions in their own behalf such as preparation, checkout, launch, and test evaluation; DOD would continue to be responsible, in both areas, for certain fundamental range functions such as scheduling of launches, flight safety, range search and sea recovery, etc. (NASA Release 63-11)

- Dr. Jerome B. Wiesner, director of President Kennedy's Office of Science and Technology, was asked in Voice of America interview whether an attempt to land astronauts on Mars in 1971 or 1973 "could be considered." Dr. Wiesner replied: "I think it could be considered, but I don't think it would be done. I think it is a very major effort, even considerably larger than the moon effort. We have made estimates of a Mars program cost and a round number like one hundred billion dollars seems to be a reasonable figure. It is obviously technically possible, [but] I doubt whether one could do it in the time period between now and 1970 without just an unbelievable crash program. I think that before the century is out we will probably have done just that, and it would be interesting to do, but I don't believe anyone can deny the vital reason for undertaking the program of the kind you are talking about." (Transcript)

- Speaking in New York on the history and the future of television, FCC Chairman Newton Minow said:

    "The year 1450 in Mainz, Germany, marked a watershed in history—the introduction of the printing press and with it the beginning of modern communication . . . .

"In these terms, we are still at the beginning of television—the year 1465, so to speak. Even Telstar, with its present technological sophistication, is but a crude vision of what will soon be the everyday, global marriage of sight and sound. I believe television is now having an impact on society as great as, if not greater than, the printing press had over the course of several centuries...." (Text, *CR*, 1/30/63, 1327–30)

January 22: Georg von Tiesenhausen, chief of future studies for NASA Launch Operations Center, told IAS meeting in New York that by 1970 U.S. would need an orbiting space station to launch and repair spacecraft; space station could double as a manned scientific laboratory. Describing station 300-ft. long and 33-ft. in diameter, von Tiesenhausen said it could be launched in two sections by Saturn C-5 vehicles and joined together in space rendezvous. (L.A. *Herald-Examiner*, 1/22/63)

- Institute of Aerospace Sciences awarded its Honors and Awards for 1962: Honorary American Fellowship to Robert R. Gilruth, Director of NASA Manned Spacecraft Center; Louis W. Hill Space Transportation Award to Dr. C. Stark Draper, head of Dept. of Aeronautics and Astronautics at MIT and Director of Instrumentation Laboratory; Sylvanus Albert Reed Award for achievement in theoretical investigation in aeronautical sciences to Walter C. Williams, Associate Director of NASA Manned Spacecraft Center; Lawrence B. Sperry Award to outstanding young man in aeronautics, Robert O. Piland, NASA Manned Spacecraft Center deputy manager of Project Apollo.

  NASA employees elected IAS fellows: John V. Becker, NASA Langley Research Center; Paul F. Bikle, NASA Flight Research Center; Dr. Hermann H. Kurzweg, NASA Office of Advanced Research and Technology; and Walter C. Williams, NASA Manned Spacecraft Center. (MSC *Space News Roundup*, 1/9/63, 1; *A-N-AF Journal & Reg.*, 1/5/63, 12; *Av. Wk.*, 1/21/63, 23)

- DOD announced establishment of Directorate for Classification Management, "designed to increase the flow of information, both general and technical, and to eliminate the unnecessary expense of protecting information which no longer warrants security classification...." (DOD Release 79–63)

- USAF's 1962 launching of six tin dipoles into orbit was disclosed by W. E. Morrow of MIT, at annual IAS meeting. The 14-in. dipoles were launched to measure effects of solar pressure, air drag, and electrical drag on small objects, so that scientists could predict how long Project West Ford copper dipoles would remain aloft. Morrow said the special tin dipoles were following their predicted orbit and were still orbiting the earth. (*Wash. Post*, 1/23/63, A1)

- James A. Van Allen, speaking at science hearing of House Committee on Science and Astronautics, said: "The spectacular nature of many space achievements, especially those involving human passengers, is such that the space race stirs the enthusiasm of all but the most obtuse or superstitious persons, and makes possible widespread support by the tax-paying public." (AP, *Wash. Post*, 1/23/63)

January 22: Rear Adm. Hugh Stirling MacKenzie was named by British Admiralty to direct Polaris project in Britain. Appointwas "first step in implementing the decisions reached by President Kennedy and Prime Minister Macmillan at Nassau" to make Polaris missiles available for Royal Navy. (AP, Wash. *Eve. Star*, 1/23/63)

January 23: International Association of Machinists (IAM) called nationwide strike against Boeing Co. for Jan. 26; President Kennedy invoked Taft-Hartley Act, naming three-man board of inquiry to determine whether national emergency existed. (AP, *Wash. Post*, 1/25/63)

- Donald L. Mallick and Harold E. Ream, NASA Langley research pilots, reported to IAS meeting on research results of simulated lunar spacecraft landing, lunar orbit rendezvous, and atmospheric re-entry made in simulated three-man spacecraft. Realistic tests demonstrated ability of research pilots to complete precise maneuvers of flight, despite necessarily cramped quarters and seven-day duration.

  LARC aerospace scientists Jack E. Pennington and Roy F. Brissenden reported on experiments to investigate astronauts' ability to rendezvous in space by visual means. Pennington and Brissenden concluded that pilot could sight target vehicle from perhaps as much as 200 mi. away if it is lighted by sun or bright beacon; he could then successfully brake his own spacecraft along an intersecting path.

  Dr. Leonard Roberts, LARC physicist, reported to IAS on implications of lunar landing if moon has, as many scientists believe, dust layer on surface. Dr. Roberts concluded that spacecraft descending to moon would erode a crater by action of its rocket exhaust. If dust exists on lunar surface, it would be blown outward and away from descending vehicle; air would not be present to recirculate dust in enveloping cloud, so astronaut's visibility should not be reduced by lunar dust.

  Other LARC scientists presented research reports covering instruments for satellites, supersonic aircraft flutter problems, and wind tunnel experiments on helicopter rotors. (Langley Release)

- Two Beacon-Arrow rockets launched from Tonopah, Nev., test range carried six flares to 56-mi. altitude, where flares burst in 6-million-candlepower brightness. Sandia Corp. conducted the test for AEC, in project developing spacecraft tracking techniques during atmospheric re-entry. (AP, *Wash. Post*, 1/25/63)

- Bell Telephone Laboratories scientist, Walter L. Brown, head of semiconductor physics research, told American Physical Society that TELSTAR I's instruments detected sudden and large increases of energetic electrons in gap or slot between Van Allen radiation belts within a few hours after Soviet atmospheric nuclear tests Oct. 22 and 28; electron enrichment of gap greatly exceeded that of radiation belts themselves. Before Oct. 21, gap was relatively free of electrons, the electrons following U.S. nuclear test in July having decayed; after Oct. 22 explosion, gap was filled. New electrons decreased at rate of 50% daily in gap's center, more slowly at gap's edges. After Oct. 28 explosion, gap was almost refilled, and decay process was repeated. Nov. 1 explosion by U.S.S.R.

added electrons to Van Allen belts but not to gap, presumably because this test differed in size and altitude from the others, Brown said. (*Wash. Post*, 1/24/63, A4)

January 23: Secretary of Commerce Luther H. Hodges told Association of Public Affairs Conference of the Chamber of Commerce of the United States, in Washington, that only one fourth of total U.S. R&D expenditure (about $4 billion) is being spent for civilian purposes. "By wrongly assuming in the first place that research and development for any purpose—space, military, or whatever—automatically fosters economic growth, we have completely missed the point that this is a highly concentrated industry, restricted by purpose, by geography, by company . . . ."

Noting that in last eight years the demand for scientists and engineers in Government R&D has increased more than 300%—10 times the increase in private industry—Hodges warned: "Unless industry itself recognizes the basic problem—that we must find ways to make engineering education possible for everyone capable of it and desiring it, and that we must put more of science to practical use—we shall not achieve the productivity rises or the new products that can lead us to faster growth." (*Wash. Post*, 1/24/63, A2)

• Dr. J. Herbert Hollomon, addressing American Meteorological Society in New York, said: "A global system of weather services and atmospheric research requires much more than the sum of separate national programs, much more than traditional cooperation or coordination. Its success depends upon a world-wide joint effort benefiting a world-wide task . . . ." (AMS Release)

• At annual dinner in New York, American Meteorological Society presented posthumously the Carl-Gustaf Rossby Award for Extraordinary Scientific Achievement to the late Dr. Harry Wexler, who had been head of U.S. Weather Bureau's Office of Research; Dr. Wexler was cited "for his contributions to knowledge of the atmospheric heat balance and dynamic anticyclogenesis, for his interdisciplinary studies in meteorology, oceanography, and glaciology, and for his outstanding leadership in international programs in the atmospheric sciences." Other awards: Charles Franklin Brooks Award for Outstanding Service to the Society, presented to Dr. David M. Ludlum "for his continuing effort, on behalf of the Society, to popularize meteorology and to stimulate the interest of young people in the field, for establishing *Weatherwise*, and for his effective participation in Society affairs over many years"; Cleveland Abbe Award for Distinguished Service to Atmospheric Sciences by an Individual presented to Dr. Lloyd V. Berkner, President of Graduate Research Center, Dallas, Tex., for "his scientific work on the upper atmosphere and for the stimulation he provided to the field of meteorology by his work as Chairman of the Committee on Meteorology of the National Academy of Sciences"; the Meisinger Award, presented to Prof. Edward N. Lorenz of MIT "for his introduction of concepts leading to a better understanding of the energetics of the general circulation, and for his recent work on low order systems indicating aperiodic vacillations from one regime of motion to another"; Award for

Outstanding Services to Meteorology by a Corporation, presented to RCA Astro-Electronics Div., "for contributions to meteorology through its role in producing the successful Tiros meteorological satellites"; AMS Award for Outstanding Contribution to the Advance of Applied Meteorology, presented to Herbert C. S. Thom, Chief Climatologist, U.S. Weather Bureau, "for a quarter century of contributions to applied meteorology, including furnished climatological estimates to the armed forces during World War II, major contributions to the problems of statistical analysis of climatological data, development of wind design data, and development of a much improved formula for calculating annual and seasonal energy use for domestic heating"; Award for Outstanding Achievement in Bioclimatology, presented to Prof. Konrad J. K. Buettner, Dept. of Atmospheric Sciences, Univ. of Washington, "for his distinguished and continuing studies on the influence of the atmospheric environment on man." (AMS Release)

January 23: USAF Minuteman ICBM traveled more than 4,000 mi. down AMR in "flawless" test flight. (UPI, *Wash. Post*, 1/24/63; *M&R* 1/28/63, 11)

- Dr. David Atlas, chief of USAF Cambridge Research Laboratories Weather Radar Branch, announced successful testing of Stradap (Storm Radar Data Processor), electronic device for short-range storm detection and forecasting. Dr. Atlas said a network of 100 long-range radars and Stradap units could provide automatic storm mapping for the entire country. (USAF OAR Release 1-63-2)

- Two major contracts for the Apollo project were awarded recently for (1) a lunar charting service by the USAF Aeronautical Chart and Information Center, St. Louis, Missouri; and (2) an exploratory study of guidance system techniques in emergency abort operation of the Apollo LEM during lunar landing, awarded to Chance Vought Astronautics Division of Ling-Temco-Vought, Inc., Dallas, Texas. (MSC Fact Sheet #96)

- Turkish Foreign Minister, Feridon Cemal Erkin, announced U.S. would dismantle Jupiter missiles deployed at NATO bases in Turkey. Land-based Jupiter IRBM was being replaced by Polaris submarine-launched missile. (*Wash. Post*, 1/24/63, A1)

- Prof. Lloyd Motz, Columbia Univ. astronomer, estimated at joint session of IAS and AMS that Milky Way Galaxy contains 200 million stars similar to our sun; if these suns also possess planets, each might possess as many as three planets inhabited by intelligent forms of life. ". . . There are in the galaxy millions of planetary systems like our solar system with similar physical and chemical properties. Hence life might have existed in many of these planetary systems, since biochemists have demonstrated that the complex organic molecules necessary to maintain life are synthesized if the physical and chemical conditions are proper." (AP, *Wash. Post*, 1/24/63, A4)

- Maj. Gen. Ben I. Funk, Cdr. of AFSSD, reported to IAS that AF Systems Command considered three fundamental projects as "space programs of prime military necessity: rendezvous and inspection, space station development, and communications satellites." (*Av. Wk.*, 1/28/63, 26f)

January 23: W. Albert Noyes, Jr., prof. of chemistry at Univ. of Rochester, told House Committee on Science and Astronautics that, to produce more scientists, emphasis should be removed from awarding of graduate fellowships and grants and should be placed on increasing high school teachers' salaries so as to attract teachers with enough scientific training to stimulate their students.

Chemist Harold C. Urey said that only about three per cent (about $15 billion) of this country's GNP was spent on education, and that amount "should be doubled." (AP, *Wash. Post*, 1/24/63, A4)

January 24: Ernest J. Sternglass, physicist at Westinghouse Research Laboratories, proposed theory at American Physical Society in New York that would offer a unified concept of nuclear and electrical forces, heretofore considered separate phenomena, and would reduce all matter to two particles—electron and positron. Sternglass hypothesized that all other particles found in atom (30 or more in recent years) are merely combinations and variations of negatively-charged electron and positively-charged positron. Simplest such grouping is single electron-positron pair, whirling near speed of light, which exhibits same properties as elementary particle known as neutral pi meson. Two such electron pairs, Sternglass said, exhibit properties of elementary particles K meson and Rho meson; other systems similarly account for all heavy mesons "within a few per cent of their observed values." Combinations of electron-positron pairs with proton can explain other elementary particles classed as hyperons. Sternglass had assumed that Einstein's special theory of relativity, which states that an object's mass increases with its velocity, governs particles' behavior; thus, if charged particles move fast enough and close enough in their orbits, they will generate sufficient force to hold atomic nucleus together. (*Wash. Post*, 1/25/63, A6)

- NASA report released on Project Mercury flight MA-8 (Oct. 3, 1962), in which Astronaut Walter M. Schirra, Jr., said: "I was also impressed with the fact that . . . [under the different lighting conditions that prevailed, the separated sustainer stage] was almost black in appearance, rather than the shiny silvery vehicle that Astronauts Glenn and Carpenter had seen at this time and that I had observed on the launching pad . . . .

"At about 10 minutes 30 seconds, I went back to flying-by-wire, low, and tracked the sustainer as it traversed down through the window, and it was a thrill to realize the delicate touch that it is possible to have with fly-by-wire, low. . . . The control system was so effective that it just amounted to a light touch and maybe a few pulses in either axis to get the response I wanted. I could point the spacecraft at anything I wanted to. I could see the sustainer and track it, but I do not believe the relative motion problem would be so easy to solve that I would be able to steam along and join up with it. Although the relative velocity was on the order of 20 to 30 feet per second, it was enough to cause a problem, particularly at a time when one is becoming acclimated to a new environment. These problems would be difficult to solve by one's own trajectory analysis, since there were no systems aboard to aid the pilot in solving the problem. I think that when we build up to the rendezvous technique, one will need more time than that just at the point of insertion to effect this rendezvous, even with proper training. The use of time while orbiting in space is only earth relative, therefore if a rendezvous is not hurried, the task should be relatively simple . . . .

"A smog-appearing layer was evident during the fourth pass while I was in drifting flight on the night side, almost at 32° South latitude. I would say that this layer represented about a quarter of the field of view out of the window, and this surprised me . . . . Seeing the stars below the glowing layer was probably the biggest surprise I had during the flight . . . ." (NASA SP-12, *Results of 3rd U.S. Manned Orbital Space Flight*, 49–55)

January 24: President Kennedy was asked at press conference if he foresaw a need for "manned strategic bomber after the current B-52's and B-47's are worn out," to which the President replied:

"Yes, there may be a need. That plane will last through 1970. We are securing, as you know, three B-70's. We have no further plans to develop at this time, but there may be a good many struggles in the globe in the late '60s or early '70s which are not subject to solution by missiles, but which may be more limited war, and where manned bombers may be very useful." (Transcript, *Wash. Post*, 1/25/63, A8)

- President Kennedy accepted "with great regret" the resignation of Philip L. Graham as an incorporator of Space Communications Corp., established by legislation to direct U.S. communications satellite activities. Graham had served as chairman of the 13-man board of incorporators. (Wash. *Eve. Star*, 1/26/63, A9)
- William G. Bade, associate mathematics prof. at Univ. of Calif., said at meeting preceding joint convention of American Mathematical Society, Mathematical Association of America, and Association of Symbolic Logic in Berkeley, Calif., that U.S. needs "perhaps three times as many [mathematics] graduates as the 250 now being produced each year." (Oakland, Calif., *Trib.*, 1/24/63)

January 25: NASA announced selection of Rohr Corp. of Chula Vista, Calif., to design, manufacture, install, and test advanced antenna system to be located at Goldstone Tracking Station. Expected to be completed in 36 months, antenna system would improve communications capability of present Deep Space Network by factor of 10; for example, when used with future deep-space probes, antenna will make possible communications over distances to the "edge" of the solar system. (NASA Release 63-13)

- Unnamed U.S. official was quoted as saying Astronaut Gordon Cooper's one-day Project Mercury flight MA-9 was set for April 2, "for planning purposes." Flight would be 18 orbits (about 27 hours) or, if all goes well, 22 orbits (about 34 hours). NASA had not yet officially announced precise launch date in April or length of flight. (AP, Wash. *Eve. Star*, 1/25/63, A2)
- Maj. Gen. Don R. Ostrander (USAF), of USAF Office of Aerospace Research and former NASA Director of Launch Vehicle Programs,

said in address to National Capitol Section of American Rocket Society, Washington, that formal definition of "basic research" as used by USAF is "effort directed toward an increase in fundamental knowledge in a science . . . .

"It is true that we have many unresolved problems in this business of basic research. On the credit side of the ledger, however, I believe that there is a general and increasing recognition at the responsible levels of government and industry that basic research provides the foundation not only for our military capability but for our economic progress as a nation. Thus, in its truest sense, basic research is everybody's business, and everyone, to a greater or lesser degree, has a responsibility toward it. I believe that this responsibility can best be exercised by a partnership of government, the civilian scientific sector, and industry. I don't believe, however, that we have yet reached a consensus as to the final terms of this partnership . . . ." (Text)

January 25: The first counter-insurgency (COIN) aircraft, a modified B-26, was rolled out in ceremonies at Van Nuys, Calif. The program is the responsibility of the Aeronautical Systems Division (ASD), Wright-Patterson AFB, Ohio. (*AFSC Operational Highlights*, 11)

- 10-day injunction by U.S. District Court Judge William J. Lindberg halted threatened IAM strike against Boeing Co. President Kennedy, after receiving report from special board of inquiry, had ordered Justice Dept. to seek the injunction on grounds that such a strike would endanger national security. (AP, *Wash. Post*, 1/26/63)
- Atlas ICBM was launched from Vandenberg AFB in routine training launch for SAC missile crews. (DOD Release 105-63)

January 26: NASA Manned Spacecraft Center announced assignment of areas of specialization for NASA astronauts. Maj. L. Gordon Cooper, flight MA-9 pilot, and Cdr. Alan B. Shepard, Jr., flight MA-9 back-up pilot, would be responsible for pilot phases of Project Mercury; Maj. Virgil I. Grissom's area of specialization would be Project Gemini; Lt. Col. John H. Glenn, Jr., would concentrate on Project Apollo; LCdr. M. Scott Carpenter's duties would cover lunar excursion training; Cdr. Walter M. Schirra, Jr., would be responsible for Gemini and Apollo operations and training. As Coordinator for Astronaut Activities for MSC, Maj. Donald K. Slayton would maintain overall supervision of astronaut duties.

Specialty areas of the 9 new flight-crew personnel: trainers and simulators, Neil A. Armstrong; boosters, Maj. Frank Borman; cockpit layout and systems integration, Lt. Charles Conrad, Jr.; recovery systems, LCdr. James A. Lovell, Jr.; guidance and navigation, Capt. James A. McDivitt; electrical, sequential, and mission planning, Elliott M. See, Jr.; communications, instrumentation, and range integration, Capt. Thomas E. Stafford; flight control systems, Edward H. White II; environmental control systems, personal and survival equipment, LCdr. John W. Young. (MSC Release 63-11; *M&R*, 2/4/63, 36)

January 26: 16-in. cannon shot a 475-lb. instrumented capsule 15 mi. into the upper atmosphere from Barbados, first launching of an altitude probe by gun and first such project undertaken by a nongovernmental agency—McGill Univ. of Canada, with support of U.S. Army. Instruments in steel-encased capsule relayed data on conditions in upper atmosphere to ground tracking stations; capsule remained aloft about three min., fell into sea nine mi. southeast of Barbados. (AP, Wash. *Eve. Star*, 1/28/'63)

January 27: Lt. Col. John A. Powers, Public Affairs Officer for NASA Manned Spacecraft Center, said Astronaut Leroy Gordon Cooper's MA-9 flight may extend to "as many as 22 orbits. If the flight lasts that long it will be a 34-hour flight." Speaking before Texas Associated Press Managing Editors' Association, Powers said NASA had not yet decided whether Cooper's flight would be last of Project Mercury; "the decision on whether another [Mercury flight] will be necessary will be made after Cooper's flight." (AP, Balt. *Sun*, 1/28/63)

- Lockheed Aircraft Corp. employees approved three-year contract, after months of negotiations over union shop proposal. Contract, which did not include union shop clause, did provide unique procedure by which Lockheed would "encourage membership in the AFL–CIO International Association of Machinists . . ." by sending letters to new employees suggesting they consider joining the union. (L.A. *Times* to *Wash. Post*, 1/28/63, A1)

January 28: Ground-breaking ceremonies for eight-story NASA structure to be used for pre-flight testing of Gemini and Apollo spacecraft were held near Cape Canaveral. $7.69-million building would be first of 40 buildings in industrial complex located in the 87,000-acre Merritt Island area NASA had acquired for its space projects. (AP, Balt. *Sun*, 1/29/63)

- President Kennedy transmitted to Congress the 1962 report on *U.S. Aeronautics and Space Activities*, stating in Preface: "The year 1962 was a period of acceleration, accomplishment, and relative progress for the United States in its space leadership drive. In both numbers and complexity of space projects, the past year was the most successful in our brief but active space history."

  In accompanying message, Vice President Lyndon B. Johnson told Congress that, during 1962, U.S. "generated a greater rate of progress in space than did the U.S.S.R. [However] . . . the records of the two countries were closely similar in regard to the ratio of space successes to space failures." (*Annual Report for 1962;* L.A. *Times*, *Wash. Post*, 1/29/63, A2)

- Reorganization at NASA Ames Research Center was put into effect by Director Smith J. DeFrance, designed to accommodate new Pioneer spacecraft program assigned to Ames. In addition to other changes, two new Assistant Directors were appointed: Assistant Director for Development, Mr. Robert M. Crane; and Assistant Director for Research and Development Analysis and Planning, Dr. Alfred J. Eggers. (Ames Memo For Staff, 1/31/63)

- NASA announced selection of Philco Corp. for negotiation of contract to develop and equip Manned Flight Mission Control Center

(MFMCC) at NASA Manned Spacecraft Center, Houston. To be operational in mid-1964, MFMCC would direct Gemini rendezvous and Apollo flights just as Mercury Control Center at Cape Canaveral directed Project Mercury manned space flights. (NASA Release 63-14)

January 28: Installation of new tracking system, Mistram (Missile Trajectory Measurement), on Eleuthera Island in the Bahamas, was announced. Mistram would provide more precise missile guidance and nose-cone impact data than presently possible on rockets fired from Cape Canaveral; it would provide accuracy required for controlling space rendezvous maneuvers in Project Gemini and for tracking return of Apollo lunar spacecraft. (AP, Balt. *Sun*, 1/29/63)

- Jean Felix Piccard, pioneer Swiss balloonist, died in Minneapolis on his 79th birthday. Twin brother of Auguste Piccard, Jean Piccard made his first flight in 1913, and on August 18, 1934, flew a *Century of Progress* balloon from Dearborn, Michigan, to an altitude of 57,579 ft. During summer of 1936, Jean Piccard developed and flew first constant-level plastic (cellophane) balloons from the University of Minnesota, and was generally credited with the development of polyethylene balloons for ONR's Project Helios which led to widespread use of plastic balloons as research tools in Skyhook, James A. Van Allen's Rockoons (balloon-launched sounding rockets), Strato-Lab, and Man High balloon research projects. (AP, *Wash. Post*, 1/29/63; *Aeronautics and Astronautics*, Appendix C.)

- M. V. Keldysh, President of U.S.S.R. Academy of Sciences, was made member of Academy of Sciences of the Czechoslovakian Socialist Republic, sixth Soviet scientist to have been so honored in recent years. (Tass, *Pravda*, 1/29/63, 6)

January 29: EXPLORER XIV, silent since Jan. 10, resumed normal transmission. (AP, Wash. *Eve. Star*, 1/30/63)

- Titan I missile successfully launched from Vandenberg AFB by SAC crew in routine training exercise, USAF announced. (AP, *Wash. Post*, 1/30/63, A1)

- NASA Aerobee 150A sounding rocket launched from NASA Wallops Station failed to boost scientific experiment to desired altitude because of launch vehicle malfunction. Experiment was to have studied spectral emission lines in upper atmosphere to help in determining "distribution of certain molecular and atomic species in the upper atmosphere." (Wallops Release 63-8)

- NASA Director of Communications Systems, Leonard Jaffee, announced NASA would attempt to launch Syncom communications satellite into synchronous orbit with Delta vehicle no earlier than Feb. 13. Previously announced for no earlier than Feb. 6, Syncom launch was postponed "to insure that the [command and control] equipment is completely checked out" aboard USNS *Kingsport*, stationed in Lagos Harbor, and on the launch vehicle at Cape Canaveral. (Tech. Background Briefing, 1/29/63)

- At an extraordinary session of the General Aviation Committee of the Fédération Aeronautique Internationale (FAI), held in Amsterdam, representatives of 14 countries agreed that there was need for a permanent organization, to be called the International Bureau of General Aviation (IBGA), which would present the views of general aviation before the International Civil Aviation Organization (ICAO). Representing the National Aero Clubs of 55 countries, FAI's IBGA would have two major objectives: (1) To gather and disseminate information pertaining to international general aviation; (2) To bring about a reduction of restrictions against an improvement in facilities for the free flow of general aviation flight operations between nations. (NAA Release)

January 30: In wake of rumors that NASA was planning three-day manned space flight for August, NASA affirmed its announced plan to end Project Mercury with MA-9 flight in April unless unforeseen problems arise in that flight. (Wash. *Eve. Star*, 1/30/63)

- NASA Administrator James E. Webb told National Transportation Institute meeting in Chicago ". . . our national defense—perhaps even our national survival—demands that we act to insure that no hostile force will be permitted to use space as an unchallenged avenue of aggression against us." (*Chicago Trib.*, 1/31/63)
- In report to House Armed Services Committee, Secretary of Defense Robert S. McNamara said USAF was conducting basic research on an "advanced hypersonic manned aircraft," follow-on project to X-20 (Dyna Soar) manned orbital space glider. McNamara described the aircraft as "an extremely advanced concept which envisages an aircraft that can take off from a conventional airstrip and fly directly into orbit and return." (AP, Wash. *Eve. Star*, 1/31/63)
- Unnamed spokesman for U.S. State Dept. confirmed reports that four unannounced objects had been launched into space between Sept. 1 and Jan. 7. "They are not our vehicles—that's all I can say," spokesman said. Presumably of Soviet origin, objects were launched between Sept. 1 and 17; Oct. 20 and 26; Nov. 1 and 5; and Jan. 1 and 7. Their existence was suspected when U.S. report to U.N. Space Registry skipped the four Greek alphabetical designations in the international system of recording objects in orbit. (Birmingham *Post-Herald*, 1/31/63, 1; UPI, *Wash. Post*, 2/1/63, A6)
- Gen. B. A. Schriever, Cdr. of AFSC, told IRE Winter Conference on Military Electronics in Los Angeles: ". . . Considering these three factors—the basic Soviet hostility, their emphasis on technology, and their achievements in space—it would be a mistake to relax our efforts in any area of technology. Space is certainly one area that calls for our best thinking and our best efforts . . . .

 "In recent years, we have been surprised more than once by Soviet technical achievements. We must not underestimate the capabilities of our opponent. It is equally important that we do not underestimate our own capabilities. We have vast natural resources, a broad and diversified industrial base, and an immense pool of scientific talent and experience. In all of these we are far more than a match for the Soviets. It is essential that we utilize all these resources wisely . . . ." (Text)

January 30: Soviet newspaper *Izvestia* reported that a Warsaw polytechnical institute had developed technique for soldering aluminum with aid of ultrasonics. Place to be soldered is first coated with a thin layer of tin, which "breaks into" the aluminum under effect of ultrasonic vibrations being produced by a generator. (*Izvestia*, 1/30/63, 2)

- In detailed explanation of decision to cancel Skybolt missile development, Secretary of Defense Robert S. McNamara told Congress that Skybolt would have been overly costly, "unsuited" to the deterrent roles proposed for it, and the poorest of U.S. missiles. (*Wash. Post*, 1/31/63, A5)
- National Aeronautic Association announced First International Aeronautical Film Festival would be held in Nice, Sept. 16–22, under auspices of FAI and sponsored by City of Nice. (NAA Release)
- Two-day strike by AFL-CIO against NRDS, Nev., nuclear test site ended; union and Reynolds Electrical and Engineering Co. agreed to continue negotiations under Federal mediation. (AP, *Wash. Post*, 1/30/63, A7)
- National Aeronautic Association and International Air Pageant, Inc., jointly announced plans for Olympics of the Air international air show, Labor Day weekend in 1964, in Southern California. (NAA Release)

January 30–31: Representatives of Canadian Defence Research Board and representatives of NASA met for preliminary exploration of scientific and technical aspects involved in proposed joint ionospheric research program. Extension of joint Alouette Topside Sounder project, proposed program would involve design and construction of four satellites in Canada, with first launching proposed for late 1964. (UPI, *Wash. Post*, 1/12/63; NASA AI)

January 31: Fifth anniversary of first U.S. satellite, EXPLORER I. Also fifth anniversary of activation of U.S. tracking network to track EXPLORER I: at the time, network included Vanguard's Minitrack stations located primarily in Western Hemisphere. Since that time, other networks were added to form truly worldwide tracking network and data acquisition system for satellites and space probes launched by U.S.: Minitrack network; Deep Space Instrumentation Facilities; Manned Space Flight Network; Smithsonian Astrophysical Observatory's Baker-Nunn Telescope-Camera Network.

Ceremonies at NASA Goddard Space Flight Center featured talks by Secretary of State Dean Rusk, NASA Administrator James E. Webb, and Astronaut Walter M. Schirra, Jr. Radio transmissions from VANGUARD I, second U.S. satellite and oldest U.S. satellite still transmitting, were heard during tour of Goddard facilities following formal ceremonies. Highlighting occasion was presentation of scrolls of appreciation to ambassadors of 16 nations that have cooperated with U.S. in establishing the international tracking networks. Field reports indicated that in at least 12 countries related special events and ceremonies were held on the same day as Goddard observances. Countries reporting special activities included Argentina, Australia, Bermuda, Brazil,

Ecuador, Great Britain, India, Iran, The Netherlands, Nigeria, Peru, and Spain. (NASA Release 63-10; UPI, *Wash. Post*, 1/27/63, A6; State Dept., USIA Field Reports—per AI/Robinson)

January 31: At ceremonies commemorating Fifth Anniversary of Tracking at NASA Goddard Space Flight Center, Vice President Lyndon B. Johnson, Chairman of National Aeronautics and Space Council, said in remarks delivered by NASC Executive Secretary Edward C. Welsh: ". . . I would like to quote briefly from the report drafted by the Senate Committee [on Aeronautics and Space Sciences] before the passage of the [National Aeronautics and Space] Act:

". . .'Space neither invites nor necessitates any re-definition of American goals and aims. Throughout our history, it has been the goal of peace and liberty that has led Americans to explore the dimensions of challenging frontiers. Our goal now remains unchanged.'

"I have chosen this paragraph to quote because it is basic to our national philosophy. In space activities, the lines of demarcation that so often separate nations in their purposes would seem to be petty. Interchange and cooperation in the exploration of a realm as vast as the solar system—indeed, the universe—should lead to better understanding among nations . . . ." (Text)

- At ceremonies commemorating Fifth Anniversary of Tracking at NASA Goddard Space Flight Center, Secretary of State Dean Rusk said: "We 17 nations have embarked on a cooperative effort to expand the knowledge of man to an unprecedented extent. We think we are creating a heritage for those who follow us in the coming generations in a great exploration for peaceful purposes . . . .

    "The world watches us in our success and our failures. And the whole world will benefit from our discoveries.

    "We seek to publish all the knowledge we uncover. We are making it available to the world scientific community. In this effort we regard ourselves as trustees for all the inhabitants of this little speck of dust in the universe . . . ." (Text)
- At Fifth Anniversary of Tracking ceremonies, NASA Goddard Space Flight Center, Astronaut Walter M. Schirra said that "the worldwide nature of the tracking network is a sign that the people of the nations throughout the world are interested in space flight and want to assist us in doing the best job possible . . . ." (Text)
- First public firing of escape rocket for Apollo spacecraft was made at Lockheed Propulsion Co. Static firing of the 150,000-lb.-thrust rocket was declared a success; NASA engineer George Lemke called test "a milestone—one of many that must be passed before we put men on the moon." (AP, Newport News *Times-Herald*, 2/1/63)
- Dr. Wernher von Braun, Director of NASA Marshall Space Flight Center, commented on scientific achievements during the first five years of Space Age: "I have always thought that these things could be done, but the rapidity with which these things took place . . . actually came quite as a surprise to me. I

thought it would take more convincing and fighting to get the public support we have received . . . .

"At the time [of orbiting EXPLORER I five years ago] you could hardly talk about space flights. Nowadays everybody talks about it . . . ." Dr. von Braun was Director of ABMA Development Operations Division which provided Jupiter C vehicle for EXPLORER I. (Huntsville *Times*, 1/31/63)

January 31: Dr. Nancy G. Roman, Chief of Astronomy and Solar Physics, NASA Office of Space Sciences, at Marymount College, Tarrytown-on-the-Hudson, New York, stated that: "There [is] a great deal of discussion of whether the United States should have women astronauts, and I am frequently asked for my own opinion on this subject. Frankly, it makes little difference to me. I believe that there will be women astronauts some time just as there are women airplane pilots, but there are so many other ways that women can contribute importantly to the space program that the fact there are no women astronauts as yet should not worry us . . . ."

NASA had over 146 women who were classified as professional Aerospace Technologists, 77 of whom were professional mathematicians whose responsibility was the programing and operation of NASA's highly complex computers and similar equipment, Dr. Roman said.

"Moreover, I should emphasize that as badly as we need scientists and engineers, we also need educated women as well as men in other fields and no one, man or woman, should go into science simply because of its glamour . . . The space program, the country and the world need young people who have been trained to think logically and clearly regardless of the area in which they have specialized." (Text)

- Thomas F. Dixon, Deputy Associate Administrator of NASA, told American Rocket Society Solid Propellant Rocket Conference in Philadelphia: "From the programs of the past two years has come conclusive evidence that large segmented [solid propellant] motors are not only feasible but highly reliable . . . .

"Also clear is the quality that is perhaps the greatest virtue of solid rockets: the relative speed with which major advances in performance can be achieved. In a brief period, the technologies of weight, thrust, burn time, and thrust vector control have moved forward rapidly . . . .

"The new DOD-NASA program carried out by the U.S. Air Force will give us a great deal of information on the technology and logistics associated with these very large high-thrust engines . . . .

"As we move deeper and deeper into space, it seems likely that we will come to rely upon solids for an increasing variety of missions for the same reasons that we have called upon them in the past—reliability, storability, ruggedness, low cost of development, and instant readiness, to mention a few . . . ." (Text)

- NASA Flight Research Center announced award of $1.325-million contract to Lockheed-Georgia Co. to purchase Lockheed Jet-Star aircraft. Aircraft would be used for research investigation of aircraft flying qualities, automatic and manual control systems, pilot instrument displays, and pilot training, with special emphasis on supersonic transport research. (FRC Release 2-63)

January 31: Reported that the National Bureau of Standards would construct a new facility for improved standard frequency and time broadcasts so as to better coordinate the global network of missile and satellite stations. 400-ft. antennas were being erected and 50-kilowatt transmitters were being built on a site near Fort Collins, Colo., to increase the coverage of the standard frequency and time transmissions of NBS stations WWVB (60 kc/s) and WWVL (20 kc/s). Both WWVB and WWVL have transmitted for several years from sites near the Boulder, Colo., Laboratories of NBS, and the high stability and long-range coverage of the lower frequencies have been established. (*Chattanooga Times*, 1/31/63; *Current News*, 2/18/63)

- Soviet newspaper *Pravda* said MARS I space probe was more than 26-million mi. from earth and successfully continuing its flight toward planet Mars. There was no mention of mid-course correction maneuver to bring the probe closer than its estimated 120,000-mi. distance from Mars in June. (Wash. *Eve. Star*, 1/31/63)

During January: Series of Gemini parachute design qualification tests was successfully completed at El Centro, Calif., for NASA Manned Spacecraft Center. Parachute system would undergo two more test series before qualification; system would be backup mechanism for two-man Gemini spacecraft recovery in case of water landings. (MSC *Roundup*, 1/23/63, 1)

- USAF Space Systems Div. requested proposals from 34 firms for development of medium-altitude communications satellite system. (*M&R*, 2/4/63, 11)
- Basic concepts of USAF five-year space plan contained in published article by Lt. Gen. James Ferguson, USAF DCS/R&T. Specific military objectives in space were: (1) to develop those military capabilities now believed required for national security; and (2) advance technology in space (i.e., "building blocks") which can be converted to military space systems. USAF was primarily interested, he said, in space from the earth's surface to the stationary orbit (22,000 mi.). (*A-N-AF Journal & Reg.*, 2/3/63, 9, 22f)
- New members of House Committee on Science and Astronautics of the 88th Congress: Democrats—Richard J. Patten (N.J.), Richard Fulton (Tenn.), Don Fuqua (Fla.), Neil Staebler (Mich.), and Carl Albert (Majority Leader-Okla.); Republicans—Donald Rumsfeld (Ill.), James D. Weaver (Pa.), Edward J. Gurney (Fla.), and John W. Wydler (N.Y.). (NASA AC; AP, *Wash. Post*, 1/25/63)
- NASA simplified its launch vehicle terminology for the Saturn space booster series: Saturn C-1 became "Saturn I"; Saturn C-1B became "Saturn I-B"; and Saturn C-5 became "Saturn V." (MA, Holmes)
- In *International Geophysics Bulletin*, NASA proposed contributions to IQSY (1964-65) were outlined. Prominently among them:

sounding rockets; ionosphere explorers and monitors; atmospheric structure satellite; Oso, Ego, Pogo satellites; Imp, Pioneer, Mariner, and Surveyor probes. (*IG Bulletin*, 1/63, 12–19)

During January: Engineers Joint Council issued report, *The Nation's Engineering Research Needs, 1965–85*, which concluded: (1) present system of allocating resources to U.S. R&D is producing imbalance in technical effort; (2) non-defense agencies of Government do not have adequate research programs relating to problems in public sector of the economy; (3) R&D by private industry is influenced heavily by Government R&D allocations; and (4) R&D efforts applied to creating new materials and products have been highly successful, but have not been matched by development of systems to utilize these products and materials efficiently.

EJC recommended: (1) periodic review of U.S. R&D allocations and "their compatibility with national goals"; (2) council and professional societies make periodic studies of technical and social problems to which engineering profession should make significant contributions, but "for which engineering programs and institutions appear to be inadequate"; (3) educational institutions and professional societies expand their programs of systems engineering "aimed at optimizing the technical engineering and social systems within which improved materials and devices perform their designed functions." (*Industrial Research*, 1/63, 5)

- Reported that General Electric scientists had succeeded in refining junction laser to point where output wavelength can be selected for optimum system performance by changing the chemical composition of the crystal. Output wavelength can be selected in the range from perhaps below 6,200 angstrom units to near 8,400 angstrom units. (*Industrial Research*, 1/63)
- Yuri Marinin, Soviet space writer, wrote in article translated by U.S. Dept. of Commerce's Office of Technical Services that "earth-orbit rendezvous" method of manned lunar flight "is regarded as the most practical." (Wash. *Eve. Star*, 1/26/63, A5)
- John C. Sanders, chief of Dynamics and Controls Branch, NASA Lewis Research Center, obtained patent for and made preliminary design of interplanetary rocket using solar rays as energy source. Rocket uses 200-ft.-wide aluminized mylar mirror to gather solar rays. (*Marshall Star*, 1/9/63, 3)
- Series of studies for Congressional Joint Economic committee reported that status of resources in U.S.S.R. force Soviet leaders to choose which of three areas to emphasize—ICBM production, antimissile defense, or space flight—with inescapable detriment to other two areas. (*M&R*, 1/7/63, 17)
- According to three Ling-Temco-Vought aerospace scientists, astronauts operating outside their capsules in outer space would have considerable difficulty estimating distances to other objects, even of known size. Results of experiments suggest that observers cannot make accurate distance estimates without the aid of some kind of artificial ranging device. The accuracy of visual observation will affect the complexity of the guidance system and the total thrust requirement for close-in maneuvering. For this reason,

man's visual performance capability must be defined as accurately as are other design data for successful space missions. (Ling-Temco-Vought, Inc.; *Space Information Digest*, Vol. 4, No. 4, 1/28/63)

During January: Dr. Douglas McKie, Univ. of London, writing in Jan. 1963 *Endeavour*, traced the history of the metric system, a subject of renewed interest in the U.K. since Parliament was considering the adoption of the metric system of weights and measures. Dr. McKie pointed out that the recent adoption of the wavelength of a line in the spectrum of krypton-86 as the basis for an international unit of length was the realization of an old French dream at the time of the adoption of the metric system—the use of an invariable natural standard. At the time of the French Revolution, Dr. McKie asserted, France had become weary of the chaos of variable systems of weights and measures and the government and the National Academy urged the States-General to adopt a national system. The States-General authorized the Academy to formulate a system of weights and measures based on measurement of an arc of the meridian as the universal standard and decimals as the notation system. This was done by 1793, but Revolutionary politics intervened, having done away with the States-General and even replaced the Academy. Not until Dec. 10, 1799, did the new standards become legal, and by then it had been discovered that the earth was not a perfect sphere and therefore that the arc of the meridian was not the invariable standard the Academy had thought. (*SIN*, Mar. 1963, 11–13)

- A new submarine cable extension 700 nautical miles in length and due for completion by late August 1963, would be laid in two links—one from Grand Turk Island AFB to Ramey AFB, Puerto Rico; the second from Ramey to Coolidge AFB, Antigua, B.W.I.—to substantially increase operational control at Cape Canaveral by providing downrange tracking and telemetry inputs faster and more reliably. The present operational system extends from the Cape to Grand Turk. (*M&R*, 1/14/63, 28)

FEBRUARY 1963

February 1: NASA announced launch responsibility for Agena and Centaur space vehicle programs had been transferred from NASA Launch Operations Center to NASA Goddard Space Flight Center's Field Project Branch at Cape Canaveral. Launches would be conducted under technical direction of Agena and Centaur project managers, NASA Lewis Research Center. Transfer would permit LOC to concentrate on manned space flight and associated launch vehicles, including Saturn series developed under management of NASA Marshall Space Flight Center. Overall project responsibility for Agena and Centaur programs had been assigned to NASA Lewis Research Center in December and October, respectively. (NASA Release 63–19)

- Unidentified space probe employing Blue Scout, Jr., booster was launched by USAF from Pt. Arguello, Calif. (*M&R*, 2/11/63, 13)
- NASA announced its first contract to study overall systems requirements for Synchronous Meteorological Satellite (SMS) had been awarded to Republic Aviation Corp. Administered by NASA Goddard Space Flight Center, contract called for four-month study to determine "technical systems needed for 24-hour surveillance of the earth's cloud cover and to identify the major scientific and engineering advances required for the ground stations." (NASA Release 63–18)
- NASA Flight Research Center announced award of $128,675 to Comcor, Inc., for service and maintenance of computer system used to simulate flights of X–15 aircraft. Computer system is capable of providing actual X–15 performance and stability conditions that may be expected in flight. (FRC Release 3–63)
- Establishment of AFSC Office of Deputy Chief of Staff for Procurement and Production was effective this date, General B. A. Schriever announced. New office combined procurement and production to form a single major staff activity. Brig. Gen. Gerald F. Keeling, formerly Director of Procurement in Office of Deputy Chief of Staff for Procurement and Materiel, was named to head the office. (AFSC Release 31–R–11)
- American Institute of Aeronautics and Astronautics (AIAA) officially came into existence, a merger, after more than a year of study and debate, of the American Rocket Society (ARS) and of the Institute of the Aerospace Sciences (IAS). AIAA is thus the leading technical society keyed to the dynamic developments of the space age. Dr. William H. Pickering, Director of JPL, became the AIAA's first President. (*Astronautics and Aerospace Engineering*, February 1963, 19)
- Morton J. Stoller, Director of NASA Office of Applications, was awarded NASA Medal for Outstanding Leadership. NASA Admin-

istrator James E. Webb, in making the award, commended Stoller for his "outstanding and dynamic leadership in planning, developing and directing a complex scientific organization whose notable achievements have significantly contributed to the preeminent position of this country in the space sciences, the development of space technology, and the practical application of such research and development." (NASA Release 63-20)

February 1: General Curtis E. LeMay, USAF Chief of Staff, in speech to Executives' Club of Chicago, quoted Gen. Douglas MacArthur's statement to the West Point Corps of Cadets in 1962:

"'You now face a new world, a world of change. The thrust into outer space of the satellite, spheres, and missiles marks a beginning of another epoch in the long story of mankind. In the five or more billions of years the scientists tell us it has taken to form the earth, in the three or more billion years of development of the human race, there has never been a more abrupt or staggering evolution.'"

General LeMay went on to say that the "trained man, whether in a manned space vehicle or in a ground surveillance control point, will be one of our most valuable assets in our national space effort—and for our survival . . . .

"We must remember that any medium . . . can be a region of danger to peace and security. In this new medium of space I believe that the military defenses of the western world must be objective, applicable and evident.

"The Air Force will use, to the benefit of military space capabilities, all scientific advances and acquisition of knowledge achieved by NASA. We don't plan to wait for a program of fallout—if we can hasten advancement or increase its utilization through collateral efforts. This we are doing, in national interest, toward advancing our considerable space testing and development of approved space programs." (*A-N-AF Journal and Register*, 2/9/63, 15; Text, *CR*, 4/6/63, A1202–1203)

• Incorporation articles for Communications Satellite Corp. were filed by board of incorporators and approved by President Kennedy. Articles of incorporation provided two series of common stock: Series I, to be issued to the public, to aerospace industries, and to noncommunications companies; Series II, to be issued to communications common carriers with permission from FCC. Shares of stock were not expected to be issued for at least six months.

Incorporators were serving as company directors until first annual meeting of stockholders—expected about six months after public offering of stock—when stockholders would elect 12 directors and President Kennedy would appoint three more. (*Wall Street Journal*, 2/4/63; AP, Newport News *Daily Press*, 2/2/63)

• Three-year tektite research program described in *Journal of Geophysical Research* by G. S. Hawkins of Boston Univ., Smithsonian Astrophysical Observatory, and Harvard College Observatory. On the basis of extensive data obtained in their research, combined with previously known data on tektites, the researchers concluded that tektites originated from terrestrial impacts by meteorites. (*Journ. of Geophys Research*, 2/1/63, in MSFC *STID*, 3/11/63, 5–6)

February 1-2: In a paper entitled "Current Developments in Space Law," Dr. John Cobb Cooper, member of the Institute for Advanced Study, Princeton University, and legal adviser to the International Air Transportation Association, spoke to the southeastern regional meeting of the American Society of International Law.

"It is now apparent that grave disputes exist between members of the United Nations as to possible limitations on the use of outer space. There is no understanding as to what constitutes authorized activities in outer space 'in conformity with international law.'

"It is to be hoped that in any final statement of basic principles as to the exploration and use of outer space some clear indication will be given of what constitutes authorized peaceful uses of outer space.

". . . . the disputes that have arisen, as indicated above, make it clear that the manner in which this undefined area called 'outer space' may be explored and used has not yet been determined. International peace certainly requires that this position be clarified." (*CR*, 3/12/63, A1349–A1351)

February 2: Responding in *Komsomolskaya Pravda* to a reader's question, Soviet radioastronomer I. S. Shklovskiy discussed question of contacting intelligent extraterrestrial beings and noted that "highly sensitive radioastronomical equipment is under development at the Shternberg State Astronomical Institute. If equipment of adequate sensitivity can be developed, and the attempt to detect artificially produced signals from the Andromeda nebula is futile, this may signify that in the regions of the universe nearest us, the intelligent life does not have adequate resources to contact us." Prof. Shklovskiy also commented on work being done by N. S. Kardashev at Shternberg Institute regarding attempts of other civilizations to communicate with us. (*Komsomolskaya Pravda*, 2/2/63, AFSS-T Trans.)

February 3: NASA Administrator James E. Webb announced appointment of Julian Scheer as Deputy Assistant Administrator for Public Affairs. Scheer had been serving as consultant to NASA Assistant Administrator for Public Affairs George L. Simpson, Jr., since November 21, 1962. (NASA Release 63–21)

February 4: Soviet interplanetary probe MARS I was nearing halfway mark on its flight to vicinity of Mars, news agency Tass announced. Tass quoted Mstislav Keldysh, president of Soviet Academy of Sciences, as saying MARS I was about 30 million mi. from earth. (UPI, Wash. *Daily News*, 2/5/63)

• Maj. J. L. Reeves, USAF veterinary officer, warned at aerospace medicine conference, SAM, that new hazard may lie ahead in Project Gemini two-man flights because of poisonous nature of Titan II launch vehicle's fuel. Poisonous quality of Titan II's fuel combination has been "known in a general way for quite some time," Dr. Reeves said, "[but] there is precious little knowledge of specific antidotes for the harmful effects of most of the materials now in use." Most dangerous component is mixture of two hydrazines, one of which induces fatty changes of the liver and the other causes "peculiar central nervous system excitement and even convulsions." (Wash. *Eve. Star*, 2/4/63)

February 4: Joint AEC-DOD-NASA report on radiation resulting from U.S. high-altitude nuclear explosion in July 1962 said: ". . . Electrons having minimum altitudes below 500 km are decaying in periods of weeks to months. Once those electrons whose trajectories dip into the denser portion of the earth's atmosphere are eliminated by air scattering and energy loss, the remaining electrons may be expected to survive for some years in the absence of pronounced magnetic storms or other perturbations of the geomagnetic field.

"Certain high altitude components of the observed electron distribution, however, were seen to decay by a factor of two or more in a period of a few weeks. Most notably observations . . . [by Bell Telephone Laboratories] have shown a pronounced decrease of intensity in regions associated with flux lines extending at the equator to distances of approximately 3 earth radii. Some more recent data indicate, moreover, that these secular changes are energy dependent. It has been noted that large time fluctuations somewhat similar to this phenomena [sic] were previously observed to occur in connection with the natural belts. The observed decrease of intensity in certain regions, and the other associated changes in the electron distribution, indicate a complicated phenomenon of considerable scientific interest. The results of the present series of observations may be expected to lead to an increased understanding of phenomena which determine the origin, intensity, and character of the natural radiation belts as they are affected by geomagnetic forces and the upper atmosphere . . . ."

On effects of artificial radiation on spacecraft solar cells, report said: "Improved types of solar cells (employing N-on-P silicon junctions), which are considerably more radiation resistant, are available and were employed on TELSTAR. With respect to manned missions in space, the shielding provided by normal capsule design effects a considerable reduction in the radiation exposure, and the artificial belt is not regarded as placing any significant restrictions on the conduct of current manned space flights . . . ." (AEC-DOD-NASA Status Report, 2/4/63)

- Astronaut M. Scott Carpenter, speaking before aerospace medicine conference at SAM, San Antonio, said that all three U.S. astronauts who had made orbital space flights had found state of weightlessness presented no problems; rather, it tended to become "addictive." (UPI, *Wash. Post*, 2/5/63, A 4)
- Senator Leverett Saltonstall (R.-Mass.) introduced in the Senate a bill (S. 656) "to promote public knowledge of progress and achievement in astronautics and related sciences through the designation of a special day [March 16] in honor of Dr. Robert Hutchings Goddard, the father of modern rockets, missiles, and astronautics . . . ." On March 16, 1926, Dr. Goddard first successfully launched a liquid-fueled rocket. (*CR*, 2/4/63, 1601)
- USAF announced it had halted all R&D on Skybolt missile. Production of the air-launched missile had been stopped Dec. 31, but research "on various components of the missile system was per-

mitted to continue pending further study of possible applications of the technology learned from the development work." (AP, *Wash. Post*, 2/5/63, A7)

February 4: Maj. Harold W. Dietz (USAF) of Vandenberg AFB told aerospace medicine conference at SAM that it was "practically impossible" for a single commanding officer of an ICBM site, even if he lost self-control in a nervous breakdown, to start a nuclear war. Major Dietz reviewed the safeguards built into the complex procedure for preparing and firing ICBM's. (*Wash. Post*, 2/5/63, A6)

- Mstislav Keldysh, President of U.S.S.R. Academy of Sciences, reported on Soviet progress in science at annual meeting of the Academy: "New successes have been achieved by our scientists in the field of space radio communications. New data have been obtained as the result of the radar observation of Mercury carried out for the first time and the repeated radar observation of Venus. Communications were made possible by reflecting telegraphy signals from the planet Venus . . .

  "Research by satellites and rockets . . . [is] opening new possibilities for understanding the powerful effect of cosmic factors on electromagnetic and hydrodynamic phenomena in the atmosphere which, in turn, decisively influence life on earth. I think that very important discoveries will be made in the near future in the field of physics of the atmosphere . . . ." (*Pravda*, 2/5/63, AFSS–T Trans.)

- *Aviation Week and Space Technology* reported that, within past 18 months, four F–1 rocket engines had been damaged or destroyed in static firings and that another 12 firings had ended with premature shutdowns. F–1 was under development by Rocketdyne for use in Saturn V launch vehicle. (*Av. Wk.*, 2/4/63, 26)

- AFSC announced improved U.S.-Canadian antiaircraft control sites, Buic (Back-Up Interceptor Control), were being built at several locations in North America. Buic system would be back-up to Sage air defense system; each site would house electronic computer, associated equipment, and display consoles to "aid Buic commander in directing and controlling his forces." (AFSC Release 31–R–9)

- First test-flights of missile warhead over continental U.S. were scheduled for next summer, it was reported. USAF Project Athena would evaluate unarmed warheads' re-entry into atmosphere in series of flights from near Green River, Utah, to White Sands, N.M. A major objective of the project would be "to obtain data that will permit defense units to tell the difference quickly between real warheads and decoys." (L.A. *Times*, *Wash. Post*, 2/5/63, A7)

- Dr. Albert J. Kelley, NASA Director of Electronics and Control, said in *Electronic News* interview that NASA's immediate need from electronics industry was greater reliability: "By far the great majority of our flight failures are due to the failure of electronic components . . . . We need more ground testing [of electronic components]."

To gain this, Dr. Kelley urged, private industry should invest more of its own funds in space electronics research: "The company that wants to get ahead and get its share of the business must pitch in with its own resources . . . . Some of the money we put in contracts is geared to steer firms to a certain phase of R&D, encouraging them to handle it independently . . . ." (*Electronic News*, 2/4/63)

February 4: President Kennedy sent message to U.N. Conference on Applications of Science and Technology in Geneva:

"The United States delegation comes to Geneva to learn, to share experiences and to probe jointly with the other delegations the great opportunity which we all share to seize upon the technological achievements of the industrialized world and reshape them for the benefit of the newly developed nations.

"We come to this task with a firm conviction that rapid and even radical progress can result if we join forces with vigor. We come with enthusiasm for a task that is the most constructive undertaking of this or any other age. And we come with a restless sense of urgency to get on quickly with a job that can mean so much to so many of the peoples of our interdependent world." (AP, Balt. *Sun*, 2/5/63)

February 4-20: United Nations Conference on Applications of Science and Technology for benefit of less developed areas (UNCAST) was held in Geneva. Heading U.S. delegation was Dr. Walsh McDermott, chairman of public health dept. at Cornell Univ. Medical Center, NYC. (Wash. *Eve. Star*, 1/8/63)

February 5: NASA Manned Spacecraft Center confirmed that Astronaut Leroy Gordon Cooper's MA-9 space flight would be delayed because of "electrical wiring problems in the launch vehicle control system which are peculiar to Atlas 130-D," the vehicle for MA-9 launch. (MSC Release 63-20)

- NASA announced it would negotiate contract with Fairchild Stratos Corp. for two meteoroid detection satellites to be launched by Saturn I vehicles. Once in orbit, the two-ton satellite would extend its two "wings" totaling 96-ft. length and providing area of more than 2,000 sq. ft. exposure to possible meteoroid hits. Aluminum skin of wings would be electrically charged; penetration of skin by meteoroid would trigger an electrical pulse which would be recorded and radioed back to earth. First of the satellites was scheduled for launch in late 1965. (NASA Release 63-24)

- House Committee on Science and Astronautics held its organizational meeting, with Chairman George P. Miller of California announcing names and ranking of committee members. (Newport News *Daily Press*, 2/6/63)

- Maj. Gen. O. J. Ritland (USAF) told news conference in conjunction with aerospace medicine conference at SAM that "this nation needs a [manned] laboratory in space" and that such a laboratory was within state of today's technology. Under USAF concept of "Mods" (Military Orbital Development Systems), he said, permanent manned stations would orbit in space and would be remanned and resupplied by spacecraft similar to Gemini. (AP, Balt. *Sun*, 2/6/63)

February 4: Adm. Sergei Gorshkov, Commander-in-Chief of Soviet Navy, wrote in *Krasnaya Zvezda* (*Red Star*) that US.S.R. had successfully fired missiles from underwater submarines during training exercises last year. He said "radical re-equipment" had provided Soviet navy with submarines and surface vessels armed with missiles capable of striking anywhere in the world. (AP, *Wash. Post*, 1/6/63, A15)

February 5: System to determine composition of moon's crust by explosions on lunar surface was described by Francis E. Lehner, senior project engineer on lunar seismology, Cal Tech. Under development for NASA by Lehner and Dr. Robert L. Kovach, geophysicist at JPL, system was designed to be sent to moon in "unmanned instrument package weighing about 50 pounds . . . .

"We propose to obtain subsurface information about the moon by touching off a series of small explosive charges and then with one or more geophones placed at some distance from the explosives detect the resulting waves that have penetrated into the crust and have been refracted again to the surface.

"A geophone converts a mechanical wave into an electrical one that may be amplified and recorded. Different kinds of rock formation, and cracks and layers of material will affect the wave patterns, which can be amplified and radioed back to earth." (CTPS, *Chicago Trib.*, 2/6/63)

February 5–6: First intra-agency technical conference on optical communications and tracking held at GSFC with representatives of NASA's field centers and installations attending. Purpose of the conference was to provide for an exchange of technical information on laser optical and tracking programs at each of the centers. (*Goddard News*, 2/25/63, 2)

February 6: NASA announced it would use TV to monitor Astronaut L. Gordon Cooper during his Project Mercury space flight MA-9. Slow-scan television monitoring equipment had been installed at Cape Canaveral and aboard tracking ship to be stationed in South Pacific; third monitor might be established in Canary Islands. Television pictures received at Cape Canaveral could later be fed into U.S. networks, it was reported. (UPI, *Wash. Post*, 2/7/63, A4)

- NASA requested industry proposals for design and construction of a "prototype six-month life support system for four men"—a 2,600-cu.-ft. cabin in which four-man crews would make simulated space flights lasting six months. Proposals were due at NASA Langley Research Center by Feb. 26. (UPI, *Wash Post*, 2/7/63, A13)
- Titan II ICBM flown 6,500 mi. down AMR with heaviest payload ever to travel that far on a U.S. rocket vehicle. This was first Titan II launch conducted by an all-USAF crew. (DOD Release 166–63; UPI, *Wash. Post*, 2/7/63, A2; *M&R*, 2/11/63, 13)
- D. Brainerd Holmes, NASA Deputy Associate Administrator and Director of Manned Space Flight, said that spectacular " 'firsts' in space are not truly accurate measurements of technical and scientific progress. The question of who is first in space will be decided by the total accomplishment and the total ability of each nation to sail this uncharted sea. Dramatic events, such as the first

flight of a man in orbit, or the first landing on the moon, make up only a part of this assessment . . . ." Holmes was addressing Sidwell Friends School Forum, Washington, D.C. (Text)

February 6: Senator Estes Kefauver (D.-Tenn.), Chairman of the Senate Antitrust Subcommittee, told the Senate that NASA proposed regulations would "waive the Government's rights to patents on taxpayer-financed research and development in most cases" and urged the proposal be killed. (AP, *Chattanooga Times*, 2/7/63)

- Speaking in New York at a special seminar at the Goddard Institute for Space Studies, Dr. Wernher von Braun, MSFC director, told scientists that benefits from America's space program would far outweigh the cost. The revenues from communications satellites alone would, by 1975, exceed the cost of NASA's program in that area. There would also be benefits on which no price tag could be placed, such as discovery of the Van Allen radiation belts and the accomplishments of the weather satellites. Dr. von Braun said that man himself is a very important link in space exploration, and that the value of placing a man on the moon cannot be debated.

  "Instrumented equipment can do the job if one knows exactly what he is seeking," he said. "However, if one doesn't know. then automated equipment is not quite good enough." (*Goddard News*, 2/25/63, 2)

- Presentation on Project Gemini, prepared by Drs. Stanley C. White and George B. Smith of NASA Manned Spacecraft Center and presented at Lectures in Aerospace Medicine series, SAM, said that present Gemini "flight program includes the performance of manned extra-vehicular operation. The accomplishment of this maneuver requires the opening of the hatch over the astronaut who will venture outside. During this period, the cabin pressure is lost, and both astronauts are dependent upon the suit for prime environmental protection.

  ". . . The program as it is now planned would start out with flights of two-days duration and move forward to flights of up to two weeks duration upon successful completion of the first phase. Ultimately, the more complex 14-day duration mission, the inter-vehicular rendezvous, and the extra-vehicular manned operation goals would be achieved. The intermediate steps offer an excellent opportunity to study the man, the vehicle, the man-machine interface, and to accomplish other bioscience experiments while the program moves toward its final objectives . . . ." (Text, MSC Fact Sheet 134)

- The consuls of 19 foreign nations were the guests of the Manned Spacecraft Center at a special program arranged to familiarize them with the Center's activities and goals in space. (MSC *Space News Roundup*, 1/8/64, 2)

February 7: Dr. Stanley C. White, of NASA Manned Spacecraft Center told aerospace medicine conference at SAM of Soviet report that Cosmonauts Nikolayev and Popovich lost an abnormal amount of calcium during their extended tandem space flights in August 1962. Dr. White said same abnormal increase of calcium in body waste occurs in bedridden patients after 10 or 12 days; in Soviet cosmonauts, condition occurred "earlier than either we or the Russians would have thought—that is, about two or three times earlier." Soviet findings had been reported to international astronautics conference in Paris in November, Dr. White said. He recommended that U.S. officials closely monitor forthcoming manned space flights to determine if this condition would constitute a problem; if it did, he said, preventive measures would be necessary because "in an extreme situation there would be demineralization of the bone, the strength of the bone would be reduced, and the bones might eventually become plastic or elastic." (AP, Wash. *Eve. Star*, 2/7/63, A3)

February 7: Senator Howard Cannon, speaking on Senate floor, said: ". . . It is my personal belief—and a belief held by many other people knowledgeable in the field—that our entire scientific effort in the decades ahead might well run aground because of an insufficient supply of trained technical personnel.

". . . I again ask this body to consider the shortage of engineers and scientists to be a matter of most vital national urgency; and I reiterate my resolve to see that the legislative branch does not default in its obligation to take action in this area." (*CR*, 2/7/63, 1949–50)

- Dr. James E. Roberts, USAF scientist, told aerospace medicine conference at SAM that permanent five-man space station could be established "in the late 1965 or early 1966 time period." Such a station, called "Mark I," could be placed in orbit by Saturn I or Titan III vehicle, he said. Mark II station, three times as big as Mark I, could replace the initial station by the end of the decade. The space stations discussed by Dr. Roberts would be primarily for scientific and engineering research. (Wash. *Eve. Star*, 2/7/63, A3)

- At banquet of "Lectures in Aerospace Medicine," School of Aerospace Medicine, San Antonio, Southwest Research Institute president Martin Goland said: "Our age is one of quickening technology, of economic opportunity accompanied by sharpened domestic and international competition, of changing social patterns and political structures. In the kaleidoscope of this environment, a major source of strength will be on the side of the nation which is creative and productive; the nation whose people are capable of exploiting for sound purposes the seemingly inexhaustible potential of science and technology.

    "Within this framework, it seems clear that we must view our space program as more than an assembly of scientific goals and technical objectives. So considerable is our intellectual and material investment in space that we cannot judiciously ignore its by-products, the opportunity to transfer knowledge and capabilities from space technology to other portions of our national effort. Each one of us must act as a catalyst in this diffusion of ideas . . . ." (Text)

- Polaris A–3 missile successfully test-flown 1,800 mi. down the AMR, first full success in seven test launchings of the advanced model. (AP, *Wash. Post*, 2/8/63, A13)

February 7: Boston Globe carried article which cited leading electronic industrialists who endorsed NASA's placement of an Electronics Center in the Boston area. Public discussion had been triggered by *Wall Street Journal* article which said that "electronic and other scientific Yankee businessmen are looking critically at a proffered Federal gift [NASA proposed electronics center]." Ephron Catlin, executive vice president of Greater Boston Chamber of Commerce, commented: "The story completely misrepresents the true feeling of the vast majority of the business and academic community in this area.

"Almost the entire area is in favor of the center, and in fact, working for it . . . ." (*CR*, 2/11/63, A663)

- Frederick R. Kappel, Chairman of the Board of AT&T, was presented Silver Quill Award of National Business Publications at 12th annual State-of-the-Nation dinner, Washington. In accepting the award, Mr. Kappel said that AT&T's TELSTAR communications satellite had "provided basic assurance that satellite communications are indeed feasible and practical, and that design and construction of commercial satellites, and ground stations to work with them, can now be approached pretty much as a straight engineering project. There is no scientific obstacle to designing a commercial system immediately, and bringing it into being as soon as the satellites could be manufactured and put into orbit, say in 2 or 3 years . . . ." (*CR*, 2/11/63, A656)
- Dr. Robert R. Gilruth, Director of MSC, was named Visiting Professor of Aerospace Engineering at Texas A&M College. (*MSC Space News Roundup*, 1/8/64, 2)
- Thomas F. Dixon, NASA Deputy Associate Administrator, said in address before faculty and student body of Vanderbilt Univ., Nashville, that for the U.S. space program, "1962 was an active and successful year . . . .

"Among the many and varied accomplishments of 1962, I would like to stress three that have very far-reaching significance.

"First, we demonstrated beyond any shadow of doubt that the United States has the capability for manned travel and manned exploration of the solar system . . . .

"Second, . . . we completed the difficult round of decisions on how to get to the moon . . . .

"Third, in the past year we more clearly defined the long-range goals of our space program . . . . Our long-range goals [are] to make America the leading spacefaring nation and to achieve pre-eminence in all major aspects of space science and space exploration . . . ." (Text)

February 8: R. J. Parks, JPL Manager of Lunar and Planetary Programs, told aerospace medicine conference that MARINER II interplanetary probe recorded unexpected temperatures on its flight past planet Venus. When spacecraft passed Venus, temperature of solar panels exceeded 250°F and temperature of batteries reached 130°F (compared to 70° at launch time), Parks said. (*Wash. Eve. Star*, 2/9/63, A5)

- First meeting of NASA-USAF board to coordinate joint participation in Project Gemini. Board included NASA Associate Administrator

Robert C. Seamans and USAF Assistant Secretary Brockway McMillan as co-chairmen; Maj. Gen. O. J. Ritland; DOD special assistant for space, L. L. Kavanau; NASA Director of Manned Space Flight, D. Brainerd Holmes; and NASA Deputy Associate Administrator for Defense, Adm. W. F. Boone. (*M&R*, 2/18/63, 9)

February 8: Hawker Aircraft, Ltd., P-1127 jet fighter completed vertical take-off and landing test aboard aircraft carrier *Ark Royal*. Spokesman for the British company said P-1127 was first aircraft other than helicopter to achieve this feat. (AP, *Wash. Post*, 2/9/63; *National Observer*, 2/11/63)

• Field Enterprises Educational Corp., publishing subsidiary of Marshall Field, was reported as having offered NASA astronauts a multimillion-dollar contract for their personal stories. Spokesman for Field Enterprises said proposition was under discussion with C. Leo DeOrsey, Washington lawyer representing original seven Mercury astronauts, and Harry A. Batten, Philadelphia advertising man representing nine Gemini-Apollo astronauts. (*NYT* [Western Ed.], 2/9/63)

February 9: First U.S. three-jet-engine airliner, Boeing Co.'s model 727, made its initial flight, with Lewis Wallick as test pilot. Aircraft was designed to cruise at 550–600 mph, carry 70–144 passengers, and operate from airports with runways as short as 5,000 ft. (AP, Wash. *Sun. Star*, 2/10/63)

• Eight senior members of the Committee on Science and Astronautics of the U.S. House of Representatives paid an informal visit to MSC. (MSC *Space News Roundup*, 1/8/64, 2)

February 10: Development and successful testing of laser demodulator by Douglas Aircraft Co. and National Engineering Co. was reported. Device converts laser beams—highly concentrated light—into electrical current to produce sounds in radio receivers and patterns of light in television tubes. (*Wash. Post*, 2/11/63, A3)

February 10–16: National Engineers' Week in the U.S. The White House released statement by President Kennedy: "Once again the observance of National Engineers' Week underscores America's vital need for the engineer and his work in the development of procedures and products which contribute to the improvement and security of our citizens.

"Trained engineers are indispensible to our efforts of meeting the recognized scientific and technological challenges of today as well as those unknown challenges of the future which we, as a nation, must be fully prepared to anticipate and meet.

"It is therefore essential to the sustained growth of American technology and to the well-being and prosperity of the people of the world that more able young men and women study engineering." (Text; NAA S&I *Skywriter*, 2/15/63)

February 10–24: FAI world soaring championships held at Junin, Argentina. (NAA Release)

February 11: Editorial in *New York Times* (Western Edition), inserted in *Congressional Record* by Senator William Proxmire (R.-Wis.), said: ". . . Energetic research and development efforts in space are vital, as a report just issued by the President makes clear, but intensive work (and expenditures) in the vast expanse of space studies is far different from—and far more important than—a mere race with the Russians. Whether the $20 billion (or $40 billion) race to the moon is justified on scientific, political, or military grounds, we do not think the matter has been sufficiently explained or sufficiently debated. We hope it will be in the present Congress." (*CR*, 2/11/63, 1989)

February 11: New NASA-DOD Contractor Performance Rating System would be "just like pulling a Dun and Bradstreet report on a contractor's performance," NASA Director of Procurement and Supply, Ernest Brackett, said in *Missiles and Rockets*. System would provide central file of past contractor performance evaluations which DOD and NASA would be required to consult before awarding future contracts. It would apply only to engineering and systems research and development contractors—not to small study contracts. Bracket disclosed the plan was first considered by committee in office of John Rubel, Assistant Secretary of Defense for Research and Engineering. "We joined the effort later," he said. (*M&R*, 2/11/63, 14)

- NASA devices for detecting life on Mars, to be ejected from Mariner interplanetary probes and soft-landed onto planet's surface, were discussed in article in *The National Observer*. Gulliver, first in series, would be 1.5-lb. dome-shaped device equipped with two 23-ft. sticky strings; upon landing, strings would shoot out of dome and then reel back inside, gathering up samples of whatever is on surface to be deposited in solution of distilled water, sterile beef broth, malt extract, vitamins, other nutrients, and radioactive carbon. If adhesive gathers microbes similar to those on earth, microbes would multiply in the solution and produce radioactive carbon dioxide which would be detected by Geiger counter in dome and radioed back to earth.

  Another device, known as Multivator, designed by Dr. Joshua Lederberg of Stanford Univ. and still under development, was designed to make 24 different biological and chemical tests of micro-organisms. Third life-detecting device is Wolf Trap, named after its designer, Dr. Wolf Vishniac of Univ. of Rochester; it would suck in samples of planetary soil or air and introduce them to broth somewhat like that of Gulliver. (*National Observer*, 2/11/63)

- NASA announced manned orbital flights were being considered for three months' duration, to study human reaction to prolonged weightlessness in space. "If men can tolerate weightlessness for three months they probably could withstand the rigors of a year-long planetary mission." (*L.A. Times, Wash. Post*, 2/12/63, B8)

- First series of tests of Gemini spacecraft back-up parachute recovery system was successfully completed at El Centro, Calif., NASA Manned Spacecraft Center announced. 20-test series checked deployment characteristics of the system and structural integrity of individual parachutes. Parachute system, primarily a back-up to paraglider recovery system, would be used for wet landings of early unmanned and manned Gemini spacecraft. Paraglider, designed for dry landings, would be primary recovery system in later Gemini two-man space flights. (MSC Release 63-25)

February 11: Writing on possible effects of subgravity on human beings, Dr. Tau-Yi Toong, consultant, USAF OAR, said: "Two major difficulties present themselves in a study of such effects in earth-bound laboratories; first it is difficult to simulate subgravity on earth for long durations and, secondly, it is impossible to scale human beings up or down. . . .

"It was pointed out to the author by Dr. Harold J. von Beckh that part of the effort of a human heart is spent to overcome gravity. It would seem that a lighter load is imposed on the heart of a man on the moon and thus his heart beat slows down. . . . The human heart might possibly degenerate to such an extent that its beat slows down further to something corresponding to the 'normal heart beat of an abnormal giant' on the earth. The situation might become even worse should he again return to the earth's gravitational field with his weakened heart. . . ." (OAR *Research Review*, 2/11/63, 3)

- NASA Argo D-8 (Journeyman) rocket launched from Pt. Arguello, Calif., the four-stage sounding rocket carrying 104-lb. instrumented payload to probe hazards of the radiation belts. Designed by Univ. of Minnesota scientists, payload reached 990-mi. altitude and relayed valuable information during its 27-min. flight, then impacted in Pacific Ocean 1,250-mi. south of Pt. Arguello. Launch was conducted by NASA Goddard Space Flight Center with support by NASA Wallops Station personnel. (Wallops Release 63-12; AP, Newport News *Times-Herald*, 2/12/63)
- USN announced successful flight-test of advanced Polaris A-3, the missile traveling 1,800-mi. down Atlantic Missile Range after being launched from land pad at Cape Canaveral. (AP, *Wash. Post*, 2/12/63, A13)
- Prof. V. Kovda, Director of UNESCO's Dept. of Natural Sciences, told UNCAST in Geneva that up to 25 per cent of the newly trained scientists in underdeveloped countries emigrate to other countries.

  Charles V. Kidd, Associate Training Director at National Institutes of Health (U.S.), told conference that as many as 20 per cent of the 43,500 scientists and engineers who moved to U.S. between 1949 and 1961 had come from Asia and Latin America. He called the migration "a national catastrophe" for emerging nations because "not only are the talents of the individuals as scientists lost, but the nucleus of people who can alone build an indigenous base for science is dissipated." (*Manchester Guardian*, *Wash. Post*, 2/12/63, A9)
- National Academy of Sciences-National Research Council announced receiving $200,000 Ford Foundation grant to support its study of U.S. utilization of scientific and engineering manpower. In accepting the grant, Dr. Frederick Seitz, President of the Academy-Research Council, said: "The study, already in progress . . . reflects the growing concern in all sectors that heightened competition for highly qualified scientists and engineers may be leading us into policies and practices that are wasteful of this scarce talent.

  "Our goal is to develop guidelines and suggest measures that will enable this vital supply of specialized manpower to work most creatively and productively." (NAS-NRC Release)

February 11: MSC announced the award of a formal contract to Raytheon Company's Space and Information Systems Division of Lexington, Mass., for industrial support to MIT, which is developing the guidance and navigation systems for Apollo, in the design and development of the onboard digital computer for the command module. (MSC Fact Sheet #96)

February 11–12: Second NASA-Industry Program Plans Conference held in Washington, featuring NASA briefings on all major NASA programs. (Advance Agenda)

- Ernest W. Brackett, Director, Procurement and Supply, at the NASA-Industry Programs Plans Conference, Washington, D.C., explained the role of small business in NASA contracting. "During fiscal year 1962 small business companies received 66 percent of the total number of contractual actions, large and small, placed by NASA. This small business share of NASA procurement amounted to approximately $125 million, or 12 percent of the total NASA procurement placed with all business firms . . . They are able to compete successfully for many of NASA's contracts, particularly in furnishing components and supplies. During fiscal year 1962, of the hundred contractors who received the largest dollar value of NASA prime contracts, 24 were small business concerns. In those procurements where small business concerns submitted bids, they were successful in receiving 57 percent of the dollar value of the awards." (*Proceedings,* NASA SP-29; Text)

February 12: NASA announced delay from a previously planned April launching date of Astronaut Leroy Gordon Cooper's MA-9 orbital space flight, because of decision to rewire Mercury-Atlas flight control system. Rescheduled date of flight was not given, but NASA Manned Spacecraft Center's Chief of Manned Flight Operations, Walter C. Williams, estimated delay would be at least three weeks—putting launch date no earlier than mid-May. (MSC Release 63–26; Wash. *Eve. Star,* 2/13/63)

- Dr. Albert J. Kelley, Director of Electronics and Control in NASA Office of Advanced Research and Technology, said problem of communications blackout during manned spacecraft re-entry may be solved by spraying water from the spacecraft as it re-enters the earth's atmosphere. Kelley told NASA-Industry Program Plans Conference that experiments conducted at NASA Langley Research Center had found that water suppressed ionized plasma, which creates shield around re-entering spacecraft and blocks radio communications. (AP, *Wash. Post,* 2/13/63, A6)

- NASA selected Marion Power Shovel Co. to design and build crawler-transport vehicle for lifting and transporting three-stage Saturn V launch vehicle mated to Apollo lunar spacecraft and associated launch equipment. Vehicle would be required to haul the 400-ft.-high, 12-million-lb. assembly a distance of a little more than two miles—keeping it within 1/10 of a degree of true level—and deposit it on Merritt Island (Fla.) launch pad. Crawler would weigh 5.5 million lbs. and would measure 130-ft. long, 115-ft. wide, and 20-ft. high. Plans called for Saturn V-Apollo to be assembled in Vertical Assembly Building at NASA Launch Operations Center's Complex 39, then transported via the crawler to launch pad. NASA planned to buy two crawlers at cost of $4–$5 million each. (NASA Release 63-27; MSC Release 63-24)

February 12: Because of weather conditions at Cape Canaveral, launching of Syncom communications satellite was postponed from Feb. 13 to Feb. 14. (UPI, *Wash. Post*, 2/13/63, A6)

- NASA announced it had awarded letter contract to RCA Astro-Electronics Div. for fabrication and test of additional Tiros weather satellites; negotiations of final contract were continuing for total purchase of seven satellites—three R&D type, two operational, and two backup. Before this order, eight Tiros satellites had been funded in R&D phase of the program; six had been placed in orbit since April 1960 and two would be launched in 1963. (NASA Release 63-29)

- Mr. Oran W. Nicks, NASA Director of Lunar and Planetary Programs, told NASA-Industry Program Plans Conference that "we anticipate the development of spacecraft capable of trips to Jupiter, Saturn, and Pluto in the 1970's."

    Dr. Joseph Shea, Deputy Director for Systems, NASA Office of Manned Space Flight, indicated manned exploration of Mars might not be attempted before the 1980's: "We consider Mars to be the most likely choice for our first [manned planetary] attempt, since its environment appears to be less hostile than that of Venus. It is also more likely to have some form of life, and is therefore of greater scientific interest." (Wash. *Eve. Star*, 2/12/63, A3)

- NASA's concern with reliability was stressed at NASA-Industry Program Plans Conference by Edgar M. Cortright, Deputy Director of NASA Office of Space Sciences. "We must now develop techniques of realizing nearly 100 percent reliability from the outset on our space projects . . . . This point of view is widely shared within the Department of Defense, and you may expect a coordinated attack on this most difficult problem." (Wash. *Eve. Star*, 2/12/63, A3)

- Soviet Power Minister Pyotr Neporozhni told news conference that two nuclear-powered stations for generating electricity would begin operating this year. Considered experimental, the stations would be located at Beloyarsk (in Ural Mountains) and at Voronezh (south of Moscow). (AP, *Wash. Post*, 2/13/63, A19)

- Chief Justice Earl Warren, speaking at ceremony commemorating 75th anniversary of Georgia Tech, Atlanta, said that law had not kept pace with science and warned that, if science were to serve peaceful purposes of mankind, it "must be given a peaceful setting in both domestic and world law." Warren said the Space Age posed many new world legal problems that man must solve or be faced with danger of self-destruction. "How high into space over his property does the right of an individual go? Does flight over national territory violate the right of a nation to the space above it? Can the privacy of our homes and offices be invaded by the use of electronic instruments that are far removed from the property? These are only a few of the problems which you men of science pose to us whose vocation it is to protect human life, human rights, and human property." (UPI, *Wash. Post*, 2/13/63, A6)

February 12: S. Fred Singer, Director of National Weather Satellite Center, told Metropolitan Board of Trade, Washington, that weather satellites would probably make possible prediction of weather trends by as much as a year. By photographing cloud formations and measuring moisture content of atmosphere, satellites should enable meteorologists to discover conditions causing weather trends. (*Wash. Post*, 2/13/63, A4)

- Eight professors of physics and geology at MIT, Harvard, and Princeton signed statement urging public to support U.S. nuclear disarmament position: "Our future depends on controlling the spread of weapons of mass destruction. The U.S. proposal includes both detection stations and on-site inspections. These together can identify almost all nuclear explosions. Only some small underground tests might go unidentified.

  "We believe that nothing can be learned from these small underground tests which can upset the balance of power.

  "In the long run, a 'no-test ban' policy would result in a great loss of national security. By contrast the U.S. proposal realistically offers us a practical first step forward." (AP, *Wash. Post*, 2/13/63, A19)

February 13: Herbert D. Rochen's appointment as NASA Deputy to Program Manager of joint AEC-DOD-NASA Snap-50 Program Office was announced by NASA. In this capacity Rochen would provide liaison between NASA programs and Snap-50/Spur (nuclear electric power plant) program. Rochen was formerly assigned to NASA Office of Advanced Research and Technology, Nuclear Systems Directorate, where he was NASA program manager for Snap-8 project. (NASA Release 63-28)

- USAF launched Atlas missile from Vandenberg AFB in "routine training launch." (DOD Release 220-63)
- J-2 rocket engine first successfully test-fired, at simulated space altitude of 80,000 ft. The one-sec. test was conducted in specially designed vacuum chamber at NAA's Rocketdyne Propulsion Field Laboratory. (MSFC Historian, 5/22/63)
- EXPLORER XVI meteoroid detector satellite recorded 16 punctures by meteoroids during its first 29 days in orbit, NASA reported. Charles T. D'Aiutolo, manager of meteoroid research programs in NASA Office of Advanced Research and Technology, said that with EXPLORER XVI "we have established conclusively that there are micrometeoroids out there which can penetrate thin surfaces." Other spacecraft had reported hits by cosmic debris, but this was first time actual punctures were recorded. If EXPLORER XVI continued as expected to report meteoroid data for a full year, it should enable scientists to determine whether meteoroids are hazardous to spacecraft—both manned and unmanned. D'Aiutolo said the satellite, which exposed 25 sq. ft. of surface to meteoroid impacts, was not large enough to provide good statistical data on larger and rarer particles in space. (Earlier this month, NASA announced plans to orbit two meteoroid-detector satellites, each with exposure surface of more than 2,000 sq. ft. See Feb. 5.) (Langley Release, 2/19/63; UPI, *Wash. Post*, 2/14/63, A1)

February 13: Invitation to European scientists to participate in ground experiments using lasers to track the NASA Polar Ionosphere Beacon Satellites (S-66) was made at Third International Symposium on Quantum Electronics, Paris, by representatives of NASA Hq. and Goddard Space Flight Center. Under proposed program, NASA would provide orbital prediction data to European participants, who would attempt to track the satellite by bouncing a laser beam from ground off corner reflectors on satellite. NASA would be conducting similar experiments in the U.S.

Lasers were expected to provide useful supplement to existing radio and radar tracking systems by providing extremely precise range and bearing data. (AI, Robinson)

- Reorganization of Apollo Spacecraft Project Office of NASA Manned Spacecraft Center was announced, with Robert O. Piland named as Deputy Project Manager for the Lunar Excursion Module and James L. Decker as Deputy Project Manager for the Command and Service Modules. Deputy directors would be responsible for cost, schedule, technical design, and production of the three-module Apollo spacecraft under overall direction of Apollo Project Manager Charles W. Frick. Other organizational changes included establishment of a Spacecraft Systems Office for the Command and Service Modules, with Caldwell C. Johnson as Manager, and a similar office for the LEM with Owen E. Maynard as Acting Manager. (MSC Release 63-27)
- "Astrovoice," device to enable voice communications with astronauts during manned spacecraft re-entry, was displayed publicly by Avco Corp. Device adds voice channel to tracking radar, which is not blacked out during re-entry ionization. Modifications to existing ground radar receivers would require only "a few hours," Avco spokesman said. (Wash. *Eve. Star*, 2/13/63)
- Senator Edward M. Kennedy (D.-Mass.), addressing Massachusetts delegates to NASA-Industry Program Plans Conference, Washington, reaffirmed his pledge to foster Government contracts for space and electronics industries in Massachusetts, and said: "It is my responsibility as a member of Congress to see to it that NASA lives up to its responsibility and make sure your case is heard. It it NASA's responsibility to see to it that the bids made by Massachusetts companies get the fullest examination and the fairest consideration . . ." (Boston *Record American*, 2/13/63)
- According to Soviet press agency Tass, MARS I spacecraft was 59,460,000 km. from Earth at 9:00 a.m. Moscow time, traveling at 12.7 km/sec as it moved away from Earth. Routine interrogations had been conducted during the past week and scientific information from MARS I had been obtained. (EOS Translation, 3/6/63, of Tass, *Pravda*, 2/13/63)
- Mirage IV aircraft, French prototype bomber, crashed near Orleans because of trouble in one of the twin jet engines. The supersonic aircraft had logged 400 flying hours during past three years. (Reuters, *Wash. Post*, 2/14/63)

February 14: NASA SYNCOM I communications satellite launched into orbit by Thor-Delta vehicle from AMR, entering a highly elliptical orbit. About five hours later, apogee-kick motor was fired for about 20 seconds in maneuver designed to place the satellite into near-synchronous, 24-hour orbit 22,300-mi. above the earth. At about the time the apogee-kick motor completed burning, ground stations lost contact with the satellite and could not confirm a synchronous orbit. NASA officials assumed that "the satellite's spin axis was misaligned at the time of the apogee motor firing. Because of this they have been unable to determine whether the satellite is damaged." Attempts to make contact with SYNCOM were continued.

SYNCOM I was to have hovered at a nearly fixed longitude over the Atlantic Ocean and traced every 24 hours a figure-8 pattern approximately 30° north and 30° south of the equator; this path would be close enough to true synchronous orbit—stationary hovering at speed equal to that of earth's rotation—to determine if synchronous orbital communications satellites were feasible. Experiments with SYNCOM were to have included telephone and teletype communications transmitted between New Jersey and Lagos Harbor, Nigeria. Syncom was NASA project, supported by DOD ground stations and communications experiments. (NASA Statement, 2/14/63; AP, *Wash. Post*, 2/14/63, A1; NASA Syncom Release, 1/29/63)

February 14: Dr. Hugh L. Dryden, NASA Deputy Administrator, addressing Cleveland-Akron sections of American Institute of Aeronautics and Astronautics (AIAA) and American Society of Mechanical Engineers (ASME), stated "a few of the social and moral responsibilities of scientists working in this field [of space science and technology]:

"(1) To search for knowledge of the facts and to conduct our work with devotion to intellectual honesty and objectivity.

"(2) To carry out our part of the national program of space exploration in such a manner as to obtain maximum benefit to the welfare of the nation and all mankind.

"(3) As citizens to lend our influence to the establishment of societies of free men in a peaceful world providing not only material benefits but also incentives for mental and spiritual growth and accomplishment.

"(4) To realize in our job and daily contacts the highest moral aspirations and ideals of which we are capable." (Text)

- Astronaut M. Scott Carpenter addressing the USAF School of Aerospace Medicine, Brooks Air Force Base, Texas, said:

"The main goal of Project Mercury is to prove the design and engineering concept and the usefulness of Man in space, but the capabilities of the Mercury system can be streched only so far. I had hoped that the flight of "Aurora Seven" would be a humble beginning in the gathering of scientific data from space about the universe in which we live. In the three short months that preceded the flight, we developed many special instruments for this purpose. . . . There is untold wealth of information to be gained as a result of space flight if we turn our talents and curiosities in that direction.

"I have worked closely with scientists from many fields, but perhaps most closely with those in medicine.

"There have been some wonderful intangibles I attach to working with these men. They are more than vein probers, brain pickers and pulse takers. I am continually challenged by their curiosity and insight, and fascinated by their research. I respect their intelligence and I treasure their friendships.

"It is this intelligence, this special education and innate curiosity scientists of all types have in common that demands they be included in future space flights.

"Perhaps one day the astronaut will be the least important person aboard as he ferries a cargo of astronomers, geologists, biologists, and flight surgeons to Mars." (Text)

February 14: Arthur S. Flemming Awards for outstanding contributions to Federal Government were presented to five administrators and five scientists at D.C. Junior Chamber of Commerce luncheon. Among the scientists were NASA's George M. Low, Director of Spacecraft and Flight Missions, Office of Manned Space Flight, and Edgar M. Cortright, Jr., Deputy Director of Office of Space Sciences. Other three scientists were Joseph F. Saunders, head of NRL Medicine and Dentistry Branch; Norman J. Doctor, physicist at Army Diamond Ordnance Laboratories; and Charles M. Herzfeld, director of DOD Ballistic Missile Defense. (*Wash. Post*, 2/11/63, A8)

- Catalytic Construction Co. of Philadelphia was selected by NASA for negotiations of contract to build, install, and test synchrocyclotron and related equipment and services for Space Radiation Effects Laboratory, to be built in Newport News, Va., under NASA Langley Research Center management. Expected to be completed in mid-1965, laboratory would be used for simulating high-energy corpuscular radiation encountered in proposed NASA space flights, and studies would include effects of radiation on spacecraft materials and components as well as means of shielding against radiation. (NASA Release 63-30)

- Representative George P. Miller (D.-Calif.) submitted article by A. A. Mikhailov, Director of the U.S.S.R.'s Pulkovo Observatory on the American astronomer, William Wallace Campbell (1862-1938) in the *Congressional Record*. Appearing in Soviet journal called *Nature*, article reviewed work of Dr. Campbell at the Lick Observatory in San Francisco, where he pioneered in stellar photography and the determination of stellar motions (1890-1923). (*CR*, 3/1/63, 1086-87)

- Dr. Franklin A. Gifford, Jr., was awarded Dept. of Commerce's Gold Medal for Exceptional Service for his "major contributions to science and administration, for extremely significant research and outstanding leadership in the study of turbulent diffusion in the atmosphere, and for highly distinguished authorship in the field of meteorology." Dr. Gifford is Meteorologist in Charge of the Weather Bureau Station at Oak Ridge, Tenn. (Dept. of Commerce Release WB 63-1)

Mid-February: AF Cambridge Research Laboratories conducted atmospheric-density experiment by dropping instrumented mylar sphere to earth from 160-mi. altitude. Aerobee rocket carried inflatable sphere to desired altitude from Eglin AFB, Fla., the sphere's telemetering equipment transmitting linear accelerometer data measuring atmospheric drag on the sphere. (DOD Release 35–63)

February 15: Unnamed USAF spokesman disclosed USAF had orbited "thermonuclear converter" aboard one of its unidentified satellites launched from PMR. Device was designed to convert sunlight into electricity with efficiency of solar cells but with more resistance to space radiation. USAF received signals from the converter for four days, according to spokesman, and "test was considered a major step in proving this concept." (AP, Balt. *Sun*, 2/16/63)

- U.S. worldwide tracking network was not able to locate SYNCOM I communications satellite; radio contact with the satellite had been lost Feb. 14, seconds after onboard rocket had fired to transfer SYNCOM I from its highly elliptical orbit into near-synchronous orbit. (AP, Wash. *Eve. Star*, 2/15/63)
- International Telephone and Telegraph Corp. (IT&T) announced it had filed application with FCC for authority to purchase stock in newly-formed Communications Satellite Corp. (UPI, *Wash. Post*, 2/15/63, C6)
- Speaking at "ground making" ceremony for Titan III launch facilities, Cape Canaveral, Col. Joseph S. Bleymaier (AFSC Titan III program manager) disclosed that 17 development tests were planned for Titan III launch vehicle, beginning autumn of 1964. Ceremony took place on dredge *Pittsburgh* anchored in Banana River, west of Cape Canaveral; dredge would move land from nearby swamps to form island in river for Titan III. (AP, Balt. *Sun*, 2/16/63)
- U.S. Army conducted successful test-firing of Nike-Zeus antimissile missile at White Sands Missile Range, the missile climbing to "a high altitude" after launch from underground cell. (DOD Release 222-62)
- Titan II ICBM exploded 18,000 ft. over Pacific Ocean after successful launch from silo, Vandenberg AFB. USAF said the missile destroyed itself as a result of automatic signal within its guidance system after 56 seconds of flight.
  Results of Titans I and II launches through February 16, 1963: Out of 47 Titan I's launched from AMR, 34 were successful, 9 were partially successful, and 4 were failures; out of 8 Titan I's launched from PMR, 7 were successful, and 1 was partially successful; out of 11 Titan II's launched from AMR, 7 were successful, 4 were partially successful; and the Titan II launched from PMR was partially successful. (*SID*, 4/22/63; AP, Wash. *Eve. Star*, 2/14/63)

February 16: President Kennedy released report by special panel of President's Science Advisory Committee, report concluding that exchange of scientific information "is an inseparable part of research and development . . . . [Scientists should] share many of the burdens that have traditionally been carried on by the professional documentalist . . . . Science can ultimately cope with the information expansion only if enough of its most gifted practitioners will compact, review and interpret the literature both for their own use and for the benefit of more specialized scientists . . . ."

In releasing the report, President Kennedy said: "Strong science and technology is a national necessity, and adequate communication is a prerequisite for strong science and technology." (NYT Co., *Atlanta Journal & Constitution*, 2/17/63)

February 16: Vice President Lyndon B. Johnson, addressing Rural Electric Cooperatives Association's annual meeting in Las Vegas, said: "In less than 5 years, our country has assumed a solid and secure place of leadership in developing the peaceful and practical uses of outer space. Where once we were, in the judgment of some, lagging behind, today we are, on the basis of clear evidence, forging ahead. Our efforts in space are in no way dictated by a policy of duplicating the activities of the Soviet. We are not engaged in a race with the Communist scientists—we are engaged in a race to extend the horizons of man's knowledge and to increase the benefits which will better the lives of all men who live on this earth . . . ." (*CR*, 2/18/63, 2299)

- Instruments onboard MARS I interplanetary probe confirmed existence of a third radiation belt, Soviet news agency Tass announced. Third radiation belt had been discovered in 1959 by Soviet rockets which had gathered particles at 50,000-mi. altitude, Tass said, and its existence was verified when MARS I recorded stream of charged particles far beyond the second radiation belt. Number of particles in this outermost belt greatly exceeded that in first two belts, according to Tass. (AP, *Wash. Post*, 2/17/63)
- Cumberland Plateau Seismological Observatory, near McMinnville, Tenn., was dedicated. Observatory was last of four U.S. observatories designed to measure earth movements as small as one-millionth of an inch and built to specifications set by 1958 nuclear testban conference in Geneva. (AP, Wash. *Eve. Star*, 2/14/63)

February 17: Scientists Institute for Public Information was founded by about 100 U.S. scientists representing 22 independent information committees at meeting in New York. Dr. Jules Hirsch, research physician and biochemist at Rockefeller Institute, was named temporary chairman of the board. (AP, *Wash. Post*, 2/18/63, A2)

- Senator Clifford Case (R.-N.J.), in interview taped for radio and television, urged public spotlight on awards of multi-million-dollar defense and space contracts to prevent "political payoffs" and other improprieties. (AP, Wash. *Eve. Star*, 2/18/63)
- Soviet Army newspaper *Krasnaya Zvezda* reported Soviet engineers had successfully tested "flying saucer" aircraft which takes off vertically, lands on cushion of air. (UPI, *Wash. Post*, 2/18/63)

February 18: Attempted launch of sodium-vapor cloud experiment from NASA Wallops Station was not successful because second stage of Nike-Asp launch vehicle failed to perform properly. (Wallops Release 63-16; *M&R*, 2/25/63)

- Internal temperatures of Mercury and Mariner spacecraft could have been reduced by 50° to 80° F—while in orbital flight but not considering effect on re-entry—if the exteriors had been painted white, according to Konrad Buettner, Prof. of Atmospheric Sciences at Univ. of Washington, Seattle, in *Missiles and Rockets*. (*M&R*, 2/18/63, 23)

February 18: County Judge Thaddeus Pruss ruled in Wisconsin that metal particles allegedly from SPUTNIK IV satellite were to be returned to their finder, Thomas Williams, Brown Deer, Wis., said to be first known test case involving space law. (UPI, Wash. *Daily News*, 2/18/63)

- NASA determination "to go ahead with its liberalization of patent licensing regulations, despite Congressional opposition," was reported in *Missiles and Rockets*, which added that NASA had "been encouraged by a still-unannounced Kennedy Administration decision to favor a similar policy throughout the government." (*M&R*, 2/18/63, 9)
- Polaris A-3 missile fired by USN from Cape Canaveral headed off course and had to be destroyed by range safety officer. (UPI, *Wash. Post*, 2/19/63)
- Six-man task group to formulate contractor evaluation group system was reported by *Missiles and Rockets* to be working under office of DOD Director of Defense Research and Engineering. Including five DOD staff members and one NASA member, group had already selected six contracts as "representative of medium and large development programs documenting contractor performance in the two major categories being studied [engineering development and operational systems development]." (*M&R*, 2/18/63, 11)
- USAF announced first silo-launching of Titan II Feb. 16 was considered partially successful; despite autodestruct 56-sec. after launch, the ICBM performed well in leaving the silo and in gaining altitude. (AP, *Wash. Post*, 2/19/63)

February 18–27: FAA, NASA, and DOD conducted a test program at Edwards AFB to determine the effects of the sonic boom on light aircraft and helicopters. Potential dangers of structural damage to aircraft other safety-of-flight factors were investigated by subjecting light aircraft and helicopters to sonic booms while on the ground, in cruising flight, in turning flight, and near-stall configuration in the case of fixed-wing aircraft. A report on test results will be issued following analysis and evaluation of data. (FAA-19)

February 19: USAF launched unidentified payload with Blue Scout booster from Pt. Arguello, Calif. (*Wash. Post*, 2/20/63, A1; *M&R*, 2/25/63, 11)

- Number of engineering students studying for doctoral degree tripled in U.S. during last seven years, said report in *Journal of Engineering Education*. (AP, *Wash. Post*, 2/20/63, B11)
- Senator John O. Pastore (D.-R.I.), chairman of the Senate Commerce Committee's Communications Subcommittee, questioned the wisdom of continued Government financing of experimental communications satellites: "We were led to believe that practically all this work would be done by the private [communications satellite] corporation." (*Wash. Post*, 2/20/63, A8)
- Dr. Hugh L. Dryden, NASA Deputy Administrator, testified before Communications Subcommittee of Senate Commerce Committee

that experiences of both TELSTAR I and RELAY I communications satellites were being "used continuously to review projects such as Syncom . . . . in an attempt to achieve the 24-hour, synchronous orbit, as well as all of our other satellite projects.

"I should like to add, finally, that the experience of Telstar and Relay to date have merely reinforced the opinion which I gave before this committee last year; that considerable research and development have yet to be performed before economic operational systems can be established . . . ." (Testimony)

February 19: Lt. Col. Robert E. Warren (U.S. Army), NASA Deputy Director of Communications Systems, reviewed launching and performance of SYNCOM I communications satellite before Senate Commerce Committee's Subcommittee on Communications. After SYNCOM was successfully launched by Delta vehicle, NASA Goddard Space Flight Center had determined apogee of SYNCOM's elliptical transfer orbit would occur at 21,700-mi. altitude, about 600-mi. short of synchronous altitude but well within allowable error. So that more nearly synchronous orbit could be achieved, signal was sent from USNS *Kingsport* in Lagos harbor, Nigeria, commanding satellite's apogee motor to fire 10 min. earlier (at 5:42 a.m.) than firing time set on onboard timer. SYNCOM received and stored signal, and at 5:42 a.m. signal was sent from *Kingsport* to execute the command. 20.1 sec. later, ground reception of satellite telemetry abruptly ceased. Because apogee motor was to have fired for 21.2 sec., Colonel Warren said it was "difficult to imagine that these two events are unrelated . . . .

"So, what can we make of all this? First we know that SYNCOM is in orbit. If the apogee motor did not fire, it is still in the elliptical orbit and will eventually be found, either by optics or by radar. If the apogee motor did fire, SYNCOM is in a very high altitude orbit, and without telemetry, there is only a slim chance that it will be found. A search has been organized which will continue until such time as it is found, or until further search seems unwarranted. If it is found, we will have another chance to try out the command system, and if it works, we could again be in business.

"Secondly, we know that one of the communications transponders worked well in the transfer ellipse.

"Thirdly, we have shown once more that the Delta launch vehicle is a reliable booster, this being its 15th successive flight with performance well within acceptable error." (Testimony)

- Bill to create Commission on Science and Technology (S. 816) introduced in Congress by Senator John McClellan (D.-Ark.). Commission would be charged with bringing about more economy and efficiency in scientific programs carried out and funded by Federal Government. (*Space Bus. Daily*, 2/20/63, 224)

February 20: Nike-Apache and Nike-Cajun sounding rockets were launched from NASA Wallops Station with scientific experiments designed to measure atmospheric temperatures, winds and diffusion rates at high altitudes. Nike-Apache's payload ejected sodium vapor trail from 25-mi. altitude to 100-mi. altitude, the vapor clouds visible for several hundred miles. Nike-Cajun's payload consisted of 12 special explosive charges ejected and detonated at intervals from 24- to 55-mi. altitudes. Experiments were part of continuing program by NASA Goddard Space Flight Center to study characteristics and composition of the upper atmosphere. (Wallops Release 63-17)

February 20: Nike-Cajun sounding rocket with experiments to measure winds and temperatures and to derive densities and pressures up to 90 km. was launched from Ft. Churchill, Canada. Twelve grenades ejected and exploded successfully at altitudes from 39 to 91.3 km. A simultaneous launching with Wallops Island was achieved. (NASA Rpt. of S. Rkt. Launching, 3/11/63)

- On anniversary of first U.S. manned orbital space flight, Kenneth S. Kleinknecht, Mercury Project Manager at NASA Manned Spacecraft Center, said:

  "Three U.S. Astronauts have accumulated 1,144.51 minutes in orbital space flights. The flights of Glenn, Carpenter and Schirra have confirmed that man can perform in a space environment and that he can enhance the mission success and flight safety by virtue of his flexibility and his capability to exercise judgment to solve operational problems.

  "The flights have also confirmed that the approach and philosophy followed in the design of the Mercury spacecraft was technically sound. Each of these flights has elevated our confidence level in the spacecraft systems and has demonstrated that, although not a simple task, the resources of the United States—that is, the National Aeronautics and Space Administration, the Department of Defense, private industry and other elements—can be unified toward a common end with unparalleled cooperation to accomplish a national objective on a minimum time schedule.

  "All these benefits derived from the Mercury Project are directly applicable to future manned space flight programs and provide a solid foundation on which to base the Gemini and Apollo programs." (MSC Release 63-30)

- In realignment of NASA Office of Manned Space Flight, two new deputy directors were appointed: Dr. Joseph F. Shea, Deputy Director for Systems, and George M. Low, Deputy Director for OMSF Programs. All major OMSF directorates had previously reported directly to D. Brainerd Holmes, Director, OMSF. In realigned structure, reporting to Dr. Shea would be Director of Systems Studies, Dr. William A. Lee; Director of Systems Engineering, John A. Gautrand; and Director of Integration and Checkout, James E. Sloan. Reporting to Low would be Director of Launch Vehicles, Milton Rosen; Director of Space Medicine, Dr. Charles Roadman; and Director of Spacecraft and Flight Missions, presently vacant. Director of Administration, William E. Lilly, would provide administrative support in both major areas. (NASA Release 63-32)

- On first anniversary of first U.S. manned orbital space flight, Astronaut John H. Glenn's Mercury spacecraft FRIENDSHIP 7 was presented to Smithsonian Institution by Dr. Hugh L. Dryden, NASA Deputy Administrator. Astronaut Glenn presented to the Smithsonian the flight suit, boots, and gloves he had worn and the tiny American flag he had carried during his three-orbit space flight.

Making the presentation, Dr. Dryden remarked: "It is appropriate that John Glenn's FRIENDSHIP 7 spacecraft should join the Wright Brothers' first airplane in this hall, dedicated to man's pioneering efforts in flight.

"These two machines . . . stand as constant reminders that mankind aspires to the stars . . . .

"As in the early days of aeronautics we can only dimly foresee the future of space exploration, what we shall find, and what benefits will come to us. We know only that we must move forward in knowledge and practical accomplishment as we strive to discover the nature of our universe . . . .

"We are surrounded here by the cherished treasures of the Nation. Today, we add still another to bring to mind the promise of the future that men of imagination and courage will always find new frontiers to conquer . . . ." (*Wash. Post*, 2/21/63, B1; Text)

February 20: NASA Manned Spacecraft Center announced award of formal contract to General Dynamics/Convair for four Little Joe II vehicles, two launchers, and launch support for testing boilerplate models of Apollo spacecraft in unmanned, suborbital flight from White Sands Missile Range, N.M. (MSC Release 63–31)

- NASA plans for new Electronics Research Center in the Greater Boston Area were outlined by Albert J. Kelley, Director of Electronics and Control in NASA Office of Advanced Research and Technology, in his address before Engineering Societies of New England, Boston: ". . . The principal function of this research center will consist of research and component technological development in the related fields of communications, data processing, guidance, instrumentation, control, and energy conversion. It will be staffed and equipped to perform original work and applied research as well as evaluation of industry and university efforts. It will sponsor NASA research and development efforts in these areas with industries and universities and act as a focal point for NASA research in these related fields . . . ." (Text)
- USAF announced Minuteman ICBM was flown on 3,000-mi. research and development flight from Cape Canaveral, the launch conducted by an all-USAF crew. (DOD Release 247–62)
- U.S. Army announced successful test-firing of Nike-Zeus antimissile missile at White Sands Missile Range. Colonel I. O. Drewry, Nike-Zeus Project Manager, said test objectives were met. (DOD Release 239–62)
- John Hawkinson, president of Television Shares Management Corp., said in St. Louis that electronics and aerospace industries provided more favorable investment prospects than any segment of U.S. business. (UPI, Chic. *Trib.*, 2/20/63)
- Appointment of Dr. Augustus B. Kinzel to NASA Industrial Applications Advisory Committee was announced by NASA Administrator James E. Webb. Dr. Kinzel was vice president for research at Union Carbide Corp. since 1955. (NASA Release 63–31)

February 20: Soviet press agency Tass reported in *Krasnaya Zvezda* that routine radio communications with MARS I interplanetary probe were conducted Feb. 13–19 and that telemetry information was received from the probe. Tass said that at 9:00 a.m. Moscow time MARS I was 67,432,000 km. from earth. (*Krasnaya Zvezda*, 2/20/63, 1, AFSS–T trans.)

- Deputy Commander of Soviet Strategic Rocket Troops, Col. Gen. Vladimir Fedorovich Tolobko, said in *Krasnaya Zvezda* that U.S.S.R. had perfected 100-megaton nuclear warheads for its missiles and had succeeded in creating antimissile defenses. (UPI, *Wash. Post*, 2/21/63, A21)
- Dr. Joseph Kaplan, U.S. Chairman of International Geophysical Year, declared greatest success of IGY was in preparing the way for other international research. Such studies would increasingly fall within scope of United Nations, he said. Dr. Kaplan was speaking before joint meeting of Peninsula Engineers Club and Institute of Aerospace Sciences in Newport News, Va. (Newport News *Daily Press*, 2/21/63)
- Special camera for optical tracking of satellites at Royal Radar Establishment, Malvern, England, described in London *Times*. Camera weighed 8½ tons, had focal length of 24 in. with aperture of f/1. According to Ministry of Aviation, the instrument had been working since last November and had photographed Discoverer and Anna satellites in orbit. A second identical camera would be operating at Malvern in several weeks. (London *Times*, 2/20/63, 15, and 2/22/63, 22)
- Report by Britain's Royal Society said number of British scientists holding doctorates that had emigrated to U.S. had doubled between 1952 and 1962 and the number was still rising. (UPI, *Wash. Post*, 2/21/63, A5)
- Jet Propulsion Laboratory said that Goldstone Tracking Station had made first radar contact Jan. 21 with planet Mars in a 12-hour per day experiment which would end early in March. 25-billion watt signal made 125-million mi. round trip Jan. 21 (the return signal being only one-billionth of one-trillionth of a watt) and indicated Mars has both rough and smooth surfaces. (JPL Release; Wash. *Eve. Star*, 2/22/63, A2; *Wash. Post*, 2/22/63, A10)

February 21: Secretary of the Air Force Eugene M. Zuckert testified to the House Armed Services Committee: "In recognition of the increased impact of space programs on our national resources and their direct effect on our future national security, Air Force space activities are closely coordinated with the National Aeronautics and Space Administration. One example is the recent agreement between the Department of Defense and the National Aeronautics and Space Administration on the Gemini program. The intent is to insure that the scientific and operational experiments undertaken as a part of the Gemini program will be directed towards satisfying both military and NASA requirements and objectives . . . .

"Our forward space development program integrates the R&D work of many years into a cohesive space program. Funding for all facets of the program is not included in the FY 1964 budget, but the program will provide for Air Force space activities for the near future.

"The Air Force has developed a solid foundation of space on which military capabilities can be built. These efforts have brought us to a point where significant military space possibilities are clearly apparent. Utilizing the technological base so far established, the Air Force must translate these technical capabilities into actual defense systems. The time required to move from a development state to operational systems is measured in years. Yet, it is the ready military capability, not the technological base, that accomplishes deterrence. Accordingly, the Air Force proposes to begin some of these conversions at once.

"The forward program aims at two basic objectives:

"FIRST, to enhance the general military posture of the United States by building a defense capability in space. Space is not only a new area of vulnerability to attack, but also affords important means of supporting the terrestrial forces of the U.S. in relation to a military threat posed by a hostile power.

"SECOND, to provide a capability within the space region for the purpose of denying to a hostile power the uninhibited military exploitation of space, and to provide a system of protection for U.S. scientific activities in space . . . ." (Text; *A–N–AF Journal and Register*, 3/2/63, 17, 37)

February 21: In testimony before the House Armed Services Committee, General Curtis E. LeMay testified: "The Soviets could be proceeding actively to develop space systems for military application. We believe that the Soviets will produce and deploy those military space systems which they find feasible and advantageous in comparison with other types of weapons and military equipment . . .

"U.S. military achievements in space will be expensive. Our objectives will be difficult to accomplish and the risks . . . large."

"However, as in all previous military progress, risks will diminish with experience and can be minimized in early stages by thoroughly planned decision points.

"We must not risk the danger of waiting for the enemy to demonstrate a capability before we undertake the development of our own. The visible threat to our National security requires a vigorous military space program." (Text; *A–N–AF Journal and Register*, 3/2/63, 16)

- General B. A. Schriever (AFSC), addressing Waco Chamber of Commerce, referred generally to aerospace activities in Texas and specifically referred to NASA Manned Spacecraft Center at Houston: "Although the buildings are not yet completed, the Center has been operating for some time. Many NASA personnel are temporarily working in facilities at Ellington Air Force Base, and an air base group has been established to provide support for personnel at the Center. This is just one illustration of the fact that Air Force and NASA efforts in space are cooperative, not conflicting. They are complementary not competitive . . . ." (Text, AFSC Release 32–R–19)

- Washington (D.C.) Academy of Sciences presented its 1962 scientific achievement awards to: the Rev. Francis J. Heyden, Professor of Astronomy and Director of the S. J. Observatory, Georgetown Univ., winner for science teaching because "by combining teaching and research, he leads others to join his adventures in science"; Dr. Marshall Nirenberg, National Institutes of Health, winner in biological sciences; Lindell E. Steele, Naval Research Laboratory, winner in engineering sciences; Dr. Bruce L. Reinhart, Univ. of Maryland and Research Institute for Advanced Study, Baltimore, winner in mathematics; and Dr. Edward A. Mason, Univ. of Maryland, winner in physical sciences. (Wash. *Sun. Star*, 2/17/63)

February 21: USAF OAR announced it would offer summer employment to six of the Youth Science Fair participants who are selected to receive USAF recognition at the 14th Youth Science Fair, International, which will take place in early May in Albuquerque, N.M. The six selected by OAR will have the opportunity to work as laboratory aides to USAF scientists engaged in basic research in OAR's in-house facilities in Ohio, Massachusetts, Colorado, and New Mexico. (OAR RREI-F-1)

February 22: Flight of balloon carrying 36-in. telescope to 80,000-ft. altitude for study of planet Mars was postponed at Palestine, Tex., after balloon developed a leak while being inflated. This was 13th time the launching had been postponed. (AP, *Wash. Post*, 2/23/63)

- NASA announced Albert F. Siepert, Director of Administration since formation of NASA in October 1958, had been appointed to new post of Deputy Director, NASA Launch Operations Center, effective about April 1. He would be succeeded by John D. Young, NASA Deputy Director of Administration since 1961. Don D. Cadle, Director of Resource Programming in NASA Office of Programs since 1961, was appointed Deputy Director of Administration to succeed John D. Young. (NASA Release 63-33)

- Brainerd Holmes, NASA Deputy Associate Administrator and Director of Manned Space Flight, said in address before Washington Day Banquet of Creve Coeur Club, Peoria, Ill.: "For centuries men have gazed at the moon; have written poems and composed songs about it; have proposed to their ladies beneath it; and now they wish to stand upon it. To do this we must hurl men some quarter of a million miles through space to rendezvous with the moon, and then return these men to earth safely. This surely is a clear goal. All men know what it means when we say 'leave the earth and go to the moon and return safely again to earth.' A further parameter is also clearly defined and that is *time*. Our goal is to accomplish this mission within this decade . . . ." (Text)

- Marshal Sergei S. Biryusov, Soviet Commander of the Strategic Rocket Force, said in article published for Soviet Armed Forces Day: "The problem of destruction of hostile rockets in flight was successfully solved in the Soviet Union . . . . It has now [also] become possible at a command from earth to launch rockets from satellites at any desirable time and at any point of the satellite's trajectory." (AP, Wash. *Eve. Star*, 2/22/63, A3; *Wash. Post*, 2/22/63, A16)

February 22: In ceremony on the 45th anniversary of the Soviet Armed Forces, Defense Minister Rodion Malinovsky said: "I maintain emphatically that in retaliation to the 344 missiles with which Mr. McNamara is threatening us, we shall deal a simultaneous blow of several times more missiles and such a tremendous nuclear yield that it will wipe off the earth all targets, industrial and administrative-political centers of the United States, will destroy completely the countries which have made available their territories for American war bases . . . ." Marshal Malinovsky said that Soviet rocket crews are required to show 95 per cent proficiency and that there are no Soviet planes equipped to drop bombs, only those equipped to fire rockets. Premier Khrushchev attended the Kremlin ceremony wearing the uniform of a lieutenant general. (AP, Wash. *Eve. Star*, 2/22/63, 1)

February 23: Unnamed NASA spokesman quoted as saying NASA had tried and failed to employ Col. Charles E. Yeager (USAF) and Maj. Robert M. White (USAF) as directors of astronaut training for Projects Gemini and Apollo at NASA Manned Spacecraft Center. Colonel Yeager is commandant of USAF Aerospace Research Pilot School, Edwards AFB, and Major White is USAF pilot of X-15 aircraft, also at Edwards AFB. (AP, Wash. *Eve. Star*, 2/23/63)

- Dr. Joseph F. Shea, Deputy Director for Systems, NASA Office of Manned Space Flight, said in address before Alabama Society of Professional Engineers in Huntsville: ". . . The lunar program is but one aspect of the technological surge which is invading every element of our society. To respond to the surge is to prosper, and maintain the U.S. leadership in the world." (*Huntsville Times*, 2/25/63)

- A one-segment solid-propellant rocket engine of the type intended as the first-stage booster power for the Titan III-C space launch vehicle was static-fired by United Technology Center. Under Air Force contract, the company is developing a single engine consisting of five such segments. Two of these five-segment engines would give Titan III-C a liftoff thrust of more than 9×10^5 kg (2×10^6 lb). They would be the largest and most powerful solid-propellant engines ever developed. (*SID*, 4/15/63)

February 24: Citing recent survey of 100 U.S. scientists who concluded top-priority scientific goal of U.S. space program should be "searching for and finding life on other planets," *Washington Post*'s *Parade* magazine said: "It would be interesting to note their answers were they asked what should be the nation's overall scientific goal—not merely the goal of our space program." (*Parade*, 2/24/63)

February 25: Testifying at Space Posture Hearings, before House Committee on Science and Astronautics, NASA Administrator James E. Webb revealed that NASA "did submit to the Director of the Budget and discussed with the President, the level of effort that might be obtained under four budget figures [$5.2 billion, $5.8 billion, $6.2 billion, and $6.6 billion] . . . . the figure recommended by the President, $5,712 million, was based on two policies established directly by the President after thorough discussion .... [1] the areas of space science ... must fall under the same kind of criteria as other scientific areas in his budget ... [2] with respect to the manned space flight activity, the budget level is set at the lowest level which will permit us to maintain target dates that give us a margin and still permit us to assure manned lunar exploration within this decade, barring some situation ....

". . . it is not a lavish budget by any manner of means; it is an austere budget set at this lowest level which would permit the maintenance of target dates that we believe are realistic ....

"When John Glenn was here last week and we presented his spacecraft to the Smithsonian Institution and found it there under Lindbergh's 'Spirit of St. Louis', Dr. Dryden used the phrase, 'An aura of obsolescence is hanging over the Friendship 7.'

"It is important to recognize that just a year ago this spacecraft was the most modern thing we had in the United States. Today it has an air of obsolescence and is now in the Smithsonian Institution." (*Space Posture Hearings*, 2/25/63, 3–18; *Posture of the National Space Program*, Rpt. of House Comm. on Sci. and Astron., 5/7/63, 5)

February 25: Dr. Robert C. Seamans, Jr., NASA Associate Administrator, presented before House Committee on Science and Astronautics an account of NASA's use of FY 1963 funds. Explaining $170 million decrease in launch vehicle costs because of lunar-orbit-rendezvous decision, he said NASA had deferred developmental phase of Nova very large launch vehicle and was conducting "further studies on the requirements and design concepts of a large launch vehicle consistent with optimum post-lunar-landing usage .... We are continuing [M-1] engine developments [for Nova], since it is a long lead time component, but have reduced the level of contractor effort to reflect the removal of a firm operational timing ...."

Speaking on space sciences, Dr. Seamans said: "Extensive study of the biological science area has demonstrated that specific recoverable satellite flights should be undertaken for the purpose of acquiring data on the effects of space environment on living matter. We are, therefore, undertaking such a bio-satellite program and have increased our estimated funding requirements in this area by $13.9 million in this fiscal year."

On communications satellites, Dr. Seamans said: "We re-examined our Communication Satellite program quite carefully in the light of the creation of the Communication Satellite Corporation and the reoriented activities of the DOD following the cancellation of the Advent project. From this programmatic re-examination we have concluded that principal NASA effort should be focused on the research and development problems associated with the synchronous altitude class of communication satellite. We have, therefore, dropped the low altitude mutliple passive satellite project, Rebound, and advanced intermediate active satellite projects from hardware development consideration at this time. As a result of these decisions, we reduced our communication satellite program by $35.2 million ...." (Testimony)

February 25: NASA announced signing of formal contract with Boeing Co. for development and production of Saturn V first stage, largest rocket unit under development in U.S. Contract worth $418,820,967, largest ever signed by NASA, called for design, development, and manufacture of 10 flight boosters and one ground test booster, as well as assistance to NASA Marshall Space Flight Center in portions of ground test program. Preliminary development of the vehicle stage, powered by five F-1 rocket engines, had been in progress under interim contract since December 1961. (NASA Release 63-37)

- Publication of *U.S. Standard Atmosphere, 1962* was announced by NASA. A detailed description of earth's atmosphere "intended to meet the needs of space age research and operations," report was prepared under joint sponsorship of USAF, NASA, and Weather Bureau, with 29 scientific and engineering organizations participating in its preparation. (NASA Release 63-35)
- 28-nation U.N. Committee on Peaceful Uses of Outer Space approved Indian progress report on plans to sponsor an international rocket base at Quilon for launchings in space above the equatorial regions. Italian delegate reported that the San Marco floating launching facility would be completed in time for use in the IQSY. (*L.A. Times*, 2/26/63)
- AFSC announced installation of 100-in. centrifuge at Central Inertial Guidance Test Facility, Air Force Missile Development Center (AFMDC), Holloman AFB. Intended primarily for calibration, testing, and evaluation of high-precision inertial accelerometer, centrifuge was capable of providing acceleration forces up to 25 g's, giving USAF "a laboratory capability for simulating accelerations common to actual missile flight." (AFSC Release 31-53-18)
- House Armed Services Committee voted 31-to-5 to authorize $363 million for RS-70 supersonic bombers in FY 1964. No RS-70 funds had been included in President's requested DOD budget. (UPI, *Wash. Post*, 2/26/63, A17)
- USAF announced award of formal contract to Martin-Marietta Corp. for design, development, fabrication, and delivery of Titan III Standardized Space Booster. Martin-Marietta was serving as systems integration contractor, with Aerojet General producing liquid propulsion systems and United Technology Corp. the segmented, solid-propellant engines. (DOD Release 256-63)
- An article in *Aviation Week*, by George Alexander, summarized the important advances in the Atlas test program: (1) 0.032-in. airframe, basically a stainless steel balloon relying on gas pressure for rigidity; (2) contaminant control in liquid oxygen systems; (3) early exploration of problems in staging: timing, hardware clearances, shifting center of gravity, flight control response and adaptability; and (4) handling of liquid oxygen to reduce geysering by using last-moment slug-filling to load the tanks. Of 87 research and development Atlas vehicles, 10 were classified as failures and 18 as partial successes. (*SID*, 5/20/63, 9; *Av. Wk.*, 2/25/63)

February 25: Federal Aviation Agency released its fourth annual report (July 1, 1961–June 30, 1962), in which was covered progress toward development of a commercial supersonic transport aircraft and an outline of required research and study to be done. (FAA Release 18)

February 26: Results of MARINER II flight towards and past Venus were detailed in NASA press conference. Analysis of microwave and infrared radiometer scans revealed Venus is covered by cold, dense clouds in upper atmosphere; preliminary estimate of Venutian surface temperature was around 800° F. Temperatures are essentially same on dark and sunlighted sides of Venus, with a cold spot in southern hemisphere. Any carbon dioxide in atmosphere above cloud layer was too small to be detected by MARINER II's instruments. High-density electron ionosphere was not detected. Radar observation on Venus from Goldstone Tracking Station established Venus' rotation rate as 230, plus or minus 40–50 days, retrograde, with 225 days as non-rotating status for Venus. This observation together with other ground measurements and data from MARINER II's magnetometer indicated Venus has very low, if any, magnetic field.

Analysis indicated solar plasma continuously flows out from the sun, velocity of observed solar winds varying from 200 to 500 miles per second (approximate) and having temperature of about 1,000,000° F. Density is normally about 10 to 20 protons and electrons per cubic inch. Because solar plasma "carries along with it lines of magnetic force which originate in the [sun's] corona, and the frequent variations in plasma velocity results in the deformation of these field lines, . . . it is difficult to deduce from the point-by-point measurement of magnetic field what the general undisturbed configuration of the solar field would be . . . . It is hoped that further data may reveal the presence of magnetic waves in space and clarify the interactions between the magnetic field and the plasma. . . . "

MARINER II data on ionized particles in space revealed a flux of about 100 million particles per sq. centimeter per sec. in the range from few hundred to a thousand electron volts. In range 0.5 to 10 mev for protons, "very few exist at times, but at other times their flux may be a number of times that of cosmic rays. In the range of energies 10 to 800 mev for protons, there is nearly complete absence of particles normally [with great number possible during solar flare]. Above 800 mev, galactic cosmic rays enter interplanetary space and these decrease in numbers quite rapidly as the energy decreases. . . . "

Data obtained from two-way doppler tracking of the probe during its flight would enable scientists to calculate the mass of Venus with probable error of 0.005 per cent (compared to probable error of 0.05 per cent calculated by astronomers); to determine precise locations of tracking stations on earth—specifically, to calculate exact location of Goldstone Tracking Station to within 20 yards, calculated to within 100 yards before MARINER II mission; to determine value of Astronomical Unit, established by radar observations as 92,956,220 mi.—50,000 mi. longer than that by classical astronomical observations; and to determine precisely the mass of the moon, now known to certainty of 0.1 per cent.

Dr. Eberhard Rechtin, Director of JPL's Deep Space Instrumentation Facility, reported MARINER II's flight "resulted in accumulation of some 65 million bits of information with an accuracy of at least one per cent and yet with the use of only three watts of radio frequency power." (NASA Releases 63-36-1, -2, -3, -4, -5; NASA Press Conf. Transcript)

February 26: Near-operational Army Pershing missile launched from tracked-launcher at Cape Canaveral and hit target 130 miles down AMR, its first complete test flight success. (AP, Wash. *Eve. Star*, 2/26/63, A3)

- Completion of M-2 lifting body's successful wind-tunnel testing and analysis was announced by NASA Ames Research Center. A wingless, maneuverable spacecraft capable of orbiting the earth and landing like an airplane, the M-2 lifting body was now ready for full-scale flight research. (Ames Release 63-7)
- NASA announced appointment of Capt. Robert F. Freitag (USN) as Director of Launch Vehicles in Office of Manned Space Flight, effective about April 1. Capt. Freitag, since 1959 Director of Astronautics in Navy's Bureau of Weapons and a recipient of Legion of Merit for work in ballistic missile programs, would replace Milton Rosen, who would become senior scientist on staff of NASA Deputy Associate Administrator for Defense Affairs, Adm. W. F. Boone (USN, Ret.). (NASA Release 63-38)
- Dr. John Billingham, British aerospace medicine specialist, was appointed assistant to Dr. George Smith of Environmental Physiology Branch in Crew Systems Div., NASA Manned Spacecraft Center. Appointment of Billingham was first direct foreign appointment to MSC. (MSC Release 63-35)

February 27: Leonard Jaffe, NASA Director of Communications Systems, testified before House Committee on Science and Astronautics: "During fiscal year 1963 the feasibility of using active communications satellites to relay television, telephone conversations, teletype, facsimile, and data over transoceanic distances has been demonstrated with Telstar and Relay . . . . Aside from these demonstrations, which have public appeal, and are necessary to show that our satellites perform as we say they do, hundreds of engineering tests have been performed, the results of which will enable the United States to design better satellites and ground stations for an operational, global communications satellite system. In addition, we have accumulated a considerable amount of radiation data, and have obtained better information on the damage to solid state components, such as solar cells, transistors and diodes, from which we will profit in the design of all kinds of satellites in the future . . . ." (Testimony)

- Nike-Apache sounding rocket launched at Wallops Station carried 65-lb. instrumented payload to 94-mi. altitude in experiment to obtain measurements of electron density profile, electron temperatures, and solar radiation in ionosphere. Test was part of continuing NASA program conducted by Goddard Space Flight Center to study ionosphere. (Wallops Release 63-20)

February 27: First Block II flight configuration of Saturn I space vehicle, SA-5, was static-fired for 33 sec. at Marshall Space Flight Center, developing full 1.5 million lbs. of thrust. All objectives of the firing were met. (*Marshall Star*, 3/6/63, 1)

February 28: Attempt to launch series of three Hopi-Dart wind research vehicles from NASA Wallops Station was canceled when first vehicle failed to perform properly. Tests, to gather wind-flow data at altitudes approaching 60 mi. for NASA Marshall Space Flight Center, would be rescheduled after analysis of cause of rocket's malfunction. (Wallops Release 63-21)

- Nike-Cajun sounding rocket launched from NASA Wallops Station in continuing series of upper atmosphere studies, the vehicle carrying payload that ejected and detonated 12 explosive charges at intervals from 22- to 51-mi. altitude. A similar experiment was launched from Ft. Churchill, Canada, at about the same time to obtain high altitude wind and temperature data. Both shots were considered successful. Experiments were being conducted by NASA Goddard Space Flight Center. (Wallops Release 63-22; NASA Rpt. of S. Rkt. Launching, 3/5/63 and 3/11/63)

- FAA Administrator Najeeb E. Halaby announced at news conference that Vice President Lyndon B. Johnson and Dr. Jerome B. Wiesner had joined interagency committee studying supersonic transport development. (AP, *Wash. Post*, 3/1/63, A2)

- Bell Telephone Laboratories announced TELSTAR I communications satellite disconnected its storage batteries and stopped communicating, apparently after misinterpreting a ground command. As in previous communications blackout in November, shutoff occurred as TELSTAR's changing orbit brought it into strong sections of radiation belt. This pattern led Bell engineers to "suspect that the continued inhibiting effects of radiation on transistors" is to blame, according to Bell spokesman Bruce Stasser. (UPI, *Wash. Post*, 3/21/63, A8)

- NASA Director of Meteorological Systems Morris Tepper told House Committee on Science and Astronautics that TIROS V and TIROS VI (launched in June and Sept. 1962 respectively) were still providing good data. Tiros data "continue to be used by the Weather Bureau for weather analysis and forecasting, storm tracking, hurricane reconnaissance, etc. The Weather Bureau disseminates its analyses and warnings widely . . . .

  "The Meteorological Soundings project has continued throughout the year as planned. The project at Goddard Space Flight Center which utilizes the larger meteorological sounding rockets continues as it has in past years with excellent results. In addition, we have initiated at the Langley Research Center a project which will develop and utilize the smaller meteorological sounding rockets. We expect to have this well underway by the end of this Fiscal Year . . . ." (Testimony)

- D. Brainerd Holmes, NASA Deputy Associate Administrator and Director of Manned Space Flight, said in testimony before House Committee on Science and Astronautics: ". . . The past year has been an exceedingly active one for our manned space flight program. It was not only the year for accomplishments both in

Project Mercury and in the initial Saturn test flights, but it was also a year for decisions in the rest of our program.

"The lunar orbit rendezvous technique was selected for Project Apollo, and our preliminary program plans were translated into firm and decisive schedules. Every major hardware element has been placed under contract and construction of the required facilities is well under way.

"In the current year we plan to carry forward all of the activities I have just described. By the end of this year much of the preliminary testing required in our program should have been completed and we should be well under way in the program designed to achieve not only a manned lunar landing in this decade, but also pre-eminence in manned space flight for this Nation." (Testimony)

February 28: White House announced that Leo D. Welch, retiring board chairman of Standard Oil Co. of New Jersey, and Joseph V. Charyk, Under Secretary of the Air Force, were being nominated by the President as incorporators of the Communications Satellite Corporation (csc). The csc announced that Welch had been selected to be chairman and chief executive while Charyk would become president and principal operating officer. Dr. Charyk became Chief Scientist of the usaf in 1959, was appointed in June 1959 as Ass't Sec'y of usaf for r&d, and in January 1960 as Under Secretary. (AP, *Wash. Post*, 3/1/63, A8)

• Testifying at Space Posture Hearings, before House Committee on Science and Astronautics, Dr. Lawrence L. Kavanau, Special Assistant for Space, DOD DDR&E, said:

"It is . . . important to understand that, within the Department of Defense itself, a 'space program' does not exist as a separate entity. Space and space-related projects are integrated, on a functional basis, throughout our program structure. The basic objective in any functional area is to develop and exploit those capabilities which will provide the maximum military effectiveness for the foreseeable future." (*Space Posture Hearings*, 2/28/63, 174)

• Harvard College Observatory reported that astronomers at Boyden Observatory at Bloemfontein, South Africa, had photographed syncom i satellite missing since February 14. The Observatory's photographs indicated syncom i probably was in orbit about 22,000-mi. high. (upi, *Wash. Post*, 3/1/63, A8)

• Air Force said that "an improved Thor booster combination of liquid and solid propellants," launched with unnamed satellite from Vandenberg AFB, was deliberately destroyed when it veered off course. Improved Thor had increased thrust from 170,000-lbs. to about 330,000-lbs.—almost double its earlier power and almost equal to the early Atlas. (upi, *Wash. Post*, 3/1/63, 1; Wash. *Eve. Star*, 3/1/63)

• usaf announced routine training launch of Atlas missile from Vandenberg AFB. (upi, *Wash. Post*, 3/1/63, A6)

• Senator Barry Goldwater (R.-Ariz.) introduced jointly sponsored resolution in the Senate to authorize presentation of an Air Force medal of recognition to Maj. Gen. Benjamin D. Foulois (ret.).

General Foulois learned to fly from Wilbur Wright in 1909 and was Chief of the Army Air Corps when he retired in December 1935. In 1907, he wrote a thesis on "Tactical and Strategic Value of Aerodynamic Flying Machines," using Jules Verne, the Bible, and Army Field Service Regulations as sources, one of the pioneering concepts of the military application of aeronautics. General Foulois served on the NACA (1929-30, 1932-35), and is currently President of the Air Force Historical Foundation. (*CR*, 2/28/63, 2975)

During February: NASA Goddard Space Flight Center plans for second-generation Oso satellite—known as Advanced Orbiting Solar Observatory, or Helios—were outlined at Philadelphia technical meeting by Goddard's Aoso Project Manager A. J. Cervenka. (*Av. Wk.*, 3/4/63, 48)

- USAF predicted Project West Ford's 250 million copper dipoles to be placed in orbit would have a very short orbiting life, a prediction based on results of six-needle orbital experiment conducted in 1962 and revealed Jan. 22. The six needles were not affected by space electricity, which had been feared would cause needles to shift from their initial orbit and then solar pressure would keep the dipoles in orbit for years. (*M&R*, 2/11/63, 21)
- California's Governor Pat Brown said in Washington that NASA Administrator James E. Webb was planning a "deliberate" effort to spread contracts throughout the nation which would result in reduction of contracts to California industries. Gov. Brown later told *L.A. Times* that Mr. Webb reassured him that NASA contracts would continue to be awarded competitively. Mr. Webb confirmed NASA policy "to spread the problems which we must solve in space work over the greatest number of able minds throughout the country, . . ." but that purpose of policy was not to penalize California but to "obtain the best solutions for the program . . . . Contracts will be let by competition alone, not on the basis of politics or geography." (*L.A. Times*, 2/16/63)
- National Academy of Sciences named a panel of scientists to advise the Federal Government on astronomy's role in space exploration during the coming decade. Headed by Dr. Albert E. Whitford, Director of Lick Observatory, NAS panel was soliciting comment from all of U.S. astronomers on astronomy's manpower and equipment needs. It hoped to compile full report this year and offer guidelines for the future of astronomy. (San Francisco *Chronicle*, 2/11/63)
- In addendum to speech delivered last Dec. 30, Dr. James A. Van Allen retracted his criticism of President's Scientific Advisory Committee concerning radiation effects of U.S. high-altitude nuclear test in July. Dr. Van Allen said his criticism should have been directed toward joint NASA-AEC-DOD committee headed by Dr. Jerome B. Wiesner, President's special assistant for science and technology and also chairman of President's Science Advisory Committee. Addendum appeared in published version of speech in bulletin of St. Louis Citizen's Committee for Nuclear Information. (NYTNS, *Chattanooga Times*, 2/14/63)

During February: Four men spent 14 days in slow-rotation room simulating slowly-rotating space station, in NASA-sponsored experiment at Navy's School of Aviation Medicine, Pensacola, Fla. "No serious disturbance of a psychological or physiological nature was encountered," and no change was found in conceptual reasoning, physical performance, perception, and sensory ability. No important changes in blood pressure, respiration, or pulse rate were observed during the experiment. (*Av. Wk.*, 3/4/63, 59)

- Dr. Dwain W. Warner, curator of ornithology at Minnesota Museum of Natural History, proposed in *Natural History* magazine that scientific satellites be used in tracking animal migrations. Dr. Warner proposed animals be tagged with transistor radios and batteries weighing ½ to 2 oz; satellite in 200-mi. polar orbit would pick up signals and transmit them to scientists who could plot animals' locations and correlate them with other data from satellites. Transistors had been successfully tested with small animals and large birds, but tracking by ground stations was hindered by horizon line; satellite with 24 ground receiving stations could provide constant contact with migratory creatures. (AP, Balt. *Sun*, 2/6/63)

- In March issue of *Physical Review*, John Linsley of MIT's Laboratory for Nuclear Science reported that an atom of high energy, presumably from some galaxy outside our own Milky Way, penetrated our solar system and crashed into the earth's atmosphere, setting off a cosmic-ray shower over New Mexico during February. Source of the atomic nucleus was reasoned because of its enormous power as the resultant cosmic-ray shower covered an area of several square miles. (AP, Wash. *Eve. Star*, 3/2/63)

- Recovery parachutes for Agena military payloads were tested with U-2 aircraft from 70,000-ft. altitudes. (*M&R*, 2/11/63, 9)

- Soviet astronomer G. A. Tikhov concluded that Martian vegetation, if it exists, is blue rather than green, because of light and temperature conditions on Mars. In translation/abstract of Tikhov's report, issued by Dept. of Commerce, he reported that "plants in the Far North on earth should have entirely different optical properties than in temperate latitudes and on mountains. Martian plants should be closest to the Arctic and mountain species, whose spectra have quite inconspicuous chlorophyll bands." (Wash. *Eve. Star*, 2/14/63)

- First issue of *Astronautics and Aerospace Engineering*, publication of American Institute of Aeronautics and Astronautics (AIAA), featured series of articles on manned orbital operations by NASA Manned Spacecraft Center personnel. (*A&AE*, 2/63)

- International Academy of Astronautics announced election of Honorary Members of the Academy: Prince Louis de Broglie, French physicist; Jacqueline Cochran, U.S. aviatrix; and Harry F. Guggenheim, U.S. philanthropist. (*Av. Wk.*, 2/4/63, 104)

- Soviet Prof. A. Martynov, director of State Astronomical Institute of Kiev, said Soviet Venus probe launched Feb. 1961 undoubtedly had reached its objective and probably landed on Venus. U.S. sources had said the probe was launched at too high a velocity to have been orbited around Venus. Martynov also said results of

Soviet radioastronomy experiments with Venus indicated Venutian surface temperatures range from 212°F on dark side and about 572°F to 752°F on the sunlighted side, conclusions differing from those of data provided by MARINER II. (*Av. Wk.*, 3/4/63, 21)

MARCH 1963

March 1: Delta Day at Cape Canaveral, the NASA Group Achievement Award presented to Delta Project Group of Goddard Space Flight Center, manager of Delta space vehicle for NASA.

At Delta Day ceremony, NASA Director of Space Sciences Dr. Homer E. Newell remarked that the "Delta Team has had 16 times at bat, and a hitting streak of 15 in a row, for a batting average of .937. That is good hitting . . . . We would wish that all launch vehicles could boast of records equal to that of Delta. The scientific community would, I am sure, wish me to express its appreciation with a hearty 'Well done!' "

Directing additional remarks to Delta Launch Team at Cape Canaveral, Dr. Newell said: "You have turned out the most cost-effective launch vehicle in the NASA stable. It has given us the highest return on the investment of any vehicle we have, and that is of fundamental interest to us.

"About 26% of all NASA attempts to place a satellite of any kind in orbit have been made with Delta, and 45% of our successful earth satellites placed in orbit have been put there by Delta. At this point in time, as new as we are in our space effort, that is an impressive record . . . ." (NASA Release 63–39; Text)

- At Delta Day ceremony, Cape Canaveral, NASA Associate Administrator Dr. Robert C. Seamans, Jr., announced NASA had ordered 14 more Delta space vehicles from prime contractor Douglas Aircraft Co., most of which were expected to be used in launching additional NASA communications satellites and meteorological satellites as well as new scientific satellites. (NASA Release 63–39)

- MARS I interplanetary probe was 78,843,000 km. (48,980,675 mi.) from earth, Soviet newspaper *Izvestia* reported. From Feb. 20 to March 1, scheduled radio contacts were made; "a series of control commands were transmitted to the station." Tass said that on March 1 "a number of corrective guidance commands were transmitted to the spacecraft. The strength of radio signals, which are being received from the interplanetary station, decreased somewhat . . . ." (*Izvestia*, 3/3/63, 5, AFSS–T Trans.; *Pravda*, 2/2/63, EOS Trans.)

- Stratoscope II tandem balloon system was launched from National Scientific Balloon Flight Station near Palestine, Tex., carrying 36-in. telescope to 15-mi. altitude to focus on planet Mars. Stratoscope II Project Manager, Dr. Martin Schwarzchild. called it "a beautiful launch." First attempt to detect life-related substances on Mars from above earth's atmosphere, project was conducted jointly by Princeton Univ. and Univ. of Calif. with Vitro Laboratories in charge of flight operations; it was sponsored by NASA, NSF, and ONR. (AP, *Wash. Post*, 3/2/63)

March 1: Patent for Mercury spacecraft parachute recovery system was awarded to its inventor, Andre J. Meyer, Jr., of NASA Manned Spacecraft Center. (*NYT* [West. Ed.], 3/2/63)
- Skylark rocket launched at Woomera, Australia, reached 123-mi. altitude with experiment to measure electron densities in upper atmosphere. Secondary experiments were measurement of distribution of positive ions, x-ray spectrum of the sun and distribution of atomic hydrogen in upper atmosphere. (*M&R*, 3/11/63, 9)
- NASA announced Ames Research Center had requested industry proposals for six-week study contracts to determine feasibility of using earth satellites as orbital, recoverable biological laboratories. Such a biosatellite project would provide means of studying animal and plant responses to periods of prolonged weightlessness. Studies would consider spacecraft systems that could be used in series of six biosatellites with flight durations of from three to 30 days. Decision on flight programs would be made after evaluation of study contracts; NASA already had received, and was evaluating, suggestions of biological experiments that could be performed. (NASA Release 63-46; Ames Release 63-9)
- U.S. Weather Bureau announced it was purchasing 11 ground stations capable of receiving cloud pictures directly from Nimbus meteorological satellites, to be launched by NASA beginning late this year. The stations would receive cloud pictures of their local areas taken by automatic picture transmission (APT) subsystem in the weather satellite. APT device would take and transmit cloud picture automatically every 208 sec., each photograph showing area of about 1,000-mi.-by-1,000-mi. Meteorologists expected to make immediate use of Nimbus cloud pictures. (Dept. of Commerce Release WB 63-3)
- Lt. Gen. James Ferguson (USAF DCS/R&D) told House Committee on Science and Astronautics that an independent USAF capability in space is essential to future security of the Nation and called for establishment of a "military patrol . . . to determine at all times what is happening in near space, whether there is a threat present, and to deal with it if necessary." (Wash. *Eve. Star*, 3/2/63, A2)
- U.S. Senate paid tributes to the late Senator Robert S. Kerr (D.-Okla.) who had been Chairman of the Committee on Aeronautical and Space Sciences. (*CR*, 3/1/63, 3156-79)
- Resignation of Thomas F. Dixon, NASA Deputy Associate Administrator since 1961, announced by Associate Administrator Dr. Robert C. Seamans, Jr. Mr. Dixon had joined NASA Sept. 1961, when he was named Director of Office of Launch Vehicles. Before that, he had been Vice President for Research and Engineering of North American Aviation's Rocketdyne Div. He was resigning to return to North American as assistant to the president. (NASA Release 63-42; NAA S&IS *Skywriter*, 3/8/63, 1)
- NASA Manned Spacecraft Center proposal to establish manned orbital space station which stressed that such a station and associated equipment could be established using rockets, equipment, and materials under development now for Project Apollo. Plan called for (1) orbiting space station with capacity for 18 crewmen; (2)

sending six men in adapted Apollo-type spacecraft to board the orbiting station; and (3) sending two more such crews within a month of each other to staff the station, docking the shuttle spacecraft at the station's three arms. Report said space station would be a national laboratory for study of space. (AP, Balt. *Sun*, 3/2/63)

March 1: USAF launched Atlas ICBM in 5,000-mi. flight down AMR, first of 20 flight-tests in Project Abres (advanced ballistic re-entry systems), purpose of which was to develop smaller nuclear warheads capable of eluding enemy radar and of reaching targets as swiftly as possible with largest possible nuclear yield.

Attached to Atlas airframe was pickaback pod carrying "space laboratory" in which were scientific experiments designed to gain information on performance of algae in space. (*M&R*, 3/11/63, 10; *Space Bus. Daily*, 3/5/63, 273)

March 1–2: The Space Vehicle Panel of the President's Scientific Advisory Committee met at MSC for briefings and presentations on the Mercury, Gemini, and Apollo programs. (*Space News Roundup*, 1/8/64, 2)

March 2: Inter-Parliamentary Union's seven-nation space subcommittee, meeting in Geneva, approved draft treaty on space law. Treaty would provide that outer space be subject to international law, including U.N. Charter, and that astronauts forced to land in foreign countries would be granted diplomatic immunity. Treaty would be presented to plenary meeting of IPU's 60-member nations in Lausanne, Switzerland, April 15–21. (UPI, St. Louis *Post-Dispatch*, 3/3/63)

- Stratoscope II 36-in. telescope, separated from tandem balloons by radio signal, landed 10 mi. north of Pulaski, Tenn.; balloons landed 14 mi. west of Pulaski. Project Manager Dr. Martin Schwarzschild, arriving at landing site, said he was "delighted" with condition of the instruments and termed the flight a success. (AP, Wash. *Sun. Star*, 3/3/63)

- FAA Administrator N. E. Halaby announced recommendation on whether U.S. should develop supersonic transport would be ready for President Kennedy by May. (AP, *Wash. Post*, 3/3/63)

- Secretary of the Air Force Eugene M. Zuckert said in address at Patrick AFB: ". . . about our future in space, let me establish two bench marks. The first is that the Air Force needs everything it can get from NASA. NASA needs us, too, as the record of how NASA puts things into space indicates, but if there weren't a NASA, the same facility and capability would have to be created some other way.

"There is reassuring precedent for the principle of having an outside-of-defense civilian agency provide the type of support we need. The case at point is the Atomic Energy Commission. Our own nuclear weapon flexibility as well as the Polaris-carrying submarine is sufficient testimony.

"The clear lesson for us in the space field is that we must put requirements on NASA to meet whatever part of our needs can be met in this way. We must utilize every possible resource to build the necessary military capability and I can assure you that NASA is ready to respond. Jim Webb, the NASA Administrator, harbors no illusions about NASA's responsibilities in support of national defense requirements.

"The second bench mark is that there is no such thing as peaceful space or military space. There is just space. The new and massive space program in a civilian agency was launched nearly five years ago, with—for reasons which seemed not unreasonable at the time—a great hullabaloo about peaceful objectives . . . .

"We have a lot to learn. The recent agreement with NASA for joint participation in the Gemini program is one way. That agreement represents an answer from both DOD and NASA to critics who said there was no place in space for military man. Our own activity directed toward manned space vehicles will increase, and with NASA's back-up, we'll attain the needed capability earlier than we would otherwise.

"A term you hear around Washington to denote the areas of cooperation between government agencies is 'interface.' Here at Patrick is one of the primary 'NASA—Air Force interfaces.' Such terms usually leave me pretty cold but this one does have some descriptive value.

"There will be plenty of problems between the Air Force and NASA, but not by any means all at the 'interface' points such as the Cape. We wouldn't either one be true to trust or tradition if there weren't. Any machine as big as the national space effort is bound to have some kind of friction. But just remember, a clutch is a friction interface. Its purpose is to join two shafts for the transmission of power.

"The power we can get will provide protection for the free world in space. This was my third point—to make sure that no aggressor can exploit space, either for expansionism on earth or interference in space with the peaceful pursuits of the free world. . . ." (Text)

March 2: Cosmic-ray shower caused by atomic nucleus so powerful that it must have come from another galaxy was reported in *Physical Review Letters*. Cosmic-ray shower was detected at Volcano Ranch recording station near Albuquerque by John Linsley of MIT Laboratory for Nuclear Science. (AP, Wash. *Eve. Star*, 3/2/63, A2)

- Officials of NASA, NSF, NAS, and other U.S. Government organizations denied allegations by Lord Hailsham, Britain's Minister of Science, that U.S. was conducting a high-pressure campaign to recruit British scientists. (AP, Wash. *Eve. Star*, 3/2/63)

- Boyden Observatory near Bloemfontein, South Africa, had confirmed location of SYNCOM I communications satellite, Harvard Univ. Observatory Director Donald H. Menzel announced. SYNCOM I was tumbling end over end in its orbital path about 19,000 n. mi. high. Boyden's unconfirmed photographs of the satellite, missing since Feb. 14, were reported Feb. 28, and NASA requested that the findings be confirmed by further observation. "Since then it has been cloudy over the Boyden Observatory until last night, when it cleared and we obtained two good plates showing images in the expected position. With this final confirmation, we have no doubt whatever of the location of the satellite. It behaved approximately as expected." (AP, Wash. *Sun. Star*, 3/3/63)

March 3: Whilden P. Breen, Jr., 35-year-old research assistant, spent his 106th consecutive day living in specially constructed isolation chamber in Univ. of Maryland's Space Laboratory building, in NASA-sponsored experiment to "obtain basic information on how man perform under certain conditions of stress, and methods to deal with such problems so that astronauts of the future [confined for prolonged periods within interplanetary spacecraft, orbiting space stations, or extraterrestrial bases] will be able to perform their duties at highest efficiency." With no direct view of outside world nor of any other person, volunteer Breen was under constant surveillance by experimenters outside and did communicate with them by teletype and voice. Subject's activities were programed automatically, with panel of illuminated pushbuttons identifying each activity for him. Experimenters said he had "responded well in the performance of his tasks . . . [and had] displayed heightened capability and richness in those creative activities which are available to him (e.g., writing and painting). He appears to be in good physical and mental condition and is eager to continue the study." (NASA Release 63–45; NASA Fact Sheet)

March 4: NASA's SYNCOM I Project Officer Lt. Col. Robert E. Warren (USAF), reported orbital characteristics of SYNCOM I, located by Boyden Observatory March 1, were: 19,767-n.m. apogee; 18,500-n.m. perigee; 1,424.8-min. period; 33.3° inclination; and 0.028 eccentricity. SYNCOM I was drifting eastward at 3.8° per day and was expected to disappear over horizon of USNS *Kingsport* within the next week. Attempts by USNS *Kingsport* to command the satellite to turn on its telemetry and communications equipment since March 1 had been unsuccessful. (NASA SYNCOM I Prog. Rpt. No. 4)

- U.S. plans for International Year of the Quiet Sun, IQSY), 1964–65, were announced by National Academy of Sciences-National Research Council (NAS-NRC), charged by President Kennedy in 1962 to correlate IQSY contributions of Federal agencies. A principal objective of IQSY would be to contrast data gathered during period of minimum solar activity with that gathered during IGY (1957–58), when solar activity was at highest level since beginning of systematic observations 200 years ago. Many IGY observations would be repeated and special experiments made possible by recent scientific advances would be added. IQSY would concentrate more intensively than IGY on the upper atmosphere and space phenomena directly affected by both the large periodic bursts of charged particles and associated magnetic fields escaping from the sun, and the continuous background activity known as "solar wind." (NAS-NRC Release)

- Water vapor and carbon dioxide on Mars were detected by Stratoscope II, project scientists announced at Palestine, Tex., press conference. Results of data gathered by balloon-mounted telescope, which observed Mars from vantage point above 98 per cent of earth's atmosphere March 1–2, were based on preliminary analysis of taped infrared-bolometer data; further analysis was expected to provide information on presence of other life-related compounds and on amount of the compounds present in Martian atmosphere. (AP, Chicago *Trib.*, 3/5/63; *Wall Street Journal*, 3/5/63)

March 4: Appointment of Walter L. Lingle, Jr., to newly created position of NASA Deputy Associate Administrator for Industrial Affairs was announced by NASA Associate Administrator Dr. Robert C. Seamans, Jr. Mr. Lingle would continue his present duties as Assistant Administrator for Management Development and, in new position, would assume responsibility for NASA's over-all relationships with industry and for development and review of NASA-wide procurement policies and procedures. (NASA Release 63–44)

- Hearings on NASA budget authorization for FY 1964 began before House Committee on Science and Astronautics. Administrator James E. Webb repeated President Kennedy's declaration of the national goal of making the U.S. "the world's leading spacefaring nation" and achieving for the U.S. "a position of pre-eminence" in space. Mr. Webb testified:

  "The mastery of space, and its utilization for the benefit of mankind, will not be determined by any single achievement. Superiority in the space environment with its great advantages and benefits will be won and very likely can be held by that nation which first fashions into a usable system all of the scientific knowledge, all of the technology, all of the experience, all of the space launch and terminal facilities, and all of the aids to space navigation required for safety and regular services.

  "These are the capabilities and resources which the United States must have, and this budget is designed to make rapid progress toward acquiring them.

  "Moving from our present position to achieve mastery of space requires that we add substantially to our scientific knowledge and to our utilization of technology. The NASA program is progressing on both of these fronts. In a complex effort such as this, conducted in a new medium about which much is yet unknown, the scientists and the engineer inevitably must work closely together and grow increasingly dependent upon one another. . . .

  "The NASA program, therefore, is designed to expand both science and technology. We are moving forward on a broad front. We have avoided becoming trapped in a narrow program—one limited, for example, to developing only the technology needed to reach the moon with state-of-the-art hardware. To do so might well be to find, some years hence, that we had won the battle and lost the war as far as ultimate and enduring superiority in space is concerned. . . ." (Testimony)

- Dr. Hugh L. Dryden, NASA Deputy Administrator, testifying on NASA's international programs before House Committee on Science and Astronautics, said that the "first substantial fruits of these programs were realized in 1962 and further significant programs were laid down for future years. During 1962,

  "—the first two international satellites, Ariel and Alouette, were successfully placed in orbit,

"—launchings of sounding rockets bearing scientific payloads were carried out in cooperation with eight countries,

"—37 countries engaged in special projects in support of our weather and communications satellite programs,

"—foreign participation continued to grow in the operation of our global tracking and data acquisition network overseas,

"—and, a new NASA international fellowship program was successfully established in our own universities."

Discussing studies of follow-on projects to manned lunar landing, Dr. Dryden listed as "obvious candidates . . . establishment of a station on the moon permitting prolonged occupancy, a manned laboratory orbiting the earth as a satellite, and manned reconnaissance of the planets . . . .

"It seems to us that an orbiting laboratory is a necessary preliminary to manned planetary expeditions . . . .

"Obviously the capsules presently in the program do not have sufficient space available for experimentation and have only a limited lifetime in orbit. Something larger, of longer duration in orbit, and with resupply capability is required. Many exploratory design studies have been made of the technological feasibility of assembling a large space laboratory in space from multiple launches with one or another of the available launch vehicles. We believe however that technical feasibility alone does not justify a project of this magnitude and cost. We are attempting to grasp the problem from the other end, that is to ask what one can and would do in a space laboratory in specific fields of science and technology with a view to establishing a realistic and useful concept. We hope that such studies will provide the information necessary to justify and support a decision to be made in time for the fiscal year 1965 budget. The program must be designed to fulfill national needs . . . ." (Testimony)

March 4: Republican study group of House Appropriations Committee said projects costing $10–$15 billion should be cut from FY 1964 budget appropriations request of $107.9 billion. Although they did not specify where the cuts should be made, Rep. Frank T. Bow (Ohio) suggested in press conference that cuts could be made in the space program and in foreign aid. (*Wash. Post*, 3/5/63; *Wall Street Journal*, 3/5/63)

• Seventh Semiannual Report of NASA (July 1–December 31, 1962) submitted by President Kennedy to the Congress. (*CR*, 3/4/63, 3192)

March 5: NASA announced agreement with Australian Government for establishment of deep space tracking facility about 11 mi. southwest of Canberra; a manned space flight and scientific satellite tracking station at Carnarvon; and a smaller station at Darwin. (NASA Release 63-47)

• Dr. Robert C. Seamans, Jr.,-NASA Associate Administrator, testifying before House Committee on Science and Astronautics, outlined status of current NASA activities, described proposed NASA program for FY 1964, and presented NASA budget estimates for FY 1964. (Testimony)

March 5: Construction was begun on Polaris Missile Facility-Pacific (PMF-PAC) at Bangor, Washington. (*Polaris Chronology*, 1955–63)
- NASA $600,000 grant for "research in aerospace physical sciences" at Univ. of Alabama Research Institute, Huntsville, was announced by Univ. president, Dr. Frank A. Rose. (*Huntsville Times*, 3/5/63)
- Insignia for cosmonauts was announced by Soviet Defense Ministry in *Krasnaya Zvezda*. Insignia represented golden wings, in center of which was world globe girdled by golden satellite orbit extending across background of unfurled Soviet flag. Below globe was removable shield with number 1, 2, or 3, according to class of cosmonaut—third class, for one space flight; second or first class, for two or three space flights respectively. (*Krasnaya Zvezda*, 3/5/63, 1 AFSS-T Trans.)
- Radio Budapest broadcast Hungarian news agency dispatch from Moscow which said U.S.S.R. was preparing "new and sensational steps in space this year, including [flight of] the first woman cosmonaut." (UPI, *Wash. Post*, 3/6/63)

March 6: Dr. Homer E. Newell, NASA Director of Space Sciences, testified before House Committee on Science and Astronautics' Subcommittee on Space Sciences that "Project Surveyor is designed to make a giant stride in lunar exploration technology beyond that of the Ranger. Surveyor will demonstrate soft landing technology, will survey various landing areas on the lunar surface, and will measure physical-chemical properties of the lunar surface in the immediate vicinity of the spacecraft . . . .

"Developments of the Surveyor orbiter were restricted during FY 1963 to permit reprogramming of funds into more critical areas in the Ranger and Surveyor landing programs. Design work has proceeded, however, and has resulted in . . . [a system making] extensive use of Surveyor lander components. In FY 1964 we will begin work on the flight hardware. The lunar orbiter will provide complete photo-reconnaissance coverage which will permit us to extrapolate to other areas our knowledge about local spots in which we have landed with Ranger and Surveyor. We will then be able to make selections of potential manned lunar landing sites. In addition, the orbiter will enable use to determine the gravity field of the moon which will affect the Apollo orbits . . . ." (Testimony)
- By this date, EXPLORER XV had transmitted 2,067 hours of data, of which 1,266 hours had been digitized and sent to the five experimenters. The radiation-investigation satellite had not been operating satisfactorily since Jan. 27 and had completely stopped transmitting Jan. 31, 1963. NASA scientists believed the satellite's attitude with relation to the sun was such that the sun was illuminating only the edge of the solar paddles; in such a case, the satellite should resume transmitting around April 1, when the solar paddles come into more favorable attitude to the sun. (NASA Proj. Explo. XV, Prog. Rpt. No. 2)
- House Armed Services Committee reported on its vote to add $363.7 million to DOD budget for development of two additional RS–70

supersonic aircraft with supporting weapon systems. Committee's report said the four chiefs of staff of the armed forces favored development of RS–70 at a "substantially higher" rate than General Maxwell D. Taylor, Chairman of Joint Chiefs, and Robert S. McNamara, Secretary of Defense. In his FY 1964 budget request, Secretary McNamara had included no funds for RS–70 beyond authority to complete three experimental models. (UPI, *Wash. Post*, 3/7/63)

March 6: The ships *Rose Knot* and *Coastal Sentry*, two mainstays of the world-wide Mercury tracking network operated by Goddard Space Flight Center, have undergone modifications for the MA-9 mission and are ready for sea duty again. (*Space News Roundup*, 1/8/64, 2)

- NASA announced Third National Conference on the Peaceful Uses of Space would be held May 1–9 in Chicago, sponsored by NASA and Committee for Economic and Cultural Development of Chicago (NASA Note to Editors)

- USAF OAR announced AFCRL was conducting survey of potential emergency landing areas for aircraft, using one-lb. device called "penetrometer" to measure weight the soil can bear. Penetrometer is carried in wing-tip tanks of an aircraft, fired at predetermined intervals over test areas, and photographed aerially after impact. If impact is greater than pre-set level, infrared bulb in tail of penetrometer lights up and is photographed. (USAF OAR Release 1-63-5)

- NASA was reported by *Space Business Daily* to be negotiating contract with Standard Oil Co. of Ohio for "a research investigation of a technique for abstracting information from non-cooperating objects in Earth-orbit." (*Space Bus. Daily*, 3/6/63, 280)

March 6–7: D. Brainerd Holmes, NASA Deputy Associate Administrator and Director of Manned Space Flight, testified before House Committee on Science and Astronautics' Subcommittee on Manned Space Flight:

"The major problem remaining in the F–1 engine program is the existence of the phenomenon known as combustion instability, which is characterized by pressure oscillations in the combustion gas inside the engine. Although combustion instability has developed in only seven of the 250 F–1 firings in the last two years, even this small incidence cannot be tolerated. Consequently, we are placing major emphasis on solving this problem. The source of instability in liquid propellant engines has been the subject of considerable research, since most engines exhibit instability in the early stages of their development. . . .

"We are confident that we will solve the instability problem on the F–1 engine, just as it has been solved for all other liquid propellant engines in use, and we are confident that we can achieve the flight rating and delivery schedules. We are, however, giving this matter our closest attention and bringing to bear the knowledge and judgment of the most qualified experts in the United States. . . .

In response to questions by Congressmen Emilio Q. Daddario (D.-Conn.) and James G. Fulton (R.-Pa.), Mr. Holmes said that both Projects Gemini and Apollo slipped about five months because of the Administration's refusal to request FY 1963 supplemental appropriations last fall. (Testimony; *M&R*, 3/18/63, 10)

March 7: NASA Director of Space Sciences, Dr. Homer E. Newell, testifying before Space Sciences subcommittee of House Committee on Science and Astronautics, described one laboratory-tested theory picturing the lunar surface as made of finely powdered sand forming "fairy castles" full of caverns and empty spaces, unable to support a weight. Dr. Newell said an equally plausible theory directly opposed this theory. ". . . This means we have to go there and find out." (AP, *Wash. Post*, 3/8/63)

- OSO I solar observatory satellite completed its first year in orbit, exceeding its estimated operating life by six months. Eleven of its 13 scientific experiments were still operating, having provided more data on behavior and composition of the sun than any single ground-based observatory and all previous rocket, balloon, and satellite flights combined. Preliminary results from OSO I would be presented at a symposium March 14. (Goddard Release)

- NASA Administrator James E. Webb, in address at Topeka, Kansas, traced the historic role of the Federal Government in animating science and pointed to the quickening pace of science and technology today. Major considerations, which "make it clear that an understanding of the space environment, and the development of the technology which will enable us to operate in space, are imperative," Webb said.

"First, the modern rocket engine, which can operate in the vacuum of space . . . has given us and other nations for the first time the means to explore and utilize the space medium. Given this ability, and the spectacular achievements already made, Americans and citizens of other nations assuredly will not remain confined to this small planet. We and they will explore space, and will learn to use it. Knowing this, we can settle for nothing less than a position at the forefront of that pioneering effort.

"Second, it is generally recognized, I believe, that our national security itself is heavily involved in the space competition. Not only our prestige but our capacity for constructive international leadership depend upon a superiority in science and technology—for economic development of national defense—that is understood and accepted. The nations of the world, seeking a basis for their own future progress and security, continuously pass judgment upon our ability as a nation to make decisions, to concentrate effort, to manage vast and complex technological programs in our own interest. It is not too much to say that in many ways the viability of representative government and of the free enterprise system, in a period of revolutionary changes based on science and technology, is being tested in space.

"Third, our national defense—perhaps even our national survival—demand that we act to insure that no hostile force will be permitted to use space as an unchallenged avenue of aggression against us.

"The fourth and most important of the major reasons for undertaking a broad national program of space research and development—the one which promises the greatest rewards for mankind—the technological applications which will flow from it, will offer vast returns on our space investment here on earth over many years ahead." (Text)

March 7: NASA Administrator James E. Webb addressed Kansas State Univ. in Manhattan, and said: "It might surprise you if I were to suggest that, in an important sense, Kansas had a part in the beginnings of the scientific age in America, almost 160 years ago. It was in 1804 that Lewis and Clark camped for three days at the junction of the Kansas and Missouri Rivers while en route to the Pacific Northwest on the United States' first important scientific exploration.

"Although the Congress authorized the investment of only $2,500 in that expedition, it established the precedent for a growing federal participation over the next 150 years in scientific research and exploration . . . .

"And what of today? In Fiscal Year 1964, Congress is considering the investment of almost $15 billion for scientific research and development and the tools and facilities with which to do it. More than $5 billion 700 million is being requested for the civilian space program alone—an activity so vast and so demanding of resources that it is something which only an entire nation could undertake . . . ." (Text)

- At Senate Small Business Committee hearings on NASA proposed patent policies, Univ. of Illinois economist Horace M. Gray charged NASA was fostering monopoly in the aerospace industry and called public safeguards in the proposed policies "mere propaganda designed to obfuscate the issue and conceal the real intent, which is to give away the public domain to private monopolists." (*Wash. Post*, 3/8/63)
- Discovery of radiation belt around planet Saturn, first reported in *Physical Review* Letters, was described by two physicists of the NRL research team, Joseph M. Bologna and Russell M. Sloanaker. From 170 radioastronomy measurements of Saturn's radiation made between July and October 1962, the researchers discovered that radiation from the planet was too intense to be heat emission, and "the only theoretical explanation is radiation from electrons spiraling in a magnetic field." Their research indicated Saturn's radiation belt circles the planet from north to south, theoretically not so likely as equatorial belts such as those of Earth and Jupiter, but Sloanaker said further studies were planned with 300-ft. radiotelescope at Green Bank, W. Va., to clarify uncertainties. (Wash. *Eve. Star*, 3/8/63)
- NASA-USAF Gemini Program Planning Board held its second meeting. (*M&R*, 3/18/63, 14)
- Mrs. Sara Bartholomae disclosed plans for $1 million Mercury Space Capsule Chapel near Los Angeles in tribute to Astronaut John H. Glenn, Jr. (UPI, *Wash. Post*, 3/8/63)

March 8: Photograph was sent from New York to London, Paris, and Rome via RELAY I communications satellite, and back to New York via transatlantic cable—in total of 11 minutes. This was first demonstration in which the three European capitals simultaneously received a picture and the first in which RELAY I sent a picture to Italy. Experiment was made by Associated Press in cooperation with NASA and the IT&T Federal Laboratories, Nutley, N.J. (AP, Wash. *Eve. Star*, 3/8/63)

March 8: Senate passed bill (S. 816) to establish a Commission on Science and Technology to conduct a two-year study of methods of eliminating duplication of research effort, to make results of research more readily available, and to determine whether a Cabinet office should be created to coordinate the Federal scientific effort. Commission would be composed of 12 members appointed by the President, Vice President, and Speaker of the House. Bill went to the House. (*CR*, 3/8/63, 1962-64; Wash. *Eve. Star*, 3/9/63, A5)

- Nike-Apache sounding rocket launched from Wallops Station with 95-lb. instrumented payload to measure "ion and electron density and conductivity in the ionosphere under undisturbed conditions," and to measure solar ultraviolet and x-ray fluxes. Rocket reached 96.7 mi. altitude and for the first time "both ionization in 'D' region and solar rays causing it" were observed. This was the very process that formed the ionosphere. (NASA Rpt. of Sounding Rkt. Launching; *Astro. and Aero. Eng.*, July 1963, 9)

- NASA announced selection of Northrop Corp. for negotiation of contract to support JPL in Project Ranger design review, reliability studies, documentation, systems testing, checkout, and spacecraft launch operations support. Contract would cover Ranger lunar spacecraft to be launched through 1964. Decision to select such an industrial contractor was among those measures adopted to strengthen the project as result of evaluations by special board of inquiry and JPL. NASA said it was "contemplated that Northrop later may be assigned complete spacecraft system responsibility for the Ranger program" beginning with Rangers to be launched in 1965; JPL would retain Ranger project management. (NASA Release 63-50)

- Three-stage Nike rocket launched from Eglin AFB, Fla., fourth in series of ten high-altitude probes to test methods of slowing reentry vehicles. Parachute was deployed when 500-lb. Cree payload reached peak altitude, and cameras in payload photographed action of parachute as it deployed and descended with payload. (AP, Balt. *Sun*, 3/9/63)

- NASA and DOD announced signing of agreement establishing areas of responsibility in nonmilitary applications of Navy's Transit navigational satellite system. Under agreement, NASA would determine suitability of Transit equipment for nonmilitary use and necessity for additional equipment. Navy would continue to be responsible for technical direction of Transit system and for R&D in support of military applications of the system, but would also provide NASA with necessary information, specifications, and equipment for meeting nonmilitary requirements. NASA and Navy would coordinate R&D "for mutual benefit and to avoid unnecessary duplication of effort." (DOD Release 315-63)

- DOD announced SOLAR RADIATION I satellite, orbited June 22, 1960, with TRANSIT II-A, was turned on again on Feb. 25, after 22 months of silence. (DOD Release 321-63)

March 8: U.S. Army announced Nike-Ajax missile accidentally ignited near Baumholder, Germany, and was propelled about 65 ft. before falling into safety embankment surrounding the site. (*Wash. Post*, 3/9/63)

- In address at Univ. of Wichita, Kans., NASA Administrator James E. Webb said: "Today, it is accepted that growing industries tend to concentrate in regions where research facilities are best. No part of the country can afford to neglect investments in advanced scientific and engineering education and in first class research facilities.

   "Modern industry, too, has much to gain from regional cooperation in support of the universities and associated research efforts. Industrial leaders are beginning, more and more, to look to the universities of their region for the most important resource of the age—ideas, scientific brainpower, and advanced technological skill, experience, and judgment.

   "It is not a question today of whether a region can already qualify—can now offer the human and natural resources required—for a particular industrial plant or government facility.

   "The question is whether the region is creatively doing what it can to equip its citizens to serve their area and their nation in a period when our prosperity and our very existence as a free people depend on scientific and technological leadership . . . ." (Text)
- Report that moon has porous surface consisting basically of silicon and aluminum oxides and that its core is very hot was made by team of Soviet radioastronomers at All Union Symposium of Radioastronomers in Gorki, U.S.S.R. (*Komsomolskya Pravda*, 3/8/63, EOS Trans.)
- Soviet embassy in Rio de Janeiro reported that Cosmonauts Pavel Popovich and Andrian Nikolayev were to arrive in Rio to represent Russia at the Brazilian air and space show on March 15. (AP, Wash. *Eve. Star*, 3/8/63, 1)

March 9: JPL released radioastronomy sky survey by Per Maltby, T. A. Mathews, and Alan T. Moffet, which mapped location and outlines of gas clouds 13 million to 5 billion light years from earth. Strongest of 24 radio sources having energy output equal to 20 billion suns (40 billion billion billion billion watts), were usually twin stars with immense but invisible gas clouds between them. "It seems likely," JPL said, "that the radio clouds are composed of material ejected from the parent galaxy." (*Wash. Post*, 3/10/63, 1)

- Two NASA Nike-Cajun sounding rockets with grenade experiments to measure winds and temperatures and derive density and pressures at altitude were launched almost simultaneously, one from Wallops Island and the other from Ft. Churchill. (NASA Rpts. of Sounding Rkt. Launching)
- In article on lasers in *Pravda*, V. Vyenikov wrote that "an auxiliary optical system can focus the laser beam in such a way that the diameter of its cross-section reaches values of 1 micron. This is approximately 60 times smaller than the thickness of a human hair. A similar beam, directed toward the Moon, illuminated on its surface a region less than 4 kilometers. This experiment

proves that theoretically it is possible to transmit energy from Earth to future participants of lunar expeditions." (*Krasnaya Zvezda*, 3/9/63, AFSS-T Trans.)

March 9: USAF Atlas ICBM exploded shortly after launch from Vandenberg AFB. Cause of explosion was not announced. (DOD Release 341-63; UPI, *Wash. Post*, 3/10/63)

March 10: Opinions of leading U.S. meteorologists on possible relationships of high-altitude thermonuclear explosions and the world's weather were quoted in *This Week* magazine. (*This Week*, 3/10/63, 7-9)

March 11: U.S.-U.S.S.R. negotiations began in Rome on technical details of a three-year agreement signed at Geneva in June 1962, for exchange of data to be gained from separate satellite launchings. Dr. Hugh L. Dryden, Deputy Administrator of NASA, headed U.S. scientific delegation, and Prof. Anatoli A. Blagonravov of the Soviet Academy of Sciences headed the Russian delegation. Joint space research program would include coordination on meteorology and communications studies, and charting of the earth's magnetic field. (AP, Wash. *Eve. Star*, 3/11/63, A3)

- RELAY I communications satellite was turned off because of severe drain on the onboard power supply, a difficulty similar to that encountered during first week after launch. Power drain was encountered March 9 after RELAY I's orbit had been in earth's shadow for five weeks and spacecraft temperatures were low. When satellite was exposed again to constant sunlight, a voltage regulator switch became too hot to operate properly and power drain occurred. NASA and RCA engineers were hopeful that onboard batteries could be recharged by power from solar cells during period of rest. Between January 3 and March 9, RELAY I had operated successfully in about 500 communications experiments, including first voice, facsimile, and teletype links via satellite between U.S. and South America. (NASA Release 63-54)

- 14-16 months' additional slippage of Rover nuclear rocket program due to engineering and materials problems in Kiwi nuclear reactor reported by *Missiles and Rockets*. Dr. Harold B. Finger, Manager of AEC-NASA Space Nuclear Propulsion Office (SNPO), was quoted as saying procurement difficulties were encountered because Rover lacked DX priority. No decision would be made to request a DX priority, he said, until results were obtained from the component test and evaluation program. (*M&R*, 3/11/63, 14-15; R-2, Stein and Morgan)

- NASA and French National Center for Space Studies (CNES) jointly announced signing of Memorandum of Understanding for a cooperative U.S.-France program to investigate propagation of VLF electromagnetic waves. First phase of the program would consist of two electromagnetic-field experiments with French-instrumented payloads to be launched from NASA Wallops Station. Second phase, to be implemented upon mutual consent that Phase I had proved the experiments to be scientifically and technically feasible, would consist of orbiting of scientific satellite, designed and built by France, with a Scout vehicle. (Memo of Understanding; NASA Release 63-49)

March 11: Unnamed Administration official confirmed cutback of USAF's X-20 program to R&D level of effort, *Missiles and Rockets* reported, and USAF proposed to NASA that X-20 be covered by same joint-management agreement as Project Gemini. (*M&R*, 3/11/63, 14)

- NASA announced signing of $387,900,000 contract with Grumman Aircraft Engineering Corp. for development of lunar excursion module (LEM) of the Apollo spacecraft. Contract provided that Grumman, on cost-plus-fixed-fee basis, would design, fabricate, and deliver nine LEM ground test vehicles and 11 LEM flight models, and also would provide certain operational support. Selection of Grumman for LEM contract negotiations was announced Nov. 7, 1962. (NASA Release 63-51; MSC Release 63-49)
- FAA announced award of $24,132 study contract to General Electric Co. Valley Forge Space Technology Center to "(1) analyze all data . . . on natural conditions at high altitudes that could be hazardous to aircraft, (2) consider SST [supersonic transport] design and operational factors, procedures, and equipment in the light of this data, and (3) recommend further paths of inquiry in this area." Study was part of broad SST research program being conducted by FAA, DOD, and NASA. (FAA Release #26)
- Bidders' conference for Voyager design-study contracts held at NASA Headquarters. Unmanned Venus and Mars probe, Voyager "is predicated on the fact that this country is serious in exploring planets in detail for a manned flight to the planets," NASA Deputy Director of Space Sciences Edgar Cortright was reported to have said. (*Space Bus. Daily*, 3/13/63, 309)
- Senator Clinton Anderson (D.-N.M.), Chairman of Senate Committee on Aeronautics and Space Sciences, questioned NASA FY 1964 request of $55 million for space communications research. He indicated he would oppose spending Federal funds to support research that would ultimately benefit Communications Satellite Corp. and would question the incorporators about possible overlapping of spending. (*Space Bus. Daily*, 3/12/63, 305)
- Senator Warren Magnuson, Chairman of Senate Commerce Committee, in nomination hearings on incorporators of Space Communications Corp., said that the corporation would have no bearing on decisions regarding NASA research in communications satellites, but that NASA would have responsibility to justify continued communications research as being in the National interest. (*Space Bus. Daily*, 3/12/63, 305)
- Winners of NASA Certificates of Outstanding Achievement at 12th annual Tidewater Science Congress, Norfolk, Va., were announced by Floyd L. Thompson, Director of NASA Langley Research Center. Awards were made to high school students for their exhibits in six fields of science and were in addition to regular Tidewater Science Congress prizes. Mr. Thompson said this marked beginning of annual NASA awards program in conjunction with Tidewater Science Congress and at regional fairs in four states. (Langley Release)
- At 17-nation disarmament conference stymied in Geneva, Soviet Ambassador Semyon K. Tsarapkin said: "The U.S. knows very

well that the Soviet Union can carry out nuclear explosions on the other side of the moon. The U.S. therefore proposed a ban on tests in outer space, but without any controls or inspections of rocket launchings or any other inspections whatsoever." This showed, Tsarapkin explained, that the U.S. demands for on-site inspections of earth tremors were made for purely political reasons. (AP, Wash. *Eve. Star*, 3/11/63, A4)

March 11: OAR issued *Basic Research Resumes for 1961 and 1962* which classified and indexed 2,500 USAF research projects. (OAR Release 3-63-5)

March 12: Dr. Lee A. DuBridge, president of Cal Tech, said in speech before Spring Recognition Dinner of Miracle Mile Association, Los Angeles: "The United States had put 118 space vehicles into orbit, whereas Russia has only placed 34. We are ahead of Russia on scientific data because of these space probes.

"Proof of this is in the papers published by the Soviet Union. There has been very little scientific knowledge gleaned from their reports in the past few years, whereas the United States has given a wealth of information on space problems.

"Through these probes we have been able to evaluate the temperature on the planet Venus. Of more important issue to us on Earth is the invaluable aid in predicting our own weather. . . ." (L.A. *Herald-Examiner*, 3/13/63)

- The Senate Committee on Commerce confirmed the nominated incorporators of the Communications Satellite Corp.: Edward P. Kaiser, David M. Kennedy, Sidney J. Weinberg, Bruce G. Sundlun, A. Byrne Litschgi, Beardsley Graham, Leonard Woodcock, Sam Harris, George Feldman, Leonard Marks, John T. Conner, George L. Killion, Leo D. Welch, and Joseph V. Charyk. (UPI, *Wash. Post*, 3/13/63)

- Reported that State Dept. officials were "dismayed" and surprised by reports from London that U.K. Government had agreed to establish independent space communications system with 12 British companies. U.S. was hoping its Communications Satellite Corp. could become the U.S. participant in a global communications system, a noncompetitive, nonduplicative, and cooperative system. State Dept. cabled London to request clarification as to whether U.K. was actually moving to establish a competitive—and potentially duplicating—space communications system. (Finney, *NYT* [West. Ed.], 3/13/63)

March 13: First stage of Saturn SA-5 vehicle was static-fired in 142.-sec. test at NASA Marshall Space Flight Center. Eight-engine cluster was fired at full thrust of 1.5 million lbs. (Huntsville *Times*, 3/14/63)

- USAF announced launch of space probe using Blue Scout, Jr., booster from Pt. Arguello, Calif. (*M&R*, 3/18/63, 11)

- RELAY I communications satellite, its power supply voltage and temperature returned to normal, responded to command signals turning on its telemetry transmitter and encoder. NASA planned to resume normal experimental operations with the satellite March 14. RELAY I had been turned off because of severe power drain encountered March 9. (NASA Release 63-56)

March 13: Results from ALOUETTE I topside sounder satellite were discussed at Goddard Scientific Satellite Symposium, held by NASA in Washington, D.C., Data from ALOUETTE showed that ionosphere is usually rough in high latitudes and smooth in lower latitudes, and that electron temperature of ionosphere increases with latitude. This evidence indicated Van Allen radiation belts, which extend to lower altitudes at higher latitudes, possibly are secondary heat source for ionosphere. Where ionospheric and radiation particles collide, ionospheric temperatures rise and F layers of ionosphere become more spread apart—causing radio waves to scatter. (AP, Balt. *Sun*, 3/14/63; *Av. Wk.*, 3/25/63, 26)

- At Goddard Scientific Satellite Symposium, results from ARIEL I scientific satellite were presented. Data from ARIEL confirmed ionospheric temperature relationship with latitude as detected by ALOUETTE I. Solar x-ray detectors found solar flares are made up of two phases—(1) heating of sun's corona above sunspot, increasing x-ray flux by factor of 10; and (2) quiet period marked by flux leveling off at accelerated level, followed by streams of electrons pushed into chromosphere, causing x-ray emissions at 500 times greater than normal. (*Av. Wk.*, 3/25/63, 26)
- Dr. George L. Simpson, NASA Assistant Administrator for Technology Utilization and Policy Planning, addressed the National Security Industrial Association Research and Development Symposium, March 13, 1963, Washington, D.C.

  "In speaking of spin-off and in looking for applications of NASA's research and development effort to the economy, we are becoming increasingly aware that we are talking about processes as well as products. Many of the useful advances will be made in extending present processes to new and more demanding uses, as well as in the establishment of higher levels of reliability and in the organization of systems." (*Space Quotes*, 9/15/63)
- Use of helium-neon gas laser for transmitting television pictures was demonstrated by scientists of North American Aviation's Space and Information Systems Div. Terming the development a "breakthrough," scientists said new system using one-inch-cube, four-pound modulator could send TV pictures from deep in space and its pictures could not be tapped by outsiders. (UPI, *Wash. Post*, 3/13/63)
- Reported that Commerce Dept. was seeking initial $8.7 million appropriation for Civilian Industrial Technology Program to aid lagging U.S. business. (*Wall Street Journal*, 3/13/63)

March 14: At hearings on NASA proposed patent policy before Senate Subcommittee on Monopoly, Committee on Small Business, Senator Wayne Morse (D.-Ore.) charged NASA Administrator James E. Webb with trying to give away "basic, substantive rights" of the public in Government financed research.

Mr. Webb cited Section 305 of the Space Act of 1958 authorizing the NASA Administrator to waive Government patent rights, then outlined present NASA patent policy, and explained how proposed policy would extend patent waiver. "A careful comparison of the proposed revision with the terms of the present regulations . . . will show, I believe, that the public interest in inventions which come out of NASA's research and development contracts is more clearly stated and would, in practice, be fully protected . . . .

"It is evident from NASA's present waiver regulations and the proposed revision that we are convinced that the public interest is often best served by permitting a contractor or subcontractor to retain commercial rights to inventions made in the course of doing research and development work funded by NASA—provided, of course, the Government acquires a royalty-free license for use of the invention for governmental purposes, and provided, also, that the invention is actually worked. NASA's policy is intended to encourage use of inventions in two ways: first, by bringing the stimuli of the patent system into play and, second, by withholding the full benefits of waiver until public realization of the fruits of the invention have been achieved. Such a policy offers many advantages over a restrictive policy which would permit contractors to retain patent rights only in the exceptional case. The economy should not be deprived of the substantial benefits of the patent system. Incentives for contractors to conceal and protect new technological developments as trade secrets rather than to disclose them as patentable inventions are not desirable in government contracting for research and development. Historically, patent protection has been one of the rewards for full disclosure and publication . . . ." (Testimony; *Wash. Post*, 3/15/63)

March 14: Preliminary results from OSO I solar-observatory satellite were presented at symposium by NASA in Washington, D.C. Dr. John Lindsay and William White, of NASA Goddard Space Flight Center, reported that the satellite had found tentative evidence that solar flares may be preceded by series of microflares whose sequence and pattern may be predictable. OSO I recorded at least four of these series during a year in orbit. (AP, Balt. *Sun*, 3/15/63)

• Dr. Jerome B. Wiesner, Scientific Advisory to the President, testified before Monopoly Subcommittee of Senate Small Business Committee on the need for flexibility in Government patent policy. "The Department of Defense feels strongly among other things that the performance of its mission requires the ability to assure the contractor at the time of contracting that he will retain commercial rights to his ideas, in order to assure the unrestrained participation by the most competent elements of American industry in the defense programs . . . . The central question . . . is whether the terms of the contract will encourage the contractor to apply his full technical background and experience to the Government. These considerations also apply to contracts of the National Aeronautics and Space Administration. The national interest requires greater uniformity between DOD and NASA patent practices since they are drawing on the services of the same sector of industrial research and development competence.

". . . I feel that a reasonable basis for framing a government-wide patent policy can be found. It seems important for the

Government to move in the direction of a more consistent policy that will eliminate the unhealthy confusion and instabilities that attend the present situation. Such a policy should provide for the Government to retain title in the range of circumstances that I listed earlier; but it is necessary to recognize that, because of the problems of the type that concern the Department of Defense and NASA, any policy, to be realistic, should enable industry to retain exclusive rights in certain circumstances . . . . The nature of the work involved, the commercial background of the contractor, and the extent to which the contractor would be expected to work the invention in the public interest would be significant factors in permitting contractors to retain exclusive commercial rights . . . ." (*CR*, 3/21/63, 4378)

March 14: Secretary of Defense Robert S. McNamara visited Boeing Co. in Seattle, Wash., and NASA Manned Spacecraft Center in Houston, Tex., for briefings on X-20 (Dyna Soar) and Gemini projects. In Houston, McNamara said to press: "In the last six months the Department of Defense has completed with NASA an agreement on joint planning for the NASA Gemini program. We want to see how Gemini and the X-20 can be fitted together to make the best program for both military and civilian purposes.

"The Gemini program provides for two men to be placed in orbit in 1964. The X-20 program proposes to place one man in orbit at a later date. While the X-20 would be in orbit less time than proposed in the Gemini program it would provide more recoverable payload because it would use a larger booster. The two programs are therefore not entirely comparable.

"But because they do have similarities and are both very costly, I believe this trip to Seattle and to Houston, Texas, to visit the Manned Spacecraft Center where the Gemini program of NASA is underway is extremely important.

"We will have spent about $350 million on the X-20 by the end of this fiscal year. We are requesting Congress to authorize an additional $125 million in the next fiscal year (1964) . . . ." (DOD Release 334-63; MSC *Space News Roundup*, 3/20/63, 1, 7; AP, Balt. *Sun*, 3/15/63)

* NASA and FAA announced they were completing plans for joint study of supersonic transport operation, the study providing data for both (1) design and operation of a supersonic transport compatible with an evolving air traffic control system; and (2) definition of requirements for air traffic control system to handle supersonic commercial air transports. Study would be based on four different SST concepts evolved by NASA Ames and Langley Research Centers and on simulation of air traffic environments of today and of the 1970's. The research would use flight simulation facilities of Langley Research Center and air traffic simulation facilities of FAA National Aviation Facilities Experimental Center; it was scheduled to run through 1964. (NASA Release 63-53: FAA Release #28)

March 15: Dr. James A. Van Allen said that artificial radiation belt caused by U.S. high altitude nuclear test last July may last for ten years. At NASA Goddard Space Flight Center's Scientific

Satellite Symposium, held by NASA in Washington, D.C., Dr. Van Allen said data from INJUN III and EXPLORER XIV satellites showed intensity at center of artificial belt had decreased only by a factor of two. As recently as last December, Dr. Van Allen had reiterated his view that the bulk of artificial radiation would disappear by summer 1963. However, based on newly available data, Dr. Van Allen said artificial radiation would decrease in intensity by factor of three or four until period of increased solar activity (around 1967) when decay rate should increase. He also said electrons in artificial belt have higher energies (up to a few mev) than natural space electrons, making them easier for scientists to study. (*Av. Wk.* 3/25/63, 27; *Wash. Post*, 3/16/63; AP, Balt. *Sun*, 3/16/63)

March 15: Aerobee sounding rocket launched 248-lb. instrumented payload to 123-mi. altitude in experiment to map night sky sources which emit photons (radiant energy) in specific wavelength interval. Impact occurred 62-mi. downrange from launch site at Wallops Station. Flight was joint project of NASA Goddard Space Flight Center and Lockheed Missile and Space Co. (Wallops Release 63-28)

* Researchers at the NASA Lewis Research Center ran a giant cryomagnet producing a field of 50,000 gauss over a one-foot-diameter test section. Although this was probably the largest magnetic field ever produced over such a volume, the cryomagnet would be enlarged from two to twelve coils with an ultimate design goal of 200,000 gauss. (Lewis Chronology, 2)
* Data from EXPLORER XII confirmed existence of low-energy proton current ringing earth in east-to-west direction, perpendicular to perpetual north-south spiraling motion along geomagnetic field lines. (*Av. Wk.*, 3/25/63, 26)
* Final parcel of giant 6,000-acre Plum Brook site near Sandusky, Ohio, was transferred from Army to NASA jurisdiction. NASA Lewis Research Center, Cleveland, Ohio, operates the Plum Brook Station. (Lewis Chronology, 2)
* Gov. Frank Clement of Tennessee signed bill authorizing construction of $2 million Tennessee Space Research Institute on site donated by USAF near Arnold Engineering Development Center, Tullahoma. Institute would be operated by Univ. of Tennessee. Mission of institute would be "to offer graduate level courses in aerospace science and engineering while carrying on research in these fields for both Government and industry." (*Nashville Tennessean*, 3/16/63; *Nashville Banner*, in *CR*, 3/18/63, A1479)
* International aerospace fair opened in São Paulo, Brazil, with U.S. exhibition of full-scale X-15 mockup, SIGMA 7 Mercury spacecraft, and other displays. (*Wash. Sun. Star*, 3/17/63)

March 16: Astronomers at Mt. Wilson and Palomar Observatories reported they had observed what they believed were five brightest objects in the universe. Joint research by astronomer groups at Mt. Wilson and Palomar, as well as Cal Tech's Radio Observatory in Owens Valley, Calif.; Jodrell Bank Observatory in England; and astronomers in Australia, was presented in Cal Tech report, which said knowledge gained about the five objects may make possible optical study of objects nearly 10 or 12 billion light years away—the edge of the visible universe. (UPI, *Wash. Post*, 3/17/63)

March 16: General B. A. Schriever (USAF) told Central Florida Medical Meeting, Orlando: "The successful development of space technology required major contributions from aerospace medicine. These contributions are twofold: first, an increased knowledge of the biological problems that will be encountered in space flight; and second, the development of equipment and techniques that will enable man not only to survive, but also to operate effectively in space. We have already made great progress through the Air Force high altitude research, through simulation of the space environment, and through the Mercury program. But additional knowledge and experience will be required to conduct extensive manned operations in space. . . .

"One of the major problems to be encountered in manned space flight is radiation, which may actually become a limiting factor on the duration of flights in a given orbit. This may come from several sources—from solar flares, from space-ambient radiation such as found in the Van Allen belts, or from nuclear devices that may be exploded in the earth's atmosphere. There are two approaches which are being taken to the solution of this problem, and they may be used in conjunction. The first is shielding, which is the current means of protection against radiation. But in space systems, where weight is of critical importance, conventional shielding methods are likely to be out of the question.

"The second method seeks to make use of a drug with antiradiation properties. Some 1,800 potential antiradiation drugs are currently being evaluated, and a number of them appear to hold great promise. Several interesting facts have already emerged from these studies. For example, cells with less than the normal supply of oxygen are protected from radiation damage. Stimulating the production of white blood cells and slowing the metabolism also protect the cells to some degree against radiation damage. Ideally, we would like to discover an antiradiation drug that can be administered orally without toxic side effects . . . ." (Text)

- Routine radio contact with MARS I interplanetary probe was made over distance of 98,863,000 km. (61,430,502 mi.), about 12,000,000 km. (7,364,400 mi.) farther than last U.S. radio contact with MARINER II Venus probe, according to Tass. (MARINER II's last contact was at 54.3 million mi.) (*Pravda*, 3/17/63, 1, AFSS-T Trans.)
- Howard M. Weiss, NASA Director of Quality Assurance and Reliability, said total cost of the two Mariner shots, the second of which brought U.S. several firsts in knowledge of planet Venus, was $47 million. Weiss was addressing American Society of Quality Control, Southern Connecticut Section, in Bridgeport. (Bridgeport *Post*, 3/18/63)

March 17: Fifth anniversary of orbiting of VANGUARD I, still orbiting the earth and still transmitting data. The satellite had made more than 19,700 orbits and had slowed about ¼ sec. from original 134 min., 27 sec. period. Present apogee: 2,447 mi.; perigee: 403 mi. Scientists estimated VANGUARD I might have a lifetime of 2,000 years. (DOD Release 365–63)

March 17: NASA announced it would sponsor study to determine interest in system of "practical" satellites to collect data from remote areas of the earth. If warranted, study would be followed by a feasibility and design study. A possible use of data collection satellite was in oceanography, using instrumented floating buoys distributed over ocean surface to measure characteristics of the sea. Satellite would interrogate each buoy, store information, then transmit it upon command. Similar application was tracking of icebergs, using transponders planted on ice formations. Study would be under cognizance of NASA Future Applications Satellites unit, which was also investigating non-military uses of navigational satellites. (NASA Release 63–55)

• Senator Clifford P. Case (R.-N.J.) questioned whether a planned space electronics research center should be located in Boston or in fact, whether one was needed at all. He said New Jersey has outstanding qualifications for such a center and it and other states should be considered a possible location—if the center is actually needed. NASA has already selected the Boston area as the site of the center. (Boston *Herald*, 3/18/63)

March 18: Dr. Wernher von Braun, Director of NASA Marshall Space Flight Center, testified before House Committee on Science and Astronautics' Subcommittee on Manned Space Flight: "The F–1 engine development program has an impressive list of accomplishments. A year after the contract was signed full-scale components were undergoing tests, and in 27 months complete engine systems testing had begun. Full thrust and full duration tests have become routine. The engine has been gimbaled during hot firing. We expect successful completion of the Preliminary Flight Rating Tests this year . . . .

"When design of the F–1 was begun in 1959, combustion instability was known to be a potential problem. Accordingly, early in the program tests of various injectors for the thrust chamber were conducted, and a design was selected which had not experienced unstable combustion. For about a year, from mid-1961 to mid-1962, engine tests were conducted without this phenomenon occurring. Then, on June 28 last year we were testing development engine #008 on the test stands in California. A test run which was scheduled for the full 2½ minutes running time was interrupted after 106 seconds of satisfactory performance by a rupture of a valve casting. The rupture was traced to combustion instability. Since this occurrence, several cases of combustion instability have taken place.

"This phenomenon is not unique to the F–1, but has occurred in the development of most liquid rocket engines . . . .

"We are presently concentrating on combustion instability and are making progress. With the contractor at Rocketdyne, we have made an exhaustive survey of all test data having a bearing on the problem . . . .

"To summarize, the F-1 engine development over the last four years has made satisfactory progress . . . ." (Testimony)

March 18: Frederick R. Kappel, AT&T board chairman, said in speech before New York Chamber of Commerce that TELSTAR II communications satellite would be equipped with instruments "far more resistant to radiation" than TELSTAR I. Like its predecessor, TELSTAR II would be launched by NASA at AT&T expense. (NYT [West. Ed.], 3/19/63)

- USAF announced launch of unidentified satellite into polar orbit from Vandenberg AFB using improved Thor-Agena booster (TAT). This was second attempt and first successful launching of the improved Thor, thrust of which was uprated from 170,000 lbs. to about 330,000 lbs. by addition of three solid-propellant rocket motors to the liquid-fueled power plant. (DOD Release 374–63)
- NASA was faced with "a sizable budget cut—up to half a billion dollars—unless a new Soviet space spectacular changes the attitude of an economy-minded Congress," according to *Aviation Week and Space Technology*. (Av. Wk., 3/18/63, 30)
- U.N. Committee on Peaceful Uses of Outer Space, in last meeting of current session, agreed to have its legal subcommittee meet at U.N. Headquarters, New York, and scientific and technical subcommittee in Geneva. Agreement ended three-week deadlock between U.S. and Soviet delegations on a meeting place. (L.A. Times, Wash. Post, 3/19/63)
- DOD announced Army Corps of Engineers, New England Div., would provide design and construction assistance to NASA on proposed Electronics Research Center in Boston area. (DOD Release 366–63)
- Four-man space laboratory described in unsolicited proposal to NASA by Douglas Aircraft Co.'s Missile and Space Systems Div. was reported in *Missiles and Rockets*. Space laboratory, which Douglas said could be operational in two years, could be fabricated mostly with existing hardware and proven equipment, would use only one new structure—the command-control module. (M&R, 3/18/63, 10)

March 18–20: American Institute of Aeronautics and Astronautics (AIAA) Space Flight Testing Conference held in Cocoa Beach, Fla. Rocco A. Petrone, Chief of Heavy Space Vehicle Systems Office, NASA Launch Operations Center, presented paper on facilities required in launching Saturn V space vehicles in Project Apollo. (*Saturn V/Apollo Launch Operations Plan*.)

March 18–21: 200 scientists from 60 countries met in Rome to plan International Quiet Sun Year (IQSY) studies in 1964–65. Congress was under auspices of International Geophysical Committee of the International Council of Scientific Unions. (NYT [West. Ed.], 3/19/63)

March 19: NASA Goddard Space Flight Center, in cooperation with NBC and RCA, accomplished first known transmission of television in color via RELAY I communications satellite. 15-min. sequence of movie *Kidnapped* was relayed by RELAY from 4,000-mi. high orbit, and was scheduled to be shown on Walt Disney's program on March 24. (AP, Wash. Post, 3/22/63, A5)

March 19: Ernest W. Brackett, Director of NASA Procurement and Supply, said in testimony before Subcommittee on Space Sciences and Advanced Research and Technology, House Committee on Science and Astronautics: " . . . NASA is trying a new contract system which we hope will accomplish the incentive objectives. The base fee in the contract will be lower than the usual fee. Criteria will be set up in the contract for evaluation of the contractor's performance. These criteria may include reduction and control of costs, excellence of performance, delivery on schedule, and other points. Periodically a NASA board, also specified in the contract, will evaluate the contractor's performance and may award some higher fee, up to a set maximum, if the facts justify an increase . . . ." (Testimony)

- Dr. Robert C. Seamans, testifying before House Committee on Science and Astronautics' Subcommittee on Applications and Tracking and Data Acquisition, outlined NASA personnel requirements for FY 1964. Dr. Seamans said some 73,000 scientists and engineers would be required to carry out the U.S. space program, and 11,300 of these would be employed by NASA. (UPI, *NYT* [West. Ed.], 3/20/63)

- NASA Administrator James E. Webb and Council of New York City President Paul R. Screvane conferred in Washington on plans for NASA exhibit at 1964 World's Fair, possibly to be a proposed 361-ft. inflatable Saturn V rocket replica to house 10,000 sq. ft. of exhibits. (NASA Release 63-58)

- Ivan A. Getting of Aerospace Corp., in address to space flight testing conference at Cocoa Beach, Fla., estimated that 90 percent of all current in-flight rocket failures could be detected and corrected by better preflight checking, including "preliminary designs, design reviews, re-design, testing, additional redesign and additional testing. The probability of a successful performance of such a product is directly related to the thoroughness of this evolutionary process . . .

  "There is an unfortunate tendency in our rush to meet end objectives and time schedules, and in our desire to save money within the program, to cut out important steps in developmental tests and ground systems tests. Not only are such economies false, but they impinge on our reputation as a nation for excellence in scientific achievement and engineering design." (AP, *Wash. Post,* 3/20/63, A2)

- Senate Committee on Aeronautical and Space Sciences ordered favorably reported the 14 nominations of incorporators of Space Communications Corp. On the Senate floor, Senator Albert Gore (D.-Tenn.) requested the Senate withhold action on confirmation of the nominations "for a reasonable time," until he could prepare "the case, which I think is a very substantial one, against the propriety, advisability, and constitutionality of the Senate allowing itself to be entrapped into the operations of a purely private corporation organized for profit . . . . " (NASA Leg. Act. Rpt. II/43; *CR,* 3/19/63, 4307; *Wash. Post,* 3/20/63)

- Dr. Abe Silverstein, Director, NASA Lewis Research Center, expressed appreciation to John W. Macy, Jr., Chairman of the

U.S. Civil Service Commission, for commission's contributions to the Lewis recruiting program. During Fiscal Years 1962 and 1963, Lewis hired 1,280 engineers and scientists from all over the U.S. (LRC Release 63-9; Lewis Chronology, 2)

March 19: Joseph G. Gavin, Jr., a vice president of Grumman Aircraft, replying to question during testimony before Manned Space Flight Subcommittee of House Committee on Science and Astronautics, said first U.S. astronauts returning from the moon would bring back about 100-(earth)-pound sample of lunar terrain. (*L.A. Times*, 3/20/63)

- Atlas booster for Project Mercury manned flight MA-9, scheduled for mid-May, arrived at Cape Canaveral from General Dynamics/Astronautics, San Diego. (AP, *Wash. Post*, 3/20/63)
- Astroscience Center, facility devoted to current and long-range space programs, opened in Chicago by Armour Research Foundation of Illinois Institute of Technology. Headed by Dr. Leonard Reiffel, Director of Physics Research at the Foundation, new center would coordinate research projects for NASA and enlarge space research activities of the Foundation. (*NYT* [West. Ed.], 3/21/63)

March 20: Joint U.S.-U.S.S.R. press conference in Rome, Dr. Hugh L. Dryden, chief U.S. negotiator, and Anatoli A. Blagonravov of the Soviet Academy of Sciences, announced signing of an accord for a joint weather satellite program and called for broader American-Russian cooperation in space activities. Agreements were reached under space cooperation accord signed by the U.S. and the U.S.S.R. last June, and after ten days of closed-door negotiations by teams of specialists. Agreement would become effective after a 60-day waiting period pending possible changes by either Government.

Under the joint weather satellite program, each nation would launch a satellite on a different orbit to collect meteorological data, while a 24-hr. cable and radio communications network would be established for exchange of information and cloud-cover pictures, such data to be made public throughout the world. Launching of Echo II later in 1963 would provide basis for communications experiments which could lead to a global communications network, including the exchange of radio and TV programs.

Both NASA Deputy Administrator Dryden and Blagonravov expressed hope that the accord just negotiated would lead to broader agreements. Blagonravov was quoted as saying that "the first step is always the hardest." Dryden was quoted regarding the Warsaw meetings later this year on probes of Mars and Venus as saying: "I hope there will be future coordination to avoid duplication." (Text; AP, *Wash. Post*, 3/21/63, 1)

- Dr. Abe Silverstein, Director of Lewis Research Center, said in testimony before Subcommittee on Space Sciences and Advanced Research and Technology of the House Committee on Science and Astronautics that Centaur liquid-hydrogen vehicle could lift 250 additional pounds of payload because of design improvements in the vehicle. Dr. Silverstein described design changes and testing program undertaken by Lewis, and said that the second and third of the planned 15 Centaur vehicles were scheduled to be launched in last half of 1963. (*M&R*, 3/25/63, 17)

March 20: John A. Hornbeck, president of Bellcomm, Inc., testified before House Committee on Science and Astronautics' Subcommittee on Manned Space Flight that safety was paramount factor in deciding on lunar landing site for Project Apollo. "Preliminary studies of this kind suggest that the characteristics of a good site for early exploration might be on a lunar sea, 10 miles from a continent and 10 miles from a post-marial crater." (UPI, *Wash. Post*, 3/21/63, A2)

- NASA Manned Spacecraft Center announced it had initiated a comparison study with Hughes Aircraft Company to determine feasibility of optical laser communications system in deep space, comparing the laser to microwave. Study, which would "outline the design of an ultimate deep space communication system, will investigate three cases: beaming the laser between the spacecraft and the earth; between the spacecraft and an earth-orbiting satellite, then relaying to earth by microwave; and beaming the laser between a spacecraft and a lunar-based laser station, then relaying the signal to earth by microwave." Study project was part of broad investigation by MSC of laser applications. (MSC Release 63–52)
- Reported that Lear Siegler, Inc., delivered to NASA the slow-scan television camera systems and receiving equipment intended for use on the forthcoming flight of Astronaut Gordon Cooper. (*Space News Roundup*, 1/8/64, 2)
- *Komsomolskaya Pravda* reported sport aircraft Yak–18P piloted by Boris Porfirov was flown to record altitude of 7,358 meters, new All-Union record for aircraft of third weight category. Former All-Union record, set 1954, was 6,311 meters. (*Komsomolskaya Pravda*, 3/20/63, 3, AFSS–T Trans.)

March 21: Soviet news agency Tass announced launching of COSMOS XIII (apogee: 209 mi.; perigee: 127 mi.; inclination to equator: 64° 58'; period: 89.77 min.). Tass stated that all systems were functioning and that onboard equipment was "intended for continuing outer space research in conformity with the program announced . . . March 16, 1962," but gave no payload weight or specific mission details. In an interview in Guinea, Cosmonauts Andrian Nikolayev and Pavel Popovich said that the Soviets plan more tandem manned orbital flights. In response to questions, they said that women sometimes have "physical superiority over men." COSMOS XIII was launched after 89 days of no announced Soviet space flight activity, longest period of space inactivity to date. (UPI, *Wash. Post*, 3/23/63, A8; *Av. Wk*, 3/25/63, 26; Tass, *Krasnaya Zvezda*, 3/22/63, 1, AFSS–T Trans.)

- USAF announced successful flight-test of Titan II from Cape Canaveral, the missile propelling a payload "twice as heavy as any other in this country's arsenal more than 6,500 miles down the Atlantic Missile Range." This was eighth successful flight test in 12 attempts. (DOD Release 390–63)
- In regular press conference, President Kennedy responded to a question about the pace of the U.S. space program relative to the Soviet

Union's: "The U.S. is making, as you know, a major effort in space and will continue to do so. We are expending an enormous sum of money to make sure that the Soviet Union does not dominate space. We will continue to do it. And we will continue to take whatever steps are necessary to prevent any action against the United States.

"The fact of the matter is the Soviet today with a nuclear weapon can reach the United States with a missile. So I would have to know in more precise detail than you [the questioner] have described the exact nature of our threat before I suggested what our counteraction would be." (Transcript in *Wash. Post*, 3/22/63, A16)

March 21: Dr. Robert C. Seamans, Jr., NASA Associate Administrator, testified on the NASA-DOD Gemini Program Planning Board before Subcommittee on Manned Space Flight, House Committee on Science and Astronautics. "The NASA-DOD Agreement on Gemini was put into effect by Secretary McNamara and Administrator Webb on January 21, 1963. This agreement represents another important example of the manner in which both the Department of Defense and the National Aeronautics and Space Administration are striving to maximize the technological productivity of national space flight programs such as the Gemini. I think it is significant that in announcing the agreement, Mr. McNamara and Mr. Webb emphasized the national character and importance of the Gemini project and reiterated their intentions to insure that Gemini is utilized in the national interest and to avoid unnecessary duplication of effort in this area as in all others . . . .

"I think it is important to point out here that increased DOD participation in the Gemini does not alter the peaceful and experimental character of the program. We are driving forward to advance the technology of manned space flight, including rendezvous and designated docking and to study and understand man's effectiveness under prolonged conditions of weightlessness. In carrying out this program, NASA continues to be solely responsible for the management of the Gemini project. . . ." (Testimony)

- Soviet scientists lost contact with MARS I interplanetary probe when it was about 66,000,000 mi. from earth; however, no mention was made of this fact by the Russians until May 16 (see May 16). (Balt. *Sun*, 5/17/63; *Av. Wk.*, 5/27/63, 24)
- NASA announced establishment of Office of Assistant Administrator for Technology Utilization and Policy Planning, with Dr. George L. Simpson, NASA Assistant Administrator for Public Affairs, filling the post. In new position, Dr. Simpson would continue to be responsible for NASA public affairs duties and would further be responsible for Office of Technology Utilization, formerly Industrial Applications Division, Office of Applications. Realignment was expected to "result in closer coordination of . . . industrial applications with [information] dissemination activities. . . ." Dr. Simpson's deputy, Julian Scheer, would become Deputy Assistant Administrator for Technology Utilization and Policy Planning. Louis B. Fong, head of Industrial Applications Div., would become Director of Technology Utilization. (NASA Release 63-60)

March 21: Dr. Leland J. Haworth of the Atomic Energy Commission was named Director of the National Science Foundation by President Kennedy. (AP, Wash. *Eve. Star*, 3/22/63, A15)

- DOD announced "clearest data yet obtained of the ionosphere and the dynamics of its formation and dissipation" had been obtained by AFCRL scientists, who matched experimental data from positive ion detector aboard a Thor rocket with theoretical data based on laboratory studies, within error of only two per cent. (DOD Release 388–63)

- Article on "The Test Ban" by Stefan T. Possony of the Hoover Institute, Stanford University, printed in the *Congressional Record*, said:

  ". . . Currently there is no feasible method of verifying [nuclear] explosions in space. If the Soviets were to launch a space vehicle and explode a nuclear device, at not too great a distance, the United States might gain a good notion of what was going on. But we would not possess the type of evidence which 'would stand up in court,' assuming that we could make public use of the evidence we have. Hence, we would be reluctant to move, just as during the Cuban crisis, Washington did not move before there was clear photographic evidence of the presence of Soviet missiles . . . .

  "Few people realize that though we are entrusting our security to ICBM's, we never have tested a full assembly of such a missile, i.e., we have not launched an ICBM together with its warhead, nor destroyed a target with an ICBM-launched warhead . . . .

  "Testing also has a bearing on space propulsion. It is generally agreed that the Orion project—which is designed to use small nuclear 'bombs' to propel a space vehicle—would allow the lifting of maximal payloads into orbit and permit the deepest penetration into the solar system. Of all propulsion systems it would provide for the most effective utilization of space. However, so far it seems impractical, because of radioactivity, to launch an Orion assembly from the ground; unfortunately, by using Orion propulsion only from orbit, much of the system's utility would be lost. An all-fusion technology would eliminate the radioactivity and thereby open the road to getting really important payloads into space.

  "Should the Soviets decide to move into space as a decisive military medium, they may employ Orion techniques even before they are able to eliminate radioactivity . . . . The brutal twin facts which we cannot circumvent are that space utilization requires maximum energy releases and that nuclear explosions are the method through which maximum energy releases can be produced— produced most cheaply . . . .

  "Plowshare explosions may be required for the building of stations on the moon . . . ." (*CR*, 3/21/63, 4365, 4367, 4368)

March 22: NASA announced RELAY I communciations satellite had achieved all of its experiments and missions. Performance of RELAY I, launched by NASA December 13, "has been uniformly excellent" throughout about 500 communications tests and demonstrations in 660 orbits December 13–March 11. Although all planned demonstrations were completed, they would be continued while the satellite remained in operation. (NASA Release 63-57)

March 22: At GSFC Colloquium, John A. O'Keefe discussed the origin and evolution of the moon, submitting his theory that billions of years ago the moon separated from the still "undifferentiated earth," thereafter was subjected to volcanic eruptions, meteorite bombardment, and ultimate cooling and transformed into a hard cinder-like material. The volcanic dust produced the comparatively smooth lunar maria. If theory is correct, O'Keefe said, the original dust has long since become firm and constitutes "no hazard" for landings of space vehicles. O'Keefe supported his theory with available evidence on tektites. (Text; *Wash. Post,* 3/23/63, 1)

- Sixth Annual Robert H. Goddard Memorial Dinner, sponsored by the National Rocket Club, Washington, D.C. In address, Vice President Lyndon B. Johnson paid tribute to the "father of modern rocketry," and said that those today who "understand the stakes of space" must help "the public to understand these stakes." He urged that communications barriers among scientists, engineers, and politicians be abolished so that public support for public policy can be obtained. "Unless and until this is done," said the Chairman of the National Aeronautics and Space Council, "the technological community cannot justifiably be impatient with those who are chosen to represent and express the public's own will."

  In addition to honoring Mrs. Goddard, the National Rocket Club presented its annual awards: the Robert H. Goddard Memorial Trophy, its premiere award, to John H. Glenn, Jr., in recognition of the first U.S. manned orbital space flight; the Nelson P. Jackson Aerospace Award to the American Telephone and Telegraph Company for development and operation of TELSTAR I; the National Rocket Club Award to the Radio-Television Industry (ABC, CBS, MBS, and NBC) for "pioneering private enterprise efforts" under the U.S. open information coverage of the space program; the Astronautics Engineer Award to Jack Norval James of Jet Propulsion Laboratory, Technical Director of MARINER II. The First Robert H. Goddard Historical Essay Award was given to R. Cargill Hall of San Jose, Calif., for his prize essay on "World-Circling Spaceships—Satellite Studies in the U.S. During the 1940's"; Honorable Mentions going to Wilfrid J. Mayo-Wells of Washington, D.C. ("Origins of Space Telemetry") and Robert D. Roach of Buffalo, N.Y. ("The First Man-Rocket Belt"). The Robert H. Goddard Scholarship was awarded to Miss Marcia S. Miner, physics major at the American University. (Program)

- DOD announced USAF was developing space structures of honeycomb material that could be compactly packaged, then expanded in space. Shelter of such material could be packaged approximately a foot square, then expanded in space to airtight structure seven ft. high and seven ft. in diameter. Geophysics Co. of America was working under contract to build the structure, developing methods of fabricating and expanding it and making the structure rigid. Other predicted uses of the honeycomb material were in solar energy collectors and re-entry vehicles. (DOD Release 405-63)

March 22: Department of Defense announced the appointment of Maj. General Leighton I. Davis (USAF), AFMTC Commander, as the DOD representative for Project Gemini support operations. (*AFSC Operational Highlights,* 11)

- USAF announced Atlas F ICBM launched from Vandenberg AFB, its re-entry vehicle impacting on predetermined area more than 4,000 miles down the Pacific Missile Range. (DOD Release 399-63)
- West German Government announced that it disapproved of its citizens assisting the United Arab Republic on military rocket projects, and said that there was no proof that any West Germans were so engaged. The Israeli Parliament had appealed to Chancellor Adenauer on March 20 to take action against German scientists assisting the U.A.R. in developing military potential. (AP, Wash. *Eve. Star*, 3/22/63, A2)

March 23: In address dedicating O'Hare International Airport, in Chicago, President Kennedy said that the U.S. "intends to be a leader in the supersonic age." Foreseeing a smaller world when the mach three airliners are developed, the President said: "Let us also see to it that it is a better world. A more just and peaceful world for our children and for their children after them."

In downtown civic address, President Kennedy spoke about automation in his remarks on labor and unemployment: ". . . underlying all of these trends is a third phenomenon, both cursed and praised; and that is technological advance, known loosely by the name of 'automation.' During the last six years, this Nation increased its manufacturing output by nearly 20 per cent—but it did so with 800,000 fewer production workers; and the gain in white collar jobs did not offset this loss. Since the second world war, the real output of the private economy has risen 67 per cent, with only a 3 per cent raise in man-hours . . . . Increasing productivity and advancing technology are essential to our ability to compete and to progress—but we also have an obligation to create an additional 2 million jobs each year for those displaced by these gains. . . ." (AP, Wash. *Eve. Star*, 3/23/63, 1, A5)

- NASA and Public Health Service announced award of contract to Massachusetts Institute of Technology for study in area of computer technology as related to biomedical sciences. Under the contract, MIT established a Center Development Office to (1) coordinate planning activities for a Regional Resource Center and (2) evaluate the potential of specially designed computer for health-related sciences. (HEW Release W93)
- USAF announced Atlas ICBM launched from Vandenberg AFB exploded shortly after liftoff. Cause of explosion was not determined. (DOD Release 414-63)
- More than 400 German scientists and technicians were reported to be working for Egyptian government on first Egyptian-made rocket and nuclear warhead. First group of German atomic scientists arrived in Egypt in 1954, Israeli sources said. (London Sunday

Times, in *Wash. Post*, 3/24/63, A17; London *Observer* News Service, in *Wash. Post*, 3/24/63, A17)

March 25: Three major U.S. television networks each broadcast seven-minute programs from Paris to New York via RELAY I communications satellite. (AP, Wash. *Eve. Star*, 3/25/63)

- Large flaming meteor visible over eastern U.S. from New York to North Carolina about 10 p.m. EST. (AP, *Wash. Post*, 3/26/63)
- NASA Administrator Webb wrote letters to the respective chairmen of the House and Senate space committees indicating that a mid-city location in Boston on land now owned by the Massachusetts Turnpike Authority might provide an alternate location for the proposed NASA Electronics Research Center. He pointed out that if authorization for the Center and its location in the Boston area were approved, NASA was going to not only carefully consider the original possibility of obtaining 1,000 acres of land as near as possible to Harvard and MIT, but also the alternate location, utilizing air-rights with multi-story structures on a smaller area. (Texts, NASA Release 62–63)
- Dr. Lyman J. Briggs (1874–1963), Director Emeritus of the Bureau of Standards and a former member of the NACA (1933–45), died in Washington, D.C. He began his government career in 1896 as a physicist in the Department of Agriculture, originating the centrifuge method of classifying soils, organizing the biophysical lab in 1906. He joined the Bureau of Standards in 1917 and became its Director in 1933; his early research resulted in improvement in the accuracy of large naval guns, the invention with Dr. Paul R. Heyl of the earth inductor compass (used by Lindbergh), and air viscosity experiments leading to improved propeller designs. NACA Report No. 207 by Briggs, G. F. Hall, and Hugh L. Dryden (on "Aerodynamic Characteristics of Airfoils at High Speeds,") issued in 1924 was considered a major contribution on airfoils at near supersonic speeds. In 1939, the White House named Dr. Briggs as chairman of the original Uranium Committee to study the military use of atomic energy, which led to the Manhattan Project and Hiroshima. For his work on the proximity fuse during World War II, he was awarded the Medal of Merit by President Truman. Dr. Briggs retired in 1945, but continued research including confirmation in 1959 that pitched baseballs could curve as much as 17 inches in 60 feet. (Wash. *Eve. Star*, 3/2/63, A5)

March 26: Vice President Lyndon B. Johnson reviewed his first 26 months in office on ABC television interview. Referring to his participation in major national decisions, he referred to pushing ahead with the large rocket booster program and the accelerated space program decisions in which he participated as Chairman of the National Aeronautics and Space Council. (*Wash. Post*, 3/27/63; Wash. *Eve. Star*, 3/27/63)

- USAF Project Stargazer balloon flight from Alamogordo, N.M., canceled because of high surface winds. Balloon was equipped with small aluminum gondola for pilot, Capt. Joseph Kittinger, Jr. (USAF), and civilian astronomer, William C. White, for a 24-hour observation with 12.5-inch gyroscopic telescope above 90 per cent of the earth's atmosphere. (UPI, Wash. *Daily News*, 3/26/63)

March 26: NASA Manned Spacecraft Center announced beginning of NASA Apollo environmental control system tests in Garrett-AiResearch Los Angeles. ECS would provide life supporting atmosphere in command module of Apollo spacecraft. Tests would include simulating prelaunch, ascent, orbital, and re-entry pressure conditions on the system. (MSC Release 63-61)

- Dr. S. Fred Singer, Director of National Satellite Weather Center, told House Committee on Science and Astronautics' Subcommittee on Applications and Tracking and Data Acquisition that reports from Tiros weather satellites were being used by Soviet scientists in their weather research. Launching of a weather satellite "is probably an immediate Soviet objective." (NASA Leg. Act. Rpt. II/46, 3/28/63; Wash. *Eve. Star*, 3/27/63)
- DOD announced Dr. Theodore von Kármán had been selected to receive the second annual Thomas D. White National Defense Award, established in 1962 by Air Force Association to be conferred on U.S. citizen for outstanding contribution to national defense. (DOD Release 420-63)
- Award of $358,076,923 contract by USAF for design and manufacture and testing of X-20 (Dyna Soar) manned space glider to the Boeing Co. was announced by Senators Warren G. Magnuson and Henry M. Jackson (D.-Wash.). Contract covered airdrop tests of the X-20 from B-52 aircraft and an orbital flight from Cape Canaveral expected in 1965. (AP, *NYT* [West. Ed.], 3/27/63)
- Britain's Postmaster General, John Bevins, dismissed as unrealistic the idea of his country's going it alone in putting communications satellites into orbit. He did not say which course the Government would choose, except that he ruled out an all-British system. A satellite system, he said, cannot be established unless other countries are willing to put ground stations on their own territories and unless they are willing to put traffic over the system. (*NYT*, Western Edition, 3/27/63)
- Ernie Smith, who made first civilian airplane flight to Hawaii from Oakland, Calif., on July 15, 1927, died in San Francisco. 2,393-mile flight took 25-hrs. 26-min. (AP, *Wash. Post*, 3/27/63)

March 27: NASA launched an Aerobee 150 sounding rocket to an estimated peak altitude of 185 mi. from White Sands, New Mexico. Objective of the experiment was to measure the absolute alpha radiation resonantly scattered by atmospheric atomic hydrogen at night. Instrumentation included eight ion-chamber nodules and two aspect photometers. (NASA Rpt. of S. Rkt. Launching, 4/4/63)

- NASA Administrator James E. Webb, in address before the Los Angeles Council on World Affairs, pointed out the unique role of California in the national space program: "During Fiscal Year 1962 and the first six months of FY 1963, 47 per cent of NASA's prime contracts over $25,000 were awarded to firms in California. If the second ranking state is excluded in each of these years, California industries received more NASA prime contract dollars than all of the other 48 states combined.

"Unless the nation decides to forego the opportunity it now has to insure a pre-eminent position in space, I foresee no possibility, certainly not in the near future, that the volume of space activity conducted in California will decline . . . It does seem likely that as our national investment in space research and development increases, California will find it more difficult to maintain its present percentage of the total . . . ."

NASA's procurement, Mr. Webb said: "In short, we are making a deliberate effort to use the self-policing forces of the market place, to avoid building government competition with industry, and also to maintain sufficient managerial and technical capability in our own organization to make certain that our contractors are giving us the reliability we must have and the taxpayers a dollar's-worth of work for every dollar we spend."

In reviewing the international aspects of the NASA program, the NASA Administrator cited the provisions of the National Aeronautics and Space Act of 1958, and said: "Some citizens, who regard military necessities as the prime justification for our space effort, find it difficult to accept the open, unclassified character of our civilian space program. Some regard basic scientific results as a national treasure, which should be kept under lock and key, and ask 'Why do we give away our secrets to the Russians?'

"The answer is that our program involves both science and technology. There are no real secrets of nature, and it would be self-defeating to try to classify the science and principles of nature. On the other hand, technological advances and inventions are the product of men and effort, and can properly be classified and kept from potential enemies where the national interest so requires.

"Accordingly, our national space program moves forward with the basic scientific research conducted by the civilian agency, available to the whole scientific community, but technological development conducted by both NASA and the Department of Defense which is shared by both agencies, and classified when defense and military interests are involved. In this manner we have protected our national security while making the maximum contribution to the advancement of science . . . ." (Text)

March 27: In a letter to Rep. George P. Miller (D.-Calif.), Dr. Jerome B. Wiesner discussed the problems of scientific manpower.

"The manyfold demands for technical manpower draw on the same pool of talent—for defense and space, for innovation of civilian products able to compete in a world market, for meeting such domestic problems as mass transportation and environmental pollution associated with a rapidly urbanized society, for medical advancement, and for the effective management of our resources in such a way that we can help upgrade the way of life of citizens throughout the world without downgrading our own. Superposed on these requirements is the need for science, mathematics and engineering teachers to meet the rapidly growing college enrollment. A sharp competition for manpower has resulted, particularly for the superior individual having advanced training.

In that competition, not all sectors of demand have succeeded in meeting their needs, and the future outlook is not encouraging.

"For the first time, we are confronted with a plurality of goals that may well outstrip our technical manpower resources.

"Faced with the same manpower problem as the Soviet Union, how, in a democratic society where each citizen must have a free choice of his or her occupation, do we meet this challenge?

"The urgent need for more individuals having advanced training is clear . . . . To keep pace with our national needs, it is the judgment of the President's Science Advisory Committee that we should and could increase the number of Ph.D.'s each year in engineering, mathematics, and physical sciences (EMP) to reach 7,500 in 1970, the number of master's or equivalent to 30,000. This annual production of 37,500 advanced degrees by 1970 should be compared with 15,000 in 1960, 11,000 in 1950.

"A substantial fraction of these new graduates, incidentally, must be plowed back into the universities for teaching, and soon . . . . With regard to students, the major barrier to continuation of full-time graduate study is financial . . . . Another barrier is inadequate graduate facilities . . . . In addition, there is very convincing evidence that some geographical areas produce a much greater proportion of doctorates than others . . . . Relatively few students are prepared to study more than several hundred miles from their homes. It would thus appear that we have a special problem in terms of meeting an unequal geographical distribution of centers of educational excellence.

"The present barriers to graduate education—limitations in student support, number of faculty, and educational facilities—must be overcome. It is to that end that the President has proposed that the Federal Government take the initiative and provide the leadership needed to meet these national needs.

"To meet goals for graduate education in the three fields of potential shortage—engineering, mathematics, and physical sciences—it is of critical importance that we increase the number of graduate students enrolling in the fall of 1964 from the 22,000 which may be expected on the basis of projected trends to some 30,000 . . . . Support would be provided within existing legislation, through NSF, AEC, and NASA fellowships, and through new NSF science development and training grant programs . . . . In virtually all of these cases, assistance to students would be accompanied by a grant directly to the college or university to help meet the true cost of education, and thus would assist the university to maintain an internal balance between fields . . . . The President also proposes Federal support for new physical facilities . . . .

"The proposed program is designed to achieve a substantial and unprecedented acceleration in the rate of production in advanced degrees in these three fields." (NASA Leg. Act. Rpt. II/86, 6/10/63)

March 27: Dr. Raymond L. Bisplinghoff, Director of NASA Office of Advanced Research and Technology, told House Committee on Science and Astronautics' Subcommittee on Space Sciences: "The essential purpose of . . . [NASA's advanced research and technology program] is to lay the foundation and to determine the building blocks which will be needed for pre-eminence in aeronautics and space in the near future . . . .

"I think that the history of technology over the past 100 years has shown that true progress in any branch of technology such as space technology requires and even demands a sharp cutting edge of advanced research. The kind of things we read about in the papers like the manned spaceflight effort, scientific probes, communications, and meteorological satellites and airliners we ride on today all rest upon a foundation of research which was done in previous decades; and the research we do today in our laboratories and universities, in industry derives vigor and freshness from the motivations of these projects and things that we are doing to benefit mankind.

"At the same time, there is really a feeding in both directions. The research derives freshness from the things that are going on in the way of projects and the projects later derive benefits from the research . . . ." (Transcript)

March 27: D. Brainerd Holmes, NASA Deputy Associate Administrator and Director of Manned Space Flight, told Manned Space Flight Subcommittee of House Committee on Science and Astronautics that current cost estimate of Project Apollo was $20 billion or less. (Transcript)

- Contract for $2,143,900 awarded to Kaiser Engineers to design a 100- by 120-ft. space environment chamber to simulate pressure and temperature conditions existing at altitudes up to 100 miles. (LRC Release 63-11; Lewis Chronology, 2)

March 28–30: First national Space, Science, and Urban Life Conference at Dunsmuir House, Oakland, Calif. 150 U.S. leaders—prominent Government figures, distinguished aerospace industrialists, eminent scholars, and leading research scientists gathered to discuss political, sociological, physical, and economic effect of space scientific R&D on the average American and his community. (NASA Release 63-62)

March 28: NASA launched Saturn SA-4 in successful flight test from Cape Canaveral, the 1.3-million-lb.-thrust rocket reaching 77-mi. altitude in last of four successful tests of first stage alone. After 100 sec. of flight, one of the booster's clustered eight engines was cut off by pre-set timer, and that engine's propellants were re-routed to the other seven, which continued to burn. Experiment confirmed engine-out capability designed into the Saturn I vehicle system by NASA Marshall Space Flight Center engineers. (NASA Release 63-61)

- Launch of NASA atmospheric structure satellite (S-6) from AMR was postponed because of difficulties in payload checkout. (AP, *Wash. Post*, 1/29/63, 1)

- NASA launched a two-stage Nike-Apache from Wallops Island, Va., with an experiment designed to measure nighttime air density and composition at altitudes of about 60 to 120 mi. The 70-lb. payload reached an apogee of approximately 118 mi.

and impact occurred in the Atlantic Ocean 114 mi. from the launch site. (NASA Release 63-30)

March 28: In speech to the North Carolina Citizens Association at Raleigh, GSFC Director Harry J. Goett reviewed the scientific exploration of space, its meaning, and future. He concluded: "From a scientific point of view it may matter little how quickly we come to understand sun-earth relationships, whether Nimbus [weather satellite] gets launched this year or next, or whether we land a man on the moon during this decade or not. But there is no doubt that our national prestige demands that the United States lead the world in the hard-won knowledge and technology in the important areas of geophysics, solar physics and interplanetary science that have been opened by the advent of the space age. We must continue an aggressive program to assure that the United States retains its leadership. . . .

"Our future in the exploration of space does not hold any simple panaceas, scientifically, technologically, and financially. But neither did the development of the United States come easy nor has the future ever been an easy one for a free people seeking truth and knowledge, and the preservation of freedom." (Text)

- Nike-Apache sounding rocket launched from NASA Wallops Station carried 65-lb. instrumented payload to altitude of 100 mi., in experiment to measure electron density profile, electron temperatures, and solar radiation in the ionosphere. Secondary objective of the flight was to check out hardware to be flown from Ft. Churchill, Can., during solar eclipse in July. (Wallops Release 63-31)
- Dr. Homer E. Newell, NASA Director of Space Sciences, addressed Space, Science, and Urban Life Conference in Oakland, California: "The budget of $3.7 billion for the current fiscal year and the requested $5.7 billion for the next fiscal year are sizable sums of money, approaching one per cent of the gross national product in magnitude. But, perhaps even more important than the absolute magnitude of these sums, is the leverage they exert or can be made to exert on other more general activities and associated resources. Although the effect of the space program on the total national scene may be quite modest, in certain localized regions it can be quite large. For example, . . . it appears at the present time that in Fiscal Year 1963 about 47 per cent of all NASA prime contracts will go to California firms . . . ." (Text)
- Defense Department announced deployment of first of three Fleet Ballistic Missile (FBM) submarines in the Mediterranean and that these three would come from those operating from Holy Loch, Scotland. (*Polaris Chronology*, 1955-63)
- Promotion of Marshal Sergei S. Biryuzov, commander of Soviet rocket forces, to Chief of the General Staff was reported in Soviet military newspaper *Krasnaya Zvezda*. (AP, *Wash. Post*, 3/29/63)

March 29: In an address to the Press and Union League Club of San Francisco, NASA Administrator James E. Webb said that the two basic policy questions repeatedly raised concerning the U.S. space program were:

(1) "Is exploration of the moon an objective of sufficient value to our drive for space leadership to warrant a total investment of as much as $20 billion in this decade, and in the next fiscal year at a rate that requires two-thirds of our [NASA] space budget?"

(2) "Is the pace at which we are proceeding and the rate of expenditure which is required to explore the moon by 1970, necessary to insure eventual pre-eminence in space, and reasonable in the light of our other pressing national needs?"

In speaking to these questions, Mr. Webb traced the evolution of the U.S. space program and the determination of the national objectives in space after the strategic surprise of SPUTNIK I. He discussed the central considerations of world leadership in science and technology, and the role of space exploration and exploitation in national affairs. He reviewed progress to date and the urgent requirement for full public understanding of the meaning and purpose of space exploration. (Text)

March 29: British Minister of Aviation Julian Amery announced U.K. had ordered a design study to determine suitable communications satellite system, the satellites to be placed in orbit by ELDO vehicle (European Launcher Development Organization). To be undertaken by Royal Aircraft Establishment at Farnborough and Signals Research Development Establishment at Christchurch, studies would consider: network of medium-altitude-orbit satellites, probably at least 12; system of stationary-orbit satellites, calling for much smaller number of satellites; use of Woomera rocket range, Australia, for launchings. U.K. was hopeful that such a system could be established in cooperation with European countries but independent of U.S. "No one country should have a monopoly of space," Amery said. He also announced that contracts to industry for a third Ariel satellite, to be launched by U.S. rockets but to be built entirely by U.K., would be awarded soon. (*Wash. Post*, 3/30/63; *Av. Wk.*, 4/8/63, 38)

• DOD announced study of emergency re-entry from manned space flight was being conducted by Armour Research Foundation under direction of AFSC Flight Dynamics Laboratory. Armour study would be based on theoretical re-entry vehicle with crew of three or less, performing in low earth orbits, re-entries, and short duration missions. Armour would provide AFSC Flight Dynamics Laboratory with design criteria and drawings for controllable emergency escape systems enabling the vehicle to land in relatively small area. (DOD Release 441–63)

• Congressman Charles A. Halleck (R.-Ind.) publicly released letter from former President Eisenhower urging that "the new spending programs in the pending budget . . . be drastically reduced . . . .

"The space program, in my opinion, is downright spongy. This is an area where we particularly need to demonstrate some common sense. Specifically, I have never believed that a spectacular dash to the moon, vastly deepening our debt, is worth the added tax burden it will eventually impose upon our citizens.

This result should be achieved as a natural outgrowth of demonstrably valuable space operations. But having made this into a crash program, we are unavoidably wasting enormous sums. I suggest that our enthusiasm here be tempered in the interest of fiscal soundness . . . ." (*CR*, 4/2/63, 5172)

March 30: In final meeting of the Space, Science, and Urban Life Conference at Dunsmuir House, Oakland, Calif., NASA Administrator James E. Webb said: " . . . we must all concern ourselves with the progress and the promise of scientific research and technological development in this country, and the effects of this activity upon society. My own work on projects concerned with civic and urban problems, has left an indelible impression of their complexity and urgency . . . .

"Today we are dealing not only with profound changes in the social and economic structure of the country and the world; we are dealing, as well, with an even more profound change in man's own conception of the boundaries and limitations of his habitable environment and his understanding of the forces of the universe . . . Man is no longer rooted to *terra firma*, either in the literal sense, or in his understanding of the powerful forces of the universe of which he is a minuscule part. Increasingly, he comprehends that the new understanding and knowledge which are being unleashed by science and technology will alter his existence in more ways, and more rapidly, than he can possibly foresee . . . .

"Man has gained mobility, his horizons have broadened not only in the geographical sense, but because he no longer feels bound by a family farm, or a traditional village industry . . . . A civilization which can move with serious purpose to gear itself for travel to the moon and the planets will not be content with old and outmoded socio-economic concepts . . . .

"Every thread in the fabric of our economic, social and political institutions is being tested as we move into space. Our economic and political relations with other nations are being reevaluated. Old concepts of defense and military tactics are being challenged and revised. Jealously guarded traditions in our educational institutions are being tested, altered, and even discarded. Our economic institutions—the corporate structure itself—are undergoing reexamination as society seeks to adjust itself to the inevitability of change."

Mr. Webb stated that two thoughts were of critical importance: (1) "The need to create favorable conditions compatible with the American private enterprise system through which the knowledge which we gain and the skills which we develop in our huge program of federally sponsored research and development . . . are identified and made available for use in the industrial stream to contribute to economic growth and our nation's power to survive in a world where second best is more severely penalized than ever before"; (2) "The necessity for developing viable relationships between science, education, industry, and government . . . ." (Text)

March 30: USAF announced 15 names of fourth class of Aerospace Research Pilots School at Edwards AFB, Calif. (*Wash. Post,* 3/31/63, 1; Wash. *Sun. Star,* 3/31/63, 1)

- Soviet news agency Tass reported astronomer Pyoter Shcheglov had discovered a "flat cloud of hydrogen shaped like a pancake" circling the earth at an altitude of 6,200 mi. Similar to a ring around planet Saturn, the cloud would be visible to spacemen on the moon with special viewing instruments. (UPI, *Atlanta Journal and Constitution,* 3/31/63)
- By this date, 53 communications companies had filed with Federal Communications Commission for authorization to purchase common carrier stock in Communications Satellite Corp. (*Av. Wk.,* 4/1/63, 30)

March 31: *Washington Post* reported that the Federal Communications Commission had proposed a compromise between astronomers and commercial television interests on channel 37 of the radio spectrum (608 to 614 megacycles). Astronomers wanted channel 37 set aside for science; TV interests wanted it for commerce. FCC proposed, article said, that 600-mile circle around Danville, Ill., would reserve channel 37 for the University of Illinois radiotelescope until January 1, 1968. Any channel 37 assigned elsewhere, such as in Paterson, N.J., would be off the air between midnight and 7 a.m. local time. (*Wash. Post,* 3/31/63, A2)

- Reported that Boeing Co., North American Aviation, and Lockheed Missiles and Space Co. had won study contracts on the proposed supersonic transport airplane. The (Seattle) *Post-Intelligencer* quoted Senator Warren G. Magnuson (D.-Wash.) as saying he and Senator Henry M. Jackson (D.-Wash.) had received confirmation of the awards from FAA Administrator Najeeb Halaby. (AP, *Wash. Post,* 4/1/63)

During March: Comet Ikeye, discovered by Japanese amateur astronomer Kaoru Ikeye, was most brightly visible in western horizon for about a week before its brightness began to fade as it drew closer to the sun. (Wash. *Daily News,* 3/12/63)

- Testimony of Defense Secretary Robert S. McNamara before House Committee on Armed Services Feb. 2 was released, in which Secretary McNamara said: "A substantial amount of funds [for FY 1964] is requested for DYNA-SOAR. I mention this only to say . . . that technology as it applies to space advances so rapidly that we must expect many changes in programs before they are completed. I personally believe that rather substantial changes lie ahead of us in this DYNA-SOAR.

  "I say this, in part, because of the GEMINI development. GEMINI is a satellite project carried on up to the present time by NASA, on which has been spent to date about $300 million, and for which they will request $300 million in 1964, toward a total program cost of $800 million; it will provide a capsule capable of carrying two men into earth orbit . . . .

  "GEMINI is a competitive development with DYNA-SOAR in the sense that each of them are [sic] designed to provide low earth orbit manned flight with controlled re-entry. DYNA-SOAR does it one way, and with flexibility, and GEMINI another . . . .

"[The NASA-DOD Project Gemini joint planning committee] . . . will insure that the military requirement for near-earth orbit is properly taken account of in the GEMINI project.

"We don't have any clear military requirement, or any known military requirement, per se. But I think we do have a requirement for environmental testing and experimentation in near-earth orbit.

"We are very much interested, therefore, in the Gemini project. When we become more familiar with it and understand its potential I suspect it will have a great influence on the future of DYNA-SOAR . . . .

"I guess that we will find that GEMINI has a greater military potential for us, even though a rather ill-defined military potential, than does DYNA-SOAR, and, moreover, that it will be available much sooner than DYNA-SOAR . . . .

"I think the DYNA-SOAR project can work out satisfactorily. The real question is, what do we have when we finish it. It will cost to complete, in total, including funds spent to date, something on the order of $800 million to a billion dollars. The question is, do we meet a rather ill-defined military requirement better by proceeding down that track, or do we meet it better by modifying GEMINI in some joint project with NASA . . . ." (Hearings on H.R. 2440, pp. 465–67)

During March: Space Orientation Center, designed for briefing Marshall Space Flight Center visitors and new employees and operated by the Historical Office, was opened at MSFC. (Marshall Historian; Huntsville *Times*, 3/6/63)

- Entire issue of *Astronautics and Aerospace Engineering* devoted to launch operations, including detailed summary articles by Dr. Kurt Debus, Director of LOC, on "Launching the Moon Rocket," Georg F. von Tiesenhausen on "Toward the Orbital Launch Facility," and Arnold W. Frutkin on "International Cooperation in Space Research."
- USN Transit navigational satellite program was classified, with no further information to be released until fully operational system is placed in orbit, *Astronautics and Aerospace Engineering* reported. (*A&AE*, 5/63, 141)
- Army Transportation Corps loaned 20 sea-going barges to NASA for use in space vehicle development program. Barges would be used initially for transporting supplies from Marshall Space Flight Center to Mississippi Test Operations; later, they would be fitted with liquid hydrogen and liquid oxygen tanks to store the propellants for fueling Saturn S–IC static-test boosters. (*Marshall Star*, 3/6/63, 1)
- Industry proposals to define advanced nuclear space propulsion systems for 1970's were submitted to NASA Marshall Space Flight Center. The mission-oriented studies would analyze post-Nerva developmental engine systems for maximum operational use and flexibility in U.S. space program. (*Av. Wk.*, 3/18/63, 53)
- Canadian government authorized four additional satellites for ionospheric research in joint U.S.-Canadian space program. (*M&R*, 3/11/63, 9)

During March: Mrs. George C. Marshall, widow of the soldier and statesman for whom NASA Marshall Space Flight Center was named, donated to the center the Executive Order naming NASA's Huntsville, Ala., facility and the fountain pen used by President Dwight D. Eisenhower for signing it. (*Chattanooga Times*, 3/10/63)

- A group of British scientists, belonging to the British Astronomical Association, was engaged in the accurate measurement of lunar craters, utilizing the Mercury computer at Manchester. By the end of this month, the group had measured the depths of more than 200 lunar craters and was in the process of studying data on numerous others. (*Space Intelligence Notes*, Oct., 1963)
- Study by Comptroller General Joseph Campbell reported that lack of "timely guidance" by DOD was largely responsible for failure of nuclear-powered aircraft project. Conceived in 1946 and declared technically feasible in 1951, project was canceled by President Kennedy in 1961 after DOD and AEC had spent more than $1 billion on the project, which did not reach the prototype stage. GAO's report said that DOD "agrees that the program suffered considerably from lack of prompt decisions and from frequent changes in emphasis and goals." (*Wash. Post*, 3/9/63; *Av. Wk.*, 3/18/63, 25)
- Dr. G. A. Tokaty, former chief of the aerodynamics laboratory of the Zhukovsky Academy and later chief rocket scientist of the Soviet Government in Germany (1946–47), published an article on "Soviet Space Technology" in *Spaceflight*, magazine of the British Interplanetary Society. Article submitted that the "Soviet Government have manifested their keen interest in the [space] field throughout the entire history of their existence [since 1917]" and reviewed this history in interesting detail to 1961. (*Spaceflight*, 3/63, 59–64)
- The development and operation of the first plastic laser was reported by the Radio Corp. of America. The experimental device is made with a fiber of transparent material of the same type used in making many familiar clear plastics. Employing a newly discovered physical effect, it produces coherent pulses of intense crimson light at the highest visible frequency yet known to have been attained by a laser. The unique laser mechanism that is employed may permit the development of plastic lasers that emit coherent light over the visible spectrum from infrared through ultraviolet. The way may now be opened for the use of still other materials that resist laser action when contained in inorganic crystals. Some of these materials may be capable of producing coherent light at yellow, green, blue, and other frequencies. (Data Supplied by Radio Corp. of America; MSFC *Space Information Digest*, 4/15/63)
- Radio and optical astronomers of the Mount Wilson and Palomar observatories announced the discovery of what probably are the brightest objects in the universe observed so far. Their brilliance, they believe, may make it possible to penetrate two to three times deeper into space than was considered possible with present-day telescopes. Five of these super-bright celestial

bodies have so far been recognized. Radio astronomers detected them by their radiations of energy in the longer wavelengths, at radio frequencies. They were identified in the third Cambridge (England) Catalogue of Radio Sources as 3C–48, 3C–196, 3C–286, 3C–147, and 3C–273. The objects are so bright that they had been thought to be nearby stars in our own Milky Way Galaxy, but there is now growing evidence that they are very distant galaxies involved in titanic explosions. The study of such distant objects is vital to cosmologists in determining the kind of universe we live in—whether it will continue expanding, stop expanding, or will eventually contract. (*NYT*, Western Edition, 3/26/63, Wm. L. Laurence)

During March: NASA scientists discovered a new "hot spot in the sky" over Wallops Island, Va., where in winter the temperatures 34 to 60 mi. above the earth reach more than 75° F and winds blow at speeds approaching 400 mph. Currently there is no existing scientific theory to account for the super-high-altitude, jet-stream-like wind and hot air that seems to blow over Washington, Baltimore, and Annapolis during the winter. Details of the discovery were distributed this month in a NASA technical note entitled "Preliminary Measurements of Temperatures and Winds Above 50 KM (31 mi.) over Wallops Island." Authors of the pamphlet are Dr. William Nordberg and Wendell Smith, of the Goddard Space Flight Center. (Balt. *Sun*, 3/31/63)

APRIL 1963

April 1: Third anniversary of orbiting of TIROS I, first U.S. meteorological satellite, which provided during its 78-day lifetime more than 19,000 usable cloud-cover photographs. TIROS I was followed by five additional successful Tiros satellites, the last two of which—TIROS V and VI—were still in operation after more than nine and six months respectively. To date, TIROS V had transmitted more than 44,000 useful cloud-system photographs and TIROS VI, more than 39,000. TIROS V and VI photographed hurricanes and typhoons in July, August, and September 1962; in August alone, they discovered two Atlantic hurricanes and four Pacific typhoons. Since April 1, 1962, there were only seven days when a Tiros satellite was not in operation. Information gathered by satellites was being used in daily operations of Weather Bureau's National Meteorological Center in weather analyses and forecasts for entire U.S. (NASA Release 63–65; Commerce Dept. Release WB 63–4)

- USAF launched unidentified satellite with Thor-Agena D vehicle. (*Pres. Rpt. on Space, 1963*, 1/27/64)
- Dr. Robert C. Seamans, Jr., NASA Associate Administrator, testified on NASA facility planning before House Committee on Science and Astronautics' Subcommittee on Space Sciences and Advanced Research and Technology. "There are several basic agency guidelines under which the Program Directors establish their facility requirements. In addition to money constraints, always present, every effort is made to meet requirements through modifications to existing facilities and the use of government owned facilities and sites in lieu of new construction and land purchase. As a result, facilities and sites which otherwise might be obsolete or unused become effective tools for the conduct of our program. I think NASA has an unusually fine record in this respect . . . ." (Testimony)
- In address to the annual meeting of the NEA's elementary school principals at Oklahoma City, NASA Administrator James E. Webb stated that every segment of society needs to reexamine its activities in the light of changes brought about by science and technology.

"In the field of education it is evident that a first and obvious demand of this age is an increasing supply of scientific and technical manpower, including scientists, mathematicians, and engineers with advanced degrees . . . .

"A second obvious demand on education is the development of a citizenry with a higher level of scientific literacy, not merely to motivate more young people toward careers in science, but to assure our nation of a body politic better qualified to understand

and cope with the problems of government and society during a period of great and rapid change.

"And, finally, achievement of our goals in space will demand the highest scholastic efforts and intellectual accomplishments in virtually every field of study.

"Space is, indeed, a new and challenging frontier, but it is a frontier of the intellect—one which challenges brain, not brawn, with creative intelligence our greatest weapon . . . ." (Text)

April 1: Selection of Lockheed California Co. and Boeing Co./North American Aviation, Inc., as contractors to conduct airframe research in supersonic transport program was announced by FAA Administrator Najeeb E. Halaby. Lockheed would perform research in three study areas under $950,000 contract; Boeing/North American, operating as a joint venture, would perform research in six areas under $850,000 contract. Under the cost-sharing agreement, government was providing $1.8 million and industries $1.6 million toward total program costs. (FAA Release 63-35)

- Norman V. Petersen, USAF FTC Technical Director, told House Committee on Science and Astronautics' Subcommittee on Manned Space Flight that if Soviets emphasize "single lunar mission, using earth-orbital rendezvous, they would have the ability to accomplish this [manned lunar landing] at the present time." Petersen estimated Soviet technique would involve launching 10-15 rockets to place lunar vehicle components in earth orbit, where they would be assembled for flight to moon. (UPI, *Wash. Post*, 4/2/63)

- V. Arsentyev of State Astronomical Institute, Moscow, was quoted as saying Soviet Cosmos satellites were planned to map the atmospheric layer between 124 mi. and 995 mi., as well as to provide specific data for space flights and spacecraft design. Cosmos satellites supplied current space weather data before and during flight of VOSTOK III and IV and were used to study micrometeoroid impact on spacecraft. (*Av. Wk.*, 4/1/63)

April 2: NASA launched EXPLORER XVII (S-6) atmospheric structure satellite from Cape Canaveral, using Thor-Delta launch vehicle (its 16th consecutive success in 17 attempts). Satellite attained orbit of 571-mi. apogee, 158.9-mi. perigee, 96.4-min. period, 57.6° inclination to the equator. The sphere-shaped satellite, measuring 35 inches in diameter and weighing 405 lbs., contained instruments (two neutral mass spectrometers, four pressure gauges, and two electrostatic probes) to measure density, composition, pressure, and temperature of the atmosphere; it was equipped to make direct samplings of atmospheric constituents such as helium, nitrogen, and oxygen. Under project management of NASA Goddard Space Flight Center, EXPLORER XVII was the first scientific earth satellite to use new pulse-code-modulation telemetry system, a solid-state system providing output power of 500 milli-watts and capable of supplying 40 separate channels of information in digital form. Useful lifetime of the satellite was estimated at two-three months. (NASA Release 63-59; NASA Project EXPLORER XVII Prog. Rpt. No. 1)

- U.S.S.R. announced successful launching of LUNIK IV probe on a 3½-day flight toward the moon. Mission of 3,135-lb. instrumented spacecraft, said to have been placed in parking orbit around the earth and then fired toward the moon, was not disclosed. Tass announcement said all onboard equipment was functioning normally, scientific stations in U.S.S.R. were tracking the probe and receiving scientific information from it. "According to data which was already processed the flight of the automatic station continues along a trajectory which is close to the calculated trajectory." At 4:00 p.m. Moscow time, LUNIK IV was reported 50,486 km. (31,350 mi.) from earth. (NYT, 4/3/63, 1, 14; *Pravda*, 4/3/63, EOS Trans.)

April 2: Dr. Robert C. Seamans, Jr., NASA Associate Administrator, testifying before House Committee on Science and Astronautics' Subcommittee on Applications and Tracking and Data Acquisition, described the four areas of NASA's applications program—communications systems, meteorological systems, future applications satellites, and industrial applications:

"The space research and development effort is so broad that there will be ideas and innovations in almost every field. Products, based upon space development and produced using techniques and materials created for space use, may become increasingly evident with time in the everyday life of the nation.

"The applications program ranges over a very broad spectrum of scientific and industrial activity. We are concerned with investigations from purely scientific interest in atmospheric physics to problems associated with commercial communications satellites. It is a very challenging area of work and it is most satisfying to our people to find that great benefits are accruing to the nation from meteorological and communications satellite developments, and now from industrial applications as well." (Testimony)

• NASA announced signing of $456.6 million contract for Project Gemini spacecraft with McDonnell Aircraft Corp. Development of the two-man spacecraft began at McDonnell in December 1961 under preliminary letter contract. Contract called for 13 flight-rated spacecraft, 12 of which would be used for space flight and one for ground testing. McDonnell would provide other services and equipment under the contract, including two mission simulator trainers, a docking simulator trainer, five boilerplate spacecraft, and three "static articles"—spacecraft for ground test evaluation in vibration and impact tests. (NASA Release 63–66; MSC Release 63–71)

• Leonard Jaffe, NASA Director of Communications Systems, testified before the House Committee on Science and Astronautics' Subcommittee on Applications and Tracking and Data Acquisition: "NASA program objectives in this area . . . [communications satellite systems] are first, to insure the full development and realization of communications satellite potentials through continued research, development and flight test and second, to assist in the early establishment of operational communications satellite systems. The fact that communications satellites are feasible has certainly been demonstrated with the Echo, Telstar, and Relay satellites. Much information and experience has been gained as a result of the last year's experiments with regard to the satellites, the space environment, and with the earth-based elements of such systems. Much remains, however, to be done before economically practical systems can be established, but a sound foundation has been laid and all of the established interests are cooperating to the fullest extent in the research and developmental programs.

"In all areas of satellite applications, NASA has many interactions with other organizations. Typical interactions already in existence in the area of communications satellites are those with the Department of Defense, and the American Telephone and Telegraph Company. Shortly, a continuing liaison will be initiated with the new Communications Satellite Corporation . . . ." (Testimony)

April 2: NASA Director of Manned Space Flight, D. Brainerd Holmes, said in response to questions by members of House Committee on Science and Astronautics' Subcommittee on Manned Space Flight that there was nothing about Soviet LUNIK IV probe "to cause us to reevaluate our [space] program . . . .

"As far as our lunar program, as far as our program for Gemini, for man's greater conquest in space, I think the rate at which we are going is reasonable . . . .

"I think it would be foolish to squander funds on an all-out program. I think it would be unnecessary to lose lives just to get there first. I think it is necessary to run a sound program . . . ." (Transcript)

• Dr. Albert J. Kelley, NASA Director of Electronics and Control, testified before House Committee on Science and Astronautics' Subcommittee on Space Sciences and Advanced Research and Technology: "As a result of our review and study on the implementation of electronics responsibilities for future space missions, in our FY 64 Budget we are requesting Congressional approval for initial funding of $5M to establish a new Electronics Research Center to be located in the Greater Boston Area. Our long range plans call for an installation buildup over a period of six to seven years yielding a staff of approximately 2,000 personnel, of whom 600–700 would be professionals, and a plant value of approximately $50M.

"The rate of Center growth will be gradual and represents an investment in the future of space flight, analogous to the investment made in the older research centers many years ago which is paying off so handsomely today. This rate of growth together with up-grading of Center personnel by continued graduate education after they are on-board is expected to allow us to meet our staffing requirements with little or no impact on scientific and engineering manpower as projected during this growth period.

"The NASA considered many potential sites throughout the country for location of the Electronics Research Center. On balance, the overall university-industrial strength and capability in electronics and guidance research in the Greater Boston Area resulted in its selection as the location for the Center. This area, noted for its past and current heavy concentration on electronic research, both in industry and universities, provides a compatible, stimulating environment for growth of NASA capabilities in this important area of technology.

"The Center will have the principal function of conducting and directing research and component technique investigations in the related fields of communications, data processing, guidance, instrumentation, and control. It will be staffed and equipped to perform original work and to work closely with industries and universities with common interests. For example, component test facilities will be provided for evaluation of industrially developed experimental components. The Center will serve to focus the increasing NASA out-of-house efforts in the burgeoning area of electronic technology. Most important, it will provide a regenerative path for the early exploitation of advanced techniques of developments for space flight, whatever their source." (Testimony)

April 2: Charles H. Zimmerman, NASA Director of Aeronautical Research, testified before House Committee on Science and Astronautics' Subcommittee on Space Sciences and Advanced Research and Technology: "Our goals are to provide the research and advanced technology which will:

"1. Enable our aircraft industry to produce aircraft which will, to the greatest practicable extent, make the convenience and speed of air transportation available, and economically feasible, to the people of this country;

"2. Enable our armed services, working with our industry, to develop military aircraft superior in performance, in utility, and in requirements for logistics and manpower support to those of any other country.

"In working toward these goals, we coordinate our efforts closely with the Federal Aviation Agency, the Department of Defense, and other government agencies . . . and with the aircraft manufacturers and their customers . . . .

"Aviation no longer offers the glamor and excitement which so strongly influenced its growth in its early years. We are not striving for speed and altitude records. Men have circled the earth at orbital speeds. The X–15 has flown practically out of the atmosphere. Our objective now is transportation, economically feasible transportation for commercial purposes, militarily superior transportation for military purposes. I would like for you to consider our program as directed toward maintenance and enhancement of our leadership in air transportation . . . ." (Testimony)

- Dr. Hugh L. Dryden, NASA Deputy Administrator, said in address at Annual Dinner Meeting of Pittsburgh Post, Society of American Military Engineers: "Space science and technology require research and development at the frontiers of almost every branch of science and technology. Any industrial nation which refuses to devote substantial effort to space exploration is incurring the hazard of future technological obsolescence of its industries, the hazard of potential loss of leadership, and the hazard of military surprise by potential enemies . . . .

"While the new technological developments offer the earliest contributions to economic development, in the long run the scientific knowledge obtained in the great unknown environment of the celestial bodies and interplanetary space may bring much greater returns. For, judging from past experience, advances in scientific knowledge are the foundation of advances in technology, which are a key factor in economic development . . . .

"We live, in the language of the old cliche, at a great moment in history. Man has begun to lift himself above the restrictions of his planet into a new and strange environment and to expand his intellectual horizons. This is no visionary activity but one essential to the strength and security of our nation and to the future of man in the universe." (Text)

April 2: Newly created NASA Management Advisory Committee on Manned Space Flight visited NASA Marshall Space Flight Center. Chairman of committee was Dr. Marvin J. Kelley, retired President and Board Chairman of Bell Telephone Laboratories; other members were Dr. Hendrick W. Bode, James McCormack, Dr. Hector R. Skifter, and Dr. Arthur E. Raymond. (Huntsville *Times*, 4/4/63)

April 3: Aerobee 150A rocket launched from NASA Wallops Station carried instrumented payload to 147-mi. altitude in experiment to flight-test components of equipment for Eogo satellite and to measure propagation of VLF signals through ionosphere. Flight was joint project of Stanford Research Institute and NASA Goddard Space Flight Center. (Wallops Release 63-32; NASA Rpt. of Sounding Rkt. Launching)

- U.S.S.R. reported LUNIK IV had covered 216,000 km. (134,150 mi.), more than half the distance to the moon, during its first 24 hours of flight. Tass announcement said several measurements had been made to establish spacecraft's trajectory but did not indicate whether the "automatic station" was on course. Telemetry data indicated onboard instruments were operating normally, and internal pressure and temperature were within prescribed limits. There still was no description of the probe's mission. Article in *Izvestia* by Soviet geologist Aleksandr Khabakov said that "a soft landing on the moon of a container with instruments and auxiliary apparatus seems to be technically feasible." (*NYT*, 4/4/63, 18; *Izvestia*, 4/4/63, EOS Trans.)

- In press conference, President Kennedy was asked: "By when do you think we will be first in space, and in view of Russia's current lunar probe [LUNIK IV], do you think we will beat the Russians with a man to the moon?"

  He replied: "I don't know. We started well behind. Quite obviously they had a tremendous advantage in big boosters and we are still behind, because obviously we haven't gotten our new boosters yet, which we won't get until 1964, '65 and '66. We will have to wait and see but I can assure you it is an uphill race at best, because we started behind, and I am sure the Russians are making a major effort. Today's indication of what they are doing makes me feel that their program is a major one, and it is not spongy, and I think that we would have to make the same ourselves.

"So I would say we are behind now, and we will continue to be behind, but if we make a major effort we have a chance, I believe, to be ahead at the end of this decade, and that is where I think we ought to be."

Questioned about his reaction to General Eisenhower's charge that the national budget could be reduced by billions of dollars and that enormous sums were being wasted in the space program (letter to Congressman Charles A. Halleck made public March 29), President Kennedy said:

". . . The United States Congress almost unanimously made a decision that the United States would not continue to be second in space. We are second in space today because we started late. It requires a large sum of money. I don't think we should look with equanimity upon the prospect that we will be second all through the sixties and possibly the seventies. We have the potential not to be. I think having made the decision last year, that we should make a major effort to be first in space. I think we should continue to do so.

"President Eisenhower—this is not a new position for him. He has disagreed with this at least a year or year and a half ago when the Congress took a different position. It is the position I think he took from the time of Sputnik on. But it is a matter on which we disagree. It may be that there is waste in the space budget. If there is waste, then I think it ought to be cut by the Congress, and I am sure it will be. But if we are going to get into the question of whether we should reconcile ourselves to a slow pace in space, I don't think so.

"This Administration has concentrated its attention since it came into office on strengthening our military. That is one of the reasons why you could not possibly put in the cut which has been recommended, nine or ten billion dollars, without cutting the heart out of the military budget. The fact of the matter is when we came into office, we had 11 combat-ready divisions and we now have 16. We increased the scheduling on Polaris, nearly double per year. We increased the number of planes on the 15-minute alert from 33 per cent of our strategic air force to 50 per cent. In a whole variety of ways—in the Navy we have added about 46 vessels, and strengthened ourselves in defense and space . . . ." (Transcript, *Wash. Post*, 4/4/63)

April 3: Dr. Edward C. Welsh, Executive Secretary of National Aeronautics and Space Council, said in speech before American Institute of Aeronautics and Astronautics in New York City that launching of LUNIK IV was an initial step "in a very well organized man-on-the-moon program." He warned that U.S.S.R. would "employ its growing space competence for military purposes if it finds such action is effective. We must be ready to handle such a threat."

Dr. Welsh disputed former President Eisenhower's recent charge that "enormous sums of money" were being wasted in the National space program. "Mr. Eisenhower is consistent. He was wrong about the space program two, three, four years ago, and he is wrong now." (*NYT*, 4/4/63, 19)

April 3: Morris Tepper, NASA Director of Meteorological Systems, testified before House Committee on Science and Astronautics' Subcommittee on Applications and Tracking and Data Acquisition. He discussed the Tiros meteorological satellite system and described the Nimbus advanced meteorological satellite, first of which would be launched in 1963, and said: "Each one of these five elements—orientation, coverage, direct local readout, lifetime and growth potential, represents a major advance over Tiros. Nimbus, including all these, represents an extremely large step forward beyond the Tiros capability. The Nimbus technical difficulties of the past were associated with the development of these five elements. We now believe that our major difficulties are behind us. . . ."

He described the meteorological sounding rocket program, and said: "In the large meteorological sounding rocket program, it is planned to fire approximately 45 per year. These will be distributed at various latitudes to measure the atmospheric structure in the various climatic regions from the Arctic to the Tropics.

"The small meteorological sounding rocket program is aimed at providing a standardized flight article which will meet the requirements at an overall minimum cost. Approximately 100 rockets will be launched per year in this program." (Testimony)

- D. Brainerd Holmes, NASA Director of Manned Space Flight, discussed the manned space flight program before Yale Chapter of Sigma Xi, New Haven, Conn., and said: "In two centuries, the United States has grown from a small collection of settlements here along the Atlantic seaboard to a great nation of almost 200 million people spread across a continent. We have dug canals, dammed rivers, built railroads, launched ships, constructed highways, and flown aircraft. We have constructed buildings, bridges, and a nationwide grid of electrical power. We have developed a system of industry, agriculture and distribution by which our people are provided with food, clothing, automotive transportation, and other necessities and conveniences in such volume that the United States has assumed an unprecedented position of wealth and power in the world.

  "Some historians explain this growth in terms of challenge and response. Under this interpretation, which appeals to me, the frontier and external danger have provided stimuli for growth. Now, however, with the frontier gone and the external danger, we hope, decreasing, we are entering a new period of change. What will now spur us to further growth?

  "Exploring space can provide such a challenge. Here is an endeavor, clearly in the American tradition, which will unleash again the urges to build and explore—and which will employ for a great national purpose the tremendous productive capacity of our economy." (Text)

- Charles W. Frick resigned as Apollo Project Manager at NASA Manned Spacecraft Center to accept a position with industry. MSC Director Dr. Robert R. Gilruth appointed Robert O. Piland as Acting Project Manager. (MSC Release 63-72)

April 4: Sixth Apollo spacecraft impact test conducted by North American Aviation's Space and Information Systems Div., Downey, Calif. Highly instrumented capsule fell from 10-ft. height at vertical speed of 18 mph into test tank containing 300,000 gals. of water. (CBS-TV "Newsnight," 4/5/63; L.A. *Herald-Examiner*, 4/4/63)

- Dr. George L. Simpson, Jr., NASA Assistant Administrator for Technology Utilization and Policy Planning, testified before House Committee on Science and Astronautics' Subcommittee on Applications and Tracking and Data Acquisition: "NASA is committed to a hard, driving effort to transfer the useful fruits of our research and development effort to the private sector of the economy in as quick and as useful a way as possible.

  "We are committed, first, because the Space Act requires it. NASA is directed to 'Provide for the widest practicable and appropriate dissemination of information concerning its activities and the results thereof.'

  "We are also committed because we are mindful that a large part of the Nation's R&D effort is occupied by the Federal agencies; and that it is essential that the maximum value be wrung out of this part of the total effort.

  "The requirements for space exploration begin in the field of anatomy and run through virtually the whole gamut of scientific and technological interest, through new power sources to zero gravity. Further, those things which are made for space travel receive no compromises from that harsh environment. Which is to say that a very large part of the total space work is at the very leading edge of the state-of-the-art; and that in another large part of the state-of-the-art must be extended.

  "It is reasonable to think that an effort of such size, range and sophistication, extending over a number of years, cannot fail to have major scientific and economic consequences: consequences that begin as scientific and technological advances and then are converted into new techniques of manufacture, new systems of organization, and new products and services. The space effort will be, we think, altogether comparable in its effect on the American economy to that of World War II. Our objective is to maximize, to facilitate, to speed up this effect. . . .

  "We are aware that past experience indicates severe problems in the transfer of government research and development efforts to private use. However, we feel that the Nation's well being requires a major effect to determine whether these problems can be solved." (Testimony)

- Alton B. Moody, Chief of Future Aplications Satellites in NASA Office of Applications, testified before House Committee on Science and Astronautics' Subcommittee on Applications and Tracking and Data Acquisition. He summarized NASA accomplishments and plans in two areas—navigation satellite systems and data collection satellite systems—and said: "Both of these projects illustrate the operating procedure adopted for Future Applications. Before a flight program will be undertaken, several questions must be answered affirmatively:

"1. Is there a requirement?
"2. Is the requirement sufficient to justify the expenditure of the funds needed?
"3. How would such a system operate?
"4. Is the design feasible and within current state-of-the-art? By answering these questions serially, we are not committing funds very far downstream and can turn back at any point if the results are unfavorable." (Testimony)

April 4: Ballute drag balloon was being built by Goodyear Aircraft Corp. for use in Gemini manned spacecraft, NASA Manned Spacecraft Center announced. Ballute would be used as part of Gemini ejection-seat-escape system, planned means of emergency escape for altitudes below 70,000 ft. The inflatable rubberized fabric structure would stabilize and slow the astronauts' fall until conventional parachutes could be deployed at lower altitudes. (MSC Release 63–67)

- Congressman George P. Miller (D.-Calif.), Chairman of House Committee on Science and Astronautics, spoke out during committee questioning of Dr. Albert J. Kelley, NASA Director of Electronics and Control, about location of proposed NASA Electronics Center in Boston area. Miller made his comments after Congressman Thomas N. Downing (D.-Va.) asked Dr. Kelley why NASA had not chosen Langley Field, Va., as site for new center: "Frankly, we're making this thing right now a question of where are we going to put it because it's going to be a plum. We'd all like this in our states. But do you want to make a W.P.A. project out of this or are we interested in the space program?" (AP, *NYT*, 4/5/63)

- Tass announced LUNIK IV was 314,000 km. (194,494 mi.) from earth at 8:00 p.m. Moscow time and would pass close to the moon's surface, indicating for the first time that the probe would not orbit or land. Speculation was that LUNIK IV might have been planned to soft-land an instrument package on the moon. Tass said radio communication with the spacecraft was good and that onboard instruments were functioning normally. (AP, *Wash. Post*, 4/5/63; *Pravda*, 4/5/63, EOS Trans.)

- NASA and AEC announced agreement to standardize the reduction ratio and size for microcopy of miniaturized forms of scientific and technical reports generated within the two agencies. (NASA Release 63–64)

- Dr. Brockway McMillan, Assistant Secretary of Air Force, gave to House Committee on Science and Astronautics a list of 10 experiments USAF would like to include as part of its participation in NASA Project Gemini: visual definition of objects in space; cooperative target inspection; angle-track-only rendezvous guidance; hybrid IR/laser guidance equipment; passive infrared star and satellite tracking; visual definition of terrestrial features; observation of missile launches; extravehicular operations; autonomous navigation; and radiation levels within the space capsule. (*M&R*, 4/15/63, 15)

- In address at Virginia Polytechnic Institute at Blacksburg, Secretary of Commerce Luther Hodges said: "Of the nearly $17 billion we in the United States are spending on research and development, only about $4 billion is spent by American industry for non-military, non-space work. And of that $4 billion, only about $1.5 billion is aimed at increasing the productivity of our economy.

"The countries of Western Europe are spending twice as large a proportion of their gross national product for civilian research and development. . . . Japan, which many people still think of as an imitator and copier of western scientific and technical developments, actually has as high a ratio of scientists and engineers working on its civilian technology as we have. . . .

"American industry must be much more alive to the need for more research and development work in the civilian sector of our economy. There must be a wider diffusion of this work throughout industry, and there must be greater industry support for the training of people capable of doing this work.

"Some 80 per cent of all our industry sponsored research is done by 300 companies, and 73 per cent in only 5 industries.

"Two of these industries, aerospace and electronics, account for 25 per cent of the industry sponsored research, but their products account for only 3½ per cent of the total production of goods and services in this country. . . ." (Text, CR, 4/25/63, 2444–2445)

April 4: Dr. James R. Killian, former president of MIT, was quoted by James Reston of *New York Times*: "Will several billion dollars a year additional for enhancing the quality of education not do more for the future of the United States and its position in the world than several billions a year additional for man in space?" (NYTNS, Chattanooga *Times*, 4/5/63)

- James J. Tart, of NASA Langley Research Center, was one of five recipients named for second annual E. H. Rietzke Achievement Awards of the Capitol Radio Engineering Institute of Washington, D.C. Cited for his work to help overcome communications blackout during re-entry ionization of space vehicles, Tart devised and built a 9,000-megacycle telemetering system for an X-band radar to enhance re-entry communications. (Newport News *Times-Herald*, 4/4/63)
- Ham, first chimpanzee to be successfully launched on (suborbital) space flight, became a permanent resident of Washington's National Zoological Park. Ham flew in Mercury-Redstone flight MR-2 down the AMR on Jan. 31, 1961. (*Wash. Post*, 4/5/63)

April 5: Tass reported LUNIK IV spacecraft was 365,000 km. (226,680 mi.) from earth at 6:00 p.m. Moscow time and was continuing on its flight toward the moon. Radio communications with the probe were good, and telemetry data indicated onboard instruments and systems were functioning normally. "Experiments and measurements, provided for by the program, are being conducted."

Sir Bernard Lovell, Director of Jodrell Bank Experimental Station (radiotelescope facility), reported receiving signals from LUNIK IV for 45 min., ceasing abruptly at 8:50 p.m. (10:50 p.m. Moscow time). Lovell said there appeared to have been some "complicated maneuvers" with the probe.

Italian astronomers reported receiving 45 min. of TV pictures showing closeup of moon's surface.

In Moscow, two scheduled lectures on LUNIK IV flight were canceled by Moscow radio; a television discussion of history of lunar research made only passing reference to LUNIK IV. (*Krasnaya Zvezda*, 4/6/63, EOS Trans.; *NYT*, 4/6/63, 2; UPI, *NYT*, 4/6/63, 2; CBS-TV "Newsnight," 4/5/63; *Av. Wk.*, 4/15/63, 38)

April 5: Lewis scientists report new and unusual use for liquid gallium, which appears ideally suited to the task of lubricating moving parts in space-bound machinery. (LRC Release 63–12; Lewis Chronology, 3)

- USAF announced launching of unidentified satellite with Blue Scout rocket from Vandenberg AFB. (AP, Wash. *Eve. Star*, 4/6/64)
- USAF memorandum charging that NASA was planning to build $77 million worth of facilities that would duplicate existing USAF facilities was detailed in *New York Times*. Interviewed regarding the memorandum, NASA Deputy Director Dr. Hugh L. Dryden said NASA had frequent coordination meetings with DOD and that "no such statement has been made to us." He described the memorandum as representing "a point of view of junior people who can't get their superiors to support their desires." On point made in memorandum about USAF's offering bioastronautics laboratories to NASA, Dr. Dryden said: "We have offered to carry half the cost of bioastronautics laboratories but this was not accepted by the Department of Defense." (*NYT*, 4/5/63)
- USN made second attempt to place operational Transit navigational satellite into orbit from Pt. Arguello, Calif., *Astronautics and Aerospace Engineering* reported. *A&AE* said spacecraft failed to attain orbit because of malfunction in fourth stage of USAF Blue Scout booster. (*A&AE*, 5/63, 141)
- John L. Sloop, Director of NASA Propulsion and Power Generation in Office of Advanced Research and Technology, testified before House Committee on Science and Astronautics' Subcommittee on Space Science and Advanced Research and Technology: "Propulsion is the key to space exploration and our push to the moon in this decade will depend completely on chemical propulsion. We see a large and continuing role for chemical propulsion in the next decade. The advent of nuclear and electrical propulsion will greatly increase our capability and enable us to use each type in the application best suited to its characteristics. We see much larger chemical propulsion systems than we have today. Some of these may use air augmentation and will be capable of ferrying large payloads economically from earth to manned satellites and between distant points on the earth. We see chemical propulsion packages carried dormant for months on long space missions but ready at the touch of a button or a signal from earth to perform their task. We see chemical propulsion systems for landing and take-off from the moon, Mars, and Venus where compact systems of high thrust per pound of weight are needed. Some may be carried as extra stored energy sources for emergency use . . . ." (Testimony)

April 5: AEC announced Dr. J. Robert Oppenheimer would be recipient of Enrico Fermi Award, AEC's highest honor, in recognition of his leadership in development of the atomic bomb and of atomic energy for peaceful purposes. Dr. Oppenheimer has headed Institute for Advanced Study, Princeton, since 1947. (*NYT*, 4/6/63, 1, 5)

- Vasily V. Parin, president of Soviet Academy of Medical Sciences and expert in space medicine, was awarded the Order of Lenin "in recognition of his great contribution to the advancement of Soviet medical science," Tass reported. (AP, Balt. *Sun.*, 4/7/63)

April 6: LUNIK IV passed within 8,500 km. (5,281 mi.) of the moon at 4:26 a.m. Moscow time, Tass reported. "The experiments and measurements which were conducted by means of the spacecraft are completed. Radio communication with the spacecraft will continue for a few more days.

"Broad experimental data, which are of great value for the solution of a number of technological problems related to the conquest of the moon, were obtained. Measurement data are now being processed and studied in scientific centers of our country which are responsible for the investigation of cosmic space." Tass still did not reveal nature of probe's mission.

Sir Bernard Lovell, Director of Jodrell Bank Experimental Station in England, was quoted as saying he believed the Soviets had failed in an attempt to land instruments on the moon. "It is hardly conceivable, after a lapse of three years, with all the improvements in techniques, that the Russians would merely set out to do again what they did with LUNIK III." (*Pravda*, 4/7/63, EOS Trans.; AP, *Wash. Post*, 4/7/63, A7; AP, Wash. *Sun. Star*, 4/7/63)

- USAF satellite (1962 Beta Kappa-1), launched Oct. 26, 1962, to study artificial radiation, had discovered evidence that radiation from U.S. Starfish high-altitude nuclear test would last up to 10 years, Project Director Dr. Ludwig Katz of USAF Cambridge Research Laboratories disclosed. The previously unidentified satellite, called "STARAD," was built at DOD request when it became apparent artificial radiation might be damaging orbiting satellites, carried nine different instruments for measuring the artificial radiation. USAF spokesmen said, in their opinion, data from this satellite were influential in Dr. James A. Van Allen's reversal of his previous position that the radiation would last only a year (see March 15). Dr. Katz said the satellite found there had been little decrease in artifical radiation's intensity since the satellite was launched. "The decay of the belt is so small that a lifetime of five to ten years would not surprise me." STARAD also detected increased radiation from Soviet high-altitude nuclear explosions of Oct. 28 and Nov. 1. It transmitted until Jan. 18 when its batteries deteriorated. (AP, Wash. *Sun. Star*, 4/7/63; *Wash. Post*, 4/7/63; *Av. Wk.*, 4/15/63, 26)

- Mining of water on the moon could provide rocket fuel for interplanetary spacecraft, thus eliminating need for more advanced rockets, John W. Salisbury of USAF Cambridge Research Laboratories said. Converting water to its components oxygen and hydrogen would provide the two fuel elements of Saturn V's upper stage. Salisbury said lunar water would be present either as subsurface ice or as' one-eighth of silicate rock formations known as olivine. USAF was conducting studies to determine most efficient way of mining water thus trapped in the moon. (NYTNS, Wash. Post, 4/7/63, A7)

April 6: Dr. Otto Struve, prominent U.S. astronomer and former Director of National Radio Astronomy Observatory at Green Bank, W. Va., died at 66 in Berkeley, Calif. (Chicago Trib., 4/10/63)

- Fourth annual Conference on Peaceful Uses of Space would be held in Boston in May 1964, NASA Administrator Webb announced. (Wash. Sun. Star, 4/7/63)

April 7: At Fourth Congress of the World Meteorological Organization, being held in Geneva, U.S. delegation circulated paper describing plans to orbit a Nimbus weather satellite in 1963, the satellite to provide daily photographic observations of the globe. (Reuters, Balt. Sun, 4/8/63)

- Senator Clifford P. Case (R.-N.J.) announced he had requested Secretary of the Air Force Eugene M. Zuckert to investigate the reported USAF memorandum asserting NASA planned to construct facilities which would duplicate those of USAF. (UPI, Wash. Post, 4/8/63)

- In nationally televised interview, Congressman Chet Holifield (D.-Calif.) said that funds for the $20-billion manned lunar program could be better spent in raising living standards on earth— on schools, hospitals, housing. Prestige value of landing a man on the moon was not worth the cost. Senator Clinton P. Anderson (D.-N. M.), Chairman of Committee on Aeronautical and Space Sciences, said that even if these funds were not spent for landing a man on the moon, the funds would not be spent on raising living standards. "If we would stop everything we're doing in space tomorrow, I think you would find there would be still plenty of people who would vote no on a program of better roads, better hospitals, better education." (CBS-TV "Washington Report"; UPI, Wash. Post, 4/9/63)

- Recipients of third annual Federal Women's Award were announced. Chosen for "outstanding contributions to the quality and efficiency of the service of the Federal Government, for their influence on major Government programs, and for personal qualities of leadership, judgment, integrity, and dedication," the six winners included Eleanor C. Pressly, Head of Vehicles Section, Spacecraft Integration and Sounding Rocket Div., NASA Goddard Space Flight Center. Miss Pressly was cited for her pioneer work in sounding rocket development and her "demonstrated organizational ability in scheduling and coordinating launchings of sounding rocket vehicles in support of upper atmospheric research." She developed the Aerobee Jr. sounding rocket, co-developed Aerobee-Hi 150, and directed improvement of Aerobee-Hi 150A, all used extensively in IGY. (FWA Release)

- Proposal by NBC to television industry to televise Astronaut Gordon Cooper's Mercury flight MA-9 in color would not be adopted, Executive Vice President of NBC News William R. Mc-

Andrew acknowledged. Proposal had been rejected, science reporter William Hines said, in February by officials of ABC and CBS networks, but NASA had been quietly trying to persuade them to accept the proposal. (Hines, Wash. Sun. Star, 4/7/63)

April 8: Attempt to launch two-stage Astrobee 1500 sounding rocket from NASA Wallops Station failed with first stage of the vehicle failing to perform properly. This was NASA's first attempt to launch the Astrobee, and purpose of test was to evaluate the rocket's performance as a NASA test vehicle. (Wallops Release 63–36)

- NASA announced selection of Avco Corp., Research and Advanced Development Div., and General Electric Co., Missile and Space Div., for negotiation of conceptual design study contracts of Voyager spacecraft. The two six-month, $100,000 contracts were to supplement in-house NASA feasibility studies to determine scientific merit and technical possibility of initial Voyager planetary flights before 1970. (NASA Release 63–70)

- Boilerplate model of Apollo spacecraft arrived at NASA Marshall Space Flight Center for dynamic and compatibility tests with the Saturn I. Boilerplate model included command and service modules and mockup of launch escape system. (Huntsville Times, 4/9/63; NAA S&ID Skywriter, 4/19/63, 1)

- Vice President Lyndon B. Johnson, in Associated Press interview, was asked: "Mr. Vice President, as chairman of the Space Advisory Council [i.e., National Aeronautics and Space Council], you were intimately concerned in drafting the 5.7-billion-dollar space budget. Former President Eisenhower says the budget is spongy, and there is no hurry about getting to the moon. In view of this recent Russian shot [LUNIK IV], what is your reaction?"

  The Vice President replied: "The Administration has submitted to the Congress what we believe is a sane and responsible proposal to insure the maximum efficiency and effectiveness of our space program. I know of no reckless desire in any responsible quarter to get to the moon at all costs and without regard to the consequences to other vital aspects of our national policy.

  "The objective of reaching the moon is one aspect of our national space program which will be of great benefit to our country, and I think to all mankind. It may be possible by appropriating more money to speed up slightly the attainment of that goal and, of course, we all know it would be mighty easy to slow that goal down. But I don't think that either alternative would be helping our country. Gaining a short period of time would almost certainly be at the expense of other crucial programs. Slowing down might bring us some one of these days face to face with the shocking knowledge that we had permanently assigned ourselves to second place.

  "So, all in all, I think the Administration has submitted a program which we can all embrace and both parties can fully support." (AP, Wash. Post, 4/8/63)

- European space organization, Eurospace, issued report recommending program of communications satellites, time-keeping and navigational satellite, and rocket development. Report said Europe's first operational communications satellite could be launched by 1968 and suggested a network of 12 communications satellites or, as alternative, two stationary satellites, in equatorial orbit; a light satellite orbiting at about 625-mi. altitude as time-keeping and navigational aid; development of launch vehicles based on existing models of European nations and ELDO (European Launcher Development Organization); and studies on equatorial launching site. Eurospace is private association of 125 members representing industries, professional bodies, and industrial associations from nine countries—Belgium, France, Great Britain, Italy, the Netherlands, Norway, Sweden, Switzerland, and West Germany. (Reuters, *Wash. Post*, 4/9/63; *NYT*, 4/9/63)

April 8: Milton B. Ames, Jr., NASA Director of Space Vehicles in Office of Advanced Research and Technology, testified before House Committee on Science and Astronautics' Subcommittee on Space Sciences and Advanced Research and Technology: "The objective of the Space Vehicle Systems Advanced Research and Technology Program is to identify and solve critical technical problems bearing on present-generation space vehicles, as well as to advance the frontiers of knowledge that will enable the development of more advanced space vehicles for future space missions. The increased complexities of the advanced space vehicle systems that will be required to accomplish our national space objectives for the next two decades attest to the need for a broad, yet thorough and timely research program to establish the technology required to insure the continued superiority of our country's space flight capability . . . ." He then reviewed the areas of activity covered in the space vehicle research and technology program— advanced space vehicle concepts, space vehicle aerothermodynamics, environmental factors and technology, space vehicle structures, design criteria, and vehicle technology flight experiments. (Testimony)

- John H. Rubel, DOD Deputy Director of Defense Research and Engineering, told House Committee on Science and Astronautics' Subcommittee on Applications and Tracking and Data Acquisition that DOD planned to initiate a communications satellite program definition study in May or June, the program based on goal of putting 24-30 operational satellites into orbit by end of 1965. "Approximately six satellites will be launched per booster, using the . . . Atlas-Agena D. The six satellites will be ejected at slightly different velocities into an approximately 6,000-mi.-altitude circular polar orbit, where they will gradually spread out. The successive clusters will be injected into orbital planes spaced 45 deg. apart." System would be developed jointly by USAF and Army, USAF being responsible for space-related activities and Army for ground stations. (*Av. Wk.*, 4/15/63, 27)
- Design for 216-ft. nuclear-powered space station had been awarded patent by U.S. Patent Office, Lockheed Missiles and Space Co. announced. Invented by S. B. Kramer and E. H. Visscher of Lockheed, prefabricated station would be assembled in orbit from cylindrical and spherical subsections. Lockheed said it would be

useful as a manned space station and observation platform. (*NYT*, 4/9/63, 10)

April 8: Polaris A-3 missile successfully flown 1,800 mi. down Atlantic Missile Range, third successful test in 11 firings. Navy reported the missile achieved all test objectives. (AP, *Wash. Post*, 4/9/63)

- District Judge David Zenoff issued temporary restraining order against 16 unions striking at Nevada atomic test site since April 4 and an injunction ordering management to "restore the status quo." Unions were protesting AEC order to contractors at the site eliminating travel pay and other fringe benefits in their contracts. Judge Zenoff scheduled a hearing for April 15. (UPI, *NYT*, 4/9/63, 53)

April 9: In its first few days of operation, EXPLORER XVII satellite had obtained data that more than tripled all previous direct measurements of the neutral gases in earth's upper atmosphere, NASA Goddard Space Flight Center announced. Satellite was traveling in elliptical orbit (578-mi. apogee, 153-mi. perigee, 96.4-min. period) and was transmitting data of excellent quality and quantity. It had been commanded more than 75 times, and each time it responded with approximately four min. of data on density, pressure, composition, and temperature of upper atmosphere gases. New communications system, utilizing special data readout station at GSFC, was providing scientific and technical data from the satellite within minutes of its transmission. (GSFC Release)

- Nike-Apache sounding rocket carried instrumented payload to 101-mi. altitude in experiment from NASA Wallops Station to measure density and conductivity of electrically charged particles in the ionosphere under undisturbed conditions. Except for failure of solar aspect sensor, experiment was considered a success. The 80-lb. payload impacted in Atlantic Ocean 77 mi. from launch site. (Wallops Release 63-38; NASA Rpt. of Sounding Rkt. Launching)

- Televised White House ceremony, with President Kennedy signing bill making Sir Winston Churchill an honorary citizen of the U.S., was transmitted to England and Europe via RELAY I communications satellite. Broadcast was viewed by millions of Britons and Sir Winston himself, and both audio and visual reception were considered perfect. (NASA Release 63-69; *Wash. Post*, 4/10/63)

- Dr. Robert C. Seamans, Jr., NASA Associate Administrator, testified on NASA's contractual arrangements with industry in systems engineering and in checkout, reliability assessment, and integration before Subcommittee on Applications and Tracking and Data Acquisition, House Committee on Science and Astronautics: "In February 1962, NASA requested . . . AT&T to provide an organization of experienced men able to employ the most advanced analytical procedures necessary to develop the factual bases needed by responsible NASA officials to make the wide range of decisions required for the successful execution of the manned space flight program.

"AT&T responded to this request by forming Bellcomm, Inc., in March, 1962 . . . .

"It is important to note that the Bellcomm organization assists and supports a technical office within the Government rather than an administrative office. This important distinction makes it possible for the Government to properly carry out its total technical responsibility utilizing a contractor for assistance, but maintaining final technical, engineering, and procurement judgments within the Government itself . . . .

"It is important to understand the difference between systems engineering, an activity that requires a relatively small organization of very highly qualified individuals, and checkout, reliability assessment, and integration, which because of the massive amount of technical detail which must be handled, require large manpower efforts.

"In systems engineering, we have developed the system concept and are now writing the overall specifications for project Apollo. You should recognize that a portion of these specifications will include the general requirements on the checkout system, definition of the requirements for reliability assessment, and technical monitoring of the program. Bellcomm supports NASA in this work.

"We have asked General Electric to provide support in areas in which the details require many people. G.E. is implementing the checkout system, beginning hardware production, and undertaking comprehensive work in integration and related data gathering and processing, which will help to insure that the overall systems engineering requirements are being met in detail." (Testimony)

April 9: LUNIK IV was Soviet Union's third attempt in 1963 to send instrumented payload to the moon, William Hines reported in Washington *Evening Star*. He quoted a Government official—outside of NASA and DOD—as saying: "We are trying to make a record to place before the world. The preparation of this record is now in progress. There has been no official statement to date regarding things that are up there which we know about and don't belong to us. But there are such things, and they should be made public—by us, if the Soviets refuse." NASA and DOD had no comment on any unannounced Soviet space launchings. (Wash. *Eve. Star*, 4/9/63, A1)

- Senator John L. McClellan (D.-Ark.) introduced in the Senate a bill (S. 1290) to establish a uniform national policy concerning proprietary rights in inventions made through expenditure of public funds. Bill was referred to Senate Committee on the Judiciary. (*CR*, 4/9/63, 5609)
- Dr. Eugene B. Konecci, NASA Director of Biotechnology and Human Research in Office of Advanced Research and Technology, outlined NASA's human factors systems program before House Committee on Science and Astronautics' Subcommittee on Space Sciences and Advanced Research and Technology and said: "This program is designed not only to do human research and development of adequate life support and protective systems for

man's survival in the aerospace environment, but to adequately determine man-machine relationships, and integrate them properly into the advanced aerospace systems. Our success and progress in the next 10 to 20 years depends upon the Biotechnology and Human research we do today . . . ." (Testimony)

April 9: First college program to train USAF astronauts had been established at Purdue Univ., USAF announced. Fourteen June graduates of Air Force Academy had been selected to begin the specialized program in Purdue's School of Aeronautical and Engineering Sciences, their studies leading to newly created Master's degree in astronautics. (*Chicago Trib.*, 4/10/63)

- Astronaut John H. Glenn, Jr., officially opened Chicago and Midwest Space Month with his televised dinner address in the Museum of Science and Industry, Chicago. Dinner marked the opening of "America in Space" exhibit in the museum, largest space display ever made available by NASA in the midwest. Space Month activities would be climaxed by third national conference on the peaceful uses of space in Chicago, May 1-9. (*Chicago Trib.*, 4/9/63)
- NASA Langley Research Center announced selection of E. W. Muller, Contractor, Inc., of Newport News for preparing the site for NASA Space Radiation Effects Laboratory, Newport News, Va. (Langley Release)
- Andrew G. Swanson, scientist at NASA Langley Research Center, was one of 45 outstanding young executives awarded an Alfred P. Sloan fellowship in executive development at Massachusetts Institute of Technology. Award was announced by Dean Howard W. Johnson of MIT's School of Industrial Management. (Langley Release)
- In speech before Public Relations Society of America luncheon meeting in Washington, MSC Public Affairs Officer Lt. Col. John H. Powers (USAF) said preparations for Project Mercury flight MA-9 were proceeding on schedule. Flight was scheduled for mid-May. (AP, *Birmingham Post-Herald*, 4/10/63)
- U.S.-U.K. agreement on purchase of Polaris missiles for use in British submarines was published in Washington. Contract was technical version of Nassau agreement signed last December by President Kennedy and Prime Minister Macmillan establishing Polaris as substitute for Skybolt weapon system. It was signed March 16 and submitted to British Parliament today. (*NYT*, 4/10/63, 3)
- Reported that Soviet cosmonaut Gherman S. Titov had been promoted from major to lieutenant colonel. (UPI, *NYT*, 4/9/63, 15)

April 10: FAA announced award of contract to Cornell Aeronautical Laboratory, Inc., an affiliate of Cornell Univ., to conduct systems analysis in supersonic transport research program. Under contract, Cornell would perform intensive effort in direct support of Joint Systems Analysis Group, composed of representatives of NASA, DOD, FAA, and the commercial airlines. (FAA Release 63-41)

- Symposium on exploration of Mars would be held in Denver June 6-7, cosponsored by NASA, American Astronomical Society, American Institute of Biological Sciences, and American Institute of

Aeronautics and Astronautics, NASA announced. (NASA Release 63-68)

April 10: USN nuclear-powered submarine *Thresher* was lost at sea with 130 persons aboard.

- First attempt to launch Polaris A-3 missile from ship (U.S.S. *Observation Island*) failed with missile's veering off course shortly after liftoff, when it was destroyed by safety officer. (*Space Bus. Daily*, 4/12/63, 474)
- Unnamed source was quoted as saying Egypt had test-fired at least two rockets near port of Suez during the past week. (AP, *Wash. Post*, 4/11/63)
- Adm. George W. Anderson (USN), Chief of Naval Operations, said in news conference that there was "no danger of radioactive contamination" from sunken nuclear-powered submarine. (*NYT*, 4/11/63, and 4/14/63)

April 11: X-15 No. 1 flew at speed of 2,864 mph (mach 4.25) and altitude of 74,000 ft. in its first test as an aerial mapping plane, a camera having been installed into the underside of the craft's nose. Piloted by Maj. Robert Rushworth (USAF), X-15's rocket engine burned a record 2½ min. Primary purpose of flight was to measure photograph distortion caused by shock waves, air flow, and other aerodynamic turbulence. Flight marked first test in advanced research program in which X-15 would photograph stars, measure meteoroids, and investigate infrared rays in space. (UPI, *Wash. Post*, 4/12/63; AP, *NYT*, 4/12/63)

- NASA announced selection of 10 experimenters for Mariner spacecraft's Mars fly-by mission, scheduled for 1964. Experiments for the probe, which would make measurements of interplanetary space as well as of vicinity of Mars, were: television, magnetometer, low-energy cosmic-ray experiment, cosmic-ray spectrum experiment, cosmic-ray ionization experiment, plasma probe, cosmic dust experiment, infrared spectrometer, ultraviolet photometers, and radio propagation. (NASA Release 63-71)
- Television broadcast of biweekly program narrated by Jacques Sallebert, French radio and TV correspondent in New York, sent direct from New York to Paris via RELAY I communications satellite for first time. Program consisted of views of lower New York City and of New York Harbor. Reception in Paris was described as considerably better than prerecorded broadcasts previously used and as comparing favorably with any TV program originating in Paris. (*NYT*, 4/12/63, 45)
- Harold B. Finger, Director of NASA Nuclear Systems Office and Manager of NASA-AEC Space Nuclear Propulsion Office, testified before Subcommittee on Space Sciences and Advanced Research and Technology, House Committee on Science and Astronautics: "The space nuclear systems program involves large amounts of funds, manpower, and facilities in NASA's program and, in addition, involves substantial effort on the part of the Atomic Energy Commission. We are convinced that the high performance potential of these systems for space propulsion and electric power generation justifies the substantial effort that we are devoting toward the early development of such systems. It is

clear that if we are to perform extensive space exploration missions beyond our first manned lunar landings, it will be necessary to have available to us nuclear propulsion systems and if we are to establish large space laboratories, lunar bases, and are to travel to distant planets, we will have to rely on the availability of large nuclear electric power generating systems...." (Testimony)

April 11: In news interview, physicist Herman Yagoda of USAF Cambridge Research Laboratories said that living organisms on planet Mars would receive 30–50 times more radiation than organisms on earth, and during a solar storm the radiation dosage on Mars is a million times what it would be on earth. "Thus, Mars with its marginal environment for sustaining primitive life forms is subjected periodically to comparatively intense radiation levels which may have profound effects on the mutation and evolution of Martian plant and animal life." (*Wash. Post*, 4/11/63)

- Reported that Senator Strom Thurmond (D.-S.C.) told closed session of Senate that U.S.S.R. was deploying antimissile missiles capable of intercepting Polaris missiles and possibly even Minuteman missiles. Senator Thurmond spoke in favor of speeding up U.S. Nike-Zeus antimissile missile development, but Senate roll call vote (58-to-16) cut a proposed $196 million authorization for Nike-Zeus from the military spending authorization bill for FY 1964. Senate Armed Services Committee had added the Nike-Zeus funds to the Administration's proposed bill. (Wash. *Eve. Star*, 4/13/63)
- Dr. Simon Ramo, of Thompson-Ramo-Wooldridge, was quoted as being skeptical of feasibility of using artificial satellites for militarily aggressive purposes, saying: "We often hear the mistaken impression that a bomb dropped from a satellite will fall to earth. Of course, it won't. It will reach the earth only if it is pushed away from the satellite with the right amount of force and with controlled direction. The whole launching operation (from a satellite) is more difficult that that of launching an ICBM from a point on earth. Also, the flight will take very much longer...." (*St. Louis Post-Dispatch*, 4/11/63)
- AEC announced it would commission scientists in several universities to study fragments of meteorite that landed last August in Upper Volta, Africa. (*Minneapolis Star*, 4/11/63)
- Scientist Dr. Charles R. Warren of U.S. Geological Survey's Military Geology Branch said in *Science* magazine that the moon is probably covered with a layer of fuzz, about an inch deep. The "skeletal fuzz consists mostly of empty space" and could be a mesh or miniature tinker-toy structure or it might resemble a loose aggregate of snowflakes or reindeer moss. Dr. Warren had made study of lunar terrain for Army Corps of Engineers. (*NYT*, 4/12/63, 29; UPI, *Wash. Post* 4/12/63)
- Next Soviet space flight would be "spectacular" and "more complicated," four Soviet cosmonauts indicated in *Evening Moscow* interview. The cosmonauts—Yuri Gagarin, Gherman Titov, Andrian Nikolayev, and Pavel Popovich—said U.S.S.R. was

busily preparing for a forthcoming space launching. (UPI, *Wash. Post*, 4/12/63)

April 12: George J. Feldman, Communications Satellite Corp., incorporator nominee, said in address on international arrangements for satellite communications: "Last July 10, when Telstar went into orbit, communications via satellite literally entered the homes of millions of men, and women. This historic event symbolized technical developments which open up unparalleled opportunities and will, in time, significantly affect the daily lives of people in many parts of the world. A new dimension has been added to communications by telephone, telegraph, radio and television. . . .

"Cooperation on the international level, bilaterally and multilaterally, is a keystone of a successful communications satellite system and it is self-evident that the wider the range of the system, the more extensive will be the need for wider participation and agreement. . . .

"The development of the communications satellite system is underway. It will involve intricate negotiation and arrangements, and there are many problems to be tackled. But technology has thrust greatness upon us and, as in some other manifestations of this technological revolution, we are summoned to the challenge. . . ." (Text, *CR*, 4/24/63, A2426)

- U.S.-Australian experiment to obtain data on the ionosphere conducted at NASA Wallops Station, with Aerobee 150A sounding rocket carrying 144-lb. payload of radio instruments 1,253 mi. high during eight-minute flight. Experiment was designed to measure intensity and spectrum of VLF radio waves above "E" region of the ionosphere. This was third U.S.-Australian attempt of the space experiment and first success, the previous two having failed to reach desired altitude because of vehicle structural failure. Scientific payload was provided by Australia's Commonwealth Scientific and Industrial Research Organization. Technical facilities, telemetry recording, and the two-stage rocket were provided by NASA Goddard Space Flight Center. (Wallops Release 63-39)

- Andrew E. Potter, Jr., Lewis scientist, described Lewis plans to launch a man-made comet. Such a comet, he said, would answer a number of questions. "If the simulated comet duplicates the behavior of an actual comet, we will have a better understanding of natural comets," he said. (LRC Release 63-13; Lewis Chronology, 3)

- Lt. Col. Stanley C. White (USAF), space surgeon and Chief of Crew Systems Div., NASA Manned Spacecraft Center, would be reassigned to Aerospace Medical Div. of Brooks AFB, MSC announced. Effective June 2, reassignment would conclude "detached duty" from USAF with NASA since October 1958. (MSC Release)

- Soviet Cosmonaut Major Andrian Nikolayev, speaking over Radio Moscow in a "Cosmonautics Day" broadcast, called U.S. Project Mercury spacecraft SIGMA 7 an "unreliable American contraption. . . . To be frank, I would not like to find myself in his [Astronaut Walter M. Schirra's] place. You have no idea how cramped and uncomfortable this capsule looks in comparison with the cabins of our remarkable Vostok cosmic ships." Nikolayev had seen SIGMA 7 during his recent visit to São Paulo, Brazil, where the spacecraft was on display. (AP, *Balt. Sun*, 4/13/63)

April 12: Prof. Eugen Saenger, Prof. of Space Research at Technical Univ. in Berlin, predicted in interview that if manned observation stations were established on the moon "no large-scale happening of any consequence anywhere in the world could go unseen." Noting that it costs several thousand dollars to orbit a pound of payload, he predicted that by 1975 cost would be only $25 per pound. Dr. Saenger's interview was published in Good Friday issue of German weekly publication *Christ und Welt*. (CDNS, *Wash. Post*, 4/15/63)

April 13: COSMOS XIV scientific satellite launched by U.S.S.R. into orbit (apogee, 318 mi.; perigee, 165 mi.; period, 92.1 min.; inclination, 48° 57' to the equator). Tass said satellite was carrying equipment "designed to continue the study of outer space," and that the satellite was functioning normally. (Tass, *Kraznaya Zvezda*, 4/14/63, 1, AFSS-T Trans.; *NYT*, 4/15/63, 25; UPI. *Wash. Post*, 4/14/63)

- Second of two Aris (advanced range instrumentation ship) ships for AMR, *Gen. H. H. Arnold*, docked at Port Canaveral after shakedown cruise from Bayonne, N.J. (*M&R*, 4/22/63, 20)
- Balloonists Don Piccard and Ed Yost, both of Sioux Falls, S.D., crossed the English Channel in hot-air balloon—first time in history such a crossing was completed. Flying mostly at 13,000-ft. altitude, they made the trip from Rye, England, to Gravelines, France, in 3-and-¾ hours. (AP, *Wash. Sun. Star*, 4/14/63, A4)
- Gas-filled balloon exploded near Herbede, West Germany, killing three persons and injuring eight. Balloon had flown with crew of four from Essen, had landed safely after flight, and was being deflated in a field when explosion occurred. (Reuters, *Wash. Post*, 4/14/63)
- Retired Admiral of the Soviet Fleet Ivan S. Isakov said in *Izvestia* interview that nuclear-powered submarine *Thresher* had been compelled to undertake "risky experiments." Admiral Isakov charged: "The sinking of the newly charged atomic reactor is sure to contaminate the Atlantic waters, all the more so since the tragedy took place in the northern current of the Gulf Stream. . . ." U.S. naval officials had repeatedly asserted that sunken reactor created no radioactivity hazard. (*NYT*, 4/14/63)

April 14: British-French supersonic transport, to be in commercial service by 1970, was described in joint statement by the two developing companies—British Aircraft Corp. and Sud-Aviation of France. Problem of sonic boom greatly influenced design of the mach 2.2 airliner, named "Concorde." "By adopting special operational techniques and by not flying at supersonic speeds below certain altitudes, disturbance can be reduced to a minimum." Airfield noise was "expected to be little more than that from current subsonic jets." (*NYT*, 4/14/63)

April 14: Reported that Marshal Nikolai I. Krylov, commander of Moscow military district, had been appointed Chief of Soviet Rocket Forces, replacing Marshal Sergei S. Biryuzov, recently named Chief of Staff of the Soviet Armed Forces. (AP, *NYT*, 4/16/63)

- Report by Legislative Drafting Research Fund of Columbia Univ., for National Security Industrial Association, was made public, the report examining statutes covering Federal compensation for deaths, injuries, and property losses caused by defense and space-related accidents. Research found victims of accidents arising from AEC activities were almost completely protected under existing laws, but that this was not the case for DOD and NASA programs. Accidents in these programs would probably be infrequent but might be "catastrophic," report said. Entitled "Financial Protection Against Risks of Major Harm in Government Programs," study was prepared under general direction of John M. Kernochan, Director of the Fund. (*NYT*, 4/15/63)

April 15: Memorandum titled "Space Race with U.S.S.R." from Dr. Edward C. Welsh, Executive Secretary of National Aeronautics and Space Council, to Vice President Lyndon B. Johnson, was made public. Dr. Welsh said U.S. space program was in the building stage this year and few space spectaculars could be expected in 1963; U.S.S.R. may attempt the following in 1963: manned orbital mission of eight days, manned orbital rendezvous mission, close-up photographs of planet Mars (with MARS I probe), and several unmanned lunar soft-landing probes. Memorandum had been written before launching of Soviet probe LUNIK IV, believed by many scientists to have been an unsuccessful attempt to soft-land on the moon. (Text; *Av. Wk.*, 4/15/63, 38)

- Hughes Aircraft Co. and Space Technology Laboratories had been selected by NASA for negotiation of contract for design, development, fabrication, assembly, and testing of four Pioneer interplanetary spacecraft. Award of contract, expected to be more than $10 million, would be made to one of these two companies in about a month. (NASA Release 63-74)

- FAA concluded two day conference (April 12 and 15) on supersonic transport aircraft with British and French aviation officials in Washington. This was first in series of meetings to develop international certification standards for supersonic transports, result of joint effort planned last February in Europe by FAA Administrator Najeeb Halaby and his British and French counterparts. Subjects discussed included:
"(1) systemworthiness standards and new areas requiring coverage, (2) flight, service and reliability testing, (3) sonic boom and noise, (4) handling qualities and performance, and (5) structural problems." (FAA Release T63-20)

- Discovery of new atomic particle, Phi-meson, was reported in *Physical Review Letters* by the two groups of researchers—Brookhaven National Laboratory, Long Island, and Univ. of California's Lawrence Radiation Laboratory, Berkeley. New particle had been predicted by Dr. Jun John Sakurai last December, based on research by team from these two institutions and on speculation by nuclear theorist Dr. Murray Gell-Mann, Prof. of Physics at Cal Tech. (*NYT*, 4/15/63, 1, 14)

April 15: Dr. Arthur R. Kantrowitz, Vice President of AVCO Corp., interviewed in *U.S. News & World Report*, said the moon was "set up as the priority target in our space program" because, at the time of the decision, "it was the most obvious goal .... But I think that, while it does provide us with a center for vigorous activity, it precluded other imaginative things that we might be doing, because it absorbs so much of our funds ....

"I am one of those who favor, as our next objective, the permanent manned station in a low orbit around the earth .... It would give us a chance to learn to live and to work in space. It would give us a chance to achieve a larger vision of the potentialities of space .... A laboratory like this would enlarge man's horizons in an important way. ..."

Dr. Kantrowitz said there was "certainly a possibility of going to the moon via a manned orbital station," and that this might be the Soviet method. Asked if U.S. should be working on this approach in addition to LOR, he said: "If we had two approaches, I would be willing to bet that it would accelerate our moon program, because there is a chance that either of these approaches will run into serious difficulties. Having two baskets for our eggs would seem to me to be a much more reasonable thing. Note that before the expensive later phases are reached by either approach a choice may be necessitated for economic reasons.

"Furthermore, we might find after we're deeper into the space program that there are more exciting things to be done than landing a man on the moon." (*U.S. News*, 4/15/63, 64–66)

- India's Defense Production Minister K. Raghu Ramiah said India would manufacture missiles with help of U.S.S.R. (*N.Y. Herald Trib.*, 4/16/63)

April 16: NASA announced it would negotiate with General Electric Co. for an extension of GE's Project Apollo support effort to provide plant and test support services at NASA Mississippi Test Facility. Estimated cost of initial one-year contract was $1.5 million. (NASA Release 63–75)

- High winds forced postponement of an attempt to send two men to 85,000-ft. altitude in Project Stargazer balloon, Holloman AFB, New Mexico. Purposes of flight were to determine at what altitude stars cease appearing to twinkle and feasibility of manned balloon observatories. (AP, Wash. *Eve. Star*, 4/16/63, A3)

- 28-nation legal subcommittee of U.N. Committee for Peaceful Uses of Outer Space opened its three-week session at U.N. Headquarters, N.Y. (*NYT*, 4/14/63, 21)

- Soviet Ambassador to U.N., Nikolai T. Fedorenko, announced Soviets had broken off private talks with U.S. on uses of space and would submit its own declaration on space exploration to the Legal Subcommittee of the Committee on Peaceful Uses of Outer Space. (*NYT*, 4/17/63, 1; *Wash. Post*, 4/17/63)

April 16: AEC announced Chairman Glenn T. Seaborg would visit U.S.S.R. in May at invitation of Andronik M. Petrosyants, Chairman of Soviet State Committee for Peaceful Uses of Atomic Energy. Primary purpose of visit would be signing of memorandum of cooperation in peaceful uses of atomic energy, calling for exchange of scientists between U.S. and U.S.S.R. and exchange of technical papers on research results. (*NYT*, 4/17/63, 6)

- Senator Clifford P. Case (R.-N.J.) said in press conference he would urge Congressional investigation to determine whether U.S. was channeling too many resources into space programs to the detriment of other Federal programs. Senator Case said he would recommend to Senator Clinton P. Anderson (D.-N.M.), Chairman of Committee on Aeronautical and Space Sciences, that the committee conduct "a pretty thorough review of the whole space program." (AP, Wash. *Eve. Star*, 4/17/63)
- Saturn I ground test booster arrived at Cape Canaveral after 10-day trip from NASA Marshall Space Flight Center, Huntsville, Ala. It would be used with S–IV stage in series of tests to check out the launch complex. (Huntsville *Times*, 4/16/63)
- The first NASA Saturn S–IV stage destined to fly in space left the Douglas Missile and Space Systems Divisions at Santa Monica, Calif., on the initial leg of its journey to Cape Canaveral. (*Space News Roundup*, 1/8/64, 2)
- Donald L. Nored, Lewis engineer, described NASA contract with Stanford Research Institute to study one of the most complex problems in high-energy rocketry—simple ignition. Preliminary results offer promise that a tiny quantity of ozone difluoride mixed with liquid oxygen will cause the lox to ignite immediately on contacting liquid hydrogen. (LRC Release 63–15; Lewis Chronology, 3)
- Federal mediators announced tentative agreement between Boeing Co. and International Association of Machinists, to be voted on by union members April 17. Agreement was reached after nine days of intensive negotiation in Washington and averted the strike of more than 40,000 Boeing employees scheduled to begin midnight April 16. (*Wall Street Journal*, 4/16/63)

April 17: Whilden P. Breen, Univ. of Maryland research assistant, emerged from isolation chamber where he had lived since Nov. 17, 1962, under continuous and highly programed experimental environment. Breen was "in excellent physical condition" and would undergo extensive medical and psychological tests during the coming week. Experiment was conducted by Univ. of Maryland Space Research Laboratory, Dept. of Psychology, and financed by NASA grant. (Wash. *Eve. Star*, 4/17/63; Univ. of Maryland Release)

- Five New Jersey newspapermen held first press conference through space, using RELAY I communications satellite in 25-min. broadcast to Rio de Janeiro, Brazil. Photo of newsmen sent via RELAY I during conference was of good quality, Rio officials said. (AP, Santa Monica *Evening Outlook*, 4/17/63)

April 17: Dr. Donald E. Gault of NASA Ames Research Center told American Geophysical Union that moon dust, from one to 10 tons a day, was falling on earth. Dr. Gault estimated that 54 per cent of dust sprayed off moon (by impact of meteoroids) enters solar orbit; 40 per cent enters earth orbit; 3 per cent impacts earth directly soon after leaving moon; and 3 per cent hits earth after a month or more. He suggested methods of finding moon dust for analysis: launching rocket to collect sample particles from upper atmosphere; and obtaining samples of sea sediment in polar regions where, he said, layers of moon dust must have been deposited and trapped. (*NYT*, 4/18/63; *Wash. Post*, 4/18/63)

- Results of optical astronomy, radioastronomy, and radarastronomy observations of planet Venus as well as MARINER II Venus fly-by data were discussed at American Geophysical Union's annual meeting, Washington, D.C. Thomas Gold, theoretical astrophysicist at Cornell Univ., told the session that there were too many incongruities in data obtained on Venus' characteristics to rule out one conclusion in favor of another. Results of the various research efforts often appeared contradictory. (Simons, *Wash. Post*, 4/18/63)

- Dr. Bruce Murray of Cal Tech reported Venus observations from Mt. Palomar Observatory during December had detected a Venus "storm" 1,000 mi. wide. The "hot spot" on Venus was observed with ultrasensitive heat detector mounted on 200-in. telescope. Dr. Murray was addressing session of American Geological Union, meeting in Washington. (Wash. *Daily News*, 4/17/63; *Wash. Post*, 4/18/63)

- NASA Administrator James E. Webb said in address before Milwaukee Press Club Gridiron Dinner:

   "From the results of the past five years of effort, it is apparent that, if we are to achieve mastery of space, we must add substantially to our scientific knowledge and to our utilization of technology. The NASA program is moving forward on both of these fronts. Essentially, our efforts in the Gemini and Apollo programs, leading to exploration of the moon, will provide us with an understanding of the limiting factors in space utilization, and the skills which will be needed to meet such future requirements in the space environment as our nation decides to undertake . . . .

   "All of these skills are needed not only in the pursuit of peaceful rewards for mankind from space exploration, but for the development of any space missions which may become necessary to our national defense, to insure that space will not be used as an unchallenged avenue of aggression against us. As these skills increase, a growing close relationship between the nation's best experts in the underlying technologies is utilized to make the best decisions on new operational systems, whether civilian or military.

   "It is also recognized that our space program must be broad enough, including considerable basic research, to give us a flexible base suitable for a continuity of effort in a direction or directions perhaps still unknown. Our broad and varied space effort permits us to lead from our great national strengths—the remarkable flexibility of our economic, social and political systems, the great variety of our talents, and the vast resources of our university-industry-government complex.

"Accordingly, we have not allowed ourselves to be trapped in a narrow competition, but have developed a national space program which embraces a wide range of activities . . . ." (Text)

- *April 17:* Soviet declaration of 11 basic principles of space activity presented to Legal Subcommittee of U.N. Committee on Peaceful Uses of Outer Space. U.S. sources said proposals were virtually unchanged from those offered last year. (*NYT*, 4/18/63, 22)
- Discovery of planet ("Barnard's Star B") in constellation Ophiuchus was reported to astronomers at Tucson, Ariz., by Dr. Peter van de Kamp, Director of Swarthmore College's Sproul Observatory, Swarthmore, Pa. Six light years from earth, planet is one-and-a-half times mass of planet Jupiter or nearly 500 times mass of earth. Dr. van de Kamp's measurements of amount of wobble it produced in motion of its sun, Barnard, determined object to be a planet. Barnard's Star B was third planet discovered outside our solar system but the most nearly planet-sized one of all. (Wash. *Daily News*, 4/18/63; *NYT*, 4/19/63, 4)
- USAF announced selection of three contractors to begin development work on large solid-fuel rocket motors: Aerojet General Corp., Thiokol Chemical Corp., and Lockheed Propulsion Co. Thiokol and Aerojet would conduct parallel programs leading to demonstration static firing of half-length 260-in. motor of about 3 million-lb. thrust; depending upon results of this work, USAF would select one of the companies to proceed with development aimed at demonstration of 6-million-lb. thrust motor. In addition, Thiokol would work toward demonstration static firing of 156-in.-diameter motor of 3-million-lb. thrust, using two segments in solid-fuel motor; Thiokol would work toward demonstration static firing of 156-in.-diameter motor of one-million-lb. thrust, in single segment, steerable through use of movable nozzles; Lockheed would work toward demonstration static firing of 156-in.-diameter motor of one-million-lb. thrust, in single segment, steerable through use of deflectors. USAF was managing the solid-propellant rocket motor development under NASA-DOD agreement. (DOD Release 493-63)
- Draft report on two bills (S. 132 and S. 152) "to create the National Weather Council and to provide coordination and central direction for an accelerated program of weather research, basic and applied," sent to Bureau of the Budget for consideration and advice prior to submission to the Congress. Proposed council "would coordinate and centrally direct meteorological and weather research and experimental programs" of represented agencies and would consist of: Chief of U.S. Weather Bureau (Council Chairman), Director of NSF, Administrator of NASA, Secretaries of Agriculture, Interior, Army, Navy, and Air Force or their respective designees. S. 132 would also include Administrator of FAA. (NASA Leg. Act. Rpt. II/55 and 56; ltr, NASA administrator to Hon. Warren G. Magnuson, 4/17/63)

April 18: NASA launched 85-lb. scientific payload to 208-mi. altitude at exact moment EXPLORER XVII atmospheric structure satellite passed over the Wallops Island, Va., launch site, an unusual "first" in NASA sounding rocket program. Launched by Aerobee 300A sounding rocket, experiment obtained temperature data on electron and neutral particles and measured ion and neutral particle densities. Data from this experiment would be compared with similar data obtained from EXPLORER XVII as it passed over Wallops Island at 198-mi. altitude during its 236th orbit of earth. Miss distance was 35,000 yards. Preliminary evaluation by NASA Goddard Space Flight Center scientists indicated data were of excellent quality. (Wallops Release 63-40; *Av. Wk.*, 5/6/63, 21)

- X-15 No. 3 flown to 92,500-ft. altitude at speed of 3,770 mph (mach 5.51) by NASA pilot Joseph A. Walker in test to investigate heat transfer rates and unusual air flow over certain parts of the aircraft. More than 400 heat-measuring devices and 25 presssure-measuring devices were mounted on the aircraft to determine how heat caused by air friction is transmitted to aircraft's surfaces and to record information concerning air flow. (FRC Release; UPI, *N.Y. Herald Trib.*, 4/19/63; *M&R*, 4/29/63, 10)

- Data from EXPLORER XVII indicated the earth is surrounded by belt of neutral helium atoms, scientists of NASA Goddard Space Flight Center said at American Geophysical Union meeting. Based on preliminary data received one day after launch, Goddard scientists said EXPLORER XVII atmospheric structure satellite had sent back more than eight hours of scientific information on physics and chemistry of tenuous gases making up the earth's atmosphere. (NASA Release 63-79)

- Physicist S. Fred Singer, Director of National Weather Satellite Center, said at American Geophysical Union meeting that the millions of tiny copper dipoles to be placed in orbit in USAF Project West Ford could remain aloft for as long as a century. Dr. Singer added that he did not regard the orbiting copper dipoles as threat to scientific observations, but he criticized secrecy surrounding first attempted West Ford launching. He urged freer and earlier public discussion of proposed space projects and suggested that U.S. invite scientists from other nations to discuss proposed U.S. scientific experiments with possible global consequences. (*Wash. Post*, 4/19/63; AP, *Balt. Sun*, 4/19/63)

- USAF X-21A experimental aircraft was flown for first time, from Northrop Corp. facility at Hawthorne, Calif., to Edwards AFB. Calif., where it would undergo extensive flight tests. X-21A is modified WB-66 type twin-jet aircraft on which were installed new laminar-flow wings, new engines relocated back on fuselage, and laminar-flow control pumping system placed in former engine pod positions. Two X-21A's would study near-elimination of friction drag in test program at Edwards. (DOD Release 547-63)

- Informed NASA sources were quoted as saying Astronaut Leroy Gordon Cooper's Mercury flight MA-9 would be delayed about a week from the May 7 planning date because of technical problems in the Atlas booster. Also revealed was name Astronaut Cooper selected for his Mercury spacecraft—"Faith 7." (UPI, *Wash. Post*, 4/19/63)

April 18: Addressing meeting of Aviation and Space Writers Association, Vice President and General Manager of Boeing Co. Aerospace Div., Lysle A. Wood, proposed: (1) orbiting weather control station that could provide a "better foreign aid program than any of our efforts to date"; (2) orbiting industrial laboratories to supplement Government space laboratories; (3) orbiting cardiovascular laboratory for studying and treating selected patients in space environment; (4) research by chemists and others for substances, compounds, and organisms "for which we do not recognize a need or desire because we cannot imagine the possibility of their existence"; and (5) orbiting vehicles as new medium of advertising. He suggested that U.S. would benefit from commercial ventures into space, and such benefits could help pay for scientific and military programs. (*NYT*, 4/19/63, 86)

- AFOAR announced USAF Cambridge Research Laboratories scientists were developing a compact, lightweight instrument to give direct readings of refractive indices of the atmosphere at microwave frequencies. (OAR Release 4-63-5)
- U.S. Army's Nike-Zeus antimissile missile successfully performed inflight maneuvers in test flight at White Sands Missile Range. Missile was launched from underground cell. (DOD Release 557-63)
- Members of International Association of Machinists Local 2061 at Cape Canaveral returned to work after one-day walkout against Boeing Co. The local had rejected Boeing's contract offer, but returned to work while awaiting results of nationwide vote on the contract. (UPI, *Wash. Post*, 4/19/63)

April 19: U.S. proposal for a major international effort to establish worldwide weather forecasting system was presented by Richard N. Gardner, Deputy Assistant Secretary of State for International Organization Affairs, at meeting of American Geophysical Union and American Meteorological Society, in Washington. Proposed system would collect and analyze weather data on regional and global—rather than national—basis; it would call for expansion of ground weather stations and launching of weather satellites to obtain global views of weather patterns. He noted that recent U.S.-U.S.S.R. agreement for coordinated weather satellite launchings would "give increased significance to this international effort." (Finney, *NYT*, 4/20/63)

- TIROS V meteorological satellite, launched June 19, 1962, entered its eleventh month of operation, surpassing previous longevity production record of TIROS II. Since launch, TIROS V had collected and transmitted more than 53,000 cloud-cover photographs of which more than 42,000 were meteorologically usable. (NASA Release 63-77)
- Titan II missile, launched by USAF from Cape Canaveral, did not reach programed range because second-stage engine cut off prematurely. (DOD Release 563-63)

April 19: Dr. Philip H. Abelson, editor of AAAS magazine *Science*, criticized in editorial the four justifications cited for manned lunar landing program: propaganda value of program had been "vastly overestimated"; possibility of military applications seemed "remote"; "technological fallout" would recover "not more than a small fraction of the cost of the moon program"; and scientific questions about the moon could be satisfied with unmanned, instrumented probes. If a scientist is not among the crewmen, "the alternative of exploration by electronic gear becomes exceedingly attractive. The cost of unmanned lunar vehicles is on the order of 1 percent of the cost of the manned variety; unmanned vehicles can be smaller and need not be returned. Most of the interesting questions concerning the moon can be studied by electronic devices . . . . Unmanned exploration could provide the basis for realistic design of manned landing craft, thus decreasing the total costs and increasing the chances of success. A reexamination of priorities is in order." (*Science*, 4/19/63, in *CR*, 6/11/63)

- Maj. Robert M. White (USAF) was awarded the Distinguished Service Medal for his achievements as pilot of X-15 rocket research aircraft. Presentation was made by Secretary of the Air Force Eugene M. Zuckert in ceremony at Edwards AFB, Calif. (AP, Wash. *Eve. Star*, 4/20/63)
- Members of International Association of Machinists employed by Boeing Co. voted to reject new labor contract that had been proposed by IAM-Boeing negotiators. Union officials said no strikes would be sanctioned until further talks with managment. (*NYT*, 4/20/63)

April 20: Preliminary test of instrumentation to be used in joint Italian-U.S. San Marco Project was made with launching of two-stage Shotput sounding rocket from NASA Wallops Station, the rocket carrying 180-lb. instrumented payload to 265-mi. altitude. Flight was first in three-phase project being conducted by Italian Commission for Space Research and NASA, to be followed by further tests of San Marco instrumentation with launching of Shotput vehicle from towable platform in Indian Ocean and to be culminated in launching of scientific satellite into equatorial orbit from the platform. Basic objective of San Marco Project was to obtain high-altitude measurements of atmospheric and ionospheric characteristics in equatorial region. (NASA Release 63-76; Wash. *Sun. Star*, 4/21/63)

- President Kennedy said in convocation address at Boston College: ". . . there is indeed an explosion of knowledge, and its outward limits are not yet in sight. In some fields progress seems very fast, in others distressingly slow. It is no tribute to modern science to jump lightly to the conclusion that all its secrets of particle physics, or molecular lift, or heredity, or outer space, are now within miraculously easy reach. The truth is more massive and less magical; it is that wherever we turn, in defense, in space, in medicine, in industry, in agriculture, and, most of all, in basic science itself, the requirement is for better work, deeper understanding, higher education.

"And while I have framed this comment in the terms of the natural sciences, I stand squarely with those who insist that at every level of learning there must be an equal concern for history, for letters, and the arts—and for man as a social being, in the widest meaning of Aristotle's phrase. This also is work for the universities . . . ." (Text, in *CR*, 4/23/63, 6494–95)

April 20: D. Brainerd Holmes, speaking before American Society of Newspaper Editors meeting in Washington, said of Project Apollo: "If we do not make these efforts, we will not be first on the moon, we will not be first in space and, one day soon, we will not be the first on earth." (*Wash. Post*, 4/21/63)

- NASA Administrator James E. Webb, speaking at American Society Newspaper Editors meeting in Washington, said purpose of U.S. space effort was "not to make a Roman holiday or a stunt, but to build a solid base of technology." Replying to charges made by Dr. Philip H. Abelson in *Science* editorial, Mr. Webb said he did not believe Dr. Abelson had given the same careful "scrutiny of facts" about manned lunar program that he would give to scientific experiment in his laboratory. Mr. Webb said he thought Dr. Abelson, in criticizing manned lunar landing effort, had exercised "editorial license" not borne out by the facts. (Wash. *Sun. Star*, 4/21/63; *Wash. Post*, 4/21/63)

- Dr. Robert Jastrow, Director NASA Goddard Institute of Space Studies, speaking at the American Society of Newspaper Editors, attempted to explain the "scientific effort" in the context of the entire space program:

  "I think, personally, that there is also a powerful general stimulus which comes from the need for developing new ideas and technologies to operate in and master a new environment. One must go back to the explorations of the late 15th Century to find a parallel to what we are about to witness in the next 30 to 50 years. Those explorations 450 years ago and their results awakened interest in the world and an intellectual ferment which were the necessary foundations for the development of a scientific revolution.

  "The ideas of a scientific revolution could not have flowered in a static society with fixed ideas of what could and could not be done. It took the explorations of the 15th and early 16th centuries to shake established notions enough to permit the birth of modern scientific thought. In a subtle and deep sense, this is one of the reasons for the interest of the scientist in the exploration elements of the space program . . . .

  "I think that the history of the last 100 years demonstrates that we need not expect a very long wait before space research will feed back into everyday affairs and play a role in them.

  "In the 19th century, the lag was rather long. The Scottish physicist James Clerk Maxwell published a treatise on electromagnetism in 1864, but not until 1901 did Marconi transmit the first wireless signal across the Atlantic.

  "More recently, in 1932, Chadwick discovered the neutron in England. Fermi set the first atomic pile critical at Stagg Field in Chicago in 1942, only 10 years from the basic discovery to a very formidable application.

"Still more recently, Bardeen and Brattain and Shockley at the Bell Labs discovered the transistor in 1948 and only six years later the first transistorized power amplifier appeared on the market . . . .

". . . the exploration of the moon has a very special role to play, precisely for the reason that would make one think the moon is uninteresting scientifically—specifically because the moon is a relatively lifeless body. It has no atmosphere, no oceans, nothing to wear away the record of the history of the solar system, and of whatever has occurred to the surface of the moon since its birth.

"On the earth, record of even tremendous surface events disappear within 10 to 50 million years, a very short time compared to the age of the planet, and mountains form and are turned over. With the action of weather, everything changes and disappears as the millions of years go by. That is probably also true on Mars and Venus which resemble the earth in this respect. But the moon has no oceans, no atmosphere, no mountain-building activity that we can discern. For this reason, the moon is a kind of Rosetta Stone of the solar system from which we can read the past. That is why it is so interesting to the scientist. He has the opportunity to find there the record of the early history of the solar system.

"To the student of the origin of the earth, the moon is even more interesting scientifically than Mars or Venus.

". . . if the moon and the planets were formed out of gas and dust in the condensation that surrounded the primitive sun, then we can expect such condensation to accompany the birth of every star. There are one hundred billion stars in our galaxy alone. Nearly every one of them we expect to have a planetary system around it. Some small fraction of those planets will be at the right distances from their suns, from their stars, and of the right size to have conditions favorable to the development of physical life as we know it.

"If that is the case, it follows that life must be a relatively commonplace thing, physical life in the universe . . .

". . . there is one small instrument on the Tiros which has not achieved as much attention as it deserves. The size of a quarter, it is a black disc which detects the infrared energy radiated from the top of the atmosphere. That is the energy which is reflected back into space from the earth. We can accurately estimate the energy coming to the earth from the sun and we can take the difference between the two and calculate the amount of energy remaining in the atmosphere.

"This energy deposited in the atmosphere constitutes the driving force that generates weather activity. With the aid of satellites, for the first time we can measure the energy difference and hope to break through from the day-to-day forecasting, which depends on immediately preceding events, to the week-long or two-week forecasts." (Text)

April 20: During prelaunch preparations at Holloman AFB, N.M., Project Stargazer balloon broke loose and soared away, finally landing 120-mi. east of Holloman. Balloon was to have carried gondola with two men, Capt. Joseph Kittinger (USAF) and astronomer William White, and their 12.5-in. telescope to study stars from above most of earth's atmosphere. USAF spokesman said sudden wind blew the moored balloon from side to side, creating static electricity which triggered a safety mechanism to release the balloon. Damaged balloon would require repairs by manufacturer before it could be re-used. (AP, Wash. *Sun. Star*, 4/21/63)

April 20: Dr. Michael J. Lipschutz, NASA Goddard Space Flight Center astrochemist, reported at American Geophysical Union meeting in Washington that 2,900-lb. iron meteorite Navititas found in Mexico in 1872 had been on earth for 800,000 years. In addition, meteorite may have traveled through space for 300 million years in asteroid belt before impacting on earth. Age was measured by examination of radioactive age of certain of meteorite's chemical elements. (Goddard Release)

- American Geophysical Union, at annual meeting in Washington, adopted resolution urging that FCC protect Channel 37 for radio-astronomy uses. FCC had received applications for operation of commercial TV on that channel in Paterson, N.J., and scientists feared that broadcasts from such station would interfere with research being performed with radiotelescope at Danville, Ill. FCC had proposed in March to exclude for five years commercial stations on Channel 37 within 600 mi. of Danville and to limit nighttime hours that stations could broadcast on Channel 37. AGU resolution asserted that the "needs of radio astronomy are not met by the limited protection" provided by FCC proposal and that the proposal "leaves unprotected nearly all of the active radio astronomy observatories in the United States." (*NYT*, 4/22/63, 24)

- In interview during annual meeting of American Geophysical Union, Dr. Lewis B. Kaplan of JPL said MARINER II's discovery of 800° temperatures of Venutian surface indicated possibility that atmosphere of Venus and clouds surrounding that planet contained organic compounds—including formaldehyde, oily hydrocarbons, and certain methyl compounds. Although large quantities of carbon dioxide were present in Venutian atmosphere, carbon dioxide alone would not account for the "greenhouse effect." Other substances—probably organic compounds—existed in Venutian atmosphere to absorb heat and radiation from planet and reflect it back to surface. (Carey, AP, Wash. *Eve. Star*, 4/20/63)

- Secretary of Defense Robert S. McNamara said in speech before American Society of Newspaper Editors: ". . . if we were to draft every scientist and engineer in the country into weapons development work, we could still develop only a fraction of the systems that are proposed . . . . This process of choice must begin with a requirement for solid indications that a proposed system would really add something to our national security . . . .

"Development costs alone on typical major weapons systems today average upward of $1 billion. Over a billion dollars was spent on the atomic airplane, which was little closer to being a useful weapon when we canceled it, shortly after taking office, than it had been half a dozen years earlier. Eighty million was spent on the Goose decoy missile, essentially a pilotless aircraft that the enemy would confuse with our B-52's . . . .

"The RS-70 is an example of a weapon which, it seems to me, fails to meet the basic requirement for a major systems development: a solid indication that the weapon, if developed, would add significantly to our national security. It happens to be a particularly expensive weapon: to develop, procure, and operate a modest force of these planes would cost us at least $10 billion. Yet considering the weapons we already have, or will have by the time the RS-70 could be operational, it is very hard to see how this weapon would add to our national security . . . .

"As weapons systems grow more complex, more expensive, and more difficult to maintain in a high state of military readiness, it is essential that we limit as far as possible the number of new systems that we bring into operation; for we want to be as sure as possible that we can depend on every system to operate when it is really needed. A basic fact of life is that under the chaotic conditions of combat you do not get anything like the efficiency of weapons systems that you get on a test range. Relative simplicity is a most desirable characteristic of a weapons system, or of a combination of systems . . . ." (Text, in *CR*, 4/24/63, 6580-83)

April 21: Dr. Jerome B. Wiesner, Science Advisor to the President, said in speech before Federation of American Scientists in Washington that the field of education was "probably the most backward field in our modern society" and that scientists were largely to blame. Pointing to "widespread lack of understanding" about science and technology, he said:

"It is surprising that we scientists and technologists in universities who, in a sense, have made major contributions to the great revolution through which we are living, have failed to apply our methods to our own profession.

"Modern teaching aids have been rarely employed, either in general education or in the universities. The process of invention has been almost ignored. The most highly developed devices in common use are still the old-fashioned textbooks, the blackboard, the written examination . . . ." (Simons, *Wash. Post,* 4/22/63)

• Dr. Homer E. Newell, NASA Director of Space Sciences, spoke at American Nobel Memorial Foundation dinner, New York:

"Because of the NASA policy of no exchange of funds, and of insisting on a mutuality of interest in entering into . . . cooperative programs, various national programs grow up with a viability that they might otherwise not have. They are not little NASA programs implanted on foreign soil. They are genuinely national programs, with strong national support, and vigorous national participation. They serve to strengthen the science and technology of the cooperating country in a way that only such direct involvement can. Moreover, in the case of scientific programs, since a universal element of these efforts is an agreement that scientific results will be made generally available to the scientific literature, the entire world community of nations benefits." (Space Quotes, 9/15/63)

April 21: Astronaut John H. Glenn, Jr., said in nationwide TV interview that U.S. could and would land a man on the moon: "We need no big, new scientific breakthroughs to complete this, we feel. We do not need the discovery of something as large as atomic power or anything like this in order to make the moon flight. Most of the technical details are known. It is now a lot of very hard and long engineering spade work and test procedure to prove out all these details and put them all together on the lunar mission." ("Meet the Press," NBC-TV, 4/21/63; AP, *Wash. Post*, 4/22/63)

- Sir Bernard Lovell, Director of Britain's Jodrell Bank Experimental Station, said in article in *New York Times Magazine:* "The challenge of space exploration and particularly of landing men on the moon represents the greatest challenge which has ever faced the human race. Even if there were no clear scientific or other arguments for proceeding with this task, the whole history of our civilization would still impel men toward the goal. In fact, the assembly of the scientific and military with these human arguments creates such an overwhelming case that it can be ignored only by those who are blind to the teachings of history, or who wish to suspend the development of civilization at its moment of greatest opportunity and drama." (*NYT Magazine*, 4/21/63, 12f)

- 22-orbit MA-9 flight of Astronaut Leroy Gordon Cooper was reported delayed from planning date of May 7 until at least May 21 because of difficulties in Atlas booster's guidance system. (Hines, Wash. *Sun. Star*, 4/21/63)

- NASA Plum Brook Research Reactor, Sandusky, Ohio, reached full power for the first time. 60,000 kilowatts of thermal energy were produced at end of two-month-long approach to power. The reactor is scheduled for basic research experiments associated with NASA's plans to develop a nuclear rocket for interplanetary exploration. (Lewis Chronology, 2; LRC Release 63-20)

- NASA announced award of study contract to Bendix Systems Division to compare behavior of soils in earth environment with that in vacuum chambers simulating lunar temperatures and pressures approaching lunar atmosphere. Results of 10-week study would assist in planning additional soil behavior research for planning possible future lunar roving vehicles. (NASA Release 63-78)

- NASA contract to High Voltage Engineering Corp. of Burlington, Mass., for laboratory study of meteoroid impact effects on space vehicle components was reported. (Boston *Sun. Herald*, 4/21/63)

- Air terminal system run by Port of New York Authority—including New York International and LaGuardia Airports in New York, Newark and Teterboro Airports in New Jersey, and two commercial heliports in Manhattan—set records in all categories of air traffic in 1962, the Authority reported. (*NYT*, 4/21/63)

- Tass announced Bruno Pontecorvo, Italian-born physicist who had been leading member of British Harwell Nuclear Research Institute before defecting to U.S.S.R. in 1950, had been awarded Lenin Prize for his pioneering research with neutrinos. (AP, Wash. *Sun. Star*, 4/21/63)

April 22: In letter to FCC Chairman Newton N. Minow, Director of National Center for Atmospheric Research Walter Orr Roberts urged that Channel 37 not be opened to television broadcasting but be reserved for radioastronomy uses. "I hope you will continue your study of this critically important question until a clear decision is reached on the public interest. If the channel is lost for research purposes, I believe that science will be significantly harmed . . . ." (Letter, 4/22/63, in *CR*, 5/1/63, 7123–24)

- U.S.S.R. placed COSMOS XV scientific earth satellite in orbit (apogee 317 km.; perigee, 173 km.; period, 89.77 min.; inclination to equator, 65°). Tass said onboard the satellite was "scientific equipment intended for continuing the space research according to the program announced by Tass on 16 March 1962." Onboard equipment was functioning normally. (Tass, *Pravda*, 4/23/63, AFSS–T Trans.)
- "Faith 7" Mercury spacecraft was installed on Atlas booster at Cape Canaveral, in preparation for Astronaut Leroy Gordon Cooper's MA–9 flight. (UPI, *NYT*, 4/23/63)
- USAF canceled Project Stargazer (manned balloon astronomy experiments) because of lack of funds. (*Chic. Trib.*, 4/23/63)
- Astronaut John H. Glenn said in speech before annual luncheon meeting of Associated Press: "As a result of the space program, tremendous and vast new areas of information appear on the horizon and are increasing daily. Consequently, this space program will undoubtedly serve as a catalyst toward an expansion of information that will dwarf anything we have known heretofore . . . .

    "What we need is a national repository for information, organized not along traditional library-type lines, but designed to utilize all the techniques at our disposal for sorting, cataloguing, analyzing, storing and making readily accessible, new information as it becomes available . . . .

    "In short, I propose that we lose no time in establishing a National Information Center. Control of such an institution could be by government, private interests, or a combination of appropriate interests . . . ." (Text)
- Three men entered compression chamber at Naval Gun Factory for eight-day experiment in living under 45 lbs. of pressure. (Wash. *Daily News*, 4/23/63)
- John L. Sloop, Director of Propulsion and Power Generation in NASA Office of Advanced Research and Technology, testified on NASA high-energy propellant program before House Committee on Science and Astronautics' Subcommittee on Manned Space Flight: "Our interest in high energy propellants dates back to the middle-forties. A large amount of research on high energy propellants was conducted by NACA. This research contributed to the development of the RL–10 engine. It also led to the choice of oxygen-hydrogen for the upper stages of Saturn I and Saturn V . . . .

    "Let me assure you, however, that we are not thinking or dealing in the past. I firmly believe there are many opportunities ahead to use high energy propellants in space exploration . . . . One phase of high energy propellant research in the 1953-1959 period is just beginning to reach fruition and offers significant potential gains in payload for existing vehicles and future vehicles. More work remains to be done . . . ." (Testimony)

April 22: Invention of edible structural material by Dr. Sidney Schwartz, Grumman Aircraft Engineering Corp. physiologist, was reported in *Aviation Week and Space Technology*. Harder than tempered masonite, the edible structure could "reduce the need for backup food in space flights," Dr. Schwartz said. (*Av. Wk.*, 4/22/63, 34)

- General Telephone and Electronics Corp. filed application with FCC for authority to purchase stock in Communications Satellite Corp. (AP, *NYT*, 4/23/63)
- Activities of Amateur Rocketeers of America (ARA) at National Amateur Missile Analysis Center (NAMAC) near Indianapolis were described in *National Observer*. NAMAC was built by ARA, national organization of high school-age rocket enthusiasts, and put into use in autumn 1962. Completion was planned for summer 1963. (*National Observer*, 4/22/63)
- National Academy of Sciences presented its awards during the 100th Annual Meeting of the Academy, in Washington. Draper Medal for investigations in astronomical physics was presented to Dr. Richard Tousey of NRL in honor of his achievements in solar spectroscopy, a principal source of information about the chemical and physical structure of the sun. Under Dr. Tousey's direction, series of high-altitude probes beginning in 1946 with use of captured German V-2 rockets produced the first detailed record of solar radiations in the far ultraviolet region of the spectrum. (NAS Release)

April 22-24: American Physical Society meeting held in Washington. At meeting, Dr. William A. Fowler, Cal Tech astronomer, presented photographic evidence he believed supported theory that gigantic galactic explosions occur regularly in the universe. With Dr. Fred Hoyle, Cambridge Univ. astronomer, Dr. Fowler contended that such explosions were only explanation for the extremely powerful radio emissions detected in select points of distant universe. (*NYT*, 4/28/63, 67)

Dr. Donald D. Clayton, Cal Tech physicist, described new method of dating creation of physical substances, at meeting of American Physical Society in Washington. Studying osmium-187 and osmium-186, Dr. Clayton concluded that iron and other heavy substances were formed 6-10 billion years before formation of our solar system, scattering into interstellar space with explosions of supernovae. These elements mixed with hydrogen clouds, from one of which the solar system was eventually formed. (*Wash. Eve. Star*, 4/23/63; *NYT*, 4/23/63, 17)

- American Institute of Aeronautics and Astronautics and NASA co-sponsored second Manned Space Flight Meeting, in Dallas, Tex. (Advance Program)

Astronaut M. Scott Carpenter described flight experiences in Project Mercury at the AIAA Second Manned Space Flight Meeting, Dallas, and concluded: "I think this philosophy sums up our attitude toward manned space flight. It is the increasing importance of man in the system that is going to characterize our space flights of the future. We will need skilled pilots to fly these missions, and we are in the process of developing those skills now.

"The era of manned space flight is just beginning, however, and Project Mercury is only the first step. We are convinced that it has been a very profitable step and that succeeding programs in manned space flight will continue to expand our knowledge of the universe, hopefully for the benefit of all mankind." (Text)

April 23: Aerobee sounding rocket launched from NASA Wallops Station carried 195-lb. instrumented payload to 125-mi. altitude in 7½-min. flight. Conducted by NASA Goddard Space Flight Center, experiment was designed to measure electron densities in ionosphere by several methods and to yield information necessary to further development of direct measurement techniques. (Wallops Release 63–42)

- Dr. Hugh L. Dryden, NASA Deputy Administrator, was elected to third four-year term as Home Secretary of National Academy of Sciences. (NAS Release)
- Illinois Governor Kerner said in address before Rotary Club in Chicago: "With a greater share in the research efforts of the space age, Illinois and its taxpaying citizens would cease to be the underwriters for the doctors of philosophy other states recruit to advance their programs." (*Chic. Trib.*, 4/24/63)
- United Arab Republic announced it planned to orbit a weather satellite this year. (Wash. *Daily News*, 4/23/63)
- At AIAA Second Manned Space Flight Meeting, Dallas, presentation by Warren J. North of NASA Manned Spacecraft Center and William B. Cassidy of Ling-Temco-Vought described Gemini launch escape system, and said: "A manual abort system will provide added operational flexibility by enabling the flight crew to choose an abort time which may reduce the possibility of aborts at high dynamic pressure; to choose optimum abort times compatible with contingency recovery areas; and to reduce the probability and the risk involved with an inadvertent abort . . . ." (Text)
- Dr. Joseph F. Shea, Deputy Director of NASA Office of Manned Space Flight, said at AIAA Second Manned Space Flight Meeting in Dallas that aerospace industry must improve quality of rocket and spacecraft components to ensure successful completion of Project Apollo. "Developing reliability of thousands of components so they all work together without error is the one main area where we need a big jump in the state of the art in the Apollo lunar landing program." (AP, Wash. *Eve. Star*, 4/23/63)
- Vice President Lyndon B. Johnson, Chairman of National Aeronautics and Space Council, said in address before AIAA Second Manned Space Flight Meeting in Dallas: "The age of space is, for all regions of America, a second industrial revolution, a revolution that is bringing a new future, a new degree of participation, a new scope of contribution for every region of our country, and for every segment of our society."

Regarding Project Apollo, he said that "to reach for the moon is a risk, but it is a risk we must take." Failure to do so would be "even riskier." (UPI, *NYT*, 4/24/63)

April 23: John W. Stone, Advanced Studies Manager of Launch Vehicles and Propulsion Directorate, NASA Office of Manned Space Flight, said at AIAA Manned Space Flight Meeting that projected Nova launch vehicle would be so costly that it would require national policy ruling—such as President Kennedy's decision to award high priority to manned lunar landing program. There would be no need for such a decision on Nova until 1968 at the earliest, he said. (*M&R*, 4/29/63, 16)

• Saturn project engineers James B. Bramlet and Robert E. Lindstrom, of NASA Marshall Space Flight Center Saturn Systems Office, said at AIAA Second Manned Space Flight Meeting in Dallas that NASA heavy launch vehicle programs are on tight schedules but are "within the bounds of our capabilities .... [and can meet requirement for a] manned lunar landing in this decade." (Huntsville *Times*, 4/23/63)

• Dr. Wernher von Braun, Director of Marshall Space Flight Center, said he doubted that Russia has developed any new booster capability despite rumors that the Soviets are about ready to launch another "space spectacular." (*Dallas Morning News*, 4/24/63)

• Report on laser research presented at meeting of American Physical Society by Robert W. Terhune, Ford Motor Co. physicist. Ford scientists had succeeded in altering laser's ultraviolet-ray frequency to other frequencies, an achievement hailed as breakthrough toward using lasers for communications. (*NYT*, 4/24/63, 18)

April 24: President Kennedy, asked in press conference if he "had any cause to reconsider" his commitment to goals of U.S. lunar program, replied:

"We looked at it, of course, when we proposed our budget for this year. We are looking at it again in relationship to next year's budget. We are also looking at it because of the concern that has been raised in the Congress and out of the Congress. I have seen nothing, however, that has changed my mind about the desirability of our continuing this program.

"Now, some people say that we should take the money we are putting into space and put it into housing or education. We sent up a very extensive educational program. My judgment is that what would happen would be that they would cut the space program and you would not get additional funds for education. We have enough resources, in my opinion, to do what needs to be done in the field, for example, of education, and to do what needs to be done in space.

"Now, this program passed almost unanimously a year ago. What will happen, I predict, will be a desire perhaps, possibly, to cut it substantially, and then, a year from now or six months from now, when the Soviet Union has made another new, dramatic breakthrough, there will be a feeling of 'Why didn't we do more.' I think our program is soundly based. I strongly support it. I think it would be a mistake to cut it. I think time will prove, even though we can't see all the answers which we will find in space, that the overall expenditures have been worthwhile. This country is a country of great resources. This program in many ways is going to stimulate science. I know there is a feeling that the scientists should be working on some other matter, but I think that this program—I am for it and I think it would be a mistake to arrest it." (Transcript, Wash. *Eve. Star*, 4/25/63, A5)

April 24: NASA announced Langley Research Center had requested industry proposals for studying manned obital research laboratory systems capable of sustaining a four-man crew in space for one year. (NASA Release 63–81; Langley Release)

- Senate Committee on Aeronautical and Space Sciences began hearings on technical and scientific aspects of U.S. space program, preparatory to hearings on NASA proposed FY 1964 budget authorizations. Testifying were NASA Administrator James E. Webb, Deputy Administrator Dr. Hugh L. Dryden, Associate Administrator Dr. Robert C. Seamans, Jr., and Director of Space Sciences Dr. Homer E. Newell. (NASA Leg. Act. Rpt. II/59)

- NASA Administrator James E. Webb, testifying before Senate Committee on Aeronautical and Space Sciences, said: ". . . It is important to note that the budget requested for the total program of NASA for fiscal year 1964 includes less than $50,000,000 for new programs. The remainder is to maintain the projects now underway on schedule and to pay the contractors for the work they have obligated themselves to do to meet these schedules.

    "The policy on which this budget is based is the mastery of space, and its utilization for the benefit of mankind. This mastery, and the relation of our position to those of other nations, will not be determined by any single achievement. Superiority in the space environment will be won and very likely can be held by that nation which first fashions into usable systems all of the scientific knowledge, all of the technology, all of the experience, all of the space launch and terminal facilities, and all of the aids to space navigation required for safety and regular operation.

    "These are the capabilities and resources which the United States must have, and this budget is designed to make rapid progress toward acquiring them . . . ." (Testimony)

- NASA Deputy Director Dr. Hugh L. Dryden discussed accomplishments of NASA's international program and some future trends of U.S. space program in presentation to Senate Committee on Aeronautical and Space Sciences. He described U.S.-U.S.S.R. agreement for cooperation in weather satellite program and mutual experiments with Echo II passive-reflector satellite, and said: "The agreement itself provides for the usual sixty-day period during which each nation may change its mind on details and suggest changes. This gives an opportunity for review by various agencies within both governments, and changes can then be made without embarrassment. For that reason, the text of the agreement will not be released until two months from March 20 . . . ."

Dr. Dryden said Soviet maneuvers in Legal Subcommittee of U.N. Committee on Peaceful Uses of Outer Space were "a small cloud on the horizon" threatening "that this effort toward scientific cooperation may be terminated." He suggested that some scientists in Soviet Academy were non-Communist Party members and were eager for international contacts and cooperation; they apparently were allowed to discuss scientific and technical matters not involving politics. But possibility existed that "political elements" might "shut off" moves toward international cooperation—evidenced by Soviet "tendency to hint that scientific cooperation may not be so easy unless there is agreement on legal matters." (Testimony; NYT, 4/28/63, 41)

April 24: Dr. Homer E. Newell, NASA Director of Space Sciences, appearing before Senate Committee on Aeronautics and Space Sciences, reviewed status of NASA space sciences programs. "During 1962, the NASA space exploration program entered a new phase of its evolution. We passed beyond the relatively simple satellites and probes which characterized our early years and are heavily engaged in the development of advanced spacecraft and launch vehicles to accomplish the more rewarding, but more difficult missions. In doing so, we met with both dramatic successes as well as some temporary delays. In general, however, the year was one of intense activity and excellent progress. . . ." (Testimony)

- Senator Margaret Chase Smith (R.-Maine) read letter from Senator Wallace F. Bennett (R.-Utah) in session of Committee on Aeronautical and Space Sciences, the letter quoting three scientists who estimated that, at present pace, 60 per cent of all physical scientists would be working as NASA employees. Senator Bennett called on Republican members of Committee to see what steps were necessary to "prevent this concentration of scientific brains in Government at the expense of our private corporations and universities." (NYT, 4/25/63)

- In AIAA speech at Dallas on "Space Policy and Space Management," Dr. Edward C. Welsh, Executive Secretary of NASC, reviewed the national objectives in space:

  "First of all, there is an over-all policy objective, which the President has expressed as the determination of this country 'to become the world's leading space-faring nation.' It is worth mentioning that this is a broad umbrella. . . . Suffice to say that the space program is not a one-project design nor a short-run episode. It is a growing, expanding, multi-project program, which I predict will become an increasingly significant part of our way of life. As important as it is, the moon project is not the whole space program; it is just one portion and one phase of an over-all program which has breadth through variety and depth far into the future.

  "A few features of this broad space policy are:

  "1. *Its objectives are peaceful* . . . the distinction as to what is peaceful and what is not peaceful is a matter of intent and is not determined by what agency of the government engages in them . . . .

"2. Our policy asserts that space travel and space exploration are subject to *international law* and that such activity should be consistent with the provisions of the United Nations charter. . . .

"3. It is our policy to seek increased *international cooperation* in, and mutually advantageous agreements for, the orderly and open conduct of space and space related activities . . . .

"4. It is also our policy that the concept and performance of our space responsibilities shall be maintained and strengthened as a *national program* rather than a series of separate and unrelated projects. . . .

"5. There are inherent risks in space exploration, particularly in manned flight, and they must be recognized . . . .

"6. While we must continue to improve on state-of-the-art competences, it is essential that we strive just as hard to make so-called quantum *jumps in technology and performance* . . . .

"7. It is a significant element of policy that progress be the joint product of government and private enterprise. . . .

"In conclusion, the national space program is an essential and dynamic feature of our economy. It benefits the many rather than the few. It strengthens the nation, both at home and abroad. It looks forward rather than backward. It highlights the contributions of the most able technologists and the most able managers and gives them an unprecedented peacetime opportunity to repay more fully the special bonus they have received by living in this great country.

"We can afford the space program. We must afford it. We can be satisfied with nothing less than first place—to the moon, to the planets, and throughout the solar system." (Text)

April 24: Evidence suggesting gigantic explosion in space 100,000 years ago was reported by Dr. Jesse L. Greenstein of Mt. Wilson and Mt. Palomar Observatories. Spectrographic study led Dr. Greenstein to conclusion that there are at least two "radio stars" or "exploding gas clouds in very distant galaxies at distances of 2 billion and 4 billion light years respectively." Gas of each cloud weighs up to 100 million times as much as the sun, and the objects are 100 times brighter than Milky Way Galaxy. Dr. Greenstein said important consequence of findings was that it showed possibility of detecting such objects twice as far away as previously thought possible. Dr. Greenstein reported his findings at National Academy of Sciences meeting in Washington. (AP, Wash. *Eve. Star*, 4/25/63)

• In press interview at NAS, Dr. Martin Schwarzchild countered scientific critics of the U.S. space program, and said: "I am for it and I think it would be a mistake to arrest it.

"The moon project is the stirring spearhead of a movement under the wing of which man will do real exciting things. . . . We must do an enormous effort of sparkling and grand and wonderful things and out of them one little pearl will come. . . .

"Energy in one program will increase the push in all others. I hope art will blossom and pure science and philosophy. The first time you stop the present momentum you will not save money but you will stop the impetus for improving education in this country." (Loory, *N.Y. Herald Trib.*, 4/25/63)

April 24: Dr. Colin S. Pittendrigh, prof. of biology at Princeton Univ., urged at National Academy of Sciences meeting in Washington that President Kennedy proclaim U.S. opposition to contamination of Mars by landing probes on that planet. He added that U.S. should make available its research on decontamination to U.S.S.R. Dr. Pittendrigh emphasized he approved decision to relax sterilization of lunar probes but declared his opposition to landing spacecraft on Mars where possibility of life exists. (*NYT*, 4/25/63)

* USAF Atlas E ICBM was launched by SAC crew from Operational Systems Test Facility at Vandenberg AFB in successful flight test. (DOD Release 585-63)
* General Thomas S. Power (USAF), CinC of Strategic Air Command, predicted at AIAA Second Manned Space Flight Meeting that "space potentially offers unique military advantages, and we must anticipate that some nation or nations will endeavor to exploit such advantages to help them attain their political objectives.... We may find that, eventually, the only really survivable command and control structure would be one employing a maneuverable command post in space." (AP, Wash. *Eve. Star*, 4/24/63)
* Nike-Zeus antimissile missile successfully performed "strenuous maneuvers" in flight test at White Sands Missile Range, U.S. Army announced. Missile was launched from underground pit and guided by ground-based radar equipment. Information gained during Nike-Zeus development would be used in Nike X antimissile missile system, advanced version of Zeus. (DOD Release 579-63)

April 25: X-15 No. 1 flown by pilot John McKay to 105,500-ft. altiude at maximum speed of 3,654 (mach 5.32) mph in test to measure distortion of photographs at extreme speeds. Camera was recessed in underside of aircraft's nose. (AP, *Chic. Trib.*, 4/26/63)

* RELAY I communications satellite was used to transmit electroencephalograms ("brain waves") from Bristol, England, to Minneapolis, Minn., in demonstration experiment conducted in connection with meeting of National Academy of Neurology in Minneapolis. (NASA Release 63-85)
* After prolonged debate, Senate confirmed nominations of the 14 incorporators of Communications Satellite Corp. (*CR*, 4/24/63, 6606f; *CR*, 4/25/63, 6698-99)
* Dr. Raymond L. Bisplinghoff, Director of NASA Office of Advanced Research and Technology, summarized scope of NASA advanced research and technology program before Senate Committee on Aeronautics and Space Sciences: "... although the cohesiveness, technical direction, and much of the research effort of the program is provided by the NASA field centers, a close coupling of the Nation's universities and industry with the centers has been established. This three-way partnership is a key to the Nation's future pre-eminence in aeronautics and space." (Testimony)
* Dr. George L. Simpson, Jr., Assistant Administrator for Technology Utilization and Policy Planning, in an interview with

Neal Stanford of the *Wall Street Journal*, April 25, 1963, stated that: "The dozens of ways that space research and development is being transmitted into industrial profits today are only a trickle preceding the deluge that will follow, for the space age is only just begun.

"When one recalls that it took 112 years to develop photography to an advanced degree, 56 years to develop the telephone, 35 years to perfect the radio, 15 years to develop radar, with television taking 12 years, the atomic bomb 6 years and even the transistor 5 years, one can appreciate not only how small the beginnings but how vast the prospects of harnessing space R&D to domestic usage." (Space Quotes)

April 25: Panel of scientists, headed by Dr. Sverre Petterssen, prof. of meteorology at Univ. of Chicago, proposed in report to National Academy of Sciences' Geophysics Research Board a program of unprecedented collaboration between scientific institutions and national governments in atmospheric research, meteorological service, atmospheric surveillance, and meteorological education "to raise the effectiveness of the meteorological services on a worldwide basis." Report was outgrowth of U.N. General Assembly resolution in December 1961 calling on "all member states, the World Meteorological Organization and other appropriate agencies" to study ways of improving atmospheric sciences. Report was basis for U.S. recommendations to World Meteorological Organization, meeting in Geneva. (NAS-NRC Release; Wash. *Eve. Star*, 4/25/63, A3; *NYT*, 4/26/63, 4)

- NASA Manned Spacecraft Center announced Cook Technological Center Div. of Cook Electric Co. had completed first of new biomedical tape recorders for use in first two-man Project Gemini space flight. Each recorder would receive and record continuously, for 30 hours, six kinds of simultaneous biomedical signals from sensors within astronaut's spacesuit. Post-flight play-back of the tapes would provide continuous analysis of astronauts' physical and mental endurance in long periods of space flight, a major goal of Project Gemini. (MSC Release 63–78)

- Dr. Harold C. Urey, Univ. of California chemist, said at press conference in conjunction with National Academy of Sciences meeting in Washington that "more effort should be made to get more first-class scientists into the space program. If NASA really is going to use only one per cent of the nation's scientists—how many among those will be first-class?" Comparing NASA with AEC, Dr. Urey said: "The atomic energy program had a very positive base. Many top scientists worked in the program. But the space program is just the opposite. NASA grew too fast." (Huntsville *Times*, 4/25/63)

- Martin-Marietta announced signing of $280 million contract with DOD for development of Titan III missile. (*Wash. Post*, 4/26/63)

- Editorial in Washington *Evening Star* quoted Vice President Lyndon B. Johnson: ". . . To reach the moon is a risk. But it is a risk we must take . . . [for] failure to go into space is even riskier . . . . Where the moon is a major goal today, it will be tomorrow a mere whistle stop for the space traveler." Editorial cited "soundness of what he has had to say about our projected lunar voyages . . ." and added: "The warning is well-timed, especially so because of what some otherwise sensible and perceptive people have been saying about the alleged nonsensicality of flying to the moon when the money could be better spent on such urgent projects as building schools, sewage plants and more and more highways right down here on earth." (Wash. *Eve. Star*, 4/25/63)

April 26: USAF launched Blue Scout rocket from Pt. Arguello, Calif., with unidentified payload. (AP, Wash. *Eve. Star*, 4/26/63)

- USAF launched unidentified satellite from PMR using Thor-Agena vehicle. (UPI, *Chic. Trib.*, 4/27/63)
- USAF successfully launched advanced Atlas ICBM with new type of nose cone more than 5,000 mi. down AMR. (*Astro. and Aero. Eng.*, June 1963, 7)
- NASA announced formation of Research Advisory Committee on Biotechnology and Human Research, chaired by Dr. Charles I. Barron, flight surgeon and president of Aerospace Medical Association. Committee would report to Dr. Raymond L. Bisplinghoff, Director of NASA Office of Advanced Research and Technology, through Dr. Eugene B. Konecci, Director of Biotechnology and Human Research in OART. (NASA Release 63-82)
- NASA Acting Director of Office of Applications, Leonard Jaffe, appeared before Senate Committee on Aeronautical and Space Sciences, reviewing NASA programs of communications systems, meteorological systems, and future applications satellites.

    Describing division of responsibility between public interest (NASA) and Communications Satellite Corp., Mr. Jaffe said that "eventually, perhaps" NASA would find way to develop necessary power for bouncing radio and TV signals off satellites to "a very large area of the earth." Such transmissions would be beyond scope of ComSatCorp. and beyond its charter as common carrier for broadcasters. Yet ComSatCorp. might benefit from NASA-developed technology, particularly in development of tremendous power supplies. Under Communications Satellite Act, Mr. Jaffe said, NASA retained responsibility to cooperate with ComSatCorp. and to advise other Federal agencies on the subject, such obligations requiring NASA to maintain independent research capability in communications satellites. (Testimony; *Wash. Post*, 4/27/63)
- Testifying before Senate Committee on Aeronautical and Space Sciences, Edmond C. Buckley, Director of NASA Office of Tracking and Data Acquisition, described basic support job that tracking and data acquisition provide to NASA's flight projects, indicated major additional resources required, and concluded that "the success of the [space flight] mission depends on our ability to communicate, to receive telemetered data, to track, and to command." (Testimony)
- Senator Wallace F. Bennett (R.-Utah) announced he had asked Senate Committe on Aeronautical and Space Sciences to investigate complaint from three scientists that NASA was using "coercion" to recruit scientists. (AP, *Wash. Post*, 4/27/63)
- Floyd L. Thompson, Director of NASA Langley Research Center, announced he had formally accepted on NASA's behalf custody of

110 acres of Government land at Oyster Point, Newport News, Va., for use as site of Space Radiation Effects Laboratory being established by NASA. (Langley Release)

April 26: NASA Langley Research Center announced award of contract to Univ. of Virginia for developing "research apparatus and techniques for laboratory studies of drag on satellites." Under one-year contract, U. of Va. would develop suitable laboratory equipment and make laboratory measurements of molecular impact forces exerted on samples of spacecraft materials. (Langley Release)

- NASA Ames Research Center selected three companies for negotiation of contracts to study solar probe project—General Electric Co. Missile and Space Div., Martin-Marietta Co., and Philco Corp. Each contract would call for four-month study providing "information in depth for the procurement of any future Solar Probe spacecraft, should the program become fully authorized." (Ames Release 63-19)

- World Meteorological Organization, meeting in Geneva, agreed to establish "world weather watch" using U.S. and Soviet meteorological satellites and a network of meteorological centers, in attempt to make possible long-range weather predictions on global scale. WWW would use world centers, regional centers, and national centers, world centers being located in Washington, Moscow, and in as-yet-undetermined site in Southern Hemisphere. Regional centers—in New York, Moscow, New Delhi, Tokyo, Frankfurt, Brasilia, Nairobi, and Melbourne—would (1) select data for analysis and forecasting on global scale and pass it on to world centers, and (2) disseminate appropriate data to national centers. Costs of program were expected to be covered by international organizations, but WMO also appealed to national governments. (Reuters, *Wash. Post*, 4/27/63)

- Dr. Philip H. Abelson, Director of Carnegie Institution's Geophysics Laboratory and editor of *Science*, criticized costs of landing men on the moon in this decade: "What we are witnessing is the expansion of a new, sophisticated form of the prewar Public Works Administration. Science is being used as a 'front' for technological leaf-raking. . . ." Dr. Abelson was speaking at Univ. of Maryland. (Wash. *Eve. Star*, 4/27/63, A1)

- U.S. and Japan exchanged notes summing up agreement to increase missile defenses in Japan. (*Chic. Trib.*, 4/27/63)

April 27: NASA Administrator James E. Webb said in interview at Princeton Univ. alumni meeting in Washington that first manned flight in Project Gemini would not be made for at least a year. He attributed the slippage to "some difficulty with the utilization of Titan II" and "some problems" with Gemini spacecraft. "No one can forecast now when we will fly Gemini."

On subject of Soviet space competition, Mr. Webb said: "With their existing rockets, they will probably be able to do a multimanned flight around the earth a year before we will do that." (Hines, Wash. *Sun. Star*, 4/28/63)

- Titan II missile launched for first time from underground silo, in USAF test from Vandenberg AFB. Re-entry vehicle landed in preselected target area and flight was described as a complete success. (UPI, *Wash. Post*, 4/28/63, A1)

April 27: Robert J. Parks, Director of JPL Lunar and Planetary Program was quoted as saying sterilization of lunar spacecraft no longer appeared necessary, because "intense ultraviolet radiation on the surface of the moon would kill any microorganisms that might be carried up from the earth.

"If a probe should dig beneath the surface, it is believed there would be no chance of contamination spreading.

"These factors, coupled with the present lack of an assured method of achieving sterilization, make it appear that efforts to sterilize Ranger VI may be unnecessary." He emphasized that planetary probes would, however, be sterilized. (L.A. Times, *Wash. Post*, 4/27/63)

- Explosions at Allegheny Ballistics Laboratory, Rocket Center, W. Va., with three men presumed dead and 10 others injured. Explosion occurred in building used for processing chemicals, destroying building and several large trailers. ABL is Navy-owned facility, primarily a research center for solid-rocket propellants, operated by Hercules Powder Co. Cause of explosion was not immediately known. (Wash. *Sun. Star*, 4/28/63)
- *Daniel Webster*, 20th U.S. Polaris missile-firing submarine, launched at Groton, Conn. (Wash. *Eve. Star*, 4/27/63)

April 28: COSMOS XVI launched by U.S.S.R. into orbit of 401-km. apogee, 207-km. perigee, 90.4-min. period, and 65°01' inclination to equator. Tass said the satellite contained "scientific equipment intended to continue the space research in accordance with the program announced . . . March 16, 1962." (Tass, *Pravda*, 4/29/63, 3; AFSS-T Trans.)

- Senator Stuart Symington (D.-Mo.), asked in televison interview about the pace of the U.S. manned lunar effort, said: "I think we might move forward a little more slowly, but I do believe that those who control space, in the years to come, will control the world." ("Meet the Press," NBC-TV, 4/28/63)

April 29: NASA Deputy Director Dr. Hugh L. Dryden, at U.S. National Committee of International Scientific Radio Union meeting in Washington, referred to recent criticism of Project Apollo by scientists who had charged that scientific returns from manned lunar landing program did not justify $20 billion cost: "They've been setting up a straw man and knocking it down. . . . No one in NASA had ever said the program was decided upon solely on the basis of scientific return. . . ." He described the manned lunar program as "a technological program, though it does have some science content," and said it, like overall U.S. space program, was designed to give this country "mastery and pre-eminence" in space. Knowledge gained through Project Apollo, in rocket techniques as well as biosciences, would help prepare U.S. "for whatever we are called upon to do for both civil and military uses" of space. (AP, Balt. *Sun*, 4/30/63)

- In editorial on "Apollo and Its Critics," *Aviation Week and Space Technology* Editor Robert Hotz said: "Like all great pioneering ventures, the Apollo manned lunar landing program is the target of a continual barrage of carping. This anti-Apollo chorus usually reaches a crescendo in the spring, when the National Aeronautics and Space Administration budget is facing significant congressionl decisions, and dwindles to a barely audible pianissimo after every successful U.S. and Russian manned space flight.

"This year the critics' chorus is louder than ever and studded with many familiar faces ranging from ex-president Eisenhower, who still wishes the space age would dry up and blow away, to an esoteric wing of the scientific community which appears to be disturbed by the fact that non-scientist types, such as engineers and pilots, have crept into the space program and are running their portion of it with vigor and enthusiasm directed toward specific goals that some scientists regard as unseemly. . . .

"The scientific community has traditionally provided the most solid opposition to the manned space flight program, but there has been a steady rate of conversion from critics to supporters during the Mercury years. This has been helped by the fact that it has become obvious that NASA does not intend to neglect scientific exploration because of the Apollo priority. As the NASA program has unfolded with the years, it has become apparent that considerable scientific exploration by instrumented satellites and probes is a necessary preliminary to extending the parameters of manned space flight. It is also evident that manned space flight itself will offer a greater opportunity for scientific exploration of the universe than could ever be possible using only remotely controlled instruments and data transmission . . . . There are a great many valid, vital reasons why this nation must continue to press its exploration of space at the maximum feasible pace.

"We intend to discuss them as the Apollo debate continues. But for the moment, there is no better summation of the purpose in this national effort than that stated by Brainerd Holmes, NASA's director of manned space flight and the chief engineer of the Apollo program, when he said in a recent speech:

"'If we do not make these efforts we will not be first on the moon, we will not be first in space and one day soon we will not be the first on earth.'" (*Av. Wk.*, 4/29/63, 17)

April 29: AEC Chairman Dr. Glenn T. Seaborg, speaking at Centennial Charter Day Convocation, Univ. of Massachusetts, said: "If we look back a number of centuries to the Middle Ages, we find that man and society in the Western World were dominated by rather static concepts. The individual was born into a world that was essentially unchanging by our standards. He was not very important. Poverty, disease, toil, and natural calamities were simply an inevitable part of life. He accepted the proffered explanations of the world and the universe, and his curiosity was not encouraged. Man was a fearful and superstitious plaything of his environment. The concept that man might improve his lot was not widely considered. Nor were the tools at hand for such a task.

"The beginning of the changes in this outlook are generally identified with the Renaissance and the Reformation. The questioning and inquisitive human spirit could not be stifled forever. Men began to question old forms and beliefs—in religion, politics, and the law; in art and literature; and in science and philosophy.

"The process of breaking out of a closed world and liberating the mind was a slow one, and it is by no means complete. There were many landmarks, established by many brave men in all walks of life, on the route to liberation. An example was the overthrow, by the evidence produced by Copernicus and Galileo, of the dogmatic concept that the earth is the center of the universe. It is difficult for us today to estimate the impact, on a closed system of thought, of the knowledge that the earth and the other planets revolve around the sun. If such an important belief could fall, were not others vulnerable? While science was opening man's eyes to the truth of his surroundings, a change was in the making concerning the concept of man himself. The themes of individual rights and freedom were being developed in the period of the enlightenment and the first crystallization of these revolutionary ideas was in the American Revolution—the first revolution of which I am to speak to you today. Beginning with the Declaration of Independence and culminating with the adoption of the Constitution, the place of the State was confined to protecting the rights of the individual and becoming the instrument for man's self-government . . . .

"The freedom to embrace new ideas, the concept that men need not accept miserable conditions of life, and the hope inherent in science, unleashed as never before the economic energies of man. Our second revolution, the industrial revolution, and the rise of science and technology, saw a beginning of the realization of . . . a huge increase in industrial and agricultural production at a cost of less total labor, the conquest of many diseases, the lengthening of life, the improvement of health . . . .

"As we now enter the early stages of the scientific revolution—our third revolution—the full flower of creative evolution is bursting upon us. Yet powerful as the forces of science and technology have already demonstrated themselves to be, relatively speaking we are only beginning to feel their influence. Our Nation and the world are committed to an accelerating cycle of knowledge-gathering and knowledge-exploitation. We cannot visualize in detail the consequences for the future. We can, however, clearly see that we can, if we have the wisdom, create an environment essentially of our own choosing. The scientific revolution can provide for the essential needs of men around the globe. Our small world will grow smaller. At the same time our perspective will be broadened unimaginably by the exploration of space. We can only guess at the distant impact of this dynamism on man, his institutions, the content of thought and the quality of life. We can be sure, however, that successful adaptation to the changing environment can be achieved, in an atmosphere of freedom, only through advancement of men generally to higher levels of education . . . ." (Text, in *CR*, 4/29/63, 6847–49)

April 29: Addressing the Senate space committee, Deputy Defense Secretary Roswell Gilpatric summed up DOD's philosophy on space:

"The United States has a single national space program. . . . Space, like other mediums, affords useful and often unique ways for achieving defense objectives. Some of these are by no means uniquely military . . . similar remarks would apply to machines and devices used in the air, on land and on sea, and space is no exception. . . . As a corollary to these observations, the totality of our military space efforts, assessed from a national rather than a departmental standpoint, greatly exceeds the efforts of the Department of Defense alone. Space systems and devices are not simply military or non-military merely because they are developed by one agency or another." (*M&R*, 5/6/63)

April 29: General Bernard A. Schriever, Commander of the Air Force Systems Command, testifying before the House Science and Astronautics Committee, said that the Air Force was reviewing its space program, at the request of Defense Secretary McNamara, to determine how the Air Force might better achieve its military mission capabilities. Stressing the importance of both the Military Orbital Development Station (MODS) and Aerospace Plane (ASP), he said that MODS was "of fundamental importance to achieving various military capabilities in space, both manned and unmanned." (*M&R*, 5/6/63)

- Lt. Gen. Thomas P. Gerrity (USAF), DCS Systems and Logistics, Hq. USAF, testifying before the Subcommittee on DOD Appropriations of the House Appropriations Committee, stated that USAF was going to modify as space boosters at least 15 of the 64 Thor IRBM's being withdrawn from the U.K. but was dismantling the 45 Jupiter missiles that were withdrawn from Turkey and Italy. Rep. Daniel J. Flood (D.-Penn.) recalled that this Subcommittee had been in the thick of the original Thor-Jupiter controversy, asked "what is the peculiar, special, and significant and distinct thing about the Thor missile which would argue that 15 at least be modified" while Jupiter missiles were being discarded as "garbage." Gen. Gerrity replied that the reason was quantity rather than quality, that USAF had had more experience with Thor, and "We have launch facilities for the Thor but do not have launch facilities for the Jupiter." Rep. Flood pressed Gen. Gerrity on "the marked physical distinction [between Thor and Jupiter] which would rush you into this conclusion?" The General did not recall a marked distinction. Rep. Flood asked "did NASA twist your arm and insist they get Thor instead of Jupiter?" Gen. Gerrity said he knew "of no such circumstances." (*DOD Appropriations Hearings*, House, 88th Congress, 1st Session, Pt. 5, 973–77)

- Four-day meeting of U.S. National Committee of the International Scientific Radio Union opened in Washington. Six of the seven technical commissions met in simultaneous sessions at National Academy of Sciences. (NAS-NRC Release)

April 30: NASA launched two Deacon-Judi meteorological sounding rockets from Coopers Island, Bermuda, first in series of tests "to measure atmospheric winds and density at altitudes above conventional balloon level up to about 300,000 feet." Data from these tests would be used in support of AEC experiment to be launched by NASA four-stage Scout vehicle from Wallops Island, Va., on planned re-entry south of Bermuda, objective of such experiment being to obtain test data on designing nuclear space reactor to break apart and disintegrate upon atmospheric re-entry. (Wallops Release 63-43)

April 30: Contract award for two Apollo mission simulators, for training astronauts from launch through lunar orbit and return to earth, was announced by NASA Manned Spacecraft Center. Link Div. of General Precision, Inc., was selected by North American Aviation's Space Systems Div. to develop and install the two spacecraft simulators, one at MSC in Houston and one at Atlantic Missile Range. (MSC Release 63-79)

- Stephen J. Grillo, Director of NASA Headquarters' Administrative Services since 1960, died.
- In speech reprinted in the *Cong. Record*, William O. Bennett, VP for research and engineering of the Bulova Watch Co., cited 145 examples of "technological fallout" from space research that have been produced in consumer byproducts. He also pointed to the development of the electronic timepiece (not developed in space research) powered by a mercury cell which was used in four families of satellites. (*CR*, 4/30/63, A2595-96)
- Dr. Glenn T. Seaborg, Chairman of AEC, testifying before Senate Committee on Aeronautical and Space Sciences, said:

    "We expect other space nuclear power units (the SNAP devices) to be available to provide continuing life support power for bases on the moon and orbiting space stations, power for unmanned missions to other planets, solar probes, or other missions using ion propulsion, and surveillance and communications satellites of any size or power range.

    "Therefore, our programs in space nuclear power units are directed toward the development of nuclear units varying in power capacity from a few watts to megawatts, varying in operational lifetimes from several days to several years, and varying in weight from less than five pounds to over a thousand pounds. . . ."

    Also testifying were Harold B. Finger, Director of NASA Nuclear Systems and Manager of Joint AEC-NASA Space Nuclear Propulsion Office; Richard L. Kirk, SNAP Program Director, AEC; and Dr. Frank K. Pittman, Director of AEC Division of Reactor Development. (Transcript)
- House Committee on Science and Astronautics' Subcommittee on Applications and Tracking and Data Acquisition resumed hearings on NASA FY 1964 budget authorization, with testimony by Dr. Joseph V. Charyk, President of Communications Satellite Corp., and Leo D. Welch, Chairman. Asked whether ComSat Corp. would conduct its own R&D or depend upon NASA-sponsored research, Mr. Welch said, "We plan to do both. We plan to contract for design studies, [and] we obviously plan to take advantage of the work done by NASA."

    Mr. Welch said space communications system may link U.S., Europe, Japan, and Latin America by 1967. Dr. Charyk said ComSatCorp. expected initial planning to take about a year and development of hardware about another three years. (NASA Leg. Act. Rpt. II/63; AP, Wash. *Eve. Star*, 5/1/63)

April 30: DOD announced U.S. would deploy ICBM's at rate of one a day during rest of 1963. Announcement came as 100th Minuteman site was turned over to SAC at Malmstrom AFB, Mont. (DOD Release 612–63)

- USAF fired Minuteman ICBM from underground silo at Vandenberg AFB "with complete success," fourth successful firing of Minuteman from West Coast in past four weeks. (DOD Release 616–63; UPI, *Wash. Post*, 5/1/63)

During April: NASA Manned Spacecraft Center issued requests for proposals for preliminary study contracts on (1) rotating manned space station to house 18 crewmen; (2) non-rotating, zero-gravity station to house 25–30 crewmen; (3) reusable logistics vehicle. (*M&R*, 4/29/63, 18)

- NASA Manned Spacecraft Center completed eighth of series of drop tests on storable gliding parachute. (*M&R*, 4/22/63, 13)
- Dr. John V. Harrington of MIT Lincoln Laboratory was selected to direct new MIT space research center, to be completed in 1965. Dr. Harrington was also named professor in departments of electrical engineering and aeronautics and astronautics. (Boston *Eve. Globe*, 4/23/63)
- Writing in *Hypnosis Quarterly*, Frank M. Frazitta said astronaut trained in self-hypnosis could alleviate pain and discomfort during space flights. In event of air-conditioning unit's failure, self-hypnosis could be used to lessen astronaut's perspiration rate or make him feel cooler or warmer. (*M&R*, 4/8/63, 21)
- NASA awarded four-month study contracts for Synchronous Meteorological Satellite to Radio Corp. of America and Hughes Aircraft. RCA would study camera resolution requirements in synchronous-orbit weather satellite and ways to modify Tiros satellite launched into highly eccentric orbit (about 22,000 mi. apogee and 200–300-mi. perigee). Hughes would study way in which a Syncom communications satellite could be modified for installation of weather satellite cameras. (*M&R*, 4/22/63, 10)
- Wind-tunnel studies to evaluate aerodynamic drag and jet engine mass air-flow characteristics with 3/10-scale model of free-flight lunar landing test vehicle were begun at NASA Langley Research Center. Model was built at NASA Flight Research Center. Data from Langley tests would be supplied to Bell Aerosystems Co., building two full-scale lunar landing research vehicles under contract from Flight Research Center. (*Av. Wk.*, 4/22/63, 77)
- Preliminary results of the first lunar contact with the new Lincoln Laboratory high-resolution radartelescope were reported at the 44th annual meeting of the American Geophysical Union by Vernon L. Lynn. The report was jointly authored with M. D. Sohigian and E. A. Crocker, also of Lincoln Laboratory. Lynn said that the first radar reflections from the lunar surface obtained at a frequency of 35,000 mc., more than four times that of previous radar probes, suggest that the moon's surface is fairly rough with respect to the dimensions of the 8.6-mm. waves employed. In addition to the advantage of being able to view a smaller area than previously possible, the new radartelescope would provide data on the absorption and reflection characteristics of the lunar

surface at a new wavelength, offering additional clues to its composition and roughness. (*Av. Wk.*, 5/6/63, 91)

During April: Alexander Kakunin, Soviet Vice-Minister for Communications, was quoted as saying U.S.S.R. would soon launch synchronous system of communications satellites for transmitting television worldwide. Using Cosmos satellites, the service would be called "cosmosvision." (*M&R*, 4/8/63, 9)

- Gaston Palewski, French Minister for Scientific Research, said in meeting of Space Council of France that France had fired more than 30 rockets in two-year period 1959–61. (AP, *NYT*, 4/24/63, 7)
- Space orbital simulator, designed to simulate conditions 200-mi. high with temperature as low as −320° F, was shipped from manufacturer, Tenney Engineering, Inc., to University of Rome. Device was believed to be first such available to Western European scientists. (*NYT*, 4/7/63)
- RAND Corp. report on communications satellites estimated that worldwide revenues from international communications would reach $210 million by 1965, $350 million by 1970, and $620 million by 1975. Large part of these revenues would come from undersea cables. (*NYT*, 4/24/63)
- F. J. Krieger, RAND Corp. specialist on Soviet astronautics, said in report for USAF Project Rand that Soviet Cosmos satellites probably are "para-military" spacecraft of two types: (1) recoverable vehicle as large as Vostok spacecraft and (2) 4,400-lb., non-recoverable vehicle launched from another site and with different booster. Krieger also stated his belief that U.S.S.R. had six attempted probe launchings end in failure between August 25 and January 4 caused by inability to get fourth stage out of parking orbit. First three were attempted Venus probes; next two, Mars probes; last one, moon probe. (RAND Memorandum RM-3595-PR, in *Av. Wk.*, 4/29/63, 22)
- Vice President Lyndon B. Johnson, speaking editorially in *Astronautics and Aerospace Engineering* magazine, April 1963, emphasized that: "As satellites extend man's vision into new dimensions, weather forecasting will make major advances as a science, until we see reliable predictions a season ahead.

  "This will have far-reaching economic benefits for all. Following are some estimates of cost-savings, assuming that we can predict weather accurately only five days in advance: $2½ billion a year in agriculture; $45 million in the lumber industry; $100 million in surface transportation; $75 million in retail marketing; and $3 billion in water-resources management. These estimates of savings are for just the United States. Worldwide benefits would be many times as great." (*A&AE*, Apr., 1963)
- Dr. Barry Commoner, Prof. of Plant Physiology at Washington Univ., St. Louis, said in speech before convention of National Science Teachers Association in Philadelphia that U.S. scientific research was becoming too "mission oriented."

  "In 1962 about one-third of the total federal obligation for basic research—and federal funds are now the major source of support for basic science—came from a single agency, the National

Aeronautics and Space Administration. In a total federal budget of about $900 million for basic research, NASA was obligated to provide $350 million, while the National Science Foundation provided under $90 million....

"We are engaged in a spectacular balancing act. Education is supported by science, science by space, and space by the man on the moon ... [The] policy of supporting education through science and science through space is dangerously unsound.... We should recognize this policy for what it is—a short-sighted, pinchpenny effort to buy a few selected fruits of the tree of knowledge, without accepting the honest responsibility of nourishing the whole living, growing organism." (*NYT*, 4/7/63; St. Louis *Post-Dispatch*, 4/7/63)

During April: William A. Hyman, internationally known trial lawyer and author of "The Magna Carta of Space—The Legal Lodestar," speaking at the Inter-American Bar Association in Panama, said:

"The Space Committee of the United Nations has failed to give due consideration and take appropriate action on the important question of sovereignty and security as they become involved in the use of space." (*CR*, 6/27/63, A4133–34)

- Dr. John R. Pierce, executive director of research and communications principles/systems divisions of Bell Telephone Laboratories, said at ceremony dedicating John Crerar Library at Illinois Institute of Technology that use of computers for document retrieval "would smother the user under a flood of information and misinformation it would produce." One of first men to suggest feasibility of communications satellites, Dr. Pierce acknowledged the value of computers for library indexing but added: "If I have painted a rather gloomy picture of the relations of computers to libraries, I have done so deliberately . . . to counteract the sadly fanciful picture that has been painted by some irresponsible people." (*Av. Wk.*, 4/29/63, 74)

- The April issue of the *Bulletin of the Atomic Scientists*, included an article by Dr. Robert E. Marshak, entitled "Reexamining the Soviet Scientific Challenge." In this article, Dr. Marshak, chairman of the Department of Physics and Astronomy at the University of Rochester, pointed out that Soviet preeminence in space does not necessarily mean that the U.S.S.R. has "achieved a decisive scientific and technological supremacy." Comparing the "closed" economic and political structure of Soviet Russia with its controlled educational system to the "open" society of the United States, he suggested that the Soviet system was geared to the more tangible results of applied science while the United States had made greater strides in pure science.

"Soviet space triumphs are due to an early and precise delineation of a major national goal in applied science and a most detailed and deliberate organization of the wherewithal to achieve it.

". . . I believe that our [the United States'] undoubted leadership in pure science will continue to be nurtured by the openness, freedom, and free enterprise spirit which characterize our society. Our accomplishments in pure science will provide a vast reservoir of ideas for important achievements in applied science, and our innate good sense and good will lead to a voluntary rational measure of coordination and control in those areas which are indispensable for the achievement of our national objectives in applied science. I believe that the Soviet methods of strict supervision and control will lead to numerous short-range breakthroughs in applied science but that the momentum will not be sustained unless there are also significant advances in their pure science. A great improvement in the quality of these achievements will depend upon their ability to fully establish the conditions of scientific freedom which are essential for highly creative work in pure science.

". . . If the Soviet Government comes to realize—and I believe that the Russian scientists already do—that scientific and political freedom go hand in hand and that it is difficult to guarantee scientific freedom without a major liberalization of Soviet society in all its aspects, there is bound to be a great efflorescence of pure science in the Soviet Union. And if a large measure of political freedom is established in the Soviet Union, it is quite likely that we shall be sending our American astronauts together with their Russian counterparts on joint expeditions to the moon and other celestial objects beyond." (*CR*, 5/27/63, 9036–38)

During April: Air Force launched an Atlas F with a Chrysler-built nose cone on a 5,000-mile flight from Cape Canaveral to the vicinity of Ascension Island in a continuation of the Advanced Ballistic Re-entry System (Abres) test program. The flight, which largely duplicated the initial Abres test of Mar. 1, was described as "highly successful." (*M&R*, 5/6/63, 11)

- Dr. Samuel J. Rabinowitz was appointed Deputy Director for Ballistic Missile Defense in DOD Advanced Research Project Agency. Dr. Rabinowitz was formerly with Columbia Univ. Electronic Research Laboratories. (*M&R*, 4/8/63, 10)
- NASA submitted the following written answer to Senate Committee on Aeronautical and Space Sciences in response to question raised by Senator Margaret Chase Smith on the contamination problem in space exploration:

"Present plans for lunar spacecraft are to use assembly techniques in clean rooms and under environmental conditions similar to surgical operating facilities. These procedures will not make the spacecraft sterile but are expected to reduce the total population of viable organisms by orders of magnitude below otherwise expected quantitites. As stated earlier, the natural environment of the Moon is believed suitably hostile to the propagation of Earth-like organisms to such an extent that any contamination will be contained in very local areas." (Committee Hearings, *NASA Authorization for Fiscal Year 1964*—Part I, 598)

MAY 1963

May 1: Testifying before House Subcommittee on Manned Space Flight, NASA Deputy Associate Administrator and Director of Manned Space Flight D. Brainerd Holmes said that "the present Gemini is a different program than the one with which we started. It had two fundamental purposes originally which we have not given up and which we still consider to be of paramount importance—to gain experience in prolonged weightlessness and to gain experience in rendezvous before Apollo . . . .

"However, as we developed the Gemini program, we found it was not wise to have only an updated Mercury spacecraft. We found as we developed the subsystems for Gemini that it was much wiser to incorporate technology known today—not something that was early developmental technology of the past. This concept would lead us step by step toward Apollo, and we would reap tremendous benefits from it. In addition . . . it became quite apparent to us that this would be a very useful Earth orbital spacecraft . . . .

"In developing the Gemini subsystems, we find that they are better and more sophisticated than initially conceived and the spacecraft is much more sophisticated than anticipated. Therefore, the costs tend to be higher and the time scales longer . . . .

"We had planned to have the first manned Gemini flight the second flight, during the first quarter of 1964. Our present plan is that the first manned flight in the series would not be until the fourth quarter of 1964. Our current plan grew in part from the advisability of having more sophisticated subsystems, which required more time, and in part from a reorientation of the program to include a second unmanned flight before the manned flight . . . .

". . . the Titan II . . . has had some developmental problems over and above what we had anticipated. However, the problems are always encountered in a developmental program. In our opinion it will be safer to use the Titan II for manned flights if we have two structural missions. We will have more time to eliminate the developmental problems . . . .

"We are planning an unmanned flight at the end of this year and about the middle of next year. We plan to have a manned flight in the last quarter of 1964 . . . ." (Transcript)

- Senator Russell Long (D.-La.) introduced bill (S. 1436) "to prescribe with greater particularity the conditions under which proprietary interests of the United States in inventions may be waived pursuant to section 305 of the National Aeronautics and Space Act of 1958." Bill was referred to Senate Committee on the Judiciary. (NASA Leg. Act. Rpt. II/64)

May 1: NASA Launch Operations Center reorganization announced. Operating under interim organization since it became operational July 1962, LOC now had 14 offices and divisions reporting directly to Office of the Director, Dr. Kurt H. Debus, and a Deputy Director and four Assistant Directors. (LOC *Spaceport News*, 5/1/63)

- NASA Ames Research Center announced award of contract to Thiokol Chemical Corp. for "hypervelocity test device," to be part of Hypervelocity Free-Flight Facility now under construction at Ames. Thiokol would design, build, test, and deliver the wind tunnel structure, to be used with light gas gun to test designs for manned orbiting spacecraft, vehicles returning from moon and planets, and possibly vehicles entering atmospheres of other planets. (Ames Release 63–20; NASA Release 63–89)
- NASA Deputy Associate Administrator for Industry Affairs, Walter L. Lingle, Jr., reviewed NASA procurement policies before National Capitol Section of American Institute of Aeronautics and Astronautics (AIAA): "It is our objective in NASA to have one overriding objective in all of our procurement policies; and that is so to make our contracts that we will achieve the highest standards in reliability and systems performance . . . ." (Text)
- Dr. Eugene B. Konecci, NASA Director of Biotechnology and Human Research, said in Los Angeles press interview that NASA planned to use balloons to float men for 30 days or more in sealed capsules at 100,000-ft.+altitudes. Such experiments would test life support systems for future manned spacecraft. (AP, Wash. *Eve. Star*, 5/1/63)
- NASA-sponsored meeting with representatives from labor and management held in Gulfport, Miss., with more than 100 men representing 30 organizations at the session to work out advance agreements on wages, hours, and working conditions for workers involved in construction of NASA Mississippi Test Facility. Meeting was "first such conference ever sponsored by the Federal government in advance of the award of a construction contract." (N.Y. *Herald Trib.*, 4/28/63)
- *New York Times* reported Australian scientists had established charred metal sphere found in western New South Wales had come from "a space vehicle of some sort." Australian Government was trying to determine original owner of the object, about 15-in. in diameter and 10-lb. in weight. (*NYT*, 5/1/63, 6)
- Titan I ICBM exploded during attempted launching at Vandenberg AFB. USAF spokesman said no one was injured and damage to silo launch facility appeared slight. (AP, *Wash. Post*, 5/2/63; DOD Release 624–63)
- A. M. Nowitsky, Director of Spacecraft Sterilization, Lockheed Missiles & Space Co., said in speech before Aerospace Medical Association meeting in Los Angeles that sterilization of lunar spacecraft was essential to prevent contamination of moon and to detect any organisms on the moon: "For millions of years the moon has been a vulnerable target in the path of extraterrestrial debris, possibly organic, bacterial and living—possibly disease-causing

and preserved in a state highly acclimated to unfavorable environmental extremes.

"If such forms do exist and are in any way dangerous or destructive, their early detection would prove vital to man's future planetary exploration and the protection of earth from mutual contamination." (Miles, *L.A. Times*, 5/2/63)

May 1: France's General Jean Thiry announced France would build nuclear test center on Mururoa Atoll, in Pacific Ocean near Tahiti. (Middleton, *NYT*, 5/2/63)

- International Association of Machinists (IAM) postponed scheduled nationwide strike against Boeing Co. after receiving telegram from President Kennedy and revised contract proposal from Boeing. Union set vote on new proposal for May 10. (AP, *NYT*, 5/3/63, 14)

- Dr. William M. Helvey, Chief of Space Environment and Life Sciences Laboratory, Republic Aviation Corp., told American Medical Association in Los Angeles that test volunteers exposed to pure oxygen in altitude chamber during four two-week periods had developed anemia but remained otherwise healthy. (Science Service, Wash. *Daily News*, 5/1/63)

- Brig. Gen. Samuel C. Phillips (USAF), Director of Minuteman Program, said at Norton AFB that the U.S. ICBM had "achieved more than was asked in all respects, including range and payload." (Dighton, AP, Wash. *Eve. Star*, 5/1/63)

- AF Weapons Laboratory at Kirtland AFB, N.M., was assigned to AFSC's Research and Technology Div. (RTD), Bolling AFB, Va. (*AFSC Operational Highlights*, 11)

- In address at dedication of *General H. H. Arnold* advanced range instrumentation ship (ARIS), Cape Canaveral, General B. A. Schriever (USAF) said:

 "Instrumentation is a key element of technology in the Space Age. It plays a crucial role in the rapid progress of our research and development efforts, through the provision of more complete and accurate information on test results. As a result, far fewer tests are required than was formerly the case . . . ." Comparing the 2,500 test firings of V-2 to the less than 50 of Thor IRBM, he said: "Better instrumentation is a major factor in this dramatic reduction."

 He recalled General Arnold's foresight "demonstrated as far back as 1917," when he and Charles Kettering worked on development of "The Bug" pilotless guided airplane—in many ways "25 years ahead of its time . . . ."

 "Throughout his career 'Hap' Arnold was a pioneer of new ideas . . . ." In 1938 he predicted development of variable-wing aircraft. In 1944 he organized Army Air Forces Scientific Advisory Group, under Dr. Theodore von Kármán, charging these scientists to " 'think about supersonic speed airplanes, airplanes that would move and operate without crews, improvements in bombs . . .; communications systems . . . television, weather, medical research; atomic energy, and any other phase of aviation which might affect the development and employment of the air power to come . . . .' " (Text, AFSC Release 35-R-36)

May 1: Ground-breaking ceremony for new astrophysics and space research laboratory in Chicago, part of Enrico Fermi Institute for Nuclear Studies, Univ. of Chicago announced. (*Chic. Trib.*, 4/28/63)

- The astronauts making the United States' first two-man Gemini endurance space flight next year would be wired for sound as never before—to a pair of small magnetic tape recorders registering their physical and mental reactions second by second throughout the flight. (*Space News Roundup*, 1/8/64, 2)

May 2: X-15 No. 3 flown by NASA pilot Joseph A. Walker to 209,400-ft. altitude at speeds up to 3,488 mph (mach 4.73), in nine-minute flight from Mud Lake, Nev. Instrumentation mounted on aircraft's tail measured radiation in ultraviolet and infrared spectra. Secondary purpose of flight was evaluation of air flow over X-15's surface conducted at high angles of attack. (FRC Release)

- U.S. and Brazil announced signing of memorandum of understanding for cooperation in satellite studies of the ionosphere. Signed by Dr. Hugh L. Dryden, NASA Deputy Administrator, and Aldo Vieira da Rosa, of Brazilian Commission for Space Activities, agreement provided for establishment of three radio-receiving stations straddling geomagnetic equator—at Belem, San Jose dos Campos, and Natal. Initial project would involve readout of telemetry data from S-66 ionospheric satellite, to be launched by NASA in 1963. Under agreement, Brazil would build, equip, and operate stations with NASA lending scientific equipment. (AP, *Wash. Post*, 5/3/63)

- In meeting of legal subcommittee of U.N. Committee on Peaceful Uses of Outer Space, Soviet delegate Roland M. Timerbaev declared that it would be "delusion" to report to the Committee that any progress had been made during 28-nation subcommittee's three-week session. Business before subcommittee had been drafting of legal code covering space exploration activities and considering of proposals concerning liability for space vehicle accidents and return of astronauts and spacecraft after emergency landings in foreign territory. (*NYT*, 5/3/63, 3)

- Three-stage solid-fueled West German rocket launched from Cuxhaven, Germany, reached 65-mi. altitude and returned to earth by newly developed parachute technique. Rocket was designed by and launched under direction of Berthold Seliger, who called test "a partial success—I meant the thing to hit 93 miles high." Seliger, who helped develop Germany's V-1 and V-2 rockets, said new rocket was developed from V-2. First major German rocket launched since World War II, it would be used in space research and possibly in development of antimissile missile. (UPI, *Wash. Post*, 5/3/63)

- Dr. Raymond L. Bisplinghoff, NASA Director of Advanced Research and Technology, said at Univ. of Minnesota Engineers' Day Convocation that "all too often" the scientist "gets the credit for the work done by the engineer." In the space program, "the boundary line between science and engineering is often so fine as to be indistinguishable . . . ." (Text)

May 2: NASA Director of Procurement Ernest W. Brackett, testifying before Subcommittee on Applications and Tracking and Data Acquisition of House Committee on Science and Astronautics, said: "During fiscal year 1962, approximately 90% of funds appropriated by the Congress were spent by contracts as compared to 64% in fiscal year 1960. In fiscal year 1964, the percentage of funds going to contractors may increase further . . . ." (Testimony)

- Speaking before a dinner of the American Association for Contamination Control in Boston, Mass., Franklyn W. Phillips, director of NASA NEO, spoke of the proposed electronics research center:

    "We have seen nothing to date that would alter our original proposal to place the [electronics] center in the greater Boston area. The reasons we made the choice in the first place are as valid now as they were then.

    "Greater Boston represents the greatest concentration of electronics know-how in the country and is the logical place for this center." (Boston *Herald*, 5/3/63)

- President of Boeing Co. William M. Allen testified in closed hearings before Senate Investigations Subcommittee that Boeing's bid to build 23 TFX aircraft was $148,000,000 less than that of General Dynamics Corp., "yet we were downgraded by reason of our lower price. It was most disillusioning and discouraging to discover that one of the principal reasons Boeing lost the award was because our fixed-price bid was too low . . . . It has been stated that our cost estimates were unrealistic, demonstrating that we did not appreciate the complexities of developing the TFX . . . ." He said Boeing developed its bid on basis of careful calculations and was fully prepared to fulfill contract commitments at quoted price of $482,000,000. (AP, *NYT*, 5/3/63, 1, 14)

May 3: First drop test of Project Apollo earth landing system successfully accomplished by Northrop Corp. for NASA Manned Spacecraft Center. Boilerplate model of Apollo spacecraft command module was released from modified C-133 cargo plane and safely landed by cluster of three Northrop Ventura ringsail parachutes. (MSC Release 63-85)

- Dr. Hugh L. Dryden, NASA Deputy Administrator, said in address before State Convention of New Jersey Society of Professional Engineers: ". . . activities in space itself, and in the complex of supporting work that goes forward here on earth, draw upon almost every branch of science and technology. In the process, the needs and interests of engineers and scientists are drawing closer together, and the lines of cleavage between the various disciplines are becoming less and less distinct . . . .

    "More than ever before, engineering has become the full and active partner of science in the exploration of this newest frontier . . . ." (Text)

- NASA requested contractor proposals for lunar base concept study, initiating 18-month broad lunar base study program to provide detailed technical data to determine feasibility of a manned lunar base. (NASA Release 63-91)

May 3: Senator J. William Fulbright (D.-Ark.) said in Tufts Univ. William L. Clayton Lecture that Project Apollo should be abandoned and money thus saved spent on down-to-earth causes such as education: "This allocation of priorities is a recipe for disaster, an unrecognized and powerful endorsement of the wrong side of the race between education and catastrophe." (*Boston Globe*, 5/3/63, 18)

- Revolutionary new helicopter, called XH-51A, was described in paper by Lockheed Aircraft Corp. engineers W. H. Statler, R. R. Heppe, and E. S. Cruz delivered at meeting of American Helicopter Society, Washington. Built for Army and Navy, experimental craft was said to be first helicopter to fly with ease and stability of an airplane because it employed rigid mounting of rotor blades to shaft and three-armed gyroscopes linking pilot control sticks with rotor. (*NYT*, 5/4/63, 38)
- Dr. Eberhard Wahl, AFSC scientist, announced development of Phototype Optical Surveillance System—electro-optical telescope capable of tracking spacecraft more than 3,000 mi. away. Telescope's three main components: 27-in. optic for gathering light, electro-optical device for measuring light, and computer-command system for controlling tracking of satellites. (AP, *NYT*, 5/4/63, 26)

May 4: In space studies symposium at Argonne, Ill., Bernard I. Spinrad, Argonne National Laboratory scientist, predicted: "On the basis of techniques which in general we now know, I believe that the larger scale exploration of the solar system will in fact be undertaken from a base on the moon." (AP, Newport News *Times-Herald*, 5/4/63)

- National Academy of Engineering was being formed as an affiliate of National Academy of Sciences, President of Engineers Joint Council Dr. Eric Walker announced. (Finney, *NYT*, 5/5/63, 60)

May 5: TIROS V ceased transmitting photographs after record 10½ months of operation. Through orbit 4,579 on May 4, TIROS V had sent more than 57,857 cloud-cover pictures, 80% of which were usable for meteorological purposes. (NASA Release 63-93)

- Estimated 200 German scientists and technicians were working in Military Factory 333, Egyptian rocket center outside Cairo, according to *Washington Post's* magazine *Parade*. (*Parade*, 5/5/63)

May 6: JPL Deep Space Instrumentation Facility (DSIF) at Goldstone, Calif., succeeded in bouncing radar signals off planet Mercury 60,500,000 mi. from earth. Traveling at speed of light, signals took 11 min. to make trip from Goldstone to Mercury and back to Goldstone. Preliminary analysis of echo patterns indicated Mercury has rougher surface than Venus or Mars but same surface reflectivity as that of moon. Mercury experiment was to be conducted 12 hrs. per day May 6–29, at which time planet would move out of antenna's range. (JPL TWX, 5/29/63; *Marshall Star*, 6/5/63, 5)

- NASA Manned Spacecraft Center announced eight NASA astronauts were working with manned space flight simulator at Ling-Temco-

Vought, Dallas, evaluating manual control and instrument display should the Apollo lunar excursion module's primary guidance system fail during descent to the moon. (MSC Release 63-81)

May 6: Interplanetary space simulation was achieved in a 6-ft.-wide, 10-ft.-high-space tank at the NASA Lewis Research Center. This included solar simulation and vacuum of 10^{-13} torr. (Lewis Chronology, 3)

- USAF announced it would attempt second Project West Ford orbiting belt of 400 million copper filaments "in the near future." First such attempt, Oct. 21, 1961, was unsuccessful because container failed to eject the needles. MIT Lincoln Laboratory, conducting experiment for USAF, said in fact sheet that "no other communication method suggested to date, by satellite or otherwise, offers comparable reliability, in terms of global coverage with virtually complete invulnerability to destruction or jamming . . . ." (DOD Release 637-63; Finney, *NYT*, 5/7/63, 29; Laurence, *NYT*, 5/12/63, E9)

- Brig. Gen. Charles H. Roadman (USAF), NASA Director of Aerospace Medicine, testified before House Committee on Science and Astronautics' Subcommittee on Manned Space Flight. Replying to press criticism of NASA plans to use pure oxygen in Project Gemini spacecraft, General Roadman said: ". . . in our investigations to date we have no physiological information that would permit us to say that 100% oxygen from a physiological standpoint is contraindicated in a two-week mission. This is not to say that there are not many physiologists who would like more information concerning 100% oxygen over a time period, which we thoroughly support . . . ." (Transcript)

May 6-9: Third National Conference on the Peaceful Uses of Space held in Chicago.

NASA Administrator James E. Webb, in keynote address May 6, cited recommendation by Iowa Summer Study that one or more scientists be included in Project Apollo lunar flights: "To those who have the responsibility for success in reaching the moon, it has appeared that the nature of the Apollo effort requires the training of astronauts who have substantial experience as test pilots in high speed jet aircraft, and to the extent possible engineering training as well.

"Meanwhile, it is apparent that the view of the scientists that trained scientific personnel should participate is valid, and that at the earliest appropriate stage in the program scientists will be included on Apollo missions. So far as we can now tell we are obligated to utilize astronauts with the maximum of test-flight experience and highly conditioned reflexes on the first flight of the most difficult of all undertakings. Should training and experience in intermediate flights indicate otherwise we will, of course, take this into consideration in determining the stage of development in manned space flight at which a scientist-astronaut will directly participate.

"In its effort to insure maximum scientific benefit from the manned space flight program NASA is doing several things:

"First, manned space flights are being used to take scientific measurements in order that they will provide results beyond crew training and technological experience.

"Second, efforts have already begun to provide scientific training for the present group of astronauts including geological field trips to enhance their capacity for scientific observation.

"Third, a manned space science working group has been established [within NASA] . . . . This group is already at work in the planning of a program for training scientists for space flight . . . ." (Text)

Dr. David H. Stoddard, Assistant Director of Medical Operations in NASA Office of Manned Space Flight, discussed human factors in manned space flight and said "we are confident that the biomedical needs of our currently approved manned space flight program can be met from within the scope of our existing technology." (Text)

Dr. Joseph F. Shea, Deputy Director for Systems in NASA Office of Manned Space Flight, reviewed U.S. manned space flight program. Describing Project Gemini, he said: ". . . the Gemini is more than an experimental spacecraft. At the end of its scheduled developmental flight program, it will be available for missions ranging from scientific and engineering experiments to the ferrying of astronauts to and from a permanent space laboratory. The end goal of Gemini is, therefore, an operational spacecraft which can be launched by relatively small boosters. We expect Gemini to be a useful space vehicle for several years . . . . The Gemini spacecraft will reach operational status by 1966 . . . ." (Text)

Capt. Robert F. Freitag (USN), Director of Launch Vehicles and Propulsion in NASA Office of Manned Space Flight, said:

"The nation's program to achieve pre-eminence in space and to insure that the United States occupies first place among the nations of this world in science, in technology and in conquest of space is critically dependent on the power of the rocket.

"You may have heard much about the 'booster gap' and the effect this situation has had on our position vis-a-vis the Soviet Union. The Soviet Union started well ahead of us in the development of large rockets, so necessary for manned flight. The Soviets have been ahead of us, and are still ahead of us in their ability to launch reliably large masses to Earth orbit.

"However, with the acceleration of our own progress and by the establishment of our program on a sound engineering basis we can, we believe, surpass the Soviets in time and clearly establish and demonstrate United States pre-eminence in manned space flight. . . ." (Text)

Reviewing NASA space program, Director of NASA Office of Programs D. D. Wyatt said: "In its broadest context the NASA program can be classified into three complementary parts. First, we have those phases of the program which are aimed at the exploration of the very nature of space . . . .

"The second major element of our program is aimed at the earliest practical utilization of space for the benefit of mankind . . . .

"Complementing our programs to understand and define space is a third major program element aimed at developing and improving the highly complex technologies which will enhance our future capabilities in space . . . ." (Text)

John E. Naugle, Director of NASA Geophysics and Astronomy Programs, said: ". . . the scientific program which we have in NASA is a logical extension of the research work which has been underway in laboratories on the earth since the time of Galileo. Just as astronomers once carried their telescopes to the top of a mountain to see better, now we carry telescopes beyond the earth on rockets and satellites. By doing this we see entirely new phenomena and enable scientists to broaden their studies in certain scientific disicplines . . . ." He then discussed scientific results from each of the six scientific satellites launched by NASA since beginning of 1962 (OSO I, EXPLORER XIV, EXPLORER XV, ALOUETTE I, ARIEL, and EXPLORER XVII). (Text)

NASA Director of International Programs Arnold W. Frutkin said in May 7 address delivered by J. H. Barnes, NASA Chief of Cooperative Programs, Office of International Programs: "What other nations are doing in space and in particular, what they are doing together, is in fact small when compared with the resources going directly into national programs in the US and the USSR. Yet, what has been accomplished provides substantial evidence of the feasibility, the benefits, and the promise of joint action by the nations in this new field . . . ." (Text)

Edgar M. Cortright, Deputy Director of NASA Office of Space Sciences, described scientific spacecraft now under development, and said: "We must continually assess our progress . . . . The flight performance of our spacecraft has been increasing dramatically . . . . The *useful* life of unmanned spacecraft launched in 1962 has exceed six months and *is still rising* because most of those intended for long life are still operating successfully.

"Our progress with launch vehicles has been equally gratifying. In 1962, 82% of all major NASA launchings were successful. This performance has been paced by the remarkable Thor-Delta which has successfully launched 16 out of 17 spacecraft. In addition, the huge Saturn has scored 4 out of 4 perfect flights. During this time period, both the Thor-Delta and the Scout payloads were increased over 60% at no increase in their $2.5 million and $1 million price tags. These dramatic increases in performance, life, and reliability of spacecraft and launch vehicles represent by far the most effective way to achieve economy in the space program . . . ." (Text)

NASA Director of Advanced Research and Technology Dr. Raymond L. Bisplinghoff said: "There are four principal areas where a continuing and driving program of advanced research is required if this nation is to achieve pre-eminence in aeronautical and space activities in the decades to come. These are energy conversion and propulsion; materials and structures; control, guidance and communications; and space sciences and the environment of space." (Text)

Louis B. C. Fong, Director of NASA Office of Technology Utilization, said on May 8 that the "basic objectives of the [NASA] Technology Utilization Program are:
- Identification of innovations
- Cataloging
- Dissemination

". . . I want to make it very explicit that NASA's Technology Utilization Program is separate and distinct from our manned space flight and space science efforts. The space exploration program stands on its own merits; our nation must occupy a position of pre-eminence. The benefits from industrial applications are not now—and never will be—the justification for the high costs of this major effort . . . ." (Text)

Harold B. Finger, Manager of AEC-NASA Space Nuclear Propulsion Office and Director of NASA Nuclear Systems, described on May 9 NASA research in and plans for nuclear propulsion: ". . . the door that was opened here in Chicago in December 1942 has in less than 20 years led us to concepts and hardware for the utilization of that new energy source in space missions that could not be generally anticipated at that time. As a result of that dramatic scientific effort, we have established major goals aimed at the early and practical utilization of nuclear energy in space. We are convinced that a substantial effort is justified by the potential performance advantages and the many applications of these systems for difficult space missions. This work will lead us not only to the development of particular hardware items, but will open new fields of rocket propulsion and power to permit us to travel freely in space." (Text)

May 6: Former Chairman of the AEC (1946–50), David E. Lilienthal, suggested a comparison of the "purple cliches about the space age, current successor to the atomic new world" with the early years of the atomic age, in Stafford Little lecture at Princeton University printed in the *Congressional Record:* "In the 1940's it was reasonable to join in General Groves' and Senator McMahon's view of the possibility of a 'new world' that peaceful applications of atomic energy might bring. Responsible men spoke of atomic power so cheap it wouldn't pay to meter it. Most of us were less exuberant by far. Yet even those with a less euphoric outlook had ample grounds to believe that the expectations of the 1940's and the early 1950's should be given a hard try, and that such an effort had a good chance of paying . . . .

"What then is the record? . . . For the country as a whole, except where the costs of producing electricity from other sources of heat are high, atomic power is not cheaper but costlier than conventional fuel . . . ." (*CR*, 5/6/63, 7321–30)

- Remarks by Charles H. Wilson on X-21A research aircraft appeared in *Congressional Record*. In experimental test program conducted by Norair Div. of Northrop Corp. for AFSC, two X-21A aircraft were "designed to test the feasibility of a suction system to maintain laminar air flow over the wings . . . . X-21A flight tests are expected to prove that up to 80 percent of airplane friction drag can be eliminated with a corresponding increase in range, endurance or payload of more than 50 percent . . . . ." (CR, 5/6/63, 7399)

May 6: Press conference with Lt. Col. K. B. Cooper of ARPA was reported in *Missiles and Rockets*. Colonel Cooper predicted Project Defender would provide in three years 95% of knowledge required to understand and interpret re-entry physics phenomena. (M&R, 5/6/63, 21)

- Speaking at American Institute of Chemical Engineers meeting in Buffalo, N.Y., engineers of Linde Co. described new process for making liquid hydrogen that would reduce its cost from $2 per pound to 35 or 40 cents per pound. (AP, *NYT*, 5/7/63, 11)
- USN announced Allegheny Ballistics Laboratory facilities destroyed by fire in April would be rebuilt "to provide support for current rocket motor development programs and those already planned for the next year." (AP, *Wash. Post*, 5/7/63)

May 7: TELSTAR II communications satellite placed in elliptical orbit (6,717-mi. apogee, 604-mi. perigee, 225.3-min. period, 42.7° inclination to equator). Thor-Delta vehicle launched from Cape Canaveral boosted the satellite into orbit for its 17th straight success, an unmatched record for U.S. satellite-launching vehicles. TELSTAR II included design changes aimed at protecting it from radiation damage which affected lifetime of its predecessor, TELSTAR I. With apogee nearly twice that of TELSTAR I, it would provide longer periods of communications between U.S. and Western Europe than did TELSTAR I. Like its predecessor, TELSTAR II was designed and built by AT&T's Bell Telephone Laboratories at AT&T expense, launched by NASA with AT&T reimbursing NASA for Delta vehicle, launching, and tracking services. Initial communications test, TV transmission from Andover, Me., to Goonhilly Downs, England, via the satellite on its fourth orbit, was successful. (NASA Release 63–83; NASA TELSTAR II Prog. Rpt. No. 2; AP, Wash. *Eve. Star*, 5/7/63)

- Aerobee 150A sounding rocket carried 153-lb. instrumented payload to 139-mi. altitude in experiment from Wallops Island to study spectral emission lines in upper atmosphere and measure their intensity as function of altitude, thus determining distribution of certain molecular and atomic species in upper atmosphere. Experiment was designed by The Johns Hopkins Univ. under NASA research grant administered by Goddard Space Flight Center. (Wallops Release 63–46)
- NASA Ames Research Center announced award of contracts to NAA Space and Information Div. and Space Technology Laboratories (STL) for studies of manned Mars landing-and-return missions. Each contract called for nine-month studies to determine requirements for possible Mars missions and to decide what research would be required during next several years to implement such a flight program. Other studies of manned Mars missions were being managed by NASA Marshall Space Flight Center and Manned Spacecraft Center, emphasizing use of Apollo-class booster and spacecraft; Ames-managed studies emphasized Mars-landing mission with no restrictions as to type of vehicle and spacecraft. (Ames Release 63–21)

May 7: Dr. Theodore von Kármán, distinguished U.S. physicist, died in Aachen, Germany, four days before his 82nd birthday. As early as 1906, Dr. von Kármán established basic principles that led to design of light and efficient aircraft structures. Among his important scientific contributions were studies of air turbulence, which influenced early aircraft design, and theory of boundary layers, which led to pioneer construction of wind tunnels. He was responsible for many scientific theories leading to development of supersonic jet aircraft and rocket engines. Dr. von Kármán in 1930 became Director of Cal Tech's Guggenheim Laboratory, which in 1944 became Jet Propulsion Laboratory. As JPL Director he pioneered in U.S. rocket propulsion, led development of Jato. Chairman of NATO's Advisory Group for Aeronautical Research and Development (AGARD) since 1951, he was recipient of more than 20 honorary doctorate degrees and was in 1963 first recipient of National Medal of Science, which cited him "for leadership in the science and engineering basic to aeronautics, for distinguished counsel to the armed services and for promoting international cooperation in science and engineering." Dr. von Kármán was born in Budapest, became U.S. citizen in 1936. (*NYT*, 5/8/63; AP, Wash. *Eve. Star*, 5/8/63, B5)

- Sir Bernard Lovell, Director of Jodrell Bank Experimental Station, England, announced his opposition to USAF's orbiting of 400,000,000 copper needles (Project West Ford) and said: "There is grave danger that projects of this sort may eventually bring the present rate of progress in astronomical research to an end." (AP, *NYT*, 5/8/63, 3)

- At Third International Conference on Atmospheric and Space Electricity, held at Montreux, Switzerland, Elden C. Whipple, Jr., of NASA Goddard Space Flight Center's Planetary Ionosphere Branch proposed that rocket be fired through thunderstorm to determine effect, if any, of such storms on earth's ionosphere. (Goddard Release)

- Rep. Ben F. Jensen (R.-Iowa) inserted in *Congressional Record* a *Washington Post* article which said that majority opinion expressed by 25 U.S. Nobel Prize Winners criticized "waste" and "inefficiency" in America's "crash program" to land man on moon by 1970. (Eisele, *Wash. Post*, in *CR*, 5/7/63, 7424)

- *Washington Post* editorial said of U.S. manned lunar program: "We may not get to the moon first, or we may not get to it at all—but that will not mean all has been in vain. In trying to reach such a spectacular goal, we are developing the rockets that will place us in the forefront of the space age, and in developing the scientific and engineering skills required to achieve this purpose we are enormously expanding the country's capacity to achieve scientific goals as yet unspecified. The skills and talents nurtured at NASA will be making their contribution to the knowledge, the comfort, the convenience and the survival of the United States long after moon landings have been accepted as a commonplace or abandoned as an impossibility. The world was altered by the voyages of Columbus, even though they failed of their planned objectives. And the world is being altered before our very eyes by a new struggle to master man's environment. Enterprises such as this are not to be judged by the ordinary criteria of scientists, economists or sociologists. They are not to be weighed on the balance against some other more practical and more appropriate or more feasible endeavor. Such heroic enterprises move by their own laws, abide by their own rules and set their own precedents and when they are over, leave humanity with its knowledge multiplied, its future expanded, its horizons widened, its outlook sharpened and its hopes uplifted by a new sense of man's unending and unlimited possibilities." (*Wash. Post*, 5/7/63; *CR*, 5/8/63, 7491)

May 7: Col. Charles Yeager, Chief of USAF Aerospace Research Pilot School, Edwards AFB, Calif., visited U.S. Capitol and was honored in tributes by Congressmen on House floor. (*CR*, 5/7/63, 7406–09)

May 8: Astronaut Leroy Gordon Cooper flew simulated launch countdown and aspects of MA-9 Project Mercury space flight at Cape Canaveral. (AP, *Wash. Eve. Star*, 5/8/63)

- Three-minute portion of U.S. color television program received in France via TELSTAR II communications satellite. NBC transmitted the program, Bell Telephone Hour, from New York to the satellite through Andover, Me., ground station. (AP, *Wash. Post*, 5/9/63)

- Nike-Apache sounding rocket launched from Wallops Island, Va., carried 63-lb. instrumented payload to 104-mi. altitude in test to measure ion concentration and composition in upper atmosphere. Flight was joint project of Lockheed Missiles and Space Co. and NASA Goddard Space Flight Center. (Wallops Release 63–47)

- D. Brainard Holmes, NASA Deputy Associate Administrator and Director of Manned Space Flight, said before Electronic Components Conference in Washington: "Today, the people of the United States are reappraising the commitment made in 1961. These discussions are natural and healthy. One of the fundamental elements of the strength of the American system is the forum of public debate. However, those of us who understand the long-term nature of major research and development programs have a duty to warn of the pitfalls involved in making frequent, drastic revisions in such programs unless it is demonstrated after thorough study that such changes are necessary. Such a course always results in excessive delays and increased costs. There is an efficient pace at which a program should proceed. To reduce this pace substantially or to pull out all stops and operate on a crash basis would result in increases in the total program costs in manpower and dollars. . . ." (Text)

- NASA Medal for Exceptional Bravery presented by Associate Administrator Dr. Robert C. Seamans, Jr., to six men who risked their lives during rescue operations following X-15 accident Nov. 9, 1962; Capt. Paul J. Balfe (USAF), helicopter pilot; John A. Gordon, NASA rocket technician; Airman 3/C Larry J. Hough, USAF physiological training specialist; Curtis C. Lyon, USAF civilian firefighter crew chief; T/Sgt. Charles L. Manes, USAF helicopter crew chief; and Dr. Lynn B. Rowe (Capt., USAF), flight surgeon. (NASA Release 63–92; FRC Release 9–63)

May 8: 16th annual McGraw-Hill *Survey of Business Plans for New Plants and Equipment—1963–66* was cited in *Congressional Record* by Senator Carl T. Curtis (R.-Neb.). Survey showed that 25% of manufacturers' 1962 sales were in new products not made 10 years ago; manufacturers expect that by 1966 13% of their sales will be in new products not now known. (*CR*, 5/8/63, 7630)

- Aetna Life Insurance Co. issued first individual life insurance policies to seven Project Mercury astronauts, astronauts' legal adviser C. Leo DeOrsey announced in Washington. (Goodman, *Wash. Post*, 5/9/63)
- Dr. Eugene B. Konecci, NASA Director of Biotechnology and Human Research, Office of Advanced Research and Technology, speaking before The Harvard Business Club at Syracuse, New York, said: "today space communism versus space democracy goes deeper than the mere competition of dumping material into orbit or the planning of landing men on the moon, or dreaming of traveling to the planets. It involves the very survival of our American way of life. The eyes of the world are upon the space race, and the Soviets using our own brand of Fifth Avenue public relations have gained a psychological advantage. Our national space objective is pre-eminence in space and not just landing a man on the moon. Although desirable and psychologically reassuring, the overriding issue is not who put the first man in orbit or who will put the first man on the moon, but which nation has been able to develop the overall research, technology, and manufacturing capability to not only explore but exploit space. . . .

    "The curtain of ignorance is thicker and more dangerous than the iron curtain." (Text)
- Editorial in *New York Times* said: "Like its predecessor, the new [Project West Ford launch] attempt is to be made in the face of protests by many scientists both here and abroad. The military hope that this experiment will lead to a long-range communication system invulnerable to destruction or jamming is apparently considered more important than the objections these scientists have raised. We disagree.

    "There is no United States right unilaterally to make changes in the space environment of this planet . . . .

    "Such unilateral decisions on our part invite the day when some unilateral Soviet act in space will make us rue the precedents we are now establishing. Politically, moreover, the plan to proceed with West Ford conflicts with our announced policy of seeking to foster international cooperation in space and to make space a zone of peace. Are the military advantages obtainable here really worth the damage being inflicted on our political posture? This issue deserves much more public debate than it has had." (*NYT*, 5/8/63, 34)
- General B. A. Schriever (AFSC) said in address to seminar participants at National Science Fair, Albuquerque: "The most challenging scientific enterprise we have today is the exploration of space. There are several reasons why this new region is of great importance to us. First of all, it is an unknown and unexplored medium, which is certain to contain many surprises. Second, it

is a tempting region for a nation that is bent on aggression. Soviet leaders have shown a keen awareness of both the psychological and military aspects of space exploration. They have boasted repeatedly that they could use their space technology for aggressive purposes. We cannot afford to leave our space frontier unguarded.

"Space exploration is also important for another reason. It greatly stimulates the advance of science by serving as a natural focus for intensive research in virtually every area of science and technology . . . .

"Today's methods of taking scientific measurements are far more accurate than those of the past, and they are improving all the time. For example, a year and a half ago we were able to measure speeds on our Holloman sled track to an accuracy of about one part in 40,000. Today that accuracy has been increased to about one part in 150,000 . . . ." (Text)

May 8: Minuteman ICBM successfully launched from underground silo, Vandenberg AFB, by SAC crew, seventh of series of Minuteman launchings down Pacific Missile Range. (AP, Wash. *Eve. Star*, 5/9/63)

May 9: 207-lb. instrumented payload carried 125 mi. high by Aerobee 150A sounding rocket launched at Wallops Island, Va., by NASA and Australian scientists. Designed and built by Australian Commonwealth Scientific and Industrial Research Organization, experiment was designed to measure VLF radio noise above E region of ionosphere and was similar to experiment conducted under nighttime conditions April 12. (Wallops Release 63-48; NASA Release 63-99)

- Titan II ICBM launched from Cape Canaveral by USAF, falling short of 5,500-mi. target because of premature second-stage engine cutoff. (DOD Release 663-63; UPI, *NYT*, 5/10/63, 14)

- Senator Margaret Chase Smith (R.-Me.) and NASA Administrator James E. Webb were co-hosts at luncheon at Senate for three women accorded national recognition for space age accomplishments—Marcia S. Miner, student at American Univ. and winner of National Rocket Club's 1963 Goddard Memorial Scholarship Award; Dr. Nancy C. Roman, Chief of Astronomy and Solar Physics in NASA Geophysics and Astronomy Program and 1962 winner of Federal Women's Award; and Eleanor C. Pressly, Head of Vehicles Section, Sounding Rocket Branch, in NASA Goddard Space Flight Center's Spacecraft Integration and Sounding Rocket Div., and 1963 winner of Federal Women's Award. (NASA Release 63-94)

- L/Gen. Thomas P. G. Gerrity (USAF), DC/S for Systems and Logistics, told American Ordnance Association in Washington that "the worst blot on the Air Force-Industry image is inaccurate estimating of costs and schedules. In these areas we simply have to do a better job at the outset of each program and we have to conduct accurate and timely reviews to be sure that program growth and cost overruns are anticipated and separately identified . . . ." (*A-N-AF Journal & Register*, 5/18/63, 38)

- Dr. Wernher von Braun, Director of NASA Marshall Space Flight Center, said in interview that plans for 360-ton, four-man spacecraft to fly from earth to Mars were in "quite an advanced stage." Emphasizing that execution of such plans was not yet approved and funded by NASA, he described nuclear-electric powered spacecraft and said it was "projected for beyond the 1970's."

May 9: A technical services contract was awarded by the Manned Spacecraft Center to Lockheed Electronics Company to provide R&D support for Apollo. The contract involves vehicle instrumentation and communication systems to be used in the early flight phases of the program. (MSC Fact Sheet #96)

- Brig. Gen. Gerald F. Keeling, AFSC Deputy Chief of Staff for Production and Procurement, said in address to American Ordnance Association meeting in Washington:

    "The fact that we will buy [for USAF] more than eight billion dollars worth of weapon systems, supplies, and services this fiscal year and will administer approximately 70,000 contracts for government purchases, which are administered by 57 field organizations, provides some idea of the sheer magnitude of AFSC's procurement management task.

    "A large majority of our funds is involved in weapon system acquisition contracts . . . . For the past two years the Air Force has undertaken a program of surveying certain selected contractors and related weapon systems. Two types of surveys are conducted. One type—called the Industrial Management Assistance Survey—looks at a particular contractor's internal organization, management and operating practices as they affect all the Air Force contracts being performed. Nine of these surveys have now been completed. The second type is called the Systems Program Management Survey and looks at all aspects of a particular aerospace system at a given point in time, including an examination, as appropriate, of the prime, associate prime and major subcontractors involved. We have now completed 20 of this second type of survey . . . .

    "Altogether, we have identified some 37 prominent deficiencies in major areas of contractor management . . . ." (Text)

May 9–10: During this night, TELSTAR II was tested from Andover, Maine, and Pleumeur-Bodou, France, ground stations. (NASA TELSTAR II Prog. Rpt. No. 4)

May 10: Astronaut Leroy Gordon Cooper (Maj., USAF) and Project Mercury launch crews conducted simulated MA-9 final countdown and launching. (AP, *NYT*, 5/11/63, 4)

- Flight of X-15 No. 3 piloted by Maj. Robert Rushworth (USAF) canceled less than 10 min. before release from B-52 mother ship because of malfunctioning hydraulic system. (UPI, *Wash. Post*, 5/11/63)

- USAF announced launching of unidentified satellite using Atlas-Agena booster launched from Pt. Arguello, Calif. (UPI, *Wash. Post*, 5/11/63)

- USAF placed into orbit two 1.5-lb. Tetrahedral Research Satellites (TRS), and a 50-lb. Project West Ford satellite, launched pickaback aboard an unidentified satellite using an Atlas-Agena B booster. TRS's were to measure solar cell radiation damage. West Ford satellite was ejected on the same day of its launch.

The satellite was to place a ring of metal fibers in a belt around the earth for passive communications experiments; experiments were successful and effect of 400 million copper filament belt on radioastronomy was found to be negligible. Natural decay of belt expected as planned. (UPI, *Wash. Post*, 5/11/63; *Press Rpt. on Space, 1963*, 1/27/64)

May 10: NASA Manned Spacecraft Center Director Dr. Robert R. Gilruth announced revision of management responsibilities, dividing operational from developmental activities within MSC projects. Walter C. Williams, Deputy Director for Mission Requirements and Flight Operations, would develop mission plans and rules, crew training, ground support and mission control complexes, and would manage all MSC flight operations; concurrently, Williams would be Director of Flight Operations in NASA Office of Manned Space Flight, with complete mission authority during flight tests of Mercury, Gemini, and Apollo. James C. Elms, Deputy Director for Development and Programs, would manage all MSC manned space flight projects, including total project planning, and would plan, organize, and direct MSC administrative and technical support activities. (MSC Release 63–88)

- NASA announced Flight Research Center had purchased three F–104 jet aircraft from Lockheed California Co. The F–104's would be modified for NASA use in many research programs, including simulated landing of X–15 rocket research aircraft. (FRC Release 10–63; NASA Release 63–98)

- Senate Republican Policy Committee made public its report questioning priority of U.S. manned lunar program. "The question is not . . . whether man will ultimately reach the moon and beyond. The question is, rather, how shall it be done, and whether other aspects of human needs should be bypassed or overlooked in one spasmodic effort to achieve a lunar landing at once . . . .

    "A decision must be made as to whether Project Apollo is vital to our national security or merely an excursion, however interesting, into space research.

    "If our vital security is not at stake, a less ambitious program may be logical and desirable." (*CR*, 5/14/63, 7916; Toth, *NYT*, 5/11/63, 1, 6)

- NASA Deputy Associate Administrator and Director of Manned Space Flight D. Brainerd Holmes, addressing Tenth Annual Conference for Engineers and Architects, Ohio State Univ., said: "The nation has a pool of about 1,400,000 scientists and engineers, which is expected to grow to 1,900,000 by 1970. At present, between 250,000 and 300,000 of these are employed in the aerospace industry. NASA employs more than 9,200 of these on its own staff and, through contract, accounts for the employment of about 32,800 others. Thus the space program occupies something over 3% of the national pool of scientists and engineers. We expect that this proportion will rise gradually, to a little more than 6% in 1968.

    "There is no evidence that this growth of the manpower requirements for space is interfering with consumer research and development. Nevertheless, as a consumer of scientific and technical manpower, we feel a responsibility to do our part to maintain and increase the supply.

"Because of this responsibility, NASA is . . . providing about 35 million dollars to universities for research projects, and 10 million dollars have been granted for university facilities. Under the graduate training grant program, 88 universities have funds for 796 fellowships. It is expected that this number will increase . . . ." (Text)

May 10: In column discussing hostility of some leading U.S. scientists to the "rotarian character" of the space program, Joseph Kraft pointed out the inherent stimulation of science: "To be sure there is something seedy about using the space program as a cover for serious scientific inquiry. Still the atmosphere of the affluent society offers only rare openings to direct public energies and resources to purposes that are not vain. For the most eminent scientists to overlook the powerful grip of space on the public imagination would be to forego an almost unique opportunity for the advancement of science. Which of them would have forgiven Galileo had he turned his back on the heavens because a donor whose heart was impure held out the telescope?" (Wash. *Eve. Star*, 5/10/63, A9)

- President Kennedy, in discussion following his address to trustees of Committee for Economic Development, said he expected U.S.S.R. "to make additional spectacular efforts" in space within next few months. (AP, Wash. *Eve. Star*, 5/10/63, A6)
- In letter to Sen. Clinton P. Anderson (D.-N.M.), Chairman of Committee on Aeronautical and Space Sciences, Sen. Margaret Chase Smith (R.-Me.), ranking minority member, recommended the committee carefully re-examine U.S. space activities and resource allocations "because I think the priority given the manned lunar program is more of an issue now than it was 2 years ago . . . .

"It is interesting to note that criticisms expressed concerning the allocation of funds within the space budget, and the amount of the total budget, are coming from members of both political parties as well as from some scientists and engineers who are not identified with either party.

"I do not, therefore, regard NASA's budget for fiscal 1964 as a partisan matter . . . ." (Letter, 5/10/63, in *CR*, 5/13/63, 7803)
- D-558-II Skyrocket dedicated in ceremony at Antelope Valley College, Lancaster, Calif., where it would be on permanent display. Designed and built by Douglas Aircraft Co. with cooperation of USN and NACA, research aircraft was first to fly twice speed of sound (mach 2.01, on Nov. 20, 1953). Antelope's Skyrocket was third model; first two were on display at Smithsonian Institution and Claremont, Calif., Air Museum, respectively. (Dedication Program)
- Dr. Raymond H. Wilson, Jr., Chief of Applied Mathematics in Research Directorate, NASA Office of Advanced Research and Technology, received $200 NASA award for invention of method for magnetic steering of spacecraft by controlled spin-damping. (NASA Release 63-100)

May 10: Establishment of Div. of Space Sciences and Applied Physics in Catholic University's School of Engineering and Architecture was announced by Rt. Rev. William J. McDonald, Rector of the Univ. (*Wash. Post*, 5/11/63)

- Nike-Zeus antimissile missile flew off course during flight test at White Sands Missile Range and had to be destroyed by range safety officer. (*Space Bus. Daily*, 5/13/63, 607)
- Soviet destroyer reported to be sailing in waters south of Bermuda, in position to observe re-entry of RDF-1 payload to be launched from Wallops Island, Va. (AP, Wash. *Eve. Star*, 5/11/63)

May 11: Astronaut L. Gordon Cooper began series of preflight medical tests at Cape Canaveral in preparation for Project Mercury MA-9 flight scheduled for May 14. With other Project Mercury officials he conducted intensive review of prelaunching activities. (Hines, Wash. *Eve. Star*, 5/11/63; Witkin, *NYT*, 5/12/63, 1)

- NASA Manned Spacecraft Center announced first drop test to qualify parachute-recovery system for two-man Gemini spacecraft was successfully completed at El Centro, Calif. Test series was scheduled for completion in 1964. (MSC Release 63-86)
- Vice President Lyndon B. Johnson, speaking before American Editorial Cartoonists' meeting in Washington, referred to complaints that excessive costs of Project Apollo and other U.S. efforts would "undermine the dollar . . . . But we are not told what would happen to the dollar—or to America—if space were defaulted to the Communist . . . ."

  "The question is which kind of philosophy, democratic or Communist, will dominate outer space?

  "I, for one, don't want to go to bed by the light of a Communist moon." (*Wash. Post*, 5/12/63; AP, Wash. *Sun. Star*, 5/12/63)
- Secretary of the Interior Stewart L. Udall announced U.S. Geological Survey had completed second in series of geological maps of the moon. Map was prepared by C. H. Marshall, astrogeologist of Flagstaff, Ariz., Observatory. (AP, *NYT*, 5/12/63, 64)
- Soviet news agency Tass announced test program of "improved types of rocket carriers for spacecraft" would be conducted May 15-July 15 in specified regions of the Pacific. (Tass, *Krasnaya Zvezda*, 5/12/63, 1, AFSS-T Trans.)
- Soviet Prof. S. Zonshayn wrote in "The Velocities that are Possible in the Universe" that only major obstacle to "relativistic space travel" was production of engine capable of approaching speed of light: "The way to the cosmos has been opened . . . . The present level of science already gives us reason to suppose that passing beyond the limits of the stars are problems which can be solved . . . ." (*Literature Gazette*, 5/11/63, in MSFC *SIN*, 5/63, 13)

May 12: Astronaut Leroy Gordon Cooper (Maj. USAF) flew simulated Project Mercury launch in FAITH 7 at Cape Canaveral in preparation for MA-9 flight scheduled for May 14. (Simons, *Wash. Post*, 5/13/63)

- Radar contact with copper dipoles launched in USAF Project West Ford confirmed they were successfully ejected into orbit from unidentified USAF satellite, MIT Lincoln Laboratory announced. Fibers were still in compact cloud circling earth every 166 min. in near-polar orbit, some 2,000 mi. high and at approximately 87° to equator; on May 13 dipoles had begun to spread in elongated cloud. Fibers were expected to spread along circular orbital path, some 40,000 mi. in circumference, to form ring around earth. Orbital ring would be used in bouncing radio signals back to earth over long distances. (DOD Release 666–63; *NYT*, 5/13/63, 1; AP, Wash. *Eve. Star*, 5/13/63, A3; Sullivan, *NYT*, 5/14/63, 18)

May 12: *Izvestia* reported about 400 spectrograms of Mars had been taken this year at observatory of Astrophysics Institute, Kazakhstan Academy of Sciences, in project to investigate properties of blue haze concealing details of Martian surface. (*Izvestia*, 5/12/63, 6, AFSS–T Trans.)

- U.S. aviatrix Betty Miller landed in Brisbane, Australia, after solo 7,100-mi. flight in twin-engine airplane from Oakland, Calif. Mrs. Miller was first woman to solo across Pacific Ocean from California. (UPI, *Wash. Post*, 5/13/63)

May 13: House Committee on Science and Astronautics, considering NASA budget authorization for FY 1964, began meeting in executive session. From March 4 to May 10, Committee had held 60 hearings with 170 appearances by NASA witnesses. (NASA AC, "Status of Leg.," 5/17/63)

- USAF selected General Electric Co. and Philco Corp. with Space Technology Laboratories (jointly) to compete for development of DOD medium-altitude communications satellite. Under two-month contracts, each contractor would propose technical approach, refine cost estimates, and develop internal management structures. Then, should DOD decide to proceed with such a comsat system employing 20–30 satellites, one of the contractors would be selected to conduct development phase of USAF portion. (DOD Release 674–63)

- Plans for modular Mars spacecraft as presented by D. M. Hammock, Assistant Chief of Spacecraft Technology Div., NASA Manned Spacecraft Center, were described in *Missiles and Rockets*. Six-man crew would undertake 400-day expedition to Mars in three-module spacecraft—Mars mission module, Mars excursion module, and earth entry module. Hammock said NASA was evaluating merits of manned space stations, permanent lunar bases, and Mars expedition as follow-on to Project Apollo, and that technology needed for Mars spacecraft would be simpler in many ways than that required for lunar base or space station. (*M&R*, 5/13/63, 34)

- USAF Titan II successfully launched from silo at Vandenberg AFB in test flight down PMR. (*M&R*, 5/20/63, 16)

- In *Barron's National Business and Financial Weekly*, article entitled "Trackless Void: The U.S. Space Program Had Better Come Down to Earth", said:

". . . Like Columbus, we have been told time and again, the United States today lives in an age of discovery. Like Columbus, the United States must not fail to seize its moment of greatness.

"While popular and plausible, the comparison strikes Barron's as unfortunate. To begin with, in contrast to the National Aeronautics and Space Administration, Columbus raised the funds for his epoch-making voyage at least partly from private sources. Far from setting out to discover America, moreover, the admiral and his financial backers merely sought to find a shorter—hence more profitable—route to the Indies; their primary aim was to make money, not history. Finally, though Washington has chosen to ignore the inconvenient fact, the conquest of the new world yielded Spain, after a short-lived fling at glory, four centuries of economic and political decay.

"The parallel to Columbus can be pushed too far, not only by friends of the space program but also by its foes. Nonetheless, the lessons of the past, coupled with some contemporary widsom on the subject, underscore the need for a sweeping reappraisal of the current U.S. space program. For example, lawmakers now suspect that NASA's skyrocketing budget (from which, according to the administration, not 1 cent can be cut without courting disaster) is designed to serve political as well as scientific ends. In turn, scientists, who know how to weigh alternatives, have grown increasingly skeptical of the overwhelming emphasis which Washington has placed upon landing a man on the moon. Thereby, they claim, the United States has ignored far less costly and more fruitful avenues of lunar exploration, while neglecting other promising ventures in space. The time has come, we submit, to ask 'Is this trip necessary?' " (*CR*, 5/28/63, A3428–3429)

May 13: W. C. Mentzer, United Air Lines Vice President, told Wings Club in New York that projected British-French supersonic airliner was a challenge to U.S. If this country accepts challenge, it must be prepared to accept "two specific obligations . . . : First, to develop an intelligent long-term program of subsidy, and second, to encourage an economic environment in which the supersonic transport eventually may thrive." (*NYT*, 5/14/63)

• Informed Nationalist Chinese sources reported by UPI to have confirmed Tokyo reports that Chinat pilots were still flying U–2 reconnaissance flights over Communist China. (UPI, *Wash. Post*, 5/14/63)

May 14: Project Mercury flight MA–9 postponed at T–13 in launch countdown because of technical failure in computer converter at Bermuda tracking station. Earlier, at T–60, countdown was halted for 129 min. to repair faulty fuel pump in gantry's diesel engine, which had failed to start.

Eight-min. recapitulation of prelaunch activities for Project Mercury flight MA–9, Cape Canaveral, was telecast to European viewers via RELAY I communications satellite shortly after flight was postponed. (NASA Release 63–104; *NYT*, 5/15/63)

• Maj. Robert Rushworth (USAF) flew X–15 No. 3 to 95,600-ft. altitude and 3,600-mph speed (mach 5.20) in test to record temperature increase rates with more than 600 sensors on X–15's skin. During the flight, Major Rushworth rolled aircraft into 90° bank to obtain the necessary data. Aircraft's exterior reached temperatures of 900° F. (FRC Release; UPI, *Chicago Trib.*, 5/15/63)

May 14: First flight test of inflatable meteoroid probe (paraglider), using Aerobee 150 launch vehicle, was conducted from White Sands Missile Range. Objective of experiment was to develop technique for determining penetration rates in thin structural materials. Preliminary results indicated Aerobee 150 followed nominal flight trajectory, but paraglider did not accomplish flight because canister in which it was contained failed to eject. Experiment was joint project of NASA Langley Research Center (paraglider) and Goddard Space Flight Center (vehicle) under direction of NASA Office of Advanced Research and Technology. (NASA Release 63-102; NASA OART/Levine)

- William S. White, editorializing in the Washington *Evening Star*, May 14, 1963, pointed out that:
"There are reformist politicians—who would never agree with the conservatives on anything else whatever—who don't want this money [for space exploration] spent simply because in their opinion not enough is being done for projects like 'urban renewal.' Such men no doubt would have screechingly stopped development of the airplane a lifetime ago if, in the meantime, the street cars in New York were in any way inadequate and the slums in Chicago or Philadelphia had not been totally replaced by public housing." (Wash. *Eve. Star*, 5/14/63)

- Scientific and Technical Subcommittee of U.N. Committee on The Peaceful Uses of Outer Space convened in Geneva. (*NYT*, 5/3/63)

May 15-16: Successful MA-9 Project Mercury flight, longest U.S. manned space flight to date (22 orbits; 34 hrs. 20 min.):
Mercury spacecraft FAITH 7, with Astronaut Leroy Gordon Cooper (Maj., USAF) as pilot, lifted off Pad 14, Cape Canaveral, at 8:04 a.m. EST. Atlas launch vehicle placed spacecraft in initial orbit of 166-mi. apogee, 100-mi. perigee, 88.7-min. period, 32.5° inclination to equator, 17,544-mph speed. Besides participation in flight procedures, astronaut ate, slept, and conducted scientific experiments. Because automatic control system (ACS) failed during 19th orbit, spacecraft was oriented manually during 22nd orbit, retrorockets were fired manually, and re-entry phase was conducted without ACS. FAITH 7 landed two miles from U.S.S. *Kearsarge* near Midway Island in Pacific. Value of trained pilot's presence in spacecraft was underscored by successful mission achievement despite ACS failure.

MA-9 provided biomedical data confirming man can survive and function during prolonged space flight with no ill effects. Experiments during flight included: ejecting from spacecraft in third orbit a 10-lb. flashing-beacon sphere (testing visibility of objects in space in preparation for future rendezvous maneuvers), which Cooper sighted during fifth and sixth orbits; photographs of zodiacal light nighttime airglow layer; horizon definition experiment, using colored filters in 70-mm. camera; radiation measurement experiment, using Geiger counters mounted on spacecraft exterior and detectors inside spacecraft and inside astronaut's space suit; tethered balloon experiment (to measure atmospheric drag) which Cooper attempted in sixth orbit but balloon did not deploy; infrared photography for meteorological purposes; television photography, which achieved historic "first" in U.S. manned space flight with direct transmission of TV pictures to tracking stations; cabin environmental temperature study, another historic "first" with cabin coolant valve and fan turned off from orbit #5 until about two hours before re-entry, astronaut relying on space suit for cooling while cabin temperature stabilized at around 96° F; HF antenna tests, another "first", involving transmissions with antenna horizontally polarized and vertically polarized; ground light experiment, with astronaut observing three-million-candlepower xenon light at Bloemfontein, Republic of South Africa; window attenuation experiment, with astronaut viewing calibrated standard light source and stars to evaluate transmission of light through spacecraft window; and white paint patch measurements, investigating changes in paint pigments during re-entry heating.

Cooper's historic flight covered estimated 593,885 mi. in 34 hrs. 20 min. Nearly 36 years before, Charles Lindbergh's historic flight from New York to Paris covered 3,610 mi. in 33 hrs. 30 min. (NASA MA-9 Press Kit; *NYT*, 5/15/63; *NYT*, 5/16/63; *NYT*, 5/17/63; Press Conf. Transcript, *NYT*, 5/20/63; *Aerospace Yearbook, 1960*)

May 15: Millions of Americans heard and watched MA-9 launching and subsequent flight activities on commercial radio and television. Estimated 8,000 people watched launching on Grand Central Station, New York, 12-by-16-ft. television screen. Attempt to telecast launching to European viewers via RELAY I communications satellite was unsuccessful because "an error on the part of the broadcast company programmers . . . resulted in the transmission of live TV monitors at Cape Canaveral which consisted of a hodgepodge of random pictures and sounds that was virtually meaningless." Videotape of launch activities and of first TV pictures from Astronaut Cooper was successfully transmitted via RELAY I later in the day. Beginning at 6:55 p.m. EDT, the three U.S. TV networks received via RELAY I a three-minute program on European reaction to Astronaut Cooper's flight. (*NYT*, 5/16/63; AP, Wash. *Eve. Star*, 5/16/63; NASA Release 63–106)

- Astronaut L. Gordon Cooper's FAITH 7 Mercury spacecraft was clearly visible to unaided eye to observers in Johannesburg, South Africa. (UPI, *NYT*, 5/16/63, 19)
- Radio Moscow wished Astronaut Cooper good luck on his space flight and said: "The Soviet people, who were pioneers in space, are hoping that this scientific experiment works out successfully. We sincerely hope that its results will serve the development of peaceful cooperation and above all cooperation between the Soviet Union and the United States in the study of the laws of the universe." (UPI, *Wash. Post*, 5/16/63)
- X–15 No. 1 piloted by John B. McKay (NASA) reached 3,856-mph speed (mach 5.57) and 124,200-ft. altitude, 156 mph faster and 26,200 ft. higher than planned because of one-degree error in flight angle. Aircraft was equipped with traversing probe, de-

vice that extends itself about three inches from X-15 surface every four seconds, to measure air flow. (FRC Release; *NYT*, 5/17/63, 19)

May 15: NASA announced from Cape Canaveral that recruiting 9-12 new astronauts would begin next month. (Simons, *Wash. Post*, 5/16/63)

- NASA announced award of $10,687,000 contract to Chance Vought Corp. for 23 Scout launch vehicles. (NASA Release 63-105)
- Chief of NASA Future Applications Satellites Capt. Alton B. Moody (USNR), addressing graduating class of U.S. Naval Academy, described nonmilitary navigation satellite system being studied by NASA. One such system could provide position determination, air traffic control, ship surveillance, search and rescue, fleet monitoring, and weather routing services. (Text)
- International Association of Machinists announced ratification of new three-year contract with Boeing Co. Agreement ended aerospace industry-labor negotiations begun last September. (*NYT*, 5/16/63, 33)
- U.K. and France were calling conference of European Postal and Telecommunications Administrations to discuss possibility of European communications satellite system, U.K. Foreign Office spokesman said. U.K. had abandoned any plan to develop its own system. Conference participants would be Austria, Belgium, Denmark, Finland, France, Greece, Iceland, Ireland, Italy, the Netherlands, Norway, Spain, Sweden, Turkey, United Kingdom, and West Germany. (*NYT*, 5/16/63, 19)
- Soviet news agency Tass reported radioastronomers at Pulkovo Observatory had discovered radiation belt around planet Jupiter. (UPI, Wash. *Daily News*, 5/15/63)

May 15-17: Second Annual National Flight Forum Symposium held at Hartford, Conn., with more than 300 representatives of business, industry, education, government, and military services attending. Symposium was sponsored by Connecticut General Life Insurance Co. (NASA Release 63-101)

Addressing the Forum, NASA Administrator James E. Webb reviewed early history of U.S. development in aeronautics, noting U.S. was far behind in aircraft development in 1915, when NACA was formed: "The wisdom of its [NACA's] establishment was amply demonstrated over the ensuing years, and led ultimately to United States supremacy in civil and military aviation . . . ." He noted U.S. was behind in 1957 when U.S.S.R. orbited SPUTNIK I, then cited formation of NASA in 1958 and President Kennedy's recommendation in 1961 of accelerated program "to give the nation a clearly leading role in the conquest of space.

"I review this history because it is necessary occasionally to remind ourselves of the past if we are to have a clear view of the present. With a warning clearly sounded in space, the nation determined, for the first time, that it would seize the opportunities offered by man's new ability to explore space, and that it would not be content merely with catching up with somebody else. Instead we set our sights on the most challenging goal within our reach, which could be achieved within a time scale competitive with the ability of others to do the same, and determined that we would explore the moon with men within this decade.

"Nothing in recent history suggests that we have reason now to change our course . . . .

"Those who view the lunar program simply as a propaganda effort fail to grasp that not only our prestige, but our capacity for constructive international leadership, our economic and military capacity for technological improvement, depend upon our ability to achieve acknowledged superiority in science and technology, and to use this capability in our own behalf and that of our allies . . . ." (Text)

Dr. Raymond L. Bisplinghoff, Director of NASA Office of Advanced Research and Technology, described NASA research in supersonic transport and in V/STOL aircraft and said: "Aeronautical and space research serves as a sharp cutting edge to advance man's knowledge in practically every technology which is fundamental to transportation devices. Although it is impossible to predict quantitatively the impact on transportation of current aeronautical and space research, past experience has shown that we may predict with confidence that it will be considerable . . . ." (Text)

Najeeb E. Halaby, FAA Administrator, said on May 16 that "there are areas—such as in pure research—where it is impossible for private industry to bear the full economic burden. For example, clearly it is up to the Government to play a role in the development of a supersonic transport, if this Nation is to provide such an aircraft . . . .

"I believe the government should act only in areas where the people 'cannot so well do for themselves' even in this increasingly complex and technical world, and that we who are charged with making decisions affecting the public must be convinced beyond any reasonable doubt that there is a need for the Federal Government to conduct the activity at all . . . ." (Text)

May 16: President Kennedy telephoned Astronaut L. Gordon Cooper aboard carrier *Kearsarge*, saying, "I just want to congratulate you. That was a great flight . . . . We are very proud of you." (UPI, *NYT*, 5/17/63, 1)

- Senate Resolution 143 resolved that "the Senate of the United States extends its profound congratulations to Major Cooper on his heroic accomplishment, and expresses its gratification to his family and to all those who have contributed to the success of his mission." (*CR*, 5/16/63, 8364–65)

- Former President Harry S. Truman said flight of L. Gordon Cooper was "one of our greatest" moments in history. . . . "I was exceedingly happy when he got down all right. That's what I was worried about.

"He made a great flight." (UPI, *L.A. Times*, 5/17/63)

- U.S.-U.S.S.R. negotiations on space cooperation resumed in Geneva with NASA Deputy Administrator Dr. Hugh L. Dryden and Soviet space expert A. A. Blagonravov heading respective teams. Attempt would be made to reach agreement on cooperative use of satellites to study earth's magnetic field. Discussion also would center on implementing agreement reached in principle last March concerning exchange of information furnished by meteorological satellites and coordinated experiments with communications satellites. (*NYT*, 5/17/63, 18; *Manchester Guardian, Wash. Post*, 5/17/63)

May 16: Prime Minister Harold Macmillan told House of Commons that British scientific criticisms of Project West Ford's orbiting copper filaments had been forwarded to U.S. Government, but he defended the project, saying it could have "very, very great significance in the whole system of deterrent defense." He also rejected Laborite Arthur Henderson's demand that he urge President Kennedy to submit plans for experimental projects to an international scientific committee. Henderson cited report by special committee of British scientists on effects of U.S. high-altitude nuclear explosion last July, report calling for international agency to control space experiments. (Lewis, *Wash. Post*, 5/17/63)

- Tass reported Soviet scientists had lost contact with MARS I interplanetary probe on March 21, when it was about 66,000,000 mi. from earth. Probe was continuing on its flight toward vicinity of Mars but telemetry data indicated difficulty in orientation system had misaligned its radio antennas. Tass said MARS I had recorded considerable change in boundaries of radiation belts around earth; "sphere of maximum intensity" in innermost belt was farther out than previously noted. Intensity of cosmic rays had almost doubled in last four years, Tass said. Important findings were made concerning stability and energy of interplanetary and solar plasma, and MARS I made possible first scientific observation of junction of earth's ionized gas envelope with outermost belt of charged particles. Other data on intensity of magnetic fields in space and distribution of meteoric particles were recorded. (*Balt. Sun*, 5/17/63; *Av. Wk.*, 5/27/63, 24)

- D. Brainerd Holmes, NASA Deputy Associate Administrator and Director of Manned Space Flight, said in address to American Ordnance Association, Washington: "The Department of Defense has not fully defined the role man should play in its space activities. However, under an agreement between Secretary McNamara and Mr. Webb, reached in January 1963, NASA will manage the Gemini program to assure the fulfillment of both DOD and NASA requirements, and the DOD will contribute funding, in an amount to be determined later . . . .

   "Although manned flight is more expensive than unmanned flight in the development phase, the increased reliability of a manned vehicle will ultimately enable us to carry out complex tasks in space more effectively and, very possibly, at less expense than with automatic equipment.

   "Thirteen years ago, the late William Faulkner made a comment that may apply to the present discussion:

   "'I believe that man will not merely endure; he will prevail. He is immortal . . . because he has a soul, a spirit capable of compassion and sacrifice and endurance.'

   "In exploring the moon and planets, I am convinced with Faulkner that man will prevail." (Text, *CR*, 5/21/63, A3220)

May 16: U.S.S.R. announced reorganization of its Academy of Sciences, designed to recentralize all basic research in physical and social sciences and to reflect trend toward specialized differentiation in the sciences. (Shabad, *NYT*, 5/18/63, 1, 9)

May 17: NASA Director of Space Sciences Dr. Homer E. Newell told Scientific and Technical Subcommittee of U.N. Committee on Peaceful Uses of Outer Space in Geneva that orbiting of USAF Project West Ford copper filaments would "create no significant interference with any other scientific activity." Dr. Newell said criticism of project by radioastronomers seemed to have been based on assumption that experiment, if successful, would be followed by similar one on much larger scale. He assured subcommittee that no further launchings would be planned until scientific findings of first experiment had been "analyzed and evaluated" and "adequate safeguards against harmful interference with any branch of scientific endeavor" had been developed. (*NYT*, 5/18/63, 9)

- Tass published Soviet Government statement declaring that U.S. was "flouting all rules of international law" and "unfolding preparations for a nuclear war" with its Project West Ford experiment. (*NYT*, 5/18/63, 9)
- Mercury Flight Operations Director Christopher C. Kraft, of NASA Manned Spacecraft Center, told reporters at Cape Canaveral that "there's a lot to be gained from flying another [Mercury] mission. We have a 92-orbit capability. But we will have to analyze the data from this flight [MA-9] to determine whether another is justified." (CTPS, *Chicago Trib.*, 5/18/63)
- Secretary of Defense Robert S. McNamara sent telegram of congratulations to Maj. L. Gordon Cooper (USAF):

    "Your Air Force leaders, Secretary Zuckert and General LeMay, and the entire Defense Department, join me in congratulating you on your record breaking achievement in behalf of our national space effort.

    "The scientific data that you have obtained will add tremendously to our store of knowledge about space.

    "You have earned the highest esteem of all Americans for your historic exploit." (DOD Release 710-63)
- Former President Eisenhower in Rochester, N.Y., hailed Astronaut L. Gordon Cooper's 22-orbit space flight as "a great step forward in the United States' space program." (AP, *NYT*, 5/18/63, 9)
- *Life* editorial said: "Now the United States can be foremost in another and greater adventure [than Columbus']—or abdicate its own national greatness by not doing enough. Unfortunately the argument about whether space is worth it has been muddied by an increasing popular tendency to talk only about the race to the moon. . . .

    "The issue is much bigger than a moon race, although of course we plan to land on the moon. The U.S. commitment to space seems a natural undertaking for the American people, who are a venturesome lot. Indeed it makes sense for military, technology, and prestige reasons. . . ." (*Life*, 5/17/63, *CR*, 5/20/63, A3160)

May 17: USAF launched Blue Scout, Jr., probe from Pt. Arguello with undisclosed mission. (*NYT*, 5/18/63; *M&R*, 5/27/63, 10)

- USAF OAR announced three USAF Cambridge Research Laboratories meteorological research groups were participating with Weather Bureau and military services' support in 1963 National Severe Storms Project to learn more about structure, development, and possible forecasting of severe storms. (OAR Release 5–63–4)
- USN test-fired Polaris A–3 missile from Cape Canaveral in successful 1,600-mi. flight. (UPI, *NYT*, 5/18/63, 8)

May 18: President Kennedy visited Nashville, Tenn., Muscle Shoals, Ala., and Huntsville, Ala. At Muscle Shoals, in address commemorating 30th anniversary of TVA, he said: "Only a national effort can explore the mysteries of outer space, harvest the products of ocean depths and mobilize the human, natural and material resources of our land. I cite these examples—not to show the growth of federal activity, for it is still small compared to the Nation's—but to show the positive side of Federal-State cooperation of which TVA is an outstanding symbol . . . ." At Redstone Arsenal he spoke briefly before boarding his plane to return to Washington:

"I know there are lots of people now who say, 'Why go any further in space?' When Columbus was halfway through his voyage, the same people said, 'Why go on any further?' 'What will he possibly find?' 'What good will it be?' And they want to stop now.

"I believe the United States of America is committed in this decade to be first in space, and the only way we are going to be first in space is to work as hard as we can here and all across the country, and support not only Major Cooper, but all those who come after him." (*Wash. Post*, 5/19/63, A8; *Marshall Star*, 5/22/63)

- In *Saturday Evening Post* article former President Eisenhower wrote:

"Let us turn to the space exploration program, which I publicly have called downright spongy. Now clearly the strong competitive spirit of the American people has been aroused by the so-called space race. But let us step back for a moment from emotion and be objective. This is what my scientific advisers and I tried to do when we started space work back in 1955 and thereafter enlarged it into a comprehensive program. We didn't and don't want to be a second-best nation, not in any important field, and certainly not in total accomplishment. But can we best maintain our overall leadership by launching wildly into crash programs on many fronts? This is where we seem to have got out of focus.

"Let me make it perfectly clear that we should have an aggressive program of research and exploration, so broadly based that in the long run there will be no question of our space leadership; but we should pursue it in an orderly, step-by-step way to enlarge systematically our knowledge of the scientific, military, and industrial potentials in space. This sort of a program would be a strong, steady push from the known into the unknowns of space science, accumulating along the way the techniques and skills to put our astronauts on the moon, among other developments. The annual cost would not include money for stunts and unnecessary contests.

"Most of the scientists who advised me on space matters—all men of great knowledge and integrity—believed in this step-by-step program as most effective for the long pull, pointing out that the things we need to learn in space are almost as endless as space itself. Proud as we may be of our astronauts and our Venus probe and other accomplishments in space, this racing to the Moon, unavoidably wasting vast sums and deepening our debt, is the wrong way to go about it, as I see it. Because it has been stated that the country's prestige is at stake in this race, the average citizen may be loath to question the huge sums now being requested. But he should . . . ." (*Sat. Eve. Post*, 5/18/63, 15–19; *CR*, 5/15/63, 8111)

May 18: General B. A. Schriever (AFSC), addressing Armed Forces Dinner in Louisville, Ky., said: "I am convinced that we must ... investigate the military potential of space systems. We simply cannot afford to leave the space frontier unguarded. Satellite systems promise to strengthen our present defenses in a number of ways. For example, they show definite potential for improved warning, communications, and navigation systems.

"We may also need to develop the ability to defend ourselves against space-borne weapons. The Soviets have threatened many times to use their space technology for military purposes. Until we know for sure that they will not carry out these threats, it is logical to pursue the development of ways to identify, inspect and rendezvous with a non-cooperative object in space . . . ." (Text, AFSC Release 35–R–41)

- U.S.S.R. launched improved models of carrier rockets for spacecraft on 7,500 mi. flight into Central Pacific. (*NYT*, 5/29/63, 3)
- USAF Minuteman ICBM launched at Cape Canaveral fell short of intended range but most test objectives were achieved. (DOD Release 722–63)
- DOD launched unidentified satellite with Thor-Agena D vehicle. (*Pres. Rept. on Space, 1963*, 1/27/64)

May 19: Astronaut L. Gordon Cooper, accompanied by NASA Associate Administrator Dr. Robert C. Seamans, Jr., NASA Manned Spacecraft Center Director Dr. Robert R. Gilruth, and MSC Deputy Director for Flight Operations Walter C. Williams, held press conference at Cape Canaveral carried on nation-wide TV. Major Cooper presented orbit-by-orbit account of his MA–9 space flight. Asked about possibility of MA–10 flight, Dr. Seamans said: "It is quite unlikely that we will have another Mercury flight," but final decision on MA–10 would be made "within a week or so." (*NYT*, 5/20/63, 24; *New York Herald Trib.*, 5/20/63)

- Presidential Boeing 707 jet airliner flew nonstop from Washington to Moscow in 8 hours 38 minutes and 42 seconds, claiming 15 speed records. Piloted by Col. James B. Swindal (USAF), airplane brought AEC Chairman Dr. Glenn T. Seaborg and his party to Moscow to conclude nuclear cooperation agreement with Soviet officials. (NYT, 5/20/63, 1, 3)

May 19: With publication of first NASA Applications Note ("Welding Tips"), NASA inaugurated new phase of its technology utilization program—disseminating information on results of NASA technological research in Applications Notes (containing innovations not thoroughly documented and evaluated) and Applications Reports (presenting comprehensive descriptions of single developments of significant industrial potential). (NASA Release 63–103)

- Atlantic Research Corp. announced development of solid rocket propellants containing beryllium to provide increased thrust. Studies leading to this development were sponsored by DOD. (NYT, 5/20/63, 46)
- Tass reported Soviet scientists had synthetically produced isotope with atomic number 256. Produced by irradiating uranium-238 with beam of accelerated neon-22 ions, isotope had half-life of eight seconds, "many tens of times" greater than expected. (Reuters, NYT, 5/21/63, 3)

May 20: Vice President Lyndon B. Johnson said at Jefferson-Jackson Day Dinner, Oklahoma City: "Space . . . has gained the lion's share of publicity—and justifiably so.

"When we talk of space and space research, there are those who raise the question that our efforts cost too much. Certainly American leadership in space is not cheap. We are now spending 20 cents per week per capita on our national space program.

"Other Americans ask if our space efforts are worthwhile.

"I can answer in terms we can all understand. While the space age is not yet 5 years old, more than 5,000 companies and research organizations have been or are now involved in our space effort. We have produced more than 3,200 space-related products, many of which are already being put to use.

"But, many prominent Republicans are questioning the value of the program. A former Republican President has referred to our moon project as a 'stunt.'

"I can answer that simply: I do not believe that this generation of Americans is willing to resign itself to going to bed each night by the light of a Communist moon . . . ." (Text, CR, 5/23/63, A3243–44)

- DOD announced first flight test of X–21A "slitwing" aircraft, built by Northrop Corp. for USAF, was successfully conducted at Edwards AFB. Inhalation of air through hundreds of tiny slits in aircraft wings showed marked reduction of drag on the plane. This was first time such laminar flow control has been attempted on aircraft wing approximating size of those of jet transports. (DOD Release 628–63)
- Two-stage sounding rocket instrumented to observe ionosphere was successfully launched to 215-mi. altitude by Japanese scientists near Kagoshima, Japan. (Reuters, Wash. Post, 5/21/63)
- Italy launched Nike-Apache sounding rocket to 126-mi. altitude for atmospheric studies. Launching was first from island of Sardinia. (UPI, NYT, 5/22/63, 34)
- NASA Ames Research Center announced award of contract to Westinghouse Electric Corp. Astronuclear Laboratory to study poten-

tial physiological radiation damage in space. One-year study would "attempt to understand the basic phenomena of biological response to cosmic radiation, and to develop the means of preserving man's well-being and insure his survival during long periods of exploration in deep space. . . ." (Ames Release 63-22; NASA Release 63-107)

May 20: A. M. Nowitzky, Head of Spacecraft Sterilization Systems, Lockheed Missiles and Space Co., said in letter to *Aviation Week and Space Technology:* "Since all known tests indicate that gaseous surface sterilization does not affect reliability, Ranger problems were definitely not due to sterilization but, instead, were ordinary design problems unrelated to the inclusion of this requirement. It is, therefore, grossly unrealistic to blame sterilization for the Ranger failures. . . ." (*Av. Wk.*, 5/20/63, 150)

- Dr. John R. Dunning, Dean of Columbia Univ. Faculty of Engineering, said in address before Design Engineering Conference of American Society of Mechanical Engineers, NYC, that "if the engineers do not aggressively claim the positions of leadership" in space effort and in exploitation of atomic energy, "this nation will pay an intolerable price for amateur improvisation." Today's engineers need knowledge of conventional engineering techniques as well as current developments in science, and some engineering schools are graduating "imitation physicists" and "disguised mathematicians" instead of men trained in the "great art" of systems design. (Sullivan, *NYT*, 5/21/63, 35)

- Robert Hotz wrote in *Aviation Week and Space Technology* editorial:

    "It is an inevitable historic trend that as new technology is transformed from the spectacular experimental stage to reliable operational use, peripheral public interest lags almost in proportion to the increase in technical significance. For example, the world press could write about little else for months after Charles Lindbergh made the first solo transatlantic aircraft flight from New York to Paris just 36 years ago this month. Last year more than two million people followed his trail across the transatlantic airway—most of them in jet transports taking about one-fourth the time of Lindbergh's flight. Yet only a few obscure statistical paragraphs in the daily press recorded this tremendously significant perspective on the trail Lindbergh blazed.

    "And so it is likely to be with manned space flight, as it passes from the experimental flight-test phase into reliable operational performance. . . . [MA-9] is less exciting than the first breathless attempt of man simply to survive short-duration space flight. But it is the type of foundation on which future capability can be built solidly. . . .

    "Mercury must go down in history as one of the most successful technical programs this nation has ever had. . . ." (*Av. Wk.*, 5/20/63, 21)

May 20-21: NASA Hq. conducted two one-day training courses dealing with conflict of interest and standards of conduct for selected Hq. personnel. (NASA Announcement 63-108)

May 20–24: Third European Spaceflight Symposium in Stuttgart, sponsored by British Interplanetary Society, French Astronautics Society, and German Society for Rocket Technology and Spaceflight. (*M&R*, 5/27/63, 14–15)

May 21: On 36th anniversary of Lindbergh's historic transatlantic flight, Astronaut L. Gordon Cooper was presented NASA Distinguished Service Medal by President Kennedy in White House ceremony. Also presented were NASA Medals for Outstanding Leadership to: G. Merritt Preston, Manager of Project Mercury Operations at NASA Launch Operations Center; Floyd L. Thompson, Director, NASA Langley Research Center; Kenneth S. Kleinknecht, Manager of Project Mercury Office, NASA Manned Spacecraft Center; Christopher C. Kraft, Director of Flight Operations Div. of NASA Manned Spacecraft Center; and Maj. Gen. Leighton I. Davis, commander of AFMTC, Cape Canaveral. NASA Group Achievement Awards were presented to: DOD Recovery Forces; and Air Force Space Systems Div.

Following ceremony, Astronaut Cooper was hailed by thousands in motorcade to the Capitol, where he addressed special joint session of Congress: ". . . I don't think I have ever been with a team that was more dedicated, or striving harder, or was more completely sold on their product than the total space effort and particularly the manned flight effort in which I am involved . . . .

"I named my spacecraft FAITH 7 for three reasons: First, because I believe in God and country; second, because of the loyalty to organization, to the two organizations, actually, to which I belong; and, third because of the confidence in the entire space team . . . .

"I think of all things that I am constantly amazed at is the public's response to this program . . . it is tremendously impressive." The parades and activities following each flight "show that Americans want to express their feelings and their confidence that we . . . can conduct peaceful research programs; that we can conduct them openly, and under the surveillance of every man, woman and child in the entire world." (*Space News Roundup*, 1/8/64, 2; *CR*, 5/28/63, A3420–21)

• At the State Department luncheon honoring Astronaut Gordon Cooper, Vice President Johnson remarked:

"In 1942, President Roosevelt called together our wartime leaders for the final decision on continuing or abandoning the Manhattan project which produced the atomic bomb. One of the most eminent leaders present heard presentation from all sides. Then, he solemnly gave President Roosevelt his verdict: 'The bomb will never go off—I speak, of course, Mr. President, as an expert on explosives.'

"In that first, uncertain spring of the space age 5 years ago, some conscientious experts took the same attitude toward Project Mercury. History has proved them grossly wrong. . . .

"Hitler once predicted the Nazis would wring England's neck like a chicken. After the Battle of Britain, Winston Churchill said to Commons, 'Some chicken. Some neck.' We have heard some say recently that the civilian space program is only 'leaf-raking.' Considering Major Cooper—considering the vast technological cooperation which made his mission a success—I would say today, 'Some leaf. Some rake.'" (Text, *CR*, 5/23/63, A3299)

May 21: U.S. and U.S.S.R. signed three-year agreement for cooperative program in nuclear studies. Signed in Moscow by Dr. Glenn T. Seaborg, Chairman of AEC, and Andronik M. Petrosyants, Chief of State Board for Peaceful Uses of Atomic Energy, agreement provided for exchanges of scientific delegations and unclassified documents as well as joint conferences and visits by research specialists in fields of controlled thermonuclear fusion, reactor techniques, and physics of high-energy particles. (Shabad, *NYT*, 5/22/63, 4)

- Nike-Apache sounding rocket launched from Ft. Churchill, Canada, carrying experiments (1) to determine upper atmospheric winds and diffusion coefficients by means of triangulation photography of sodium vapor trail, and (2) to determine atmospheric temperature from photoelectric measurement of emission line width. The flight was part of an international effort to measure winds at widely separated geographical locations. (NASA Rpt. of S. Rkt. Launching, 6/14/63)

- Presidential Boeing 707 jet airliner flew nonstop from Moscow to Washington in 9 hours 53 minutes 52 seconds, claiming 15 speed records, the same number it had claimed for the outbound flight over the same route on May 19. (AP, *NYT*, 5/22/63, 18)

- Dr. L. R. Shepherd, President of British Interplanetary Society, urged European Space Research Organization (ESRO) and European Launcher Development Organization (ELDO) combine to form a European space agency to "develop the technology and logistics of an economically viable space transportation system." Suggesting it would be wasting time and money for Europe to attempt to duplicate U.S. communications satellite systems, he said European agency should concentrate on advancement of transportation techniques—including construction of orbital stations and establishment of lunar base capable of constructing and supplying space vehicles. (Wilcke, *NYT*, 5/22/63, 37; *M&R*, 5/27/63, 14–15)

May 21–24: Aviation/Space Writers Association met in Dallas, Tex., more than 500 writers from throughout U.S. and Canada. (MSC *Space News Roundup*, 5/16/63, 1)

Dr. Lawrence L. Kavanau, Special Assistant for Space, DOD DDR&E, said: "None of us harbors any illusions about the expense of developing and operating military space systems. Any and all military efforts in this field will be costly. Of the two, development costs may well prove to be the relatively minor one. Building and supporting an operational military space force may involve costs far out-distancing the investment in its research and development. Military mission capabilities in space will not ipso facto evolve merely because we are able to attain an experimental capability in the medium. The fundamental point here is that military needs, by definition, require the capacity to conduct reliable repetitive space operations . . . .

"One of the important aspects of the [NASA–DOD] Gemini agreement is the stipulation that neither agency will initiate any major new programs or projects in the field of near-earth orbit manned space flight without the consent of the other. That paragraph was inserted for a specific purpose. The concept of a Manned Orbital Space Station, Manned Orbital Development System, or similar names by which such a facility has been called has been proposed and considered in the Department of Defense for at least five years . . . a limited space station project need not represent a national commitment as far-reaching and as large in scope as the manned lunar landing program. It should, nevertheless, require a national, rather than a single agency decision, to get it under way. There is no question that, if a decision is made to proceed, the Department of Defense should be a full partner in this enterprise all the way, and may even qualify for management of the national project to satisfy both NASA and DOD needs. This, of course, requires a Presidential decision if and when the project is approved." (Text, DOD Release 718–63)

Col. James W. Little, Director of Pilot Training, USAF Air Training Command, said: "We envision that the basic preparation for space flight taught now by the Edwards AFB Aerospace Research Test Pilot Course will be taught eventually to all pilot trainees in the Air Training Command. Student training will be conducted into the suborbital area above 200,000 feet instead of below 50,000 feet as it is at present . . . ." (A-N-AF Journ. & Reg., 6/1/63, 2)

William Littlewood, American Airlines vice president, said: "The United States is woefully deficient in funds and efforts to solve the essential research problems, to gain the basic knowledge, to develop the potentials of atmospheric flight, as to safety, comfort and efficiency, or to compete successfully and for long with the foreign advantages in this field . . . .

"Whereas all efforts and expenditures of NACA were devoted to aeronautic research, including large assignments in safety and reliability, we now find aeronautic research relegated to a fifth echelon level in NASA and reduced to a negligible portion of the total budget and to a fraction of the former NACA effort." (AP, NYT, 5/22/63, 65)

May 21: Rep. A. J. Multer (D.-N.Y.) said at Briefing Conference on National Patent Policies and Practices, Washington: "Among the many subjects around which differing opinions swirl is that of Government patent policy, particularly as respects ownership of patent rights on inventions resulting from governmentally financed research . . . .

"I have introduced in the 88th Congress two bills of major importance in this field, H.R. 701 and H.R. 2349. Both bills were also introduced in the last, the 87th Congress.

"The purpose of H.R. 701 is to attempt to bring some order out of chaotic and conflicting Government patent policies to the end that a uniform practice as respects ownership of patents derived from Federal contracts, and grants, research, and otherwise, would be carried out by executive agencies.

"This bill has been referred to Subcommittee No. 3 of the House Judiciary Committee . . . .

"The second bill, H.R. 2349, would amend the existing patent code to permit patent holders to bring civil actions against Government contractors who infringe their patents while carrying out Government contracts.

"This bill is also in the Judiciary Committee but not yet referred to a subcommittee . . . ." (Text, *CR*, 5/23/63, 8743)

May 21: Scientists and engineers from major NASA centers convened at Lewis Research Center to discuss future manned planetary missions. (LRC Release 63-30, Lewis Chronology, 4)

May 22: New York City honored Astronaut L. Gordon Cooper with tickertape parade. At city luncheon honoring Major Cooper, NASA Administrator James E. Webb, and the entire Project Mercury team, Mayor Robert Wagner presented Major Cooper and Mr. Webb with city's Medal of Honor and scroll acclaiming MA-9 space flight. Mayor Wagner said: ". . . It is remarkable how such a feat not only unites our own country in prideful pleasure, but also gives satisfaction to the rest of the world—to friend and foe alike. This is not man against man, but man against the elements, man against the forces of nature, man armed with the power of knowledge . . . ." (*NYT*, 5/23/63; *CR*, 5/29/63, A3487)

- At regular press conference, President Kennedy was asked: "Did the astronauts raise with you . . . their desire for another Mercury flight? Do you have any opinion yourself, tentative or otherwise, as to the desirability of another Mercury flight?"

President Kennedy replied: "I think they feel that it's worthwhile. I haven't discussed it with Mr. Webb. NASA should make the judgment and will make the judgment, and I would not intervene. But they do feel that a flight is useful and that the experience of Major Cooper has indicated that the time between the last Gemini flight as scheduled—the Mercury flight and the new Gemini flight, which is a period of almost 18 months—they feel may represent a gap which could be filled very usefully by another Mercury flight.

"This will be a matter which I think they're going to be talking about this week with Mr. Webb and which I will discuss with him next week. But the final judgment must be NASA's." (Transcript, *NYT*, 5/23/63, 18)

- NASA launched Scout rocket from Wallops Island, Va., with RFD-1 (re-entry flight demonstration-1) payload for AEC. 480-lb. mockup nuclear reactor traveled 800-mi. suborbital trajectory and reentered southwest of Bermuda, but initial recovery attempts were not successful. Purpose of flight was to obtain data for designing space nuclear reactors that disintegrate upon atmospheric re-entry. This was first of series of AEC flight tests to evaluate safety of aerospace nuclear power systems in operation. (Wallops Release 63-49; AP, Wash. *Eve. Star*, 5/23/63)

- Two Nike-Apache sounding rockets launched from Ft. Churchill, Canada. The flights were part of a series of experiments to measure winds and temperatures of the upper atmosphere. (NASA Rpt. of S. Rkt. Launching, 6/14/63)

May 22: Maj. Robert S. Sowers (USAF), Capt. Robert MacDonald (USAF), and Capt. John T. Walton (USAF) named by General Curtis E. LeMay (USAF) as recipients of 1962 MacKay Trophy "for the most meritorious flight of the year." The fliers established new transcontinental speed records on March 5, 1962, in B-58 Hustler. (DOD Release 724-63)

- COSMOS XVII launched into orbit by U.S.S.R. (apogee, 488 mi.; perigee, 161 mi.; period, 94.82 min.; inclination, 49°02' to equator). As with previous Cosmos satellites, Tass said scientific equipment "intended for continuing the study of interplanetary space . . . was mounted on board." (Tass, *Izvestia*, 5/23/63, 1, AFSS-T Trans.)
- FAA Administrator Najeeb E. Halaby announced at Aviation/Space Writers Association meeting in Dallas that he had submitted recommendation to President Kennedy on U.S. prospects for supersonic transport aircraft development. (Jackson, *L.A. Herald-Examiner*, 6/22/63)

May 23: Arcas rocket instrumented to measure electrical characteristics of lower ionosphere launched from Birdling's Flat on the Canterbury Plains, New Zealand. First sounding rocket launched in cooperative U.S.-New Zealand program, Arcas with its launcher was supplied by NASA; payload was designed and built by Univ. of Canterbury staff; rocket was assembled and launched by Royal New Zealand Air Force. Two more Arcas rockets would be launched in the joint Univ. of Canterbury-NASA project, sponsored by National Space Research Committee of the Royal Society of New Zealand. (NASA Release 63-110)

- Sodium-vapor experiment to measure high-altitude winds and diffusion rates was launched on Nike-Apache sounding rocket from Wallops Island, Va. Sodium vapor trail, ejected from 27 to 127-mi. altitudes, was visible for several hundred miles from launch site. (Wallops Release 63-51; AP, *Wash. Post*, 5/24/63)
- Astronaut L. Gordon Cooper greeted by crowds of thousands in Houston as he returned home for first time since his 22-orbit MA-9 space flight. (UPI, *NYT*, 5/24/63)
- NASA announced plans to award contracts for support services in Launch Operations Center Merritt Island Launch Area (MILA) in four broad categories: base services; launch support services; administrative and management services; and communications. (NASA Release 63-111)
- Senator Clinton P. Anderson (D.-N.M.), Chairman, Senate Committee on Aeronautical and Space Sciences, speaking on the Senate floor stated:

    "The decision to try to land a man on the moon by 1970 does not constitute a crash program . . . . Failure to decide on a schedule for attempting to reach the moon is equivalent to setting no deadline at all. Work is done best—even by the technical community—which is not exempt from the laws of human nature—when a specific goal is set. An objective of landing on the moon in the 1960's . . . permits us to work at a challenging pace; yet, does not absorb a greater percentage of the nation's resources than our country can afford. This is only one percent of the gross national product at current rate . . . .

"Project Apollo does not fit the description of a crash program. The wartime Manhattan project was a crash program. We used all the resources we could to develop the atomic bomb in the shortest period of time, regardless of cost. We maintained duplicating operations over a 3-year period in the hope that one would work—and one did, in time to help win a war. Apollo is no such project. If it were, we would not have had the controversy over whether earth-orbiting rendezvous, lunar-orbiting rendezvous, or a direct landing was the best way to get a man to the moon. A crash project would have gone into operational stages of all three alternatives to a moon landing . . . .

"The shortage of scientific manpower has not been aggravated in a measurable way by Apollo. Those critics who fear that it has, have not established their case. For example, in 1970, it is predicted that U.S. scientists and engineers will total 1,995,000, of which only 114,700 or less than 6 percent, will work for NASA and its contractors. It should be noted that a substantial portion of those are merely moving over from research and development work on aeronautics and missiles . . . NASA is enlarging its efforts to increase the national supply of scientific and engineering personnel. Last fiscal year, the agency provided graduate science and engineering fellowships to 100 students at 10 universities. In its budget proposal for fiscal 1964, NASA requests authorization for 1,250 fellowships covering 110 universities.

"Additionally, the agency has provided funds for 150,000 square feet of research space at five universities; it is asking for funds to provide 400,000 square feet of laboratory space at 11 universities." (*CR*, 5/23/63, 8961–66)

May 23: Dr. George L. Simpson, NASA Assistant Administrator for Technology Utilization and Policy Planning, testified before Foreign Operations and Government Information Subcommittee of House Committee on Government Operations, appearing at Committee's invitation to "discuss NASA's responsibilities with respect to the Satellite Situation Report and related procedures and reports . . . .

"NASA has been publishing the Satellite Situation Report since . . . [Feb. 17, 1961]. It has included all unclassified information made available to it by NORAD except for a period from January to April 1963 when it was thought that each agency, NASA and DOD, would handle its own reporting. NASA has resumed publishing all unclassified information received from NORAD . . . .

"In summary, NASA was never assigned the responsibility for reporting satellite tracking information but did agree at DOD's request, when the DOD's Space Tracking effort became operational, to disseminate all unclassified information on satellites to the scientific community using the information made available to it by NORAD.

"There apparently has been some variation in the type and scope of the information which Defense has made available to us. We have released all unclassified material received from NORAD without modification . . . ." Under recent NASA-DOD agreement, NASA's *Satellite Situation Report* lists information on Soviet satellites when such information is: (1) cleared for release by Assistant Secretary of Defense for Public Affairs, (2) officially reported to U.N. Registry and confirmed by NORAD tracking data, and (3) publicly announced by Soviet government and confirmed by NORAD.

Subcommittee members joined Chairman John E. Moss (D.-Calif.) in accusing NASA of "unjustified subservience" to DOD policy. Chairman Moss said that "official space information has dwindled to the point where a true perspective of where we stand, in relation to the Russians, scarcely exists so far as the general public is concerned . . . . [DOD policy] clearly has resulted in the withholding of information vital to public understanding of United States and Russian space activities." Asked if any Soviet space failures in recent months had been kept secret by U.S. authorities, Dr. Simpson replied that his answer could be given only in executive session. (Testimony; Transcript)

May 23: Arnold W. Frutkin, NASA Director of International Programs, outlined progress in NASA international programs before Third European Space Flight Symposium, Stuttgart, and said: "I believe that the essential meaning of this brief review of our own programs is that international pooling of efforts in the space field can be highly successful if it is rigidly directed to solid scientific and technical objectives of mutual interest, requiring significant rather than token contributions by both sides and thoughtful planning. Perhaps even more important is the clear evidence that these successes engender growing interest and greatly expand the basis for succeeding cooperative enterprises.

"We are quite aware of our great indebtedness to European science and technology. We believe that the freest possible association of our respective programs, where they complement and support each other, is the soundest way to proceed if we are to continue to benefit in this way and if we are to reciprocate generously. I believe that in the long run cooperation between NASA for the United States and ESRO and ELDO for Europe affords the best means to expand the hard core of cooperation in space research, and that such collaboration will have the greatest potential for further extension to forward-looking nations everywhere in the world . . . ." (Text)

- In paper presented at Lunar Surface Materials Conference, Boston, NASA scientists Curtis W. McCracken (GSFC) and Maurice Dubin (Hq.) asserted moon is covered with dust from four inches to three feet deep. Basing estimate on data gained from spacecraft and ground observations of interplanetary dust particles, McCracken and Dubin said particles traveling at hypervelocities impact moon at such speed that both particle and its impact point are destroyed, creating lunar dust. They supported theory that dust layer is porous cobweb-like structure. (NASA Release 63–109)

- Warren Rayle, Lewis Research Center engineer, described work he is doing in watching for "ball" lightning, a phenomenon occurring during thunderstorms and appearing as a round, glowing object which may move slowly or hang in the air. Lewis scientists engaged in research with plasma propulsion devices think ball lightning may be much like the plasmas they work with in the laboratory. (LRC Release 63-31, Lewis Chronology, 4)

May 23: Senator John L. McClellan (D.-Ark.) inserted in *Congressional Record* editorial from *Industrial Research* regarding S. 816, bill to establish Commission on Science and Technology: "We simply doubt that it would be possible to pull together all Federal scientific activities under a single roof; and failure to do so would mean that the agency or department would be little or no improvement over the existing Office of Science and Technology. It is highly unlikely that the National Bureau of Standards could be taken from the Department of Commerce; the National Institutes of Health from the Department of Health, Education, and Welfare; the Bureau of Mines from the Department of the Interior; the research and engineering activities from the Department of Defense; the nuclear laboratories from the Atomic Energy Commission; etc., without impairing the effectiveness of these governmental bodies. However, we are willing to be convinced. But there is not much chance of change unless someone—such as the proposed Commission—takes a good, hard look at the ongoing Federal science programs." (*IR*, 5/63, in *CR*, 5/23/63, 8786-87)

- Heinz Koelle, Director of MSFC Future Projects Office, was awarded Hermann Oberth Medal at third annual European Space Vehicle Congress, Stuttgart. (*Marshall Star*, 5/29/63, 1)

May 24: Sodium-vapor experiment launched on Nike-Apache sounding rocket from Wallops Island, Va., ejected sodium-vapor trail from 26- to 122-mi. altitudes. Pink and reddish vapor clouds were visible for several hundred miles from launch site. This was last experiment in current sodium-vapor series, which included similar experiment May 23 and unsuccessful attempt earlier May 24 (payload failed to operate properly and no sodium vapor was ejected).

Data from these tests would be correlated with information obtained from similar experiments conducted from other sites around the world. Eight countries (Algeria, France, Italy, Canada, Pakistan, Argentina, Japan, and the United States) were now participating in the international sodium program for obtaining high altitude wind measurements and diffusion. (NASA Release No. 63-51, Wallops Releases 63-51 and 63-52)

- Soviet press agency Tass announced COSMOS XVIII satellite had been placed in orbit (apogee, 187 mi.; perigee, 130 mi.; period, 89.44 min.; inclination, 65°01' to the equator). Onboard equipment was functioning normally. (Tass, *Komsomolskaya Pravda*, 5/25/63, 1, AFSS-T Trans.)

- Dr. Hugh L. Dryden, NASA Deputy Administrator, and Dr. Anatoli A. Blagonravov of Soviet Academy of Sciences announced U.S. and U.S.S.R. had reached agreement on coordinated launching of satellites for measuring earth's magnetic field. Agreement was subject to final approval by NASA and Soviet Academy of Sciences; agreement was expected to be approved in about a month. (*NYT*, 5/25/63, 25)

- USAF announced all radio experiments conducted with orbiting Project West Ford needles were successful. Tests included coast-to-coast radio tests in which signals were bounced off 2,000-mi.-high cloud. Needles were now stretched over 11,000-mi. section of polar orbit, lengthening at rate of 1,000 mi. per day. Scientists expected in six weeks cloud would lengthen to form closed path about 40,000 mi. in circumference. (Toth, *NYT*, 5/25/63, 6)

May 24: Titan II ICBM launched from Cape Canaveral successfully placed re-entry payload in target area more than 6,500-mi. down AMR, third such success for Titan II. This was the 1,400th missile to be launched from the cape. (DOD Release 755–63; *CR*, 1/8/64, 2910)

- NASA Manned Spacecraft Center announced award of contract to Westinghouse Electric Corp. for design and delivery of main drive system of flight acceleration facility. Main drive motor will provide power to rotate 50-ft. arm, to end of which gondolas would be affixed for men or equipment. (MSC Release 63–91)
- Chief of NASA Recruiting and Examining Programs Madison B. Smith said at meeting of Federal Personnel Associations of New York and New Jersey, Massena, N.Y., that recent study of 3,348 scientists and engineers hired by NASA over 15-month period showed about one third of group came from industry. Bulk of those hired from industry were from firms with personnel cutbacks. Another third of new scientists and engineers came from other Government agencies. "The rest of the 3,348 were mainly new college graduates. About 30, or less than one percent, were from faculties." (NASA Release 63–113)
- Communications Satellite Corp. Chairman Leo D. Welch and President Joseph V. Charyk left Washington, D.C., for 10 days of discussions with Canadian and European communications officials. (*Wash. Post*, 5/24/63)
- Development of new helium isotope, Helium 2, was reported by Dr. O. M. Bilaniuk of Univ. of Rochester and Dr. Rodolfo J. Slobodrian of Argentine Atomic Energy Commission. Finding opened "whole new vistas for the possibility of holding other experiments of this kind," Dr. Bilaniuk said. (AP, *NYT*, 5/16/63, 18)
- New York *Journal-American* reported as many as five Soviet cosmonauts had died in unsuccessful space flights. Newspaper quoted informed Congressional and space authorities. (N.Y. *Journal-American*, AP, Balt. *Sun*, 5/25/63)
- USAF Minuteman ICBM launched from Vandenberg AFB silo in successful flight. (DOD Release 767–63)

May 25: Panel of scientists, headed by Dr. Norman F. Ramsey and convened by President's Science Advisory Committee and AEC's General Advisory Committee, recommended U.S. build atom accelerators with much greater energies than those now being used. For a "deeper understanding of the fundamental nature of matter," panel proposed: construction of proton accelerator with approximately 200-bev energy at Lawrence Radiation Laboratory; beginning design studies at Brookhaven Laboratory on proton accelerator with proton energies from 600 to 1,600 bev; construction of "storage rings" to be used with existing Brookhaven accelerator. (Finney, *NYT*, 5/26/63, 1, 45)

May 25: Award of patent for manned orbital spacecraft designed to glide into land or water landing was reported in *New York Times.* Patent was obtained for NASA by Alan B. Kehlet, Dennis F. Hasson, and William W. Petynia. (*NYT*, 5/25/63, 13, 18)

- West German scientists outlined plans for German space program at Third European Space Flight Symposium, Stuttgart. Program includes: communications satellite to be placed in polar orbit; high-energy rocket propulsion stage; space transport system; space probe rocket with paraglider landing system to allow recovery and reuse. (Wilcke, *NYT*, 5/26/63, 13)
- U.S.S.R. launched improved models of carrier rockets for spacecraft on 7,500-mi. flight into Central Pacific. (Tass; *NYT*, 5/29/63, 3)

May 26: Eight U.S. scientists issued statement supporting U.S. program to land men on moon and countering recent criticism that instruments would be more valuable than men in lunar exploration. "This criticism raises important issues which underlie the United States space effort." Statement cited "overwhelming" Congressional approval of President Kennedy's proposal for a manned lunar landing within this decade, presented on May 25, 1961.

"Was this support tendered for scientific reasons primarily, or was it motivated by a broader concern with national interests and national goals?" Project Apollo cannot be assessed on scientific merits alone, but even if it could be, "situations are bound to arise in which the human performance is indispensible." Furthermore, "the momentum and significance of the lunar program are derived from its place in long-range United States plans for exploration of the solar system," at heart of which is role to be played by man. The eight scientists were: Dr. W. Maurice Ewing, Director of Lamont Geological Observatory; Dr. Robert Jastrow, Director of NASA Goddard Institute for Space Studies; Dr. Joshua Lederberg, Chairman of Genetics Dept., Stanford Univ. School of Medicine; Dr. Willard F. Libby, Director of UCLA Institute of Geophysics and Planetary Physics; Dr. Gordon J. F. MacDonald of UCLA Institute; Dr. Lyman Spitzer, Jr., Director of Princeton Univ. Observatory; Dr. Harold C. Urey, Prof. of Chemistry, Univ. of Calif.; and Dr. James A. Van Allen, Chairman of State Univ. of Iowa Physics Dept. (Sullivan, *NYT*, 5/27/63, 1, 13)

- D. D. Wyatt, Director of NASA Office of Programs, received honorary doctorate in engineering from Univ. of Missouri School of Mines and Metallurgy. In commencement address, he told graduating class: "If you think you can rest on what you have already learned you will find yourself overtaken and pushed aside by the eager young graduates of tomorrow. Engineering, the application of the sciences, is no more static than the sciences themselves, and they are bursting at the seams of discovery . . . .

"No area of human activity better focuses and dramatizes the meaning of today than does this boldest, most challenging arena of space . . . ." (Text)
- Dr. Wernher von Braun, Director of NASA Marshall Space Flight Center, was awarded 1963 American Citizenship Award at 11th

annual German-American Day festival, North Bergen, N.J. (*Marshall Star*, 5/29/63, 7)

May 26: Reported that Russian trawlers and other Soviet vessels regularly appeared off of NASA's Wallops Station to observe rocket launchings. (Perkinson, Balt. *Sun*, 5/26/63)

May 27: Senator Clinton Anderson (D.-N.M.) replying on the floor of the Senate to Senate Republican Policy Committee's criticism of Project Apollo (May 10), cited buildup of U.S. scientific and technological base: ". . . insofar as our national security is concerned, the development of this basic structure for space power, and the scientific knowledge and technical skill required to enable men safely to explore the moon is probably of greater consequence than the lunar landing itself . . . .

"The decision to try to land a man on the moon by 1970 does not constitute a crash program. The decision-making process does not allow for vacuums; failure to decide on a schedule for attempting to reach the moon is equivalent to setting no deadline at all. Work is done best—even by the technical community—which is not exempt from the laws of human nature—when a specific goal is set. An objective of landing on the moon in the 1960's was chosen as the one that would permit us to work for a lunar landing in a rapid, yet efficient manner. This permits us to work at a challenging pace; yet, does not absorb a greater percentage of the Nation's resources than our country can afford. This is only 1 percent of the gross national product at current rate.

"Once having been made—in 1961—the decision is subject to review and modification every year at budget time and can be modified within limits . . . . NASA is not immune from congressional appropriations review . . . .

"Administrator Webb has announced that NASA, in this decade, will accomplish all the programs now planned, including lunar landing for under $35 billion. This is less than two-thirds of the budget requested by the Department of Defense for fiscal year 1964 alone—a small price for the experience, technology, industrial base, and facilities which may be required for national security . . . .

"Mr. President, there is ample room in the current debate on the space program for divergent views and criticism and for constructive suggestions. But are we to believe that if we cancel the Project Apollo, the savings will be so translated into new hospitals, modern classrooms, and better diets for the ill nourished? . . . .

"We are not faced with an 'either-or' proposition. We can afford to have an effective, logical, and successful space program and we can afford these other efforts to improve well-being here on earth. Mr. President, we cannot afford not to go to the moon . . . .

"Any slowdown in our Nation's space programs would certainly afford great comfort to our enemies and spread dismay among our friends overseas. To slacken the space effort after having accepted publicly on a national basis the Soviet challenge would open us to accusations of lack of firmness and resolution regarding national goals . . . .

"In 1958, President Eisenhower declared:

" 'There is the factor of national prestige. To be strong and bold in space technology will enhance the prestige of the United States among the peoples of the world and create added confidence in our scientific, technological, industrial, and military strength.' " (*CR*, 5/27/63, 8961–8965)

May 27: First USAF F–4C high-speed tactical fighter plane test-flown at Lambert Field, Missouri. F–4C was modified USN Phantom II aircraft. (DOD Release 759–63)

- Proposed NASA-USN program of high-altitude, long-duration balloon flights was outlined in *Missiles and Rockets*. Pending DOD approval, program would be conducted by NASA OART and USN BuWeps and would call for three types of flights: three-man, three-day flight using basic gondola and existing balloon of 10-million-cu.-ft. class; 14-day flight, possibly with larger crew and expanded equipment; and 30-day flights with six- or seven-man crew. Program objectives: to check out space station equipment and components and to study biomedical and psychological factors. (*M&R*, 5/27/63, 18)

- Washington *Evening Star* proposed appointment of "a devil's advocate" for U.S. space program. He would be "a member of Government, well qualified in science and engineering and well advised on the doings of lawyers and publicists . . . charged with building the strongest possible case against every space proposal—before it becomes sanctified as a line item in the Federal budget . . . .

    "This functionary . . . might well save the country a good deal of money and enhance the progress of the space program at the same time . . . ." (Editorial, Wash. *Eve. Star*, 5/27/63, A12)

- *New York Times* reported USAF was conducting Project Forecast—a series of self-analysis studies "designed to assure the integration of available scientific knowledge with the Air Force requirements for new systems to support future national security posture." Official sources said USAF was "stepping up its continuing efforts in a series of intensive studies" because of "accelerated rate of technological progress." Ordered April 17 by Secretary of the Air Force Eugene M. Zuckert and USAF C/S General Curtis E. LeMay, studies were overall responsibility of General Bernard A. Schriever, AFSC Commander. (Raymond, *NYT*, 5/27/63, 1)

- Wade St. Clair of Welch, W. Va., joined NASA Hq. to develop the agency's educational radio programing. (NASA Release 63–96)

- Westinghouse Electric Corp. said its scientists had developed method for lubricating equipment in space. Dry lubricant, tungsten disenlenide, "is held in tiny microscopic pockets throughout a matrix of silver or copper . . . . The dry lubricant continuously transfers to the metal surfaces over which the bearing metal rubs." (AP, Balt. *Sun*, 6/5/63)

May 28: Letter from David Sarnoff, RCA Board Chairman, to Leo D. Welch, Communications Satellite Corp. Chairman, was disclosed by Drew Pearson in *Washington Post*. Referring to new technology of transoceanic transistorized cables, Sarnoff said:

"In my view, the transistorized cable does not alter the fundamental premise that early development of an operational communications satellite system is vital to our national prestige and to continued U.S. leadership in both communications and space technology.

"Nevertheless, a second premise—that the facilities of a global satellite system will be needed to meet the expected growth of commercial demand for oversea communications services beyond 1965—now needs careful reevaluation in terms of the possible effects upon the [Communications] Satellite Corporation of this latest technological breakthrough. . . .

"One transistorized cable, laid across the North Atlantic, could connect areas possessing nearly 90 percent of all the world's present telephones . . . .

"One new transistorized cable will have a capacity almost double that of all existing transoceanic cables. . . .

"We should now consider an amendment to the present law which would enable the satellite corporation to acquire through purchase the international telephone and telegraph facilities and operations, both present and planned, of all American carriers.

"Unification of all the Nation's international communication facilities and operations—satellites, cables, and radios—would in my opinion assure the satellite corporation's commercial success, best serve our overall national interests, provide the most flexible and economical service to the public, and maintain America's leadership in world communications . . . ." (*Wash. Post*, 5/28/63; Letter, 4/2/63, in *CR*, 5/28/63, 9207-09)

May 28: Eric Sevareid, in Washington *Evening Star* column titled "If Columbus Had Never Sailed," discussed the debate on landing men on the moon:

"It was two years ago this week end that President Kennedy sent his special message to the Congress in which he said that he himself believed we 'should go to the moon.' It has taken two years to develop the beginnings of a national debate on the question, among Congressmen, scientists and editorialists.

"I say the beginnings of a debate because, on its public plane at least, the argument has not yet come into its true focus. The true question is not whether we should try to land men on the moon—the nature of this political world as well as the nature of men's curiosity and the unquenchable spirit of science make it inevitable that we try—but how we go about it . . . .

"This is only the beginning. Anyone has only to let out his imagination a short notch to see the ultimate possibilities—to see humanity's push into space transforming this society, dominating its intellectual pursuits, absorbing its resources, altering the training of its youth and its moral and religious concepts, upsetting the priorities for the its social and humanitarian efforts on terra firma.

"Those who scold the worriers say that to cancel the moon voyage would be as if Ferdinand and Isabella had cancelled Columbus' voyage which opened the New World. They are more often right than they know. What is at stake are not only the new marvels to be found, but also the profound transfiguration of the source of the search.

"After the voyage of Columbus the Old World was never the same, in political, economic, military, social, religious or intellectual terms. After the first men walk upon the moon, Old Earth will never be the same and the change will begin in the two societies, Russia and America, now competing for the cataclysmic honor of commencing the alteration." (Wash. *Eve. Star*, 5/28/63, A13)

May 28: Stanford Research Institute report, prepared under Aerospace Industries Association grant, criticized "industry-Government aerospace relationship" as inefficient, full of overregulation and ineffective administration by Government and "over-management" by industry. Report said its findings were of national significance because aerospace industry's output represents three per cent of gross national product and even greater share of manufacturing volume and employment. (Hines, Wash. *Eve. Star*, 5/29/63, A4)

- A. Adzhubey, Chief Editor of *Izvestia*, wrote an open letter to William R. Hearst, Jr., publisher and owner of the *New York Journal-American*, denying the New York newspaper's report that several cosmonauts had died in space flight.

  One of the cosmonauts reported dead, Petr Ivanovich Dolgov, "actually did die but not in 1960 as the newspaper claims, but in 1962, . . . when together with Major Yvgeniy Andreyev, he made an unprecedented parachute jump from the stratosphere." Col. Dolgov and Maj. Andreyev received the title "Hero of the Soviet Union" for their test jump from the stratosphere balloon "Volga."

  The other "Soviet cosmonauts" reported missing by the *Journal-American* were: A. Belokonev, I. Kachur, A. Grachev, and G. Mikhaylov. These men were not cosmonauts, said Adzhubey, but instead technical workers "who test various kinds of equipment, instruments, and clothing used both in space flight and in high-altitude aviation." None of them made space flights and they are all alive and well. (*Izvestia*, 5/28/63, 2, ATSS-T Trans.)

May 29: X-15 No. 3 flown by Joseph A. Walker (NASA) to maximum speed of 3,858 mph (mach 5.52) and maximum altitude of 92,000 ft. in flight test near NASA Flight Research Center to investigate heat rates at relatively high speeds and low angles of attack. As secondary mission, pilot performed several maneuvers to check control methods while X-15's automatic damping system was turned off. (NASA Release 63-119)

- NASA announced problems with two connectors to electrical amplifier in FAITH 7 Mercury spacecraft had caused loss of automatic control system during MA-9 space flight, forcing Astronaut L. Gordon Cooper to re-enter from orbit with manual control. Premature lighting of .05 g panel light during 19th orbit had indicated that automatic control system would be inoperative until spacecraft descended to point of feeling .05 g. Then, in final orbit, trouble in inverter caused loss of automatic control system altogether. Postflight examination of FAITH 7 circuitry traced .05 g signal and inverter failure to independent electrical connectors that failed to function due to moisture; in both cases, electrical insulation of the connectors broke down. Correction of problems for future spacecraft would include tighter control of moisture and increase of insulation of electrical components. (NASA Release 63-118)

May 29: NASA announced award of four-month study contract to RCA Astro-Electronics Div. for engineering evaluation of whether Tiros meteorological satellite could be modified to increase its cloud-cover photography capability. RCA would evaluate Tiros "wheel configuration," calling for two TV cameras mounted in satellite to point in opposite directions; satellite would be launched into polar orbit spinning on its side, pointed towards earth at all times. Such satellite could provide complete earth coverage every three days. (NASA Release 63-115)

- Maj. L. Gordon Cooper received USAF pilot-astronaut wings from General Curtis E. LeMay, USAF C/S, in ceremony at the Pentagon. (CTPS, *Chic. Trib.*, 5/30/63)
- A U.N. space subcommittee addressed a resolution to the United Nations Committee on the Peaceful Uses of Outer Space emphasizing the "urgency and importance of the problem of preventing potentially harmful interference with peaceful uses of outer space." The warning was contained in a diluted version of a resolution in which the Soviet Union had hoped to see a direct criticism of the recent United States launching of millions of small copper wires in a communications experiment. Anatoli A. Blagonravov, the Soviet delegate, had denounced the experiment as a danger to other space studies, including flights by manned satellites. This was denied by Dr. Homer E. Newell, the U.S. representative. In the resolution adopted at the close of a two-week session, the 28-nation subcommittee said that "certain experiments conducted in space may affect present or future scientific activities." There was no reference to the U.S. "needles" or to any other experiment. (*NYT*, Western Edition, 5/31/63)
- NASA announced award of $97,000 study contract to Sylvania Electric Products, Inc., for preparation of plan to use satellites for collection of weather and oceanographic data from unmanned weather stations, buoys, and balloons. Data collection satellite would interrogate each station, store information on magnetic tape, then transmit it to ground retrieval station upon command. (NASA Release 63-116)
- Grumman Aircraft Engineering Corp., contractor to NASA's Manned Spacecraft Center to build the lunar landing vehicle in which two U.S. astronauts would descend to the moon's surface, named Space Technology Laboratory, a division of Thompson-Ramo-Wooldridge Corporation, to develop the lunar descent and landing engine. The lunar excursion module (LEM) of Project Apollo would detach from the Apollo spacecraft orbiting around the moon and slowly descend to the lunar surface. (MSC Release 63-92)
- Titan II exploded after about 50 sec. of flight in USAF launch, Cape Canaveral. Cause of explosion was not yet determined. This was first Titan II explosion in 16 launches from Cape Canaveral, 9 of which were completely successful and 6 of which fell short of intended range because of second-stage difficulties. (DOD Release 780-63)

May 29: Nike-Zeus antimissile missile successfully intercepted reentry target vehicle launched by Titan I ICBM in recent tests, DOD announced. Nike-Zeus was fired and controlled from Army test installation on Kwajalein Island in the Pacific. (DOD Release 772-63)

- USAF Minuteman flew more than 4,000 mi. in last launch conducted by Boeing Co. employees at AFMTC, Florida. Future AMR launchings would be conducted by USAF personnel. (DOD Release 780-63)
- U.S. Army announced it was testing flexible-wing gliders for hauling cargo to troops in field. Built by Ryan Aeronautical Co., gliders were installed on aluminum cargo containers carrying up to 1,100 lbs. and towed by helicopters, then released to deliver cargo. (*NYT*, 5/29/63, 15)

May 30: Dr. Glenn T. Seaborg, AEC Chairman, said in news conference ending 10-day tour of Soviet atomic energy installations that prospects of building joint atomic reactor facility with U.S.S.R. would be negotiated under agreement signed in Moscow last week. Choice of site and cost of project were among initial problems to be solved. Dr. Seaborg and his colleagues were first foreign delegation to visit new Soviet reactor development and testing center at Melekess and 70-bev accelerator under construction at Serpukhov. (Shabad, *NYT*, 5/31/63, 2)

- Chief of the U.S. 14th Coast Guard District in Honolulu, Capt. George D. Synon, was quoted as confirming that a U.S. cutter had entered the Soviet rocket-test zone "by accident" earlier in the month. This was in reply to accusation by the Soviets that the U.S. had observed rocket tests on May 18 and 25. Capt. Synon stated that the cutter "did not do any observing." (AP, *Wash. Post*, 5/31/63)

May 31: Federal Radiation Council reported radiation fallout would be higher in 1963 than in 1962 and would deposit twice as much strontium-90 as in 1962 and four times as much as in 1961. Fallout would be "far short of figures which would cause concern or justify countermeasures," and would diminish in 1964 and 1965 if atmospheric testing was not resumed. (Grigg, Wash. *Eve. Star*, 6/1/63)

- Dr. Frank J. Malina, U.S. astronautics pioneer, elected by International Academy of Astronautics to succeed the late Dr. Theodore von Kármán as Director of the Academy until its annual meeting next autumn. Dr. Malina had served as Academy's Deputy Director since 1961. (IAA Release No. 14)

During May: Paul D. Lowman, Jr., geologist in NASA Goddard Space Flight Center's Theoretical Div., asked whether lunar base, permanent manned space station, or manned Mars exploration would "give the greatest scientific return in the shortest possible time," replied: "A lunar base, very definitely. The moon will be more useful scientifically than Mars because it has no atmosphere, and because, being much closer, we can send back information and samples at a much higher rate. Compared with the space station, the moon has the advantage of providing a large, stable platform with usable raw materials; in addition, the moon has a stratigraphic record of its history and of meteoritic material which has accumulated on it. These advantages are in a sense a one-way street. We can make many of the observations proposed for the space station from the moon as well, because the moon has essentially a space environment. But we cannot examine the lunar rocks from a space station; we must get down on the surface for that . . . ." (Goddard Release)

During May: NASA awarded $115-million contract to Rocketdyne Div. of North American Aviation, Inc., for continued development of F-1 engine. New contract was follow-on to letter contract which had initiated F-1 development. (*M&R*, 5/6/63, 13)

- U.N. Legal Subcommittee of Committee on the Peaceful Uses of Outer Space concluded second session of this year. Summary report described session as "a very useful and constructive exchange of views . . ." and recommended subcommittee delegations continue trying to reach agreement before next meeting of full committee next fall. Reason for subcommittee's failure to resolve even non-controversial general principles: Western delegations could not agree to Soviet demands that Soviet package proposal (April 17) be adopted in entirety and that international treaty incorporating these principles be drawn up and ratified by individual states. (*Av. Wk.* 5/20/63, 129)

- Conclusions on where the armed forces "are going" in the rest of this decade were outlined recently by Gen. William F. McKee, USAF Vice Chief of Staff.

    ". . . we must watch our own and the Soviet's space programs carefully. This is probably the area of greatest demand on our vision as a nation. For it is here that one of us, probably, will find the key to the strategic superiority of the 1970's. Secretary McNamara says, 'I think there is no clear requirement for military purposes for manned operations in space as we look at the future today.' However, Mr. McNamara said more: '. . . But, should one develop it could come so suddenly in a field in which the lead-time for the developmental technology is so long that I believe we must anticipate that possibility today . . .'.

    "I, too, think we must anticipate that today. While this is a national effort, it is also an absolute obligation of those of us responsible for our national security." (*Air Force Information Policy Letter*, June 1963)

- Link Div. of General Precision, Inc., awarded $9.5-million contract to develop two Apollo mission simulators for training astronauts at NASA Manned Spacecraft Center and Cape Canaveral. Contract was awarded by North American Aviation, Inc., prime Apollo spacecraft contractors. (*Av. Wk.*, 5/6/63, 25)

- First meeting of NASA Advisory Committee on Biotechnology and Human Research held at Ames Research Center. (*Av. Wk.*, 5/6/63, 103)

- U.S. Army Corps of Engineers officially announced opening of its Canaveral District office at Merritt Island, Fla., first district organized solely for missile and space construction. (*M&R*, 5/13/63, 45)

During May: X-15 pilot Maj. Robert M. White (USAF) awarded Villanova University's Mendel Medal for scientific achievement in honor of his record-breaking flight July 17, 1962. (*A-N-AF Journ. & Reg.*, 6/11/63, 6)

- Aerospace Medical Association awarded Lyster Award to Maj. Gen. M. S. White (USAF), Air Training Command Surgeon at Randolph AFB, Tex. (*A-N-AF Journ. & Reg.*, 6/1/63, 5)
- USAF completed its 30-day test of regenerative atmosphere system developed by Boeing Airplane Co. Three airmen occupied 1,115-cu.-ft. chamber for full duration of test, relying on oxygen-recovery system to replenish normal four-hour air supply. (*Astronautics and Aerospace Engineering*, 5/63, 137)
- Announced that the Air Force launched six research rockets designed to study the structure of the earth's atmosphere by photographing chemical trails released by the rockets. The flights, launched from Eglin AFB, Florida, on May 15, 17 and 21, reached the ionosphere, 60–90 mi. above the earth. Each rocket was fired between sunset and sunrise. Luminescent chemicals were released and photographed by ground-based cameras. The high altitude studies are being made of turbulence and wind shears. (OAR Release #6-63-5)
- Bell Telephone Laboratories received patent for improved system of controlling satellite attitude. Invented by C. C. Cutler, system operates by two-way radio which receives signals from satellite for measuring its alignment; ground computer analyzes attitude data and figures any necessary attitude changes, relays them to instruments on satellite which makes corrections in satellite's position. (Jones, *NYT*, 5/11/63, 28)
- NASA Administrator James E. Webb sent a report to the Congress on the construction of facilities for the University of California:

  ". . . the University's geophysics building is being expanded by the addition of wings for geology and chemistry laboratories. Accordingly, the National Aeronautics and Space Administration plans to provide funds for the construction of a 62,000 gross square foot addition to that building complex in the amount of $2,000,000. The addition will be known as the space science building." (Leg. Act. Rpt. II/84)
- Igor I. Sikorsky International Trophy for 1962 awarded to Sikorsky Aircraft Div. of United Aircraft Corp. for flight of Sikorsky S-61 helicopter Feb. 5, 1962, setting world speed record of 210.6 mph over 19-km. straight-line course. Lt. Robert W. Crafton (USN) and Capt. Louis K. Keck (USMC) were pilot and co-pilot respectively. (*Av. Wk.*, 5/13/63, 23)
- Inflight testing of infrared communications between two moving aircraft conducted at AFSC Aeronautical Systems Div., Wright-Patterson AFB. With communications carried over extremely narrow infrared beam, system is difficult to jam, intercept, or detect. (AFSC Release 31-105-24)
- Richard N. Thomas, professor of astrophysics and chairman of Faculty Committee for Astrophysics of Dept. of Physics and Astrophysics, Univ. of Colorado, was awarded Dept. of Commerce

Gold Medal for Exceptional Service. (*Astronautics and Aerospace Engineering*, 5/63, 9)

During March, April, and May: Total of 227 NASA appearances at 70 hearings before the House Committee on Science and Astronautics, involving 246 hours. This involved 546 man-hours of primary witnesses, not including backup witnesses. (NASA Leg. Act. Rpt. II/88)

JUNE 1963

June 1: Astronaut John H. Glenn, Jr., returned to U.S. after vacationing in Japan with his family since shortly after MA-9 manned space flight, during which he had been stationed aboard Pacific tracking ship *Coastal Sentry.* Glenn's candidness and openness during visit was said to have "boosted U.S. image" in Japan. (AP, Wash. *Sun. Star*, 6/2/63)

• In address on "Prelude to Independence" at Williamsburg, Va., Barbara Ward of *The Economist* (London) said: "If the men of 1776 had attempted the opening up of America with the testy impatience of so many politicians today, they would barely have crossed the Adirondacks . . . .

"Seen in the aspect of not a single decade but of the 50 or 60 years needed for full modernization, economic assistance to our crowded world could seem as challenging as our space ventures and far more rewarding. . . ." (Text, *Wash. Post*, 6/2/63, A8)

• Sen. Robert C. Byrd (D.-W. Va.), addressing Alumni of Concord college, said: "Our marvelous technological developments, our flights into outer space, have demonstrated that there is virtually no limit to the development of the human mind. Today the moon is within range, tomorrow the stars—after that, who knows? Human knowledge is a pyramid whose topmost point stretches into infinity . . . .

"Who develops the mind of the astronaut, and his co-workers? That is the job of the teacher, of education. That pyramid of knowledge . . . . rests upon a base of solid, fundamental learning which makes all of our splendid achievements possible . . . ." (Text, *CR*, 6/6/63, A3655-56)

June 2: In commencement address at MacMurray College (Jacksonville, Ill.), Vice President Lyndon B. Johnson said: ". . . We are more masters of the future than of the past but we allow the hand of the past to rest heavily upon us . . . .

"At today's pace, the sum of human knowledge is being doubled every decade. On a relative basis, the class of 1963 may well be three centuries of learning ahead of the class of 1933. But we are asking a generation capable of going to the moon to live in a world ordered and fixed by generations which scarcely knew the model T—and that makes no real sense . . . ." (Text, *CR*, 6/4/63, 9438-39)

• Japanese newspaper *Yomiuri* reported Japanese Space Development Council was preparing to build navigational satellite for launching by U.S. within five years. Plan was to be submitted to Prime Minister Hayato Ikeda for approval. (*L.A. Times, Wash. Post*, 6/4/63)

June 2: NASA Associate Director of Plans and Program Evaluation Addison M. Rothrock said in keynote address, Southern States Work Conference: "Please educate our youth well. Teach them of the power that inevitably comes with knowledge of our universe and the people who inhabit it. Teach them sciences so they can achieve triumphs in science and engineering. Teach them the humanities so they will use these triumphs for man's betterment. And teach them to blend these two great branches of learning . . . ." (Text)

June 3: NASA requested proposals for a broad range of experiments applicable to life-support systems for men and equipment in space. Experiments would be conducted by NASA on spacecraft in orbit and on aircraft with from 15 sec. to 5 min. of zero gravity. NASA Office of Advanced Research and Technology would evaluate industry and university proposals. (NASA Release 63–117)

- In commencement address at Univ. of Pittsburgh, NASA Administrator James E. Webb said: "In the world you enter today, man is no longer rooted to terra firma, to his native land, or even to his city block, either in the literal sense or in his understanding of the powerful forces of the universe, of which he is such a minuscule part. Increasingly man comprehends that the new understanding and knowledge that are being unleashed by science and technology will alter his existence in more ways and more rapidly than he can possibly foresee . . . .

  "Forty years elapsed between the Wright brothers and the first supersonic flight. Only 10 years were required to reach 4,000 miles per hour in the X–15. Applying this knowledge, we now see a shrunken and bipolarized world, within which we can travel anywhere in a few hours, but cannot solve without force and violence man's relation to man. And all of this was achieved using the atmosphere up to twice the height of mountains.

  "By the end of this decade, man will have stood on the moon and looked outward to the vast universe and beyond, and learned in the process vastly more about the earth itself . . . ." (Text, CR, 6/12/63, A3792)

- In Wells College (Aurora, N.Y.) commencement address, NASA Director of Manned Space Flight D. Brainerd Holmes said: "How about women in space? The answer is yes, but not immediately. At the moment, however, all of the qualified pilots are men.

  "In a larger sense, of course, women are already in space. The astronaut does not fly alone. He relies on the thousands of people in his supporting team on the ground. These include women as well as men. On the staff of the National Aeronautics and Space Administration, there are about 150 women scientists and engineers. Many more are employed by the contractors throughout the country who perform the bulk of our work. In this rapidly growing program, there are many interesting career opportunities for those of you who complete the necessary preparation . . . ." (Text)

June 3: Students in Washington, D.C., area began helping Weather Bureau in Project Skywatch—drawing cloud formations, recording wind speed and direction, temperature and relative humidity and precipitation, in project to identify specific cloud formations photographed by Tiros meteorological satellites. About 50 students were participating in project. (Commerce Dept. Release WB-63-6)

- NASA Director of Advanced Research and Technology Dr. Raymond L. Bisplinghoff said in address before Society of Aerospace Material and Process Engineers in Philadelphia: "There is no area of technology more important to our space effort than materials and their processing. Virtually every device that is employed in space is limited in either performance or efficiency by its constituent materials. The space program is literally paced by our ability to use and process materials . . . ." (Text)

- In *Aviation Week and Space Technology* editorial, Robert Hotz said: "The current debate has done much to explain to the public and Congress the real aims of the space program, which are the development of a broad capability in space technology that eventually can be applied to whatever national purpose or need that may arise in the future. Gradually, the point that the manned lunar landing Apollo program is simply the best possible focal point for this effort in its present phase is emerging from the verbal pyrotechnics of the current debate. It is becoming more readily apparent that Apollo is not just an 'over-Niagara-Falls-in-a-barrel' type of dare-devil stunt, but a well-conceived program of broad exploration across the whole spectrum of space technology as we know it today.

    "The solid technical success of the Mercury program, particularly in the six- and 22-orbit flights of Cdr. Schirra and Maj. Cooper, has done much to make this point clear. It should be obvious now that Mercury, Gemini and Apollo spacecraft are really the Nina, Pinta and Santa Marias of space exploration, and are simply paving the way for whatever useful purposes this nation eventually decides to utilize space. . . ." (*Av. Wk.*, 6/3/63, 17)

- Maj. Robert M. White (USAF) received 1962 General Thomas D. White USAF Space Trophy from Secretary of Air Force Eugene M. Zuckert for his outstanding contributions to U.S. aerospace research through piloting X-15 rocket research aircraft. USAF Space Trophy is presented annually by National Geographic Society. (DOD Release 789-63)

June 3-11: Sixth annual meeting of 18-nation COSPAR (Committee or. Space Research, International Council of Scientific Unions) held in Warsaw. U.S. scientists, headed by Dr. Richard W. Porter and participating through NAS Space Science Board, delivered 70 papers at Fourth International Space Science Symposium held in conjunction with COSPAR meeting. (NAS Release)

Dr. Richard W. Porter, head of U.S. delegation, summarized U.S. space activities from approximately Jan. 1 to Dec. 31, 1962: more than 100 high-altitude research balloons; approximately

846 synoptic rockets (by Meteorological Rocket Network from six North American sites); and more than 100 scientific sounding rockets. U.S. had announced launching of 19 scientific and scientific-applications satellites and space probes. (Porter, Report to COSPAR)

Head of Soviet delegation A. A. Blagonravov summarized Soviet space activities of 1962: launching of interplanetary probe MARS I; 71 geophysical and meteorological sounding rockets; 12 unmanned (Cosmos) satellites; and two manned spacecraft. Elaborating on MARS I, he said that at 106,000,000 kilometers from the earth "a defect in the probe's orientation system resulted in a violation of the directionality of the probe's antenna to earth, making it impossible to maintain further radio contact with the probe." (*NYT*, 6/4/63, 10)

Soviet scientist S. V. Vernov reported on June 4 that cosmic-ray bombardment of earth from outside the solar system had doubled in last five years. Findings were based on more than 4,000 Soviet scientific balloon flights since 1957. (Underwood, *NYT*, 6/5/63)

On June 5, NASA Goddard Space Flight Center offered world scientists the design of small rocket payload and ground telemetry station suitable for ionospheric research. GSFC scientists Siegfried J. Bauer and John E. Jackson said payload's "versatility, simplicity and relatively low cost should make it an ideal tool for the investigation of the many problems of the ionosphere by the international scientific community, especially during the IQSY [International Year of the Quiet Sun]." (GSFC Release)

Prof. Anatoli Blagonravov, head of Soviet delegation, announced U.S.S.R. would conduct unspecified number of manned space flights in 1963, would launch two unmanned probes toward Venus and Mars, and would continue orbiting satellites in Cosmos series. (UPI, *Wash. Post*, 6/6/63)

Dr. Wilmot N. Hess, Chief of NASA Goddard Space Flight Center Theoretical Div., said that U.S. high-altitude nuclear explosion July 9, 1962, as well as Soviet high-altitude nuclear explosions Oct. 22, Oct. 28, and Nov. 1, added temporarily to radiation particles in Van Allen belts. Lifetime of electrons and protons trapped by earth's magnetic field had been "one of the most important and perplexing problems in radiation belt work for several years . . . [and] for the first time, this past year we have obtained definite information on trapped electron lifetimes . . . ." (GSFC Release; AP, Wash. *Eve Star*, 6/6/63)

U.S. and U.S.S.R. delegations on June 6 announced readiness of their countries to cooperate in IQSY space experiments. Also at COSPAR, Japanese scientists reported plans to use U.S. Relay communications satellites for television broadcasts of Olympic Games in Tokyo in 1964. (UPI, *Wash. Post*, 6/7/63)

Dr. W. R. Adey of UCLA Brain Research Institute on June 7 challenged Astronaut L. Gordon Cooper's report that he sighted smoke rising from individual houses in Tibet during his MA-9 orbital space flight. Dr. Adey asserted weightlessness produces feeling of euphoria: "Under such conditions one is inclined to make judgments that are not critical, because normal controls on judgment are not operative . . . . Unrecognized disorders of visual perception and of judgments arising therefrom may be of considerable importance in space research."

Dr. George T. Hauty of FAA Civil Aeronautics Research Institute reported on experiments showing that marked reduction of sensory stimuli could induce slowing of subject's thought processes and even hallucinations. (*NYT*, 6/8/63, 26; *Electronic News*, 6/10/63)

B. E. Welch of USAF School of Aerospace Medicine reported that results of six experiments showed oxygen toxicity would not be problem with pure-oxygen atmosphere in 17-day manned space flights. (NAS Release)

Dr. William F. Neumann of Univ. of Rochester Medical Center said astronauts under weightlessness up to two weeks could be expected to lose some calcium from their bones, probably equivalent to calcium loss of bedridden patient over 6–7 weeks: "For flights longer than two weeks, particularly space-capsule flights of indefinite length, grave questions of the continued normalcy of bone structure and bone development arise." (*NYT*, 6/8/63, 26)

Soviet scientist V. V. Sharnov reported that moon was covered with slag-like material, confirmed by Soviet radioastronomical observations. Speculating on surfaces of Mercury and Mars, he said former should resemble that of moon and latter was "mostly covered with dust. The fact that this material remains uncemented is a result of humidity." Radar contact with planet Mercury was made at distance of 54 million mi. Radar contact with planet Venus permitted refinement of value of astronomical unit obtained in 1961 and estimate of Venus rotational period. (Underwood, *NYT*, 7/9/63, 81; *Av. Wk.*, 6/10/63, 31)

Dr. D. E. Graveline of USAF School of Aerospace Medicine reported provision of artificial gravity in manned spacecraft was being considered for prolonged manned space flights. Artificial gravity may be required because re-entry after a month or more of weightlessness could be dangerous to human beings—in prolonged weightlessness, natural processes which attune the circulatory system to gravity forces would decay, and body could not make necessary readjustment to g forces encountered in re-entry (*NYT*, 6/8/63, 26; AP, Balt. *Sun*, 6/9/63)

R. S. Young of NASA Ames Research Center reported June 8 on experiments providing evidence that certain earth organisms survive and grow under simulated Martian environment. Experiment simulated equatorial summer day-night cycle on Mars with presence of water; further experiments were being conducted to determine minimum water and nutritional requirements for these organisms and to find their shortest growing period. If experiments show terrestrial organisms can survive "under the most rigorous laboratory simulation" of Martian environment, there would be "decisive evidence" of need for spacecraft sterilization. (NAS Release)

Dr. Carl Sagan of Harvard Univ. Observatory said June 9 that astronomer on Mars, using most advanced equipment known today, would probably be unable to prove existence of life on earth.

Of 80,000 photographs of earth taken from "an American Tiros satellite, only one could be interpreted as showing biological life on earth." (AP, *NYT*, 6/10/63)

June 3–6: NASA 16 astronauts received jungle survival training at USAF Caribbean Air Command Tropic Survival School, Albrook AFB, Canal Zone, including classroom instruction and three-day field practice in local jungle. (MSC Release 63–96)

June 4: "First conclusive detection" of water vapor on Mars was announced by Cal Tech. Mars' water supply was found to be minimal—perhaps as little as half a cu. mi. compared with millions of cu. mi. on earth. Finding was made by analysis of spectrum of light from Mars gathered by Mt. Wilson and Mt. Palomar Observatories in California. (Hill, *NYT*, 6/5/63, 17)

- Video tape of President Kennedy's address before World Food Congress in Washington was transmitted to European television viewers via RELAY I communications satellite. (NASA Release 63–121)
- George J. Feldman, a director and incorporator of Communications Satellite Corp., said in address before Federal Bar Association in Washington:

    "The execution of this national program [developing a commercial comsat system] and the task faced by the corporation is difficult, complex, and truly formidable and should be viewed by the objective with understanding. This is the uncharted region of technology, international relationship and agreements, and national support . . . .

    "Up to this point the United States has been the innovator and leader in this field, with great benefit to our national prestige, and international standing and with the satisfaction of pioneering a part of the future. Having taken this pioneering step, we have now assumed our share of responsibility and the latter is the operative word . . . ." (Text, *CR*, 6/20/63, A3948–49)
- Pan American World Airways President Juan T. Trippe announced Pan Am had ordered six Concorde supersonic airliners being developed jointly by British Aviation Corp. and Sud-Aviation of France. BOAC also reported ordering six of the 1,500-mph jetliners; Air France was believed to have ordered six, but report was unconfirmed. (Carter, *NYT*, 6/5/63, 34)
- Historian Arnold Toynbee, in Washington to address World Food Congress, said in news interview that U.S. and U.S.S.R. together controlled 80 per cent of world's technical power but both nations were neglecting humanitarian applications. He cited the "moon race" as example of both nations' "escapism" from basic problems of mankind. (Richmond *News Leader*, 6/12/63)
- Sen. Edward V. Long (D.-Mo.) and 13 other Senators proposed "freedom of information" bill to require Government agencies to publish in *Federal Register* all policy statements, rules, and procedures, and to make available for inspection all final opinions, statements, and interpretations. (AP, Wash. *Eve. Star*, 6/4/63)
- USAF launched Atlas E ICBM on successful 4,000-mi. test-flight from Vandenberg AFB. (DOD Release 810–63)

June 5: NASA awarded contract for designing and fabricating advanced four-man, six-month, life-support system to General

Dynamics Corp.'s Astronautics Div. (GD/A). 18-month contract called for research, design, fabrication, testing, and delivery to NASA Langley Research Center of fully operating prototype life-support system for space flights. (NASA Release 63–126; Langley Release)

June 5: USAF conducted successful test-firing of "second generation" Minuteman ICBM from Cape Canaveral, its re-entry vehicle impacting target area near Ascension Island. (DOD Release 820–63)

- NASA announced selection of Boeing Co. and Douglas Aircraft Co. for final negotiations leading to manned orbital research laboratory study contracts. NASA Langley Research Center would negotiate definitive three-month study contracts with the two companies. (NASA Release 63–120; Langley Release)
- In Air Force Academy commencement address, President Kennedy said: "It is my judgment that this Government should immediately commence a new program in partnership with private industry to develop at the earliest practical date the prototype of a commercially successful supersonic transport superior to that being built in any other country of the world . . . . Neither the economics nor the politics of international competition permits us to stand still in this area." (*Wash. Post*, 6/6/63)
- President Kennedy witnessed successful firings of six Army missiles at White Sands Missile Range: Honest John, Little John, Sergeant, Hawk, Nike-Hercules, and Nike-Zeus. (*M&R*, 6/10/63, 12)
- In commencement address at Mundelein College, Chicago, AEC Chairman Dr. Glenn T. Seaborg said:

  "Today, science has become the dynamic factor in our society. There are a great many intelligent and dedicated men and women in the world using their highest talents and intelligence to try to solve complex problems about our universe . . . .

  "Answers to these questions will serve not only to satisfy our human curiosity, but, of course, will have enormous practical implications. Curiosity about our universe is part of our human inheritance, and the scientist is, after all, a person with a highly skilled and intelligently directed sense of curiosity . . . ."

  Discussing modern technology's solutions to some old problems, he said: "Perhaps a lesson to be gained from these accounts is the fact that there is a unity of knowledge and knowledge seeking. There is a continuum that extends from theories of the origin of the universe through the rise of life on earth to the evolution and history of man, and on to the completion of the cycle with such an abtruse field as high energy particle theory. There is no knowledge that is isolated from the total fabric . . . ." (Text, *CR*, 6/6/63, A3667–68)
- USAF CH–3B helicopter landed at LeBourget Airport near Paris after 35-hr., 5-stop flight from Otis AFB, Mass. (UPI, *NYT*, 6/6/63, 4)

June 5–6: India's President Sarvepalli Radhakrishnan toured Cape Canaveral and witnessed two successful test launchings: USAF Minuteman ICBM on June 5 and USN Polaris A–3 missile on June 6. (AP, *NYT*, 6/7/63, 2; *M&R*, 6/17/63, 12)

June 6: U.S. Ambassador to U.N. Adlai E. Stevenson said in letter to U.N. Secretary General U Thant that U.S.S.R. had failed to register six space vehicles which achieved orbit last Aug. 25, Sept. 1 and 12, Oct. 24, Nov. 4, and Jan. 4 (1963). Stevenson repeated charges that U.S.S.R. also had concealed four high-altitude nuclear tests in autumn 1961. Letter was in response to Soviet Ambassador Nikolai Fedorenko's asking U.N. Secretary General to circulate Tass article criticizing U.S. Project West Ford, which Stevenson defended. (*L.A. Times, Wash. Post,* 6/7/63)

- Unnamed NASA spokesman quoted as saying ECHO I's orbit was "holding up fine" and could last another year or more. Although partially crumpled, balloon satellite launched Aug. 12, 1960, was still visible to naked eye. (UPI, *Wash. Post,* 6/6/63)

- Sen. Margaret Chase Smith (R.-Me.), speaking on the Senate floor, said Committee on Aeronautical and Space Sciences would "provide a forum for [non-NASA] witnesses who are recognized as experts in their respective fields. They will have an opportunity to analyze our space objectives and to air their views, pro and con, on the manner in which we are seeking to achieve national goals . . . .
  "Why is this particular year [FY 1964] of unusual significance?
  "First, there is a tremendous increase in NASA's budget request and there seems to be no ceiling in sight for succeeding fiscal years . . . .
  "Second . . . NASA's funds are largely committed to landing a man on the Moon and returning him safely to Earth during this decade. The priority given this project, which has been variously estimated to cost between $20 billion and $40 billion, raises the question of adequate funding of other space activities . . . ." (*CR,* 6/6/63, 9690)

- U.K. Office of the Minister for Science announced contract had been awarded to British Aircraft Corp., Ltd., for development and construction of UK-3 (S-53) satellite. NASA would launch the experimental satellite into orbit in about four years. (AP, *Wash. Post,* 6/7/63)

- Scientists at the NASA Lewis Research Center succeeded in driving an experimental solid-propellant rocket motor into unstable combustion. This new experimental technique makes possible a planned research investigation of combustion instability problems. (Lewis Chronology, 5)

- USN test-fired Polaris A-3 missile from land pad at Cape Canaveral in successful flight down Atlantic Missile Range. (*M&R,* 6/17/63, 12)

- DOD announced two single-engine Cessna 180 airplanes of USN Arctic Research Laboratory made deepest known light-plane penetration of Arctic Ocean Basin when they landed May 24 at North Pole. From May 21 to June 2, nine sites were occupied on pack ice so that scientists could collect scientific data on gravity and on ice conditions. (DOD Release 814-63)

- National Aeronautic Association president Martin M. Decker announced National Frank G. Brewer Trophy for 1962 would be awarded to Dr. Merlyn McLaughlin, president of College of Osteopathic Medicine and Surgery of Des Moines, Iowa. (NAA Release)

June 6–7: Symposium on exploration of Mars held in Denver, with objectives of assessing effort necessary for manned exploration of Mars, reviewing planning for such mission, estimating timetable, and defining scientific value. Symposium was sponsored by NASA, American Astronomical Society, American Institute of Biological Sciences, and American Institute of Aeronautics and Astronautics. (*Marshall Star*, 6/5/63, 1, 8)

Eugene M. Zuckert, Secretary of the Air Force, speaking at Symposium on the Exploration of Mars of the American Astronautical Society, discussed the two-fold task of the Air Force: ". . . . first, to protect the Nation's exploration and research in space as effectively as the Air Force contributes to protection of our national activities on the Earth's surface, and second, to act as the Department of Defense agency which helps insure the National Space Program's success through provision of Air Force facilities and services to the overall effort.

"The question of whether there is a need of an operational capability for United States military forces beyond the atmosphere may already have been answered for us in a number of public statements by Soviet military and political leaders . . . . Their major new book on *Military Strategy* . . . says, "Soviet military strategy acknowledges the need to study the use of space and space vehicles to reinforce the defense of the socialist countries. . . . It would be a mistake to allow the imperialist camp to gain any superiority in this area.'" (Text)

Speaking at a symposium on the Exploration of Mars in Denver, Colo., Wolfgang E. Moeckel, chief of LRC's Electromagnetic Propulsion Division, said: "The potential of electric propulsion for Mars exploration is as great or greater than that of other systems now under development and it becomes even more attractive for the exploration of the farther reaches of the solar system." (LRC Release 63–35, Lewis Chronology, 4–5)

Highlights of parellel studies on three classes of Nova launch vehicles, conducted by General Dynamics/Astronautics and Martin-Marietta Corp. for NASA Marshall Space Flight Center, were presented by Andrew Kalitinsky, GD/A Program Director. (*Av. Wk.*, 6/10/63, 34)

Dr. H. O. Ruppe of NASA Marshall Space Flight Center's Future Projects Office estimated manned Mars mission would require about $9 billion for interplanetary spacecraft, $5 billion for operating costs (including procurement of Nova launch vehicles), and $2 billion for remaining production and procurement costs.

Dr. Ruppe presented paper on "Vehicle Design for Earth Orbit to Mars Orbit and Return," discussing results of Project Empire (Early Manned Planetary-Interplanetary Roundtrip Expedition) contract studies by Lockheed Aircraft Corp., Ford Co. Aeronutronics Div., and General Dynamics/Astronautics Corp. Results of studies indicated: only marginal capabilities exist for all-chemically propelled space vehicles; even with nuclear-powered upper stages, manned roundtrip flight would be difficult task; because of eccentricity of earth's orbit, flights during 1975–80 period may be technically or economically unfeasible; manned Mars program would require unusually high degree of coordination within entire National space program. (*NYT*, 6/7/63, 11; *Marshall Star*, 6/12/63, 1, 4)

Harold B. Finger, Manager of NASA-AEC Space Nuclear Propulsion Office and Director of NASA Nuclear Systems, said: "[Manned Mars mission] will be a very major undertaking exceeding the Apollo program in total cost and difficulty. For that reason, we must realistically recognize that one of the factors determining the timing of such a mission will be the availability of funds and manpower. We must recognize that we cannot undertake such a mission until the Apollo mission begins to phase out.

"If, as will probably be the case, rendezvous in earth orbit and some orbital assembly or propellant transfer will be required to perform the Mars landing missions, an earth orbiting space laboratory will probably precede the accomplishment of the actual planetary missions. The cost of such a system would also be high but it would probably be a necessary stepping stone to the planetary missions . . . .

"The unknowns in such a mission will require the accumulation of a great deal of basic information and . . . substantial development efforts for all of the systems and facilities that would be required in such a mission. I think it is to be expected that the time required for development of these systems puts the accomplishment of a manned Mars landing mission off until late in the seventies and more probably into the early eighties . . . .

"I expect that large chemical rocket booster stages, in combination with nuclear rocket upper stages, will be used to accomplish the Mars landing mission. I expect also that such vehicles and propulsion systems will be used to perform the preliminary missions of manned fly-by trips around Mars and trips into orbit around Mars. It is conceivable that such missions could start in the latter part of the 70's . . . ." (Text)

A. James Shiner of NASA Man-System Integration delivered paper prepared by him and Dr. Eugene B. Konecci, Director of NASA Biotechnology and Human Research: "A manned orbital research laboratory . . . becomes a distinct requirement to obtain human design requirements for a Mars vehicle system, whether that system be nuclear or chemical. The manned orbital research laboratory will help answer many of our questions about the space environment, and man's ability to not only survive, but to perform useful functions in prolonged space missions. The manned orbital research laboratory under study by the Langley Research Center will have to operate for prolonged periods of time to obtain the required human research and biotechnology data needed for design of advanced space systems. A manned orbital research laboratory will eventually be used to simulate Martian voyages as well . . . ." (Text)

Secretary of the Air Force Eugene Zuckert predicted need for military patrol in space "to determine at all times what is hap-

pening in near-space and to deal with any threat there . . . .
[Space patrol] would provide on-call protection for U.S. space
activities, both scientific and military, in the event of hostile enemy
actions." (*Rocky Mountain News* [Denver], 6/7/63)

Dr. W. M. Sinton, Lowell Observatory astronomer, said in
June 7 address that possible vegetation on Mars would probaby
resemble cactus, especially barrel or saguaro cactus. (AP, Balt.
Sun, 6/8/63)

Dr. Gerard de Vaucouleurs of Univ. of Texas described recent
Mars observations from Pic du Midi in French Pyrenees. Studying spotted patterns of dark areas, Pic du Midi found seasonal
darkening of Martian surface was actually a filling in of spaces
between dark spots on the planet. Dr. de Vaucouleurs found it
difficult to reconcile large-scale changes of dark areas with recent
findings that Mars was extremely dry. (Sullivan, *NYT*, 6/8/63)

June 7: France's President Charles de Gaulle opened 25th International Air Show at LeBourget Airport, visiting U.S. space exhibit
as well as other exhibits. (*NYT*, 6/8/63, 35)

- President's Science Advisor Dr. Jerome Wiesner announced establishment of national standard reference data system, integrating
  data compilation activities of National Bureau of Standards,
  DOD, AEC, NASA, National Science Foundation, and other agencies.
  (*CR*, 6/12/63, A10127)

- DOD Distinguished Public Service Medal presented by Secretary of
  Defense Robert S. McNamara to John H. Rubel, Deputy Director
  of Defense Research and Engineering. (DOD Release 824-63)

During First Week of June: Technical progress reports on supersonic transport study contracts were made by Boeing Co's Transport Div. and Lockheed-California Co. to NASA Langley Research
Center. Final reports on the nine-month contracts would be made
in November. (NASA Release 63-129; Langley Release)

First Week in June: Dr. Barry Commoner, biology professor of
Washington Univ., St. Louis, referred to conclusion of Space
Science Summer Group that search for extraterrestrial life
"should be acclaimed as the top-priority scientific goal of our
space program," and charged that chances of finding extraterrestrial life were so small that they did not justify multibillion dollar
cost. Dr. Commoner's criticism came in commencement address
before Hahnemann Medical College in Philadelphia. (Finney,
NYT, 6/9/63)

June 8: Vice President Lyndon B. Johnson said in commencement
address at Univ. of Maryland: "We are not reaching for prestige in
space—we are reaching for peace. And this is considerably more
urgent than many realize or others will yet admit . . . .

"In exploring space, we are exploring the environment of the
earth itself. We are finding paths through a new dimension—and
historically civilizations have stood or fallen according to their
ability to move through a dimension.

"We seek to make space an instrument for peace and the development of mankind. But if we abandon the field, space can
be preempted by others as an instrument for aggression.

"And in a world of competing social systems, we would be naive
indeed if we failed to recognize how bleak the future would be

if this new dimension became the realm of tyranny. For your children, and your children's children, an iron curtain would be drawn across the pathway to the stars . . . ." (Text, *CR*, 6/18/63, A3867)

June 8: Soviet Academician Nikolai Kozyrev of Pulkovo Observatory wrote in British magazine *Nature* describing evidence of volcanic processes on the moon. Ejection of gases, including molecular carbon, was observed from (1) central peak of Alphonsus crater on Nov. 3, 1958, and Oct. 23, 1959; and (2) Aristarchus crater in Nov. and Dec. 1961. (*Nature*, 6/8/63, 979–80)

June 9: In commencement address at Univ. of Delaware, NASA Administrator James E. Webb said: ". . . For the first time in the history of mankind the opportunity to leave the earth and explore the solar system is at hand. Only two nations, the United States and the Soviet Union, have the resources with which to exploit this opportunity. Were we, as the symbol of democratic government, to surrender this opportunity to the leading advocate of the Communist ideology, we could no longer stand large in our own image, or in the image that other nations have of us and of the free society we represent.

"Like it or not, therefore, the hard facts of international life leave no doubt that we are in a contest with the Soviet Union. Like it or not, it is a fact of international life that space achievement is and will be a symbol of national character and national strength . . . ." (Text)

- Floyd L. Thompson, Director of NASA Langley Research Center, was awarded honorary Doctor of Science degree by College of William and Mary. On previous day, he was awarded honorary Doctor of Science degree by Univ. of Michigan. (LARC Release, 6/10/63)
- From 1950 to 1960, U.S. professional and technical work-force grew by 2,400,000, according to Columbia Univ. Bureau of Applied Social Research. In 1950 professionals and technical workers made up 8.3% of the total employment, and in 1960, 11.3%. Increase was attributed to rapid growth of government activities and industries involved in government-sponsored work. (Lissner, *NYT*, 6/19/63, 75)

June 10: Demonstration of transatlantic typesetting was conducted between Chicago and Goonhilly Downs, England, via ground computer and RELAY I communications satellite. Originating from Chicago conference of American Newspaper Publishers Association Research Institute, dispatch was received in final typeset form. In second phase of demonstration, articles were similarly sent from London and Rio de Janeiro and were received in Chicago. (Wehrwein, *NYT*, 6/11/63)

NASA announced selection of Space Technology Laboratories, Inc., for negotiation of contract for design and fabrication of four Pioneer interplanetary space probes. First launching in series was scheduled for early 1965 with Delta launch vehicle. (NASA Release 63–131; Ames Research 63–24)

- Selection of two new X–15 pilots, Capt. Joe H. Engle (USAF) and Milton O. Thompson (NASA) was announced by Paul F. Bikle,

Director of NASA Flight Research Center, and Brig. Gen. Irving L. Branch (USAF), Commander of AF Flight Test Center. Pilots would fill vacancies left by NASA research pilot Neil Armstrong, now a NASA astronaut, and Maj. Robert White (USAF), recently assigned other USAF duties. (NASA Release 63-127; FRC Release 12-63)

June 10: In American Univ. commencement address stressing action for world peace, President Kennedy said: ". . . Genuine peace must be the product of many nations, the sum of many acts. It must be dynamic, not static, changing to meet the challenge of each generation. For peace is a process—a way of solving problems . . . ." He announced his agreement with Prime Minister Macmillan and Chairman Khrushchev that "high-level discussions will shortly begin in Moscow looking toward early agreement on a comprehensive test ban treaty," and declared that U.S. "does not propose to conduct nuclear tests in the atmosphere so long as other States do not do so. We will not be the first to resume . . . ." (Text, *CR*, 6/10/63, 9870)

- Rep. George P. Miller (D.-Calif.) said on House floor:

    "Fulfillment of our diverse programs to which we are already committed may be threatened unless strong action is taken to increase the output from our graduate schools, particularly of more engineers, mathematicians, and physical scientists who combine high talent and high training.

    "For example, in 1950 we turned out 2,000 Ph.D.'s in these 3 fields; in 1960, we graduated 3,000, an increase of 50 percent. But in 1970 we need a minimum of 7,500, an increase in the decade of 150 percent . . . .

    "The study conducted for the President [by President's Science Advisory Committee] suggests that the manpower requirements of this decade can be met only if more students studying science and engineering . . . are encouraged to continue on to full-time graduate study.

    "The major barrier to such advanced study is financial . . . .

    "Another barrier is limited graduate facilities and faculty . . . .

    "Both because the Federal Government is a major consumer of scientific talent, and because the problem is urgent, the President proposes a program with Federal leadership, but involving a close relationship with State and local governments, industry and the universities.

    "Fellowship and training grant subventions are proposed in budgets especially of the National Science Foundation, Atomic Energy Commission, and the National Aeronautics and Space Administration, and extension and expansion of the National Defense Education Act.

    "These proposals would represent a deliberate first step whereby the Federal Government accepts the Nation's responsibility to assure adequacy of technical manpower resources to meet its commitments . . . ." (*CR*, 6/10/63, 9950-54)

- In hearings on U.S. space program, Senate Committee on Aeronautical and Space Sciences heard testimony of Dr. Philip Abelson,

Carnegie Institution of Washington's Geophysical Laboratory; Dr. C. S. Pittendrigh, Princeton Univ. biology dept.; Dr. Simon Ramo, Thompson-Ramo-Wooldridge, Inc.; Dr. Harold C. Urey, Univ. of Calif. prof. of chemistry; and Dr. Polycarp Kusch, Columbia Univ. physicist.

Dr. Abelson charged Project Apollo was harming future expansion of scientific knowledge and "taking away from our national security"; he declared that "manned space exploration has limited scientific value and has been accorded an importance which is quite unrealistic . . . ."

Dr. Pittendrigh testified he approved of lunar exploration program chiefly because of its value toward eventual exploration of Mars. He cautioned that manned Mars landing should be made only after it is assured there would be no contamination by terrestrial organisms.

Dr. Ramo expressed his approval of Project Apollo, but advised a "calm watch" to assure such a project would not begin to "dominate" U.S. technology and economy.

Dr. Urey, taking issue with Dr. Abelson's suggestion that Apollo costs were hindering U.S. science and education, said that "if we should decide not to do the space program we will probably do less for education than we would otherwise . . . ."

Dr. Kusch testified he opposed high priority for Project Apollo because "we have great national needs that far outweigh the exploration of space." He cited such necessities as conservation of natural resources. (NASA Leg. Act. Rpt. II/86; Hines, Wash. Eve. Star, 6/11/63)

June 10: USAF decision to recover manned spacecraft at Edwards, Holloman, and Wendover Air Force Bases was reported in *Missiles and Rockets*. All mission command and control would be exercised by Satellite Test Center, Sunnyvale, Calif. Unnamed USAF spokesmen said first program operating under this arrangement would be X-20 (Dyna Soar) in its air-launched phases. (*M&R*, 6/10/63, 34)

- USAF proposal that X-15 research aircraft program be extended to hypersonic speed range was reported in *Aviation Week and Space Technology*. Aircraft would be modified to attain new goals of 400,000-ft. altitude, 2,400° F skin temperatures, and 6,000-mph speed (about mach 8). (*Av. Wk.*, 6/10/63, 38)

- Quality of Soviet electronic equipment seen by recent U.S. scientific delegation to U.S.S.R. was described by Dr. Manson Benedict in *Electronic News* as very good. Very little of it is transistorized, and "most of it is similar to what we had five or six years ago." Chairman of AEC General Advisory Committee and head of MIT nuclear engineering dept., Dr. Benedict was among delegates accompanying AEC Chairman Dr. Glenn T. Seaborg on tour of Soviet scientific facilities May 19–30. (Connolly, *Electronic News*, 6/10/63)

June 11: NASA announced granting of three patent waivers to Ampex Corp. (for two inventions) and Space Technology Laboratories, Inc. Action brought to 46 the total of NASA's waivers of U.S.

commercial rights to inventions developed under NASA contracts. (NASA Release 63-128)

June 11: Senate Committee on Aeronautical and Space Sciences heard testimony on U.S. space program from U.S. scientists: Dr. Lloyd V. Berkner, President of Graduate Research Center of the Southwest; Dr. Lee DuBridge, President, Cal Tech; Dr. H. H. Hess, Chairman NAS Space Science Board; Dr. Martin Schwarzschild, Princeton Univ. Dept. of Astronomy; and Dr. Frederick Seitz, NAS President.

Dr. Berkner testified: "Men everywhere see in the conquest of space the peaceful demonstration of the superiority of one of the two competing systems of economic organization—capitalism vs. communism.... [The nation gaining space superiority] will have won the equivalent of a war in demonstrating the superior viability of his system in the eyes of the world...."

Dr. DuBridge testified that factors such as national security, prestige, and advancement of scientific knowledge all were strong reasons for proceeding with Project Apollo.

Dr. Hess outlined NAS Space Science Board proposal for "reserve corps" of young scientists to participate in U.S. manned space program. Under proposed plan, NASA would select about 500 volunteer scientists who would each year "do a period of active-duty training, say six to eight weeks, having to do with co-pilot activities on a space flight." About two years before projected mission, 20 reservists would begin fulltime training and further screening at NASA Manned Spacecraft Center. "Maybe five would be chosen, and maybe one would fly....

"Train astronauts to be scientists? I think this would be virtually impossible. It would take 10 years." (NASA Leg. Act. Rpt. II/87; Averill, *L.A. Times*, 6/12/63; Hines, Wash. *Eve. Star*, 6/12/63)

June 12: Testifying before Senate Committee on Aeronautical and Space Sciences, NASA Administrator James E. Webb announced NASA had decided not to conduct another Project Mercury flight: "We will proceed at once to a very careful, thoughtful but vigorous reorientation and realignment of the NASA organization, to move on with Gemini and with Apollo...." (Transcript)

- Project Mercury officially ended. Initiated in autumn of 1958, Mercury had achieved its goal of placing manned spacecraft in orbital flight around earth, investigating man's survival ability and performance capabilities in space, and recovering man and spacecraft. Mercury-Redstone suborbital flights paved the way toward ultimate goal, with Astronaut Alan B. Shepard becoming first U.S. man in space in FREEDOM 7 (May 5, 1961), followed by Astronaut Virgil I. Grissom in LIBERTY BELL 7 (July 21, 1961). First U.S. manned orbital space flight was achieved with Mercury-Atlas MA-6 flight, Astronaut John H. Glenn in FRIENDSHIP 7 (Feb. 20, 1962). This three-orbit space flight was followed by: MA-7, three-orbit flight of Astronaut M. Scott Carpenter in AURORA 7 (May 24, 1962); MA-8, six-orbit flight of Astronaut Walter Schirra in SIGMA 7 (Oct. 3, 1962); and MA-9, 22-orbit space flight of Astronaut L. Gordon Cooper in FAITH 7 (May

15–16, 1963). Project Mercury logged 34 orbits of manned space flight with 52 hrs. 23 min. of flight time.

June 12: NASA Manned Spacecraft Center announced award of six-month study contract to Hamilton Standard Div. of United Aircraft Corp. for establishment of environmental control and life support requirements for manned earth-orbiting space station. Duration of 24-man station would be from one to five years. (MSC Release 63–100)

- D. Brainerd Holmes announced his resignation as NASA Deputy Associate Administrator and Director of Manned Space Flight, effective some time next autumn. Holmes had joined NASA autumn of 1961, would return to industry "within the period of two years which was understood to constitute his obligation for government service at the time of his appointment." (NASA Release 63–133)
- Ford Motor Co.'s Aeronutronic Div. was selected by NASA Manned Spacecraft Center to investigate requirement for Mars excursion module (MEM)—unit designed to ferry astronauts between orbiting spacecraft and surface of Mars. (MSC Release 63–98)
- Appointment of Robert H. Charles as Special Assistant to NASA Administrator was announced. Consultant to Administrator since September 1962, Mr. Charles would work closely with NASA Deputy Associate Administrator for Industry Affairs in development and negotiation of NASA-industry contractual relationships. (NASA Release 63–134)
- At breakfast meeting of Republican Congressmen in Washington, Former President Dwight D. Eisenhower said that anyone who would be willing to spend $40 billion to get to the moon was "nuts." (Finney, *NYT*, 6/13/63)
- Rep. Roman C. Pucinski (D.-Ill.), speaking on House floor about his proposed bill (H.R. 1946) to establish National Research Data-Processing and Information Retrieval Center, said: "I firmly believe that the related activities in data storage and retrieval . . . must be carried on very much as it is today by the many scientific societies, abstracting societies, and universities. But the finished product of this widely dispersed effort must ultimately find its way to a central center if we are to recapture the full contribution of the world's technological explosion . . . ." (*CR*, 6/12/63, 10127)
- Atlas-Agena B launch vehicle exploded shortly after launch from Pt. Arguello, Calif. (AP, *Wash. Post*, 6/13/63; *M&R*, 6/24/63, 10)
- USAF launched Thor-Agena D space vehicle with unidentified payload from PMR; launch occurred several hours after another booster exploded. (*Pres. Rpt. on Space, 1963*, 1/27/64; *NYT*, 6/12/63)
- USAF conducted routine training launching of Atlas D ICBM from Vandenberg AFB, Calif. (DOD Release 844–63)
- President's Award for Distinguished Civilian Service presented to: Dr. Alain C. Enthoven, Deputy Assistant Secretary of Defense for Systems Analysis; David D. Thomas, Director of FAA Air Traffic Service; and Dr. Fred L. Whipple, Director of Smith-

sonian Institution Astrophysical Observatory. (*Av. Wk.*, 6/10/63, 36)

June 12: Friends and associates of the late Dr. Edward R. Sharp, former director of Lewis, dedicated the Edward R. Sharp Medical Library at Southwest Community Hospital, Berea, Ohio. (LRC Release 63-39, Lewis Chronology, 5)

June 13: Responding to former President Eisenhower's criticism of U.S. manned lunar program, Senator Edmondson (D.-Okla.) said on floor of the Senate: "We stand in the position of leader for the free world as we battle for superiority in this challenging field. We cannot afford to do less than we are now doing.

"The national goals set by the President of the United States and ably administered by the distinguished James E. Webb deserve the united support of Americans, for it is this program which assures this country preeminence in space."

Rep. Charles H. Wilson (D.-Calif.) said on floor of the House: "Surely the critics of this program do not wish our country to be left behind in the space race? . . .

"In our own day and age, brave men are still exploring uncharted oceans. I have no doubt that these journeys, too, will be well justified and that once again in time the expense will be repaid many times over.

"I suggest we study history a little more fully to prepare ourselves for the future. I regret that General Eisenhower will be pictured to our great grandchildren in future textbooks as being shortsighted and narrowminded . . . ." (*CR*, 6/13/63, 10168, 10248)

* House Committee on Science and Astronautics' Subcommittee on Space Sciences voted to delay NASA proposed electronics research center until NASA provided detailed study justifying need for center and for its location in the Boston area. (Finney, *NYT*, 6/14/63, 12)
* Morton D. Stoller, Director of NASA Office of Applications, died at 46 in Bethesda, Md. Formerly an engineer with NACA, Stoller had been with NASA since its creation Oct. 1, 1958. He received NASA Medal for Outstanding Leadership in February 1963. NASA had previously announced appointment of Dr. Robert F. Garbarini to succeed Stoller effective July 1. (NASA Release 63-95; *Wash. Post*, 6/14/63)
* Dr. Harold Brown, Director of DOD Research and Engineering, testified before Senate Committee on Armed Services' Preparedness Subcommittee that past performance—including "questionable" performance on Skybolt missile—of Douglas Aircraft Co. was basis for DOD's selection of Bell Aerosystems Co. for development of X-22 V/Stol aircraft. (UPI, *Wash. Post*, 6/14/63)
* NASA Marshall Space Flight Center employee Dave M. McGlathery became third Negro registered at Univ. of Alabama, first at Huntsville Extension Center. (Smith, *NYT*, 6/14/63, 16)
* Lewis Research Center engineers reported on studies involving simulated micrometeoroid penetration of future spacecraft. Using specially-built .22 caliber rifles that fire at velocities up

to 10 times the speed of normal bullets, engineers were puncturing different types of fuel tanks to duplicate what might happen in space to a spacecraft struck by a micrometeoroid. (LRC Release 63-42, Lewis Chronology, 5)

June 14-19: Manned orbital flight of VOSTOK V conducted by U.S.S.R., with Lt. Col. Valery F. Bykovsky onboard. Launched from Baikonur, VOSTOK V completed 81 earth orbits in 119 hrs. 6 min.; initial orbital path 146-mi. apogee, 112-mi. perigee, 88.27-min. period, 64°58′ inclination to the equator. Tass stated objectives of flight were: to continue studies of influence of various factors of space flight on human organism; to conduct extensive biomedical research of conditions of prolonged space flight; and to make further improvements and adjustments in piloting of spacecraft. Cosmonaut Bykovsky landed by parachute in Republic of Kazakhstan, shortly after landing of Cosmonaut Valentina V. Tereshkova and VOSTOK VI (see June 16-19). (Tass, *Izvestia*, 6/18/63, 1, AFSS-T Trans.; *NYT*, 6/15/63-6/20/63, 1ff)

June 14: Television photographs of Lt. Col. Valery F. Bykovsky (U.S.S.R.) in orbiting spacecraft were viewed by U.S. television audience. Soviet-bloc TV system Intervision's broadcast of photographs from space was received in Helsinki, where telecast was put onto Western Europe TV system Eurovision. Telecast was taped and edited in London, then sent to Goonhilly Downs for transmission via TELSTAR II communications satellite to receiving station in Andover, Me.; there it was sent to New York, where all three U.S. networks picked it up. (*NYT*, 6/15/63, 35)

- NASA Goddard Space Flight Center announced series of sounding rocket tests had confirmed association of Sporadic-E disturbances with presence of wind shears in altitude regions measured Nov. 7, Nov. 30, and Dec. 5, 1962. Under NASA contract, Geophysics Corp. of America scientists measured velocity of wind movements (using Nike-Apache rockets with sodium vapor trails) and ionospheric phenomena (using Nike-Cajun with Langmuir probe electrical equipment) at nearly the same time. Experiments confirmed theory of Australian scientist J. D. Whitehead that action of upper atmosphere wind pulls electrons from above and below into thin cloud-like layers, causing Sporadic-E layers that often interfere with radio signals' being reflected off higher F layer of ionosphere. (GSFC Release)

- In letter to Speaker of the House, President Kennedy described his proposed program of U.S. supersonic transport development. "A successful supersonic transport can be an efficient, productive commercial vehicle which provides swift travel for the passenger and shows promise of developing a market which will prove profitable to the manufacturer and operator.

    "It will advance the frontiers of technical knowledge—not as a byproduct of military procurement, but in the pursuit of commercial objectives.

    "It will maintain the historic U.S. leadership in aircraft development.

"It will enable this country to demonstrate the technological accomplishments which can be achieved under a democratic, free enterprise system.

"Its manufacture and operation will expand our international trade.

"It will strengthen the U.S. aircraft manufacturing industry—a valuable national asset—and provide employment to thousands of Americans . . . ." (Letter, *CR*, 6/17/63, 10304)

June 14: NASA announced signing contract with AC Spark Plug Div. of General Motors to fabricate, assemble, and test navigation and guidance system for Apollo command module. (NASA Release 63-136)
- Deputy Secretary of Defense Roswell L. Gilpatric testified before Senate Committee on Armed Services' Preparedness Investigating Subcommittee on X-22A development contract award to Bell Aerosystems Co. rather than USN-recommended Douglas Aircraft Co. "I made the best judgment I could in the light of a very long experience and knowledge with the aerospace industries . . . . I gave adequate consideration to all the factors" before selecting Bell because of its "experience and past performance." (Raymond, *NYT*, 6/15/63, 1; *Av. Wk.*, 6/24/63, 36)
- Invention of retrometer, device relaying light beam over which voice messages are transmitted, by NASA Langley Research Center scientist N. E. Thomas was reported in Newport News *Times-Herald*. (Newport News *Times-Herald*, 6/14/63)
- NASA announced German Federal Minister for Scientific Research Hans Lenz and Director of Ministry's Space Research Div. Max Mayer had concluded two weeks of discussions on space research and survey tour of NASA and NASA-contractor facilities. (NASA Release 63-135)
- U.S. Army announced Nike-Zeus antimissile missile successfully intercepted re-entry target vehicle launched by Atlas ICBM in recent test. Nike-Zeus was fired from Kwajalein Island in the Pacific, Atlas from Vandenberg AFB, Calif. (DOD Release 858-63)

June 15: Official opening of 1963 hurricane season. TIROS VI meteorological satellite, launched September 18, 1962, was still operational and U.S. Weather Bureau said additional meteorological satellites may be orbited by NASA before end of season, Nov. 15. (Commerce Dept. Release WB 637)
- USAF announced launching of unidentified satellite using Blue Scout booster from Pt. Arguello, Calif. (UPI, *Wash. Post*, 6/16/63; *M&R*, 6/24/63, 10)
- USAF launched five satellites—LOFTI IIA, SOLARAD IV, and three unidentified satellites—using one Thor-Agena D launch vehicle. LOFTI IIA was to conduct low frequency communications experiments and SOLARAD IV was to measure solar radiation. (*Pres. Rpt. on Space, 1963*, 1/27/64)
- In May 6 testimony before House Armed Services Committee's Subcommittee on Military Appropriations released today, Director of Defense Research and Engineering Dr. Harold Brown said USAF Midas satellite program had been reoriented and cut back substantially because of technical difficulties. DOD recommenda-

tion of cut in Midas funding for FY 1964 was based "on the conclusion that the way the program was going it would never produce a reliable, dependable system." Midas program had cost $423 million through FY 1963. "Of that, I would say about half has been wasted. Half has been spent on gaining data which are, I think, quite necessary for any system of the kind that we may develop in the future . . . ." During past 10 years DOD has initiated and then terminated 61 projects at cost of more than $6 billion, he testified. (AP, Wash, *Sun. Star*, 6/16/63)

June 15: GSFC *Satellite Situation Report* identified three Soviet space launchings not previously released by either the U.S. or U.S.S.R.: Sept. 12, 1962 (1962 Alpha-Phi 1-5), with fragments re-entering from Sept. 12 through Sept. 17; Nov. 4, 1962 (1962 Beta-Xi 1-5), with fragments re-entering from Nov. 5 through Jan. 19, 1963; and Jan. 4, 1963 (1963 1A-1C), with fragments re-entering from Jan. 5 through Jan. 11. (GSFC *Sat. Sit. Rpt.*, June 15, 1963)

- Following radio broadcast received in London:

    "Attention all transmitters working on the frequencies between 20,000 kilocycles and 20,012 kilocycles.

    "You are interfering with the reception of signals from the Vostok V space ship. You are asked to stop working on these frequencies." (AP, Wash. *Eve. Star*, 6/15/63)

- In *Saturday Evening Post*, Sen. Howard W. Cannon (D.-Nev.) wrote:

    "A perilous notion grips Washington—that through some sort of gentleman's agreement with Russia we can quarantine space, can keep it from becoming a theater of war. Wishfully the nation gears its space program to peaceful purposes, blind to the fact that the Russians have no such inhibitions. While we dream, the Russians bend every effort to develop military space systems. Quite openly they are striving for a decisive advantage in that newly penetrated region.

    " 'Peaceful purposes,' the Administration admits, means 'keeping the peace' as well as engaging in scientific exploration and in commercial activities in space. President Kennedy has said we must insure that no other nation gains a position in space that would threaten our security. And we are orbiting numerous unmanned space vehicles which could have significant military purposes. But the disturbing fact is that we continue to put overwhelming emphasis on nonmilitary space programs and limit military efforts to a small list of defensive possibilities . . . ." (*Sat. Eve. Post*, 6/15/63, 10, 12)

- Reported that USAF Vice c/s General William F. McKee had dispatched message to USAF commanders urging "personnel attention and active support" in recruiting volunteers for forthcoming NASA astronaut selections. (*A-N-AF Journ. & Reg.*, 6/15/63, 4)
- NASA Deputy Administrator Dr. Hugh L. Dryden received honorary Doctor of Science degree from Northwestern Univ. (Letter)
- Henry F. Auter, Chief of Electrical Systems Engineering Branch of Test Div., NASA Marshall Space Flight Center, appointed Deputy Manager of MSFC Mississippi Test Operations. (*Huntsville Times*, 6/15/63)

June 16–19: Flight of VOSTOK VI with Lt. Valentina V. Tereshkova, first woman in space, conducted by U.S.S.R. Launched from Baikonur, VOSTOK VI completed 48 earth orbits in 70 hrs. 50 min.; initial orbital path was 141-mi. apogee, 112-mi. perigee, 88.3-min. period. Shortly after entering orbit, Tass reported, Lt. Tereshkova established radio communications with Lt. Col. Bykovsky in VOSTOK V; the two cosmonauts then radioed joint message to Premier Khrushchev. Tass stated flight of VOSTOK VI was being made "to continue the study of the effect of various space-flight factors on the human organism, including a comparative analysis of the impact of these factors on the organisms of a man and a woman; to carry out new medical-biological research and to further improve and perfect the systems of piloted spaceships in conditions of simultaneous flight." During initial orbit VOSTOK V and VI came within about five km. (three mi.) of each other, apparently closest distance achieved. Cosmonaut Tereshkova, whom Tass described as having never piloted an aircraft, landed by parachute in Kazakhstan. After she and Cosmonaut Bykovsky landed, Tass issued statement on flight results: "New valuable data have been obtained about the influence of different factors of a space flight of long duration on the organisms of man and woman. Rich factual material necessary for further perfection of the systems of piloted space ships has been obtained." (Tass, *Izvestia*, 6/18/63, 1, AFSS–T Trans.; *NYT*, 6/17/63–6/20/63, 1ff)

June 16: Results of six-year experimental study of 25 generations of mice exposed to large radiation doses were reported in *New York Times*. Conducted by Los Alamos Scientific Laboratory, study disclosed large doses of radiation over long period of time have an "amazingly small" effect on mice: no evidence of malformations was found, and there appeared "to be no threat to genetic extinction of the species." (Plumb, *NYT*, 6/16/63)

- AF Cambridge Research Laboratories terminated flight of 54-ft. mylar balloon after 28-day journey across Pacific Ocean. Launched May 19 from Chico, Calif., superpressure balloon had reached point 500 mi. south of Kamchatka (Siberia) peninsula when AFCRL destroyed it by radio command from Chico. This was third successful balloon flight in AFCRL series to evaluate superpressure design. (*OAR Research Review*, 9/9/63, 2)

June 17: Aerobee 150 sounding rocket launched to peak altitude of 113 mi. from White Sands Missile Range, N.M. Purpose was to obtain positive information on gyro drift and general control characteristics. The aspect instrumentation on the rocket included two cameras, two Adcole digital sun sensors, a fine tracking Ball Brothers System and a Whittaker three-axis gyro package. All of the instruments worked exceptionally well. Good control, stabilization, and maneuver accuracy was demonstrated, but the target accuracy was less than expected. (NASA Rpt. of S. Rkt. Launching, 7/2/63 & 7/19/63)

- Commenting on Soviet Cosmonaut Valentina Tereshkova's space flight in VOSTOK VI, Chairman of Senate Committee on Aeronautical and Space Sciences Clinton Anderson said: "I'm not surprised. There isn't any reason why women can't go up into space...."

Chairman of House Committee on Science and Astronautics George P. Miller said: "It doesn't surprise me that the Russians want to pull another spectacular by putting a woman in space. It shouldn't interfere with our scientific program. I don't want to downgrade their achievement, but it doesn't mean we have to follow suit ...." (AP, Wash. *Eve. Star*, 6/17/63)

June 17: NASA Associate Administrator Dr. Robert C. Seamans, Jr., testifying before Senate Committee on Aeronautical and Space Sciences, observed Soviet manned spacecraft VOSTOK V and VI were traveling in two different orbital planes: "It appears unlikely that sufficient fuel could be taken aloft by either of these Vostok spacecraft to effect a rendezvous." (Transcript)

- U.S. Geological Survey's lunar mapping project, mapping 8.5 million sq. mi. of moon's surface, was described in Washington *Evening Star*. Undertaken for NASA, project would provide basic geological information needed for placing instruments and men on the moon. (Henry, Wash. *Eve. Star*, 6/17/63)

- Addressing Washington, D.C., meeting of American Red Cross, NASA Administrator James E. Webb said:

  "Project Apollo offers us our greatest opportunity to overcome the lead in manned space exploration now held by the Soviet Union. It gives us assurance that if an American is not first on the Moon, he certainly will not be far behind. And most important of all, it demonstrates to all the world that we in the United States have the capability and the determination to carry on those things which we have declared seriously that we intend to do." (Text)

- NASA $150 awards to Dr. Karlheinz Thom and Joseph Norwood, Jr., for invention of plasma accelerator, were announced. Accelerator is capable of controlled ignition under very low gas pressures. (NASA Release 63–137)

- USN Polaris A-3 missile fired from U.S.S. *Observation Island* in test said to be first success at sea. (*M&R*, 6/24/63, 10)

- Tass announced postage stamp commemorating VOSTOK V space flight had been issued by Soviet Ministry of Communications. (Reuters, *NYT*, 6/18/63, 10)

June 17–20: American Institute of Aeronautics and Astronautics held its summer general meeting in Los Angeles. (*Av. Wk.* [Space Calendar], 5/6/63, 9)

Walter C. Williams, NASA Manned Spacecraft Center Associate Director and Project Mercury Operations Director, summarized Project Mercury, emphasizing that primary lesson taught by the manned space flights was value of attention to detail. Discussing biomedical results, he said orthostatic hypotension first noted in Astronaut Walter M. Schirra (MA-8, six orbits) was more pronounced in Astronaut L. Gordon Cooper (MA-9, 22 orbits). Lowered blood pressure was believed to be cause of Cooper's temporary dizziness when he emerged from Mercury spacecraft: "The effects seem to be post-flight, and continued from 18 to 24 hours." (*Av. Wk.*, 7/1/63, 71)

Astronauts Walter M. Shirra, Jr., and L. Gordon Cooper received the AIAA Astronautics Award. (*Av. Wk.*, 7/1/63, 31)

Rombus (Reusable Orbital Module—Booster and Utility Shuttle) concept described June 18 by Philip Bono, Chief Advance Project Engineer of Future Space Systems, Douglas Missile and Space Systems Div. Single-stage manned space cargo carrier, capable of orbiting many hundreds of tons and reusable at least 20 times, could be developed by late 1970's. Eight strapped-on liquid-propellant tanks would be jettisoned and parachuted to earth after providing initial boost; Rombus would be recovered from orbital mission with use of parachutes, main engine retrothrust, and extended landing legs. Bono's paper was based on study performed for NASA Marshall Space Flight Center and was one of many large vehicle concepts under study by industry and government. (DAC Release 63–122)

Speaking on June 19, Dr. Stanley Deutsch, Chief of Systems Research and Analysis in NASA Office of Advanced Research and Technology, declared Astronaut L. Gordon Cooper could not have sighted houses, boat, and train during his MA-9 orbital space flight "because of the visual angle." Dr. Deutsch speculated the visual impressions might have involved hallucinations or transfer of stored impressions. "We need a great deal of study in the area of perception [in the weightless condition]." Interviewed at meeting, Astronaut Cooper said: "I'm not changing my mind one bit, I know what I saw." (*Av. Wk.* 7/1/63, 31)

June 18: X-15 No. 3 flown by Maj. Robert A. Rushworth (USAF) to 223,700 ft. altitude and 3,539 mph (mach 4.97). Principal research objective was gathering vertical-tail pressure data. (X-15 Project Office, NASA Hq.)

- NASA Manned Spacecraft Center announced recruitment of astronauts had been broadened in current recruitment call for volunteers: change in criteria for selection, eliminating need for test pilot school certificate; and soliciting letters of recommendation from industries, professional groups, military services, and other organizations throughout the country. (MSC Release 63–102)

- Dr. Eugene B. Konecci, NASA Director of Biotechnology, said in speech before National Rocket Club that Lt. Tereshkova's orbital space flight established feasibility of sending scientists not trained as pilots to moon and nearby planets: "I think it is significant that she is not a pilot. This points to the importance the Soviets attach to gathering biomedical information on the effects of space flight and may be a preparation for flying scientists who are not trained pilots . . . ." (UPI, *NYT*, 6/19/63, 3; *Av. Wk.*, 6/24/63, 33)

- Members of recent U.S. Mt. Everest expedition, William F. Siri, Univ. of California physiologist, and Prof. Maynard M. Miller, Michigan State Univ. scientist, said in New Delhi interview that laboratory study of rock and snow samples from Mt. Everest might provide clues to life on Mars. Siri said environment on Mars is "just a little bit worse" than on Everest's peak, citing Everest's thin air, low temperatures, lack of liquid water, and intense solar radiation. (AP, *Richmond News Leader*, 6/18/63)

June 18: ComSatCorp incorporator George J. Feldman said at Conference on Space Science and Space Law, Univ. of Oklahoma: "The reality of a viable [satellite] communication system remains to be accomplished with the accompaniment of hard work, hard and yet imaginative intelligence, and cooperation, nationally and internationally . . . .

"The choice of system . . . in turn is of importance to the international users. How do they participate? Do they own stations? Do they furnish personnel? What about their own research and development? And in addition to the economics there is the political national interest and national pride in participation . . .

"The communication satellite has arrived of experimental age. It will be brought into being as a working system. The United States has no monopoly on its creation. There are others who are capable of doing it. There is strong worldwide interest. The corporation is beginning to tackle the formidable problems in earnest enthusiasm, and I may say in the spirit of the kind of dedication that accompanies a pioneering effort of responsibility. The expert group being assembled can make a substantial contribution of national benefit, as it researches and develops its program. It is time to give this effort the understanding required, for much is at stake for all . . . ." (Text, CR, 7/2/63, A4188-89)

June 19: TIROS VII (A-52) meteorological satellite placed in orbit with Thor-Delta launch vehicle launched from Cape Canaveral. Initial orbital data: 405-mi. apogee, 388-mi. perigee, 97-min. period, 58.2° inclination to equator. On satellite's first orbit, Command and Data Acquisition Station at Wallops Island, Va., obtained direct pictures from Camera 2 showing cloud vortex over Newfoundland and set Camera 1 to read out pictures on next orbit. First pictures were transmitted within an hour to Cape Canaveral, Fla.; NASA Goddard Space Flight Center, Greenbelt, Md.; and National Weather Satellite Center, Suitland, Md. In addition to two wide-angle TV cameras, TIROS VII carried infrared sensors and electron temperature probe. Orbiting marked 18th straight successful satellite-orbiting by Thor-Delta launch vehicle. (NASA Release; Wallops Release 63-59; UPI, *NYT*, 6/20/63, 3)

- Experiment to study behavior of liquid hydrogen when exposed to radiant heating and zero gravity conditions was launched to approximately 120-mi. altitude by Aerobee sounding rocket from NASA Wallops Station. Preliminary analysis of telemetry data indicated only partial success of the experiment, provided by NASA Lewis Research Center. (Wallops Release 63-60)
- Letter from Harold S. Geneen, IT&T President, to Sen. Warren G. Magnuson, Chairman of Senate Commerce Committee, and Rep. Oren Harris, Chairman of House Committee on Interstate and Foreign Commerce, was made public. Replying to RCA Chairman David Sarnoff's April 2 letter to Leo D. Welch, Communications Satellite Corp. Chairman, and to Former FCC Chairman Newton N. Minow's May 31 letter to President Kennedy, Geneen said he "must respectfully disagree with General Sarnoff's suggestion

that the merger [of international communications carriers] should be compulsory and include all international communications facilities, both voice and record, under the Communications Satellite Corporation." Reasons against such merger: it would run counter to historical Government opposition to monopoly in any field; IT&T believed that private enterprise should not abandon field to which it has made substantial contribution; and IT&T believed that ComSatCorp was established as "a private corporation and intended . . . to be a carrier's carrier" rather than competition for carriers using its services. "The real problem, as we see it, is rather to strengthen competition than abandon it . . . .

"We have also urged that the international record communication carriers be permitted to merge, as a means of offsetting American Telephone and Telegraph's dominant position in international voice communications . . . ."

AT&T also issued statement opposing RCA Chairman Sarnoff's proposal to create single communications company: ". . . The present arrangement . . . is a good one. It is based on the principles of good service and sound economics and is working well. We see no reason to experiment with substitutes of doubtful merit . . . ." (Smith, *NYT*, 6/20/63)

June 19: FAA Administrator Najeeb E. Halaby told AIAA meeting in Los Angeles: "First, technically, a superior supersonic transport is feasible.

"Second, it is an expensive aircraft to develop, but not fantastically so, and certainly not beyond the resources of the greatest industrialized nation in the world.

"Third, there is a market for an American transport of perhaps 200 to 250 first-round orders. If we do not develop an American SST, then the entire market, which could be something like 400 planes during perhaps a decade of first-round buying, will fall to our overseas competitors. Among the purchasers would be the airlines of the U.S., which would be forced to purchase planes produced overseas in order to remain competitive.

"Fourth, in the event that we did not proceed with SST development, the Nation would default its position of preeminence in international commercial aviation—with all that would entail in practical terms and in less measurable terms of confidence and prestige . . . ." (Text)

- U.S.S.R. MARS I passed planet Mars, but with its communications systems dead as they had been since March 21. All subsequent attempts to restore communication had failed. (AP, Wash. *Eve. Star*, 6/26/63)

- NASA announced RCA Service Co. had been selected to negotiate contract for operation and maintenance of satellite data-acquisition stations—two near Rosman, N.C., and two near Fairbanks, Alaska. One station at each location would be operational by Sept. 1963; other two stations were expected to be operational in 1964. Facilities would be first high-capacity data-acquisition stations, supporting Oao, Pogo, Ego, and Nimbus satellites. (NASA Release 63–138).

June 19: NASA Deputy Administrator Hugh L. Dryden received international aspects of the U.S. space program at a luncheon meeting of the National Capitol Section of the AIAA. Dr. Dryden was awarded the First Annual von Kármán Memorial Citation and a letter from Vice President Johnson was read which said, in part: "A few more Hugh Drydens—appropriately placed about the globe—and we would have a world of peace, honor, and progress." Preceding Dr. Dryden's address Mr. Robert F. Packard of the State Department reviewed U.S. policy with regard to space. (Letter; Citation; Text)

- Brig. Gen. Gerald F. Keeling (USAF) said in address before Fifth Annual USAF—Industry Conservation Conference: "Department of Defense procurement from the aerospace industries for air weapon systems, and research and development, is now running about $20 billion a year....

  "The Air Force share ... totals $10.6 billion a year, which is more than half the DOD total....

  "With expenditures of this magnitude, both the DOD and Air Force have reason to be concerned that conservation, cost reduction, and value engineering be practiced...." (Text, AFSC Release 36–R–47)

June 19–July 7: First nationwide display of space research innovations with possibilities for industrial applications was held at 1963 Chicago International Trade Fair. Exhibit was sponsored by NASA Office of Technology Utilization and presented by Illinois Institute of Technology Research Institute (IITRI). (NASA Release 63–122)

June 20: Dr. Edward C. Welsh, Executive Secretary of National Aeronautics and Space Council, said in speech before Conference on Space Science and Space Law at the University of Oklahoma:

"The impact of the space budget ... is greater than its size would indicate. For example, a major portion of it goes for research and development, with the attendant flow of new processes, new inventions, new products, and new ideas which stimulate many sectors of the economy. So far, during the past six years we have spent less than $10 billion on all aspects of space development and exploration. This has averaged less than 20 cents per capita per week. Contrary to the careless thinking of some critics, this money has been spent for materials, jobs, facilities, etc., right here on earth rather than on the moon or some distant planet. In other words, its stimulating effect has resulted in a greater demand for consumers' goods and greater savings and investments.

"It is unrealistic to conclude that, if this money were not going into space work, it would necessarily flow into other essentials such as slum clearance, education, and cancer research. The facts are that we have been spending more for those essentials since there has been a space program; the space program contributes to such progress; and it is not likely that the Congress would vote more money for such other projects if there were no demand for space funds. This is not an either/or proposition...."

He attributed to the space program major credit for such benefits as: "comprehensive knowledge and understanding" of earth-sun relationships, interplanetary space environment, origins of solar system and life itself; improved communications, weather forecasting, and navigation; improvements in metals, alloys, and ceramics, and "substantial progress in the whole field of electronics . . . ."

In "peaceful application of space competence to national security" he listed:

"1. Support of our terrestrial armed forces . . . in communications, meteorology, navigation, and mapping;

"2. Increase in our alertness against pending dangers and our knowledge of potentially hostile territories through early warning and observation capabilities; and

"3. Competence to examine and neutralize, if necessary, hostile and threatening spacecraft.

". . . Of even greater importance is the potential impact the space program can have on world peace through substituting competition in space exploration for competition in building nuclear striking power. If sanity prevails, the path to the stars can be the path to peace." (Text)

June 20: Dr. Robert L. Barre, Scientist for Social, Economic, and Political Studies, NASA, said at University of Okla. Conference on Space Science and Space Law:

"The demands of our space goals require a very great increase in functional and intellectual specialization, in both kind and degree. The effort to drive scientific knowledge and technological capability ever further into the unknown has the effect of differentiating skills and areas of knowledge and of enhancing their uniqueness on a scale not previously experienced. The ever increasing complexity of man-machine problems, particularly in regard to man's daring to venture into the uncharted realm of space, produce demands for reliability, durability, mechanical precision, exactness and speed of human interaction which, if uncontrolled, could be corrosive of the human and social characteristics this nation has for some time held in high esteem. These demands require integrative processes, operating through man, his institutions and his tools which allow much narrower tolerances of error of variation than those acceptable to a mass production society." (Text)

- Titan II ICBM launched from Vandenberg AFB concrete silo site in successful USAF test flight. (AP, *Wash. Post*, 6/21/63)
- Aerobee 150 sounding rocket with flight test instrumentation duplicating that intended for installation in the solar-pointed section of the S-17 satellite spacecraft (Oso–B) launched to 119.2 mi. from White Sands Missile Range, N.M. Results undetermined until recovery of payload effected. (NASA Rpt. of S. Rkt. Launching, 6/24/63)
- Sir Charles (C.P.) Snow, British scientist and author, said in New York interview that pace of U.S. lunar program was "probably a mistake," an "all-out" effort in race with U.S.S.R. Most

dramatic scientific achievements of next decade, he said, would be in molecular biology. (Sullivan, *NYT*, 6/21/63, 11)

June 20-21: In hearings conducted by House Committee on Interstate and Foreign Commerce, FAA Administrator Najeeb E. Halaby outlined details of proposed Government-industry project to develop supersonic transport. (NASA Leg. Act. Rpts. II/93 and II/94)

June 21: Addressing Conference of National Organizations in Atlantic City, NASA Deputy Director of Manned Space Flight Dr. Joseph F. Shea said that "now that the costs are rising as predicted, the country is reappraising the commitment [to land men on the moon in this decade]. Such discussions are entirely proper. Indeed, the resolve with which we carry out our national activities is frequently strengthened [by the] tempering in the heat of debate . . . ." (Text)

- At 39th annual joint meeting of Virginia Engineering and Technical Societies Dr. Floyd L. Thompson, Director of Langley Research Center, announced that whenever the U.S. decided to go ahead with a manned orbiting laboratory, LARC would be ready. "Although there is no NASA approved flight project for a manned space station at the present time, Langley will be in a position through its studies to provide many of the answers to design and operational problems when such a program is given the go signal." (Newport News *Daily Press*, 6/22/63)
- U.S. television audiences witnessed first public appearance of Pope Paul VI via RELAY communications satellite. (CBS TV)
- USAF announced $500,000 study contracts awarded to General Dynamics/Astronautics, North American Aviation, and Douglas Aircraft Co. for investigation of Aerospace Plane (Asp) concepts. Powered by air-breathing ramjet-like engine, Asp would take off from conventional runway, then travel through space propelled by liquid-oxygen/liquid-hydrogen rocket engine, orbit, and return to land at conventional airfield. (DOD Release 891-63; *M&R*, 7/1/63, 9)
- Establishment of Supersonic Transport Authority within Federal Aviation Agency was recommended by advisory group of airline and manufacturing experts headed by Gen. Orval R. Cook (USAF, Ret.). Report of advisory group, released by FAA, recommended Authority direct U.S. program to develop supersonic transport airliners. (AP, *NYT*, 6/22/63, 34)
- USN launched Polaris A-3 missile more than 1,500 mi. from U.S.S. *Observation Island*, second such test in one week. (UPI, *NYT*, 6/22/63, 35)
- Edmond E. Bisson, Lewis scientist, installed as president of American Socity of Lubrication Engineers. (LRC Release 63-45, Lewis Chronology, 5)

June 22: Soviet Premier Khrushchev formally welcomed Cosmonauts Valery F. Bykovsky and Valentina V. Tereshkova in ceremony before crowds in Moscow's Red Square. Later, in Kremlin reception, the cosmonauts were awarded Hero of the Soviet Union and Pilot-Astronaut citations as well as Order of Lenin and Gold Star medals. (Topping, *NYT*, 6/23/63, 1, 8)

June 22: EXPLORER XIV was eclipsed by the moon—a possible first in satellite history. Satellite was some 70,000 km. from the earth with apogee portion in the sun's direction when a 19-min. shadow occurred between approximately 0151 and 0210 UT. (GSFC Historian)

- J. C. Garrett, president and founder of the Garrett Corp., died of a heart attack in Beverly Hills, Calif. One of the pioneers in research on pressurization of aircraft, Mr. Garrett engineered the pressurization system on the B-29 bomber. His company developed the life-support system used in the Mercury spacecraft. (*NYT*, 6/24/63, 27)

June 23: NASA Daytona Beach Operation established as integral part of NASA Launch Operations Center (LOC), Cocoa Beach, Fla. (NASA Circular 267A)

- A travel ban by NATO countries on East German scientists has been modified by a "trial relaxation" that would permit East German scientists to attend scientific meetings in NATO countries if they are part of an all-German delegation along with West Germans. The meeting of the International Union of Geodesy and Geophysics, scheduled for August 1963 in Berkeley, Calif., was almost withdrawn from the U.S. because of the ban on East German scientists. Intervention by the National Academy of Sciences achieved the compromise relaxation. (*NYT*, 6/24/63, 1)

June 24: President Kennedy sent to Congress a request to add $60,000,000 to the FY 1964 budget to finance detailed design studies on the supersonic transport aircraft. In his June 14 letter to Congress on the aircraft, the President had set a total ceiling of $750 million for Government support of the project, with $250 million additional to come from the aircraft industry, and with part of the Government investment to be recovered through royalty payments. (*NYT*, 6/25/63, 55)

- Soviet Premier Khrushchev, opening session of World Woman's Congress in Moscow, warned that third world war would "be a rocket nuclear war of extermination, . . . [destroying] much of what we generally call civilization." Cosmonaut Valentina Tereshkova was stellar attraction at Communist-dominated congress. (AP, Wash. *Eve. Star*, 6/24/63 A-20)

- Reported that NASA Administrator Webb and Defense Secretary McNamara were considering proposal that USAF perform military experiments during first 12 Gemini space flights, *Aviation Week and Space Technology* reported. (*Av. Wk.*, 6/24/63, 25)

- NASA announced selection of Lear Siegler to negotiate contract for design study of "psychophysical information acquisition processing and control system" (Piapacs). Proposed system, mounted in pilot's garment and headgear, would replace present system of sensors attached to pilot and would permit continuous acquisition of psychological and physical data. NASA pilots would test system in flights of aircraft including F-104, Jet Star, and X-15, at NASA Flight Research Center. (NASA Release 63-139; FRC Release 13-63)

- NASA and DOD negotiation of agreement giving NASA authority to purchase its own Atlas-Agena launch vehicles was reported by *Mis-*

siles and Rockets. NASA had signed agreement, and DOD approval was "expected shortly." (*M&R*, 6/24/63, 9)

June 24: John Stack, pioneer of aircraft development over 34 years with NACA and NASA, now director of engineering for Republic Aviation Corp., interviewed on supersonic transport development in *U.S. News and World Report.* Asked how much had been done on U.S. supersonic transport to date, he said:

". . . You could almost take it off the shelf today. The technology is there. It's just a matter of pulling it together and putting it into the airplane you want. We were doing research on the problems of supersonic transport down at Langley Research Center all through the late 1950's. We knew then that it was feasible. There was no fundamental scientific obstacle.

"In December of 1959, at the request of E. R. Quesada who was head of the Federal Aviation Agency and President Eisenhower's aviation adviser, we made a presentation on development of a supersonic transport. Mr. Quesada recommended that work get underway. He didn't get the money he asked for, but supersonic-transport development was established as a national policy.

"The Kennedy administration came in, in 1961, and Najeeb Halaby replaced Mr. Quesada at the FAA. He pushed ahead with the development program. Congress appropriated a total of $31 million for the project in 1961 and 1962. This year, 1963, was to be the prototype year.

". . . as you know, the day after Pan American Airways announced that it was placing an order for six of the British-French Concorde supersonic transports President Kennedy said there will be a supersonic transport program.

". . . If we wait another year [for Congressional appropriations], we will be in a serious position in this competition. . . ." (*U.S. News*, 6/24/63; *CR*, 7/9/63, 11621–22)

- Dr. Glenn T. Seaborg, Chairman of the Atomic Energy Commission, was unanimously confirmed by the Senate for an additional five-year term. Dr. Gerald F. Tape was also confirmed to finish out the remaining three years of Dr. Leland J. Haworth's term on the Commission. Dr. Haworth had resigned to become Director of the National Science Foundation. (*NYT*, 6/25/63, 48)

- USAF contract award to Avco Corp. for production of re-entry vehicles was announced by DOD. $1.3 million contract was awarded by AFSC Ballistic Systems Div. (DOD Release 901–63)

- Address by Dr. Glenn T. Seaborg, AEC Chairman, before Conference on Higher Education in Chicago inserted in *Congressional Record* by Rep. Barrett O'Hara, in which he said:

"The most important source of creative evolution in the future will continue to be, as it has been in the past, fundamental scientific research—the kind of investigations that give us a better understanding of the true nature of our world . . . .

"We have inherited a revolutionary philosophical concept of man as an architect of his environment, a concept that seemed to flower only about two centuries ago. Men gained confidence in this idea when their economic energy and inventions brought unparalleled modifications of life through the industrial revolution.

In the period starting some two decades ago, which may be designated as the beginning of the third revolution, Western man became irrevocably dedicated to the concept of creative evolution. The future of the third revolution—of man's power to mold the world to his liking—is almost unlimited." (Text, *CR*, 6/24/63, A3993-95)

June 24: Maj. Gen. Leighton I. Davis, Commander of AFMTC, in an interview with *Electronic News*, stated that the recent Soviet tandem space flight added urgency to a strong U.S. military space program: "The thinking that goes with the exploitation of space for peaceful purposes is inconsistent and incompatible with the thinking necessary to exploit the military and defense aspects of space." Mercury, Gemini, and Apollo are "not technical weapon systems with any military potential." They require "long checkout time, lack quick reaction time, and don't have the degree of maneuverability in space needed for a weapon system." (*Electronic News*, 6/24/63)

- Medical investigation of the 4,565 general aviation accidents in the U.S. in 1961 was inadequate, a group of USAF doctors reported in the *Journal of the American Medical Association*. Physicians should autopsy the dead and examine survivors with the aim of making flying safer, the medical officers asserted, but this is done only for military and commercial airline accidents—not for general aviation, where often only a single investigator from FAA or CAB visited the crash. (*NYT*, 5/24/63, 6)

June 24–July 10: Fourth Annual Aviation Education Workshop, sponsored by Mississippi State Dept. of Education, various aviation interests, military services, and NASA, was held at Univ. of Southern Mississippi, Hattiesburg. Robert F. Thompson, Head of NASA Manned Spacecraft Center's Flight Operations Recovery Branch, was principal speaker June 26, with addresses on "Tracking, Search and Rescue Procedures" and "Past and Future Programs of Manned Space Flight." (MSC Release 63–104)

June 25: X–15 No. 1 was flown by NASA pilot Joseph Walker to 3,911 mph (mach 5.51) and 111,800 ft. at Edwards AFB, Calif. During the flight a new measuring instrument, the traversing probe, popped in and out of the wing, measuring the whole profile of boundary layer over surface of the X–15. Surface temperature rose to 1,000° during the flight, the 86th for the X–15. (UPI, *Chic. Trib.*, 6/26/63; NASA Hq.)

- U.S.S.R.'s most recent cosmonauts, Jr. Lt. Valentina V. Tereshkova and L/Col. Valery Bykovsky, held press conference in Moscow. Both insisted that tandem flight had gone off according to schedule, both as to duration and maneuvers. Lt. Tereshkova, the first woman into space, added another fact to her official biography—that she had had pilot training prior to entering the space program. (*N.Y. Herald Trib.*, 6/26/63)

- House Committee on Science and Astronautics began full committee hearings on proposed NASA budget for FY 1964 and authorized expenditure of $3.9 million for NASA proposed electronics research center. However, Committee deleted references to Boston area as site in FY 1964 budget authorization and also directed NASA to first undertake detail study of "geographic location, the need for, and the nature of the proposed center." (NASA Leg. Act. Rpt. II/95; Quincy [Mass.] *Patriot-Ledger*, 6/25/63; Boston *Traveler*, 6/26/63)

June 25: NASA announced receiving from eight nationally known artists 60 sketches made during Astronout Cooper's MA-9 mission under NASA Artists' Cooperation Program. Sketches were preliminary to finished paintings, which would be exhibited and from which reproductions would be made. MA-9 was first event which NASA invited artists to cover; other NASA activities of historic interest would be recorded by artists. (NASA Release 63-140)

- Langley Research Center issued invitations to bid to 39 contractors for construction of the laboratory building of the NASA Space Radiation Effects Laboratory at Newport News, Va. With cost of the laboratory building estimated at $3 million, this is the last major contract to be awarded for the $12.3 million facility. (LARC Release)

June 26: House Science and Astronautics Committee cut $134,248,600 from the scientific exploration of space portion of the NASA authorization for FY 1964. With a previous cut of $259,122,000 from the manned space flight portion, total reduction in the $5.7 billion request was now $483 million. (*NYT*, 6/27/63, 1)

- Development and construction of frictionless platform by Martin Co. for NASA Manned Spacecraft Center was announced by MSC. MSC Crew Systems Div. would use simulator to test and evaluate space suits, stabilization devices, tethering lines, and space maintenance tools. (MSC Release 63-108)

- House of Representatives passed (410-1) and sent to the Senate DOD authorization bill for FY 1964 in the amount of $47,082,009,000, second largest in peacetime history. Amount was $1,932,228,000 less than requested by the Administration. Major House changes in R&D requests by the Administration: slash of $100 million (to $43.1 million) in the MMRBM program, because of lack of program definition: crease to $125 million (from $98 million requested) for USAF's X-20 "or Mach 3 aircraft programs," together with a strong endorsement of the X-20; no money voted (none requested) for the RS-70, thus apparently ending the House's long struggle to get DOD to increase the scope of the RS-70 program. (*NYT*, 6/27/63, 1; *M&R*, 7/1/63, 34)

- Postmaster General J. Edward Day, speaking at a luncheon in Washington honoring the 25th anniversary of the CAB, said: "Today, the 11 domestic trunkline carriers serve more than 600 points over routes in excess of 185,000 miles in length. Twenty-four smaller carriers provide service to some 900 smaller communities over routes in excess of 88,000 miles. Three carriers provide domestic all-cargo service to 38 points over routes in excess of 14,000 miles and another 15 carriers are licensed to provide supplemental services. Fifteen carriers are authorized to provide service to a total of 428 points outside the United States over routes in excess of 320,000 miles." (*CR*, 6/27/63, A4098)

- Charles Michels, LRC scientist, described Lewis work with a "coaxial plasma-gun," being studied as a method to propel future interplanetary spacecraft. Michels spoke before the American Physical Society in Cleveland. (LRC Release 63-44, Lewis Chronology, 5)

June 26: USAF launched Thor-Agena D from Vandenberg AFB, Calif., with unidentified satellite and a hitchhiker radiation monitoring satellite that separated in orbit on July 1, 1963. (UPI, *Wash. Post*, 6/27/63; *Pres. Rpt. on Space*, 1/27/64)

- FAA proposed new rule of certification of manned free balloons, establishing minimum airworthiness requirements. (FAA Release T 63-33)
- Reorganization of NASA Marshall Space Flight Center's Michoud Operations was announced by Michoud General Manager Dr. George N. Constan. (*Marshall Star*, 6/26/63, 7)

June 27: X-15 No. 3 piloted by Maj. Robert A. Rushworth (USAF) to 285,000-ft. altitude and 3,425-mph speed (mach 4.89) in 13-min. flight. X-15 was launched from B-52 mother ship over Delamar Lake, Nev., its engine burning 80 sec. Maj. Rushworth qualified for military astronaut wings by piloting the rocket research aircraft to altitude higher than 50 mi. Purpose of flight was to familiarize 38-year-old Major Rushworth with comparatively high-altitude flight. (FRC Release 14-63; NASA Release 63-142)

- About ⅓ of the U.S.—1,072,000 sq. mi.—was brought under FAA's positive separation control system for high-speed aircraft flying on instruments and over 24,000-ft. altitude. Latest addition was 285,000 sq. mi. of the Midwest when new Air Traffic Control Centers were activitated at Kansas City and Denver. Most of the U.S. was scheduled to be under the system by the end of 1963. (FAA Release 63-61)
- NASA announced three personnel appointments to Headquarters: Earl D. Hilburn, Vice President and General Manager of Curtiss-Wright Corp. Electronics Div., as Deputy Associate Administrator (for non-manned space flight centers); Robert F. Garbarini, Chief Engineer of Sperry Gyroscope Co. Air Armament Div., as Director of NASA Office of Applications; and Robert W. Long, President of Long Construction Co., as consultant on construction matters to Dr. Robert C. Seamans, Jr., NASA Associate Administrator. (NASA Release 63-141)
- Sen. Ernest Gruening (D.-Alaska) criticized NASA on the Senate floor for its indifference to training and use of women astronauts in the manned space program. Also he inserted in the *Congressional Record* a recent article in *Life* by Clare Boothe Luce, in which she said: "The astronaut of today is the world's most prestigious idol. Once launched into space he holds in his hands something far more costly and precious than the millions of dollars' worth of equipment in his capsule; he holds the prestige and the honor of his country . . . . But the astronaut is also something else: he is the symbol of the way of life of his nation.

"In entrusting a 26-year-old girl with a cosmonaut mission, the Soviet Union has given its women unmistakable proof that it believes them to possess these same virtues. The flight of Valentina Tereshkova is, consequently, symbolic of the emancipation of the

Communist woman. It symbolizes to Russian women that they actively share (not passively bask, like American women) in the glory of conquering space." (*CR*, 6/27/63, 11228-30)

June 27: USAF launched Minuteman ICBM from AMR in 4,000-mi. flight down Atlantic Missile Range in successful test. (DOD Release 929-63)

- Centaur space vehicle arrived at Lewis to begin series of environmental tests in the center's space power chamber. (LRC Release 63-49, Lewis Chronology, 6)
- Senate unanimously confirmed Gen. Curtis E. LeMay (USAF) to be Chief of Staff, USAF, for a one-year term, beginning July 1, 1963. (*CR*, 6/27/63, 11204-5)
- Senate unanimously passed a joint resolution authorizing a special medal to be struck in honor of Maj. Gen. Benjamin D. Foulois (Ret.), one of the first military pilots in the U.S., chief of the Air Service of the American Expeditionary Force in World War I, and Chief of the Air Corps from 1931 to 1935. (*CR*, 6/27/63, 11203)

June 28: NASA concurred in Grumman Aircraft Engineering Corp. selection of Radio Corp. of America as subcontractor for Lunar Excursion Module (LEM) electronic subsystems and engineering support. (NASA Release 63-143)

- Rocket-boost testing of the Apollo spacecraft would begin during the summer of 1963 from White Sands Missile Range, N.M., NASA announced. The first flight tests would see dummy Apollo capsules launched to relatively low altitudes by the Little Joe II solid rocket to subject sections and the assembled vehicle to maximum combinations of g forces and aerodynamic resistance that would be encountered in the launch phase of normal flights. (*NYT*, 6/29/63, 35)
- Mercury astronaut M. Scott Carpenter, accompanied by NASA astronomer Dr. Jocelyn R. Gill, would observe solar eclipse over central Canada on July 20 from a specially equipped DC-8 jet, flying above most of atmospheric haze at 42,000 ft. During the flight Dr. Gill would point out to Astronaut Carpenter various scientific phenomena likely to be encountered in future manned space flights, and Carpenter would attempt to photograph airglow caused by zodiacal light. Scientists from NASA Ames Research Center also would be onboard to photograph rays of the sun's corona, seeking particular detail in sun's polar regions to trace solar magnetic fields. Flight would be sponsored by Douglas Aircraft Co., National Geographic Society, and 11 other organizations. (*NYT*, 6/28/63, 1; NASA Release 63-148)
- Astronaut L. Gordon Cooper arrived in home town of Shawnee, Okla., for weekend celebrations in his honor, his first visit there since Project Mercury MA-9 space flight (May 15-16). (AP, *Wash. Post*, 6/29/63)
- Speaking at Third International Technical-Scientific Meeting on Space at the 10th International Congress on Electronics, Nuclear Energy, Radio, Television, and Cinema, in Rome, NASA Director of Space Vehicles (OART) Milton B. Ames, Jr., reviewed content and objectives of U.S. space program, describing NASA's space research program: "[NASA] was created to carry out a program

to explore space in the best interests of men everywhere. The entire world is affected, directly or indirectly, by this effort to investigate, explore, and make use of outer space. This tremendous activity affects economy, science, technology, education, and human welfare . . . ." (Text)

June 28: USAF successfully launched a Minuteman ICBM from Vandenberg AFB, Calif., the eighth successful Minuteman launch from the West Coast base. (*NYT*, 6/29/63, 8)

- Dr. Leland J. Haworth was sworn in as the Director of the National Science Foundation, succeeding Dr. Alan T. Waterman, who was retiring. Dr. Haworth is a leading authority on high-energy physics, a former director of Brookhaven National Laboratory, and AEC Commissioner. (*NYT*, 6/29/63, 8)
- Editorializing on cuts made by the House Committee on Science and Astronautics in the NASA FY 64 appropriations, *New York Times* said: "Now for the first time it is taking a closer look at NASA's activities and is making a belated attempt to reduce the waste and duplication uncovered there.

  "This approach should have been utilized all along. Instead, the committee's largesse and laxity encouraged NASA to act as if there were no limit to what could be spent in the skies." (*NYT*, 6/28/63, 28)
- USAF launched unidentified satellite for geophysics research purposes from NASA Wallops Station, Va., using Scout booster. (AP, *Wash. Post*, 7/7/63; GSFC *Sat. Sit. Rept.*)
- Aerobee 150 sounding rocket launched to 126 mi. altitude from White Sands Missile Range, N.M. Payload contained experiments to study the solar corona and extreme ultraviolet radiation—studies which would later apply to the Oso-B (S-17) experiment. (NASA Rpt. of S. Rkt. Launching, 7/10/63)

June 29: Dr. Robert R. Gilruth, director of the Manned Spacecraft Center, was the principal speaker at the Space Age Symposium held in Shawnee, Okla., in conjunction with the city's homecoming celebration for Astronaut L. Gordon Cooper. (*Space News Roundup*, 1/8/64, 2)

- USAF launched unidentified satellite using Thor-Agena launch vehicle from Vandenberg AFB, Calif. (AP, Wash. *Sun. Star*, 6/30/63, 4)

June 30: During FY 1963 NASA awarded 8% of its total contract value to small business, lower than in previous years: FY 1960, 17%; FY 1961, 15%; FY 1962, 12%. (*M&R*, 1/29/64, 9)

- Based on net value of awards, NASA Marshall Space Flight Center awarded almost one third of NASA's FY 1963 contracts to industry ($949.8 million). Second highest was Manned Spacecraft Center ($737.2 million), followed by Western Operations Office ($412.3 million) and Goddard Space Flight Center ($303.5 million). (*M&R*, 1/20/64, 9)
- Space Systems Division (SSD) recorded its 200th Thor launch. The figure includes the use of the Thor as a space booster—Thor-Agena, Thor-Delta, Thor-Able, and Thor-Able-Star—and as an intermediate range ballistic missile (IRBM). (AFSC Release #312-R-114, Atch #2, 1/1/64, 2)

June 30: Estimated 10,000–15,000 persons visited NASA Marshall Space Flight Center on "Family Day." (*Huntsville Times,* 7/1/63)

During June: Dr. Edward C. Welsh, Executive Secretary of the National Aeronautics and Space Council, writing in *Ordnance,* said: "It should be unnecessary to point out that the space race is between the United States and Soviet Russia, not between NASA on the one hand and the Department of Defense or any of its services on the other. Yet, I believe the point needs to be made.

"We are trying to build national competence and international good will and cooperation through our space program, not agency competence or intragovernmental rivalries. We are trying to build an effective image of a nation devoted to peace, to the improvement of mankind's well-being, and to the furthering of man's abilities and knowledge.

"To do so we must be able to discourage aggression, which is the antithesis of our objectives. To fail to develop strength for peace through competence in space would, in a sense, be to disarm by default." (*Ordnance,* May–June 1963, 658)

- NASA announced that it would recruit 10 to 15 new astronaut trainees this summer. (MSC *Space News Roundup,* 1/8/64, 2)
- Study of Soviet astronautics from 1957 through 1962 (RM–3595–PR) by F. J. Krieger of the RAND Corp., under USAF contract, concluded that the Soviet space program is an integral part of the Soviet military program. Krieger listed four main categories of Soviet effort: (1) earth-orbiting man in space; (2) lunar scientific exploration; (3) interplanetary exploration; and (4) Cosmos earth satellites, paramilitary in nature.

  Cosmos satellites have come in two sizes—large ones similar to the Vostoks, launched from Tyuratam cosmodrome east of the Aral Sea, orbiting at around 65°, and brought back to earth; and smaller ones launched with smaller rockets from Kapustin Yar cosmodrome north of the Caspian Sea, orbiting at about 49°, and not recovered.

  Krieger counted 10 Soviet attempts at interplanetary flight, of which 2 have been partially successful (one passing Venus, another Mars, but both with communications failure) and six have failed to get out of earth orbit. (*M&R,* 6/10/63, 16)
- U.S.S.R. would use earth-orbit rendezvous for an early circumlunar manned flight with present-generation hardware, according to article in *Pravda* by "chief designer" of Vostok spacecraft. Author indicated that the tandem flight of L/Col. Valery Bykovsky and Jun. Lt. Valentina Tereshkova was successful preliminary to such flight, which he said would last 8 to 12 days. Re-entry, he indicated, was the most difficult part of the moon flight. (*M&R,* 7/1/63, 29)
- Bill (H.R. 6866) to provide each House of Congress with a three-man scientific staff was commented on approvingly by *Science,* publication of the AAAS: ". . . Congress is showing interest in sound scientific dissent against official policy set by the executive agencies—the lunar landing program is an example—and is looking increasingly to the scientific community for help in seeing the other side of the question." (*CR,* 6/26/63, A4064)

During June: MSFC officially changed the name of its Mississippi Test Facility to Mississippi Test Operations (MTO). (MSFC Hist. Off., MHM-7, Draft of MTO Chapter, 2)

- Details of plan for $250,000,000 research center near Pittsburgh disclosed by Dr. Edward Litchfield, Univ. of Pittsburgh chancellor and Oakland Corp. board chairman. Oakland Corp. was specially formed to undertake development of center, which would include nuclear reactor and instrumentation complex, computer center, and data bank, in addition to theaters, school, restaurant, and Carnegie Museum. (*NYT*, 6/23/63, 39)

- Editorial in *Industrial Research* on "Myopia in Space": "Contained in the concept of space exploitation is at once an economic substitute for war, an opportunity for peace on earth, an answer to overpopulation, and, more important, an unprecedented expansion of everything we know! The exploration of space and the planets presents a future so wonderful as to make its accomplishment imperative at almost any cost. It is the next logical move in man's quest for new knowledge and a better way of life." (*CR*, 6/27/63, A4105)

- Dr. Eugene Fubini succeeded John H. Rubel as Deputy Director, DOD Research and Engineering. Rubel resigned to return to private industry. (*Av. Wk.*, 6/10/63, 25)

- Patent #3,093,346 was awarded to seven engineers for design of Mercury manned spacecraft. (Jones, *NYT*, 6/15/63, 26)

- In *Interstellar Communication*, anthology of articles on man's search for intelligent extraterrestrial life published by W. A. Benjamin, Inc., Dr. Freeman J. Dyson proposed that highly advanced civilization elsewhere in Milky Way galaxy may be using gravity machines instead of sunlight as principal power source. If a civilization on planet which orbits twin stars fired vehicle from planet toward twin stars so as to circle approaching star, vehicle would return with far more energy than initially; vehicle could then be recovered in manner exploiting the energy. Dr. Dyson is theoretical physicist at Institute for Advanced Study, Princeton, N.J. (Sullivan, *NYT*, 6/9/63)

- Engineer Nathan Price described in *Science Digest* his Air Space Transportation (ASTRA) system proposed for the 1970's: transcontinental passenger rockets would be vertically launched on predetermined courses from vertiports in the middle of cities. (*NYT*, 6/22/63)

- A working experimental model of a one-man space scooter was successfully flight-tested at NAA's Space and Information Division. The jet-powered platform was designed by Jack Bell, director of Lunar and Planetary Systems, who made the first flight, and John W. Sandford, research specialist in Advanced Systems. According to Bell, "precision control of platforms can be learned in a few minutes' flight time." The vehicle hovers, rotates, or moves in any direction desired. An extremely versatile piece of equipment, the lunar scooter could permit moon explorers to traverse flat, rough terrain where walking is difficult and where surface vehicles would stall; as a "crater hopper," it would permit a space-suited moon explorer to span or descend into fissures or crevasses; as a "lunescape climber," it would serve as a vehicle to scale steep cliffs. As a "shuttle bus," it would transport crew members between orbiting space vehicles, or it might be used to support man in assembling advanced space stations in earth orbit. Finally, its potential for use in military maneuvers on earth is considerable. (S&ID *Skywriter*, 6/28/63)

JULY 1963

July 1: NASA announced selection of Boeing Aircraft Co. to negotiate contract for four-month study of lunar base concept. Study would be first phase of broad lunar base study program to determine feasibility of such a project following Project Apollo. (NASA Release 63-145)

- NASA Administrator James E. Webb, quoted in press interview, expressed his hope that Congress would continue its record of consistently taking "the necessary action to support the space program . . . As the space program builds up toward pre-eminence for the United States, a position we are rapidly approaching, the space agency is endeavoring to obtain the maximum space value for each dollar it spends and has made an excellent record in this regard . . . . [But] the successful and expeditious carrying out of the programs recommended in the President's budget will require the $5,712,000,000 which he recommended . . . ." (Hines, Wash. *Eve. Star*, 7/1/63)

- Letter from RCA Chairman David Sarnoff to Sen. Warren Magnuson, Commerce Committee Chairman, and Rep. Oren Harris, Interstate and Foreign Commerce Committee Chairman, made public. Repeating his view that commercial communications carriers be consolidated under Communications Satellite Corp., Sarnoff charged that ITT proposal for separate, competing carriers (advocated in June 18 letter to Congress) would "turn back the clock" and be a "disservice to the public and an exercise in futility." He urged Congress undertake public hearings aimed at providing "powerful impetus" to search for unified U.S. communications policy. (*Wall Street Journal*, 7/1/63)

- Reported that DOD cost study of Titan III launch vehicle concluded total development cost of Titan III "exclusive of mission payload adaptations" would be $808.3 million. Study concluded the military space booster would repay its development costs over three or four years of operation—conclusion based on predictable and expectable payload launchings within booster's capability. (*Space Bus. Daily*, 7/1/63)

- International Academy of Astronautics announced 1963 Daniel and Florence Guggenheim International Astronautics Award would be awarded to Prof. Marcel Nicolet, Director of Centre National de Recherches de l'Espace in Belgium. Prof. Nicolet's achievements in aeronomy and planetary atmospheres included that of accurately predicting existence of earth's helium belt, later verified by space probes. Award would be presented during IAF XIVth Congress in Paris next autumn. (IAA Release No. 15)

- A. O. Tischler, NASA Launch Vehicles and Propulsion Assistant Director (for Propulsion), quoted as saying combustion instability problem of F-1 rocket engine was now "under control," in *Missiles and Rockets*. Installation of mechanical isolator separated oscillations in engine pump from oscillations in fuel system. "Since that time the engine has been tested under the most severe conditions and there has been no evidence of combustion instability." (*M&R*, 7/1/63)

July 1: Hitchhiker radiation monitor satellite on DOD unidentified satellite launched June 27, 1963, with Thor-Agena vehicle, was ejected and fired kick motor to attain higher apogee. Satellite measuring magnetically trapped electrons and protons of all significant energy levels showed relation between solar flares and low-energy particles in solar plasma. No distinct division between inner and outer Van Allen belts was found, but instead a gradual transition. (*Pres. Rpt. on Space, 1963,* 1/27/64)

- USAF Minuteman ICBM "met all its test objectives" in 5,000-mi. flight from silo launch, Cape Canaveral. (UPI, *Wash. Post*, 7/2/63)
- U.S. Army Nike-Zeus antimissile missile performed successfully in development test at White Sands Missile Range, N.M. (*M&R*, 7/8/63, 10)
- Dr. Robert W. Buchheim became USAF Chief Scientist, replacing Dr. Launor F. Carter. Dr. Buchheim was formerly head of RAND Corp., Aero-Astronautics Dept. (DOD Release 690-63)
- The 6555th Aerospace Test Wing, Patrick AFB, Fla., was placed under the command of the Space Systems Division (AFSC), Los Angeles, Calif. The Wing was previously assigned to the Ballistic Systems Division, Norton AFB, Calif. (*A-N-AF Journal and Register*, Jan., 1964)
- Restructuring placed the AF Materials Laboratory, the AF Avionics Laboratory, the AF Aero-Propulsion Laboratory and the AF Flight Dynamics Laboratory at Wright-Patterson AFB, under the operational control of RTD, Bolling AFB. (*AFSC Operational Highlights*, 12)
- FBM program had at this date as its objective the deployment of additional submarines carrying the Polaris A-2 missiles and the deployment in 1964 of FBM submarines carrying Polaris A-3 missiles. Thirty-five SSBN's and four tenders were authorized with long leadtime items authorized for an additional six SSBN's. (*Polaris Chronology*, 1955-63)

July 2: 50-lb. payload of ionospheric measuring instruments launched with Argo D-4 sounding rocket from Wallops Station, Va., into orbital path of ALOUETTE satellite. Preliminary data indicated measurements were made in upper ionosphere within two minutes of soundings taken from ALOUETTE. Payload reached peak altitude of 590 mi. Purpose of experiment was to obtain measurements of ion and electron temperatures and densities; data from payload instruments would be compared with similar data transmitted simultaneously by ALOUETTE. Analysis of data later confirmed that ALOUETTE I data were valid. (*Pres. Rpt. on Space, 1963*, 1/27/64, 128; Wallops Release 63-63)

- NASA announced selection of Genisco, Inc., to negotiate development, fabrication, and assembly contract for man-carrying motion generator for research into guidance and psycho-physiological problems of space flight. Motion generator would be major component of NASA Ames Research Center's Space Guidance Research Facility, expected to be operational in about two years. (NASA Release 63–146; Ames Release 63–26)

July 2: JPL spokesman said tests showed deterioration of washer could have caused short circuit and overheating in RANGER V lunar probe, which failed to generate power after launch Oct. 18, 1962, and passed within 450 mi. of moon Oct. 21. (Wash. *Eve. Star,* 7/2/63)

- In his public statement to President Kennedy at the Vatican, Pope Paul VI said: "These past few years have been impressive developments in the exploration of space to which the United States has made notable contributions.

  "May these undertakings take on a meaning of homage rendered to God, Creator and Supreme Law Maker. Because they augur so much for the benefit of mankind, may they be indicative of true and peaceful progress which would bring men together in a closer relationship of universal brotherhood . . . ." (Text, *NYT,* 7/3/63, 3)

- Soviet Premier Khrushchev addressed East Berlin rally, and said: ". . . the Soviet Government expresses its willingness to conclude an agreement banning nuclear tests in the atmosphere, in outer space and under water . . . ." (*NYT,* 7/3/63)

- Dr. Seth B. Nicholson, leading solar astronomer, died at 72 in Los Angeles. He had been staff member at Mt. Wilson and Palomar Observatories for 42 years. (AP, *NYT,* 7/3/63)

July 3: With President Kennedy's return to Washington from Europe, NASA communications satellite RELAY I marked end of its busiest programing period. RELAY was "booked solid" during past weeks to cover President's trip, death of Pope John XXIII, and election of Pope Paul VI. During its six months of operation, RELAY I had been used for 85 public communications demonstrations, including transmission of television, voice, radio-photo, and teletype. (NASA Release 63–144)

- Selection of Republic Aviation Corp. for negotiation of Advanced Orbiting Solar Observatory (Aoso) contract announced by NASA. Under $5.5 million contract, Republic would complete Phase I of Aoso development, including detail design, systems engineering, reliability assessment, limited hardware development of critical systems and components, and trade-off analysis. (NASA Release 63–147)

- NASA Manned Spacecraft Center announced ejection seat escape system for Gemini two-man spacecraft had successfully undergone first high-speed rocket-sled test, at Naval Ordnance Test Station, China Lake, Calif. Both dummy astronauts were safely recovered after ejection at nearly 600 mph from boilerplate spacecraft, simulating emergency ejection during boost phase of Gemini mission. Tests simulating ejection before launch were also being conducted, consisting of firing ejection seats from 150-ft. tower and parachuting dummy astronauts to safe landing. (MSC Release 63–111)

July 3: Addressing First World Conference on World Peace Through Law in Athens, NASA Communications Systems Director Leonard Jaffe described role of satellites in communications and said: ". . . Today, approximately 300 telephone channels are available across the North Atlantic.

"Notwithstanding these recent and radical improvements in trans-Atlantic telephone service, we can predict the need for still more channels. Every time communication channels have increased, they have been loaded to capacity almost at once—new means for communication have never caused the abandonment of the old ones. During 1960 there were more than three million overseas telephone calls, and reliable estimates indicate a sevenfold increase in overseas calls during the next decade. Such an increase in traffic will overtax the current and planned undersea cables, and high frequency radio facilities. And our current communication facilities have *no* capability for transmitting television. . . .

"Communications satellites should soon be providing a substantial increase in our capabilities for world-wide high quality common carrier communications, at least over those routes where the volume of traffic exceeds present or forecast capabilities, and is also great enough to justify the necessarily large investment in a communications satellite system. Eventually, we hope that communications satellite technology will be improved to the point where service can also be provided to areas not served by other means, either because of their isolation or because the traffic volume does not justify an investment in more conventional communication means.

"The objective of the United States in this area was voiced by President Kennedy in a public statement, not too long ago, in which he said: 'There is no more important field at the present time than communications, and we must grasp the advantages presented to us by the communications satellite to use this medium wisely and effectively to insure greater understanding among the peoples of the world.'" (Text)

- Bochum Institute for Satellite and Space Research, Germany, announced reception of 15-min. radio transmission on frequencies normally used by Soviet spacecraft. Commercial radio station in Paris broadcast that U.S.S.R. had placed a man into space but did not elaborate. Soviet news agency Tass made no mention of space launching. (UPI, *Wash. Post*, 7/4/63)
- Effective date of appointment of Dr. Eugene G. Fubini as Assistant Secretary of Defense, with specific duty of serving as Deputy Director of Defense Research and Engineering. (DOD Release 7/8/63)
- National Aeronautic Association President Martin M. Decker announced five Americans would receive 1962 awards from Fédération Aeronautique Internationale: Paul F. Bikle, Director of NASA Flight Research Center, Lilienthal Medal for his record-making glider flights; Donald L. Piccard of Raven Industries, Montgolfier Award for best international balloon performance in 1962; and (jointly) Mrs. Grace M. Harris of Kansas City Aero

Club, Philip S. Hopkins, National Air Museum Director, and W. W. Millikan, Northrop Corp. Washington representative, the FAI Paul Tissandier Diploma for outstanding service to aviation in general and sporting aviation in particular. (NAA Release)

July 3: First Minuteman missile wing declared operational in ceremony turning over last of the Montana sites to Strategic Air Command at Malmstrom AFB, Great Falls, Mont. (DOD Release 952-63)

- USAF launched Atlas ICBM from Vandenberg AFB in successful routine training launch. (DOD Release 987-63)
- Polaris A-3 missile launched from land pad on successful 1,500-mi. flight down Atlantic Missile Range. (AP, Wash. *Eve. Star*, 7/4/63; *M&R*, 7/15/63, 10)
- Army's Nike-Zeus antimissile missile performed successfully in White Sands Missile Range test studying precise control of the missile by ground equipment under "severe" conditions. (DOD Release 954-63)

July 4: NASA Marshall Space Flight Center announced study contract award to Space Technology Laboratories, with objective of identifying and defining essential design requirements for an operational nuclear space propulsion system in the 1970's. Study would be based on Saturn V or Nova-class first-stage booster. MSFC also let conceptual design studies to Douglas Aircraft Co. and General Dynamics/Astronautics for parallel studies of launch vehicles having chemical-propulsion first stages and nuclear-propulsion upper stages, to be used in mid-1970's. Chemical-nuclear rocket configuration would have one-million-lb. orbital capability. (*Huntsville Times*, 7/4/63; *Space Bus. Daily*, 7/8/63, 33)

- NASA Manned Spacecraft Center announced two types of orbital manned space stations would be studied to compare concepts for 24-man operational station—one concept by Lockheed Co. California Div. and the other by Douglas Aircraft Missiles and Space Systems Div. Each station would be designed to remain in orbit at 200- to 300-mi. altitude for about three years, with resupply and crew changes every three months. (*L.A. Times, Wash. Post*, 7/6/63)

July 5: Minuteman ICBM fired from Vandenberg AFB down the Pacific Missile Range in routine training launch. (DOD Release 987-63)

July 6: In the wake of Washington reports that Lt. Col. John H. Glenn might be candidate for U.S. Senate from Ohio, the NASA astronaut said in *Houston Post* interview that his plans were to remain with NASA and that he had given no thought to a political career. (*Houston Post*, 7/7/63)

- In study of 26 U.S. colleges and universities for Carnegie Foundation for Advancement of Teaching, the institutions unanimously concluded that Federal aid to education was "highly beneficial." Study reported that 28 Federal agencies supplied funds to the 26 institutions in period covered (total of 42 agencies support U.S. colleges and universities at rate of $1,760,000,000 per year). Most influential area of Federal support was research; without Federal funds for research "facilities in many instances would shrink. Many research efforts would have to be abandoned completely. Others would be sharply curtailed." (Amer. Council on Educ., *Educational Record* [special issue "Partners in Search of Policies"], 7/6/63; *NYT*, 7/7/63, 28)

July 6: Eight Soviet women parachutists, led by T. Voinova, established new world record in group jumping for accuracy of landing from 600-meter (1,968.5-ft.) height. Average deviation from center of circle was 22.15 ft.—almost 200% better than previous record. (Tass, *Krasnaya Zvezda*, 7/7/63, 3, AFSS-T Trans.)

July 7: TELSTAR II was successfully tracked for 18 min. by ground station northeast of Tokyo. Satellite microwave beacon signal was turned on by ground station at Andover, Maine. No communications tests were conducted, but tracking exercises between Tokyo and Andover were to continue through July 20, in preparation for communications experiments to be conducted "during the next period of mutual visibility, beginning in April 1964." (Bell Lab. Release, 7/9/63 in *SID*, August 1963; Reuters, *NYT*, 7/8/63, 36)

- USAF announced launching of unidentified satellite June 28 from NASA Wallops Station, using Scout booster. Goddard Space Flight Center *Satellite Situation Report* listed the satellite as "research satellite for geophysics." (AP, *Wash. Post*, 7/7/63; GSFC *Sat. Sit. Rpt.*, 6/30/63)

- Dr. E. C. Welsh, Executive Secretary of the National Aeronautics and Space Council, spoke before the Georgetown University Forum:

    "In brief I would list some solid reasons for the moon trip:

    "1. [There is] No other place so near in space where we can test the equipment and the men for future space travel.

    "2. A clear objective gives impetus, order, and efficiency to a program.

    "3. The lunar project requires the development of powerful rocket engines, sophisticated spacecraft, trained astronauts, tracking systems, and capability to protect man against the multiple hazards of space.

    "4. Success in this venture gives prestige essential at negotiating tables.

    "5. It will give an impetus to our standard of living, to education and employment, and to new methods and materials for the productive process.

    "6. There will be substantial defense spin-offs, in addition to the political and economic benefits which also tend to deter aggression. For example, rendezvous technique, life-protective measures, control and guidance systems, improvement in rockets, etc.

    "7. The moon is an excellent platform for mounting astronomical instruments, without atmospheric handicaps.

    "8. The moon can be a relay point for communications and a refueling point for space travels." (Text)

- Astrophysicist Dr. F. Curtis Michel (Capt., USAFR) said in AP interview:

"... NASA knows the requirements for the first space flights much better than I do.

"But I personally believe it can't be very long before we must send scientifically trained men into space. We are going into space because we expect to find the unexpected, and only a man trained in science will know what he's seeing when the unexpected comes along." The 29-yr.-old scientist, with 500 hrs. jet flying time, would begin new job next Sept. as Assistant Prof. of Space Science, Rice Univ., in Houston. Rice President Dr. Kenneth Pitzer had described Dr. Michel as "a possible candidate as scientist-astronaut," adding he was one of few men in U.S. with such qualifications. (AP, N.Y. Herald Trib., 7/7/63)

July 7: General Electric Co. said four men would spend one month inside full-scale model of space station at GE's Space Technology Center, in test of man's reactions and performance in simulated space-flight conditions. Experiment would begin in September. (AP, *NYT*, 7/8/63, 25)

- Brig. Gen. Frank Purdy Lahm (U.S. Army, Ret.), second Army pilot officially to fly Army's first airplane in 1909, died at 85. Taught to fly by Wilbur Wright, he served during World War I as air chief of Second Army in France, later became first commander of first U.S. Air Corps Training Center, at Randolph Field, Tex. (AP, *NYT*, 7/9/63, 31)

July 8: Field Enterprises Educational Corp. announced it was withdrawing its $3.2 million contract offer to 16 U.S. astronauts for their personal stories. "It is with deep regret that we are withdrawing our bid. We knew there would be a vast number of subtle and complex problems to resolve before we could arrive at a viable contract.

"Following our most recent conversations with [NASA] officials, we have reached the conclusion, most reluctantly, that further negotiations would be futile." (UPI, *NYT*, 7/9/63, 1)

- NASA Launch Operations Center awarded two contracts for construction at the Merritt Island site adjacent to Cape Canaveral: Ingalls Iron Works, $11,500,000 contract for three launching towers; and American Bridge Div. of U.S. Steel Corp., $23,534,000 contract for work on Saturn V vertical assembly building. (*NYT*, 7/10/63, 5)

- Dr. Nello Pace, physiology prof. at Univ. of Calif., sworn in as consultant on life science programs to NASA Administrator James E. Webb. (NASA Release)

- Jr. Lt. Valentina Tereshkova's confirmation that she landed by parachute separately from her VOSTOK VI spacecraft was reported in *Aviation Week and Space Technology*. (*Av. Wk.*, 7/8/63, 26)

July 9: 164-lb. payload sent to 127-mi. altitude with Aerobee 150A sounding rocket from NASA Wallops Station in experiment to obtain nighttime electromagnetic noise and propagation data. Included in payload were three sweeping receivers and a broadband receiver of the type to be included in Eogo satellite (Eccentric Orbiting Geophysical Observatory) next year. Preliminary telemetry evaluation indicated all experiment objectives were met. (Wallops Release 63-65)

July 9: X-15 No. 1 flown by Joseph A. Walker (NASA) to 226,400-ft. altitude and 3,631-mph speed (mach 5.07), the X-15 engine burning for 84 sec. Mounted in aircraft's fuselage was traversing probe, retracting every four sec. and measuring air pressures at varying distances from aircraft skin. (FRC Release; FRC X-15 Flight Log; NASA Hq. X-15 Proj. Off.)
- House Committee on Science and Astronautics cut $95.5 million from NASA proposed FY 1964 budget, based upon recommendation of Committee's Subcommittee on Applications and Tracking and Data Acquisition. Today's reduction brought Committee's total reduction to $488.88 million, cutting requested $5.712 billion authorization to $5.223 billion. (*NYT*, 7/10/63, 12; UPI, *Wash. Post*, 7/10/63)
- Dr. Samuel D. Estep, of the faculty of Univ. of Michigan Law School, in a paper "Some International Aspects of Communications Satellite Systems," offered a series of assumptions on the Communications Satellite Corp., among them: no profits for at least 8 or 10 years; interdependent with international communications companies, U.S. Government, and foreign governments; for next decade, will transmit only normal TV relay (not broadcast), and the telephone, and data relay already provided by overseas telephone, radio, and telegraph facilities; for some years the primary use of the system will be in the high-density traffic between U.S. and Europe; receiving facilities and operations in other parts of the world will probably have to be Government subsidized. (*CR*, 7/9/63, 11572–79)
- Advertising executive Harry A. Batten, representing nine NASA astronauts, said that withdrawal of Field Enterprises Education Corp. contract offer for astronauts' personal stories did not mean NASA disapproved this type contract. Rather, he said NASA believed astronauts should conclude such a contract to protect their own and their families' privacy. He predicted astronauts would reach story-contract agreement with some other firm. (AP, *Houston Post*, 7/10/63)
- DOD named Dr. Albert C. Hall, Vice President and General Manager of Martin Co. Space Systems Div., as Deputy Director of Defense Research and Engineering for Space Technology. Post was formerly that of Special Assistant to the Director (Harold Brown), held by Dr. Lawrence L. Kavanau. (DOD Release 969–63; *Av. Wk.*, 7/15/63; *Wash. Eve. Star*, 7/10/63)
- Plum Brook Station, NASA Lewis Research Center, awarded a $450,000 contract for completion of nuclear rocket and dynamic control test stand. The 200-ft.-high stand was to be used for testing second-generation nuclear rocket components. (Lewis Chronology, 3)

July 10: NASA Administrator James E. Webb issued statement to press:

"The record shows that Congress has consistently taken the necessary action to support the space program, and I am confident that this Congress will continue this record. Although the report of the House Committee on Science and Astronautics is a major milestone on the road to approval of the 1964 NASA

Budget, it is still too early in the consideration of this budget by Congress to draw firm conclusions as to what the final result will be. The House subcommittees have given the NASA programs a most careful review, accumulating data from hearings which will total more than 4,000 pages of testimony. Their devotion and attention to details has been impressive, but in my opinion the overall result of their actions is an inadequate level of support for a program that is urgently needed, has achieved a high level of success, and is now giving this nation the promise of early preeminence in all phases of space exploration and use. In the areas reduced, NASA will present a strong case for restoration as the legislation proceeds to enactment.

"After its many details were brought under close scrutiny, and the possibility of postponement of many of these examined, the Committee has endorsed the program and the way it is being carried out but decided to defer approval until subsequent budgets in a number of important areas. My view is that the necessary resources should be authorized this year, rather than have important segments postponed with the consequent introduction of uncertainty at many levels in the government-industry-university team now engaged in this urgent national effort.

"At the reduced level of funding recommended, we will certainly have to slow up or postpone a number of programs and direct a reduction in the scheduled level of effort in the plants of our contractors where 90% of the work of the space program is done. Unless an adequate level of support is restored, momentum already attained with great effort and difficulty will be lost and the stretch-out required will add to the final cost of the work required to carry out the program." (Statement)

July 10: "Town Meeting of the World"—international live telecast via TELSTAR II communications satellite—presented on CBS-TV. The one-hour forum featured former President Dwight D. Eisenhower from Denver; former British Prime Minister Sir Anthony Eden, Earl of Avon, from London; Dr. Heinrich von Brentano, majority leader of West German Bundestag, from Bonn; and Jean Monnet, France's "Father of the Common Market," from Brussels. Program was to have been first live transatlantic viewing, but French government refused permission to use Pleumeur-Bodou ground station needed to relay telecast throughout Europe. (*NYT*, 7/11/63, 3)

- Univ. of Chicago scientists led by Dr. Peter Meyer, physicist of Enrico Fermi Institute for Nuclear Studies, left Chicago for Churchill, Manitoba, where they would conduct experiments testing theory that electrons and positrons reaching earth are created in collisions of atomic nuclei in space. Experiments would involve sending 185-lb. magnet, plus instrumentation and cameras, to 120,000-ft. altitude with 40-story-high balloons, to record and analyze particles in area affected by field near the earth's magnetic pole (where particles can pass with minimal interference). (*Chic. Trib.*, 7/8/63)

- National Airlines President L. B. Maytag, Jr., told National Aerospace Education Council in Miami Beach that mach 2 commercial airliner would be far more economical and utilitarian than mach 3, usable only in transoceanic flight because of altitude requirements. Mach 2 aircraft could make transcontinental U.S. flights faster than mach 3 aircraft, which would have to cruise at 12-mi. altitude in order to prevent sonic boom damage. Mach 2 could be built of aluminum, but mach 3 airliner would have to be built of titanium and steel, Maytag said. (UPI, *Wash. Post*, 7/11/63)

July 10: Camera equipped with special synchronization system making it capable of taking 8 million pictures per sec. was reported in Washington *Star*. Synchronization system was devised by Pvt. 1/C R. Carey of Army Engineer Research and Development Laboratory, Ft. Belvoir, Va. (Henry, Wash. *Eve. Star*, 7/10/63)

- West German Cabinet approved West German participation in two European rocket and space projects—development of three-stage rocket by 7-nation ELDO and development of space probe by 11-nation ESRO. (*NYT* [West. Ed.], 7/11/63)
- Supersonic Transport Advisory Group was dissolved by Federal Aviation Agency following completion of task for which it was formed in Nov. 1961. (FAA Release 63-63)

July 11: NASA Ames Research Center announced seven NASA astronauts were conducting simulated manned spacecraft missions at Ames centrifuge facility. Tests involved subjecting astronauts to space flight stresses they might encounter during future manned space flights. (Ames Release 63-29)

- Senators Warren G. Magnuson and Henry M. Jackson announced three NASA study contracts awarded to Boeing Co.: $199,920 contract to continue studying methods of recovering Saturn first-stage rocket boosters for reuse; $106,000 award to continue studying large clustered solid-propellant rocket motors; and $84,108 contract to study vehicle-integrated rocket powerplant with air augmentations. Contracts were let by NASA Marshall Space Flight Center. (*Seattle Times*, 7/11/63)
- Communications Satellite Corp. issued its first request for contract proposals—proposals for study of multiple access system for communications satellites. (ComSatCorp Release)
- Soviet news agency Tass announced improved types of rocket carriers for spacecraft were launched successfully from Soviet Union to central Pacific during June and July, and area was no longer off-limits for navigation of ships and aircraft. (Tass, *Komsomolskaya Pravda*, 7/11/63, 3, AFSS-T Trans.)
- FAA announced month-long series of tests would evaluate use of helicopters in addition to ground fire-fighting methods to increase survival chances of aircraft passengers in crash and fire situations. Tests would begin in September. (FAA Release 63-39)
- USAF announced Minuteman ICBM was successfully launched from Vandenberg AFB in routine launch. (DOD Release 997-63)

July 12: Dummy S-IV stage for Saturn I rocket arrived at Cape Canaveral from Douglas Aircraft Co. via converted Boeing 377 Stratocruiser, marking first use of aircraft for transporting major NASA rocket components from West Coast. (UPI, *NYT*, 7/14/63, 64)

July 12: Columbia Univ. School of Journalism announced $131,000 award from NASA to evaluate NASA's dissemination of space news and to study problems and competence of news media in reporting and interpreting space news, to determine whether better ways can be found to improve public understanding of space sciences and space exploration. Study would take 3 years, with comparable funds expected for the second and third years.

Commenting on the grant, Sen. Clinton P. Anderson, Chairman of Senate Committee on Aeronautical and Space Sciences, said: "It seems a strange expenditure, especially when we are desperately trying to maintain their budget . . . ."

Rep. George P. Miller, Chairman of House Committee on Science and Astronautics, said: "It sounds like the agency is carrying out the Space Act by trying to find ways of better interpreting to the public the great technological developments in space."

Unnamed NASA spokesman said the grant was "a legitimate study fulfilling our Congressional mandate to keep the public informed." (Toth, *NYT*, 7/13/63, 21)

- NASA announced signing $35,200,018 contract with International Business Machine Corp. for computer complex design, equipment, and associated services for Integrated Mission Control Center at NASA Manned Spacecraft Center. (NASA Release 63–151)
- USAF launched unidentified satellite using Atlas-Agena D launch vehicle from Pt. Arguello, Calif. This was the 100th launch of an Agena space vehicle, the first having been launched on February 28, 1959. (UPI, *Wash. Post*, 7/13/63; *M&R*, 7/22/63, 11; *A-N-AF Journal and Register*, January 1964; *CR*, 2/18/64, 2909)
- U.S. Army announced Nike-Zeus antimissile missile fired from Kwajalein Island had successfully intercepted target from Atlas ICBM fired from California 5,000 mi. over Pacific Ocean. (DOD Release 988–63)

July 13: NASA Director of Electronics and Control Dr. Albert J. Kelley made public two studies conducted by NASA on need for electronics research center and letters from NASA Administrator James E. Webb to Rep. George P. Miller and Sen. Clinton P. Anderson. Letters and studies pointed out that electronics difficulties comprised greatest single cause of failures in spacecraft and rockets during past five years. NASA had no highly developed skill in design of electronic components, and new center was expected to fill that need. (Loory, *N.Y. Herald Trib.*, 7/14/63)

- Federal agencies announced plans to support new center for computer technology and biomedical research in Cambridge, Mass., area. Center would be managed by 12 universities and institutions in New England. $2,800,000 grant from National Institutes of Health was made for first-year support of center, and next year's Government grant of a "comparable" amount would be made by NASA. Government grants would continue for at least seven years and would cover cost of establishing center and operating it for that period. (Toth, *NYT*, 7/14/63, 1)

July 13: Whirlpool Corp. won NASA contract for development of food supply, personal hygiene items, and waste disposal system for Project Gemini manned space flights. (AP, *NYT*, 7/14/63)

July 13-19: "Women in Aviation Week" proclaimed by Maryland's Governor J. Millard Tawes, spotlighting All-Woman Transcontinental Air Race from Bakersfield, Calif., to Atlantic City. (Wash. *Eve. Star*, 7/3/63)

July 14: Nike-Apache sounding rocket carried instrumented payload to 107.5-mi. altitude in flight from NASA Wallops Station to measure electron density and temperature with Langmuir probe and solar radiation with ion chambers and Geiger counter. This flight was control experiment for series of six similar rockets to be launched July 20 during solar eclipse. (NASA Rpt. of Sounding Rkt. Launching)

July 15: Dr. George M. Knauf (Col., USAF Ret.) NASA Acting Director of Space Medicine, was awarded Legion of Merit by USAF Surgeon General Maj. Gen. O. K. Neiss in ceremony at NASA Manned Spacecraft Center. Citation accompanying medal covered Dr. Knauf's contributions to aerospace medicine from Jan. 1, 1962, when he became NASA Deputy Director of Space Medicine on assignment from USAF, to Oct. 31, 1962, when he retired from USAF. He continued in this NASA position in civilian capacity, becoming Acting Director of Space Medicine June 1. (MSC Release 63-118)

- Little Joe II launch vehicle left General Dynamics/Convair plant en route to White Sands Missile Range, N.M., where test launch would be made next month. Purpose of flight would be to qualify Little Joe II launch vehicle for later flights with boilerplate Apollo payload and Apollo spacecraft built to production standards; Little Joe II tests would provide engineering information for use on manned Apollo orbital flights. (MSC Release 63-115)
- Two French Berenice rockets have reached mach 12 speed and 60,000-ft. altitudes at Ile du Levant launch center. (*M&R*, 7/15/63, 9)
- By this deadline, NASA Manned Spacecraft Center received 271 applications for astronaut openings. 71 applicants were military pilots recommended for possible astronaut duty; remaining 200 were civilians, including three women. (*Av. Wk.*, 7/22/63, 325)
- Sen. Wallace F. Bennett (R.-Utah) inserted in *Congressional Record* a letter from physics prof. John H. Gardner, Brigham Young Univ., criticizing "the trend in NASA and in the Department of Defense toward the development of inhouse capability for the performance of basic research and other scientific tasks which can and, under our private enterprise system, should be performed by the private sector of the economy with a secondary role being played by the university . . . .

  "It is far from evident that the creation of new Government laboratories in competition with and at the expense of existing laboratories will help us to achieve our goals in space any faster. Appropriate utilization of the private scientific resources of the country would seem to be faster and more in keeping with our traditions. The statement that NASA has too much money is frequently heard among scientists and it is indicative of the widespread fears and misgivings scientists have about the ominous changes taking place in the kind of role they play in our society. These fears and misgivings may underlie much of the growing opposition to our lunar landing goal . . . ." (Letter, *CR*, 7/15/63, 11884)

July 15: Egyptian army test-fired two ground-to-air antiaircraft missiles in "remarkably successful" maneuver witnessed by President Nasser. Rockets' range was not disclosed, and launch site was unidentified. (AP, Wash. *Eve. Star* 7/16/63)

July 16: TELSTAR II communications satellite went dead during its 450th orbit, and subsequent efforts to reactivate the satellite by radio signal were not successful. Cause of TELSTAR II's failure was not known, but AT&T said that telemetry data had given "no indication that radiation damage has caused the satellite to fail." Efforts to turn on the satellite's communications equipment were continuing. (Osmundsen, *NYT*, 7/18/63; Hines, Wash. *Eve. Star*, 7/18/63)

- House Committee on Science and Astronautics met in executive session and ordered NASA FY 1964 authorization bill (H.R. 7500) favorably reported to the House. (NASA Leg. Act. Rept. II/107)

- Sir Bernard Lovell, director of Jodrell Bank Experimental Station, held press conference after returning from three-week tour of Soviet space tracking observatories. Lovell said that "there is a great deal of discussion in the Soviet Academy as to whether it will ever be worthwhile getting a man on the moon . . . ."

  "I think, at the moment, the Americans are racing themselves . . ." [for manned lunar landing], adding that he favored manned lunar exploration.

  He announced agreement on tracking deep space probes reached between tracking station in Crimea and Jodrell Bank facility, and agreement on three astronomical programs in which Soviet and Jodrell Bank telescopes would be synchronized. (Farnsworth, *NYT*, 7/17/63; AP, *Wash. Post*, 7/17/63)

- Addressing National Rocket Club in Washington, Rep. George P. Miller, Chairman of House Committee on Science and Astronautics, said: "To me, the most important indirect benefit we are gaining [from the space program] is probably the least tangible, the least measurable, and yet the most far reaching in effect insofar as our future national growth is concerned. I am referring to the impact our space program is having upon our educational programs and institutions. The fact that NASA places demands upon and draws from almost every academic discipline has produced a stimulus in universities and colleges that is unprecedented in peacetime . . . .

  "What is important to understand is that scientific research and development is for the first time in our peacetime history being organized within one agency to accomplish a national goal. The achievements of our immediate space objectives are, of course, of paramount importance. But, I have little patience with the critics who cry out that our tax money is being thrown away in the exploration of space, or that going to the moon will result in the neglect of our problems here on earth.

"It should be obvious to those critics, if they take the time for a short historical review, that many years of experience have shown that the indirect payoff in human terms for technical innovation is many times more valuable than the original investment . . . ." (Text)

July 16: Dr. Milton Clauser, former Vice President of Space Technology Laboratories, and Dr. J. P. Ruina, former Director of DOD Advanced Research Projects Agency, joined Communications Satellite Corp. as consultants. (ComSatCorp. Release)

- Five men entered Boeing Co. space chamber, simulating quarters in manned space station or manned lunar base, for 30-day engineering test of integrated life-support system. Designed and built for NASA Office of Advanced Research and Technology, system included all elements of life support necessary for 150 man-day space mission. Associated with life-support equipment were specific crew tests simulating problems of space flight. Members of test crew: R. H. Lowry, Boeing Chief of Bioastronautics; Maj. Edward Westlake (USAF) of Air Force Systems Cmd.; Roger Barnicki, NASA Flight Research Center X-15 personnel equipment specialist; Charles Proctor, Boeing biochemist and food specialist; and Richard Farrell, Boeing psychologist. (NASA Release 63-155)

- Disputing *New York Times* editorial (June 28) which approved House Science and Astronautics Committee's rejecting NASA lunar orbiter for special survey of the moon, Assistant Director of Univ. of Calif. Institute of Geophysics and Planetary Physics Gordon J. F. MacDonald wrote that "no other program funded or contemplated will provide critical data on the moon's gravity field and on the radioactivity of the lunar surface. . . .

  "[NAS Space Science Board] . . . noted that the surveillance of the moon on a planetary scale is of first importance both to the manned landing and to the long-range scientific investigation of the moon . . . .

  "The full value of [Ranger and Surveyor] limited observations can be realized only if they are tied in to a less detailed but broader coverage of the kind that can be provided by the orbiter . . . .

  "The long-range manned and unmanned exploration of the moon will be delayed by failure to fund during the current fiscal year the orbiter program. This failure is another expression of the inability of the scientific community adequately to present their case to Congress." (Letter, 7/1/63, *NYT*, 7/16/63, E8)

- 51 British scientists working in U.S. and 17 in Canada have been offered government appointments in Britain, Parliamentary Secretary for Science Denzil Freeth told House of Commons. British government board interviewed 171 candidates in U.S. and 65 in Canada. (Reuters, *Wash. Post*, 7/17/63)

- Westland Co. announced Bell Aerosystems Co. had obtained Western Hemisphere rights to manufacture Hovercraft, developed by Westland in England. (UPI, *NYT*, 7/17/63)

July 16: LRC Plum Brook reactor was started for first six experiments. One, under contract to the Lockheed Aircraft Corp., would study the effects of neutron bombardment on small samples of various alloys in a cryogenic environment. (LRC Release 63–57, Lewis Chronolgy, 6)
- USAF launched Titan ICBM in routine training test from Vandenberg AFB. (DOD Release 1018–63)
- USAF launched Minuteman missile from Cape Canaveral, the ICBM exploding about five sec. after lift-off. Cause of malfunction was not yet determined. (DOD Release 1018–63)

July 17: President Kennedy, asked in press conference whether reports that U.S.S.R. was not striving for manned lunar landing would affect Project Apollo, said:

"Well, in the first place, we don't know whether the Russians are—what their plans may be. What we are interested in is what their capabilities are. While I have seen the statement of Mr. Lovell [July 16] about what he thinks the Russians are doing, his information is not final. Their capacity is substantial; there is every evidence that they are carrying on a major campaign and diverting greatly needed resources to their space effort. With that in mind, I think that we should continue. It may be that our assumption or the prediction . . . that they are not going to the moon might be wrong a year from now, and are we going to divert ourselves from our effort in an area where the Soviet Union has a lead, is making every effort to maintain that lead, in an area which could affect our National security as well as great peaceful development? I think we ought to go right ahead with our own program and go to the moon before the end of this decade.

"The point of the matter always has been not only of our excitement or interest in being on the moon, but the capacity to dominate space, which would be demonstrated by a moon flight. [This] I believe, is essential to the United States as a leading free world power. That is why I am interested in it and that is why I think we should continue, and I would be not diverted by a newspaper story."

Asked about the possibility of cooperating with U.S.S.R. in a joint lunar mission, he replied:

"We have said before to the Soviet Union that we would be very interested in cooperation. As a matter of fact, finally, after a good many weeks of discussion, an agreement was worked out on an exchange of information with regard to weather, but we have never been able to go into more detail. The kind of cooperative effort which would be required for the Soviet Union and the United States together to go to the moon would require a breaking down of a good many barriers of suspicion and distrust and hostility which exist between the communist world and ourselves.

"There is no evidence as yet that those barriers will come down, though quite obviously we would like to see them come down. Obviously, if the Soviet Union were an open society as we are that kind of cooperation could exist, and I would welcome it. I don't see it as yet, unfortunately." (Text. *Wash. Post*, 7/18/63)

July 17: National Aeronautics and Space Council considered NASA-DOD coordination in Project Gemini and space stations. Vice President Lyndon B. Johnson, NASC Chairman, said after the meeting: "Coordination is a natural and effective enemy of duplication.... I am encouraged by the coordination already being exhibited between these two agencies in these important fields, but continuing attention is required if this country is to fulfill its destiny as the leading space-faring nation." (NASC Release)

- In letter to Chairman Miller of House Committee on Science and Astronautics, NASA Administrator Webb expressed NASA views on H.R. 5171, bill to "authorize the Administrator of the General Services Administration to coordinate and otherwise provide for the economic and efficient purchase, lease, maintenance, operation, and utilization of electronic data processing equipment by Federal departments and agencies."

  Mr. Webb wrote that the "scientific and technical use and management of computers cannot be separated from the responsibility for conducting the national space program. The bill would necessarily have the effect of fragmenting and watering down that responsibility. Accordingly, I see no alternative but to recommend that NASA be exempted from the terms of the bill . . . ." (Letter, *CR*, 7/18/63, 12273)

- NASA announced six-month contract with Space Technology Laboratories, Inc., for study of and recommendations for NASA management of manned space flight program. Study was associated with NASA organizational changes necessitated by completion of Project Mercury and focusing on Projects Gemini and Apollo. Included in study would be ways of improving relationship of NASA centers and industrial contractors involved in manned space flight program. (NASA Release 63-156)

- In roll call vote (245-144), House of Representatives approved $511,000 in Interior Dept. budget for FY 1964 to plan National Air Museum. (Wash. *Eve. Star*, 7/17/63, B1)

July 18: X-15 No. 1 flown by Maj. Robert Rushworth (USAF) to 3,925-mph speed (mach 5.63) and 104,800-ft. altitude, the X-15 engine burning for 84 sec. In test to determine stability of the rocket-powered research aircraft, automatic stability controls were turned off, and in spite of yawing the X-15 remained successfully on course. This was first such test of X-15 without lower tail fin. (FRC Release; FRC X-15 Flight Log; NASA Hq. X-15 Proj. Off.)

- House of Representatives passed H.R. 5171 to authorize automatic data processing coordination by General Services Administration. No amendment was passed to exempt NASA from bill's provisions. (NASA Leg. Act. Rpt. II/109; *CR*, 7/18/63, 12277)

- National Federation of Business and Professional Women's Clubs, convening in Dallas, adopted resolution calling upon NASA to give qualified women equal opportunity for astronaut-training selection. (*Wash. Post*, 7/19/63)

- NASA announced signing $980,488 contract with Pratt and Whitney Aircraft Div. of United Aircraft Corp. to conduct research in use of fluorine/hydrogen and fluorine-oxygen/hydrogen as rocket propellants. Objective of 13-month research project was to demonstrate feasibility of a complete engine system using the two mixtures. (NASA Release 63-158)

July 18: In ceremony at NASA Hq., Francis M. Rogallo and Gertrude S. Rogallo were awarded $35,000 for their invention of flexible-wing paraglider. Other awards: Walter K. Victor and Dr. Eberhardt Rechtin, $5,000 for deep space communication system: William J. Alford, Edward C. Polhemus, and Thomas A. Toll, $3,000 for inventions leading to application of variable-sweep wing to supersonic aircraft; Noah S. Davis and Andrew J. Kubica, $1,500 for decomposition unit; Robert V. Hess, $1,200 for Hall-current plasma accelerator; Curt P. Herold, $1,000 for multiple quick disconnector; and Elden C. Whipple, Jr., $1,000 for method and apparatus for determining orientation of a space vehicle. (NASA AB, Memo, 7/11/63; NASA Release 63-154)

- NASA announced withdrawal of controversial research grant to Columbia Univ. School of Journalism at the University's request because of an apparent misunderstanding of its scope and purpose. NASA had awarded the grant for studying "ways of making scientific advances of the space program more readily available" to news media. (Simons, *Wash. Post*, 7/19/63)
- Polaris A-3 missile launched from Cape Canaveral land pad in successful flight down AMR. (*M&R*, 7/29/63, 28)

July 19: X-15 No. 3 flown by NASA pilot Joseph A. Walker attained record altitude of 347,800 ft. (65.9 mi.) and speed of 3,710 mph (mach 5.50). Flight was planned as 315,000-ft. altitude build-up for pilot Walker in preparation for later 350,000-ft. attempt, but X-15's engine burned 85 sec. instead of planned 83 sec. Nitrogen-filled balloon was to have ejected from X-15's tail during ascent to obtain altitude density measurements, but electrical ejection signal did not function properly. This was 90th flight of X-15. (FRC Release 25-63; FRC X-15 Flight Log; NASA Hq. X-15 Proj. Off.; NASA Release 63-160)

- Aerobee 150A sounding rocket launched from NASA Wallops Station with 230-lb. instrumented payload to measure intensity of light from stars. Equipped with four scanning photoelectric spectrophotometers, payload reached 115-mi. altitude, impacted in Atlantic Ocean 68 mi. downrange. (Wallops Release 63-68)
- NASA announced appointment of James T. Dennison and Dr. Thomas P. Murphy as special assistants to NASA Assistant Administrator for Technology Utilization and Policy Planning, Dr. George L. Simpson. Dennison, formerly Director of Research and Engineering at Dennison Manufacturing Co., was assigned overall responsibilities in technology utilization. Dr. Murphy, formerly staff assistant to NASA Administrator, was assigned responsibilities in policy planning, socio-economic studies, and research projects. (NASA Release 63-159)
- Astronaut John H. Glenn, Jr., repeated in Washington press conference that he was not planning a career in politics. (Hines, Wash. *Eve. Star*, 7/20/63)
- USAF OAR announced formation of Scientific Advisory Group, composed of 12 leading U.S. scientists: Dr. Joseph Kaplan, chairman; Dr. Oliver G. Haywood, Jr.; Prof. Robert J. Havig-

hurst; Prof. Henry Houghton; Dr. John P. Howe; Dr. Mark Kac; Dr. Carl Kaplan; Dr. Nathan L. Krisberg; Dr. Gerard P. Kuiper; Dr. David B. Langmuir; Prof. Leonard Schieff; and Prof. Frederick Seitz. Group would serve as advisory body to OAR Commander, Maj. Gen. Don R. Ostrander. (OAR Release 7-63-4)

July 19: USAF launched three unidentified satellites and a Tetrahedral Research Satellite (TRS) from Pt. Arguello: a single unidentified satellite from a Thor-Agena D vehicle and the other two unidentified satellites with the TRS from an Atlas-Agena B launch vehicle. The 1.5-lb. TRS was to measure solar cell radiation damage. (*Pres. Rpt. on Space, 1963*, 1/27/64; *M&R*, 7/29/63, 28)

- Sen. Barry Goldwater, speaking before Air Force Historical Foundation at Maxwell AFB, criticized U.S. space policies: "By choice of official policy, so far, we are choosing to relegate the military function of space to a secondary position. We have delayed and debated to the point where it must be said that the United States today has no fully defined and effective military space program at all.

    ". . . We have, in fact, policy declarations stating that we will not orbit weapons of mass destruction in space unless forced to do so by the hostile action of others. In short we say that we will not utilize the military potential of space until such time as it may be too late. . . .

    "Just what is this race in space? By our actions we are clearly saying that it's a race for the Moon. We are moonstruck. And, to be sure, the Moon is most romantic . . . . But, while our eyes are fixed upon it, we could lose the Earth or be buried in it . . . ." (Text)

- Mrs. Betty Miller was awarded medal for exceptional service to aviation by FAA Administrator Najeeb E. Halaby, for being first woman to fly alone across the Pacific. 7,400-mi. flight from California to Australia was completed in 54 hrs., 8 min. flying time April 30–May 12. (AP, *NYT*, 7/20/63)

- Gen. Lucius D. Clay (USA, Ret.) named recipient of 1964 John Fritz Medal, citing him for "his distinguished service to the engineering profession, the nation, and to the world." (*NYT*, 7/19/63)

July 20: Eclipse of the sun visible across Canada and Northeastern U.S. NASA joined other scientists and astronomers in scientific studies during the eclipse, with emphasis on ionosphere and on sun's corona.

At Churchill Research Range, USAF OAR facility located at Ft. Churchill, Can., six Nike-Apache sounding rockets equipped with instruments to measure electron density, electron temperature, and solar radiation in ultraviolet and x-ray regions were launched for NASA Goddard Space Flight Center; Aerobee 150 sounding rocket equipped with spectrophotometric instruments to measure absolute intensity of spectral features in ultraviolet region was launched for Johns Hopkins Univ.; and Canadian Black Brant sounding rocket with instruments to measure variations in D and E layers of ionosphere was launched for USAF Cambridge Research Laboratories. GSFC and AFCRL scientists said preliminary results indicated collected data confirmed previous predictions of composition of the ionosphere.

At White Sands Missile Range, GSFC project for eclipse was Aerobee 150 sounding rocket equipped with coronagraph to photograph sun's corona, ultraviolet spectrometer aimed at sun's center, and camera to photograph Lyman-Alpha rays.

At Wallops Station, GSFC project was measurement of electron and neutral particle temperatures in the ionosphere using instrumented payload launched by Aerobee 300A sounding rocket to 207-mi. altitude.

At Pleasant Pond, Me., GSFC team photographed eclipse with specially made instrument for photographing stars and comets near the sun.

DC-8 flying observatory flew north from Edmonton, Can., to meet eclipse at Great Slave Lake, following path of moon and flying above much of atmospheric-haze layer. Sponsored by Douglas Aircraft Co. and National Geographic Society with 11 other participating organizations, expedition included NASA Astronaut M. Scott Carpenter and NASA astronomer Dr. Jocelyn R. Gill studying various scientific details of the phenomenon. Also onboard were Sheldon Smith and Ray Torrey of NASA Ames Research Center, photographing rays of sun's corona. (NASA Release 63-148; MSC Release 63-113; *Goddard News*, 7/15/63, 1, 8; OAR *Research Review*, Vol. II, No. 11; DOD Release 978-63; Wallops Release 63-70; *M&R*, 7/29/63, 24)

July 20: Five men locked in Boeing Co. test chamber on simulated 30-day space mission were forced to cut the test short after 104 hrs. because hole developed in reactor tank (comparable to septic tank). (AP, *Cleveland Plain Dealer*, 7/21/63)

- NASA launched Scout rocket with flight experiment to test re-entry heat shield material for spacecraft, but Scout veered off course and had to be destroyed before it left vicinity of Wallops Station launch area. Cause of malfunction was being investigated. (Wallops Release 63-69)

- Solid-propellant rocket motor for Titan III test-fired for first time at United Technology Center, burning for about two min. and producing more than a million lbs. of thrust. Two such motors would be used in first stage of Titan III. (*M&R*, 7/29/63, 15)

- Cathode of experimental ion engine set endurance record of 1660 hours operation. Electric propulsion engineers at LRC estimate they reached 16 per cent of the cathode life required for a round-trip Mars mission. (Lewis Chronology, 7)

July 22: U.S. Navy announced it had successfully permanently stabilized an orbiting satellite through use of 100-ft. boom. Difference in gravitational forces acting on satellite and on end of the boom aligned satellite so that its face will be permanently pointed toward earth. As satellite was tumbling through space with its face toward earth, boom was released from canister, causing entire assembly to rock through 40° arc. Weight attached to boom by 40-ft. spring bobbed yo-yo fashion, slowing oscillation of boom and satellite until nearly all sway was removed. Satellite would remain stabilized indefinitely. Navy announcement said this was first successful stabilization of satellite without fueled mechanical controls. (DOD Release 1034-36)

July 22: Sen. Mike Mansfield (D.-Mont.) paid tribute to X-15 pilot Joseph A. Walker and his record flight of July 19, and said that "there can be no question that Mr. Walker and all the members of the X-15 team are pioneers of the space age. The solid contributions of the X-15 program have been many and valuable. They include information on aerodynamic heating, the behavior of the boundary layer, the effect of noise levels on aircraft structures, measurement of friction drag on aircraft skin, new control systems, new piloting techniques, new landing gear designs and new ways of measuring landing gear loads, better instrument displays, and many other improvements to 'performance, efficiency, or safety of air and space vehicles.' In addition, during the current series of tests, the X-15 is being used to carry scientific experiments into space. One of these experiments will use a camera which may give us the first meaningful photographs of stars taken from outside the earth's atmosphere. Speed and altitude records may be the incidental contributions of the remarkable X-15 program. More significant in the long run is the information and experience it has provided and continues to provide about this strange new sea of space upon which man has embarked . . . ." (*CR*, 7/22/63, 12354)

- NASA announced extension of contract with Research Triangle Institute, Durham, N.C., for abstracts and evaluation of literature on reliability and quality assurance. (NASA Release 63-161)
- NASA Manned Spacecraft Center announced $1,000 incentive award by NASA Inventions and Contributions Board to Matthew I. Radnofsky and Glenn A. Shewmake of MSC Crew Systems Div. for design of one-man life raft used in Project Mercury landings. (MSC Release 63-119)

July 23: 238-lb. instrumented payload to measure intensity of light from stars was launched with Aerobee 150A sounding rocket from NASA Wallops Station to altitude of 110 mi. Telemetry data obtained during flight would be compared with data from similar experiment conducted July 19. (Wallops Release 63-71)

- Dr. George E. Mueller, Vice President for R&D of Space Technology Laboratories, named to new position of NASA Deputy Associate Administrator for Manned Space Flight, succeeding D. Brainerd Holmes who would resign Sept. 1 as Director of Office of Manned Space Flight and Deputy Associate Administrator for Manned Space Flight Centers. (NASA Release 63-162)
- First test-firing of twin-engine system which would power Titan III upper stage, engines producing 16,000 lbs. of thrust in 284-sec. test at Aerojet-General Corp., Sacramento. Engines were started and stopped three times in the firing. (*M&R*, 7/29/63, 19)
- Addressing commencement of Parks Air College of Univ. of St. Louis, NASA Director of Advanced Research and Technology Dr. Raymond L. Bisplinghoff discussed research and development

required to build an "economically attractive supersonic transport airplane." He pointed out that this nation's "total manned aircraft flight time beyond Mach number 2.5 is less than four hours, all of which has been amassed by NASA's Flight Research Center at Edwards, California.

". . . we are undertaking a challenging and difficult development, but one that we can handle by a reasonable stretching of today's technology." (Text)

July 23: Vice President Lyndon B. Johnson, addressing Governors' Conference at Miami Beach, said:

"In the world of today, we must recognize that the winds and the waves are with us. This is the time for us to steer straight and true toward the horizons we have for so long hoped to reach.

"But we should also recognize that favoring winds are rising at home. Our record in space is a prime example. We started late. We moved too slowly at first.

"But the record shows now that where the Soviet Union has successfully launched 43 payloads into orbit, the United States has launched 144.

"Our ratio of sucesses to failures in launchings prior to 1961 was only 1 to 1—but it has now reached the ratio of 5 to 1.

"The Soviet continues its lead in the ability to launch heavier weight into space. But we are pulling far ahead in developing practical uses of space with weather, communication, and navigation satellites.

"We are not racing for prestige in space. We are engaged in a deadly serious race for peace—and for control of our own destiny. This is a race free men must win . . . ." (Text, *CR*, A4759–60)

- X–19 Vtol aircraft, first of three vertical-rising aircraft being built under Army-Navy-Air Force program, unveiled at Caldwell, N.J. X–19 was designed to combine agility of helicopter with speed of fixed-wing airplane. (Hudson, *NYT*, 7/24/63)

July 24: William E. Stoney, Jr., appointed Chief of Spacecraft Technology Div. of NASA Manned Spacecraft Center's Office of Engineering and Development. Stoney previously was chief of Advanced Vehicle Conceptual Studies in NASA Hq. Office of Advanced Research and Technology. (MSC Release 63–20; MSC Space News Roundup 8/7/63, 6)

- USAF announced orbiting on June 27 of 176-lb. HITCHHIKER satellite to measure distribution and energies of radiation particles in earth's magnetic field and upper atmosphere. Ejected from another satellite (unidentified) launched from Vandenberg AFB June 27, HITCHHIKER was propelled by its own engine into elliptical orbit July 1: 208-mi. perigee, 2,568-mi. apogee, 132-min. period, 82° inclination to the equator. (DOD Release 1042–63)

- Paul Bikle, Director of NASA FRC, piloted glider from Hailey, Idaho, to Swift Current, Saskatchewan, Canada, for an unofficial national soaring distance record of 545 mi. (AP, Wash. *Eve. Star*, 7/25/63, 2)

- Manned Spacecraft Center was scheduled to complete its move into the new Clear Lake complex by July 1, 1964. Complete plans for the move had been laid out in a Master Move Plan. (*Space News Roundup*, 1/8/64, 2)

July 24: USAF launched Minuteman ICBM from underground silo on test flight down Atlantic Missile Range. (DOD Release 1055-63)

July 25: U.S.-U.K.-U.S.S.R. nuclear test-ban treaty prohibiting nuclear testing in the atmosphere, in space, and under water was initialed in Moscow by negotiators W. Averell Harriman, U.S. Under Secretary of State for Political Affairs; Viscount Hailsham, British Minister for Science; and Andrei A. Gromyko, Soviet Foreign Minister. Treaty was subject to formal signing and parliamentary ratification by U.S. Senate, British Parliament, and Supreme Soviet. (Topping, *NYT*, 7/26/63, 1)

- House Committee on Science and Astronautics reported H.R. 7500, NASA FY 1964 authorization bill, to House. Committee's authorization recommendation was $5.2 billion. (NASA Leg. Act. Rpt. II/113)
- Secretary of the Air Force Eugene M. Zuckert testified before Subcommittee on Investigations of Senate Committee on Government Operations that his own personal study of General Dynamics and Boeing Co. contract proposals for TFX convinced him that General Dynamics provided the most practical design and the most realistic cost estimate. (Testimony)
- Maj. Robert A. Rushworth (USAF), X-15 pilot, was presented military astronaut wings by Air Force Chief of Staff Gen. Curtis E. LeMay. (*Wash. Post*, 72663)
- AT&T spokesman said cause of TELSTAR II's failure was still not known. "We're still sending command signals, but we still have no clue to what happened." (Doan, N.Y. *Herald Trib.*, 7/26/63)
- Senate Committee on Aeronautics and Space Sciences voted (6 to 5) to delete nearly $5 million requested in NASA budget for FY 1964 for proposed electronics research center. Vote was tentative, pending final committee action on NASA FY 1964 request. (*NYT*, 7/26/63)
- Aeronautics press briefing held at NASA Hq., with presentations by officials from Hq., Ames Research Center, Flight Research Center, and Langley Research Center. NASA Director of Office of Advanced Research and Technology, Dr. Raymond L. Bisplinghoff, said:

  "It is important to point out that we have chosen to confine NASA's aeronautical program to research and advanced technology activities leading to new aeronautical vehicle concepts, but not including the very costly development of new aircraft. In this respect, NASA's aeronautical activities differ from those in space where new vehicles are developed, constructed and operated. It is our belief that in general the development and construction of new aircraft should be the responsibility of the user since only he can make the compromises necessary to produce aircraft which can play their proper role in civilian or military applications. We believe this view to be consistent with the advanced state of aeronautical technology, the sophisticated aircraft industry which now exists and the nation's use of commercial and military aircraft . . . ." (Text)

July 25: In aeronautics press briefing, NASA Administrator James E. Webb replied to reporter's question on cost of supersonic transport development: "We can do a great deal for $1 billion. I think we could probably build a flying prototype for less than that." In President Kennedy's proposed plan for supersonic transport development, $750 million would be spent by Government (FAA, DOD, and NASA) and remaining $250 million by industry. (*L.A. Times, Wash. Post,* 7/26/63)

- Lt. Col. John Powers (USAF), MSC Public Affairs Officer, announced in Washington he had conferred with NASA Administrator James E. Webb and Deputy Administrator Dr. Hugh L. Dryden about "the changing nature of the manned space flight program." In accordance with this change, brought about by the end of Project Mercury, NASA would "realign our organization and our people, including myself." (West, *Houston Post,* 7/26/63)

July 26: NASA SYNCOM II communications satellite launched into orbit with Thor-Delta launch vehicle from AMR, entering elliptical orbit (140-mi. perigee, 22,548-mi. apogee). Five hrs. 33 min. after launching, apogee-kick motor onboard fired for 21 sec., placing SYNCOM II in orbital path ranging from 22,300-mi. to 22,548-mi. altitude and adjusting its speed to near-synchronous 6,800 mph. Traveling in slightly lower than synchronous orbit and at less than synchronous speed, satellite began drifting eastward at rate of 7.5° per day. Ground signals would attempt to reverse drifting so that satellite would attain synchronous position over Brazil.

Communications tests during SYNCOM II's ascent into orbit were successful, including reception and transmission of "The Star Spangled Banner," a voice message, and a teletype transmission. Once in synchronous position, SYNCOM II would provide telephone, teletype, and photo facsimile communications between Lakehurst, N.J., and Lagos Harbor, Nigeria. (NASA Release 63–152; *NYT,* 7/27/63; *N.Y. Herald Trib.,* 7/27/63)

- In televised speech to the Nation on the nuclear test-ban treaty, President Kennedy said: "The treaty initialed yesterday . . . is a limited treaty which permits continued underground testing and prohibits only those tests that we ourselves can police. It requires no control posts, no on-site inspection and no international body.

"We should also understand that it has other limits as well. Any nation which signs the treaty will have an opportunity to withdraw if it finds that extraordinary events related to the subject matter of the treaty have jeopardized its supreme interests; and no nation's right to self-defense will in any way be impaired. Nor does this treaty mean an end to the threat of nuclear war. It will not reduce nuclear stockpiles; it will not halt the production of nuclear weapons; it will not restrict their use in time of war.

"Nevertheless, this limited treaty will radically reduce the nuclear testing which would otherwise be conducted on both sides; it will prohibit the United States, the United Kingdom, the Soviet Union and all others who sign it from engaging in the atmospheric tests which have so alarmed mankind; and it offers to all the world a welcome sign of hope . . . ." (Text, *NYT*, 7/27/63, 2)

July 26: U.S. Army announced successful test firing of Nike-Zeus antimissile missile at White Sands Missile Range. (DOD Release 1059-63)

- Hanson W. Baldwin reported in *New York Times* that during latest Soviet nuclear tests "one high-altitude explosion destroyed two incoming missiles. In one American high-altitude test in the Pacific, a fairly small nuclear detonation high above the earth caused fission—an atomic chain reaction—in a nose cone 150 miles above the earth and 800 miles from the explosion.

  "This phenomenon, called neutron flux, travels great distances in a virtual vacuum. Thus it may be able to neutralize fissionable material in incoming warheads. This is a technique about which the Russians are believed to know more than American scientists . . . ." (Baldwin, *NYT*, 7/26/63)

- FCC released letter to Communications Satellite Corp., indicating ComSatCorp. directors should make definite plans for stock issue. Pointing out that ComSat Act of 1962 required ComSat Corp.'s directors be divided among stock-owning public, communications companies, and presidential nominees, FCC said that "undue delay in the establishment of the corporation may force the present [appointed] Board of Directors to engage in activities and to make decisions which should be left to the representatives of the owners of the corporation."

  ComSatCorp. Chairman Leo D. Welch told press that there was "no basis for the concern expressed by the Commission," and ComSatCorp. would issue stock "at as early a date as the directors determine to be compatible with the public interest and the carrying out of the purpose and objectives of the Act." (Clayton, *Wash. Post*, 7/26/63; Toth, *NYT*, 7/27/63)

- USAF announced routine training launch of Atlas ICBM from Vandenberg AFB. (DOD Release 1080-63)

July 27: Eastward drifting of SYNCOM II communications satellite was reversed by ground command firing of hydrogen peroxide jet controls onboard the satellite. SYNCOM II was now drifting westward at rate of 5° per day. After drifting into desired position, satellite would be stopped by onboard control jets so that it would be in synchronous position of 24-hour orbit. Command signals to SYNCOM II were sent by U.S. Army Satellite Communications Agency station aboard USNS *Kingsport* in Lagos Harbor, Nigeria. (NASA Release; AP, Wash. *Sun. Star*, 7/28/63; AP, *Wash. Post*, 7/28/63)

- NASA Deputy Administrator Dr. Hugh L. Dryden, at District Convention of American Legion, Dept. of Virginia, in Roanoke, said: "Those who view the lunar program simply as a 'propaganda' effort fail to grasp that not only our prestige, but our capacity for constructive international leadership, our economic and military capacity for technological improvement, depend upon our ability to achieve acknowledged superiority in science and technology, and to use this capability in our own behalf and that of our allies.

"With a billion people already allied against us, and the uncommitted and emerging nations weighing events that will affect their own future welfare, the United States must present the image of a can-do nation with which they can confidently align their futures.

"Thus the program, in addition to its other aims, becomes literally also a matter of national necessity. The world has come to regard space exploits as a measure of a nation's strength. And alliances and loyalty are given to the strong.

"We must master this new environment—just as we have had to master the land on which we make our homes, the oceans that carry our ships, and the air that sustains us—to guard against the day when mastery of space might mean world domination . . . ." (Text)

July 27: House Rules Committee announced it was considering a "complete, full and thorough investigation" of Federal expenditures on research, to determine: what departments and agencies conduct research, at what cost, and with what results; amounts being spent on scholarship grants and grants for research to colleges and industry; and what facilities, if any, coordinate the various research programs, including grants to colleges and scholarship grants. (*NYT*, 7/28/63, 1, 49)

- Six Republican members of House Committee on Science and Astronautics recommended establishment of a special Congressional committee, composed of members of Armed Services Committee and Science and Astronautics Committee, to re-evaluate the national goals in space. The Congressmen said U.S. was ignoring "the main thrust of the Soviet space aim, which is to dominate inner space through the ability to exercise control over the surface of the earth." Statement was included in a Committee report on NASA FY 1964 authorization bill. (AP, *NYT*, 7/29/63, 12)

- Writing in *Washington Post*, Howard Simons said: "There is no absolute guarantee that U.S. experts, using scientific means alone, can detect every nuclear detonation in . . . [the air, under water, or in space]; but the over-all U.S. detection capability is regarded as good enough to make cheating in those environments a very risky business, indeed.

  "Generally speaking, this detection system provides an excellent probability for detecting and identifying any and all atmospheric nuclear blasts larger than a few kilotons from ground level up to 6 miles above the earth's surface . . . .

  "Detection and identification from 6 to 15 miles above the earth's surface is not as effective, but considered to be adequate.

  "From 15 miles out to 100 million miles, detection becomes easier—with ground-based and satellite-based detection systems—unless a cheater were to go to complex and costly lengths to hide the blast behind a shield in space . . . ." (*Wash. Post.* 7/27/63)

- USAF Minuteman ICBM launched from Vandenberg AFB in routine training test by Strategic Air Command crew. (DOD Release 1080-63)

July 28: NASA scientists expanded on reports that future astronauts might use crayons and paper to record sights during space flights, saying that Astronaut Walter M. Schirra, Jr., was particularly interested in applying his sketching talents to space flights. Cdr. Schirra was apparently disappointed with some photographs taken during his MA-8 space flight, finding that such phenomena as the moon's halo did not show in photographs taken from his spacecraft. (Witkin, *NYT*, 7/28/63, 30)

- "Outstanding Unit Award" given to 6593rd Test Squadron at Hickam AFB, Honolulu, for "its record in the development and application of aerial techniques for the recovery of space capsules returned from orbiting satellites." USAF announcement said more than 70 per cent of satellites launched last year with recovery as goal were actually retrieved by the squadron; in a more recent series, 88 per cent were recovered. Maj. Gen. Ben I. Funk, AFSSD Commander, said recovering capsules from space has reached "the point where recovery is accomplished on a routine basis." (UPI, *Wash. Post*, 7/29/63)
- AFSC reported it was experimenting with gelatin for structural uses in space. In simulated space vacuum chamber, gelatin pressed into fiberglass cloth becomes rigid, AFSC said. Gelatin was said to have excellent resistance to ultraviolet radiation. (AP, *NYT*, 7/28/63, 13)
- FAA Administrator Najeeb Halaby announced appointment of Gordon M. Bain as Deputy Administrator for Supersonic Transport Development. Formerly Assistant Administrator for Appraisal, Bain would head FAA organization charged with overall responsibility for Government-industry development of supersonic transport aircraft. (FAA Release 63-69)
- General Electric Co. announced USAF Titan II "carries a re-entry vehicle which would 'significantly increase' this country's ability to penetrate antimissile defenses." (Witkin, *NYT*, 7/29/63)

July 29: NASA announced SYNCOM II communications satellite was performing "extremely well," drifting westward at rate of 4.5° per day. Orbital data: 22,800-mi. apogee, 22,110-mi. perigee, 1,454-min. period (slightly more than 24 hours). SYNCOM II was expected to be on station above Brazil in about three weeks. (AP, *NYT*, 7/30/63)

- RCA announced RELAY I communications satellite had operated successfully for 203 days, setting record for performance and durability by a communications satellite. RCA, which built satellite for NASA, said RELAY I had traveled estimated 156 million mi. and carried out more than 1,350 communications experiments and demonstrations. (AP, Wash. *Eve. Star*, 7/29/63)
- President Kennedy appointed Robert M. White to succeed Francis W. Reichelderfer as Chief of U.S. Weather Bureau. White was president of Travelers Research Center, Inc., an independent nonprofit organization engaged in R&D in environmental and mathematical sciences. (Simons, *Wash. Post*, 7/30/63)
- Rep. James Fulton (R.-Pa.) introduced bill to establish in NASA an inspector of programs and operations (H.R. 7770). Bill was

referred to Committee on Science and Astronautics. (NASA Leg. Act. Rpt. II/114)

July 29: 347-acre Government site in Newport News, Va., presented to Commonwealth of Virginia for Virginia Associated Research Center (VARC), established by Univ. of Va., VPI, and College of William and Mary. VARC would operate the adjacent Space Radiation Effects Laboratory under NASA contract.

Speaking at luncheon following site presentation ceremonies, NASA Deputy Administrator Dr. Hugh L. Dryden discussed role of physicist today:

"It seems clear to me that man's newly won ability to leave the confines of earth and explore the frontiers of space offers a challenging new opportunity for scientists of all disciplines, and for physicists in particular . . . .

"The need . . . is for our creative physicists to think of simple devices, or at least small ones, to make the critical physical measurements that will heighten our knowledge of the characteristics of the universe around us. The task is not easy, but the challenge is great, and the opportunities for new findings are, it seems to me, almost unlimited . . . .

"Today space exploration is the great motivating development that can serve to promote the over-all growth of science and enhance its role in the national life . . . ." (Text)

• Sen. Jack Miller (R.-Iowa) inserted in *Congressional Record* the May 10 Republican policy committee staff study examining U.S. priorities in space as well as article from current issue of *Reader's Digest* entitled "We're Running the Wrong Race with Russia" which said:

"The Russians have recently achieved an increase in the effectiveness of nuclear explosions so devastating that it dwarfs all previous records. Confirming this, our own atomic experts have warned that such stupendous forces, let loose above us from a satellite, could, in a few seconds, literally cremate a large part of the United States. There would be nothing left below, no man, beast, vegetation, buildings, nothing at all but the glare of white-hot cinders.

"In the face of such a dire threat, it must come as a stunning shock to all thoughtful Americans to learn that the United States has no top-priority programs beamed at preventing, anticipating, or deterring such a terrifying prospect. Our top priorities are all tied up with steamrollering through the moon shot. Many billions of public money, together with the rich cream of scientific talent, are lavished on an effort that could well be aborted in midstride by Soviet enterprise and realism.

"There is a crying need for a reappraisal of our space aims, for more specific public information, especially since statements from those highly placed in Washington are so contradictory that they blur rather than clarify our understanding . . . ." (*Reader's Digest*, 8/63, *CR*, 7/31/63, 13035-39)

• *New York Times* editorialized that "there is a need for a thorough review of all the activities and outlays carried on by NASA.

"This would not be necessary if Congress had been exercising ordinary supervision over our space efforts. But, goaded by Mr. Kennedy's determination to beat the Soviet Union in the race to the moon, Congress has permitted NASA to lead a charmed life, providing what amounted almost to a blank check for its operations and a free hand to its managers. NASA, it seemed, could do no wrong. . . .

"There is no doubt that false starts and dead ends are inevitable in exploring the unknown frontiers of space. But NASA's effectiveness will be enhanced by tighter controls over spending and greater coordination of its activities." (*NYT*, 7/29/63)

July 29: Rep. Donald Rumsfeld (R.-Ill.) inserted in *Congressional Record* Dr. James D. Atkinson's "An Approach to American Strategy" from the book *National Security, Political, Military, and Economic Strategy in the Decade Ahead*, in which Dr. Atkinson said:

"Space military capabilities would broadly appear to be:

(1) Intelligence operations: observing, recording, detecting.
(2) Direction: of land, sea, air, or combined operations.
(3) Offensive actions: against land, sea, or air targets; against other space vehicles.
(4) Defensive actions: neutralizing attacking missiles; counteracting other space vehicles.
(5) Psychopolitical operations in advance of or in support of military actions. . . .

"The transferrence of classical military operational concepts to space remains to be accomplished and the next decade will see this taking place apart from the obvious projections of well-understood military operations. Space offers intriguing new possibilities for the exercise of power at low intensity levels, but with enormous strategic import . . . ." (Text, *CR*, 7/29/63, 12852)

July 30: Speaking on Senate floor, Sen. E. L. Bartlett (D.-Alaska) proposed establishment of Congressional Office of Science and Technology "as a creature of Congress, responsible only to the Congress.

"This office would have a small, highly skilled, permanent, professional staff and a large body of consultants, available to the Congress to furnish advice, evaluations, and reports . . . . It would be the task of the permanent staff to ask questions for Congress, to answer questions of Congress, and to assist the Members in handling the scientific matters which come before them. It would also be the responsibility of COST to report to the appropriation committees whatever achievements, happenings or failures in the scientific world appear to be of importance . . . ." (*CR*, 7/30/63, 12896–98)

- At Second Space Industry Assistance Symposium in Houston, sponsored by MSC and Houston Chamber of Commerce in cooperation with Rice Univ., Dave W. Lang, Chief of MSC Procurement and Contracts Div., said that average basic salary for all MSC employees was $8,500. "The industrial support payroll [of MSC] stands at $27 million per year locally and is expected to reach $45 million next year. The Clear Lake site itself is expected to cost nearly $150 million by the time it reaches completion. The majority of these expenditures will provide income for Houston area firms and individuals through prime and subcontracts to Houston area firms, purchase of materials through local suppliers, and employment of local skilled and unskilled labor." Since locating in Houston, MSC had awarded more than $12 million in contracts to Houston area firms. In addition, MSC attracted some 300 business representatives per month from out-of-town. "If these out-of-town visitors spend a minimum of $15 per day, it adds up about $100,000 more in revenue each year to the growing impact of MSC."

Lang also remarked on the influence of MSC on local universities:

"The reason for the high interest in education is apparent when you see that out of 1213 scientific and technical personnel, 1050 have bachelor of science degrees, 140 have masters, and 23 doctorates." (Text; NASA-MSC Fact Sheet #202)

July 30: USAF launched Blue Scout, Jr., probe from Cape Canaveral to altitude of more than 8,000 mi. 50-lb. instrumented payload was designed to conduct measurements of the ionosphere. (DOD Release 1080-63)

- NASA would support pilot program to develop adult education course on the space program in Pawtucket, R.I., E. E. Collin of NASA Education Services said in New York *Herald Tribune*. NASA $2,500 contract with Rhode Island State Dept. of Public Instruction would finance 10 two-hr. presentations covering overall space exploration program, each given by expert in the subject area. (Loory, N.Y. *Herald Trib.*, 7/30/63; Collin, NASA AFEE)

- NASA Manned Spacecraft Center held Second Industry Assistance Symposium at Rice Univ., Houston. (MSC Release 63-114)

- Secretary of Commerce Luther H. Hodges released letter of commendation to Francis W. Reichelderfer, retiring as Chief of Weather Bureau, which said: ". . . When you took your post 25 years ago, the Bureau established its first network of radio-sonde upper air observations. On your departure you are leaving a legacy of the world's largest and most sophisticated weather system, one that is already effectively using such modern tools as satellites and computers.

  "During your tour of duty, your leadership and inspiration guided meteorologists throughout the world to work toward the common goal of a truly global weather system. Your plans to expand your international activities will, I am sure, contribute much to world cooperation and advancement in atmospheric sciences . . . ." (Commerce Dept. Release G-63-142)

- Atlas ICBM fired from Vandenberg AFB by Strategic Air Command combat crew in successful 7,000-mi. test. (DOD Release 1080-63)

July 31: National Aeronautics and Space Council met "to examine the interrelationship between the military and non-military aspects of the space program, with special attention to the Apollo project." Following the meeting, Vice President Lyndon B. Johnson, NASC Chairman, said: "I am impressed with the solid competence which we are building in space—a competence which can be used to further our national security, our prestige, our scientific knowledge, and our standard of living. The benefits flowing from NASA's lunar program, for example, contribute a foundation for a wide range of capabilities, including those of national defense. This is a coordinated national program." (NASC Release)

July 31: NASA announced Paul Haney, Public Affairs Officer for NASA Office of Manned Space Flight, would replace Lt. Col. John A. Powers, MSC Public Affairs Officer, effective Sept. 1. Colonel Powers would become special assistant to Dr. Gilruth. (AP, Wash. *Eve. Star*, 7/31/63; NASA Release 63-167)

- NASA Administrator James E. Webb said in address before National Association of Counties, Denver: "The question which now confronts the Congress, and the nation, is whether our progress [in the space program] has been so great that we can afford to be complacent; that we can begin to rest on our laurels with the comfortable assurance that a slower pace will still win the race.

  "The clear record of success which has come from five years of consistent, expanding effort has moved us well along the road to leadership in space. Those of us who have worked with this program have confidence that, given a continuation of that effort on a sustained basis, and at a level consistent with what our national resources and other national requirements permit, we will establish pre-eminence in all fields and we will achieve the national goal of exploring the moon within this decade.

  "Meanwhile, no one can have assurance that if we depart from the policies which have served us so nobly during our first five years in space, there is not grave danger that we will remain second best for a long time to come, or forever. We are in space because we believe that great scientific and economic benefits will result from our efforts. But, we are also in space because it is intimately related to our strength and security in a world in which men have not yet learned to live in peace with each other . . . ." (Text)

- Article by NASA scientists Dr. Homer E. Newell and Dr. Robert Jastrow in *Atlantic Monthly* inserted in *Congressional Record* by Sen. Clinton P. Anderson, Chairman of Committee on Aeronautical and Space Sciences. Replying to scientists critical of pace set for landing men on the moon, the authors said:

  "This question requires a further exploration of the motives underlying the U.S. space effort. Is it primarily a scientific program, or is it motivated by a broader concern with the national interest? Looking back to the overwhelming support given the new space program by the Congress in 1961, it seems clear that this support was not tendered for scientific reasons primarily, but came from a deep-seated conviction that the expanded program will make an important contribution to our future strength and security. We believe that this is the reason why the people have supported the enlarged space program. That brings us to the point on which we take issue with some of our scientific colleagues, who complain, 'The scientific exploration of the moon has been accorded a secondary priority in the lunar program.' This remark is based on the premise that science should have top priority in the space program. However, while science plays an important role in lunar exploration, it was never intended to be the primary objective of that project. The impetus of the lunar project is derived from its place in the long-range U.S. program for the exploration of the solar system. The heart of that program is man in space, the extension of man's control over his physical environment. The science and technology of space flight are ancillary developments which support the main thrust of manned exploration, while at the same time they bring valuable returns to our economy and our culture. The science which we do in space provides the equivalent of the gold and spices recovered from earlier voyages of exploration. It is the return to the taxpayer for his investment in his nation's future. But the driving force of the program is not in scientific research alone, valuable though that may be in the long run. Thus, the pace of the program must be set, not by the measured patterns of scientific research, but by the need for a vigorous response to the national challenge . . . ." (Text, *CR*, 7/31/63, A4874–76)

July 31: USAF launched Thor-Agena D with unidentified satellite from PMR. (*Pres. Rpt. on Space, 1963*)

During July: NASA Marshall Space Flight Center selected General Dynamics/Ft. Worth and Douglas Aircraft Co. to conduct parallel studies of manned Mars exploration flight "in the unfavorable time period" (1975–85). Under seven-month contracts, the companies would outline the mission and development program for the Mars flight, identifying spacecraft and booster systems and selecting most promising mission profile. (Space *Bus. Daily*, 7/5/63; *L.A. Times*, 7/9/63; *Marshall Star*, 7/10/63, 8)

- NASA Manned Spacecraft Center awarded $100,000 contract to Boeing Co. for six-month study of logistics spacecraft to be used for ferrying men and equipment to and from orbiting space station. Studies would be of two wingless lifting bodies: M-2, developed by NASA Ames Research Center, and HL-10, originated at NASA Langley Research Center. (MSC *Space News Roundup*, 7/24/63, 8)

- Raytheon Co. announced development of one-lb. laser capable of sending 10 voice messages for more than one mile over infrared beam. New laser used mixture of neon and xenon gases. (*NYT*, 7/28/63)

- USAF announced it was establishing three-man Gemini liaison and support office at NASA Manned Spacecraft Center. Under Col. William A. Stellenwerf, office would deal primarily with Titan II launch vehicle for Gemini and recovery requirements. (*Av. Wk.*, 7/15/63, 31; *A-N-AF Journ. & Reg.*, 7/13/63, 16)

AUGUST 1963

August 1: MARINER II interplanetary space probe completed its first orbit of the sun, after traveling approximately 540,000,000 mi. (aphelion, 113.8 million mi., occurred June 18, 1963; perihelion, 65.5 million mi., occurred Dec. 28, 1962). Launched Aug. 27, 1962, the spacecraft passed within 21,648 mi. of Venus Dec. 14, 1962, and provided 111 million bits of information on Venus and interplanetary space. (UPI, *NYT*, 8/2/63, 7; MSC *Space News Roundup*, 8/7/63, 1, 2)

- House passed bill to authorize $5,203,719,400 for NASA in FY 1964 (334–57). House bill would authorize $4,013,175,000 for research and development; $508,185,000 for administrative operations; and $682,359,400 for construction of facilities. House cut $34.4 million from the bill as reported from Committee on Science and Astronautics: Amendment introduced by Rep. Richard L. Roudebush (R.-Ind.) cut $24.4 million from Committee's recommendation for facility, training, and research grants, the amendment intended "to stabilize" NASA's university program at FY 1963 level. Another amendment cut $10 million from advanced design development. Amendment offered by Rep. John W. Wydler (R.-N.Y.) to cut $3.9 million authorization for electronics research center was defeated (111–64). (*CR*, 8/1/63, 13070–13129)

August 1–November 1: Project Stormfury, experiment to investigate ways of reducing destructive force of hurricanes, conducted jointly by Weather Bureau and Naval Weather Service. Experiment involved seeding hurricane clouds and cumulonimbus clouds (not associated with hurricane winds) with silver iodide to learn if energy patterns can be changed. Results would be compared with similar experiment conducted during 1961 Hurricane Esther. (DOD Release 1077–63)

August 1: Senate Committee on Aeronautical and Space Sciences ordered H.R. 7500 favorably reported (pending receipt from the House). With amendments, bill would authorize $5,511,520,400 for NASA in FY 1964. (NASA Leg. Act. Rpt. II/117; *N.Y. Herald Trib.*, 8/2/63)

- Commenting on action of Senate Committee on Aeronautical and Space Sciences, Committee's Chairman Sen. Clinton P. Anderson, (D.-N.M.) said Committee approved NASA appropriations requests whenever they "made sense in terms of National objectives. On the other hand, some projects failed to get committee authorization because they are in need of further explanation . . . .

   "If an Apollo landing on the moon were to be the end of the line, then the attitude of the Congress toward NASA expenditure will be one thing. But if space is here to stay, and we believe it is, then our attitudes would be something else again. When we are asked for funds to cover a variety of advanced research projects and new engines and vehicles beyond those occasioned by Apollo, it is hard to be enthusiastic in supporting these projects in the absence of clearly stated national goals." (*Missile/Space Daily*, 8/2/63)

August 1: In report to Bureau of the Budget, NASA commented on bill (S. Res. 50) to establish Senate Select Committee on Technological Developments. NASA recognized that "there may be substantial advantages derived both for the Senate and for the general public from the establishment of a committee such as this," which would study and investigate responsibilities of Government departments and agencies for scientific and technical developments and the effect of such activities upon U.S. scientific, technical, and economic progress and upon structure of the U.S. economy. (NASA Leg. Act. Rpt. II/118)

- In Washington press conference, President Kennedy was asked about relative status of antimissile missile programs of U.S. and U.S.S.R. in light of last year's Soviet nuclear tests "in which very large warheads were detonated." The President replied:

    "I don't think that the problem is solved by the explosion of a large megaton bomb. The problem is really one, as you know, of discrimination, of being able to prevent saturation, of having to protect many targets while the adversary can select a few . . . .

    "The problem of developing a defense against a missile is beyond us and beyond the Soviets technically, and I think many who work in it feel that perhaps it can never be successfully accomplished because the whole problem, as you know, is to have 100 objects flying through the air at thousands of miles an hour [and] to be able to pick them out. And if you can do that there is an advantage, it still seems to me, to the offense because they can pour in 200, or 300.

    "And therefore, the problem is not the size of the bomb but rather the problem of discrimination, the problem of selectivity, targeting and all the rest.

    "On those matters we can continue to work. But I must say those who work the longest are not particularly optimistic that a scientific breakthrough can be made . . . ." (Transcript, *NYT*, 8/2/63, 10)

- Speaking at the Fourth International Space Science Symposium in Warsaw, Poland, Dr. R. S. Young of NASA's Ames Research Center, said that certain earth bacteria not only survive but grow satisfactorily in a laboratory-created Martian environment. Knowledge of growth of earth organisms on Mars is important for two reasons, Dr. Young said. First, the extremely complex problem of decontaminating earth spacecraft depends upon proof that earth organisms do not present a hazard to other planets. Second, as much information as possible about survival on Mars or on other celestial objects is needed for designing instruments on spacecraft to detect and analyze such life if it exists. (*The Cleveland Press*, 8/1/63)

August 1: Australian Minister of Supply Allen Fairhall announced U.S.-financed space tracking station would be built near Canberra. The 60-sq.-mi. complex would be built by Australia and staffed by Australian scientists and technicians. (*NYT*, 8/2/63, 8)

- On his third anniversary as president of USAF's Aerospace Corporation, which itself was three years old last June 25, Ivan A. Getting reported the organization now has annual budget of $22,600,000 and 4,500 total personnel (1,617 scientists, engineers, and technicians.) During year just completed, Aerospace Corp. provided guidance on $1 billion in space systems contracts and $3 billion in ballistic missiles and components. (Aerospace Corp. Release; *NYT*, 8/4/63)
- First technical description of U.A.R.'s Pioneer rocket appeared in Cairo newspaper *Akher Saa*. Article said Pioneer multistage rocket has range of 625 mi., is powered with liquid oxygen and another unnamed liquid, and is intended as space research vehicle. (AP, *Houston Post*, 8/2/63)

August 2: Second Shotput suborbital sounding rocket flight in U.S.-Italian San Marco project was launched from NASA Wallops Station, the 177.5-lb. payload reaching 183-mi. altitude in flight to test instrumentation for San Marco satellite. Another purpose of launching was to determine whether Shotput despin difficulties of first San Marco launching had been corrected, and flight officials indicated the vehicle performed satisfactorily.

Shortly after the Shotput launch, a related experiment was launched from Wallops Island using Nike-Cajun sounding rocket. Three mylar balloons were ejected from Nike-Cajun and tracked by radar, to provide air density data to aid in evaluating San Marco experiment.

A third Shotput launching later this year would continue testing instrumentation performance and launch techniques for San Marco project, which would culminate in launching of satellite into equatorial orbit from platform off east coast of Africa. San Marco is cooperative project of Italian National Research Council and NASA. (NASA Release 63-172; Wallops Release 63-75; NASA Rpt. of Sounding Rkt. Launching)

- After 10 months of satisfactory operation in orbit, EXPLORER XIV scientific satellite, with its six experiments to measure energetic particles and magnetic fields in space, was still functioning. During the 10 months, data acquisition stations received 6,003 hrs. of data from EXPLORER XIV, of which more than 4,500 hrs. were digitized and 3,200 hrs. were sent to the six experimenters. (NASA EXPLORER XIV Prog. Rpt. No. 5)
- Joint U.S.-U.K.-Australian Project Dazzle to study behavior of objects re-entering atmosphere would be conducted at Woomera Rocket Range in Australia, DOD announced. U.K. would provide Black Knight rockets and re-entry payloads; U.S. would provide instrumentation for test. (DOD Release 1088-63)
- Sweden successfully launched U.S. Army Nike-Cajun rocket from Kronogård rocket range in test to explore "bright night clouds." (*M&R*, 8/12/63, 11)

August 2: USN Sparoair research rocket launched from F-3B jet at 30,000-ft. altitude, the two-stage rocket climbing to 350,000-ft. altitude while its instrumented payload measured ultraviolet radiation of the stars above 99% of earth's atmosphere. Payload impacted Pacific Ocean 70 mi. west of Pt. Mugu and no recovery was attempted. (AP, *NYT*, 8/7/63)
- In response to inquiries about rumors that Astronauts M. Scott Carpenter and Walter M. Schirra were ill with space-flight aftereffects, unnamed NASA Hq. spokesman quoted Dr. Charles A. Berry, Chief of MSC Medical Operations Office, as saying: "Neither Carpenter nor Schirra has any difficulties at all, and they are in excellent health.
  "In fact, none of the astronauts who have made space flights has suffered any ill effects, and the entire group is in top physical condition." (UPI, *Houston Chronicle*, 8/3/63, 9)
- Atlas booster, first stage of Centaur space launch vehicle, arrived at Plum Brook Station to begin series of structural tests in E-Stand. (LRC Release 63-64, Lewis Chronology, 7)

August 3: Patent to Boeing Co. engineers Joseph H. Doss and Gary A. Graham for method of suspending astronaut in his spacecraft with netting and cables was reported in *New York Times*. Inventors claimed system would give astronaut more mobility than couches used in Mercury spacecraft. (Jones, *NYT*, 8/3/63, 21)
- Photographs of members of Leningrad Cosmonaut Club featured in Soviet press. School children members of clubs said to have flown in jet aircraft, trained in pressure and isolation chambers, and executed parachute jumps; they have studied radio operation and were familiar with astronomy, space medicine, and "cosmonautics." (*Komsomolskaya Pravda*, 8/3/63, 4, AFSS–T Trans.)

August 4: First public demonstration of communications exchange via synchronous satellite, when two U.S. wire services and Nigerian newsmen exchanged news stories of about 300 words each via SYNCOM II communications satellite, hovering 22,823 mi. over Western Africa. Photographs of President Kennedy and Nigerian Governor General Dr. Nnamdi Azikiwe also were exchanged. Transmissions were made from NASA station at Lakehurst, N.J., and USNS *Kingsport* communications ship in Lagos Harbor, Nigeria. (AP, Wash. *Eve. Star*, 8/5/63; NASA Release 63-171)
- Los Angeles Chamber of Commerce issued report urging selection of Edwards AFB as U.S. Space Recovery Center. Report was based on 12-month study of potential recovery sites throughout U.S. for all types of space vehicles. Chamber General Manager Harold W. Wright said: "We hope that the National Aeronautics and Space Administration will seriously consider the Edwards facilities for its use and not attempt to duplicate similar facilities elsewhere . . . .
  "The federal government has already poured millions and millions of dollars into the highly-instrumented Pacific Missile Range, the electronic recovery control center at Sunnyvale and the highly developed facilities in the Edwards-Wendover, Utah-Holloman-White Sands, N.M., complex.

"This provides the nation's space effort with the best available land recovery site."

NASA Hq. spokesman, queried about report, stated Project Gemini landing operations were still under study, "with a possibility we will require no large, fixed base for recovery of the spacecraft." First series of Gemini flights would employ water-landings, and there was possibility that all Gemini flights would depend on water recoveries. In any case, "NASA is well aware of Edwards," spokesman said, and the agency would give Chamber's study careful consideration. (*L.A. Times*, 8/5/63)

August 5: NASA announced SYNCOM II communications satellite, now drifting westward over Atlantic Ocean at 22,800-mi. altitude, would be stopped when it reached desired position at 55° west longitude. At this location SYNCOM II would be lowered into precise synchronous orbit, so that it would appear to trace elongated figure-8 pattern along 55° meridian to points 33° north and south of the equator, period 1,463 min., perigee 22,221 mi., apogee 22,912 mi., drift rate (degrees per day westward) 6.8. (NASA Release 63-168; GSFC Historian)

- TIROS VI, still orbiting the earth and still providing excellent quality photographs, exceeded durability record of 302 days set by its predecessor TIROS V. Launched Sept. 18, 1962, TIROS VI provided 58,589 pictures (several hundred more than TIROS V), of which about 90 per cent were usable. TIROS VII, launched June 19, 1963, continued to operate in conjunction with TIROS VI. (NASA Release 63-174)

- Static test-firing of first Saturn S–IV flight stage conducted by Douglas Aircraft Co. at Sacramento, the cluster of six RL-10 engines generating the full 90,000 lbs. of thrust for more than one minute. The stage was slated for Saturn SA-5 launch vehicle, to be flight-tested later this year. (*Marshall Star*, 8/7/63, 1)

- Some historic achievements of NASA Langley Research Center, NASA's installation specializing in basic aeronautical and space research, reported in press: invention of world's first transonic wind tunnels; discovery of the "area rule," a design principle regarded as key to practical supersonic flight; invention of automatically inflatable satellites, leading to development of world's first communications satellite (Echo); development of first all-solid fuel launch vehicle (Scout) to place U.S. satellite in orbit. (W. Warwick, R.I., *Pawtuxet Valley Times*, 8/5/63)

- Richard Tereselic, Lewis engineer, described hydraulic press intensifier which he designed at a considerable savings. The device cost approximately $4,000 and can obtain pressures up to 100,000 pounds per square inch. A similar system purchased commercially would cost $60,000. (LRC Release 63-62, Lewis Chronology, 7)

- At Grand Kremlin Palace in Moscow, treaty banning nuclear weapon tests in the atmosphere, in outer space, and under water was signed by Secretary of State Dean Rusk for the United States, Foreign Secretary the Earl of Home for U.K., and Foreign Minister Andrei A. Gromyko for U.S.S.R. (*NYT*, 8/6/63, 12)

August 5: On test ban treaty, *U.S. News and World Report* said:
"Nuclear test ban does not mean disarmament. Nuclear weapons still will be the weapons of future war. Weapon testing will not come to a full stop.

"Tests will continue underground. France will go on testing in the air. Red China will test in the air, too, if and when she gets the bomb ....

"Test agreement, now being entered into, favors Russia. Russia, behind in smaller nuclear weapons, can catch up by testing underground. United States, behind in bigger weapons and in an antimissile missile, will be hindered in her effort to catch up by the bar against testing in the atmosphere ....

"Just as long as Russia is a closed country, ruled by a dictatorship, possessed of weapons capable of destroying United States, there can be no real disarmament, no real end to the arms race without great danger ...." (*U.S. News*, 8/5/63, in *CR*, 7/31/63, 13060)

- Rep. Melvin R. Laird (R.-Wis.) introduced resolution (H. Res. 473) to create select committee for investigating expenditures for research programs conducted by or sponsored by Government departments and agencies; proposal was referred to House Committee on Rules. (NASA Leg. Act. Rpt. II/119)
- French plans to establish tracking network, to be integrated later into European network, reported in *Missiles and Rockets*. French network would include tracking stations in Canary Islands and South Africa, four telemetering stations in Africa. (*M&R*, 8/5/63, 9)
- USAF launched Minuteman ICBM from Cape Canaveral underground silo, but malfunction caused missile to fall short of expected target area. (*M&R*, 8/12/63, 11)

August 5–9: Desert survival training for NASA's nine Gemini/Apollo astronauts conducted at Stead AFB and Carson Sink, Nev., the astronauts participating in lecture sessions, field demonstrations, and actual survival practice. (MSC Response to Query)

August 6: U.S.S.R. announced COSMOS XIX had been placed in orbit (519-km. apogee, 270-km. perigee, 92.2-min. period; 49° angle of inclination to equator). (Tass, *Pravda*, 8/7/63, 1, AFSS–T Trans.)

- Tracking and data acquisition operations ceased for OSO I (Orbiting Solar Observatory), launched March 7, 1962. (GSFC Historian, 8/14/63)
- In hearings on H.R. 7381, bill to modernize dual-compensation and dual-employment laws, conducted by House Committee on Post Office and Civil Service, NASA Administrator James E. Webb testified:

"The National Aeronautics and Space Administration is strongly of the opinion that the proposed legislation should be enacted (1) to permit the Government to capitalize on certain skills, competence, and experience in the areas of aerospace science and technology possessed by a sizable group of military personnel who have completed their service careers, and (2) to correct certain inequities in the present statutes pertaining to the Government employment of these personnel.

"Many of the abilities and skills and much of the knowledge valuable to NASA are possessed by certain military personnel who have been closely associated with missile and military space programs and aeronautics. Collectively, this body of men, educated and skilled in aerospace science and technology, represents a reservoir of knowledge and experience which constitutes a national asset of inestimable value. Unless this limited group of specialists can be attracted to remain in the Government, this asset will not be available directly to the Government's space program. As of August 5, 1963, 227 active duty military officers are detailed to NASA by the Department of Defense . . . ." (Testimony)

August 6: License granted by the Atomic Energy Commission for operation of low pressure, low power (100 kw) Mock Up Reactor to Plum Brook Station of NASA Lewis Research Center. The MUR would be used to determine the effects of experiments on the radiation flux in the core of its high-power twin, the Plum Brook 60-megawatt reactor. Also, experiments requiring low-level nuclear bombardment would be run in the MUR. (Lewis Chronology, 3)

- NASA announced preparations for dual launching of satellites with Scout launch vehicle late this year. Under contract to NASA Langley Research Center, State Univ. of Iowa would provide Injun satellite containing instruments to record corpuscular radiation streaming into earth's upper atmosphere from space. The other satellite, a 12-ft. inflatable sphere to measure air density, was being built by LARC. The two satellites would be launched into near-polar orbit from PMR. (NASA Release 63-170)
- Communications Satellite Corp. announced selection of American Telephone and Telegraph Co., Hughes Aircraft Co., and Radio Corp. of America to study multiple access techniques for communications satellites. Each company would study different approach and provide ComSatCorp with specific data on existing types of equipment for engineering information so that ComSatCorp could choose design of first operational commercial comsat system. (ComSatCorp Release)
- Weather conditions forced postponement of NASA pilot Joseph A. Walker's 68-mi. altitude flight in X-15. Flight was made Aug. 22. (UPI, *NYT*, 8/7/63)
- DOD plans to orbit its first two experimental satellites for detecting nuclear tests in outer space reported by *New York Times*. The Project Vela satellites, under jurisdiction of DOD's Advanced Research Projects Agency, were expected to be capable of detecting unshielded nuclear tests as small as 10 kilotons and as far away as 180,000,000 mi. from earth. To be launched this fall, satellites would orbit earth at 50,000–60,000 mi. (*NYT*, 8/7/63, 12)
- Raytheon Co. announced appointment of D. Brainerd Holmes as senior vice president, effective Oct. 1. (*NYT*, 8/6/63, 8)
- Dr. M. Samuel White named Federal Air Surgeon of Federal Aviation Agency by FAA Administrator Najeeb E. Halaby. (FAA Release 63-73)

August 6: Second anniversary of Titov's orbital space flight in VOSTOK II, *Pravda* using occasion to announce the cosmonaut and his wife were expecting the birth of a baby. (UPI, Wash. *Daily News,* 8/7/63)

August 7: By its 1,842nd orbit, RELAY I communications satellite had been used in 930 wideband experiments; 409 narrowband experiments; 95 demonstrations (TV and narrowband). Transponder had been operated for 190 hrs. over period of 454 operations. (GSFC Historian, Memo, 8/14/63)

- Senate began debate on H.R. 7500, NASA authorization bill for FY 1964. In opening statement, Sen. Clinton P. Anderson, Chairman of Senate Committee on Aeronautical and Space Sciences, said:

    "I am fully aware of the many opinions which have been espoused as to the priority of funding between our military and civilian space programs, and the concern expressed in many quarters with respect to the high authorization of expenditures for the civilian portion of our space program. I should only like to say at this point that our administration has declared as a national goal the landing of a man on the moon by 1970. NASA has been given the responsibility by the administration of carrying out this national objective and while I, like others, would disagree with some of the decisions that have been made by Mr. Webb, the Administrator, and his associates, I, personally, feel, and I believe the other members of the committee would generally agree, that NASA deserves unqualified credit for the manner in which it has embarked upon this formidable task . . . .

    "The United States is in space to stay. NASA is dedicated to the objective of making our Nation preeminent in space, and has, indeed, to date compiled an enviable record and taken a long step toward the accomplishment of this objective . . . ." (*CR*, 8/7/63, 13710–16)

- Under Secretary of the Air Force Brockway McMillan said in letter to Sen. Clinton P. Anderson, Chairman of Senate Committee on Aeronautical and Space Sciences:

    "Our assessment of the NASA programs in the light of your questions has strengthened our conviction that the potential joint value of the NASA and Defense Department programs can be more fully realized by closer collaboration in the early conceptual phases, to insure that the objectives of each agency are clearly recognized at each successive stage of program evolution. Our point is illustrated by recent experience in developing experiments to be sponsored by the Defense Department on the Gemini flights. Although the concerted joint effort, under the guidance of the Gemini program planning board, has been fruitful, it has become obvious that an earlier beginning of intensive collaboration might well have allowed greater results. Looking to the future, we hope to apply this lesson in the establishment of an orbital space station program. We view a space station as an essential preliminary to an operational space defense system, and as a valuable source of experience which will apply to NASA objectives. As this program is in the early conceptual phase, we think it possible to begin now to fully exploit its high potential to serve both civilian and military needs. As you know, the Defense Department has proposed to NASA that there be collaboration in studies leading to the definition of the space station program. . . ." (Letter, *CR*, 8/9/63, 13900)

August 7: In letter to FCC Chairman E. William Henry, ComSatCorp Chairman Leo D. Welch said ComSat directors agreed that stock be issued to the public "at the earliest practicable date" and were planning stock issuance not later than early 1964. Replying to FCC contention that stockholders should have voice in ComSat policy making (see July 26), Welch asserted that FCC was invading management functions of the directors and argued that the corporation's initial studies were necessary to give potential investors "a responsible presentation of material facts in compliance with the applicable Federal and state securities laws." (N.Y. *Herald Trib.*, 8/8/63; Clayton, *Wash. Post*, 8/8/63)

- A City of Los Angeles Commendation award was presented to Air Force Space Systems Division by Los Angeles mayor, The Honorable Samuel Yorty. The citation, in recognition of SSD's contribution to the national space effort, was accepted by General Ben I. Funk, Commander. (*A-N-AF Journal and Register*, Jan., 1964)

August 8: With launching of Nike-Cajun sounding rocket from Kronogård Range, Sweden and U.S. completed series of sounding rocket experiments to study noctilucent clouds near Arctic Circle. Sponsored by NASA and Swedish Committee for Space Research, program included launchings of Arcas rockets with payloads to measure winds and Nike-Cajun rockets with payloads to make direct cloud samplings during 1961 and 1962. Four Nike-Cajun rockets with rocket grenade payloads were successfully launched during summer 1963, these experiments measuring upper atmosphere temperatures, winds, pressure, and density and measuring changes in size of artificial cloud particles created by smoke puffs from the payloads. Experimenters were scientists from Institute of Meteorology, Univ. of Stockholm; NASA Goddard Space Flight Center had responsibility for U.S. coordination in the project. (NASA Release 63-179)

- Senate debate of H.R. 7500, NASA authorization bill for FY 1964, continued. Sen. Stephen Young (D.-Ohio) said:

"The argument has been made that it is unwise to spend billions of dollars to explore outer space when we have so many unsolved problems remaining here on earth. In my judgment, however, and I am sure most of my colleagues agree, elimination of expenditures for space exploration would not necessarily result in the transfer of those funds to other forms of research. Moreover, an investigation of the funding of all forms of Federal research reveals that no real conflict exists between space exploration and research in other fields.

"In fact, this investigation provides persuasive evidence that, rather than interfering with other forms of research, our space effort has stimulated a more favorable climate for research in general, so that progress is being made simultaneously in a great many areas totally unrelated to space . . . ." (*CR*, 8/8/63, 13751)

August 8: Urging passage of $5.5 billion NASA FY 1964 authorization bill, Sen. Stuart Symington said: "It should be made clear that the United States is not committed to an all-out race, a crash program, so as to be first to the moon.

"The moon is no finish line in a race. It is but a proving ground, where we will test and demonstrate that competence in space technology which our security should insure . . . .

"[There is] a proper division of effort between military and civilian space programs, between space science and space technology, between current programs and advanced research in support of future programs, between the space program and other possibilities for substantial Federal contributions to research and development." (AP, *NYT*, 8/9/63)

- NASA Manned Spacecraft Center announced drop-test of parachute recovery system for two-man Gemini spacecraft had been conducted successfully, the drop made from 20,000 ft. into water at Salton Sea, Calif. (MSC Release 63-125; UPI, *NYT*, 8/10/63)
- Mackay Trophy presented to crew of SAC B-58 for record flight of Mar. 5, 1962, which established new Los Angeles-to-New York flight time of 2 hr., 58.7 sec. Administered by National Aeronautic Association, trophy was presented by Under Secretary of AF Dr. Brockway McMillan in ceremonies at Carswell AFB, Tex. (see May 22). (DOD Release 1119-63)
- Nike-Zeus antimissile missile fired at White Sands Missile Range in series testing components for Nike X. (DOD Release 1121-63)

August 9: Voice and teletype messages exchanged via SYNCOM II communications satellite between ground station at Paso Robles, Calif., and communications ship *Kingsport* in Lagos Harbor, Nigeria. The test spanned 7,700 mi., greatest surface distance ever spanned between two points on earth via a communications satellite. (NASA "SYNCOM II Fact Sheet"; N.Y. *Herald Trib.*, 8/10/63)

- Senate passed NASA authorization bill (H.R. 7500) providing $5,511,520,400 in FY 1964, a restoration of $307,801,000 cut by House action Aug. 1. Senate bill would authorize $4,225,275,000 for research and development; $539,185,000 for administrative operations; and $747,060,400 for construction of facilities. Senate adopted amendment to bar, under certain circumstances, any authorized funds for R&D for exclusive benefit of any person providing satellite communication services (other than Government agency) and amendment requiring that, in addition to study of location for Electronic Research Center, there would be written notice to Administrator that study committee had no objection to selected location. Bill was ordered to House-Senate conference committee. (*CR*, 8/9/63, 13877-889, 13893-912)

August 9-11: TIROS VI and VII meteorological satellites observed Hurricane Arlene approximately 600 mi. northeast of Bermuda, Typhoon Bess approximately 100 mi. west of Japan, and Typhoon Carmen approximately 500 mi. east of the Philippine Islands. (GSFC Historian Memo, 9/4/63)

August 9: NASA Langley Research Center announced contract awarded to Basic Construction Co. for building to house synchrocyclotron of NASA's Space Radiation Effects Laboratory at Oyster Point, Newport News, Va. (LARC Release)

- Letter from Sir Bernard Lovell, Director of Jodrell Bank Experimental Station, to NASA Deputy Administrator Dr. Hugh L. Dryden reporting on his recent visit to U.S.S.R. was inserted in *Congressional Record* by Sen. Joseph S. Clark (D.-Pa.). Sir Bernard quoted M. V. Keldysh, president of Soviet Academy of Sciences, as saying U.S.S.R. had rejected, at least for present, any plans for a manned lunar landing, because of insurmountable problems of radiation in space. According to Sir Bernard, Keldysh said "that the manned project might be revived if progress in the next few years gave hope of a solution of their problems, and that he believed the appropriate procedure would be to formulate the task on an international basis. He stated that the Academy believed that the time was now appropriate for scientists to formulate the task on an international basis (a) the reasons why it is desirable to engage in the manned lunar enterprise and (b) to draw up a list of scientific tasks which a man on the moon could deal with which could not be solved by instruments alone. The Academy regarded this initial step as the first and most vital in any plan for proceeding on an international basis . . . ."

  NASA Administrator James E. Webb's reply to this letter also was inserted, in which he said:

  ". . . With regard to space research and exploration, as you know, our present relationships with the Soviet Union have developed directly from the correspondence between President Kennedy and Chairman Khrushchev on specific possibilities of cooperation in this field. Dr. Dryden's discussions with Academician Blagonravov over the past year or so have followed within this framework. There is already a current agreement between the Soviet Academy of Sciences and NASA which represents the first fruit of these early efforts.

  "Accordingly, if the Soviet Academy is indeed interested in the matters you describe in your letter, we will look forward to the possibility of further explorations by Dr. Dryden and Academician Blagonravov as to their views and desires . . . ." (Letters, *CR*, 8/9/63, 13903)

August 10: Soviet government announced Cosmonaut Valentina Tereshkova would visit Prague Aug. 15 at invitation of Czech government. (AP, Wash. *Sun. Star*, 8/11/63, A6)

- Company began final razing of World War II TNT manufacturing facilities at Plum Brook Station. At the height of production during WW II, 12 TNT production lines were in operation and 6,000 employees were at work. The 6,000-acre site now is the home of a nuclear research reactor and numerous rocket engine test stands. (LRC Release 63-65, Lewis Chronology, 7)

August 11: Dr. Edward P. Ney, Univ. of Minnesota physicist, said in *New York Times* interview that sightings by Astronauts Glenn, Carpenter, and Cooper during their orbital Mercury space flights represented the "first direct observation" of what probably is cosmic dust in the airglow layer of about 48- to 54-mi. altitude. Major Cooper reported that light of two stars in Big Dipper had been extinguished by airglow layer, and Dr. Ney said this was "strong evidence for the existence of dust" which could block out the light. (Toth, *NYT*, 8/12/63, 17)

August 11: Sen. Clinton P. Anderson (D.-N.M.), Chairman of Senate Committee on Aeronautical and Space Sciences, said in television interview that "with very good luck" the U.S. could land a man on the moon by 1968 or 1969, but he personally thought U.S. lunar landing would be in early 1970's. Senator Anderson said he doubted reports that U.S.S.R. was not racing U.S. to the moon. (CBS-TV "Washington Reports," AP, *Wash. Eve. Star*, 8/12/63)

- Questionnaire regarding Government agencies' policies and practices in preparing and releasing information to the Congress and the public was sent to all Federal agencies by House Foreign Operations and Government Information Subcommittee. (AP, *NYT*, 8/12/63)

August 12: Goddard Space Flight Center engineers directed radio signal sent from Lakehurst, N.J., to SYNCOM II communications satellite 22,000 mi. in space, signaling hydrogen peroxide jets onboard to fire and slow satellite's drift westward from rate of about 7° to about 2° per day. With aid of additional maneuvers signaled from ground, SYNCOM II was expected to attain synchronous orbit within three days, appearing to hover over 55° west position and prescribing figure-8 pattern 33° north and south of equator. (AP, *NYT*, 8/12/63)

- TELSTAR II communications satellite resumed operating for first time since July 16. Satellite received and transmitted sound and television picture test sent by Bell Telephone Laboratories on TELSTAR II's 622nd orbit. Cause of the satellite's restoration, like its silence, was not known. (AP, *Wash. Post*, 8/13/63; *NYT*, 8/13/63, 12)

- S-IV stage for the Saturn SA-5 launch vehicle completed its static-test program with a full-duration firing lasting 479 sec. (*Av. Wk.*, 8/26/63, 33)

- Third birthday of ECHO I balloon satellite, launched into orbit Aug. 12, 1960. World's first artificial passive communications satellite, the 100-ft. inflatable mylar sphere demonstrated use of radio wave reflection for global communications. ECHO I was still orbiting the earth, having traveled some 425 million mi., and was still usable for communication.

    NASA announced G. C. Schjeldahl Co. had been selected to build Echo II satellite, scheduled to be placed in orbit this winter. Under $362,000 contract Schjeldahl would build three models, one for static-inflation tests, one for orbital flight, and one for backup. ECHO II would be larger and more rigid than its predecessor. (NASA Release 63-183; Rep. Joseph E. Karth, *CR*, 8/12/63, A5097; AP, *NYT*, 8/13/63)

- First anniversary of first tandem manned space flight, which began Aug. 12, 1962, when VOSTOK IV piloted by Lt. Col. Pavel R. Popovich was launched into nearly same orbit as that of VOSTOK III piloted by Maj. Andrian C. Nikolayev, launched into orbit less than 24 hrs. earlier. At one point the two spacecraft were within 6.5 km. (4 mi.) of each other, according to Soviet newspaper *Pravda*. Upon landing Aug. 15, Maj. Nikolayev in VOSTOK III had completed 64 earth orbits, 1,645,000 mi.; and 95 hrs., 25 min. flying time. Colonel Popovich in VOSTOK IV had completed 48 earth orbits; 1,242,000 mi.; and 70 hrs., 29 min. flying time.

August 12: NASA selected Bendix Eclipse-Pioneer Div. and RCA Data Systems Div. to provide components and equipment for Saturn IB and Saturn V guidance and checkout systems. Bendix and RCA already were providing similar services in Saturn I launch vehicle system. (NASA Release 63-182)

- Sir Bernard Lovell, Director of Britain's Jodrell Bank Experimental Station, said in *U.S. News and World Report* interview that U.S.S.R. did not have a priority program for landing a man on the moon. The president of the Academy of Sciences [Mstislav Keldysh] said that the Soviet Academy saw certain insuperable difficulties.

  "First, they did not at the moment see how it was possible to protect any lunar voyages from the lethal effects of solar radiation.

  "Secondly—and mixed up with this—is that they did not think it was economically feasible with present techniques to land sufficient material on the moon to give the necessary protection against solar radiation and, at the same time, to enable a scientific program to be carried out and also to give the chap a reasonable chance of getting back to earth.

  "Most important of all, it seemed to me, is this: They have decided that with a soft landing of instruments on the moon they can extract nearly all the scientific information they want long before there would be a chance of getting a man there to do it . . . .

  "I told the president of the Academy quite frankly that I didn't agree with this sentiment, because I'm a very firm believer in the importance of getting a man on the moon not only for scientific reasons but also because it is a great challenge facing the human race . . . ."

  Sir Bernard said U.S.S.R. was concentrating on two objectives in space exploration: (1) "soft landing of instruments on the moon . . . and, subsequently, other planets" and (2) "assembly of a platform in space . . . . The cosmonaut program is related . . . to this effort to erect a big platform in space and put a telescope on it . . . ." (Interview, *U.S. News*, 8/12/63)

- Washington *Evening Star* editorialized that NASA "should lose no time in exploring the possibilities suggested by Sir Bernard [Lovell] in his letter to James Webb, NASA's head. The Russians may be sincere about the matter; anyhow, we should try to find out whether they are or not." (Wash. *Eve. Star*, 8/12/63)

- NASA announced three-year $400,000 grant to Univ. of Washington for research in new, advanced ceramics and the improvement of existing materials for use in the space program and for industrial applications. (NASA Release 63-180; AP, Wash. *Eve. Star*, 8/12/63)

August 12: NASA Director of International Programs Arnold Frutkin discussed U.S.-U.S.S.R. cooperative meteorological satellite program in *Aviation Week and Space Technology* interview. Beginning late next year, NASA would launch Nimbus meteorological satellites into polar orbit and U.S.S.R. would launch polar orbiting satellites. Paths of two satellites would be phased so that they are 90° apart, so that one satellite of each pair will cross each spot on globe every six hours. (*Av. Wk.*, 8/12/63)

- Experiments to determine behavior of dust on moon's surface described by Dr. John W. Salisbury of AFCRL Space Physics Laboratory in *OAR Research Review*. Conducted in lunar simulation chamber in which effects of low pressure, radiation, and temperature on physical properties of probable lunar surface materials were studied, experiment demonstrated for first time that silicate powders, like metals, adhere to one another in ultrahigh vacuums. Particles also adhere to glass and metal surfaces. Experiment director Salisbury concluded "properly designed vehicles will not sink away into a sea of lunar dust, should one exist. The tendency of dust particles to stick to all metal and glass surfaces will, however, provide an unusual difficulty in the operation of camera lenses, portholes and mirrors." (*OAR Research Review*, 8/12/63, 3-4)

- In address before National Editorial Association's airplane division, inserted in *Congressional Record* by Rep. Garner E. Shriver (R.-Kans.), Boeing Co. president William M. Allen discussed proposed Government-industry program to develop supersonic transport aircraft in light of past progress in U.S. aircraft development. "The first large expansion of passenger air transportation in 1931 employed trimotored airplanes each of which provided on the average, in actual operation, about 3,500 seat-miles of transportation per day. These airplanes cost $75,000 each. In rapid jumps we have been able successively to increase the capability of the product until now, in the form of the 707 or the DC-8, it is delivering 400,000 seat-miles per day, which is 114 times the output of the 1931 trimotor. The jet airplane cost is $6 million, or 80 times the cost of the trimotor.

    "Each step of progress which has made this change possible has been larger than the one before, and the jump to supersonic is by far the largest of all. The development cost has mounted accordingly . . . ." He cited Boeing's financing in 1935 of prototype four-engine airplane which became the Flying Fortress, at cost of $660,000, and Boeing's designing and building in 1952 of first U.S. jet transport, at cost of nearly $20 million. "A prototype supersonic transport of the type regarded now as most practical is estimated to cost a possible $150 million . . . .

    "Already Boeing has expended in the neighborhood of $15 million on supersonic transport study . . . . But the big costs are still ahead in supersonic prototype development . . . .

    "It becomes evident that there is valid reason for the initiative which our Government is taking in developing means to finance the program and that the continued cooperation of Government, the manufacturing industry and the airline industry is required in working out a program that can be practically accomplished with the economic capability of industry and will serve the national interest.

"In carrying out this program and undertaking to meet the intensive competition which we will face from abroad, we feel it is important, as Federal Aviation Agency Administrator N. E. Halaby has recognized, that we retain the advantages of the private enterprise system of contractual responsibility which has proved so effective in the past . . . ." (Text, CR, 8/12/63, A5106–07)

August 12: Transfer of Pacific Missile Range from USN to USAF would be accomplished in six months, *Missiles and Rockets* reported. USN would retain Pt. Mugu portion and continue to operate range instrumentation ships; Pt. Arguello would be merged with Vandenberg AFB, which would be transferred from Strategic Air Command to AF Systems Command. New division to be created under AFSC would operate PMR, AMR, and Satellite Test Center (composed of Sunnyvale facilities and Hickam AFB recovery group). (M&R, 8/12/63, 12)

- U.S. Weather Bureau was considering purchasing Japanese weather observation rockets from Prince Motors, Ltd., of Tokyo. (M&R, 8/12/63, 9)

August 12–16: International Symposium on the International Geophysical Year held in Los Angeles and attended by 300 scientists from 32 countries. Sponsored by Geophysics Research Board of NAS-NRC, symposium featured discussions of discoveries made and advances achieved since beginning of IGY in 1957. (NASA AFC, *Consolidated Calendar*, 6/15/63; Wash. *Eve. Star*, 8/12/63)

- Conference on artificial satellites held at Virginia Polytechnic Institute, Blacksburg, Va., sponsored by VPI in cooperation with National Science Foundation and NASA Langley Research Center. 13 NASA scientists and engineers were among the featured speakers. (NASA Release 63-173)

Dr. Homer E. Newell, NASA Director of Space Sciences, said in introductory lecture:

"Those who argue that we should dispense with the frills of science and space exploration, and concentrate on the necessities of military development, forget that we can't really say what the military necessities in space will be. Our crystal ball is not that good, and it would be foolhardy to pretend that it is. We do not wish to develop a Maginot line in space, only to have it flanked by forces of greater flexibility. We need to develop in a broad way our space capability so that we shall have the ability to move in any direction required by future events to meet any threats along whatever lines may develop.

". . . Out of this broad activity in space will come the ability of the United States to use space and to operate in space either as it may chose to do voluntarily or may find itself compelled to do in its own defense. The development of our ability to operate in space, including manned space flight, gives to our country another dimension in which to meet the challenges—both opportunities and threats—of the future. We can do engineering in space, advance our science in a way that cannot be accomplished at the surface of the earth, and extend the range of practical applications for the benefit of man. And, if necessary, we can thwart the attempts of any enemy to use space against us.

"In this day and age we cannot afford to ignore this last point. In our own self interest, and for the safety of our country, we cannot permit others to develop space capabilities that we cannot match, and that may, therefore, be used disastrously against us.

"This is a capability we must have to ensure our survival in the space age as the independent, self-determining nation that our forefathers set us up to be, and that we have always insisted on being.

"This is the capability that we shall have from the development of the ability to investigate scientifically with satellites and space probes, and from space applications, from the ability to perform manned space flight and manned space operations, from the vast complex of manufacturing and assembly plants, launchings complexes, tracking and telemetering facilities, and from the invaluable experience that this initial stage in the space program will give us.

"This is the most significant point about the present era in space. This is the most important aspect of the present activity in space.

"We are now laying the groundwork for whatever role we may have to play in space in the future. We are ensuring that no one will ever be in a position to use space against us while we, helpless and frustrated through lack of the necessary space capability, have to take what comes . . . ." (Text)

Dr. Wilmot N. Hess of Goddard Space Flight Center told Conference August 14 that U.S. high-altitude nuclear explosion in 1962 provided new information for determining path and rate of decay of high-energy electrons, thus providing better understanding of the Van Allen radiation belts. "While the nuclear explosion shortened the lives of some satellites and increased the intensity of radiation around the globe, it had a valuable side effect.

"To students of the Van Allen belt, it was almost like having a controlled experiment in space . . . ." (AP, *NYT*, 8/15/63, 11)

ANNA I geodetic satellite's flashing beacons had resumed after more than two months of blackout, Richard B. Kershner of Johns Hopkins Univ. Applied Physics Laboratory disclosed in Aug. 15 press conference. In Blacksburg for Artificial Satellites Conference, Kershner said: "We have no explanation for the lights coming back on two weeks ago. We don't like to believe in space gremlins, but we've reached the point where that's as good an explanation as any." (AP, *Chic. Trib.*, 8/16/63)

Dr. Floyd L. Thompson, Director of NASA Langley Research Center, addressed banquet. Dr. Thompson called on educators attending the conference to join with NASA and other research organizations to develop new and more effective means for accelerating effective distribution and assimilation of newly acquired research information. He cited prospective establishment of Virginia Associated Research Center (VARC) adjacent to the LARC Space Radiation Effects Laboratory at Newport News, Va., as a new and imaginative device to promote flow of knowledge between a major laboratory and three outstanding Virginia institutions of higher learning. (LARC Release)

Christopher C. Kraft, Jr., Chief of NASA Manned Spacecraft Center's Flight Operations Div., said Aug. 16:

"The first U.S. manned spaceflight program [Project Mercury] was designed to (1) put a man into earth orbit, (2) observe his reactions to the space environment, and (3) bring him back to Earth safely at a point where he could be readily recovered. All of these objectives have been accomplished, and some have produced more information than we expected to receive . . . ."

Discussing the stringent reliability requirements of manned space flight: "The smallest mistake in a man-rated system can bring totally unexpected results. The unexpected is the rule in the unknown, and if man is going to live in the region beyond our atmosphere, he is going to live under new rules or not at all. We have been aware of these rules . . . but they have not been brought to our attention so vividly as they have in the manned flight program . . . .

"The manned space flight program has added greatly to our knowledge of the universe around us and demonstrated that man has a proper role in exploring it. There are many unknowns that lie ahead in space, but we are reassured because we are confident in overcoming them by using man's capabilities to the fullest . . . . We now depend on man in the loop to back up automatic systems rather than using automatic systems alone to insure that the mission is accomplished . . . . We have arrived at what we think is the proper mixture of that formula. Man is the deciding element; but we cannot ignore the usefulness of the automatic systems."

Kraft told news conference that two engine problems were encountered in the Titan II booster for Project Gemini. Titan II's first stage generated chugging oscillation which could cause "the pilots to have their eyeballs shook out." Second stage engine was producing only about 90% of desired thrust. Both problems apparently were associated with turbine drive mechanisms (MSC Release 63–126; AP, *Chic. Trib.*, 8/17/63)

Gerald M. Truszynski, NASA Deputy Director of Tracking and Data Acquisition, discussed satellite data recovery and tracking system for manned satellite program. He described the manned space flight network used in Project Mercury and the augmentations planned for support of Project Gemini, concluding:

". . . while certain equipment augmentations will be effected, many of the techniques learned during the Mercury program, particularly in terms of procedures required to assure positive Network support, will be utilized in Gemini and in this way, an ever-increasing background of know-how will be available for the extremely complex missions which will be encountered in the manned lunar landing program shortly to be upon us."

August 13: Spokesman for Westinghouse Defense Center, which was preparing program definition study of nonmilitary navigational satellite under NASA contract, predicted navigation by satellite could become reality "in three or four years . . . . Using electronic sensors, the system would determine automatically the position of each ship or aircraft with respect to known references and relay this information to a ground station. This station would then compute the navigation 'fix' and relay this information to the user via the satellite . . . ." (*Space Bus. Daily*, 8/13/63, 230)

August 13: Sen. E. L. Bartlett (D.-Alaska) introduced in Senate a bill (S. 2038) to establish a Congressional Office of Science and Technology (COST) to advise and assist members and committees of the Congress on matters relating to science and technology. Bill was referred to Senate Committee on Rules and Administration. (NASA Leg. Act. Rpt. II/125)

• Speaking on floor of Senate, Sen. Barry Goldwater (R.-Ariz.) advocated deployment of Nike-Zeus and development of more advanced antimissile missiles:

"Imagine a world, treaty or no treaty, in which the Soviets had achieved a substantial lead in development of an anti-ICBM. Where then would be the sword and shield of peace? What then would be the position of the Soviets to back their aggressions, and what would be our ability to stop them?

"President Kennedy is fully confident that we can put a man on the moon—no mean feat—and is willing to spend $5 billion a year to do it . . . .

"I have talked enough to competent scientists and engineers who believe that a good anti-ICBM, capable of protecting the American people, would be no tougher task. But of course it would require an equal enthusiasm, money, and effort as our moon program does to accomplish. How can we fail to apply anything less than our best efforts to produce it? . . ." (*CR*, 8/13/63, 14043)

August 14: New information on solar wind, obtained from analysis of data from MARINER II interplanetary probe, released by National Academy of Sciences. Data showed continuous flow of solar wind from sun due to expansion of solar corona. Velocity of solar wind is between 300 and 800 mi. per sec., indicating great fluctuations in coronal expansion. Temperature of solar wind ranges around 500,000°. On approximately 20 occasions, velocity of solar wind increased from 20 to 100% in periods of a day or two; these fluctuations correlated with magnetic disturbances on earth. (Henry, Wash. *Eve. Star*, 8/14/63)

• At JPL briefing of NASA Hq. scientists, it was reported that Ranger 6 would be ready for flight to the moon by about Thanksgiving and Mariner 3 would be available for flight past Mars by fourth quarter of 1964. (Hines, Wash. *Eve. Star*, 8/14/63)

• AEC announced discovery of "anti-xi-zero" antiparticle by team of physicists from Yale Univ. and AEC's Brookhaven National Laboratory. AEC said discovery confirmed fundamental theory of physics which states that for every elementary particle there must be an antiparticle. Anti-xi-zero antiparticle filled in the final gap in family of elementary particles and their antiparticles, AEC said. (Carey, AP, *Nashville Tennessean*, 8/15/63)

August 14: "A Report Covering Evaluation of Areas Considered for Proposed NASA Electronics Research Center," by Robert C. Sellers & Associates, inserted in *Congressional Record* by Rep. John W. Wydler (R.-N.Y.). Report concluded:

"1. NASA's statistics on graduate educational capability in the Boston area do not stand up as against that of New York [metropolitan area].

"2. NASA's inference that the industrial complex in the Boston area offers potential benefits does not stand up in light of the current backlogs in that area without expansion of existing organizations, as against New York's Long Island area.

"3. In reviewing NASA's statements, the Senate and House committees should also ask: After seeing overexpansion's disruptive effects in other areas of the Nation, is it good planning to create a new situation when other areas can support their needs with minimal expansion?" (Seller Report, *CR*, 8/14/63, 14203-04)

- Brig. Gen. Joseph Bleymaier, Titan III program director, quoted as saying USAF planned to fly at least 10 X-20 (Dyna Soar) spacecraft, the first two to be unmanned and remaining eight carrying astronauts. (*Space Bus. Daily*, 8/14/63, 238)

- Reported that a U.S. military satellite, launched by Agena, had established a record 46 days of operation under continued attitude stabilization. (*Space Bus. Daily*, 8/14/63)

- President Kennedy named Eugene R. Black, former president of International Bank for Reconstruction and Development, as special adviser on financing development of U.S. supersonic transport airplane. Stanley deJ. Osborne, chairman of Olin Mathieson Chemical Corp., would assist Black as deputy adviser. (*Wall Street Journal*, 8/15/63; CTPS, *Chic. Trib.*, 8/15/63)

- Missile tests for Project Dazzle, a defense research program to develop antimissile missiles, involving cooperation between the United States, Britain, and Australia, would be conducted at Australia's big Woomera rocket range, using the British-developed Black Knight rocket.

    Project Dazzle would involve identifying and tracking missiles re-entering the earth's atmosphere from space. The program would include basic research into the physical phenomena associated with the re-entry of objects from space. Australian and British natural scientists would be responsible for analyzing the results recorded by an instrumentation system provided by ARPA. (*U.S. Naval Institute Proceedings*, November, 1963; *The Christian Science Monitor*, 8/14/63)

- Dr. Albert J. Kelley, NASA Director of Electronics and Control, said in address before Fourth International Electronic Circuit Packaging Symposium:

    "Electronics and its associated disciplines constitute the brain, nerves and senses of flight vehicles. It is estimated that about 40 per cent of our booster costs, 70 per cent of our major spacecraft dollars, and 90 per cent of our tracking and data acquisition funds go into electronics. Unfortunately, in the same context approximately 90 per cent of our flight failures, not to mention flight delays, arise from electronic failures. Success in this area is obviously a major factor in over-all mission accomplishment since a dead or unintelligent vehicle is useless if it cannot measure data and send it back to earth . . . .

"NASA electronic research must satisfy the following requirements:

"Undertake, under contract and in-house, advances in the electronic art to meet the requirements of future space flight.

"Guide industry and university sponsored research, in space electronics, into channels useful to both NASA and the sponsoring industry.

"Provide facilities to verify the performance of advanced electronic equipment developed for space use, irrespective of source of funding.

"Conceive, develop, ground test, flight test and interpret data for NASA flight experiments designed to prove advanced electronic equipment 'in-use.' "

Dr. Kelley then described the functions of the proposed Electronics Research Center, which would be "the focal point of our electronics research efforts." (Text)

August 14: GSFC scientist reported U.S. nuclear blast above Pacific in 1962 disclosed key data on travel pattern and life of high-energy particles in Van Allen belt with accuracy to a distance of 5,000 mi. (*A&AE*, Oct. 1963, 7)

- FAA Administrator Najeeb E. Halaby predicted in news conference that U.S. supersonic transport aircraft "will be in actual air service not more than six months after the [British-French] Concorde starts carrying passengers." Prototype of Concorde was expected to be flying by end of 1966. Mr. Halaby said U.S. should be able to test-fly supersonic transport by 1967 that would be "better, faster and longer-ranged with superior handling and capacity characteristics." (UPI, *NYT*, 8/15/63, 48)

- FCC Chairman E. William Henry predicted in address to American Bar Association's Committee on Communications, meeting in Chicago, that a single communications satellite eventually may equal carrying capacity of all telephone cable circuits between U.S. and rest of the world. Last year 5 million telephone calls were made from U.S. to 176 countries. Existing facilities probably will be used to capacity by 1965, he said, and $500-million international communications system will be in operation by 1970. New transistorized cable, with capacity of almost six times present capacity, will be available. (Wolters, *Chic. Trib.*, 8/15/63)

August 15: SYNCOM II communications satellite was successfully maneuvered into synchronous position 55° west longitude, over Brazil and South Atlantic Ocean. The satellite was slowed from westward drift speed of 1.2° per day to nearly zero drift by firing of onboard nitrogen jets. Previous firings of hydrogen peroxide jets occurred Aug. 11, when SYNCOM II was slowed from 7° drift per day to 2.7° drift per day, and Aug. 12, when satellite was reduced to 1.2° drift. Firings were directed by engineers at NASA Goddard Space Flight Center, and actual command was executed from ground station at Lakehurst, N.J. SYNCOM II was now stationed about 22,300-mi. altitude and traveling at speed of about 6,800 mph, matching earth's rotation speed of 1,040 mph at equator to keep it on station. It was hovering in figure-8 pattern 33° north and south of equator. NASA Administrator James E. Webb called completion of the positioning maneuvers the culmination of "one of the outstanding feats in the history of space flight." (NASA "SYNCOM II Fact Sheet"; NASA Release 63-185)

August 15: Dr. Wernher von Braun, Director of NASA Marshall Space Flight Center, announced appointment of Robert B. Young as MSFC Director of Projects and Industry Operations. Presently Vice President and General Manager of Aerojet General Corp.'s Sacramento Plant, Young would assume his MSFC position Nov. 1. (NASA Release 63-184)

- NASA Administrator James E. Webb said in letter to Sen. Thomas Kuchel (R.-Calif.): "It is not NASA policy to direct or instruct its prime contractors as to where to place their subcontracts or the identity of subcontractors." Statement was in response to Senator Kuchel's request for comment on charges that NASA was ordering contracts be awarded in specified locations. NASA provisions in prime contracts regarding subcontracting policy: selection of subcontractors be made "on a competitive basis to the maximum practical extent"; subcontracts be awarded to small business firms in labor-surplus areas "to the maximum practical extent"; NASA must approve contractor's plan to manufacture more than $1 million worth of space work outside of contractor's own plants; and NASA reviews contractor's basis for selecting proposed subcontractor "and the degree of competition obtained by the prime contractor." (MacDougall, *Wash. Post*, 8/16/63)

- Dr. T. L. K. Smull, NASA Director of Grants and Research Contracts, said in commencement address, Univ. of Georgia:

  "Today we are dealing not only with profound changes in a social and economic structure of our country and of the world; we are dealing as well with even more profound changes in man's own conception of the boundaries and limitations of his habitable environment and of his understanding of the forces of the universe.

  "Therefore, the opportunities available to you graduates in the next decade are, in a very real sense, boundless. The challenges of tomorrow are not just for the scientist, the engineer, and the technician working toward scientific developments, but for all men and women in all professions—for every one of you. Even if science is not your field, you must understand what it implies simply because science will continue affecting your lives with even greater impact as each year passes . . . .

  "Never before has a civilization seized upon and exploited a new technology at the rate man is now exploring the new dimension of space. In the 2,060 days since the Russians launched Sputnik, NASA has conducted 96 major vehicle launchings, an average of one every 22 days. In this period, the USSR has announced about 35 successful space flights.

"NASA's impressive accomplishments will be dwarfed by United States activities during the rest of this decade as our understanding of the nature of space, the sun, and the map of the Universe is greatly enhanced.

"In an age of exploding knowledge in all of the physical sciences no area of human activity better focuses and dramatizes the meaning of today than does this boldest, most challenging arena of space . . . ." (Text)

August 15: FAA issued to industry the Request for Proposals establishing performance objectives for U.S. supersonic transport airplane. Initial designs were to be submitted to FAA by Jan. 15, 1964. (FAA Release 63-76)

- Rep. R. C. Pucinski (D.-Ill.) introduced in House an amendment to Federal Aviation Act of 1958 to make illegal any civil supersonic aircraft within U.S. that generated sonic boom overpressures exceeding 1.5 lbs. per sq. ft. on ground directly beneath its flight path. (*CR*, 8/15/63, 14271-76)

- 85-ft.-tall research balloon drifted out over Pacific Ocean at 90,000-ft. altitude and was reported "hopelessly lost" by balloon flight center at Palestine, Tex. Balloon was sent aloft Aug. 13 to gather meteor particles, and had been intended to land at Pecos, Tex. (UPI, *Wash. Post*, 8/16/63)

- Findings of Trendex survey of public attitude on space exploration, conducted by Hicks and Greist, Inc., advertising agency: "There seems to be a definite relationship between age and attitude toward the space program. Generally, there is an increasingly negative opinion about the program with increase in age . . . ." (Kaselow, N.Y. *Herald Trib.*, 8/15/63)

- USAF successfully fired Titan I ICBM from Vandenberg AFB, a routine training launch by SAC crew. (DOD Release 1142-63)

August 16: NASA announced it and Soviet Academy of Sciences had approved Memorandum of Understanding providing for implementation of cooperative space agreement reached June 1962 in Geneva. Memorandum outlined procedures for U.S.-U.S.S.R. exchange of meterological data from weather satellites operated by each country, joint communications experiments using Echo II passive reflector satellite, and contributions of geomagnetic data from special respective satellites to World Magnetic Survey in 1965. Memorandum of Understanding was drafted at meetings between U.S. negotiators, led by NASA Deputy Administrator Dr. Hugh L. Dryden, and Soviet negotiators led by Academician Anatoli A. Blagonravov, Chairman of Commission of Exploration and Utilization of Outer Space of Soviet Academy of Sciences. (Memorandum of Understanding; NASA Release 63-186)

- At International Geophysical Year Conference in Los Angeles, Dr. W. G. Beynon of Great Britain said that plans for another IGY of international cooperation during 1968-69 peak sunspot activity were being discussed by Committee on Space Research (COSPAR). (Wash. *Eve. Star*, 8/16/63)

August 17: NASA Ames Research Center announced laboratory synthesis of Adenosine Triphosphate (ATP) molecule, universal medium of energy exchange in earth's life-forms. ATP was most complex molecule yet produced in a laboratory, according to Dr. Cyril Ponnamperuma of Ames Exobiology Div., who performed the experiment by shining ultraviolet light on nine combinations of chemical mixtures. Dr. Ponnamperuma said the nine different mixtures may have resembled composition of earth's oceans at the dawn of life four billion years ago. "Such experiments are lending significant support to the theory that biological molecules, which are the prerequisites of life, could have appeared by the interaction of forces and materials which existed on the earth before life did." Collaborating in the ATP synthesis were Dr. Carl Sagan of Smithsonian Astrophysical Laboratory, on the theoretical level, and Ruth Mariner of Ames. (AP, Wash. *Sun. Star*, 9/18/63)

- Centaur stage was fired for first time with both RL-10 engines equipped with flight-type propellant supply systems. Conducted at General Dynamics/Astronautics, San Diego, test ran for planned 30 sec. in steam-diffusion chamber to simulate space vacuum. (*M&R*, 8/26/63, 12)
- NASA Deputy Assistant Administrator for Technology Utilization and Policy Planning, Julian Scheer, said Lt. Col. John A. Powers, former Public Affairs Officer at NASA Manned Spacecraft Center, would become technical consultant to NASA's exhibits program at Hq., effective Sept. 1. Colonel Powers would supervise construction of spacecraft models used in NASA exhibits. (AP, Wash. *Sun. Star*, 8/18/63)
- NASA-DOD agreement on Manned Orbital Research and Development System (MORDS) was signed, which required AACB to coordinate all study contracts. (Joint NASA-DOD Release)
- AFSC announced the cooperative, tri-service program conducted to build a new family of VTOL. The three services were sharing equally in funding, support, and evaluation of the X-19, XC-142A, and the X-22A experimental aircraft. (*AFSC Operational Highlights*, 12)

August 18: According to results of survey by National Industrial Conference Board, about two per cent of U.S. working population and nearly one third of its scientists and engineers are engaged in research and development. Survey also showed total R&D spending, including both private industry and Government, is more than $17 billion per year. In 1960, more than one half of overall financing came from Federal government. Defense, atomic energy, and space projects accounted for 57% of all resources devoted to R&D in 1960. (AP, *Wash. Post*, 8/19/63)

- Rep. Gerald R. Ford (R.-Mich.), Chairman of House Republican Conference, named Rep. Charles S. Gubser (R.-Calif.) head of 10-member study group to make a "constructive and positive examination" of aeronautics and space fields. (UPI, *NYT*, 8/19/63)

August 19: House Committee on Appropriations' Subcommittee on Independent Offices began hearings on NASA appropriations for FY 1964. NASA Administrator James E. Webb said in opening statement before the executive meeting:

"I urge that your Committee approve the total amount of appropriations authorized. I should like to point out that even if the full amount requested by the President were authorized and appropriated, we face in these funds and in the plants of the contractors a similar situation to that with our own Centers where they requested 2,000 more people than we were able to approve under the budget restriction. In our manned space program we will be short between one hundred and two hundred million dollars out of the three billion dollars requested by the President. We are prepared to operate with a tight belt, we are prepared to operate without contingencies. However, I do want to point out to the Committee that we are still going to have a serious problem if we encounter any major difficulties in the form of a catastrophe on the launching pad or with a test station.

"I should also like to say to the Committee that after the Authorization Bill has been enacted and after the Appropriations Bill has been passed, we are going to prepare an over-all operating plan. We plan to show it to both the House and Senate Space Committees and we will be happy to show it to the Appropriations Committees. With this re-examination and the basis for committing the funds actually appropriated, we will be in a position where we will know what our situation is and whether we will have to ask for a supplemental appropriation . . . ." (Testimony, NASA Leg. Act. Rpt. II/128)

- Under questioning by the House Appropriations Subcommittee on Independent Offices, NASA Deputy Administrator Dr. Hugh L. Dryden discussed the future space programs to follow the manned lunar landing program:

"There are three principal candidates for what might be called the next program. These are the manned orbital laboratory; the further exploration of the moon; and finally, man's interplanetary travel.

"One difficulty is the definition. MERCURY is a one-man orbital laboratory; GEMINI is a two-man orbital laboratory; APOLLO is a three-man orbital laboratory; and all will stay up for the duration of the trip to the moon.

"If you are going further, you come to a question, Do you want something for 3 men, 6 men, 12 men, or 20 men?

"One of our problems is to define the requirements and both the Department of Defense and ourselves are making studies. We have agreed there should be only one project for the time being because this will be a 2 or 3 billion project. We have agreed we will not proceed unilaterally. A decision, of course, has to be made as to what, if anything, in the 1965 budget will be in this field or another . . . .

"As to the further exploration of the moon, the definition is a very variable one. You might leave an astronaut out and, figuratively speaking, take camping equipment along for a few days. Or you might convert a whole lunar bug to a freight carrier to extend the length of stay. You can go from that to a lunar base where you try to build a station, as in the Antarctic, that man can occupy for some time ....

"All of these things run into the billions of dollars and certainly Congress will be involved in it.

"There are people who favor one or another of these projects. I personally favor the manned orbital laboratory as being the next project. What we find on the Moon may change our minds on this, but from present knowledge I feel priority should go to a manned orbital laboratory. Some of this equipment has been in space for a period of a year or more. That is the length of a trip to Mars or Venus and back.

"As you know, the matter of making major decisions of this kind in Government is a complicated one and I cannot say at this time when the hardware will be recommended by the administration. One reason for saying the manned orbital laboratory has priority is the possible military application." (Hearings . . . *Independent Offices Appropriations for FY 1964*, Part 3, 242–243)

August 19: New evidence verifying existence of long-lived solar plasma streams in interplanetary space was published in article by four GSFC scientists in *Physical Review Letters.* Evidence was based on data gathered by EXPLORER XII energetic particles satellite: On two occasions the satellite observed solar events producing solar plasma streams. Indication that they were long-lived was revealed when the plasma streams caused geophysical disturbances more than 20 days after the streams were created. (GSFC Release G-14-63)

- National Geographic Society announced two scientists with astronomical instruments were recently lofted by plastic-skinned balloon to 15-mi. altitude in experiment to record radio sounds from stars. National Geographic Society outlined other recent scientific contributions by man's oldest means of flight, including balloon-lifted telescopic camera which made unprecedented photographs of sunspots and another instrumented payload which obtained man's clearest observations of planet Mars. Balloon-carried instrumentation now probing upper atmosphere is obtaining valuable data on astrophysics, meteorology, and aeromedicine. (*Chic. Trib.*, 8/20/63)
- Communications Satellite Corp. announced award of contract to Univac Div. of Sperry Rand Corp. for computer services to facilitate ComSatCorp's comparative global satellite communications system studies. (ComSatCorp Release)
- *Missiles and Rockets* quoted NASA spokesman on letter from Sir Bernard Lovell in which the British astronomer indicated U.S.S.R. wanted to explore possibilities of cooperating with U.S. in lunar exploration program:

    "[The letter] . . . represents the personal impressions of a private individual on a matter of the greatest obscurity, namely the Soviet space program. . . .

    "It doesn't seem wise to give the letter any undue importance, although we were glad to get it . . . .

"We already have a channel of communication, namely the discussions between Dr. Dryden and Soviet Academician Blagonravov. It is reasonable to expect that channel would be used for a serious proposal." (*M&R*, 8/19/63, 21)

August 19: Tremendous atmospheric explosion in Kimberley Mountains of Western Australia was reported by observers, who saw trail of smoke and ball of orange flame and heard the explosion. Australian Weather Office spokesman said cause was unknown but suggested possibility of artificial satellite's exploding during atmospheric re-entry. (AP, *Tulsa Daily World*, 8/21/63)

- DOD plans for consolidation and expansion of the three U.S. missile ranges would be closely reviewed by National Aeronautics and Space Council to see if they duplicate NASA plans, *Aviation Week and Space Technology* reported. (*Av. Wk.*, 8/19/63, 30–31)

- Nat Welch, Federal representative on Southern Interstate Nuclear Board, suggested to Southern Governor's Conference ways to attract and develop nuclear and space industries in their states:

  "Suggestion 1. Be a champion of education . . . . Your greatest contribution in your State is to see that educational budgets are substantially increased . . . .

  "Suggestion 2. Capitalize on the nuclear and space industries in your State. Industry begets industry . . . .

  "Suggestion 3. Exploit your new breakthroughs . . . [such as, for Louisiana] NASA's new Michoud plant . . . .

  "Suggestion 4. Build a solid legislative base now for the growth of nuclear and space industries . . . .

  "Suggestion 5. Work with your congressional leaders . . . .

  "Suggestion 6. Romance your unique advantages . . . . Carefully inventory your industrial advantages . . . .

  "Suggestion 7. Encourage the entrepeneur as well as the scientist . . . .

  "Suggestion 8. Lend a helping hand to your existing industries . . . .

  "Suggestion 9. Recognize and appreciate [achievements of industries in your State] . . . .

  "Suggestion 10. Do your homework well . . . .

  "Your efforts in promoting the growth of nuclear and space industries will help not only your State but the entire region . . . .

  "Your efforts will also substantially contribute to the attainment of national goals. As President Kennedy points out, if we can increase our growth rate from 3 to 5 percent a year this will substantially aid in providing employment for all of our people, in increasing Federal revenues, and in creating a more dynamic economy." (Text, *CR*, 8/20/63, A5312–13)

- On 10th anniversary of first successful Redstone missile launching, Army fired Redstone rocket at White Sands Missile Range in successful test by Redstone Field Artillery unit. (*M&R*, 8/26/63, 14)

- USN launched Polaris A–3 from land pad at AMR in "very successful" flight test. (*M&R*, 8/26/63, 14)

August 19: Informants said in Cairo that Egyptian scientists were interested in obtaining a satellite tracking station from U.S. (AP, *Wash. Post*, 8/20/63)

August 20: Asked in press conference about U.S.-U.S.S.R. collaboration in manned lunar flight program, President Kennedy said: "Well, we haven't had any success in reaching any agreement. The kind of agreement, to be really meaningful, would require a good deal of inspection on both sides, and there's no evidence as yet that the Soviet Union is prepared to accept that. All we've gotten was an agreement to exchange weather information. We haven't anything more substantial." (Transcript, *NYT*, 8/21/63)

- Deputy Secretary of Defense Roswell L. Gilpatric said in testimony before Senate Committee on Appropriations' Subcommittee on DOD Appropriations: "Our goal is to establish, at an early date, a developmental communications satellite system consisting of a minimum number of simple, reliable satellites together with a minimum number of widely spaced ground stations, which can be tested and evaluated as a 'package' within the framework of the established military communications network. This program would demonstrate the technical and operational characteristics of such a system as well as its compatibility with the existing communications complex. The ability to transmit signals between a satellite and two ground stations has already been demonstrated. But the ability of a 'system,' comprising a number of satellites and ground stations, to provide a reliable means of military communication has yet to be explored . . . ." (Testimony)

- Dr. Eugene G. Fubini, Assistant to DOD Director of Defense and Engineering, said in address to National Rocket Club in Washington that DOD viewed X-20 (Dyna Soar) and Gemini projects as "insurance" against uncertainties of future military manned space missions. Dr. Fubini expressed hope that DOD-NASA cooperation in Gemini and X-20 would provide some answers to help clarify man's possible role in military space missions. (*M&R*, 8/26/63, 12, 14)

- Army Corps of Engineers announced start of construction of world's largest structure, the Vertical Assembly Building for Saturn V launch vehicle. VAB is being built on Merritt Island, adjacent to Cape Canaveral, Fla. As design and construction agent for NASA's manned space flight facilities, Corps was managing construction contracts in Merritt Island for NASA Launch Operations Center. (DOD Release 1141-63)

- Sen. Clinton P. Anderson, Chairman of Senate Committee on Aeronautical and Space Sciences, inserted in *Congressional Record* an art'cle from *Missiles and Rockets*, which said:
"The United States has gained more knowledge of outer space than the Russians because we've done a better overall job with our unmanned satellite program. We are using unmanned vehicles to help predict weather around the world, to improve navigation, to make possible international television and other communications, and to keep track of what the enemy is doing to establish missiles and other weapon systems aimed at us.

"Not only have we learned more about the planet Mars through unmanned space research, but we've gained much knowledge of the vast reaches of space in between and beyond.... We've pursued unmanned space research down far more streets and alleys than have the Soviets.

"For the most part, our unmanned satellites are yielding knowledge that will assure success with manned flights into space in the future. For that reason, if no other, it is easy for the unmanned satellite program to be overlooked, its significance lost in the international competition of manned flights. This is a pity, for what we do in unmanned space exploration today is the very foundation on which we will build future manned space exploration.

"In saying this we are not being partisan to the claim so often heard these days that we ought to forget about the manned program and explore space only with unmanned vehicles. We regard the controversy as a mere tempest in a teapot that will prove meaningless in time . . . ." (CR, 8/20/63)

August 21: USAF fired Titan II ICBM in 5,800-mi. flight down the Atlantic Missile Range, the missile carrying malfunction-detection system similar to the one to be incorporated in Titan II Gemini booster. The Titan II also carred scientific pod containing instruments to study radiation in the missile's exhaust plume. (*NYT*, 8/22/63, 22; *M&R*, 8/26/63, 14)

- S-1 stage for Saturn SA-5 launch vehicle arrived at Cape Canaveral, having left NASA Marshall Space Flight Center, Huntsville, Ala., by barge Aug. 10. (*Av. Wk.*, 8/26/63, 33)
- Gordon M. Bain, FAA Deputy Administrator for Supersonic Transport Development, told representatives of 11 foreign airlines that U.S. could build a safe, commercially profitable supersonic transport airliner. He stressed that "we don't think we are lagging far behind" British-French development of Concorde. (*NYT*, 8/22/63)
- NASA announced it would negotiate contract with Pratt and Whitney Div. of United Aircraft Corp. for continuation of RL-10 rocket engine development. Contract would complete the RL-10 qualification program; it would specify improvements to be incorporated in the liquid-hydrogen engine as well as certain support services to be provided to Centaur and Saturn vehicle systems. RL-10 engines were being used to propel the Centaur upper-stage launch vehicle and the S-IV stage of Saturn I launch vehicle. (NASA Release 63-189)
- NASA selected General Electric's Missile and Space Div. to develop recoverable biosatellite (Bios) system. Based on study contracts awarded to GE, Lockheed Missile and Space Co., and Northrop Corp. in April, new contract called for development, assembly, and testing of six 1,000-lb. flight spacecraft plus ground test vehicles. Biological experiments for the flights had not yet been defined. (NASA Release 63-190)

August 22: X-15 No. 3 flown by NASA pilot Joseph A. Walker to record 354,200-ft. altitude (67.08 mi.), the aircraft attaining maximum speed of 3,794 mph (mach 5.58) at 167,000-ft. altitude.

Purpose of flight was to obtain data on stability and control in extreme ranges of flight, especially during re-entry from high altitude with X-15's ventral fin removed. X-15 carried, for first time, an altitude predictor using an internal computer to provide instantaneous measurements of aircraft's climb angle and rocket engine's energy; these measurements were then relayed to pilot through use of an altimeter displaying the predicted altitude. Immediately after launch from B-52, X-15 went into steep climb about 44° above horizontal, its engine burning full thrust for 85.5 sec. On upward trajectory pilot Walker applied speed brakes to prevent excessive velocity. For about three min. he experienced near weightlessness. This was 91st flight of X-15 rocket research aircraft, 25th in X-15 by Walker. (NASA Release 63-191; FRC Release)

August 22: President Kennedy pressed button at White House unveiling by remote control the first USAF C-141A Starlifter jet transport at Marietta, Ga. General Joe W. Kelly, MATS Commander, said the 550-mph aircraft "promises to be the fastest, most efficient and flexible airlift system in the world." (DOD Release 1146-63; *A-N-AF Journal & Register*, 8/24/63, 5)

- Dr. Lee A. DuBridge, President of Cal Tech, addressing Western Electronic Show and Convention banquet in San Francisco, noted that "many people talk about the space program as a scientific program. . . . [Actually] the engineers of the world ought to have credit." Dr. DuBridge said scientific bases of space program were laid from 60 to 300 years ago, but enormous advances in engineering and rocketry in last 50 years had made space accomplishments possible. "The scientists [of today] are working in their laboratories on things that may lay the groundwork for application 50 years from now." (Davies, *NYT*, 8/23/63, 38)

August 22-26: 310 nuclear physicists from 22 countries exchanged ideas on high-energy accelerators at conference held in Dubna, U.S.S.R. (AP, Wash. *Eve. Star*, 8/28/63)

August 23: SYNCOM II communications satellite relayed its first live telephone conversations, a transmission between President Kennedy and Nigerian Prime Minister Sir Abubaker Tafawa Balewa and other messages between U.S., Nigerian, and U.N. officials. Arranged by USIA, the demonstration program originated from the White House and Voice of America studios in Washington and from ground station aboard USNS *Kingsport* in Lagos Harbor, Nigeria. (*Goddard News*, 8/26/63, 3; *Av. Week*, 9/2/63, 24)

- U.S.-Canada agreement for cooperative testing of communications satellites launched by NASA was announced by NASA and Canada's Dept. of Transport. Each cooperating national agency would provide a ground station to receive and transmit television and multichannel telephone and telegraphic signals via communications satellites, according to Memorandum of Understanding signed in April and made operative by exchange of notes today. (NASA Release 63-194)

- Senate-House conference committee on NASA FY 1964 authorization bill (H.R. 7500), meeting in executive session, agreed to file conference report on differences between Senate-passed and House-passed versions of the bill. Conference committee's bill would authorize total of $5,350,820,400 for NASA in FY 1964. (NASA Leg. Act. Rpt. II/132)

August 23: USAF awarded two contracts in its program to develop technology for large solid-fueled rocket motors: (1) $5 million contract to Thiokol Chemical Corp. for demonstration static-firing of half-length 260-in.-diameter motor of approximately 3 million lbs. thrust and for demonstration static-firing of 156-in.-diameter, two-segment motor of 3 million lbs. thrust; and (2) $3 million contract to Aerojet General Corp. for demonstration static-firing of half-length 260-in.-diameter motor of approximately 3 million lbs. thrust. Based on results of these contracts, USAF will select one contractor to continue development of "full-length" 260-in.-diameter motor. (*Wall Street Journal*, 8/26/63; *Space Bus. Daily*, 8/28/63, 307)

- Rep. George P. Miller, Chairman of House Committee on Science and Astronautics, announced formation of nine-man Subcommittee on Science, Research and Development. Chaired by Rep. Emilio Q. Daddario (D.-Conn.), Subcommittee would have following objectives: overall evaluation of scientific research and development throughout the country; strengthening of Congressional sources of information and advice on science and technology; achievement of most effective utilization of scientific and engineering resources of U.S. to accomplish national goals; and Congressional overseeing of the National Science Foundation. (NASA Leg. Act. Rpt. II/133)

- NASA Manned Spacecraft Center announced the 271 applicants for astronauts had been screened to 30. Of these, 10–15 would be selected in late October. (AP, *Wash. Post*, 8/29/63)

- Request for proposals to provide services and materials necessary for operating and maintaining X–15 "High Range" issued by NASA Flight Research Center. Range consisted of terminal station at Edwards AFB and two up-range stations at Beatty and Ely, Nev. (*Space Bus. Daily*, 8/22/63, 280)

- German Ministry of Defense ordered 140 Arcas sounding rockets from Atlantic Research Corp., through Bureau of Naval Weapons, to be used for meteorological soundings during IQSY. This same week Atlantic Research Corp. was issued patent for the seven-foot-long, solid-fueled Arcas. (*NYT*, 8/24/63; *M&R*, 8/26/63, 9)

- Deputy Secretary of Defense Roswell L. Gilpatric, in letter to Sen. Richard B. Russell, Chairman of Senate Committee on Armed Services, detailed the four safeguards to maintain national security in connection with test ban treaty: "The conduct of comprehensive, aggressive, and continuing underground nuclear test programs . . .; The maintenance of modern nuclear laboratory facilities and programs in theoretical and exploratory nuclear technology . . .; The maintenance of the facilities and resources necessary to institute promptly nuclear tests in the atmosphere should they be deemed essential to our national security . . .; The improvement of our capability . . . to monitor the terms of the treaty . . . ." (Letter)

August 23: Balloon flight with instrumented payload to study oxygen conditions in stratosphere was conducted successfully from Palestine, Tex., John Sparkman, manager of National Center for Atmospheric Research's Balloon Flight Station, announced. Balloon rose to 108,000 ft.; its 680-lb. instrument package was recovered in good condition 40 mi. north of San Angelo, Tex. Three similar attempts had failed when balloon split on two occasions and weather forced postponement on a third. Project was sponsored by Massachusetts Institute of Technology. (*Houston Post*, 8/24/63)

- Donald T. Gregory, Technical Assistant to Director of NASA Manned Spacecraft Center, addressed nearly 1,000 Dade County educators at workshop for new teachers in Miami, Fla. Noting U.S. requires about 60,000 engineers yearly—while only 33,700 graduated in 1963—Gregory emphasized the importance of teaching all children about scientific and technological events of the space age and stimulating their interest and knowledge in science and engineering at an early age. (MSC Release 63-139)

August 24: USAF SAC crew conducted routine training launch of Atlas ICBM from Vandenberg AFB, Calif. (DOD Release 1174-63)

August 25: USAF launched unidentified satellite from PMR using Thor-Agena D launch vehicle. (*Pres. Rpt. on Space, 1963*, 1/27/64)

August 26: Rep. George P. Miller (D.-Calif.) submitted to the House the conference report and statement on bill (H.R. 7500) to authorize FY 1964 appropriations to NASA. As reported, bill would authorize total appropriations of $5,350,820,400. Report was submitted to Senate August 28. (NASA Leg. Act. Rpt. II/133)

- Office of Construction established in NASA Hq., with nationally recognized construction contractor Robert W. Long as Director. Reporting to NASA Deputy Associate Administrator for Industry Affairs, Walter L. Lingle, Long would review NASA's construction projects and future construction plans to assure their timeliness, efficiency, and economy. Since June 27, 1963, Long had served as consultant for construction to Associate Administrator Dr. Robert C. Seamans; before that appointment he had been managing partner of Long Construction Co., Kansas City, Mo. (NASA Release 63-192)

- NASA announced publication of "Handbook on Space Radiation Effects to Solar Cell Power Systems," prepared for NASA by Exotech, Inc., of Alexandria, Va. The handbook and the recently published transcript of Photovoltaic Specialists Conference, held in Washington earlier this year, were expected to help industry produce more durable, radiation-resistant solar-cell systems for satellites. (NASA Release 63-187)

- Legislative Manager of U.S. Chamber of Commerce Theron J. Rice said in public letter that CofC recommended NASA's FY 1964 budget request be cut by $1.2 billion "and that a total of $4,512,000,000 be appropriated for NASA programs for Fiscal 1964. We believe this amount will adequately provide for a continuation of our national space program and will enable the space agency to meet the established goal of a manned lunar landing by the end of this decade. The amount we recommend will give the space agency an approximate 25% increase in appropriations over Fiscal 1963, which is the maximum growth rate we believe can be sustained and supported within the economic as well as scientific and technical capabilities of the nation . . . ." (Text, *Av. Wk.*, 8/26/63, 21)

August 26: President Kennedy appointed Willis M. Hawkins, Jr., vice president and general manager of Lockheed Aircraft Corp., to position of Assistant Secretary of the Army for Research and Development, replacing Finn J. Larsen who resigned July 31. (Wash. *Eve. Star*, 8/27/63)

• Rep. Joseph E. Karth (D.-Minn.), in *Missiles and Rockets* interview, advocated selling space boosters to foreign countries. Noting Scout and Thor boosters already were available to European countries, he said Atlas also should be made available, under these conditions:

"One, that they pay for the vehicle—I don't think we should give it away.

"Two, we would need to launch it for them. . . . Third, whatever secrecies surround the launch vehicle and/or other devices, all of this would be handled by U.S. personnel . . . .

"In my opinion, other nations are not as interested in advancing scientific exploration as they are in embarking on programs to advance their own economic position. And in this instance I say it's fine for us to cooperate by selling to them a booster or whatever other paraphernalia they need.

"As these countries become more involved in space programs, it may be they will be desirous of buying Saturns or Centaurs. We don't necessarily need to stop at Atlas. I think the whole high-thrust field of boosters is a tremendous market with great potential.

". . . we should do this through an agency of the U.S. government, probably through the Defense Department or NASA. Whatever profit might be involved does not go to NASA, does not go to DOD—it goes to private industry, to those who make the Atlas or the Saturn or Centaur. This money gets back into the economy. It provides jobs. It stimulates the U.S.'s gross national product, causing our overall economy to grow and expand . . . ." (*M&R*, 8/26/63, 21)

• *Missiles and Rockets* quoted Moscow radio as announcing that U.S.S.R. would attempt orbital rendezvous of two spacecraft "in the very near future." Soviet announcement said earlier tandem flights of Vostok spacecraft were part of the rendezvous program but attempts to link them during flight had not been planned. (*M&R*, 8/26/63, 9)

• U.S. Army announced Nike-Zeus antimissile missile, fired and controlled from Kwajalein Island in mid-Pacific, recently made successful interception of target vehicle launched by Titan I ICBM from Vandenberg AFB, Calif. (DOD Release 1170–63)

• General B. A. Schriever, AFSC Commander, received the General Henry H. "Hap" Arnold Aviation Gold Medal Award at national convention of Veterans of Foreign Wars, Seattle. (AFSC Release 38–R–65)

August 27: The Rev. Dr. O. Frederick Nolde, Director of Commission of the Churches on International Affairs, said in report to World Council of Churches' Central Committee that a joint U.S.-U.S.S.R. lunar exploration program would "avoid the menace of military use" of space by either country. Dr. Nolde's proposal was part of an analysis of nuclear test-ban treaty which he said would provide a "new and more promising atmosphere" for peace. (Price, N.Y. *Herald Trib.*, 8/28/63)

- Thermonuclear researchers at LRC reported on devices able to measure temperatures in excess of 400,000 degrees. In recent tests, Lewis apparatus recorded an electron temperature of 20 volts, or about 420,000° F, in an experimental helium plasma. (LRC Release 63-71, Lewis Chronology, 8)

August 28: NASA launched Little Joe II booster rocket with dummy payload simulating size and weight of Apollo command module and escape tower in flight test at White Sands Missile Range, N.M. The LJ-II attained 24,000-ft. altitude and mach 1.1 speed, impacted about 47,000 ft. downrange instead of planned 35,000 ft. because destruct system did not respond to ground command signal. This was first launching of LJ-II and was a flight qualification test to verify performance of LJ-II before using it to test Apollo command module. 100 sec. of telemetry was recorded and five of the six test objectives were achieved. (NASA Release 63-193; *Av. Wk.*, 9/2/63, 20)

- Senate and House passed H.R. 7500, bill to authorize NASA appropriations for FY 1964. Senate and House both adopted the conference committee's bill authorizing $5,350,820,400 total; of which $4,119,575,000 was for research and development, $713,060,400 for construction of facilities, and $518,185,000 for administrative operations. (*CR*, 8/28/63, 15298-303, 15268-270)

- Hughes Aircraft Co., prime contractor for SYNCOM II communications satellite, announced the synchronous-orbiting satellite had accumulated more message time than all other communications satellites combined. SYNCOM II was launched July 26, maneuvered into 24-hr. station Aug. 15. (*Space Bus. Daily*, 8/28/63, 306)

- House Resolution 504 to create a select committee to investigate expenditures for research programs conducted by or sponsored by departments and agencies of the Federal Government was reported to House floor from House Committee on Rules. (*CR*, 8/28/63, 15297-98)

- Portions of the March on Washington for Jobs and Freedom, telecast by nationwide networks within U.S., were transmitted via TELSTAR II communications satellite to Europe. (CBS-TV, 8/28/63; *Wash. Post.* 8/30/63, 5)

- 13 security guards at NASA Wallops Station, Va., set up picket lines in strike for higher wages. (AP, Wash. *Eve. Star*, 8/28/63)

- USAF SAC crew conducted routine training launch of Atlas ICBM from Vandenberg AFB, Calif. (DOD Release 1188-63)

August 29: In ceremony celebrating establishment of new Australian tracking stations near Carnarvon and Canberra, NASA Deputy Administrator Dr. Hugh L. Dryden presented scale model of 85-ft.-diameter tracking antenna to Australian Ambassador Sir

Howard Beale. Carnarvon station was being established for Project Gemini and would also monitor Ogo and Oao unmanned satellites. Canberra station would be a portion of Deep Space Network which tracks space probes to moon and planets. (NASA Release 63-195)

August 29: Monorail rocket sled attained mach 5 speed along missile test track at Holloman AFB in recent test, AFSC announced, highest known speed ever reached by a monorail sled. Fired in two stages with Cajun and Cherokee rockets, the test vehicle reached 3,753 mph before being stopped near end of seven-mile track. Attachment of "flow separation spike," similar to needle on nose of jet fighter aircraft, reduced air resistance and thus increased speed. To brake the sled, spike was flipped away from front of sled when peak speed was attained, thus increasing air drag by a factor of four. Sled was then caught by several lengths of tied-down nylon restraining straps. Major test objectives were accomplished. (AFSC Release 37-74-72)

- At meeting of American Institute of Biological Scientists in Amherst, Mass., Max D. Lechtman of Magna Corp. presented report by team of researchers predicting use of certain bacteria in life-support systems of future manned spacecraft. One means of supplying oxygen to astronauts on long-duration flights would be electrolysis of water—breaking down water into its oxygen and hydrogen components by electric current. Microbes such as H. eutropha and H. ruhladi which absorb hydrogen would be valuable in removing the hydrogen. (*Wash. Post*, 8/30/63, A11)

- Senate Foreign Relations Committee approved nuclear test-ban treaty by 16 to 1 vote. Senate debate on treaty was scheduled to begin Sept. 9. (Marder, *Wash. Post*, 8/30/63, 1)

- USAF launched Thor-Agena D booster rocket with two unidentified satellites from Vandenberg AFB, Calif. (*Wash. Eve. Star*, 8/30/63; *M&R*, 9/9/63, 13; *Pres. Rpt. on Space, 1963*, 1/27/64)

- Special Subcommittee of House Committee on Armed Services, established to investigate research and development in the Armed Services, met in executive session. (NASA Leg. Act. Rpt. II/136)

- Minuteman ICBM launched from Vandenberg AFB silo on 5,000-mi. flight down Pacific Missile Range. Launching was conducted by SAC crew. (*M&R*, 9/9/63, 13)

August 30: NASA announced its Langley Research Center had issued requests for proposals for instrumented lunar-orbiter probes which would be launched by 1966 to secure topographic data on the moon's surface. Photographic lunar-orbiters would team with Ranger hard-landing spacecraft and Surveyor soft-landing spacecraft in gathering data preparatory to Apollo manned lunar landing. To be launched by Atlas-Agena launch vehicle, lunar-orbiter would contain camera system capable of obtaining pictures not closer than 22 mi. above moon's surface. (NASA Release 63-196; LaRC Release)

- Marvin W. Robinson resigned position as Deputy Director of NASA Office of International Programs to serve United Nations as Scientific Secretary in Secretariat of Committee on the Peaceful Uses of Outer Space. (NASA Announcement 63-186)

August 30: Reporting on experiments with mice in simulated space flights up to 90 days, Chicago researcher Bernard Miezkuc said at annual meeting of American Institute of Biological Sciences that resistance to infection is critically low on first, fourteenth, thirtieth, and possibly sixtieth days of flight. Apparently physical and psychological stress of early flight could account for lowered resistance of the mice on first day, he said. Then, after temporary recovery, mice build up resistance during flight, with lapses occurring at periods of about two weeks, one month, and two months. (Carey, AP, *Wash. Post,* 8/31/63)

- USAF launched Titan I ICBM from Vandenberg AFB in flight some 5,000 mi. down Pacific Missile Range. (UPI, *Wash. Post,* 8/31/63; *M&R,* 9/9/63, 13)
- Kaj Aa. Strand, former Chairman of Northwestern Univ. Astronomy Dept., succeeded retiring Gerald M. Clemence as Scientific Director of U.S. Naval Observatory. (Wash. *Eve. Star,* 8/27/63)
- Nike-Zeus antimissile missile, fired from Kwajalein Island in the Pacific, scored its ninth "intercept" of Atlas ICBM target fired from Vandenberg AFB, Calif. (*M&R,* 9/9/63, 13)
- DOD announced the direct communication link between Washington and Moscow was now operational. (DOD Release)
- USN launched Polaris A-3 from land pad at AMR, in successful night flight to gather data on guidance, propulsion, and thrust vector control and to evaluate re-entry body flight dynamics. (*M&R,* 9/9/63, 13)

August 31: Results of industrial survey made by Univ. of Denver Research Institute under NASA contract were released, survey based on questionnaires of 988 companies. Results indicated: (1) transfers of technology from Government-sponsored research and development to private industries was the most important contribution of space-related programs to civilian sector of nation's economy; (2) space-related R&D stimulates both basic and applied R&D of improved materials, processes, and techniques, as well as new test and laboratory equipment; (3) space-related R&D output contributed to cost reduction in both commercial manufacturing and fabrication methods; (4) institutional and economic barriers exist and must be surmounted in the transfer process; (5) time lag exists between development of new or improved technology whether through industrial or Government-sponsored research and its ultimate utilization. NASA Office of Technology Utilization would use this information to determine ways of accelerating transfer of NASA's technological advances to civilian/industrial community. (NASA Release 63-197; Text)

- Rescheduling of first manned Apollo flight—a three-man earth-orbiting flight—from early 1965 to late 1965 or early 1966 reported by John W. Finney of *New York Times.* Unnamed NASA official attributed 9- to 11-month delay to technological difficulties inevitable in such a complex undertaking. (Finney, *NYT,* 9/1/63, 1, 27)
- NASA Marshall Space Flight Center Director Dr. Wernher von Braun announced reorganization of MSFC, designed to strengthen the Center's dual function of performing in-house R&D and monitoring industrial contracts. In Office of the Director, Deputy Di-

rector Dr. Eberhard Rees was designated Deputy Director-Technical, and Deputy Director H. H. Gorman, Deputy Director-Administrative. Hans Maus was designated Director of newly formed executive staff replacing MSFC Central Planning Office. In second major group, MSFC's nine technical divisions were formed in new Research and Development Operations, with Hermann Weidner as Director. Third major group was new Industrial Operations, comprised of MSFC elements concerned with work performed by industrial contractors, and headed by Robert Young. (*Marshall Star*, 9/4/63, 1, 3; N.O. *Times-Pic.*, 9/1/63)

August 31: Research balloon launched by MIT scientists from National Center for Atmospheric Research, Palestine, Tex., was lost when it sprang a leak at several thousand feet. Its payload of instruments, designed to obtain data on cosmic rays, was parachuted safely to ground. (UPI, *Wash. Post*, 9/1/63, 2)

- Concluding its two-week meeting at Berkeley, Calif., International Union of Geodesy and Geophysics elected as president Joseph Kaplan, physicist of UCLA, to succeed Vladimir V. Beloussov, Soviet seismologist. During the meeting, the 3,000 delegates laid plans for Upper Mantle Project to explore rock layer of the earth and for International Hydrological Decade (1965–1975) to study earth's water resources and for projects during International Quiet Sun Year. (AP, *Wash. Post*, 9/2/63)

During August: Mockup model of Gemini two-man spacecraft shipped from McDonnell Aircraft Corp. to NASA Launch Operations Center, Cape Canaveral, for tests of ground support equipment at the launch complex. (AP, Wash. *Eve. Star*, 8/22/63)

- NASA Director of Aeronautical Research Programs Charles H. Zimmerman resigned to accept position with Army Material Command's research and development directorate. Since June 1962, Zimmerman directed NASA's aeronautical research programs including X–15, V/Stol research, and supersonic transport study. (*Av. Wk.*, 9/2/63, 24)

- In first altitude test of AEC's Project Sand (Sampling Aerospace Nuclear Debris), Honest John-Nike rocket carried nuclear sampler payload to 36-mi. altitude where it was jettisoned and parachuted back to earth. Launch was conducted from Tonopah Test Range, Nev., by Sandia Corp. for AEC. Project Sand was designed to determine amount of nuclear debris in atmosphere 23 to 43 mi. above earth and to develop techniques for predicting its dispersal in this region. (*Space Bus. Daily*, 8/27/63, 300)

- In introduction to U.N. Annual Report, Secretary General U Thant said:

". . . the development of cooperation in outer space exploration and use continued in an encouraging manner, especially in the scientific and technical field.

"The Scientific and Technological Subcommittee of the Committee on the Peaceful Uses of Outer Space, at its second session held in Geneva in May 1963, agreed upon a series of new and revised recommendations concerning the exchange of information, encouragement of international programs, education and training, potentially harmful effects of special experiments and the organization of international sounding rocket facilities. The meeting of the subcommittee provided once more the occasion for the scientists of the two leading space powers to continue their private talks on cooperative space programs.

"The World Meteorological Organization, the International Telecommunication Union and the United Nations Educational, Scientific and Cultural Organization continued to participate actively in the field of peaceful exploration and use of outer space.

"No agreement was reached on legal problems relating to outer space, but a valuable exchange of views took place in the Committee and its legal subcommittee . . . ." (Text, *CR*, 8/27/63, 15099–103)

During August: The *Atlantic Monthly* published an article "Why Land on the Moon?" by Dr. Robert Jastrow, Director, Goddard Institute for Space Studies, and Dr. Homer E. Newell, Director, Office of Space Sciences, NASA.

"These are the specific values of space exploration: the benefits of basic research, economically valuable applications of satellites, contributions to industrial technology, a general stimulus to education and to the younger generation, and the strengthening of our international position by our acceptance of leadership in a historic human enterprise. The current discussion of these values of the space program has served the United States well in directing its attention to questions of national purpose. But, however we may try to break the program down into its elements and to attempt a detailed balancing of debits and credits, the fact remains that the space effort is greater than the sum of its parts. It is a great adventure and a great enterprise, not only for the United States but for all humanity. We have the power and resources to play a leading role in this effort, and it is inconceivable that we should stand aside.

"Scientific administrators ask . . . can we afford the cost of the space program in technical manpower? Their concern is heightened by the fact that federal activities in defense, space, and atomic energy together consume nearly half of the science and engineering talent available in the United States. But is the space agency the major consumer of trained manpower within this federal complex of technical agencies? In actuality, NASA will be using six percent of the national manpower pool in science and engineering through its contracts with private industry, plus an additional one percent in government laboratories. If the space program has substantial value, this is not an overwhelming drain." (Text)

- Dr. Simon Ramo, Vice Chairman of Board of Thompson-Ramo-Wooldridge, Inc., said in *Air Force and Space Digest* article: "The landing of men on the moon needs to be understood by the American people merely as a specific framework on which is built a broad research program to explore many areas of science. Space is new and special; it is only now that we have the ability to enter a region of the universe previously denied to us, except for viewing from a distance through the insulation of our atmosphere. Nevertheless, without a definite task, a broad space program of this size would be very inefficient, would wander and become redundant, unwieldy, and confused. We need to agree that what counts most is advancing science on a broad front . . . .

". . . the addition of a human passenger and observer to a total space effort is a necessary ingredient. An ideally laid-out program . . . requires emphasizing broad research aims in all of the sciences . . . . How can we justify a large national space-research program whose true objective is scientific discovery across all the spectrum of science if we are going to leave out the life sciences . . . .

"It is essential then that the public should understand that the program they are buying is not just a man-on-the-moon program. Like an iceberg, the greater part of the program will be unseen. Space exploration will be lengthy and continuing, requiring patient backing as a long-term venture . . . ." (*AF & Space Digest*, 8/63, 49f)

During August: Peak power output of 50w from pulsed gas laser demonstrated by Orlando Div. of Martin-Marietta Corp. At higher power levels the company predicted application of the laser in optical radar system. (*Av. Wk.*, 8/19/63, 34)

- Astronaut John H. Glenn, Jr., received 1962 Gold Medal of French Association Pour l'Etude et la Recherche Astronautique. Colonel Glenn accepted the award "on behalf of myself and the thousands of persons here and abroad who have made these space flights possible." (*Av. Wk.*, 8/12/63, 109)
- COSPAR re-elected M. Maurice Roy, member of French Academy of Sciences and professor at Ecole Polytechnique in Paris as President for 1963–66. (*A&AE*, 8/63, 116)
- S. Fred Singer, Director of National Weather Satellite Center of U.S. Weather Bureau, was awarded first Astronautics Space Research Medal of British Interplanetary Society. (*A&AE*, 8/63, 119)

SEPTEMBER 1963

September 1: Theory of "hypersonic organosynthesis," creation of organic molecular compounds by giant explosions, offered as a clue to origin of life on earth by Adolph R. Hochstim, researcher of Institute for Defense Analysis and National Bureau of Standards. Hochstim said 1,000-yd.-diameter meteorite would have exploded with tremendous force upon impact, transforming molecules of earth's atmosphere, its water, and meteorite itself into heavy organic compounds. Since theoretically such compounds could be formed by meteorite impact "on any planet which contains any type of atmospheric and liquid phase," Hochstim warned, space probes "sent to a planet on a crash landing may on impact produce complex chemical compounds which will need to be discounted when searching for a sign of life in the vicinity of the planet." (Myler, UPI, *Wash. Post*, 9/1/63, A3)

- In speech to American Psychological Association, Dr. Philip H. Abelson, Director of Carnegie Institution's Geophysical Laboratory and editor of *Science*, repeated his view that there were no predictable economic advantages to the exploration of the moon or possible subsequent flights to Mars. Much of the justification for manned lunar program was national prestige, he said, but "the half of the world that is undernourished could scarcely the expected to place a higher value on landing on the moon than on filling their stomachs." (Barbour, AP, Wash. *Eve. Star*, 9/2/63)

- With launching and recovery of 120-ft. balloon, Balloon Research Center at Palestine, Tex., completed its cosmic-ray study project. Today's balloon floated to more than 10,000-ft. altitude to collect cosmic-ray data. (UPI, *NYT*, 9/3/63, 4)

September 2: Two flight tests of M-2 wingless lifting body conducted by NASA Flight Research Center, with Milton O. Thompson as test pilot. Towed to 13,000-ft. altitude by C-47 aircraft, M-2 glided each time to landing on Rogers Dry Lake bed in a few seconds over three minutes. These were sixth and seventh airtow flight-tests of M-2 in series to investigate man's ability to control M-2 during low-speed operations, particularly during landing phase. (FRC Release 17-63; *M&R*, 9/9/63, 14)

- Results of NASA research project reported: 18 species of earth bacteria survived in "Mars jars" in which atmospheric conditions of Mars were simulated. One type enlarged itself by four or five times in the Martian environment. (*M&R*, 9/2/63)

- AFSC announced establishment of field office at NASA Manned Spacecraft Center, headed by Col. Daniel D. McKee (USAF), to serve as central AFSC point of contact at MSC. (AFSC Release 38-R-76)

September 2: Recently approved program by NASA-USAF Project Gemini Program Planning Board called for about 12 USAF experiments to be included in later Gemini flights, John Finney reported in *New York Times*. Board was reported to have concluded it could make no recommendations on USAF pilots participating in Gemini until DOD clarified its position on military requirements for a manned space flight program. (Finney, *NYT*, 9/3/63, 1, 15)

- Reported that one example of space flight experience vitally benefiting other Government projects was adoption of Mercury spacecraft and landing maneuvers to the escape and survival projection system for F–111 (TFX) tactical fighter aircraft. In emergency escape from F–111, entire two-man crew compartment and section of aircraft's wings are separated from aircraft; drogue parachute then deploys and pulls out large Mercury-type parachute which lowers compartment to a landing. Like Mercury spacecraft, ejected compartment contains environmental life-support system and necessary survival equipment. (Cook, Wash. *Daily News*, 9/2/63)
- Results of Harvard Business School poll of U.S. industrialists on how they viewed America's space program, published in *Harvard Business Review*, indicated that industrialists expect a considerable payoff from space in terms of tangible benefits to our planet and think that the space program is "a great energizing force on our society." (Greenough, Boston *Morning Globe*, 9/2/63)

September 3: Rep. Emilio Q. Daddario (D.-Conn.), Chairman of Subcommittee on Science, Research and Development of House Committee on Science and Astronautics, sent letter to all Committee members and to Panel on Science and Technology, saying: "The Subcommittee will explore plans and programs to accomplish the best orientation and use of science and its resources.... [For example], what standards may effectively be applied to determine the need and the priority for scientific research, how much may be evaluated as conforming with national scientific policy and goals, and what proper levels may be established for basic, applied and developmental research." (NASA Leg. Act. Rpt. II/17)

- NASA Goddard Space Flight Center announced Belgian astrophysicist, Dr. Francois V. Dossin, working at GSFC on a National Academy of Sciences fellowship, discovered faint comet about 5° from sun during July 20 solar eclipse. Dr. Dossin made seven camera-plate exposures of comet from Pleasant Pond, Me., during 60 sec. of total eclipse. He used blue-green filter to bring out the light of carbon molecules in the comet. Microscopic examination of developed plates showed a diffuse image emitting the light of molecular carbon. (GSFC Release)
- National Science Foundation announced U.S. and U.S.S.R. would conduct cooperative cosmic-ray study in Antarctica, on IQSY project. From one station, VHF radio signals would be beamed at ionosphere; some of the signals would be scattered downward by ionosphere and recorded at other stations. By studying changes in reception of radio signals, scientists can take indirect measure-

ments of cosmic rays. Antarctic region is especially suited for such measurements because cosmic rays are guided by earth's magnetic lines of force which converge in polar regions. (Finney, *NYT*, 9/3/63)

September 4: Aerobee 150 sounding rocket launched from Ft. Churchill, Can., with nuclear emulsion payload to study very-low-energy cosmic ray heavy nuclei. Payload reached 150-mi. altitude, was recovered from an inland lake approximately 90 mi. from launch site. Instrumentation and nuclear emulsions were in excellent condition, but analysis of nuclear emulsions would require considerable time. Experiment was project of NASA Goddard Space Flight Center. (NASA Rpt. of Sounding Rkt. Launching)

- NASA Manned Spacecraft Center announced issuing requests for proposals for vacuum chamber to generate space environmental conditions for use in thermochemical tests. Chamber would permit MSC engineers to study function of propellant equipment, heat rejection, and energy collecting devices during long periods of space flight. (MSC Release 63-144)

- According to Soviet lunar-exploration timetable published in Hungarian magazine *Lobogo*, U.S.S.R. planned first manned circumlunar flight for 1964-65, first manned lunar landings for 1966-68, and first "temporary laboratory" on moon for 1968-70. Article said manned Vostok flights had "virtually accomplished 80 per cent of a space rendezvous maneuver, lacking only the coupling of the two spaceships." Next step after orbital rendezvous should be trip to moon. Abstract of article, which appeared in June 26 issue of *Lobogo*, was published in translation by Joint Publications Research Service of Dept. of Commerce's Office of Technical Services. (Schmeck, *NYT*, 9/4/63)

- Urging a "breakthrough" by CAB and FAA "toward more economical operations in 18-passenger to 24-passenger" range of commercial aircraft, Sen. Thomas J. McIntyre (D.-N.H.) pointed out that in 1958 "the passage of the Federal Aviation Act provided the basis for such [design] studies. . . . Originally, of course, such studies were the province of the National Advisory Committee for Aeronautics, set up as early as 1915. Senator Ralph Flanders said, in 1955:

    "'There is not an airplane flying in this country today—certainly not a military one—the design of which does not depend to some degree upon scientific investigations of the NACA.'"
(*CR*, 9/4/63, 15444)

- Gemini astronauts would learn how to "apprehend" another vehicle traveling in space on a special trainer in a darkened hangar-high structure now being erected at Manned Spacecraft Center's Clear Lake home. (MSC *Space News Roundup*, 1/8/64)

September 5: NASA announced SYNCOM II communications satellite achieved perfection of orbit probably unparalleled by any other previous satellite. At more than 22,000-mi. altitude, orbit varies from absolute circle by no more than 4.5 mi. Orbital period—23 hrs., 55 min., 54 sec.—is only 10.09 sec. shorter than mean sidereal day. Virtually stationary over 55th meridian of west longitude,

SYNCOM II is drifting eastward so slowly that it would take nearly a month to move a single degree. Drifting will be corrected periodically, and the communications satellite can be expected to remain on station and operating for more than two years. (Wash. Eve. Star, 9/5/63)

September 5: At NASA Hq., D. Brainerd Holmes, former Director of NASA Manned Space Flight, was presented NASA Medal for Outstanding Leadership by NASA Administrator James E. Webb. (NASA Release 63-199)

- Dr. Edward Teller, Univ. of Calif. physicist, declared at symposium on "open space and peace" held at Stanford Univ. that surveillance from space of every part of the world should be made legal. Noting that "open skies" policy was first proposed in 1955, he stressed that today the security of U.S. and Free World depend in "a most vital fashion on information obtained from Russia." Surveillance from space should be conducted "publicly, freely and in the interest of the peace of our nation." (Davies, *NYT*, 9/6/63, 10)
- Boeing Co. announced rendezvous and docking in space can be achieved by pilot with little, if any, assistance from instruments. Findings were based on three months of experiments with rendezvous-docking simulator in preparation for Project Gemini. (Macomber, *San Diego Union*, 9/5/63)

September 6: President Kennedy signed NASA authorization bill for FY 1964 (PL. 88-113), authorizing NASA $5,350,820,400 for the fiscal year. (NASA Leg. Act. Rpt. II/141)

- NASA and DOD announced agreement for NASA use of USAF-developed Agena launch vehicles, replacing Feb. 1961 agreement under which NASA used Atlas-Agena and Thor-Agena vehicles for Ranger, Mariner, and Alouette projects. Signed by NASA Associate Administrator Dr. Robert C. Seamans, Jr., and AFSC Vice Commander Lt. Gen. Howell M. Estes, Jr., agreement provided: USAF will be responsible for design, engineering, and acceptance testing of basic Atlas and Thor vehicles and Agena D stages, with NASA buying these from USAF; NASA will have responsibility for all its mission modifications to basic vehicles; NASA will launch its Atlas-Agena vehicles from Complex 12 at AMR and USAF will conduct all Agena launch operations from PMR; technical and administrative control of 11 remaining NASA Agena B stages transferred from USAF to NASA. Atlas-Agena vehicles for Project Gemini were covered by separate NASA-USAF agreement. (NASA Release 63-198; DOD Release 1212-63)
- NASA Manned Spacecraft Center announced signing of definitive contract valued at $7,658,000 with Kollsman Instrument Corp. for guidance and navigation equipment for Apollo spacecraft's command and service module. (MSC Release 63-147)
- NASA Director of Space Sciences Dr. Homer E. Newell said in address at dedication of Stedman Hall of Science, Central Methodist College, Fayette, Mo.:
    "We are now laying the groundwork for whatever role we may have to play in space in the future. We are developing the ability of the United States to use space and to operate in space either as

it may choose to do voluntarily or may find itself compelled to do in its own defense. This is the capability that we shall have from the development of the ability to investigate scientifically with satellites and space probes, from space applications, from the ability to perform manned space flight and manned space operations, from the vast complex of manufacturing and assembly plants, launching complexes, tracking and telemetering facilities, and from the invaluable experience that this initial stage in the space program will give us.

"This is the most significant point about the present era in space. This is the most important aspect of the present activity in space . . . ." (Text)

September 6: USAF launched Atlas-Agena D from Pt. Arguello with unidentified satellite. (*M&R*, 9/16/63, 11–12)

- USAF announced launching two Atlas ICBM's from PMR. (AP, Wash. *Eve. Star*, 9/7/63)

September 7: Second birthday of Michoud Operations, NASA facility in Louisiana used for fabrication of Saturn I, Saturn IB, and Saturn V rocket stages. To date, NASA had awarded $739 million in contracts which will be carried out primarily at Michoud, and some $41 million had been spent on rehabilitation and modification of plant and construction of new facilities. More than 7,800 persons are now working at Michoud, most of them employees of NASA contractors. (*Marshall Star*, 9/11/63, 1, 2)

- French plans to orbit four-pound "Satmos" satellite in 1964 revealed by unnamed French Government officials. To be launched by Army's National Office of Space Research using four-stage Bérénice rocket, satellite would provide data on re-entry from outer space. (UPI, *NYT*, 9/8/63, 42; *M&R*, 9/16/63, 12)

- Cal Tech reported new study of meteorite records going back 260 years has been undertaken by geochemist Harrison Brown, who would use electronic computers to map meteorite patterns. (*A&AE*, Nov., 1963, 11)

September 8: USAF announced establishment of worldwide network of 19 riometer (relative ionospheric opacity meter) stations for monitoring cosmic noise and absorption of galactic radiation in the ionosphere. (NANA, *Denver Post*, 9/8/63)

- In news interview, John W. Locke, aerospace engineer for Booz, Allen Applied Research, Inc., reported on conclusion of six-month study of feasibility of launching rockets from aircraft, study conducted for USAF Special Weapons Center with cooperation of USAF Office of Aerospace Research. Suggesting that cost of U.S. space launchings could be reduced 40 per cent by employing aircraft instead of launching pads, Locke said:

"Results from the study indicate the B–58 can be used as the prime launch vehicle. A four-stage rocket like the small version of the Scout missile, weighing 14,000 pounds, will provide the desired performance with minimum development effort.

"The air launching of even larger space vehicles is both feasible and acceptable and may be done economically." (Copley News Service, N.O. *Times-Picayune*, 9/8/63)

September 9: Nike-Apache sounding rocket launched from NASA Wallops Station carried 70-lb. instrumented payload to 106-mi. altitude, payload impacting in Atlantic Ocean 87 mi. downrange after 6 min., 40 sec. flight. Payload contained instrumentation to determine altitude and intensity of electric current systems in the ionosphere. New method of altitude determination during rocket flight was also tested. Flight was in preparation for electrojet program to be conducted from India later this year. (Wallops Release 63–81; NASA Rpt. of Sounding Rkt. Launching)

- Pratt & Whitney Aircraft Co. announced successful demonstration of new method of cooling high-pressure hydrogen rocket-engine chamber. P&W spokesman said that "successful demonstration of cooling during this company-sponsored firing was a research breakthrough on high-pressure hydrogen rocket engines.... The high pressure concept is the next step in increasing the performance of hydrogren fueled rocket engines." P&W is producer of RL–10 liquid hydrogen rocket engines for Saturn S–IV stage and Centaur upper stage booster. (*Space Bus. Daily*, 9/9/63, 356; *Hartford Courant*, 9/9/63)

- Clarence A. Syvertson, NASA Ames Research Center scientist, said in paper presented at Space Rendezvous, Rescue and Recovery Symposium at Edwards AFB that future spacecraft based on M–2 lifting body design could satisfy requirements for maneuvering re-entry vehicles on missions from near-earth orbit and on return from manned Mars missions. With lift-drag ratio of over one, M–2 lifting body allows use of maneuvering technique to solve critical heating problems of re-entry. Choice of M–2 rather than spacecraft covered with heavy ablative material could save considerable spacecraft weight. (Ames Release 63–34)

- James T. Koppenhaver resigned as Director, Office of Reliability and Quality Assurance, NASA. Deputy Director John E. Condon was appointed Acting Director of the Office. (NASA Announcement 63–198)

September 10: FAA announced six industries had notified FAA of their intent to submit proposals for design competition in national program of Government assistance to industry for development of supersonic transport airplane. (FAA Release 63–81)

September 11: Aerobee 150A launched from NASA Wallops Station carried instrumented payload to approximately 102-mi. altitude in experiment to study behavior of liquid hydrogen in conditions of radiant heating and zero gravity. Radiant heat measurements were as predicted, and there were 4 min. of zero-g time. Experiment was conducted for NASA Lewis Research Center, responsible for NASA development of liquid-hydrogen-propelled rocket engines. (Wallops Release 63–83; NASA Rpt. of Sounding Rkt. Launching)

- At luncheon meeting with NASA Deputy Administrator Dr. Hugh L. Dryden, Soviet Academician Dr. Anatoli A. Blagonravov reported to have suggested that it might be desirable to discuss possibility of U.S.-U.S.S.R. cooperation in manned lunar expedition. (Bartlett, Wash. *Eve Star*, 9/17/63; Simons, *Wash. Post*, 9/18/63; Finney, *NYT*, 9/18/63, 11)

September 11: House passed (336–0) H. Res. 504 to create select committee to investigate expenditures for research programs conducted by or sponsored by Government departments and agencies. Committee would submit its final report by Dec. 1, 1964. Speaker of the House appointed nine Representatives as members of the Committee, with Rep. Carl Elliott (D.-Ala.) as Chairman. Rep. Elliott, principal sponsor of the resolution, pointed out that 90 per cent of Government research is conducted by or sponsored by five agencies: DOD, NASA, HEW, AEC, and National Science Foundation; ". . . more than 75 per cent of these research and development dollars are spent on projects conducted by private research organizations, institutions and universities under Federal grants and contracts." (NASA Leg. Act. Rpt. II/143; *CR*, 9/11/63, 15867ff)

- Speaking at Symposium on Space Rendezvous, Rescue and Recovery at Edwards AFB, NASC Executive Secretary Dr. Edward C. Welsh said:

  ". . . It is worth keeping in mind that irrefutable facts and events demonstrate that the Soviets are conducting a vigorous space effort, and the evidence is strong that this program includes a manned lunar trip.

  "Let me emphasize that we should be making a vigorous effort to go to the moon and that we should continue to do so even if the Russians delete such an objective from their space program."

  Dr. Welsh listed some major reasons for manned lunar project:

  "(1) There is no other place so near in space where we can test the equipment and the men for future space travel.

  "(2) Such a clear objective as that of the lunar project gives impetus, order, and efficiency to the program . . . ." Other outstanding reasons he listed were national prestige, economic spin-off, defense spin-off, and scientific advancement. (Text)

- NASA Manned Spacecraft Center announced construction of permanent Center facilities at Clear Lake was 75 per cent complete. First large personnel move was scheduled for October. (MSC Release 63–150)

- In ninth month of strike against Florida East Coast Railroad, members of Order of Railroad Telegraphers picketed NASA construction site at Merritt Island. Union alleged NASA was accepting building materials delivered by railroad line. 865 construction workers from 14 craft unions refused to cross picket lines; next day, 1,500 workers were absent. (*Av. Wk.*, 9/16/63, 38)

- USAF announced routine launching of Atlas ICBM from Vandenberg AFB. (UPI, *Wash. Post*, 9/12/63)

- USAF announced plans to test fuel cell in space by sending new type of electrical power system into orbit in a satellite within next few months. Working model of the fuel cell, developed by Allis-Chalmers Manufacturing Co., was displayed at AFA convention in Washington. (Finney, *NYT*, 9/12/63, 14)

- In formal statement of policy, Air Force Association declared Senate ratification of nuclear test ban treaty would entail "unacceptable" risks to security of U.S. and Free World. Statement also said:

"One area of military technological potential in which this nation obviously is making less than a maximum effort is military space.... We must exploit military space to extend and preserve our deterrent strength, to control and inhibit those who would use their power aggressively to dominate life on earth. All of our national space efforts, including purely scientific exploratory programs, must be measured first against this yardstick. The national interest must be the prime purpose of every national program...." (AFA Statement)

September 11: A $3.5-million spaceship simulator to train crews for space flight, space intercepts, and orbital rendezvous at AFFTC, Edwards AFB, has been ordered by AFSC, the command revealed at the Air Force Association meeting, Washington, D.C. (*AFSC Operational Highlights,* 13)

September 12: Department of Labor's second annual William F. Patterson Memorial Award presented to NASA Langley Research Center "in recognition of apprenticeship program excellence." (*Newport News Daily Press,* 9/12/63)

• Secretary of the Air Force Eugene M. Zuckert said in letter to Air Force Association President J. B. Montgomery regarding AFA's Statement of Policy on nuclear test ban treaty: "... I recognize the right of the Association to express itself but the resolution is so immoderate and based on such misinformation as to disregard completely the best interests of the United States, as well as of the Air Force...." He also canceled his appearance at AFA reception in his honor. (Letter/DOD Release)

• General Curtis LeMay (c/s USAF) said in address to Air Force Association, Washington, that USAF was "making a concerted effort to maintain a mixed force of manned aircraft and missiles and, for the longer term, vehicles that could operate in space. The Secretary of Defense ... has directed that we continue our studies of follow-on manned, strategic vehicles that can counter threats and survive in the varied plateaus of aerospace operations...." (Text)

General B. A. Schriever (USAF) said that during the next "ten to fifteen years aircraft technology promises to make major advances which will provide significant improvements in range, speed, and versatility." As examples, he cited use of laminar flow control which "promises to make possible a significant increase in the subsonic range and endurance of large aircraft"; technical progress in such areas as "aircraft configuration, high temperature structural materials, and propulsion systems" which will "make feasible the development of a great variety of advanced types of aircraft, ranging from conventional design to V/Stol, variable geometry, and paraglider configurations." (Text, *A-N-AF Journal & Reg.,* 9/21/63, 18)

Lt. Gen. T. P. Gerrity (USAF) told AFA Industrial Associates' Luncheon that "there is a recognition of our future mixed force needs and new manned strategic systems are currently under consideration" and spoke of threat of a "manned aircraft gap." (*A-N-AF Journal & Reg.,* 9/21/63, 19)

September 12: Sunflower turboalternator failed after 4,328 hours—longest operating time to date. Sunflower is a 3-kilowatt solar-collector power system being developed for NASA by Thompson Ramo Wooldridge. (Lewis Chronology, 8)

September 13: SYNCOM II and RELAY I linked Rio de Janeiro and Lagos, Nigeria, in 20-min. voice conversation, first operation employing both communications satellites in single communications circuit and world's first three-continent telephone conversation. Signal began from USNS *Kingsport* in Lagos harbor, then to SYNCOM II, which sent it to Lakehurst, N.J., ground station, then by overland wire to Nutley, N.J., ground station, then to RELAY I overhead which sent it to Rio de Janeiro ground station. NASA Goddard Space Flight Center engineers monitoring the conversation declared quality of transmission to be good. (GSFC Release G-17-63; AP, *Wash. Post*, 9/14/63; N.Y. *Herald Trib.*, 9/14/63)

- NASA issued procedures on decontamination of lunar-landing spacecraft and planetary-landing spacecraft, based on recommendations by NAS-NRC Space Science Board made public this date.

  Because sterilization procedures employed on previous Ranger lunar spacecraft—particularly heat cycle—were suspected of directly contributing to Ranger mission failures, NASA lowered decontamination requirements for current Ranger spacecraft. It was now only required that microbes on Rangers be "reduced to a minimum." Regarding lunar probes, Space Science Board said: ". . . The lunar surface with its high temperatures, intense ultra-violet radiation, paucity of moisture, and high vacuum is a most unfavorable environment for proliferation of terrestrial organisms . . . ." It recommended U.S. ". . . minimize contamination to the extent technically feasible . . . [and develop] a sterile drilling system to accompany an early Apollo mission to return an uncontaminated sample of the lunar subsoil . . . ."

  NASA procedures for Mars landing missions required that spacecraft be sterilized and that this decontamination be effectively achieved without jeopardizing spacecraft reliability. Regarding Mars probes, Space Science Board said: ". . . The contamination of Mars through the impacting of nonsterile probes from the Earth could destroy an opportunity to carry out a meaningful search for life forms on Mars with remote detectors. This opportunity is unique and its loss would be a catastrophe . . . ." It recommended U.S. ". . . accord the highest priority to the prevention of the biological contamination of Mars until sufficient information has been obtained about possible life forms there so that further scientific studies will not be jeopardized . . . ." (NASA Release 63-200; NAS-NRC Release; NAS-NRC *Space Probe Sterilization*)

- Astronaut M. Scott Carpenter disclosed seeing twin tails on one side of sun's corona during solar eclipse July 20. The twin-tailed corona did not appear in photographs Carpenter made of the eclipse while flying in DC-8 jetliner with other scientific observers. (O'Leary, *Houston Post*, 9/14/63)

September 13: NASA Manned Spacecraft Center had begun training NASA's 16 astronauts in techniques of water and land parachute landings. In preparation for safe parachute landings in low-altitude abort during Project Gemini launching, astronauts were towed to heights of 400 ft. suspended under pre-inflated canopy, then cut loose for free descent to earth. Towing was by power boat for water descents and truck for land descents. (MSC Release 63-151)

- NASA Lewis Research Center awarded a $48,131,315 contract to Aerojet-General Corp. for redefinition of Snap-8 (Systems for Nuclear Auxiliary Power) development program. Snap-8 is designed to provide 35 kilowatts of on-board electric power to future spacecraft. (Lewis Chronology, 8)
- USAF announced most promising and practical method for protecting spacecraft against meteoroids appeared to be the "particle wall" concept: a layer of tiny metallic particles trapped by an electrostatic field would be used to cover surface of the spacecraft, absorbing impact of space particles. Study of this concept was completed by Fundamental Methods Associates, for AFSC Research Technology Div. (DOD Release 1233-63)
- Air Force Association presented Astronaut L. Gordon Cooper the David C. Schilling trophy for the year's most significant aerospace achievement in field of flight—his 22-orbit Project Mercury space flight MA-9. (MSC *Space News Roundup*, 9/18/63, 8)

September 14: U.S.-Scandinavia approval of Memorandum of Understanding for testing of NASA-launched experimental communications satellites announced by NASA and Scandinavian Committee for Satellite Telecommunication. Vice President Lyndon B. Johnson, on official tour of Scandinavia, received in Copenhagen the Danish Government's note of approval, making the Memorandum effective; Norway had approved Memorandum in note dated Sept. 11 and Sweden, in note dated July 25. Under agreement, Scandinavian Committee would provide ground station to receive multichannel telephone or telegraph signals transmitted from U.S. via orbiting communications satellite. (NASA Release 63-205)

- Discussing the National Space Program before Iowa City Daily Press Association in Des Moines, NASA Associate Director of Plans and Program Evaluation Addison M. Rothrock said:

"The prime objective of . . . [the manned space flight] program is to land men on the moon before the end of the decade. The scope covered by the program is much broader than this. It is in fact the research, development and operation of a series of research manned spacecraft that will determine for us man's ability to operate in space. By focusing on the moon-landing we have set a goal that insures we develop all basic phases of the operation of manned spacecraft. I cannot emphasize this too strongly.

"By the time our three astronauts leave for the moon-landing we as a nation will have accumulated 2000 flight hours in earth orbit of manned craft. To indicate to you the significance of this number—we have as of now less than five flight hours of manned aircraft at speeds in excess of Mach 2.5. . . . In these 2000 hours we will have orbited the earth some 1300 times. Men will have gone to and from craft in orbit. We will have rendezvoused and inspected craft and transferred men and material between craft. We will have maneuvered the craft both in regards to the attitude of the craft as has been done with Mercury and the Orbiting Solar Observatory and in regards to changes in the flight path as has been done with Mariner II and Syncoms. These are the things that must be done and will be done in the development, operation, and functioning of manned craft in space. . . ." (Text)

September 14: Univ. of Tennessee announced USAF had provided 400 acres at Tullahoma, Tenn., as site for a Univ. of Tennessee space institute. State government has appropriated $1,250,000 for construction of first buildings. (*NYT*, 9/15/63, 54)

- U.S. and U.S.S.R. were both awarded three gold wings (first prize) at first World Festival of Aeronautic and Space Films, sponsored by Fédération Aeronautique Internationale and held in Deauville. Among U.S. winners were "The Mastery of Space" and "The X-15 Story." (*NYT*, 9/17/63, 35)

September 15: NASA announced EXPLORER XVI meteoroid-detection satellite, launched Dec. 16, 1962, had ceased transmitting usable experimental data on July 25, 1963. Throughout its 7½ months of transmitting useful data, EXPLORER XVI fulfilled all its primary objectives. The satellite reported these results (one mil equals 1,000th of an inch): 44 penetrations of one-mil beryllium-copper; 11 penetrations of two-mil beryllium copper; 6 penetrations of one-mil stainless steel; one penetration of two-mil copper; and one penetration of three-mil copper. There were no penetrations of five-mil beryllium-copper or six-mil stainless steel. One cadmium-sulfide cell was penetrated in such a way that sunlight saturated it and rendered it inoperable. More than 15,000 meteoroid hits were recorded by microphone impact sensors which covered about one-tenth of total exposed experiment surface. These sensors did not measure penetration. (NASA Release 63-203; LARC Release)

- Third command and data acquisition station in Tiros meteorological satellite CDA system became operational, the Fairbanks, Alaska, station joining those at Wallops Island, Va., and Pt. Mugu, Calif. CDA stations receive cloud-cover photographs and other data from orbiting Tiros satellites, relay them to Weather Bureau's National Weather Satellite Center, Suitland, Md., for analysis. (NASA Release 63-206)

- Dr. W. E. Morrow, Jr., of MIT's Lincoln Laboratory, told International Scientific Radio Union in Tokyo that Project West Ford had "demonstrated that a significant communications capacity between large ground terminals can be provided by a dipole belt so diffused as to be barely detectable by optical astronomers and virtually undetectable by radio astronomers." He disclosed that only about half of the 400 million released dipoles had proved effective as signal reflectors, but the project "has already fulfilled virtually all its major objectives." (Simons, *Wash. Post*, 9/16/63; *Space Bus. Daily*, 9/16/63, 388)

September 16: AFSC announced it launched more than 60 space experiments in FY 1963, encompassing following areas of study: aurora borealis; infrared spectrum of Agena engine plume; variations in earth's magnetic field; effect of space environment on materials; impedance of near space; air density at satellite altitudes; earth's albedo (whiteness); altitude of ozone layer; radio waves from outside solar system; emissions from upper atmosphere; thermal electricity; and space radiation. Spacecraft were launched with Atlas-Agena and Thor-Agena launch vehicles from Vandenberg AFB and USAF facilities at Pt. Arguello. (*Space Bus. Daily*, 9/17/63, 400)

- Addressing National Conference on Citizenship in Washington, Astronaut John H. Glenn, Jr., said:

  "No one can foretell with certainty the tremendous import this broad technological revolution may have. But its implication is really much more broad and deep than a single reference to space, for space accomplishments are but one fruit of a greater advancement. There have been matching broad advances in many technical fields that have affected, and will continue to affect every facet of life from our homes, to automobiles, to medicine, to buildings, to roads, and to education. History has shown that such surges of advancement are times of greater opportunity to mold the future . . . ." (Text, *CR*, 9/16/63, A5799)

- In a speech at the 1963 Corporate Leadership Dinner, Detroit, Michigan, Robert W. Sarnoff, Chairman of the Board of NBC, said:

  "Within the past quarter century alone, scientific research has more than doubled the inventory of human knowledge.

  "The simple fact that we have shortened the distance between research and product development has itself altered and enriched American life immeasurably, and has created a potential for raising the living standards of people everywhere . . . .

  "Historically, man's adaptation to new circumstances has proceeded by fits and starts, aided by a generous allotment of time in which to accomplish massive transformation. But now science and technology are pumping a new form of quick-change fuel into the lifestream of civilization, and the time is past when we might count upon years of grace in which to alter concepts and methods to suit a differing environment. Today the unprecedented rate of change we experience—and can anticipate—is a fact unique in human experience, calling for more rapid and drastic accommodations than have ever before been required." (Text)

- Two contracts were awarded by NASA Lewis Research Center to study possibility of "floxing" the Atlas booster. If feasible, the mixture of liquid fluorine and liquid oxygen might increase the Atlas payload capability for earth orbit by as much as 88%. (Lewis Chronology, 4)

September 17: Opening of U.N. General Assembly transmitted via RELAY I and SYNCOM II to Europe and Africa. (NASA Release 63–207)

- Addressing International Northwest Aviation Council in Edmonton, Alberta, Deputy Director of NASA Office of Educational Programs and Services James V. Bernardo said: "The destinies of all nations are inextricably entwined with aeronautics and space exploration. Aviation—or perhaps more appropriately, aerospace activities—will continue, on an ever-increasing scale, to have profound effects upon the lives of all people.

"The frontiers of science and technology are limitless. The years ahead will bring about not only tremendous changes, but also an increased and unquenchable thirst for knowledge of the unknown beyond our earthly atmosphere. These years ahead will also witness, I'm sure, another great revolution in education, one which will give us a continuity and sequence in science and mathematics, from the kindergarten through the college graduate school. As the revolution takes place, great care must be exercised by the architects of the curriculum that we achieve a necessary balance. We must teach our young people—since the future lies with them—the physical sciences so that they can achieve our national goals in science and engineering. But we must also teach them the humanities so that they can use these achievements for man's betterment. The two great branches of learning must be blended into the best possible educational program." (Text)

September 17: AEC announced installation of experimental Snap-type reactor in its Western Test Station, Idaho, as first step in new series of reactor-destruction tests. Test series would simulate conditions of "several conceivable, though unlikely, accidents that could destroy [such] a reactor." (AP, *NYT*, 9/18/63, 16)

- Dr. Robert R. Gilruth, NASA Manned Spacecraft Center Director, said in speech before National Rocket Club in Washington that suggestions of U.S.-U.S.S.R. joint lunar landing program were impractical. Such exchange of manpower, knowledge, and funds could be of some benefit, he said, "but I tremble at the thought of the integration problems we could expect . . . . I really have mixed emotions, but I'm speaking only as an engineer, not as an international politician . . . ." (*NYT*, 9/18/63, 11; NRC Release)

- The Ballistic Missile Early Warning System (BMEWS), developed by the Electronic Systems Division, became fully operational with turnover of the third station at Fylingdales, England. The other two are located at Clear, Alaska, and Thule, Greenland. (AFSC Release 312-R-114, 1/1/64, 3; *A&AE*, Nov., 1963, 11)

- NASA signed contract with North American Aviation for studies of Apollo spacecraft modifications to permit its being used as a space science laboratory for missions up to one year. Contract was awarded by Manned Spacecraft Center. (*Space Bus. Daily*, 10/14/63, 76)

- Field Enterprises Educational Corp. and *Life* Magazine announced signing of four-year contracts with 16 NASA astronauts for their personal stories; Field Enterprises acquired newspaper syndication and book rights, *Life*, the magazine rights. (AP, *NYT*, 9/18/63, 15)

- National Capitol Section of AIAA announced it was sponsoring construction of full-scale copy of Wright brothers' airplane which made man's first powered flight. Replica would be presented to Kill Devil Hills, N.C., Wright Memorial Museum at 60th anniversary program on Dec. 16. (AIAA/NCS Release)

September 17: USAF launched Titan I ICBM with dummy nuclear warhead from Vandenberg AFB. (*M&R*, 9/23/63, 10)

September 17–19: Technical session on supersonic transport aircraft held at NASA Langley Research Center, with representatives of Boeing Co. and Lockheed California Co. reporting on results of their feasibility studies performed under LARC contract. Contractor reports indicated two of the four AST concepts studied appeared to be most promising for commercial design: (1) design using principle of variable-sweep wing to provide good flying characteristics at low speeds needed for landing and take-off as well as for efficient supersonic cruise—concept evolved by scientists and engineers of LARC; (2) design using fixed delta wing mounted well aft on long fuselage with canard control surfaces—concept evolved out of research conducted by NASA Ames Research Center. (NASA Release 63–210; LARC Release 9/19/63)

September 17–23: Symposium on applications of the Theory of Functions in Continuum Mechanics, sponsored by International Union of Theoretical and Applied Mechanics, held in Tbilisi, U.S.S.R. Dr. Adolf Busemann, Research Staff Scientist of NASA Langley Research Center, represented the Center and NASA at the symposium. (LARC Release)

September 18: One-year anniversary of orbiting of TIROS VI meteorological satellite, its year-long operational lifetime setting new record for weather satellites. On July 31, 1963, TIROS VI discovered first hurricane (Arlene) of 1963 season in tropical Atlantic; altogether, TIROS VI photographed two hurricanes in Atlantic, two tropical storms in eastern Pacific, eight typhoons in central and western Pacific, as well as sand storms in Saudi Arabia and ice conditions in southern and northern hemispheres. Along with TIROS V it supported Mercury space flights of Astronauts Schirra and Cooper. National Weather Satellite Center issued about 600 weather advisories around the world based on some of the 63,000 cloud-cover pictures from TIROS VI. (NASA Release 63–209; Commerce Dept. Release WB 63–11)

- Sen. Hubert Humphrey (D.-Minn.) said on Senate floor that U.S. "should take the initiative, at this session of the United Nations General Assembly, to create a constructive and conciliatory climate which could lead to a serious discussion between the United States and the Soviet Union of the basic political issues of the cold war." Of the five issues he proposed, two dealt with space exploration:

"We should make clear our willingness to cooperate with the Soviet Union and with other nations in the field of outer space. The whole question of exploration of outer space and of the law that will govern outer space are ripe for further discussion and bold new actions . . . ."

Commenting on *New York Times* report that Soviet Academician Anatoli A. Blagonravov had "suggested that it might be advisable to discuss the possibility of cooperating between the United States and the U.S.S.R. in the assault on the moon," Senator Humphrey urged: "Despite the technical problems that might be involved in such a cooperative venture, it is my view that such proposals on the part of Soviet officials should not be hostilely rejected. Rather they should be given careful consideration." (*CR*, 9/18/63, 16536–37)

September 18: GSFC selected two companies for negotiation of contracts pertaining to Nimbus weather satellite. $252,000 contract to General Electric Co. called for development of operating procedures for Nimbus control center as well as training of personnel to operate the center. $165,000 contract to RCA Electron Tube and Semiconductor Div. required contractor to furnish solar cells for Nimbus satellites and Nimbus operational system. (GSFC Release G–19–63)

- USAF launched Thor rocket with delta-wing re-entry payload from AMR, first flight test in Project Asset (Aerothermodynamic-Elastic Structural Systems Environmental Tests). Payload reached 35-mi. altitude before re-entering atmosphere at 10,900 mph speed and landing about 1,000 mi. southeast of Cape Canaveral. Recovery efforts were not successful. Good data on temperatures and pressures of the payload were obtained during the 20-min. flight; the various materials of the payload were subjected to re-entry heat up to 4,000°F. (AP, *NYT*, 9/19/63, 8)
- Army Corps of Engineers awarded NASA contract to Westinghouse Astronuclear Laboratory for estimate of power needs of a manned lunar base and design of nuclear generator that could be sent to the moon for such a base. Data would be used by NASA in evaluating feasibility of lunar base. (DOD Release 1255–63)

September 19: SYNCOM II 24-hour communications satellite used to relay oceanographic data from research vessel *Geronimo* in Gulf of Guinea off Africa to National Oceanographic Data Center in Washington, which compared the data with its records and sent back to the *Geronimo* the deviations to correct errors. Demonstration via SYNCOM II was performed to determine practicability of providing research ships quickly with information to correct errors. Line of transmission: from *Geronimo* to *Kingsport* in Lagos harbor, to SYNCOM II some 22,300 mi. above Atlantic Ocean, to ground station at Lakehurst, N.J., along ground lines to NODC, and return. (NASA Release 63–212)

September 19–22: Conference on Outer Space, sponsored by St. Louis Univ. and Columbia Univ., held in St. Louis. 75 persons from 13 states represented business, industry, education, professions, science, labor, Government, and the press at the third Midwest Assembly. (*St. Louis Post-Dispatch*, 9/8/63)

NASA Administrator James E. Webb said in Sept. 19 address: "... as a nation we are going through a new and vital experience in achieving the mastery of space. We have undertaken to perform the largest job of research, of development, and of manufacture ever mounted by this or any other nation. There is required here the mobilization of the best of science, of engineering, of industry, and of government. The technical job is in itself staggering. Yet for American society perhaps the greater significance lies in the fact that the objective of this program is fully as much to gain knowledge as to secure concrete objectives; that it is being carried on in peacetime without the impetus of war; that, while insuring national security, its central purpose is to secure the peaceful uses of space for all mankind; that NASA is a civilian agency; and that it necessarily will require many years of sustained and costly effort. We have in fact set ourselves a hard challenge of national purpose and will: to sustain such a program through all of the troubles and difficulties that are inevitable in a research and development effort, an effort whose failures as well as its successes are set before the world to see. This is a difficult undertaking; but if we succeed in these ways, then we will have accomplished much and learned more as a free society . . . ." In interview following his address, Webb commented on that day's proposal by Soviet Foreign Minister Andrei Gromyko for summit meeting of leaders of the 18 nations participating in Geneva disarmament conference. Webb said Gromyko's U.N. speech was the most recent example of growing awareness of U.S. advances in space technology. ". . . The image of America as a can-do nation in space has been increasing rapidly, and the Russians cannot fail to be influenced by that image." (Text; *St. Louis Post-Dispatch*, 9/20/63)

September 19: USAF announced two 1½-lb. Tetrahedral Research Satellites (TRS) had been ejected into orbit from an unidentified orbiting satellite earlier in 1963. Twin satellites were second and third of series; first TRS was launched in 1962. Primary purpose of TRS experiments was investigation of radiation damage to spacecraft solar-power systems and evaluating methods of protecting the solar cells from Van Allen belt radiation. Valuable data obtained by the five radiation-damage experiments in each satellite were transmitted back to earth and received by NASA Minitrack telemetry network, cooperating with USAF in the project. (DOD Release 1260–63; TRS Fact Sheet)

- Quoting President Kennedy's Inaugural Address, in which he said, "Let both sides seek to invoke the wonders of science instead of its terrors. Together let us explore the stars . . .," *New York Times* editorial stated:

    "When the President spoke these words, there was far more reason to think his idea an impractical one than there is today. Of course there are difficult problems still in the way of serious Soviet-American cooperation on a manned space flight. But the financial, political and scientific advantages to both sides of a joining of forces in this area would be so great that every opportunity should be seized to explore this possibility with the Soviet Government. Moscow's feelers in this direction should be followed up, not rebuffed out of hand . . . ." (*NYT*, 9/19/63, 26)

- USAF announced issuing Requests for Proposals for studies of manned orbital space station. From the bidding industries USAF will select three contractors to conduct parallel four-month studies "to define characteristics from which a space station could be designed to demonstrate and assess quantitatively the utility of man in space for military purposes." (DOD Release 1261–63)

September 20: Addressing U.N. General Assembly, President Kennedy listed results of U.S.-U.S.S.R. negotiations:

"We have, in recent years, agreed on a limited nuclear test-ban treaty, on an emergency communications link between our capitals, on a statement of principles for disarmament, on an increase in cultural exchange, on cooperation in outer space, on the peaceful exploration of the Antarctic, and on tempering last year's crisis over Cuba . . . .

"Finally, in a field where the United States and the Soviet Union have a special capacity—the field of space—there is room for new cooperation, for further joint efforts in the regulation and exploration of space. I include among these possibilities a joint expedition to the moon.

"Space offers no problem of sovereignty; by resolution of this Assembly, the members of the United Nations have foresworn any claims to territorial rights in outer space or on celestial bodies, and declared that international law and the U.N. charter will apply. Why, therefore, should man's first flight to the moon be a matter of national competition?

"Why should the United States and the Soviet Union, in preparing for such expeditions, become involved in immense duplications of research, construction and expenditure? Surely we should explore whether the scientists and astronauts of our two countries—indeed, of all the world—cannot work together in the conquest of space, sending some day in this decade to the moon, not the representatives of a single nation, but the representatives of all humanity.

"All these and other new steps toward peaceful cooperation may be possible. Most of them will require on our part full consultation with our allies, for their interests are as much involved as our own, and we will never make an agreement at their expense.

"Most of them will require long and careful negotiations. And most of them will require a new approach to the cold war—a desire not to 'bury' one's adversary but to compete in a host of peaceful arenas, in ideas, in production, and in service to all humanity.

"The contest will continue, the contest between those who envision a monolithic world and those who believe in diversity, but it should be a contest in leadership instead of destruction, a contest in achievement instead of intimidation. Speaking for the United States of America, I welcome such a contest. For we believe that truth is stronger than error, and that freedom is more enduring than coercion. And in the contest for a better life, all the world can be the winner . . . ."

In addition to urging U.S.-U.S.S.R. negotiations, President Kennedy urged U.N. to work on five major projects, among them "a global system of satellites [which] could provide communication and weather information for all corners of the earth." (Text, *Wash. Post,* 9/21/63, A10)

- Commenting on President Kennedy's proposal for joint U.S.-U.S.S.R. lunar exploration negotiations, NASA Associate Administrator Dr. Robert C. Seamans said in Houston press conference that the proposal did not mean Soviet cosmonauts would fly in

Apollo spacecraft. Dr. George E. Mueller, NASA Deputy Associate Administrator, compared a cooperative moon program to cooperative exploration of Antarctica—U.S. and Soviet scientists explore the continent together but, he said, "they get there in different ships." (Burnett, Wash. *Sun. Star*, 9/22/63)

September 21: Upper stage (S–IV) for Saturn SA–5 rocket arrived at Cape Canaveral aboard modified Stratocruiser aircraft. To be test-flown late this year, SA–5 would be first Saturn testing live upper stage. (*Marshall Star*, 9/25/63, 1)

- TIROS VII meteorological satellite discovered Hurricane Debra, fourth hurricane of season, headed north in Atlantic southeast of Bermuda. (UPI, *Chic. Trib.*, 9/23/63)
- Unnamed White House spokesman disclosed that President Kennedy first mentioned possibility of joint U.S.-U.S.S.R. manned lunar exploration to Soviet Premier Khrushchev during their June 1961 meeting in Vienna. Premier Khrushchev neither accepted nor rejected the proposal at that time. (Simons, *Wash. Post*, 9/22/63, A12; *NYT*, 9/22/63, 1)
- British and European press headlined President Kennedy's proposal for a joint U.S.-U.S.S.R. lunar expedition, while Soviet news media ignored that portion of the President's address. (UPI, *Wash. Post*, 9/21/63, A10; Reuters, *NYT*, 9/22/63, 23; UPI, *NYT*, 9/22/63, 34)
- NATO announced U.S. Prof. Courtland D. Perkins had been elected chairman of NATO's Advisory Group for Aeronautical Research (AGARD). (Reuters, *Wash. Post*, 9/22/63, A15)

September 21–22: Worldwide press reaction to President Kennedy's U.N. speech of Sept. 20, proposing joint manned lunar landing effort:

". . . The very reasons which impelled the powers to reach a test ban treaty—a combination of common sense and a recognition that costs were going up by geometric progression—should now bring them to combine a space program." (*London Times*)

". . . The horizons opened by the President . . . were vast." (*Paris Aurore*)

". . . Benefits from such a joint effort militate in favor of this second major step to develop the improved international relationship already stemming from the agreement on the limited nuclear test-ban treaty." (*NYT*)

". . . Will excite world-wide attention . . . eminently sensible suggestion . . . national advantages to be derived by the first nation to the moon are great, in a political and propaganda context, but they are lesser than the advantages that all mankind will derive from the extension of human knowledge that is going to be involved in the effort to reach the moon . . . ." (*Wash. Post*)

". . . Stirred the imagination of the world . . . tend to open up the closed society of the U.S.S.R." (*Wash. Star*)

". . . Kennedy envisages an incredibly bold scientific advance [and] accepts as possible a degree of collaboration between two opposing systems . . . which would have seemed incredible only a few months ago." (*London Daily Express*, *NYT*, September 22, 1963)

September 23: SYNCOM II communications satellite relayed transmission of speech and teletype between Fort Dix, N.J., and moving ship *Kingsport* about 40 mi. west of Lagos, Nigeria. This was first such transmission via a communications satellite to a moving ship at sea. This was first in series of experiments designed to test shipboard equipment and reception in fringe areas. (NASA Release 63-213)

* USAF announced launching unidentified satellite with Thor-Agena D launch vehicle from Vandenberg AFB. (AP, *NYT*, 9/25/63, 14)
* President John F. Kennedy wrote a letter to Rep. Albert Thomas (D.-Tex.):

    "I am very glad to respond to your letter of September 21 and to state my position on the relation between our great current space effort and my proposal at the United Nations for increased cooperation with the Russians in this field. In my view an energetic continuation of our strong space effort is essential, and the need for this effort is, if anything, increased by our intent to work for increasing cooperation if the Soviet Government proves willing.

    "As you know, the idea of cooperation in space is not new. My statement of our willingness to cooperate in a moon shot was an extension of a policy developed as long ago as 1958 on a bipartisan basis, with particular leadership from Vice President Johnson, who was then the Senate Majority Leader . . . . Our specific interest in cooperation with the Soviet Union, as the other nation with a major present capability in space, was indicated by me to Chairman Khrushchev in Vienna in the middle of 1961, and reaffirmed in my letter to him of March 7, 1962 . . . . So my statement in the United Nations is a direct development of a policy long held by the United States Government.

    "Our repeated offers of cooperation with the Soviet Union have so far produced only limited responses and results . . . . But as I said in July of this year, there are a good many barriers of suspicion and fear to be broken down before we can have major progress in this field. Yet our intent remains: to do our part to bring those barriers down.

    "At the same time, as no one knows better than you, the United States in the last five years has made a steadily growing national effort in space . . . .

    "This great national effort and this steadily stated readiness to cooperate with others are not in conflict. They are mutually supporting elements of a single policy. We do not make our space effort with the narrow purpose of national aggrandizement. We make it so that the United States may have a leading and honorable role in mankind's peaceful conquest of space. It is this great effort which permits us now to offer increased cooperation with no suspicion anywhere that we speak from weakness. And in the same way, our readiness to cooperate with others enlarges the international meaning of our own peaceful American program in space.

    "In my judgment, therefore, our renewed and extended purpose of cooperation, so far from offering any excuse for slackening or weakness in our space effort, is one reason the more for moving ahead with the great program to which we have been committed as a country for more than two years.

"So the position of the United States is clear. If cooperation is possible, we mean to cooperate, and we shall do so from a position made strong and solid by our national effort in space. If cooperation is not possible—and as realists we must plan for this contingency too—then the same strong national effort will serve all free men's interest in space, and protect us also against possible hazards to our national security. So let us press on . . . ." (Letter, 9/23/63)

September 23: Dr. Hugh L. Dryden, NASA Deputy Administrator, said in *Missiles and Rockets* interview that Soviet Academician A. A. Blagonravov had informed him that "once the two nations have landed instrumented payloads on the Moon discussions of cooperation in a manned lunar-landing program should begin." Dr. Dryden added that "it's the first time they have indicated interest in a joint effort.

"The climate has changed. It is somewhat more favorable." (*M&R*, 9/23/63, 14)

- USAF announced launching Titan II ICBM from underground silo in test flight from Vandenberg AFB. (AP, *NYT*, 9/25/63, 14)
- France's Centre National d'Etudes Spatiales announced France would send a cat into space from rocket launching site in Sahara. (Reuters, *Wash. Post*, 9/24/63)
- First child born to space traveler—Tanya Titov, daughter of Maj. Gherman Titov (VOSTOK II orbital space flight) and wife Tamara. (AP, *NYT*, 9/25/63, 8)

September 24: Subcommittee of House Committee on Appropriations voted to approve $5.1 billion appropriations bill for NASA in FY 1964, $250,820,400 less than NASA authorizations bill. Rep. Olin E. Teague (D.-Tex.), Chairman of Manned Space Flight Subcommittee of House Committee on Science and Astronautics, said President Kennedy's proposal for U.S.-U.S.S.R. manned lunar flight was "bound to hurt" chances of full appropriations being passed by House and criticized idea of joint project as impractical. (Finney, *NYT*, 9/25/63, 1, 15)

- First public showing of guidance and navigation system for Apollo manned space flights, the briefing conducted at MIT's Instrumentation Laboratory, in charge of the system design and development for NASA. (*NYT*, 9/25/63; Wash. *Eve. Star*, 9/25/63)
- AFSSD briefed prospective bidders on two-phase program definition of a standardized space guidance system (SSGS), aimed at developing single system capable of providing guidance and control for launch vehicle, for orbital injection, and for actual spacecraft mission performance. Phase One would study various missions and analyze hardware requirements; Phase Two would investigate means to develop hardware based on Phase One concepts. (*M&R*, 9/9/63, 16; *M&R*, 9/30/63, 18)
- After lengthy debate, Senate voted (80-19) to ratify nuclear test-ban treaty, initialed by U.S., U.K., and U.S.S.R. on July 25 in Moscow and prohibiting nuclear tests in the atmosphere, in space, and under water. (*CR*, 9/24/63, 16909)

September 24: In his *Washington Post* column, Walter Lippmann said:

"The President has made his suggestion of collaboration in going to the moon at a time when there is some improvement in U.S.S.R.-U.S.A. relations. It happens also to be a time when there is a growing doubt among American scientists and among the people generally about the commitment to put an American man on the moon by the year 1970.

"The President's proposal at the U.N. is, it seems to me, excellent even if the joint effort proves to be technically and politically impracticable. It is excellent because it may offer an honorable way to correct the mistakes of our original commitments about going to the moon.

"There were two big mistakes. One was the commitment to put a man, a living person rather than instruments, on the moon. The other mistake was to set a deadline—1970—when the man was to land on the moon . . . ." (*Wash. Post*, 9/24/63)

- Senate passed (77–0) Defense appropriations bill for FY 1964, totaling $47,339,707,000. This was $258 million more than passed by House and $1.4 billion less than Administration request. Bills would be sent to conference committee for compromise. (*CR*, 9/24/63, 16972; *NYT*, 9/25/63, 1)
- At International Atomic Energy Agency conference in Vienna, AEC Chairman Dr. Glenn T. Seaborg said agricultural and industrial applications of radioactive isotopes in U.S. had approximately doubled in past five years and that "we look for perhaps an increased trend in the next five years." (Underwood, *NYT*, 9/25/63, 14)
- U.S. Army conducted successful test-firing of Pershing ballistic missile from Black Mesa, Utah, to impact point within White Sands Missile Range, N.M. Test was part of current overland series conducted under near-tactical conditions. (DOD Release 1282–63)
- DOD launched unidentified satellite from PMR employing Thor-Agena D launch vehicle. (*Pres. Rpt. on Space, 1963*, 1/27/64)

September 25: NASA launched two-stage Aerobee 150A sounding rocket from Wallops Island, Va., with 185-lb. payload of U.S.-Japanese instrumentation to measure electron density and temperature in the ionosphere. Payload was lofted to 139-mi. altitude and no recovery operation was involved. Purpose of experiment was to compare simultaneous measurements made by two different methods: instruments supplied by NASA Goddard Space Flight Center and radio-frequency resonance probe developed by Japan's Radio Research Laboratory. (Wallops Release 63–85; NASA Release 63–211)

- NASA Administrator James E. Webb, speaking before annual meeting of Texas Mid-Continent Oil and Gas Association in Houston, quoted President Kennedy's proposal for U.S.-U.S.S.R. negotiations toward joint lunar exploration:

"The President's statement has captured the imagination of the peoples of the world. In itself, cooperation in space activity between the great antagonists of the Cold War is a thrilling prospect. But the significance of this possibility is not limited to space; rather, it lies equally in the fact that cooperation in space is one more step toward cooperation on Earth, toward the banishment of the fear of the annihilation of life as we know it.

"The President's statement received weight and conviction from the fact that the United States has a powerful space capability; and from the fact that this Nation is well on its way to achieving preeminence in this new environment.

"This national strength in space represents a massive and productive effort on the part of the United States during the past five years, an effort whose potential value is only now beginning to emerge . . . .

"While proceeding with the development of the launch vehicles and spacecraft of the future, more than 90 per cent of it under contract with American industry, we have also learned the essentials of space operations, and achieved a degree of reliability which has produced many specific achievements in space.

"In 1958, the United States had five successful flights, but for each success we had two failures. By 1961, out of 54 flights, the success ratio was 83 per cent. In the first eight months of this year every NASA launch has succeeded, with the exception of one small Scout rocket launched from Wallops Island . . . ." (Text)

September 25: Eugene Wasielewski, Associate Director of NASA Goddard Space Flight Center, in address before Scientific Research Society of America in Paoli, Pa., emphasized accomplishments of NASA during its first five years:

"While it is easy for me to say that NASA has, as an organization, launched over 50 satellites, of which 30 were put into orbit by Goddard . . ., I am sure that you must all realize the tremendous research and development effort that was required to put these spacecraft into orbit.

"Certainly Goddard . . . owes much to the successful development of the Delta launch vehicle which has given us 20 successes in 21 shots . . . ." (*Goddard News.* 10/7/63, 6)

- Louis B. C. Fong, Director, Office of Technology Utilization, addressing the American Management Association, Inc., said:

"NASA's activities in space operate across all of the interfaces in the total transition process which starts with basic research and advances through applied research, development, engineering design, test and fabrication to ultimate production. Industry must be ready to determine in which phase it should act and direct space technology to an industrially oriented goal.

"Resistance to new ideas and new technologies is part psychological; a good deal of it is practical when new expenditures have to be justified to stockholders in terms of an upward sales curve. Often it is economic, since to implement a new technology may result in a tremendous impact upon a way of life of a major industry—e.g., oil vs. coal, transistors vs. tubes, solid state physics vs. conventional circuit design, diesel vs. steam engines, etc. . . .

"Companies must organize to accept this data; must be geared to use what is helpful; must work to break down the barriers between one division and another. For economic survival, the time lag can no longer be accepted as standard operating procedure in the laboratory-to-consumer cycle . . . .

"The accepted business patterns of the past must change if you singularly as a company and we collectively as a national business enterprise are to stay in the race." (Text)

September 25: Sixty Eastern and Western scientists meeting at 11th Pugwash Conference on disarmament and world security endorsed President Kennedy's proposals for international scientific cooperation in space and other areas. The scientists also proposed U.S., U.K., and U.S.S.R. conduct joint study of seismic phenomena, with hope that this study would lead to banning of underground nuclear tests. Conferees endorsed appeal of Soviet Foreign Minister Andrei A. Gromyko for agreement barring orbital atomic weapons. (Binder, *NYT*, 9/26/63, 11)

- Study memorandum identifying U.S. "stamp lag" inserted in *Congressional Record* by Sen. Frank E. Moss (D.-Utah). In the memo Frank Ballard, principal engineer of Sperry Utah Co., noted Communist bloc has issued at least 18 postage stamps identifying Soviet space successes, whereas U.S. has issued only two stamps associated with its space program, and these in limited editions. ". . . In the face of the existing world struggle, it is imperative, even urgent, that we utilize all available media of communication to advertise the forces of democracy. The U.S. 'stamp lag' is self-evident. We must engage in a vigorous program calculated for optimum utilization of the vast potentialities of the U.S. postage stamp. As a start, I should like to suggest a chronological documentation of the entire space program beginning with the suborbital flight through individual Mercury orbital flights. Gemini and Apollo . . . ." (Study Memorandum, *CR*, 9/25/63, 17044–45)

- Communist China's press agency Hsinhua announced that more than 1,000 Chinese and foreign scientists met in Peking to celebrate establishment of center for the World Federation of Scientific Workers devoted to the "advancement of science in Asian, African and Latin American countries." (*NYT*, 9/26/63, 5)

September 26: NASA announced first television experiments via SYNCOM II communications satellite had been conducted. Test pattern signals sent Sept. 23 were followed by TV pictures Sept. 24 and 25; because of bandwidth limitations, no audio was sent. Officials of NASA Goddard Space Flight Center said transmissions were of good quality. Transmissions originated at Fort Dix, N.J., ground station, were sent to SYNCOM II 22,300 mi. above the earth, and retransmitted to AT&T ground station at Andover, Me. (NASA Release 63–216)

- Rep. Albert Thomas (D.-Tex.) made public a letter from President Kennedy written in reply to the Congressman's request for clarification of the President's position on U.S. lunar landing goals in light of his proposals for U.S.-U.S.S.R. joint program. In his letter President Kennedy said:

"This great national effort and this steadily stated readiness to cooperate are not in conflict . . . ." Rather, they are "mutually supporting elements of a single policy.

"In my view an energetic continuation of our strong space effort is essential, and the need for this effort is, if anything, increased by our intent to work for increasing cooperation if the Soviet Government proves willing . . . .

"If cooperation is not possible—and as realists we must plan for this contingency, too—then the same strong national effort will serve all free men's interest in space, and protect us also against possible hazards to our national security. So let us press on." (UPI, *NYT*, 9/27/63, 2)

September 26: James T. Dennison, of NASA Technology Utilization Office, said in paper delivered at annual meeting of National Association of Business Economists, Cleveland:

". . . We suggest that the greatest contributions of our government's aerospace programs to the business economy have been the psychological boost of setting for ourselves an utterly impossible goal and then buckling down to reach it. And this adventure has come at a time when there were those who cried that our democratic capitalistic economy was sluggish, that we were being outdone by overseas and overland competitors, that gloom was here and that doom was close at hand.

"The challenge of the exploration of space, of putting our fellow man safely into a desperately hostile environment, has provided a spur to great segments of our American business economy. And that spur will be keeping us on the jump until we get those men to the moon and back, and, then perchance, take off for Mars." (Text)

- On his 66th birthday, Pope Paul VI recognized 175th anniversary of Georgetown Univ. by a special address broadcast to U.S. via TELSTAR II communications satellite. (*NYT*, 9/27/63, 9; NBC-TV, 9/26/63)
- USN launched Polaris A-3 test missile from surface test ship U.S.S. *Observation Island* using completely new eject launch system. (DOD Release 1301-63)

September 26–October 1: XIVth International Astronautical Congress held in Paris, with more than 1,000 delegates from 34 countries attending. 80 papers covering wide range of scientific and technical subjects were presented. (Program)

Dr. Eugene B. Konecci, Director of NASA Biotechnology and Human Research, and Chairman of IAF Bioastronautics Committee, presented his "Bioastronautics Review." Dr. Konecci discussed U.S. bioastronautics programs, results, and plans, and compared U.S. program with that of U.S.S.R. Summing up Project Mercury MA-9 flight, he said:

"The performance of Gordon Cooper proved in detail man's integration with the operation of the spacecraft systems. This accomplishment of the Mercury program is of major significance as it does indicate that much more dependence can be placed on man as a reliable operating portion of the man-spacecraft combination . . . . Once again, the recent flight of MA-9 has proved that man can adapt very rapidly to a 35-hour period in a new environment. His senses and capabilities are little changed in space. At least for the duration of Cooper's mission (a total of 34+ hours) weightlessness affecting the normality of his judgment, orientation and other human functions in space was no problem . . . ." (Text)

Dr. Konecci also reviewed Russian progress in biotechnology. He said the Soviets had demonstrated the reliability of their spacecraft and their "high regard for life of the cosmonauts, since, to our knowledge, 'not a single life has been lost in the U.S.S.R. space effort.' The Soviets concede that, in the long run, human intellect and sensations are the only forces capable of learning the secrets of the universe." (Text)

Four papers by Soviet scientists presented at IAF Congress indicated biological effects of cosmic radiation were one of the most complex and most urgent problems of bioastronautics. Papers reported on radiation effects on seeds, fruit flies, animal tissue, and bacteria as well as on cosmonauts Gagarin, Titov, Nikolayev, and Popovich, who experienced no harmful effects. (Abstracts)

At IAF Congress in Paris, Dr. Charles Draper of MIT told *Missiles and Rockets* that joint lunar landing was possible within five years if progress continued at present rate and if U.S. guidance were used with a Soviet rocket. (*M&R*, 9/30/63, 26)

Lt. Col. Yuri Gagarin said in address Sept. 28 that Soviet scientists were working on problems of rendezvous and coupling of at least two spacecraft in orbit. Cosmonaut Gagarin acknowledged that rendezvous technique was "extremely complex" and that a number of problems remain in areas of "communications, optics and maneuvers." He said Soviets would dock several individual vehicles in earth orbit to construct space platform from which man would be sent on flight trajectory toward the moon.

Le Journal du Dimanche reported Cosmonaut Gagarin's prediction that U.S.S.R. would land a man on the moon before the U.S. (*Av. Wk.*, 10/7/63, 30; *Le Journal du Dimanche*, 9/29/63, 22)

Soviet General Nicolai Petrovich Kamanin, Deputy Chief of Staff of Soviet Air Force and head of cosmonaut group, said at IAF luncheon that joint U.S.-U.S.S.R. manned lunar program would achieve a manned lunar landing more quickly and cheaply than separate efforts by the two countries. After the luncheon he told newsmen that a "more favorable political evolution" was necessary before a joint lunar expedition could be planned. (Reuters, *Wash. Post*, 9/30/63)

Edgar M. Cortright, Deputy Director of NASA Office of Space Sciences, said in IAF address:

"One of the most exciting technological aspects of space exploration has been the development of automated spacecraft. Most of the scientific exploration of space and the useful applications of space flight thus far have been made possible by automated spacecraft. Development of these spacecraft and their many complex subsystems is setting the pace today for many branches of science and technology. Guidance, computer, attitude control, power, telecommunication, instrumentation, and structural subsystems are being subjected to new standards of light weight, high efficiency, extreme accuracy, and unsurpassed reliability and quality . . . ." He then reviewed automated spacecraft already flown or currently under development by NASA. (Text)

September 26: Dr. Max Tishler, President of Merck, Sharp & Dohme Research Laboratories, received the 1963 Chemical Industry Medal of the American Section, Society of Chemical Industry, in Houston. In his acceptance speech Dr. Tishler discussed growth of Federal support of basic research, concluded: "Industrial research—once devoted exclusively to the promotion of economic growth—has become predominantly a Government-dominated institution for protecting the Nation's security and exploring the solar system.

"The question is being asked with increasing persistence whether we are devoting a disproportionate amount of our scientific and technical resources to the Government's objectives. It has been pointed out that between 1954 and 1961 three-quarters of the increase in scientists and engineers engaged in research and development was absorbed by the defense and space programs, and that the programs on the drawing boards of these same agencies will require as many scientists during the next decade as all our institutions of higher learning will graduate during the same period . . . . How far and how long can this situation progress without damaging our country's capacity to maintain its broad leadership in research?

"The Defense Department and the space agency have countered these concerns with predictions of spinoffs for the civilian economy from developments financed by them. So far, the evidence for this contention is unpersuasive. Whether or not the prophecy will come true does not alter the unrelenting fact that too little attention is being paid to the long-range effect on the civilian economy and to the manner in which we have been using tax money to divert our scarce scientific and technical resources into such crash projects as the race to the moon. Our Government has been prone to act as if these resources are either unlimited or can be expanded indefinitely by appropriating dollars. Neither proposition is correct . . . ." (Text, *CR*, 1/15/64, A143–145)

September 27: NASA EXPLORER XIV satellite progress report indicated no usable scientific data had been obtained from the scientific satellite since mid-August. In its ten months of operation since launch into highly elliptical orbit Oct. 2, 1962, EXPLORER XIV sent back more than 6,500 hours of data from the six onboard scientific experiments to chart boundaries of earth's magnetosphere, measure particle population and energies of electrons and protons, and determine how magnetic fields influence these particles. 3,700 hours of data had been processed through computers and scientific analysis was continuing. (NASA Explorer XIV Prog. Rpt. No. 6)

- Press briefing on contracting procedures for Lunar Orbiter spacecraft held at NASA Hq. Contract for five Lunar Orbiters would be first major NASA contract on basis of incentives, with contractor sharing cost savings and/or penalties. Objectives of incentive contracting were to assure greater reliability of products and meeting of scheduled deadlines. (Press Briefing Transcript)

September 27: USAF launched Scout vehicle from Vandenberg AFB with undisclosed payload. (UPI, *Chic. Trib.*, 9/28/63)
* USAF announced awarding $3,000,000 original definitive contract to Lockheed Propulsion Co. for research and development of solid propellant motors for the large solid booster program. (DOD Release 1304-63)

September 27-October 10: NASA displayed models of manned and unmanned spacecraft in lobby of NASA Hq. building, part of fifth anniversary observance. (NASA Release 63-214)

September 28: First birthday of ALOUETTE I ("topside sounder") satellite, built by Canada and launched by NASA into orbit Sept. 28, 1962. ALOUETTE was still functioning well and transmitting ionospheric data to 13 telemetry stations around the world.

NASA launched four-stage Javelin (Argo D-4) rocket with instrumented payload to 645-mi. altitude from Wallops Station, in experiment to compare rocket-borne measurements of ion and electron temperatures and densities with similar data obtained by ALOUETTE. Measurements were taken by the rocket payload in the ionosphere 10 min. before ALOUETTE passed through the region and made its measurements. Data were telemetered to ground receiving stations and no recovery operation was involved. Flight was follow-on to similar experiment conducted July 2, results of which indicated close agreement in measurements. During the first year of operation, ALOUETTE I orbited the earth 4,981 times, executed 12,900 commands, and provided 2,060 hours of telemetry transmissions. (GSFC Historian; NASA Release 63-211; Wallops Release 63-87; *Goddard News*, 9/23/63, 1)

* Aerobee 150A sounding rocket launched from NASA Wallops Station with U.S.-Japanese experiment to measure electron temperatures and densities in the ionosphere by two different methods: Langmuir probe, supplied by NASA Goddard Space Flight Center, and radio-frequency resonance probe, developed by Radio Research Laboratory, Tokyo. 185-lb. payload reached 141-mi. altitude and transmitted approximately 8 min. of telemetry before impacting in Atlantic Ocean about 71 mi. from launch site. Data obtained from the daytime experiment were compared with data obtained from similar experiment conducted at night, three days earlier. (NASA Releases 211 and 218)
* Navy SKMR-1, air-cushion Hydroskimmer, developed by Bell Aerosystems for speeds up to 80 mph, went through its paces at Lake Erie. (*A&AE*, Nov., 1963, 11)
* Unidentified USN satellite placed in orbit with Thor-Able-Star launch vehicle launched from Vandenberg AFB. Satellite was first to be completely powered by nuclear generator, a 27-lb. Snap-9A (Systems for Nuclear Auxiliary Power) which will produce 25 watts of power continuously for five years. Press sources unofficially identified the satellite as TRANSIT V-B. The Navy later disclosed that a second satellite weighing 120 lbs. had been launched pickaback style along with the 160-lb. nuclear-powered satellite. This sun-powered satellite carried several radiation detectors and six transistors in a test of means of protecting these devices from radiation damage. (Finney, *NYT*, 9/15/63, 75;

Hill, *NYT*, 10/1/63, 1, 76; *Av. Wk.*, 10/7/63, 37; *Wash. Post*, 1/10/64)

September 28: Article by commentator Stanislav Kondrashov, first published mention in U.S.S.R. of President Kennedy's proposal for U.S.-U.S.S.R. cooperation in lunar exploration, appeared in Soviet magazine *Za Rubezhom*. Article said:

"American propaganda . . . is pushing to the forefront the President's idea of sending a joint Soviet-U.S. team to the moon.

"In lavish headlines American newspapers describe the grandiose character of the project, although it seems it is somewhat premature. . . .

"Leaving aside judgment of the President's lunar project, it should be noticed that the hard emphasis on it is hardly worthwhile.

"It distracts attention from joint earthly exploits directed at attaining peace and reduction of world tension. . . ." (UPI, *Wash. Post*, 9/29/63)

- Discussing President Kennedy's proposal for U.S.-U.S.S.R. lunar cooperation, James J. Haggerty, Jr., said in *Army-Navy-Air Force Journal and Register* article:

  "The big question . . . is how far along are the Soviets in a moon program and which mode of approach did they select?

  ". . . if the Soviets have adopted one of the approaches we have discarded, any cooperative arrangement would require program reorientation on the part of both nations of such magnitude that the only imaginable result is complete chaos, compounded of the basic difference in approach, the language barrier, differing technological philosophies and engineering standards.

  "Even if the U.S.S.R. is already proceeding on the same lines as the U.S., i.e. the lunar orbit rendezvous mode, there are different methods of attaining the same end and it is extremely unlikely that their systems development parallels our own in every detail.

  "It is possible of course that the cooperative program consist of joint funding, a single lunar launch base and a common pool of astronauts, with the technical differences being resolved through the use of one spacecraft/launch vehicle combination without altering current developmental lines. But which one? Who gets the program management? One can see a political and technical argument over the 'Apollostok' (or should it be 'Vostapollo?') which would leave the moon safe from assault for at least a couple of decades." (*A-N-AF Journal & Reg.*, 9/28/63, 14)

- Communications Satellite Corp. set 1966 as target date to start commercial operation of global satellite communications network. (*A&AE*, Nov. 1963, 11)

- Sermon by Dr. Duncan Howlett, Washington Unitarian minister, was quoted in *Washington Post*. On June 18, 1961, Dr. Howlett said:

  "Why can we not boldly propose now a cooperative moon-shot, built and manned by Russian and American scientists? Why can we not send a capsule to the moon with a Russian and an American inside it; and when they get there let them claim the moon—not for Russia, not for the United States, but for humanity? . . .

"We can claim space for humanity or we can make it a battle ground to which to extend the ancient quarrels of earth. Which shall it be? There is but one choice before us. We can try to make the exploration of space a cooperative venture. We cannot succeed unless we try . . . ." (Glaser, *Wash. Post*, 9/28/63)

September 28: The International Academy of Astronautics, at regular meeting, elected Dr. Charles Stark Draper to succeed Dr. Frank J. Malina as President of the Academy. Dr. Draper is one of the world's leading space engineers, engaged in the solution of problems of guidance and control of space vehicles. His latest and largest undertaking is the design of the guidance-navigation system to be used aboard the Apollo spacecraft, which is to carry man to a Moon landing and a safe return. He is Head of the Department of Aeronautics and Astronautics at the Massachusetts Institute of Technology, as well as Director of the MIT Instrumentation Laboratory, in which capacity he leads a team of 1,800 scientists, engineers and technicians. (IAF Release 17, 10/1/63)

- Trevor Gardner, former Assistant Secretary of the Air Force for R&D (1955-56) and advocate of increased funding for ICBM development, died at his home in Washington. (AP, *NYT*, 9/30/63, 29)

- The Iven C. Kincheloe Award was presented by Mrs. Dorothy Kincheloe to the seven original Mercury astronauts as the highlight of the Seventh Annual Awards Banquet of The Society of Experimental Test Pilots held at Lancaster, Calif. The award is presented annually for the recognition of outstanding professional accomplishment in the conduct of flight testing. (*Space News Roundup*, 10/16/63)

September 28-29: Open house at NASA Wallops Station, Va., in observance of NASA Fifth Anniversary (Oct. 1). Approximately 8,000 persons visited the facility. (Wallops Releases 63-84 and 63-89)

September 29: Argo D-4 sounding rocket launched from Wallops Island, Va., in experiment to measure vertical distribution of ionospheric parameters simultaneously with overhead passage of ALOUETTE I topside sounder satellite. The rocket reached 644.6-mi. altitude and all experiments functioned normally, but telemetry system failed at approximately 12 min. so about 7 min. of re-entry data were not obtained. (NASA Rpt. of S. Rkt. Launching)

- Vice President Lyndon B. Johnson, Chairman of National Aeronautics and Space Council, interviewed in *Washington Post* magazine, *Parade*. Asked "with all our needs on earth, can we afford to spend $20 billion to go to the moon?" he replied:

"We can't afford *not* to spend it. Only the U.S. and the U.S.S.R. have the resources for extensive space exploration. If we are to lead the free world and insure our own security, we must be first in space. This does not mean that we must neglect other urgent needs. We have ample resources to explore space and do the other things, as well."

Asked "What is our ultimate destiny in space?" he replied:

"I don't know, nor does anyone else. Columbus didn't find what he was looking for, but I think we're all pretty glad that he took that voyage. Einstein, when he produced the formula $E=MC^2$, didn't know that it would change the course of history.

"I am sure of one thing—the benefits which will flow from our venture into space will be beyond anything any of us could imagine.

"Until now, in space, no shot has been fired in anger. Thank God. My hope is that, in the years ahead, the conquest of space will encourage peaceful co-operation among nations and become a substitute for war.

"In the hostile environment of space there are challenges all mankind must share. We—all nations, that is—should go out there together, hand in hand." (*Parade, Wash. Post,* 9/29/63)

September 30: Francis W. Reichelderfer retired as Chief of U.S. Weather Bureau, succeeded by Robert M. White. (*Wash. Post,* 10/1/63)

- AEC announced that "signals from a Department of Defense Satellite launched recently from Vandenberg AFB, by a Thor-Able-Star missile are being transmitted successfully with electricity from a nuclear power source developed by the AEC." The device, SNAP 9-A was boosted into space by a launch crew of the Air Force Space Systems Division's 6595th Aerospace Test Wing. (*A-N-AF Journal and Register,* Jan. 1964; *A&AE,* Nov. 1963, 11)
- Tiros meteorological satellite discovered Hurricane Flora in the Atlantic off northern coast of South America. (UPI, *Wash. Post,* 10/1/63)
- NASA Flight Research Center announced refinement of tracking data showed Joseph A. Walker reached record altitude of 354,200 ft. in his X-15 flight of Aug. 22, 3,200 ft. higher than preliminary figure previously announced. (NASA Release 63-219)
- AFSC announced the development of a new pressure suit by ASD, Wright-Patterson AFB, for X-20 Dyna-Soar pilots. The suit would permit more freedom of movement than its predecessors and could be worn for 36 hours without discomfort. (*AFSC Operational Highlights,* 13)
- Address on Research and Development and the Federal Budget, by BOB Executive Assistant Director William D. Carey, inserted in *Congressional Record* by Rep. John W. McCormack. Carey said:

"... Government's part in the research and development business has now reached the point where it commands attention because of its sheer size and propensity for growth. From here on, we will have to be more choosy in what we do, and better prepared to supply answers to questions about marginal costs and benefits. The budget this year for research and development is a husky $15 billion. Its growth potential dwarfs anything else in the budget. Someone has figured out that the doubling time for research and development as a fraction of national income is only 7 years, and that if this continued for 30 years research and development would rise to one-half of the national income.

"... it is my view that the difficulty here is ... one of organizing research about research, of developing more adequate insights into cost-benefit relationships, of illuminating our value analysis so that we can with greater confidence strike a balance between being first in high energy accelerators and being first in education and in decent living and job opportunity. I do not think that Government alone can reach these answers, but perhaps Government can—and indeed I believe it must—be as proportionately lavish in stimulating this kind of intellectual inquiry as it has been in endowing science and technology ...." (Text, CR, 9/30/63, A6108-09)

September 30: Columnist David Lawrence wrote that many billions of space research dollars could be more beneficially used for developing the vast uninhabitable areas of the earth to accommodate bigger populations. (Wash. *Eve. Star*, 9/30/63)

- In *Missiles and Rockets* editorial, William J. Coughlin offered five answers to President Kennedy's rhetorical question, "Why, therefore, should man's first flight to the Moon be a matter of national competition?" Coughlin's reasons: "Because it is important to the survival of the United States that we develop the science and technology which gives us the capacity to send a manned expedition to the Moon and back ... ; because there is national prestige at stake ... ; because it is important to our military security ... ; because we are engaged in an economic war with the Soviet Union ... ; [and] because we believe in finishing what we start." (*M&R*, 9/30/63, 86)

During September: NASA Research Advisory Committee on Missile and Space Vehicle Structures issued its conclusions regarding space structures problems of the 1970's. Committee considered structural problems arising from need for larger spacecraft, reusable spacecraft, higher re-entry velocities, integration of nuclear and/or electric propulsion systems in spacecraft, and greater spacecraft maneuverability. Report stressed multipurpose minimum-weight structures emphasizing maximum efficiency for lowest material cost. (*Av. Wk.*, 9/16/63, 71)

- In Washington ceremony, Astronaut John H. Glenn, Jr. (Lt. Col., USMC) presented U.S. flag he carried during orbital space flight in FRIENDSHIP 7 to Marine Corps Assistant Commandant Lt. Gen. Charles H. Hayes. Flag would be placed in Marine Corps Museum. (*A-N-AF Journal & Reg.*, 9/28/63, 8)

- Thirty NASA astronaut candidates underwent physical examinations at School of Aerospace Medicine. Final selection would be made next month. (*M&R*, 9/2/63, 12)

- NASA Manned Spacecraft Center awarded $30,500 contract to Whirlpool Corp. for design and development of feeding system for four astronauts for long periods of time, first hardware fabrication for space station simulation facility. (*M&R*, 9/9/63, 9)

- JPL awarded Avco Corp. Research and Advanced Development Div. an $83,000 contract for study of entry and landing of ballistic capsules on Mars and Venus. (*M&R*, 9/2/63, 12)

- NASA issued "Reliability Program Provisions for Space System Contractors," set of guidelines to strengthen and unify reliability and reliability assurance methods in spacecraft and launch vehicle development. This was companion to and consistent with previous NASA publications "Quality Program Provisions for Space Systems Contractors" and "Inspection System Provisions for Suppliers of Space Materials, Parts, Components and Services." (NPC 250–1).

During September: NASA awarded $39,000 grant to Amateur Rocketeers of America (ARA) to produce educational and safety materials kits for young people interested in rocketry. (*Marshall Star*, 10/2/63, 10)

- Hughes Aircraft Co., NASA contractor for Syncom communications satellite, outlined design features of military synchronous communication satellite for USAF Space Systems Div. and Aerospace Corp. (*Av. Wk.*, 9/9/63, 23)
- Laser for space communications was patented by AF scientist Janis A. Sirons. (*A&AE*, Nov., 1963, 11)
- Soviet press reported refined flight parameters for VOSTOK III and VOSTOK IV manned space flights: VOSTOK III apogee, 234.6 km.; perigee, 180.7 km.; period, 88.33 min. VOSTOK IV apogee, 236.7 km.; perigee, 179.8 km.; period, 88.39 min. (*Space Bus. Daily*, 9/9/63, 357)
- Reduction of France's 1964 civilian space budget from $70 million to $48 million announced in Paris by General Aubiniere, President of Centre National d'Etudes Spatiales, who explained reduction was part of France's economic austerity program. (*M&R*, 9/30/63, 18)
- In an article in the September issue of *Aerospace* magazine, Vice President Lyndon B. Johnson pointed out some of the many advantages to be gained from the lunar project:

    ". . . The lunar project has forced us to develop many competences which have military as well as non-military significance. These are competences which we would have been slow to develop were it not for this national moon objective. For example, rendezvous technique so basic to our moon project is essential to detecting and examining other spacecraft which may be hostile. Life protective measures are essential to a useful police force in space for maintaining the peace. Powerful rockets, reliability of space equipment, development of control and guidance systems, experience with manned spacecraft, etc., are all spin-offs from the lunar project, which help build our defense capability." (Text, *Aerospace*, Sept. 1963)
- Dr. Edward C. Welsh, Executive Secretary of National Aeronautics and Space Council, compared U.S. and U.S.S.R. space achievements in *General Electric Forum:*

    "1. U.S. has put about 4 times as many payloads into earth orbit (approximately 130).

    "2. U.S.S.R. has put substantially more weight into earth orbit with its smaller number of payloads.

    "3. U.S. space applications in weather, communications, and navigation have been impressive. U.S.S.R. has potential in these fields but not accomplishments.

    "4. Both nations have obtained remarkable amounts of scientific data, with no measurable advantage to either country.

"5. U.S.S.R. has operational more powerful rockets than U.S., while the U.S. is developing more powerful ones than any known to be in U.S.S.R.

"6. U.S.S.R. is ahead in manned space flight.

"7. Since the race began, ratio of successes to failures has been comparable. In 1962, U.S. had one payload launching failure for each five successes." (*GE Forum*, 7-9/63, 23)

During September: NASA Deputy Administrator Dr. Hugh L. Dryden said in *General Electric Forum* interview: ". . . There is no question that setting definite technical tasks before people does motivate them. For instance, war efforts generally provide goals and accelerate progress—jet propulsion, electronics, miniaturization, and other technological advances. Along the same line, one of the major byproducts of space research has been a revitalization of education, getting people interested in scientific knowledge again—and particularly important, getting young people interested in science and technology." (*GE Forum*, 7-9/63, 20)

- Results of Research Institute of America's survey of more than 1,400 businessmen on U.S. goals and national policy, published in *General Electric Forum*, revealed 21% considered present expenditure on space fully warranted, 41% considered it probably fully warranted, 27%—probably not, and 11%—not warranted. Other replies:

  To "In years immediately ahead, what *should* be level of space expenditure, in relation to GNP?" 6% said "sharply higher than present expenditures, 16%—a little higher, 38%—remain at about the present ratio, 24%—a little lower, and 16%—sharply lower.

  To "Assuming that national policy in the immediate future corresponds to your answer above, do you believe that the U.S. will exceed the Russians in space achievements by 1970?" 38% said probably will exceed the Russians, 44%—will approximately equal them, and 18%—probably will not equal them.

  Asked to identify a statement closest reflecting their opinion on what space policy should be, 66% checked the statement, "Our prime objective should be to sustain a prudent and orderly program of scientific progress in space achievements with little regard for who leads in this or that particular aspect of space technology."

  Asked about the military role in the U.S. space program, 16% replied it should be greater, 67%—about the same, and 17%—smaller. (*GE Forum*, July-Sept. 63, 31-39)

OCTOBER 1963

October 1: Fifth anniversary of NASA. Ceremonies throughout the week included open house at Wallops Station, Va., Sept. 29 and 30; Fifth Anniversary Honor Awards Ceremony in Washington; Project Mercury Summary Report Conference at Manned Spacecraft Center, Houston, Tex., Oct. 3 and 4; and NASA Fifth Anniversary Banquet in Washington Oct. 5.

In its fifth year of space operations (October 1, 1962–October 1, 1963), NASA launched 12 orbital, deep space, and manned space flights, of which 10 were successful, 1 partially successful, and one unsuccessful. The X-15 rocket research aircraft set a new altitude record of 350,000 ft. and began its follow-on flights involving space research. The Mercury program added a 6-orbit flight and was concluded with a 22-orbit flight. Among other successful flights were four scientific satellites, two communications satellites, the first and second synchronous-orbit communications satellites, and another weather satellite. The fifth scientific lunar probe (RANGER V) was unsuccessful and led to a reworking of the remaining Ranger probes.

Project Apollo flight testing was begun.

In its five years of space operations, NASA had launched a total of 68 orbital, deep space, and manned space flights, of which 39 were successful, 9 partially successful, and 20 unsuccessful, for an overall average of $2\frac{1}{2}$ successes for every failure. The rate of improvement ranged from 1 success per failure in 1958–59 and 1959–60, to 2 to 1 in 1960–61, to 5 to 1 in 1961–62, to 12 to 1 in 1962–63. (HHR–14)

- At NASA Fifth Anniversary Honor Awards Ceremony, held at Smithsonian Institution, 23 individuals were singled out for outstanding contributions to the civilian space program. Vice President Lyndon B. Johnson stated that U.S. space policy was clear: "With or without cooperation from any other country, we are going to the moon, and we are going to make that trip as soon as we can. . . . We would like to do this through international cooperation. Leaders of both parties have sought cooperation under two administrations. The moon represents our major space goal for this decade, and if cooperation is possible we are willing to share this goal as well as others."

NASA Administrator James E. Webb summarized the five years of space achievements and looked ahead: "From the first two United States satellites, Explorer I and Vanguard I, launched early in 1958, we learned that the earth was slightly pear-shaped rather than being the sphere it had been thought to be. We also learned that the earth was surrounded by a zone of radiation, called the Van Allen Belt after its discoverer Dr. James Van Allen.

"In addition to these first satellites of scientific significance, we have orbited highly successful weather and communication satellites, probed the area of Venus, put men into orbit and brought them safely back to earth, and sent thousands of sounding rockets into space to study the earth's environment. From these and other experiments installed in more than 100 satellites and deep space probes, we have succeeded in vastly expanding knowledge and understanding of our own planet earth and of the solar system of which it is a part . . . .

"Looking ahead through the remainder of this decade, we have set our sights on a truly global weather satellite system, advanced communications and navigation satellites, space observatories for astronomical and solar studies, and a manned expedition to explore the moon . . . ."

Awards were presented by Mr. Webb, NASA Deputy Administrator Dr. Hugh L. Dryden, and Associate Administrator Dr. Robert C. Seamans, Jr. Eight cash awards totaling $12,200 were divided among 14 NASA employees for inventions and scientific and technical contributions: Robert L. Trimpi, LARC, $3,000; Charles H. McLellan, LARC, $2,000; James H. Schrader, LARC, $1,500; Jesse M. Madey and Xopher W. Moyer, GSFC, $1,500 shared; James B. Newman, FRC, $1,200; Hershel M. Nance, MSFC, $1,000; Lee B. Malone, Charles E. David, and Harold R. Lowery, MSFC, $1,000 shared; and Frank L. Clark, Charles B. Johnson, Wayne D. Erickson, and Roger I. Buchanan, LARC, $1,000 shared.

The NASA Medal for Exceptional Scientific Achievement went to Dr. Dean R. Chapman, ARC, for his research on tektites; to Dr. Ernst D. Geissler, MSFC, for contributions to Mercury-Redstone, Saturn, and Nova boosters; and to Dr. John C. Houbolt, formerly of LARC, for work on lunar orbit rendezvous.

The NASA Medal for Outstanding Leadership was presented to Charles J. Donlan, LARC; and Dr. Walter Haeussermann and Dr. William A. Mrazek, MSFC.

The new NASA Public Service Award, given to persons not employed by the Government, was presented to Jack N. James and Robert J. Parks, JPL, and John F. Yardley, McDonnell Aircraft Corp., Cape Canaveral. (NASA Releases 63-217, 215; *Wash. Post*, 10/2/63; Webb, Text; Program, Honor Awards Ceremony)

October 1: NASA Administrator James E. Webb appeared on Voice of America, "Press Conference USA." Speaking of President Kennedy's proposal for a joint U.S.-U.S.S.R. lunar program, Mr. Webb said: ". . . after the ratification of the test ban and one day after Mr. Gromyko proposed that we agree not to put into orbit multi-megaton weapons, the President followed with a suggestion that had five elements. Most people have centered on only one of them." The five elements were (1) measures against war by accident or miscalculation; (2) measures against surprise attack; (3) measures to curb the nuclear arms race; (4) exchange of information and persons; (5) U.S. consent to Gromyko's proposal to negotiate an agreement not to orbit large nuclear weapons.

"And then he added as a sixth item, we should also explore, now mark you, he said *explore*, possibilities for cooperation in manned space flight and then set a clear indication that he was not thinking of a limited exploration, that rather he was thinking of a broad exploration . . ., that we could even go step by step to the planning of a joint expedition."

Speaking of the preparations for the President's U.N. speech, Mr. Webb said: ". . . I myself personally attended a meeting in the White House called by Mr. Arthur Schlesinger, perhaps a month ago, to consider things that the President might want to discuss at the United Nations. And I did myself on the day before Mr. Gromyko's speech have a long talk with the President, maybe 35 or 40 minutes, about the whole space program in which we discussed this. And I think I should say that he had his office phone me in St. Louis on Thursday afternoon after Mr. Gromyko made his speech to read me the language that he then decided he would use, because it seemed such a natural follow-on to Mr. Gromyko's proposal." (Text)

October 1: In a ceremony coincident with NASA's fifth anniversary, Dr. Floyd L. Thompson, Director of NASA's Langley Research Center, presented a model of a Mercury space capsule to the new municipal aerospace park in Hampton, Va. (Newport News (Va.) *Times-Herald*, 10/1/63, 13)

- George Low, NASA's Deputy Director of Manned Space Flight, speaking at an AIAA-NASA symposium on interplanetary exploration in Palo Alto, Calif., spoke of two areas of possible U.S.-U.S.S.R. cooperation on manned lunar flight: (1) a series of jointly programed space probes to survey the lunar surface prior to a manned flight; and (2) exchange of information on problems of re-entering the earth's atmosphere at 25,000 mph. (San Francisco *Chronicle*, 10/2/63)

- A Mars Excursion Module (MEM) would be the best means of landing men on Mars, according to Temple Neuman of Philco Corp.'s Aeronutronic Div., which had just completed a study of the subject under NASA contract. The MEM would carry two men, house and feed them for 40 days, and enable them to bring back 800 lbs. of equipment and samples, for a gross vehicle weight of about 55,000 lbs. (*Space Bus. Daily*, 10/2/63, 12)

- Aerobee 150 sounding rocket was fired by NASA from White Sands Missile Range, N. M., the 235-lb. payload to take observations of ultraviolet light and the sun's corona. (GSFC Hist.; AP, *Wash. Post*, 10/2/63)

- Definitive contract for research and development of a paraglider system for landing of Project Gemini spacecraft was awarded by MSC to Space and Information Systems Div. of North American Aviation, Inc. The contract set paraglider R&D at $20,015,100, of which $10.8 million had already been spent under a letter contract issued about one year ago. The other landing system being developed for Gemini is by parachute, similar to the one used in Project Mercury. Paraglider would be designed for land landings, parachute for water landings, although both systems have eventual capability for land and water landings. (MSC Release 63–159)

October 1: Marcel Nicolet of the Belgium National Space Research Committee was the winner of the Guggenheim International Astronautics Award at the banquet of the International Astronautics Federation in Paris. Nicolet had, in early 1961, suggested the existence of the helium layer between the oxygen-nitrogen atmosphere and the hydrogen upper atmosphere, later confirmed independently by R. E. Bourdeau, E. C. Whipple, P. C. Donnelly, and S. J. Bauer, using EXPLORER VIII ion trap data.

Cosmonaut Yuri Gagarin attended the XIVth IAF banquet dressed in civilian clothes. (Cf. R. Jastrow, "Results . . . ," 12/18/61)

- Robert F. Six, president of Continental Airlines, speaking to an aviation group of the Los Angeles Chamber of Commerce, proposed that the U.S. join with France and Great Britain to develop a mach 2.2 commercial airliner and that at a later date those countries join the U.S. in developing a stainless-steel—titanium mach 3 airliner: "If our country can propose a joint effort with the Russians to reach the moon, we certainly should be able to entertain the idea of joint development of a supersonic transport with two of our best friends." (*Wall St. Journal*, 10/2/63)
- First transpolar nonstop flight from Capetown, South Africa, to McMurdo Sound, Antarctica, was made in a ski-equipped C-130 Hercules aircraft by Rear Admiral James R. Reedy (USN). The 14-hr.-31-min. flight covered 4,700 mi. and crossed the entire Antarctic continent, inaugurated Operation Deep Freeze 64, the Navy's logistic support of U.S. Antarctic research. (DOD Release 1313-63)
- Value of weather Tiros satellites in giving advance warning of tropical storms was underscored by Rep. Paul G. Rogers (D.-Fla.): "Yesterday's discovery of Hurricane Flora is a prime example. . . . Through the efforts of the U.S. Weather Bureau, using data supplied by a weather satellite, warnings were issued a bare 3 hours before the 110-mile-an-hour winds hit the islands [of Trinidad and Tobago]. Because of the 22-mile-an-hour forward speed of the storm, little or no warning would have been possible without the quick work of U.S. weathermen and their new ally, the satellite. While the damage was extensive, untold lives were saved and property damage prevented by the advance warning." (*CR*, 10/1/63, 17456)
- Polaris A-2 was successfully fired by the Navy from the U.S.S. *Andrew Jackson*. (*M&R*, 10/7/63, 20)
- Nuclear test ban treaty had caused cutbacks of two DOD space nuclear propulsion projects—Orion and Pluto, according to F. C. Diluzio, staff director of the Senate Aeronautical and Space Sciences Committee, speaking at a press conference in Albuquerque, N.M. Project Orion, conceived in 1957 by Dr. Stanislaw Ulam, would have provided propulsion through a series of controlled atomic explosions. Project Pluto would employ a nuclear reactor to propel a missile through the lower levels of the earth's atmosphere at supersonic speeds. (Wash. *Eve. Star*, 10/2/63)

October 1: The *Astrophysical Journal* reported that explosion of galactic core of M–82, galaxy some 10-million light years from earth, was photographed by 200-in. Mt. Palomar telescope last spring. (*A&AE*, December 1963, 5)

October 1–3: Youth Science Congress, sponsored by NASA and the National Science Teachers Association, was held at GSFC. Feature event was presentation of 25 award-winning research papers by high school students from Washington, D.C., Md., Del., Pa., and N.J. (GSFC Release G–20–63)

October 2: The Russian people heard of President Kennedy's U.N. proposal for joint U.S.-U.S.S.R. lunar exploration for first time. *Pravda* reprinted without comment Walter Lippmann's column praising the President's proposal. (AP, *Wash. Post*, 10/2/63)

- Sen. J. W. Fulbright (D.-Ark.), Chairman of the Senate Foreign Relations Committee, stated on the Senate floor that he favored space exploration but considered the present large space budgets "an unwarranted diversion of resources from the Nation's most pressing needs." He favored President Kennedy's proposal for joint U.S.-U.S.S.R. lunar exploration for two reasons: "First, it would greatly reduce the costs of space exploration, releasing funds for important domestic programs, such as education, employment, urban renewal, and conservation of resources; second, it would further reduce world tensions by opening up a significant new area of Soviet-American cooperation." (*CR*, 10/2/63, 17598)

- Talks at U.N. between East and West on agreement not to orbit large nuclear weapons were termed promising by Secretary of State Dean Rusk and Soviet Foreign Minister Andrei A. Gromyko. Said to be the most important discussions since the negotiations on the nuclear test ban treaty, the talks also dealt with exchange of observers between East and West to reduce possibility of surprise attack and with nondissemination of nuclear weapons. (*NYT*, 10/3/63, 1; AP, *Wash. Post*, 10/3/63)

- USAF and Martin Co. engineers reportedly adopted a tight engineering review system in the number and types of changes in the Titan II missile to the man-rated Titan II booster for Gemini flights. System is intended to improve reliability by holding changes down to fewest possible and seeing that each of these contributes to overall reliability. Modifications that have been made include the necessary structural changes to accept the Gemini capsule, redundancy in certain key systems, and a malfunction detection system that would alert the flight crew to possible trouble. (*Space Bus. Daily*, 10/3/63, 20)

- Dr. Karl G. Guderley, senior scientist at USAF's Aerospace Research Laboratories, Wright-Patterson AFB, Ohio, was presented DOD's highest civilian award, the DOD Distinguished Civilian Service Award. Award was for contributions to transonic flow theory and other problems in aerodynamics and mathematical physics. (OAR Release 10–63–1)

- *Space Business Daily* reported that Dr. Robert C. Seamans, Jr., NASA Associate Administrator, "said flatly this week that he would resign rather than become involved in a technical bog-down caused by U.S.-Soviet co-operation on a lunar flight." In its Oct. 4 issue, the same publication said that NASA had officially "denied" the report as being "flatly untrue in language, concept and fact" and asserted that Dr. Seamans was " 'enthusiastic' " over the possibility of a joint program. NASA spokesman told the publication that Dr. Seamans had been addressing a group on Capitol Hill, had spoken of technical obstacles to lunar cooperative program, and had said " 'in a joking manner' " that the problems would be so difficult "as to require NASA to find someone to replace him." (*Space Bus. Daily*, 10/2/63, 9; 10/4/63, 26)

October 3: First official Soviet comment on President Kennedy's proposal for U.S.-U.S.S.R. lunar cooperation came from Russian news agency Tass. While not identifying the proposal with the President, Tass said: "Maybe it is too early to discuss now the question what is better—to combine a Soviet rocket with an American spaceship or to include an American in the Soviet space crew, but first steps in cooperation cannot but give satisfaction." A few hours later, Cosmonaut Pavel Popovich made an unscheduled talk on Moscow Radio, calling for a "common effort by the peoples of all states of our planet" to solve the earthly difficulties that hinder man's flight to the stars. (*Wash. Post*, 10/4/63)

- Rep. Thomas M. Pelly (R.-Wash.) charged that President Kennedy had completely reversed U.S. space policy in his U.N. speech of Sept. 20 by shifting the objective of the manned lunar landing from one of international competition and world prestige to one of cooperation with the Russians. He opposed the new proposal but said it offered justification to cut back the manned spaceflight program and substitute cheaper unmanned flights. (*CR*, 10/3/63, 17666)

- Vice President Lyndon B. Johnson quoted in article in *Washington Daily News* as saying of the manned lunar landing program: "To default would be as catastrophic as if we had defaulted exploration of the atom." (Text)

- In commenting to the Senate Committee on Government Operations on the bill (S.1577) authorizing the General Services Administration to "coordinate and otherwise provide" automatic data processing equipment for all Federal agencies, NASA noted that on June 30, 1963, NASA had 118 general-purpose digital computer systems, of which only seven were used full time for administrative purposes. The rest were used in science and engineering tasks. (NASA Leg. Act. Rpt. II/158 cont'd)

- Thomas J. Watson, Jr., chairman of the board of International Business Machines Corp., speaking at the 46th annual meeting of the American Council on Education in Washington, D.C., expressed concern about the current course of education. "That concern has its origin just 6 years ago tomorrow. It was the day that the Soviet Union put Sputnik I in orbit. Do you remember the thoughts which flooded in on us that day? Thoughts of embarrassment—dismay—concern. . . . We could even try to say that satellites were just stunts, that Sputnik was a silly hunk of iron. But try as we would to whistle in the dark,

Americans finally had to reach the conclusion that the Russians had outdistanced us at least temporarily in the exploration of space, a pursuit which in our century might well become as important as the exploration of the oceans had been in the time of Columbus and Magellan."

Mr. Watson pointed out that the impact of sputnik on education "has thrown the sciences and the humanities badly out of balance. . . . In the fiscal year of 1961, of all Federal funds for basic research, the physical and biological sciences got 97 percent.

"There are many traditional reasons for reemphasizing the humanities—which today are more important than ever.

"In the first place, a thorough grounding in the humanities is vital training for many leaders who can wisely manage people as well as sufficiently manage machines. This need exists in universities, in public service, and in business.

"Secondly, as our planet continues to shrink, we will more and more have to become citizens of the world. We must acquire a better understanding of the language, history, and culture of people of other lands. With present day mass-destruction weapons, this understanding is not just desirable—it's vital.

"Finally, the greater our skills in the humanities—in literature and the arts—the greater our capacity for a constructive use of leisure time, which is bound to increase as machines lift old burdens from men's shoulders and minds.

"On this sixth anniversary of the flight of sputnik, we can be sure of one thing; no second sputnik will come along to jolt us into action, and do for American education in the humanities what the first Sputnik did for American education in the sciences." (*CR*, 11/12/63, A7010–A7012)

October 3: General Accounting Office (GAO) released report on NASA's Centaur booster which was sharply critical of both Government and contractor performance. Done at the request of the House Committee on Science and Astronautics and submitted to them on March 29, 1963, but not released until Oct. 3, the report charged that poor management and program slippage had added $100 million to the cost of the program. The most significant charge in the report was that NASA and the contractor—General Dynamics/Astronautics—had in their possession in 1960 data from which could have been predicted the failure of the first test flight in 1962. The only immediate reply came from Julian Scheer, NASA's Deputy Assistant Administrator for Technology Utilization and Policy Planning: "On Sept. 30, 1962, we moved the Centaur project from Marshall Space Flight Center at Huntsville to the Lewis Research Center because we knew we had problems. Still, we hope to fly Centaur by the end of this year." (*Wash. Post*, 10/4/63; *Space Bus. Daily*, 10/4/63, 25)

- Reporting under a NASA contract to study requirements for a manned mission to Mars in "the unfavorable (1975–1985) time period," Douglas Aircraft Co.'s Missile and Space Systems Div. proposed a nuclear-chemical four-stage vehicle carrying a seven-man crew on a 400-day trip. Mission would begin out of 300-mile earth orbit: first stage would boost vehicle into Mars trajectory; 140 days later, second stage would put vehicle into 300-mile orbit of Mars, where for 20 days the crew would study the planet; third stage would provide escape from Mars and 240-day trip back to earth; fourth stage would provide retrothrust for re-entry. (*Space Bus. Daily*, 10/4/63, 27)

October 3–4: Project Mercury Summary Conference held at MSC, Houston, Tex. Two days of papers were presented on various aspects of Project Mercury as a whole and on the final, 22-orbit flight by Astronaut L. Gordon Cooper in particular.

The most controversial point raised during the conference was statement on the large amount of faulty or careless manufacture and assembly of Mercury spacecraft. On the MA-9 backup spacecraft, for example, some 720 discrepancies were recorded, of which 526 were attributable to unsatisfactory workmanship and 444 of these required specially scheduled time to correct. (MSC Releases 63–153–158; *Project Mercury Summary*)

October 4: Sixth anniversary of launching of SPUTNIK I, the opening of the Space Age.

- First flight-rated model of Gemini spacecraft (No. 1) delivered by the contractor to MSC Test Conductor at AMR, Paul C. Donnelly, the spacecraft to be used for pre-flight checkout procedures leading to the first Gemini mission. This spacecraft will structurally simulate weight, center of gravity, and aerodynamic form of the final Gemini spacecraft. (MSC Release 63–160)

- NASA selected three companies for negotiations to lead to a contract for installation, operation, and maintenance of a technical communication system in NASA's Merritt Island Launch Area (MILA). Eventually the system would link some 50 buildings on the 87,000-acre site, including Complex 39, where Saturn V will be launched. This system would be separate from the normal administrative telephone system with its dial telephone exchange. (NASA Release 63–221)

- Weather Bureau stopped its portion of the funding of the Nimbus meteorological satellite because the current Nimbus offers an operational lifetime of no more than one year. Weather Bureau would continue using improved Tiros satellite while cooperating with NASA in development of a follow-on satellite with a lifetime of several years. NASA said it would continue the development of the Nimbus R&D models. (NASA Release 63–220; Commerce Release; *Wash. Post*, 10/4/63, 2)

- An aircraft somewhat similar to the U-2 reconnaissance plane was seen by reporters on the flight line at Air Force Flight Test Center, Edwards AFB, Calif. Air Force officials identified the aircraft as an RB–57F, a drastically modified version of the British Canberra jet light bomber. Chief modifications visible were two extremely long tapered wings similar to those of the U-2, with two very large engines mounted on them. Two smaller J–60 engines were mounted on the fuselage. The aircraft is used in Project Peewee, an environmental study of navigation and weather equipment at high altitude. (AP, *NYT*, 10/6/63)

October 4: NASA Administrator James Webb spoke in Washington before the American Council on Education on the relationship of university activity to the space effort:

"Dr. Frederick Seitz, President of the National Academy of Sciences and head of the Department of Physics at the University of Illinois, noted recently that he could 'think of no aspect of university activity relating to science or technology which is not involved in a fundamental way in the space effort' . . . .

"NASA [sponsors] a pre-doctoral training program with the ultimate objective of assisting in the production of 1,000 Ph. D.'s a year . . .

"I think I can say categorically that the universities with which NASA works are not becoming so reliant on federal research funds that it is interfering with their traditional role of undergraduate instruction. Although the ratio of graduate to undergraduate effort in many universities may be increasing, those with which we work are not reducing their undergraduate instruction role. In fact, we know that all the major schools are most concerned about keeping up with the undergraduate demand.

"It is worth emphasizing that NASA is attempting to foster a broader base of competence in graduate research by awarding training grants to institutions rather than to individual scholars. Thus the university as an institution is given the opportunity to select its own candidates for instruction on its own campus." (Text)

- In an article in *Pravda*, V. Golobachev reported that U.S.S.R. was filing with the Fédération Aeronautique International claiming new records for the tandem space flight of Cosmonauts Valery Bykovsky and Valentina Tereshkova June 14–19, 1963. U.S.S.R. was claiming eight new records for the tandem flight: for VOSTOK V (Bykovsky), duration of orbital flight, 119 hours; distance, 3,326,000 km. (2,063,377 mi.); these represented four new records, since they were records not only for orbital flight but for any type of flight. For VOSTOK VI (Tereshkova), four records claimed for women in orbital flight: duration, 71 hrs.; height, 231 km. (144 mi.); distance, 1,971,000 km. (1,224,720 mi.); maximum weight orbited, 4,713 kg. (10,390 lbs.). Total maximum thrust for each flight was listed as 600,000 kg. (1,322,760 lbs.). (*Pravda*, 10/4/63, AFSS–T Trans.)

- Federal Communications Commission (FCC) ordered that TV channel 37 in the U.S. be reserved for exclusive use of radio astronomy until Jan. 1, 1974. FCC said it would urge Mexico and Canada to take similar action and would recommend world wide reservation of the channel at the world conference on space communications convening in Geneva on Oct. 7. (Wash. *Daily News*, 10/5/63)

- Sen. Clinton P. Anderson (D.-N.M.), Chairman of the Senate Committee on Aeronautical and Space Sciences, wrote a letter to NASA Administrator James E. Webb asking for an explanation of the statements made at the summary conference on Project Mercury that there had been poor workmanship on the Mercury spacecraft, according to William Hines, in the Washington *Evening Star*. He was also reported to have inquired about a remark attributed to Dr. Robert Gilruth, Director of MSC, that industrial shortcomings included outright falsification, citing an instance where a component used " 'black iron where there should have been stainless steel.' " (Hines, Wash. *Eve. Star*, 10/4/63)

October 4: Atlas F ICBM was launched by USAF from Vandenberg AFB, but blew up during liftoff. (UPI, *NYT*, 10/8/63, 28)

October 5: Rep. Thomas M. Pelly (R.-Wash.), minority member of the House Committee on Science and Astronautics, asked that the Committee investigate alleged waste in the space program to determine how much money has been wasted, by whom, and whether some projects have fallen so far behind that money earmarked for them in the FY 1964 budget can be cut out because it cannot be spent within the fiscal year. He requested that the investigation be undertaken immediately so that it can be finished before Congress takes final action on the NASA budget. (Troan, Wash. *Daily News*, 10/5/63)

- NASA Fifth Anniversary Banquet held in Washington, D.C., sponsored by Aerospace Industries Association, American Institute of Aeronautics and Astronautics, and Aviation/Space Writers Association. Dr. James Doolittle, last Chairman of the NACA, served as Master of Ceremonies. After speaking of the tangible achievements of NASA, Dr. Doolittle said: "Just as well known, to this audience at least, are the less tangible achievements of NASA's last five years. I think, however, that they bear reemphasizing.

"I refer first of all to the rapidly growing competence and effectiveness of the Government-industry-university space team—an American space team that simply did not exist five years ago. Today, the team is a valuable source of national strength, prestige, and security." Other factors mentioned were "openness of the NASA program," "inspiration which the astronauts of our space program have given to our youth," and that difficult achievement, "a balanced program." (*CR*, 10/9/63, A6327)

Senator Clinton P. Anderson (D.-N.M.), Chairman of the Senate Committee on Aeronautical and Space Sciences, called on the banquet guests to assist in getting a $5-billion-plus budget passed by Congress. He added: ". . . let's remember that there would be no basis for the modest space research agreements the United States has with the Soviet Union if this country had not demonstrated to the Russians the great strength of our space program. We bargain—and must continue to bargain—from a position of strength and those who criticize space spending should not overlook this fundamental point." (*CR*, 10/9/63, A6333)

Dr. Edward C. Welsh, Executive Secretary of the National Aeronautics and Space Council, made five main points about the space program: "the space successes of this country have been truly remarkable"; "This is a national program"; "This is also an international program"; "This is a growing and permanent program"; and "This is no time for complacency about our space competence." (Text)

NASA Administrator James E. Webb took time in his remarks to comment on the impression that NASA was dissatisfied with industry in the space program. He suggested that it was faulty reporting to pick one page of criticism out of a 443-page report that essentially spoke of industry's "'great job.'" (*Wash. Post,* 10/6/63)

October 5: On the eve of its fifth anniversary, NASA had run into a patch of "attack by partisan politicians, budget-cutters, those who want the military to have a stronger role in space, mismanagement sleuths, and persons confused by President Kennedy's bid for a joint U.S.-U.S.S.R. moon venture," Howard Simons claimed in the *Washington Post.* He cited NASA's budget troubles, with the $5.7 billion request already cut to a $5.3 billion authorization, another $200 million already cut by a House Appropriations Subcommittee, and even further cuts threatened in the full Committee. NASA Administrator Webb, Simons said, had blamed budget problems on House Republicans, charging that they "'are following a political line trying to make the President's program look foolish—'Nuts,' as Eisenhower called it. They have moved to the position that military activities in space are the only ones worthwhile.'" Under mismanagement charges, Simons listed the GAO report on the Centaur program and the NASA charges of substandard workmanship by industry in the Mercury program, which in itself implied, according to Simons, a NASA admission that it had not supervised industry effectively. (*Wash. Post,* 10/5/63, 4)

- Speaking on Cuban television from Havana, Soviet Cosmonaut Valentina Tereshkova said that she was a member of the team of Russian cosmonauts chosen to make the trip to the moon and that Yuri Gagarin headed the team. She said she would propose that a Cuban woman be included on the team. The Russian plan for a moon flight was described by Miss Tereshkova: the manned capsule would be launched into earth orbit, where it would rendezvous with an unmanned tanker spaceship; then another rocket would be launched from earth and tow the manned-vehicle-tanker combination to the moon. (AP, *Wash. Eve. Star,* 10/5/63)

- A change in the political climate in the last few months had NASA in trouble on its fifth anniversary, according to John Finney of the *New York Times:* The trouble really began when scientists began to question the validity of the manned lunar landing. "The approach being urged by the scientific critics is a step-by-step program calling first for lunar investigation by instruments and then a manned program that would not divert resources from other fields of scientific research." It was this approach that was being followed until May 1961, when President Kennedy declared the high priority goal of a manned lunar landing in this decade.

    "Increasingly, the attack on the space budget is taking on a partisan tinge, and the Administration in turn is reacting in a partisan manner. It is significant, for example, that in the House Appropriations subcommittee the votes to cut the space budget were strictly along partisan lines with the exception of Clarence Cannon . . . ." (Finney, *NYT,* 10/6/63)

- Aerospace manufacturers on the West Coast indicated they had no interest in a study commission made up of representatives of the companies, the two big unions—United Auto Workers and

International Association of Machinists—and Government. Such a commission had been recommended last year during wage negotiations and endorsed by Prof. George W. Taylor, of the Univ. of Pa., who was President Kennedy's representative. The proposal was endorsed by the two unions on Oct. 4. The companies were reported to fear that such a commission would be prelude to industry-wide bargaining, which they oppose, and that Government representatives would be partial to labor. (Langguth, *NYT*, 10/7/63, 22)

October 5: In Rome, Pope Paul VI received several hundred delegates from 30 countries attending the Sixth International Congress of Aeronautical and Space Medicine. Among those received was Prof. Vasily Parin of the U.S.S.R. (*NYT*, 10/6/63, 25)

October 6: The Hall of Science building at the New York World's Fair will not be ready to open for at least a month after the Fair opens on Apr. 22, 1964, the *New York Times* reported. Contests over whether the site should become a permanent science museum, the belated discovery that pilings had to be driven to support the foundations, and the resulting increased costs of the building have all delayed construction. (Sullivan, *NYT*, 10/6/63, 1)

- MSC is experiencing an increasing manpower shortage as it moves into high gear on Project Apollo, John Finney of the *New York Times* reported. Particularly scarce are top-flight engineers needed to oversee large systems and subsystems. MSC hopes that its manpower ceiling will be raised this year from the present 3,500 to around 5,500. (Finney, *NYT*, 10/6/63, 71)

- Construction of the new NASA Manned Spacecraft Center at Houston, Tex., was 75% complete. The first contingent of 80 people would move into the new quarters over the weekend, with general occupancy scheduled for June 1964. (MSC Release 63–166; Finney, *NYT*, 10/7/63)

- Dr. Hugh L. Dryden, NASA Deputy Administrator, speaking in Cleveland, Ohio, before the eighth annual Inter-Church World Peace Seminar, said that possible U.S.-U.S.S.R. cooperation on space exploration would bog down in the Soviet's extreme security measures. Their security measures toward their space program were as stringent as that of the U.S. in regard to nuclear weapons, Dr. Dryden said. (Text in *CR*, 10/8/63, A6325)

- D. Brainerd Holmes, former NASA Director of Manned Space Flight and now a senior vice president of Raytheon Co., said in a TV interview on ABC that the manned space flight program needed a single manager with great freedom of action. The only way to run a large program, he said, "is to put the project manager in such complete charge . . . that he makes policy with very little approval above." (*NYT*, 10/7/63, 23)

- COSMOS XIX transmissions have been picked up by the Sohio Research Center near Cleveland, Ohio. The Center said that Russian satellites normally have an on-off switch which the Russians can activate for transmission as the satellite passes over Russian territory. They theorized that the switch might have stuck, causing the satellite to transmit continuously. (*Cleveland Plain Dealer*, 10/6/63)

October 6: "Informed sources" in Moscow said that Soviet Cosmonauts Andrian Nikolayev and Valentina Tereshkova were engaged to be married. (UPI, *NYT*, 10/7/63)

- Editorial in the *Washington Post* called the weaknesses in quality control mentioned in the final report on Project Mercury "shocking," calling for "sweeping revision of control standards previously existing in the aeronautics industry." Tying this into the GAO report on Centaur, the editorial concluded, "The NASA report on Mercury and the GAO report alike suggest it may be necessary to proceed with more deliberation, caution and control." (*Wash. Post*, 10/6/63)
- Announced that transistorized sensory system, Myoelectric, to provide astronauts with automatic control during high g was patented recently by two Sperry Gyroscope engineers. (*A&AE*, December 1963, 5)

October 7: House Appropriations Committee voted $5.1 billion appropriation for NASA in FY 1964, $612 million less than the Administration request and $250 million less than previously authorized by Congress. NASA Administrator James E. Webb said the 1970 target date for landing a man on the moon could not be met unless Congress restored the $250 million cut, either by putting it back now or by approving a supplemental request in Jan. 1964. Furthermore, Webb said, the FY 1965 budget would have to make up the "shortfall" from the original Administration request for $5.7 billion.

Rep. Clarence Cannon (D.-Mo.), Chairman of the Appropriations Committee, said he had wanted the NASA budget cut further to $4.9 billion, that the Subcommittee on Independent Offices had wanted $5.3 billion, and that the $5.1 billion represented a compromise. He predicted House approval, said: "Although it's only a token cut, it shows the attitude of the Committee and the House towards the space budget." Rep. Albert Thomas (D.-Tex.), Chairman of the Appropriations Subcommittee on Independent Offices, said the cut had nothing to do with President Kennedy's proposal for a joint U.S.-U.S.S.R. manned lunar landing program. "There was simply one big thing. Everybody wants a tax cut, and many people want a reduction in spending. Of all the spots, this was a good one to cut." (*NYT*, 10/8/63, 1)

- NASA testimony before a subcommittee of the House Committee on Appropriations on Aug. 19, 1963, indicated that at that time NASA officials saw little prospect for U.S.-U.S.S.R. cooperation in manned lunar exploration, William Hines, of the Washington *Evening Star*, quoted from testimony released today. Dr. Hugh L. Dryden, NASA Deputy Administrator and chief U.S. space negotiator, said: "In my discussions with Blagonravov, the question of cooperation in manned space flights was mentioned in the early part, but was pretty well dropped as something that was not practical . . ." because among other reasons, of "the very high degree of classification the Russians attached to their space program. . . ." Summing up, Dr. Dryden said: "I do not foresee in the near future any prospects that the Russians, any more than ourselves, will see a clear way to co-operate in a manned lunar landing."

NASA Administrator James E. Webb agreed: "Up to now, there has been no evidence whatever that we will gain anything significant from it." (Hines, Wash. *Eve. Star*, 10/7/63)

October 7: In *OAR Research Review*, Louis Block of AFCRL described instrumentation pods used on Titan II missile launched from Cape Canaveral on Aug. 21, 1963. Two 3-ft. by 1-ft. cylindrical pods were released into the missile exhaust plume, one at 300,000 ft., the second at 600,000 ft. Both carried four infrared spectrometers to scan the spectral region of the plume and a 3-axis magnetometer to record the orientation of the pod as it fell through the flame. On the Titan II were mounted two radiometers and a spectrometer looking down the plume from outside the flame. Purpose of the experiment was to learn characteristics of infrared spectral plume of particular missiles, to learn more about Titan II performance in terms of fuel combustion mixture ratios and engine efficiency, and to gain data helpful in evaluating effect of telemetry blackout occurring during missile launch. (*OAR Research Review*, 10/7/63, 10)

- National Science Foundation appropriations of $323 million were voted by the House Appropriations Committee, a cut of $265 million from the $589 million requested by the Administration. No funds were approved for new programs. Specifically eliminated were two new programs to improve the Nation's scientific manpower by (1) encouraging new "centers of scientific excellence" in universities that had unrealized potential, and (2) encouraging more students to go into engineering, mathematics, and physical science. (*NYT*, 10/8/63, 26)
- Rep. George P. Miller (D.-Calif.), Chairman of the House Committee on Science and Astronautics, inserted the Goddard Historical Prize Essay for 1962—"Early U.S. Satellite Proposals," by R. Cargill Hall—into the *Congressional Record*. Chairman Miller pointed out that "in space affairs we cannot afford to be shortsighted or wrong" and asked that proposals of 1946–49 be considered in the light of the current discussion on the U.S. space program: "What if the United States had launched a satellite in 1952 or even 1954? Would history have been different? . . . Would not the image of the U.S. in the eyes of the rest of the world have been spared the historical blemish of Sputnik, the appearance that we were not the scientific and technological leader in the world? . . . If we would remain free and maintain peace on a small and troubled planet, America had better make sure that superior space science and technology are on our side."

  Prize historical essay reviewed proposals by the U.S. Navy's Bureau of Aeronautics and the U.S. Air Force's RAND Corp. for for the launching of earth satellites, proposals turned down by Department of Defense officials as not having "scientific or military utility." Goddard Historical Essay Competition is sponsored annually by the National Rocket Club, Washington, D.C. (*CR*, 10/7/63, A6277–6284)
- FAA announced that its air traffic control system would be expanded to cover most of Continental Air Defense Command (CONAD) airborne interceptor operations that are made under instrument flight rules (IFR). Improvements were expected in air safety

from having the same traffic control system in charge of both air defense operations and civil air traffic. (FAA Release 63-85)

October 7: Missiles and Rockets quoted Harrison Storms, president of North American Aviation's Space and Information Systems Div., prime contractor for Apollo program, on proposals that a joint U.S.-U.S.S.R. lunar program would enable U.S. to proceed at more leisurely pace and less expensively: "When is soon enough? I don't see anything blocking us now. . . . If you put it off, it will cost more money. You must have the level of effort capable of developing the systems to do the job. If you go below that level, it costs money. . . . When you begin to lower your effort it's not the least capable man who leaves first, it's the most capable. You lose critical talent and you may actually need the same or more manpower to reach the same goal at a later date." (*M&R*, 10/7/63, 86)

- Sen. Joseph S. Clark (D.-Pa.), speaking on the Senate floor with regard to various aid-to-education bills before the Senate, asserted a connection between excellence of education and loss of jobs to automation. As a case in point he inserted in the *Congressional Record* two *New York Times* articles on NASA's findings of careless or inadequate workmanship in Project Mercury. "In both articles," the Senator said, "the point is made that workmen on the projects do not have the necessary technical skill to complete them, in many instances, without the defects which resulted not only in serious failures of the end product but also in running up a substantial expense to the Government." (*CR*, 10/7/63, 17847)

- Many U.S. airline officials have grave doubts about the economic feasibility of supersonic transport aircraft, the *Wall Street Journal* reported. Costing up to four times as much as today's $6 million subsonic jets and carrying no more and probably fewer passengers, with high fuel costs and crew costs, the aircraft are supposed to be able to do more work in less time thanks to their high speed. But many doubt that the one will balance the other. (*Wall Street Journal*, 10/7/63)

- X-15 No. 1 flew to an altitude of 77,800 ft. and a speed of 2,834 mph (mach 4.21), as Capt. Joe H. Engle (USAF) made his first flight in the research rocket aircraft. (NASA Release 63-224; *Chic. Trib.*, 10/8/63)

- It would be two or three years before the U.S.S.R. tried to land a man on the moon, according to Soviet scientist Prof. Leonid Sedov in an interview with the Brussels newspaper *Le Soir*. He said the Russians do not yet have a definite system for landing a man on the moon but did have the technical ability to do it. (UPI, *Wash. Post*, 10/8/63)

- British enthusiasts are still actively pursuing attempts to develop a man-powered aircraft, the National Geographic Society reported. A British industrialist has offered a $14,000 prize to the first man to propel himself in the air over a mile-long, figure-8 course, keeping at least 10 ft. above the ground. Some 25 teams, including some of Britain's top aeronautical engineers, are competing. Best effort to date is over one half mile at 19 mph, flown by a young man whose furious pedaling turned a propeller on a slender-winged, featherweight craft. (*NYT*, 10/7/63)

October 7: The aerospace industry "could better be defined as the national survival industry," Stuart H. Clement, Jr., of Hayden, Stone & Co., Inc., investment bankers, told the New York Society of Security Analysts in New York. "It provides national defense but, through space exploration, it may also provide a substitute for mutually suicidal nuclear war." The aerospace industry is the fastest growing in the economy and has the broadest future, he said. (*NYT*, 10/8/63)

- NASA announced it would negotiate with Lear-Siegler, Inc., of Anaheim, Calif., for an extension of the company's present $3,753,059 contract for design, manufacture, and installation of equipment for testing subsystems of the Saturn V launch vehicle at MSFC. The additional work would include design, manufacture, and installation of instrumentation for a dual liquid-hydrogen test position and an acoustical-model test position at MSFC. (NASA Release 63–222)

- World Conference on Space Communications opened in Geneva with delegates from 70 countries present. U.N. Secretary General U Thant was to have addressed the delegates via Telstar communications satellite, but a technical problem prevented the broadcast. Principal purpose of the conference was to reallocate radio frequencies to provide growth room for future satellite communications needs. (*NYT*, 10/8/63, 16)

- Dr. Hugh L. Dryden, NASA Deputy Administrator, received Harry E. Salzberg Memorial Medal for his "distinguished contribution to the field of transportation" and became the 1963 Salzberg Lecturer at Syracuse Univ. (Ltr. of Invitation, 4/18/63)

- Astronaut John H. Glenn, Jr., would decide during the week whether he would seek the Senate seat now occupied by Sen. Stephen Young (D.-Ohio), according to George Clifford of the Washington *Daily News* after an interview with C. Leo DeOrsey, adviser to the astronauts. (Clifford, Wash. *Daily News*, 10/7/63)

- NANA reported Lufthansa sources in Germany claim that U.S.S.R. is farther along on supersonic transport development than either Britain-France or the U.S. Being designed and built by the famous Ilyushin and Tupolev teams, the Russian entry is reported to be in the same speed class as the British-French Concorde—mach 2.2 (1,500 mph)—and carry about the same passenger load of 100, and to land at quite low speeds, probably through the use of variable-sweep wings. First version of the supersonic transport would be a freight carrier, not only because it can be introduced with a lower safety factor than a passenger version but because the increased friction between Soviet Russia and Communist China dictates a fast priority freight system to eastern Russia. (NANA, Wash. *Eve. Star*, 10/7/63)

- Atlas D ICBM was launched from Vandenberg AFB, Calif., but exploded shortly after launch. (UPI, *NYT*, 10/8/63, 28)

- Lewis awarded study contracts totaling $576,000 to General Dynamics/Astronautics and Rocketdyne Division of North American Aviation to investigate the possibilities of using "flox"—a mixture of liquid fluorine and liquid oxygen—as the oxidizer in the Atlas space launch vehicle. Preliminary studies indicate the combination could result in an 88 per cent increase in payloads for 100-mile high orbits. (LRC Release 63–78, Lewis Chronology, 9)

Early October: FAA Deputy Administrator Gordon M. Bain said that NASA's studies of a supersonic commercial air transport (SCAT) were not in accord with FAA thinking—that they derived too much from the B-70 work of some years ago and that they recommended the technical desirability of a mach 3 aircraft, while FAA leaned toward a mach 2.5 aircraft as the one economically feasible for airline operation. (*Av. Wk.*, 10/14/63, 38)

- Army announced $213,385,000 contract for continued development of the Nike-X antimissile system, awarded to Western Electric Corp. Largest single missile contract in Army history, the award calls for development and testing but does not provide for production or deployment. (*M&R*, 10/7/63, 18)

October 8: EXPLORER XIV energetic particles satellite had definitely ceased useful transmission after almost 10 months of successful operation. Scientists at GSFC said trouble began in August, when the satellite's transmitter failed to modulate—translate instrument signals into telemetry code—properly. Intermittent modulation had occurred since then, but little useful data had been received. The satellite signal is still useful for position reference.

EXPLORER XIV, launched on October 2, 1962, had an initial elliptical orbit of 61,500-mi. apogee and 173-mi. perigee. After 10 months, gravitational influence of the sun and moon had changed the orbit to 60,220-mi. apogee and 1,280-mi. perigee, with a period of about 37 hours. Some 6,500 hours of data were received from the satellite. While not all the data had been analyzed, Dr. L. Cahill, Univ. of New Hampshire, said a number of new insights had already emerged, among them being: earth's magnetosphere, as shown by mapping charged particles, flared away from the earth in an ogival—pointed arch—shape; confirmation that the vector magnetic field changes gently from a dipole configuration to a radial field at increasing distance on the night side of the earth near the equatorial plane; and further evidence probably supporting EXPLORER VI's finding of a ring current flow on the night side of the earth. (NASA Release 63–223)

- NASA announced that Dr. Joseph F. Shea, Deputy Director (Systems), Office of Manned Space Flight in NASA Hq., had been named Program Manager, Apollo Spacecraft, at MSC. He would be responsible for the development of the Apollo command and service modules and for the lunar excursion module (LEM). George Low, Deputy Director (Programs) in the Office of Manned Space Flight, would assume the added responsibility for the Systems organization. (NASA Release 63–226; MSC Release 63–163)
- In an editorial entitled "Bungling in Space," the *New York Times* referred to Congressional criticism of the space program, NASA's criticism of industry performance in Project Mercury, GAO criticism of industry and NASA in the development of the Centaur booster rocket, and the "aroma of pork-barrel" the Administra-

tion had allowed to befog the space program: "The disgraceful maneuvering to establish multi-million-dollar space projects—whether in Texas or Massachusetts—reflects on the Administration, the Congress and the country and undermines popular support for the program itself. No wonder the expenditure of such vast sums has been meeting so much well-founded public resistance." (*NYT*, 10/8/63, 40)

October 9: President Kennedy's news conference. Asked about developments on his proposal for U.S.-U.S.S.R. cooperation on manned lunar landing, the President replied: "We have had no indication that the Soviet Union is disposed to enter into the kind of relationship which would make a joint exploration of space or to the moon possible. But I think it is important that the United States continue to emphasize its peaceful interest and its preparation to go quite far in attempting to end the barrier which has existed between the Communist world and the West and to attempt to bring as much as we can the Communist world into the free world of diversity which we seek . . . ."

Asked whether the question of verification had come up in U.S.-U.S.S.R. talks on an agreement to ban the orbiting of nuclear weapons, President Kennedy said: ". . . there is not an agreement. The United States has stated it would not put weapons in outer space. We have no military use for doing so, and we would not do so. The Soviet Union has stated that it does not intend to. We are glad of that. There is no way we can verify that, but we are glad to hear the intention . . . .

"This is a matter, it seems to me, that can best be handled not through any bilateral agreement but as a General Assembly matter, because other countries may have the same capability, and I think every country should declare that they are not going to put atomic weapons in the atmosphere, which could threaten not only the security of a potential adversary but our own security, if for some reason the weapons should miscalculate and descend upon us . . . ." (*Wash. Post*, 10/10/63, 14)

- NASA announced the details of a major reorganization of NASA Hq., to become effective Nov. 1, 1963. Designed to consolidate authority and responsibility for major program management and for direction of field centers and to realign headquarters management of agency-wide supporting services, the reorganization provided the following: Associate Administrator for Manned Space Flight (Dr. George Mueller) directing the manned space flight program and the related centers, MSFC, MSC, and LOC; Associate Administrator for Advanced Research and Technology (Dr. Raymond Bisplinghoff), directing the advanced research and technology program and the related centers, ARC, FRC, LARC, and LRC; and the Associate Administrator for Space Sciences and Applications (Dr. Homer Newell), directing the scientific explorations of space and peaceful applications program and the related centers, GSFC, Wallops Station, PLOO, and administering the contract for JPL.

These program managers would report to the Associate Administrator, Dr. Robert C. Seamans, Jr. Dr. Seamans acquired a deputy, Walter L. Lingle. Also reporting to Dr. Seamans would be four staff functions: Deputy Associate Administrator for Industry Affairs (Earl D. Hilburn); Deputy Associate Administrator for Administration (John D. Young); Deputy Associate Administrator for Programming (D. D. Wyatt); and Deputy Associate Administrator for Defense Affairs (Adm. Walter F. Boone (USN, Ret.)).

Several agency-wide support functions would report directly to the Administrator: Assistant Administrator for Public Affairs (Julian Scheer); Assistant Administrator for Technology Utilization and Policy Planning (Dr. George Simpson); Assistant Administrator for Legislative Affairs (Richard Callaghan); Assistant Administrator for International Programs (Arnold Frutkin); Executive Director of the Policy Planning Board (Paul Dembling); General Counsel (John A. Johnson). In addition to his assistant administrator duties, Dr. Simpson would also become Associate Deputy Administrator, coordinating the staffs reporting to the Administrator and Deputy Administrator. (NASA Release 63–225)

October 9: Nike-Apache sounding rocket launched at twilight from Wallops Station, Va., with 100-lb. payload to test theories of free radical formation in comets. At 95-mi. altitude a flame was produced that left a faintly luminous cloud of combustion products visible from earth for 10 to 20 seconds. The cloud reflected sunlight in the way a comet does. Conducted by Lewis Research Center, the studies would test theories that natural comets are composed of the materials existing at the time of formation of the solar system. A second test was scheduled for the following night. (Wallops Release 63–90; *Space Bus. Daily*, 10/11/63, 68)

- NASA officials testified before the Subcommittee on NASA Legislative Oversight of the House Committee on Science and Astronautics on the summary report on Project Mercury that had accused industry of poor workmanship. Dr. George Mueller, Director of Manned Space Flight, and his Deputy, George Low, and Walter C. Williams, Deputy Director for Mission Requirements and Flight Operations, MSC, agreed that the bulk of the 444-page report was "a recital of success," with only one 10-page section listing deficiencies. The committee agreed with the witnesses and considered the matter closed. (*Space Bus. Daily*, 10/10/63, 60)
- Sen. Hubert H. Humphrey (D.-Minn.) spoke on the Senate floor about the results of the Senate Committee on Government Operations' five-year fight for more comprehensive, integrated technical information programs in Government. The Senator said that useful programs were now at last underway in all major Federal agencies; he particularly cited the Science Information Exchange, which as of June 30, 1963, had a roster of 75,000 research proposals and projects in the life sciences and physical sciences. (*CR*, 10/9/63, 18128)
- By this date, engineering design work on the Apollo spacecraft was more than 75 per cent complete and work had begun on fabrication of the stainless steel skin and molding of honeycomb aluminum inner walls of the command module at the North American Aviation plant at Downey, Calif. (Finney, *NYT*, 10/13/63, 80)

October 9: AEC memo, "Electromagnetic Effects on World Communications," described the effects of high-altitude, high-yield nuclear explosions on worldwide communications, according to *Space Business Daily*. Data had been accumulated from high-altitude nuclear explosions in Project Argus, 1958, Operation Hardtack, 1959, and Starfish Prime, July 9, 1962. The AEC report said that a 50-megaton blast at 50-mi. altitude would black out communications within a radius of 2,500 miles for as long as a day. Such effects had been an area that opponents of the nuclear test-ban treaty had argued the U.S. had too little information about and that the knowledge was critical for ICBM penetration and anti-missile defense. (*Space Bus. Daily*, 10/9/63, 50, 51)
- A "flying carpet" escape system from orbital space stations has been advanced by Douglas Aircraft Co. The escape system would be a saucer shape that would expand into a blunt-nosed, cone-shaped vehicle 25 ft. across at its base. Passing through the atmosphere, the vehicle would act as its own brake. Re-entry heating problems would be met by using fabrics woven with filaments of nickel-based alloys that can withstand 1,600-degree temperatures. (*Space Bus. Daily*, 10/9/63, 52)
- Air Force Systems Command and Air Force Logistics Command announced a joint program of data management applicable to all Air Force contracts from the two commands. Program calls for tight data control by project officer, selection of minimum essential data, line-iteming each data point in the contract. (AFSC Release 310-R-87)
- An editorial in the *Washington Daily News* supported the idea that a slower program would not save money in the long run.

  ". . . today the moon and space programs are in high gear and almost $5 billion in prime contracts have been placed with American industry. Plants are tooled up, scientists are on the job, research centers are going up and near Houston even a new city is being built.

  "We are inclined to agree with Dr. Wernher von Braun, NASA Director James Webb and others that a major slash in space funds at this stage would not save money in the long run because a stretched-out lunar landing program would come to an even greater total cost. Mr. Webb, for example, estimates an extra $3 billion for the additional overhead costs alone of a three-year stretch-out.

  "It was judged two years ago by the President and Congress to be in the national interest to place a man on the moon—and before the Russians, if possible. If the reasons were valid then, and we think they were, they are valid today." (Text)

October 10: Nike-Apache sounding rocket was launched at twilight from NASA Wallops Station in the second of a series of comet experiments. At 95-mi. altitude the 100-lb. payload produced flames visible from earth and combustion products that lingered after burnout as a faint luminous cloud somewhat similar to a comet. The cloud effect was visible for a surprising five minutes, compared with less than one half minute the night before. The simulated comet was photographed from three ground positions. (Wallops Release 63-92)

October 10: Aerobee 150 sounding rocket was launched from White Sands Missile Range, N.M., but sustainer engine failed to ignite, destroying vehicle and payload containing a GFSC nebular spectra experiment. (GSFC Historian)

- House passed H.R. 8747, the Independent Offices Appropriations for FY 1964, which included the $5.1 billion for NASA recommended by the Appropriations Committee. Among the amendments accepted was one which would require Senate approval of any agreement that would use NASA funds in support of joint lunar exploration with Communist countries. Among amendments rejected were ones which would have (1) reduced NASA R&D funds by another $700 million and set limitations of $597,400,000 on Project Apollo and $124,400,000 on lunar and planetary exploration; (2) deleted the facility construction funds for the proposed NASA electronics center ($3,990,000); and (3) reduced the amount for development of a supersonic transport aircraft from $60 million to $500,000.

    An involved debate preceded passage of the bill, most of it concerned with the NASA program. Rep. Clarence Cannon (D-Mo.), Chairman of the Appropriations Committee, was one of those opposing passage of the NASA funding. He argued that "the appropriation reported out by the committee has been provided largely because of commitments. It is impossible to escape the conviction that the project [manned lunar landing] must eventually be abandoned."

    Rep. Albert Thomas (D.-Tex.) led the support of the NASA program, arguing that the NASA budget had already been cut $1.1 billion from that ($6.2 billion) originally presented by NASA to the Bureau of the Budget. Another in support of the NASA appropriation was Rep. George W. Andrews (D-Ala.), who read into the record a letter from Dr. Wernher von Braun, Director of MSFC. Dr. von Braun argued that a program as big and complex and with such long lead times as the space program must have sustained, consistent support year after year. He stated that of the various major expenditures—rockets, capsules, facilities, etc.—only the Lunar Excursion Module was being designed principally for the manned lunar landing. "All other elements of these programs create what we as a nation most urgently need independent of our lunar landing effort: A national manned spacefaring capability, to sail the new ocean, as the President said." (*CR*, 10/10/63, 18260–18309)

- A. Scott Crossfield, first pilot of the X-15 and now an executive with North American Aviation, Inc., said in an interview that he thought the X-15 "can fly twice as high as its present record and probably can go several hundred miles faster than it has been flown . . . ." The increased altitude ". . . would be more difficult than the extra speed and it would create re-entry problems too." (Tulsa *Daily World*, 10/11/63)

- The President's memorandum on patent policy for work performed under Government contract was read into the *Congressional Record*. Written as guidance on patent policy for all Government agencies the memorandum provided: (1) the Government would normally acquire rights to inventions made in the course of work on a Government contract, except when the head of the agency concerned certified that the public interest would best be served by some other arrangement; (2) in cases where the Government contract calls for building on existing knowledge in a field of technology directly related to field in which the contractor has an established nongovernmental commercial position, the contractor would normally acquire rights to any inventions. (*CR*, 10/10/63, 18320-21)

October 10: Nobel Peace Prize for 1962 went to Dr. Linus C. Pauling, of the California Institute of Technology. It was the second Nobel Prize for Dr. Pauling, who won the chemistry award in 1954 for his theory describing the fundamental nature and behavior of molecular bonds. He had been a controversial figure in the U.S. for some years because of his outspoken opposition to nuclear testing. (*NYT*, 10/11/63, 1)

- The Collier Trophy for 1962 was presented to the seven Project Mercury astronauts by President Kennedy in the White House Rose Garden:
  "I hope this award, which in effect closes out a particular phase of the program, will be a stimulus to them and to the other astronauts who will carry our flag to the moon and, perhaps, some day, beyond." (AP, *NYT*, 10/11/63, 18; NAA Release 10/7/63)

- Department of Commerce asked the National Inventors Council, previously concerned with routing the ideas of independent inventors to DOD, to broaden its activities and advise the Commerce Department on all steps the Federal government can take to spur invention and application of inventions. Dr. J. Herbert Holloman, Assistant Secretary of Commerce for Science and Technology, said application of new ideas was the biggest problem: "There is immense resistance to new methods. Innovation often comes from an invader—from a foreigner to a company, an industry or a country. For example, polymers came from the chemical industry to the textile industry. Fluorescent lighting came from Europe. The transistor was not developed by a radio-tube manufacturer." (Jones, *NYT*, 10/11/63, 53)

October 11: NASA Deputy Associate Administrator Earl D. Hilburn, speaking before the Midwest Regional Conference of the Society of American Military Engineers in Omaha, Neb., described NASA's steady increase in facilities and techniques for ground simulation of the space environment in the testing of spacecraft: "To date, . . . it has been estimated that the monetary cost of the spacecraft environmental test programs has been less than 10 percent of the irrecoverable cost of the spacecraft launch. It is estimated that the cost of placing a Thor-Delta launched spacecraft into orbit is $10 million. Included in the 10 percent figure I have just cited is the prorated cost of spacecraft test facilities distributed over about 20 launches and 10 years of time." (Text)

October 11: Douglas Aircraft Co. graduated first class from on-the-job training program for potential executives. The six trainees, with no previous managerial experience, were assigned full time to departments completely different from their prior jobs with the company. Acting as staff assistants, they sat in on all policy conferences and later reviewed with the manager in private his reasons for the decisions made during the conference. W.E. Maschal, director of management systems for the Aircraft Division, said the program differed from other manager rotation plans "in its emphasis upon teaching potential managers the decision-making process from the excutive point of view." (*Space Bus. Daily*, 10/11/63, 68)

October 12: Joseph A. Walker, NASA's senior X-15 pilot and holder of the world altitude and speed records for research aircraft, was awarded the Christopher Columbus International Prize for Communication. The award is made by the city of Genoa, Italy, each year on Columbus Day. U.S. Ambassador G. Frederick Reinhardt accepted the award on behalf of Mr. Walker. (NASA Release 63-229; AP, *NYT*, 10/14/63, 9)

October 13: NASA Administrator James E. Webb, interviewed by the *New York Herald Tribune*, acknowledged that parts of Project Apollo were behind the schedule as written two years ago. He singled out on-board power—"we have counted on fuel cells . . . . To say that the fuel cell is behind is not an adequate statement, but to say that problems have been encountered in solving the problems of enough on-board electrical energy to do the job would be an accurate statement"—and the Titan II booster for the Gemini program—"We must know what happens to men and equipment under a weightless condition for a week or two. We had hoped to get this information at an earlier date in making our plans for Apollo and future flight missions. We still will get it early enough to do the job within this decade." (Ubell and Loory, *N.Y. Herald Tribune*, 10/13/63)

- Editorial in the *New York Times:* "The House of Representatives, moved by sheer frustration and a kind of primitive emotionalism, occasionally acts in such a way as to undermine its reputation for rational lawmaking. The vote to prohibit the use of moon-project funds in any cooperative lunar attempt with the Soviet Union is a case in point. This restrictive proposition was presented without advance notice, received virtually no debate on its merits, and was then adopted with barely half the membership voting . . . . The Congress of the United States should not relieve Premier Khrushchev from making the decision on cooperation, or noncooperation, with the United States in the exploration of space." (*NYT*, 10/13/63, 8E)

October 14: The concensus of a number of visual experts consulted by NASA is that Astronaut L. Gordon Cooper did in fact see houses, roads, and vehicles from his orbiting FAITH 7 capsule. Dr. John A. O'Keefe of GSFC dismissed the theory that the atmosphere had acted as a magnifier of the objects. Maximum atmospheric magnification would be 1.00002, or a rise of 8.5 ft. Normal human visual acuity is normally considered to be one minute of arc, but under good combinations of illumination and contrast objects of ½ minute of arc or even smaller can be seen. From 100 miles altitude, this would work out at 30 ft. on the ground. Line or ribbon objects can be seen much more easily, down to one-sixth or less the diameter of circular objects. (*M&R*, 10/14/63, 14)

October 14: The lunar landing research facility was nearing completion at Langley Research Center. A gantry structure 400 ft. long and 250 ft. high with a 50 ft. clearance would suspend a model of the Lunar Landing Module (LEM), sustaining ⅚ of the model's weight to simulate the ⅙ lunar gravity and enable astronauts to practice lunar landings. (*Av. Wk.*, 10/14/63, 83)

- NASA Hq. had tentatively approved Project Luster, designed to capture lunar dust that was in orbit around the earth, *Missiles and Rockets* reported. Proposed by Ames Research Center, the project is based on the assumption that meteorite impacts on the moon above a certain velocity should throw out more than the meteorite's own weight in dust. Some of this lunar dust should find its way into earth orbit. Ames would launch an Aerobee 150 sounding rocket to attempt to trap some of the dust. (*M&R*, 10/14/63, 9)

- X-15 flight to test a modification of the vertical tail section was postponed. The pilot, Maj. Robert Rushworth (USAF), called off the flight 15 min. before he was to drop away from the B-52 mother aircraft. A malfunction had developed in the X-15's inertial guidance system. (AP, *Wash. Post*, 10/15/63)

- MSC awarded $100,000 contract to North American Aviation, Inc., for a study of what modifications would be necessary to the Apollo spacecraft to use it as a space science laboratory in orbital flights lasting up to one year. (*Houston Post*, 10/15/63)

- Edward Z. Gray was appointed Director of Advanced Studies, Office of Manned Space Flight, NASA Hq., succeeding Dr. William A. Lee, who was joining the Apollo Spacecraft Program Office, MSC. Mr. Gray had been with Boeing Aircraft Co. for 24 years, most recently as Development Program Manager of Advanced Space Systems, Aero-Space Div. (NASA Release 63-228)

- Total manpower working at NASA's Michoud Operations, of which only 244 were NASA employees: 8,463. Contractor employees were: Boeing, 4,136; Chrysler, 3,207; Mason-Rust, 769; Rocketdyne, 13; and Telecomputing Services, 94. (*Marshall Star*, 10/23/63, 4)

- MSC selected Hamilton Standard Div. of United Aircraft Corp. to develop detailed theoretical analysis of mechanics involved in rocket-belt transportation of one man plus a small load of supplies or tools in a zero-gravity or lunar surface environment, including capability to ascend or descend steep lunar crater walls. (*Space Bus. Daily*, 10/15/63, 81)

- FAA exhibited a device intended to detect bombs carried aboard aircraft. The device, displayed at Dulles International Airport, Washington, was based on the principle of the geiger-counter, would depend on the willingness of explosives manufacturers to mix small quantities of radiative material in their products. (FAA Release)

October 14: Astronauts on the five-day manned lunar landing trip would not only pilot the Apollo but would make repairs to malfunctioning systems out of a stock of spare parts carried along for that purpose. This would be one means of upgrading the reliability of the entire system and reduce the need for redundancy of systems, saving weight and space. (Finney, *NYT*, 10/15/63, 22)

- Dr. Jerome B. Wiesner, Director of the White House Office of Science and Technology, appeared before a Senate Appropriations subcommittee to request restoration of $245,000 that the House had cut from his office's $1,025,000 budget request. If the cut were not restored, he said, he would have to cut back his 29-member staff just as the one-year-old organization was beginning to "realize some of the potential foreseen for it as a new innovation in the Federal science structure." (*NYT*, 10/15/63, 25)

October 15: Seventeen of the 18-member U.N. Disarmament Committee presented a resolution to the U.N. Political Committee of the General Assembly banning the orbiting of nuclear weapons and other weapons of mass destruction. This was a follow-up to the Oct. 3 announcement of such an agreement among the U.S., U.S.S.R., and U.K. (Brewer, *NYT*, 10/16/63, 1)

- NASA launched an Aerobee rocket to an altitude of 123.5 mi. over White Sands Missile Range, N.M., with a 245-lb. payload to study solar radiation. Three telescopes with automatic camera attached to each took x-ray photos of sun at exposures ranging from $3/10$ sec. to 100 sec. Equipment was designed to obtain an optical image of the sun from its own low-energy x-rays and would be used in future scientific satellites. GSFC cooperated with American Science and Engineering Corp. of Cambridge, Mass., in the experiment. The payload was recovered 55 mi. downrange. (*Wash. Post.* 10/16/63; AP, *NYT*, 10/16/63, 2; NASA Rpt. of S. Rkt. Launching, 11/14/63)

- MIT announced that a laser had been used to observe meteoritic dust at heights of 35 to 85 mi., with the dust seeming to concentrate at 50 to 75 mi. It was suggested that the lower layer might be related to the noctilucent clouds, since last year's rocket launchings from Sweden suggested that noctilucent clouds found at that altitude could be ice-covered particles of meteoritic dust. The higher layer might be at the level where meteors break into dust on entry into the earth's atmosphere. (*NYT*, 10/16/63, 2)

- Dr. Frederick Seitz, President of the National Academy of Science, was the opening witness in hearings on the Government's role as a patron of science and technology, conducted by the subcommittee on Science, Research and Development of the House Committee on Science and Astronautics. Dr. Seitz said that the period of almost unlimited growth of Federal investment in science and technology was coming to an end, would have to level off and force some "hard decisions" on choices, and expressed the fear that basic science would suffer in competition with more glamorous hardware projects. For the most part he thought U.S. science was well ahead of Russia, but noted remarkable progress on the part of West Germany and Japan. (Finney, *NYT*, 10/16/63)

October 15: Second Town Meeting of the World telecast by CBS-TV via TELSTAR II satellite. The subject was "The Christian Revolution" and linked in a discussion church leaders in New York, London, and Rome. Program was only a partial technical success, since the audio signal for the speakers in Rome was missing for most of the program. (CBS-TV)

October 16: Twin Vela Hotel satellites were launched by DOD from Cape Canaveral on an Atlas-Agena vehicle, according to newspaper reports. The two satellites, designed to detect nuclear explosions in space to a distance of 100,000,000 mi., were launched into an elliptical orbit of 57,000-mi. apogee, 230-mi. perigee. Some 18 hours after launch, one of the two satellites received a signal from Vandenberg AFB which activated a rocket motor at apogee to kick the satellite into circular orbit at 57,000 mi. On Oct. 19, the same was done for the second satellite, so that both would orbit at 57,000 mi. but always be on opposite sides of the earth. Reports also indicated that a third satellite, known as "Pygmy" and weighing 2½ lbs., was included in the launch, would remain in elliptical orbit returning radiation readings across the depth of the Van Allen Belt. This function was said to be completely independent of the purpose of the twin Vela Hotel satellites. (Witkin, *NYT*, 10/18/63, 13; *Wash. Post*, 10/18/63; *N.Y. Herald Tribune*, 10/18/63; DOD Release 1396-63)

- B-58 with USAF crew flew nonstop from Tokyo to London, completing the 8,028-mi. flight in 8 hrs. 35 min. for an average speed of 938 mph. The B-58 was refueled in midair five times. Previous record of 17 hrs. 42 min. was set in 1957 by U.K.'s W. Hoy, who flew a Canberra at an average speed of 335.7 mph. Crew members on the B-58 were Maj. Sidney G. Kubesch, Capt. Gerald R. Williamson, and Maj. John O. Barrett. (*Av. Wk.*, 10/21/63; *NYT*, 10/22/63, 8)

- Political Committee of the United Nations General Assembly unanimously approved a 17-nation resolution to prohibit the orbiting of nuclear weapons and weapons of mass destruction in space. Committee chairman, Carl W. A. Schurmann of the Netherlands, said: "We are all so happy about this resolution that it will not be necessary to take a formal vote." Previously, only the United States and the Soviet Union, under the limited test ban treaty, had agreed to such a matter in principle. Resolution would go to the General Assembly next day. (Brewer, *NYT*, 10/17/63; *Wash. Post*, 10/17/63)

- U.S.S.R. Ambasador Anatoli Dobrynin told the U.N. that his country had a plan for manned lunar flight prior to 1970. This was said to be the first official acknowledgment on a political level that such a plan existed. (*Space Bus. Daily*, 10/21/63, 116; *Chic. Trib.*, 11/27/63)

- Maj. Donald K. Slayton (USAF) one of the original seven Mercury astronauts and the only one of them who was grounded during the course of the project, submitted his letter of resignation to the USAF and would join NASA as a civilian pilot qualified for space flights. (Finney, *NYT*, 10/17/63)

October 16: Dr. Jerome B. Wiesner, the President's Scientific Adviser, testified before House Science and Astronautics subcommittee on the Federal role in science and technology. He indicated that the tremendous surge in development of military weapon systems that had been the principal cause of the 15-to-20 per cent increase in the Federal science and technology budget each year for the last several years would level off. Now the effort would shift from military systems to bringing technology "to bear on serving the collective needs of our people." He was concerned lest without the defense motive the Nation slip back into indifference toward support of science, particularly the basic research that provided the new ideas for further advances. He also said that he not only supported the U.S.-U.S.S.R. cooperative lunar landing proposal but that he had often commended the idea to the President. (*NYT*, 10/17/63; *Space Bus. Daily*, 10/17/63, 98)

- FAA Administrator Najeeb E. Halaby and Civil Aeronautics Board Chairman Alan S. Boyd testified before the Senate Commerce Committee Subcommittee on Aviation on the U.S. plans for development of a supersonic transport (SST). They agreed that the U.S. must and would win the competition with the U.K.-France Concorde. Airlines were even now holding back on purchase commitments of the Concorde because the U.S. entry promised to be faster, carry more passengers, and have more range, even though it would be a year or more later in schedule. Halaby said three problems were still unsolved: the sonic boom level, materials to withstand the heat of mach 2.5 flight, and producing a better aircraft at a price competitive with the Concorde. (*Wash. Post*, 10/17/63)

- Soviet Cosmonauts Lt. Col. Yuri A. Gagarin and Junior Lt. Valentina Tereshkova visited the U.N. in New York. After a standing ovation by the General Assembly, Col. Gagarin listed five areas in which U.S.-U.S.S.R. space cooperation was possible: exchange of scientific knowledge, tracking, rescue of downed spacemen, global weather study, and an international radio communications system. Manned space flight cooperation required "certain clarification," he said. (*NYT*, 10/17/63)

- AFOAR announced that its scientists at AFCRL had developed a precision pressure gauge that would greatly improve accuracy of measuring altitude in high-altitude flight. Called the hypsometer, the new device senses altitude by measuring the boiling point of a liquid (butylbenzene). Testing in a series of balloon flights, hypsometer proved 10 times as accurate at 120,000 ft. as previous altimeters (240 ft. error as opposed to 2,600 ft.). (OAR Release 10-63-2)

- Dr. Glenn T. Seaborg, Chairman of the AEC, was named recipient of the 1963 Franklin Medal for the advancement of knowledge in the physical sciences. The medal is presented annually by the Franklin Institute of Philadelphia. Dr. Seaborg was co-winner of the Nobel Prize for chemistry in 1953. (*NYT*, 10/17/63)

- NASA said it was investigating "possible improprieties" in reports that cars leased to NASA by a contractor, Management Services, Inc., at Cape Canaveral were being sold after two or three years use for as little as $50. Some of them had been sold to NASA employees, the reports said. (*NYT*, 10/17/63)

October 17: U.S.-France Aerobee 150A sounding rocket was launched from Wallops Station, Va., in cooperative program between NASA and the French National Center for Space Studies (CNES) to study very low frequency (VLF) radio wave propagation in the ionosphere. The French were developing a technique to measure simultaneously electronic and magnetic VLF field strength and electron densities. If successful the technique would be used in a scientific satellite. The 197-lb. payload rose 116 mi. and transmitted seven min. of data before impacting 59 mi. downrange. This was repeat of unsuccessful attempt two days earlier when Aerobee 150A failed on launch. (Wallops Release 63-95)

- DOD launched two unidentified satellites and Tetrahedral Research Satellite (TRS II) using Atlas-Agena D launch vehicle. 3-lb. TRS II was to measure charged particle intensity in Van Allen belts. (*Pres. Rpt. on Space, 1963*, 1/27/64)

- NASA and DOD announced joint agreement to coordinate studies and actions concerning a manned orbital research and development station. The Aeronautics and Astronautics Coordinating Board (AACB) would be the means by which the joint appraisals would be carried out. If the eventual findings were that such a program was required, the NASA Administrator and the Secretary of Defense would submit a joint recommendation to the President, including a statement as to which agency should have project responsibility. If the President agreed to proceed with the program, a joint NASA-DOD board would be set up to establish objectives and approve specific experiments. (NASA Release 63-231; DOD Release 1380-63)

- NASA announced that TIROS VI weather satellite was no longer sending usable cloud-cover pictures, after 13 months of successful operation and some 67,000 pictures. (NASA Release 63-232)

- By acclamation, the United Nations adopted a resolution barring weapons of mass destruction from outer space. A request for a rollcall vote by Mexico was ignored. Thus, the attitudes towards East-West agreements of Cuba, Albania, and France were not revealed on this resolution. (Fleming, *L.A. Times, Wash. Post*, 10/18/63, A-3)

- Sen. J. W. Fulbright (D.-Ark.), Chairman of the Senate Foreign Relations Committee, spoke on the Senate floor on the NASA FY 1964 appropriations bill: "There is, I believe, a dangerous imbalance between our efforts in armaments and space on the one hand and employment and education on the other. The proposed appropriation for NASA, in my opinion, reflects this imbalance. I believe that is should be substantially reduced. I further believe that any funds which are withheld from the space program should be reallocated to programs of education and employment which are before Congress this year.

  "The question before us, as I have said, is not whether we should or should not send a manned rocket ship to the moon but whether the project is so vital and so urgent as to warrant the indefinite postponement of other national efforts." (*CR*, 10/17/63, 18777)

October 18: COSMOS XX was launched into earth orbit by U.S.S.R. A scientific satellite for "continuing the study of space," COSMOS XX was stated to have an apogee of 311 km. (193 mi.); perigee of 206 km. (128 mi.); period, 89.55 min.; and inclination, 65°. (*Komsomolskaya Pravda*, Oct. 19, 1963, 1, AFSS-T Trans.)

- NASA announced at MSC the selection of 14 astronauts for the Gemini and Apollo projects, bringing the total number of NASA astronauts to 30. New astronauts were: Maj. Edwin E. Aldrin, Jr., Capt. William A. Anders, Capt. Charles A. Bassett II, Capt. Michael Collins, Capt. Donn F. Eisele, Capt. Theodore C. Freeman, and Capt. David R. Scott, all USAF; Lt. Alan L. Bean, Lt. Eugene A. Cernan, Lt. Roger B. Chaffee, Lt. Cdr. Richard F. Gordon, Jr., all USN; Capt. Clifton C. Williams, Jr., USMC; and R. Walter Cunningham and Russell L. Schweickart, civilians. NASA announced another contingent of astronauts would be named in the fall of 1965. (*NYT*, 10/19/63, 9)

- NASA officials testifying before a subcommittee of the Senate Appropriations Committee offered sharply revised estimates on the cost of a manned lunar landing. Under questioning from senators, NASA Administrator James E. Webb agreed that NASA would spend much of the $20 billion normally stated as the cost of manned lunar exploration even if there were no program to land a man on the moon. When asked how much the manned lunar landing would cost over and above other space exploration, Mr. Webb estimated less than $3 billion. NASA Associate Administrator Dr. Robert C. Seamans, Jr., estimated $2 billion; NASA Associate Administrator for Manned Space Flight Dr. George Mueller estimated $1 billion. Mr. Webb and Dr. Seamans had included the cost of unmanned investigation of the lunar surface, while Dr. Mueller was counting only the cost of the Lunar Excursion Module (LEM). (*NYT*, 10/19/63, 1)

- France successfully launched a cat into space onboard a Veronique rocket and brought it to a safe landing. (Reuters, *Wash. Post.* 10/19/63, 4)

- Dr. Edward Teller testified before the House Science and Astronautics subcommittee studying the role of the Federal government in science and technology. Dr. Teller suggested that the Congress devote its main effort to the development programs, where most of the Federal money is being spent, but leave the selection of basic research projects largely to the scientists because this kind of research "deals in the unexpected." Asked to comment on Dr. Wernher von Braun's statement of the value of the manned lunar landing, Dr. Teller answered: "Von Braun is a competent technician and a superb salesman. Far be it from me to disagree with him." (Toth, *NYT*, 10/19/63, 25)

- NASA announced it would negotiate a contract with MIT, AC Spark Plug Div. of GM, Raytheon Co., Kollsman Instrument Co., and Sperry Gyroscope Co., for the development, fabrication, and testing of the guidance and navigation system for the Lunar Excursion Module (LEM) of the Apollo spacecraft. These same companies already are developing the guidance and navigation system for the Apollo command and service modules; the LEM system would be as compatible as possible and have as many interchangeable parts as possible with the other systems. (NASA Release 63-234)

October 18: Air Force biomedical support of Projects Gemini and Apollo included: at School of Aerospace Medicine, Brooks AFB, Tex., studies of physiological effects of exposure to various atmospheric pressures and gases, special medical examinations of flight candidates, and thermal balance and water requirements for wearing the Apollo spacesuit; at Aerospace Medical Research Laboratories, Wright-Patterson AFB, Ohio, studies of effect on men of predicted impact velocities for Apollo landing, nutrient requirements, weightless orientation flights for astronauts, and thermal evaluation of prototype Gemini spacesuit; at Aeromedical Research Laboratory, Holloman AFB, N.M., studies on effects of deceleration forces on Apollo pilots during recovery. (*Space Bus. Daily,* 10/18/63, 108)

- A 3,000-man strike virtually halted construction at the Nevada Test Site, one week after labor and management had signed an agreement designed to stop such walkouts. Labor Union Local 872 set up a picket line that was honored by employees of the prime contractor, Reynolds Electrical and Engineering Co. (AP, *Wash. Post,* 10/19/63)

- Soviet scientists in Geneva for the meeting of the World Conference on Space Communications presented a background paper which stated that the further man ventured into space the more obvious it was that certain problems could only be solved by collective, peaceful cooperation among nations. (Reuters, *Wash. Post,* 10/18/63, 23)

October 19: In convocation address at the University of Maine, President Kennedy outlined U.S. foreign policy. Referring to the Cuban crisis of October 1962, the President said:

"A year ago it would have been easy to assume that all-out war was inevitable—that any agreement with the Soviets was impossible—and that an unlimited arms race was unavoidable. Today it is equally easy for some to assume that the cold war is over—that all outstanding issues between the Soviets and ourselves can be quickly and satisfactorily settled—and that we now have, in the words of the psalmist, an 'abundance of peace so long as the moon endureth' . . . .

"The fact of the matter is, of course, that neither view is correct. We have, it is true, made slight progress on a long journey . . . . We have concluded with the Soviets a few limited enforceable agreements or arrangements of mutual benefit to both sides and the world . . . .

"In times such as these, therefore, there is nothing inconsistent about signing an atmospheric nuclear test ban, on the one hand, and testing underground on the other; about being willing to sell to the Soviets our surplus wheat while refusing to sell strategic items; about probing their interest in a joint lunar landing while making a major effort to master this new environment; or about exploring the possibilities of disarmament while maintaining our stockpile of armaments. For all of these moves, and all other elements of American and Allied policy toward the Soviet Union, are directed at a single, comprehensive goal—namely, convincing the Soviet leaders that it is dangerous for them to engage in direct or indirect aggression, futile for them to attempt to impose their will and their system on other unwilling peoples, and beneficial to them, as well as all the world, to join in the achievement of a genuine and enforceable peace . . . .

"It is in our national self-interest to ban nuclear testing in the atmosphere so that all our citizens can breathe easier. It is in our national self-interest to keep weapons of mass destruction out of outer space—to maintain an emergency communications link with Moscow—and to substitute joint and peaceful exploration for cold war exploitation in the Antarctic and in outer space . . . .

". . . For without our making such an effort, we could not maintain the leadership and respect of the free world . . . ." (Text, AP, *Wash. Post*, 10/20/63, A7)

October 20: Soviet Cosmonauts Yuri Gagarin and Valentina Tereshkova visited the Berlin Wall at the Brandenburg Gate. Later in interview on East Berlin Radio-TV, Cosmonette Tereshkova stated:

"There are married women among our female cosmonauts. A friend of mine is married and has children and she has been prepared for a space flight in the same way as I was." Visit of cosmonauts to Berlin coincided with East German Parliamentary elections. (Reuters, *Wash. Post*, 10/21/63, A3)

- DOD was conducting a major R&D program for ground-to-space weapons capable of destroying hostile satellites, according to Richard Witkin of the *New York Times*. DOD Director of Research and Engineering Dr. Harold Brown had been quoted as saying that antisatellite defense systems were thought "important enough to put substantial work into" and that ground-based weapons seemed cheaper and more efficient than orbital weapons. (Witkin, *NYT*, 10/20/63, 1)

October 21: NASA announced it had selected a fact-finding committee to study various geographic locations for the proposed NASA Electronics Research Center. Once the committee completed its study, the NASA Administrator would make the final decision and report his findings to Congress in conformance with P.L. 88–113. (NASA Release 63–233)

- NASA Administrator James E. Webb, testifying before the Senate Commerce Committee Subcommittee on Aviation, pointed out the long history of NACA-NASA contribution to aircraft development and its continuing research contribution to the supersonic transport. Denying that such a transport would be economically unfeasible, Mr. Webb said that further improvements were attainable during the aircraft's development: "Reduce fuel reserves by 17%; Increase airplane lift-to-drag ratio by 7%; Reduce engine specific fuel consumption during cruise by 3%; Reduce engine weight by 15%; Reduce airframe structure weight by 5%. If all of these gains can be achieved simultaneously the payload can be increased 34% for a given weight aircraft designed to fly 4,000 statute miles." (Text)

October 21: National Academy of Sciences began its centennial series of scientific meetings in Washington. (*NYT*, 10/22/63, 23)

- Sen. Clinton P. Anderson (D.-N.M.), Chairman of the Senate Committee on Aeronautical and Space Sciences, speaking at the Women's National Democratic Club in Washington, said the space age was "one of the great movements in human history . . . . Our ability to maintain the peace and deter aggression depends on keeping ahead—or at least abreast—of any other nation in space technology." (Wash. *Eve. Star*, 10/22/63)
- NASA spokesman was quoted by *Missiles and Rockets* as saying that NASA would switch many of its satellite programs from the Delta launch vehicle to the Thrust Augmented Delta (TAD). TAD would use the Delta second stage with the USAF-developed Thrust Augmented Thor (TAT), which uses three strap-on solid rockets to raise thrust from 170,000 lbs. to 330,000 lbs. and provide 200 lbs. increase in payload in a 300-mi. orbit. Many NASA programs had been pushing the payload limit of the Delta. Those benefiting from the new booster were reported to be Syncom, the Bios satellites, Oso, Tiros, and the Interplanetary Monitoring Probe (Imp). (*M&R*, 10/21/63, 15)
- NASA Administrator James E. Webb, speaking to the New York section of the AIAA in New York, pointed out that the U.S. manned space program involves technical requirements "about 20 times the complexity of the Minuteman program." For this reason, among others, he said leaders of the space program had recognized from the beginning that leadership could be achieved "only through a sustained, steadily accelerating effort." He quoted Dr. Hugh L. Dryden, speaking on Apr. 26, 1958, when he said that without this big, long-term national commitment "I'd just as soon we didn't start"; Dr. Wernher von Braun in 1958 telling the House Select Committee studying the Space Act that the biggest problem was "how we can eliminate this lethal and wasteful hot-and-cold blowing that has plagued all our missile projects in the past. I think this lack of steady determination and unwavering support over the years has hurt us more than anything"; and Dr. T. Keith Glennan, NASA's first Administrator, speaking at the first NASA budget hearings before the House Committee on Science and Astronautics: "If our space programs are to be run on an off-again, on-again basis, zigging and zagging with the turn of every new year, then we'd better spend our money buying telescopes to watch the Russian pioneers in space." (Text)
- FRC completed first phase of a program to determine how much visibility a pilot needs to fly an aircraft through all phases of flight. The problem is that the present standards of large glassed areas for pilot visibility would be virtually unacceptable in vehicles capable of orbital flight and re-entry. FRC's approach to the problem was to configure a U.S. Army L–19 light aircraft with two wide-angle monocular telescopes mounted with their axis convergent to an angle of 55° toward the pilot. Exposing only four inches of glass, the two telescopes afforded the pilot 140° visual field horizontally, 90° vertically, with a 40° horizontal overlap. Forty flights were made with the system, testing all phases of

flight. The second phase of the program would employ a high-speed jet aircraft with a similar optical system. (FRC Release 25-63)

October 21: NASA awarded a $236,520 grant to the Lowell Observatory, Flagstaff, Ariz., to provide a building for processing, storage, and research of lunar and planetary photography. This would assist Lowell Observatory in fulfilling the request of the International Astronomical Union in August 1961 that Lowell establish a Western Hemisphere repository of lunar and planetary research materials corresponding to the Eastern Hemisphere center at Meudon, France. (NASA Release 63-235)

- Rep. Joseph E. Karth (D.-Minn.) refuted charges of waste and inefficiency in the space program, reminded members of a time three years before when similar statements were being made about the Mercury program. As evidence, he read into the record an article by James Baar in *Missiles and Rockets* of Aug. 15, 1960, entitled "Is Mercury Program Headed for Disaster?": "NASA'S Mercury manned satellite program appears to be plummeting the United States toward a new humiliating disaster in the East-West space race.

"This is the stark conclusion that looms in the minds of a growing number of eminent rocket scientists and engineers as the Mercury program continues to slip backward." (*CR*, 10/21/63, A6547-48)

October 22: President John F. Kennedy addressed the centennial meeting of the National Academy of Sciences in Washington: "... it took the First World War to bring science into central contact with government policy, and it took the Second World War to make scientific counsel an indispensable function of government....

"Every time you scientists make a major invention, we politicians have to invent a new institution to cope with it—and almost invariably, these days, it must be an international institution. I am not just thinking of the fact that when you people figure out how to build a global satellite communication system, we have to figure out a global organization to manage it. I am thinking, as well, that scientific advance provided the rationale for the World Health Organization and the Food and Agriculture Organization—that splitting the atom leads not only to a nuclear arms race but to the establishment of the International Atomic Energy Agency—that the need for scientific exploration of Antarctica leads to an international treaty providing free access to the area without regard to territorial claims—that the scientific possibility of a World Weather Watch requires the attention of the World Meteorological Organization—that the exploration of international oceans leads to the establishment of an Intergovernmental Oceanographic Commission.... The ocean, the atmosphere and outer space belong not to one nation or to one ideology but to all mankind.

"... ours is a century of scientific conquest and scientific triumph. If scientific discovery has not been an unalloyed blessing, if it has conferred on mankind the power not only to create but also to annihilate, it has at the same time provided humanity with a supreme challenge and a supreme testing. If the challenge and the testing are too much for humanity, then we are all doomed. But my own faith is plain and clear. I believe that the power of science and the responsibility of science have offered mankind a new opportunity not only for intellectual growth but for moral discipline—not only for the acquisition of knowledge but for the strengthening of nerve and will." (CR, 10/28/63, 19279–80)

October 22: After a technical lecture on the chemical structure of matter, Dr. Linus Pauling, 1963 Nobel Peace Prize winner, attacked the U.S. lunar program before the National Academy of Sciences meeting in the Department of State auditorium: "I believe that it [the lunar program] is a pitiful demonstration. Something is wrong with our system of values when we plan to spend billions of dollars for national prestige." Pauling claimed that for the same investment it would be possible "to answer 1000 interesting and important questions about the human body for every one question answered about the moon." (Simons, Wash. Post, 10/23/63, 1)

- Boilerplate Apollo capsule underwent eighth drop test of the 3-parachute landing system. At El Centro, Calif., U.S. Naval-Air Force Parachute Facility dropped the capsule from a C-133 transport aircraft from 13,000 ft. in a test of the parachute combination under low-altitude abort conditions. (MSC Release 63-178)
- GFSC announced negotiations with Republic Aviation Corp. for Phase I contract for Advanced Orbiting Solar Observatory (Aoso). Aoso would be launched into a 300-mi. near-polar orbit for observations of x-rays, gamma rays, and ultraviolet emissions of the sun. Phase I calls for one-year development of systems engineering and detailed design of the satellite. (GSFC Release G-22-63)
- NASA awarded $2 million contract to Lockheed Missiles and Space Co. for a vehicle systems test complex to be used for full checkout of Agena target vehicles to be used in Project Gemini rendezvous flights. (Space Bus. Daily, 10/23/63, 133)
- Lewis Research Center tested a newly developed seal for liquid fluorine pumps. This seal eliminates the explosion inherent in previous seals for liquid fluorine transfer systems. (Lewis Chronology, 10)
- U.K.'s newest commercial jet aircraft, the BAC 111, crashed in a field in southern England during a test flight, killing all seven crewmen. The British Aircraft Corp.'s twin-jet, 74-passenger 111 was planned as a replacement for the prop-jet Viscount, bringing jet service on short-haul routes. Prior to the crash, the aircraft had had some 80 hours of flight test. (NYT, 10/23/63, 66)
- Aviation executives criticized Government plans for the supersonic transport in testimony before the Senate Commerce Committee Subcommittee on Aviation. C. R. Smith, president of American Airlines, objected to the provision that industry pay one quarter

of the development costs, wanted Government to pay all the costs and defer decision on how to recapture some of the costs until after the prototype had flown. John Stack, vice president and director of engineering of Republic Aviation Corp. and former NASA Director of Aeronautical Research, complained that competition with the British-French Concorde would force U.S. industry to set up management and assembly lines prematurely. Stack urged that a manufacturer be chosen to begin work on a prototype next spring and that the one-year-long second design phase be eliminated. He also supported the steel-and-titanium mach 3 concept as more efficient than an aluminum mach 2+ aircraft presently favored by FAA. (Clark, *NYT*, 10/23/63, 9)

October 22-25: DOD's Exercise Big Lift successfully airlifted an entire 14,500-man armored division from bases in the U.S. to Germany in less than 72 hours. This was the first large-scale transatlantic test of the C-135 jet transport fleet recently acquired by the Military Air Transport Service (MATS). (DOD Release 1339-63; Olsen, *NYT*, 10/23/63, 1)

October 23: Dr. Walter C. Williams, Deputy Director for Mission Requirements and Flight Operations, MSC, was named NASA's Deputy Associate Administrator for Manned Space Flight Operations. He would supervise manned space flight operations at MSC, MSFC, and LOC. During manned space flight missions he would have full authority and responsibility for conduct of the flights. His appointment would become effective Nov. 1, 1963. (NASA Release 63-237)

- NASA Administrator James E. Webb, speaking before the Michigan Industry-University Research Conference in Detroit, urged a steadily supported space program based on public understanding of the total purpose of the space effort: "If this debate is to serve the best interests of the country, it is essential that every citizen of influence, whether editor, businessman, scientist, professor, government official, or military man, study and analyze all the principal reasons for a well-balanced, fast-paced space effort. The scientist who thinks only of scientific results or the businessman who looks only for economic results or the writer who seeks only for the spectacular or controversial will not be making his most constructive contribution to the process of national decision making . . . ." (Text)

- NASA Administrator James E. Webb, speaking at Iowa Bankers Association 77th Annual Convention in Des Moines, said: "Let me list the principal reasons why our investment in manned space flight is sound:

"1. The current level of effort is vital if we are not to settle for second position in space . . . .

"2. Almost all our advances in space technology will add directly or indirectly to our ability to deter aggression . . . .

"3. The scientific capital we shall acquire in the lunar program will long serve the nation . . . .

"4. The facilities we are building and the technology we are acquiring are tangible assets that will extend benefits to our children and grandchildren . . . .

"5. The dollars that are putting us into space are being spent right here on earth . . . .

"6. We are clearly demonstrating to the world the ability and determination of our democratic society to organize whatever large-scale scientific and industrial effort is required to meet critical national and international needs in time of peace as well as war . . . .

"7. To meet the stringent demands of the space environment . . . engineers are developing new materials . . . certain to pay off dividends, many of which we cannot foresee, some of which may be of revolutionary nature . . . .

"8. We are rapidly building up a large pool of highly trained and creative scientists, engineers, and technicians . . . .

"9. We are currently devoting about one per cent of the country's income to the civilian space effort. If the inducement to hard work and inspiration of participating directly or indirectly in space exploration spurs our people to increase production of goods and services by only one per cent, the space program will pay for itself.

"10. Our swift progress has now brought us to the point where our space power can be employed in peace-making as well as peace-keeping." (Text)

October 23: Ranger 6 launch rescheduled by NASA from last quarter of 1963 to first quarter of 1964. JPL had discovered defects in some of the diodes in the same shipment as 242 diodes used in the assembled lunar spacecraft. Gold used in the bonding process was found to be flaking inside the diode capsules; under space conditions these minute flakes could cause malfunction. The launch was postponed so that the diodes could be replaced. (NASA Release 63–238)

- Dr. Albert C. Hall, DOD's new chief of space development, cited the Saint and Midas programs as representative of large-scale R&D efforts that only proved "the state-of-the-art was not equal to the task," according to *Space Business Daily*. Midas, he said, was a system "that could not compete with contemporary systems on a cost effectiveness basis." (*Space Bus. Daily*, 10/24/63, 137)
- New camera tracking theodolite telescope developed by the U.S. Army was undergoing final test evaluation at White Sands Missile Range, N.M. Called GORID (Ground Optical Recorder for Intercept Determination), the telescope would track and film hypersonic, high-altitude missiles while simultaneously recording missile flight data on magnetic tape in computer format. (DOD Release 1396–63)
- Army Pershing missile exploded shortly after launch at Ft. Wingate, N.M. (AP, *Wash. Post*, 10/24/63)
- Second wing of 150 Minuteman ICBM's was activated and turned over to Strategic Air Command. This wing was at Ellsworth AFB, S.D. First Minuteman wing was activated at Malstrom AFB, Mont., in July 1963. (DOD Release 1413–63)

October 24: NASA press conference in Washington on results of EXPLORER XVI meteoroid hazard experiments. Principal conclusion expressed by Dr. Fred L. Whipple, Director of the Smithsonian Astrophysical Observatory, was that in its 7½ months of useful lifetime EXPLORER XVI demonstrated that most of the meteoroids in space were bits of "fluff" in existence since the formation of the solar system some 4.6 billion years ago and not, as feared, small rocks that would cause major impact damage to satellites. While the rock type meteoroids do exist, they are so rare as to make collision chances very remote. (NASA Release; Myler, *Wash. Post*, 10/25/63)

October 24: Dr. Jerome B. Wiesner, Director of White House Office of Science and Technology, testified before the House Subcommittee on Science, Research and Development on the NASA-DOD agreement on joint study and recommendations on National Orbital Space Station (NOSS): "We have finally made a good start. Now, we are going in the right direction." He also said that much of NASA's R&D was work that DOD would have to do for itself if not done by NASA, that DOD's R&D budget was leveling off and would continue to do so because the U.S. was no longer behind U.S.S.R. in weapon systems such as the ICBM. (*Space Bus. Daily*, 10/25/63, 145)

- NASA announced it would negotiate with Linde Co. and Air Products and Chemicals, Inc., for final selection of the contractor to supply liquid hydrogen to the NASA test sites at Mississippi Test Operations and MSFC for the Saturn launch vehicle test program. Supplier would have to have the capability of producing as much as 115 million pounds of liquid hydrogen from April 1965–Dec. 1970. (NASA Release 63–239)

- Vice President Lyndon B. Johnson defended the space program as a major bulwark of U.S. posture. In a speech to the Massachusetts Associated Industries in Boston, the Vice President said: "If we want to maintain the credibility of our claim to the superiority of a free political system—and a free private enterprise system— we cannot seriously entertain the thought of precipitating now so massive a disillusionment as would follow a political default on our commitments in space exploration." (AP, *Wash. Post*, 10/25/63)

- DOD confirmed that Australia would buy 24 TFX aircraft from the U.S., with delivery beginning in 1967. In the meantime, the U.S. would make available two squadrons of B-47 aircraft to replace Australia's obsolete Canberra bombers. (DOD Release 1415–63)

- DOD officials see very few applications of laser for military purposes in the near future, according to *Space Business Daily*. Use of laser within the earth's atmosphere would be hampered by absorption and diffusion of many frequencies even though the light is coherent. Greatest use seems to be in space, although size of power source may be a problem here. NASA was said to be taking more interest in laser for space applications, perhaps raising R&D investment to $7 or 8 million per year. (*Space Bus. Daily*, 10/25/63, 147)

October 25: DOD launched two unidentified satellites on one Atlas-Agena D launch vehicle (*Pres. Rpt. on Space, 1963*, 1/27/64)

October 25: The Air Force accepted the first Gemini Launch Vehicle from the Martin Company. The booster, a specially modified Titan II, would be used in the NASA Gemini program. (*A–N–AF Journal and Register*, Jan. 1964; AFSC Release 312-114, Atch. 2, 1/1/64, 2)

- Weather Bureau and U.S. Navy announced first tentative results from Project Stormfury, the cloud-seeding exeriments performed on hurricanes to determine whether the energy patterns of the storm can be changed. Hurricane Beulah was seeded with silver iodide particles on Aug. 23 and 24, 1963. The Aug. 23 seeding apparently missed the eye of the storm; winds continued to increase. The Aug. 24 seeding did enter the eye of the storm; the eye appeared to degenerate and reform in a wider and less violent circle. Scientists cautioned that such an effect might have occurred even without seeding and that several more hurricanes would have to be seeded before any firm conclusions could be drawn. (WB Release 63-13)
- Lewis announced final details of contract to General Dynamics Astronautics for construction of second Centaur launch pad at Cape Canaveral. Pad would give Centaur a capability of one launch a month after completion in late 1964. (LRC Release 63-88, Lewis Chronology, 10)
- Albert J. Evans was appointed Acting Director of Aeronautical Research, NASA Hq., following the resignation of Charles H. Zimmerman, who left NASA to become Chief Engineer, Army Materiel Command. (NASA Announcement 63-232)
- Secretary of Defense Robert S. McNamara ordered the Navy to proceed with the construction of the conventional-powered aircraft carrier (CVA 67) authorized by Congress in FY 1963. Navy had delayed construction to plead for nuclear propulsion in the carrier; this would have cost an additional $180 million. (DOD Release 1424-63)
- House Post Office and Civil Service Committee reported 1,248 computer systems were in use in the Federal government. The Committee attributed the computers with holding down Federal payrolls. (UPI, *NYT*, 10/26/63, 7)

October 26: Premier Nikita Khrushchev, in an interview with Communist newsmen published in *Izvestia*, said: "At the present time we do not plan flights of cosmonauts to the moon.

"I have read a report that the Americans wish to land a man on the moon by 1970. Well, let's wish them success.

"And we will see how they will fly there, and how they will land there or to be more correct 'moon' there.

"And most important—how they will get up and come back.

"We will take their experience into account.

"We do not wish to compete in sending people to the moon without thorough preparation.

"It is obvious there would be no benefit from such competition.

"On the contrary it would do harm since it would lead to the destruction of people.

"It is a joke in our country to say 'Who is impatient on earth, let him go to the moon.'

"For us it is good enough on earth. But if we are to talk seriously we will have to work a lot and prepare well in order to complete a successful flight of man to the moon." (UPI, *NYT*, 10/27/63, 1)

October 26: NASA Administrator James E. Webb, asked for his reaction to Premier Khrushchev's statement that U.S.S.R. was not competing to land a man on the moon, said it was "useful to have an authoritative statement" on Soviet plans but that "we will have to think a long time before modifying our program.

"We will not want to change the status of our program until we determine precisely what the statement means and what is in our national interest." (*NYT*, 10/27/63, 12)

- Administration reactions to Premier Khrushchev's statement that Russia would not attempt to land a man on the moon indicated strong reservations on Russian motivations. Dr. Edward C. Welsh, Executive Secretary of the National Aeronautics and Space Council, said the Russian withdrawal announcement was "a very wise pronouncement. Whether they planned to go to the moon or not, they have sensed some slackening of support for the moon program here and have made a strategic move to capitalize on this situation." Another official said: "All during the moratorium on nuclear testing the Soviets secretly prepared for an intensive series of tests. How can we be certain that they won't do the same thing in space?" (*Wash. Post*, 10/27/63)
- The Rosman, N.C., tracking and data acquisition facility was dedicated by NASA. A key station in NASA's Satellite Tracking and Data Acquisition Network (STADAN), the 85-ft.-diameter parabolic antenna at Rosman would be used to track and receive the large flow of telemetered data from the large orbiting observatories and would relay the data by telephone and radio to GSFC for processing and analysis. (NASA Release 63–240)
- AEC announced an underground nuclear test of 12 kt had taken place at Fallon, Nevada, some 1,200 ft. underground. Part of Project Shoal, the test was made in conjunction with ARPA in the effort to improve instrumentation for distinguishing between underground nuclear tests and natural earthquakes. (AEC–DOD Release F–221)
- Polaris A–3 missile was launched from a submerged submarine for the first time. The submarine U.S.S. *Andrew Jackson* was cruising submerged about 30 miles off Cape Canaveral; the Polaris flew successfully down the AMR. (DOD Release 1428–63)

October 27: A roundup of Congressional opinion on Premier Khrushchev's statement of Oct. 26 that the Soviets did not plan to land a man on the moon and its possible effect on the U.S. manned lunar landing program indicated that most Congressmen thought it would result in some reduction in funds for the U.S. program. (*Wash. Post*, 10/28/63)

- Editorial in *New York Times* commenting on rising importance of National Academy of Sciences: "The President [of the U.S.] hinted at but did not elaborate on the need for greater teamwork between the natural and social sciences. The discoveries of the physical scientists create needs for new institutions, as the President noted, but the efficacy of these new institutions depends in large part upon utilizing the insights and knowledge of the social scientists. As the natural scientists themselves participate now more often in the shaping of public policy, they must feel increasingly the need for guidance from the economist, the sociologist, the political scientist, the historian.

"Sir Charles Snow has recently stressed the importance of social scientists in meeting the human problems caused by the scientific revolution. But that importance is still too little recognized. As it begins its second century the National Academy of Sciences could help upgrade the social sciences by opening its membership to social scientists, thus helping to diminish the present invidious distinction between talented workers in these two vital areas." (*NYT*, 10/27/63)

October 28: NASA Administrator James E. Webb, speaking to the Chamber of Commerce, Billings, Mont., noted that President Jefferson's Louisiana Purchase in 1803, which brought much of the state of Montana within the boundaries of the U.S., was criticized at the time as a "prodigal waste of 'the public monies.'" Similarly the space program had been criticized. But Mr. Webb predicted that "the decisions for a dynamic national space program made by the Congress, by President Eisenhower, and by President Kennedy during the past five years will earn a place in history that in significance will go far beyond Jefferson's Louisiana Purchase and many of the other major decisions which have characterized turning points in our national history." (Text)

- NASA's Langley Research Center asked 36 companies for proposals on Project Scanner, a spacecraft that would establish the radiation characteristics of the Earth's horizon and develop horizon-scanning techniques. Program would involve two flight models complete with re-entry heat shield, plus one backup. (*Space Bus. Daily*, 10/29/63, 165)

- Final approval for contracting the military Medium Altitude Communications Satellite (MACS) was being delayed by Secretary of Defense Robert S. McNamara pending a reply from ComSatCorp as to whether a civilian comsat could handle military traffic under wartime conditions, according to *Aviation.Week*. It was anticipated that the answer would be that several of the military requirements—jam-resistant channels, protection against attack, transportable receiving stations—would not be met by the civilian system. (*Av. Wk.*, 10/28/63, 25)

- Development of the Apollo spacesuit and backpack was entering the manned test phase of the prototypes. Under a 3-phase program, Hamilton-Standard Div. of United Aircraft Corp. and International Latex Corp. were now well into the second phase. The first phase had been design; the second called for development of prototypes of three suit designs, each building on test data from its predecessors. Prototypes of first suit were finished in August; prototypes of the second suit were nearing completion. The third phase, to be concurrent with the second phase, would include environmental and physiological tests. (*Av. Wk.*, 10/28/63, 49)

October 28: Rep. James Weaver (R.-Pa.) spoke on the space race following Premier Khrushchev's statement that Russia was not sending a man to the moon: "This 'moondoggle' has been a wrong objective in a race with the Soviet Union that never existed and has resulted in a space gap which threatens our national security.

"Therefore, it becomes imperative that we immediately realine [sic] our space objectives to national security and eliminate this gap. I call upon James E. Webb, Administrator of the National Aeronautics and Space Administration, to carry out this objective. If he cannot immediately reevaluate and report frankly and honestly to this Congress that this has been accomplished, then I feel it is our responsibility to demand his resignation. The Congress and the public can no longer tolerate public relations gimmicks and double-talk concerning the space program and our space gap when our national security is threatened." (*CR*, 10/28/63, 19365)

October 28-29: Ames Research Center held regional contest of highschool scientists, with the three winners to compete in the National Science Youth Congress finals in Washington, D.C. National Science Youth Congress is sponsored by NASA and National Science Teachers Association. (Ames Release 63-42)

October 28-30: Symposium on the Physics of Solar Flares was held at GSFC, sponsored by NASA and the American Astronomical Society. (GSFC Hist.)

October 29: X-15 No. 1 was flown by NASA's Milton O. Thompson for the first time. Flight reached maximum speed of 2,712 mph (mach 4.10) and 74,400-ft. altitude. Thompson was the ninth pilot to fly the X-15 and the only civilian among the potential pilots of the X-20 (Dyna-Soar). (FRC Release 26-63; UPI, *Chic. Trib.*, 10/30/63; Martin, NASA Hq.)

- X-15 No. 2, damaged in a landing accident in Nov. 1962, would be returned to flight status early in 1964 with a number of improvements, NASA announced. Under a USAF contract with North American Aviation, Inc., the damaged X-15 was being modified to take another 13,500 lbs. of propellants. This would increase the rocket engine burning time from 86 sec. to 145 sec., which theoretically would add some 1,300 mph to maximum speed capability. There were no plans for extending altitude capability because it would increase re-entry stresses. Other modifications included strengthened nose wheel and landing skids, dropable wing tanks, provision for installation of two 50-gal. liquid-hydrogen tanks in the fuselage midsection, to be used for advanced propulsion experiments, and windshield changes and ablative material additions to cope with the higher heat accompanying higher speed. (FRC Release 27-63)

- Speaking at the dedication of the Douglas Space Center in Huntington Beach, Calif., Vice President Lyndon B. Johnson emphasized the peaceful purpose of the U.S. in space research and compared our policy of cooperation in space to the secretive Soviet program:

"Those who say—as some are inclined to do—that our purposes and the purposes of communism in space exploration are the same misread and misunderstand the history and meaning of our times. In 1957 when the Soviets placed the first sputnik in orbit, the Communist rulers of Russia refused to consider sharing the fruits of space research with other nations—refused to consider committing themselves to developing space for peaceful purposes alone. In that same year, we of the United States clearly stated our own national policy and purpose. We committed and dedicated ourselves to sharing the fruits of space research with all mankind—and to the sole objective of developing the uses of space for peaceful purposes.

"It is significant that while the Communist efforts are conducted behind a curtain of secrecy, we of the United States enjoy the cooperation, support and alliance of more than 60 nations for our space program. It is significant that while we of the United States can and do propose joint endeavors for peaceful purposes, the Soviet both rejects our proposals and devotes itself to attempts to dissuade us from continuing on the programs we have established.

"America's commitment to the exploration of space for peaceful purposes—and for the good of all mankind—is a firm commitment. We will not retreat from our national purpose. We will not be turned aside in our national effort by the transparent maneuvers of those who would attempt to divert us.

"Our national purpose in space is peace—not prestige.

"Our foremost objective is not to send a man to the moon but to bring a greater measure of sunlight into the lives of men on earth." (CR, 11/19/63, A7163)

October 29: Rep. J. Edward Roush (D.-Ind.) spoke on the space race: "Mr. Speaker, the announcement that the Soviet Union is withdrawing from the race to the moon has produced some disturbing reactions here in the United States. Here we have a breakdown in the Communist system. Here we have an opportunity to exert real leadership and show what a free economy can do. And we have those who are now advocating that we abandon our efforts to explore the universe. I am perturbed. Why must we do something just because the Soviet Union does it? Why should we refuse to do something just because the Soviet Union refuses to do it?" (CR, 10/29/63, 19384)

- Addressing the Joint Meeting of Service Clubs, Bozeman, Montana, NASA Administrator James E. Webb said:

  "To take a clear lead in rocket power, or even to match the larger rockets the Russians may be building now, we must carry on with the development of the Saturn V. This giant rocket is needed to do extensive maneuvering in space near the earth and has the power to send the Apollo spacecraft on its journey to the moon and on prolonged flights in earth orbit.

  "The United States will inevitably need space power of this order even if we had no plans for going to the moon. We do not know now what future military requirements for rocket power may arise." (Text)

- Project Gemini was running into serious problems in its onboard power systems, according to John Finney of the *New York Times*. The new fuel cell source for electric power was to have provided power for up to two weeks at only 1/6 to 1/10 the weight of batteries required to produce equal wattage. But fuel cell development has run into problems of leakage, inadequate lifetime, and poor thermal control. While such problems could probably be engineered out by the time fuel cells were needed for Apollo, the Gemini timetable might require batteries to be substituted for the fuel cells. This in turn might cut Gemini missions down to two days. NASA's new Associate Administrator for Manned Space Flight Dr. George E. Mueller was said to favor beginning development of backup systems in some of the areas of new technology such as fuel cells but to be hampered in this approach by lack of money. (Finney, *NYT*, 10/30/63, 26)

October 29: FAA Administrator Najeeb E. Halaby and his Deputy Administrator Gordon Bain testified on plans for supersonic transport before Senate Commerce Committee Subcommittee on Aviation. They said the U.S. would cancel its development of the supersonic transport if events proved the U.S. could not produce an aircraft competitive in price with the British-French version. But they argued that estimates previously presented to the Subcommittee on costs of the aircraft were too high. Aviation spokesman had estimated cost of the U.S. aircraft at from $30 million to $40 million per aircraft, compared with the $10 to $12 million estimated for the heavily subsidized European aircraft. FAA said U.S. aircraft should not cost more than $20 million. (AP, *NYT*, 10/30/63, 26)

* USAF launched two unidentified satellites from Vandenberg AFB, Calif., on a Thrust-Augmented Thor-Agena D booster. (UPI, *Wash. Post*, 10/30/63; *Pres. Rpt. on Space, 1963*, 1/27/64)
* Atlas missile was launched from Cape Canaveral in test of new slim-silhouette warhead intended to present less image on enemy radar screens. Atlas went out of control 2½ min. after launch and fell into the Atlantic far short of its intended goal. This was the sixth successive failure of Atlas missiles, the previous five at Vandenberg AFB, Calif. (AP, Wash. *Eve. Star*, 10/29/63)
* USAF awarded contract to Goodyear Aerospace to study and develop a post-Echo passive communications satellite. Ten ft. in diameter, the satellites would be a wire grid covered with a plastic film. Once inflation in orbit had occurred, the plastic film would dissipate under ultraviolet radiation from the sun, and the circular wire grid would be left. It was thought that greater reflectivity could be achieved with the wire grid than with solid-skin spheres such as Echo and that they would have a more stable orbit since they would not be as subject to the solar wind. (*Space Bus. Daily*, 10/30/63, 172)

October 30: NASA announced it was dropping four manned Apollo orbital flights employing the Saturn I booster and accelerating the all-systems manned Apollo flights employing the Saturn I-B booster. Saturn I would have been able to orbit only the command and service modules of Apollo. Saturn I-B can launch the entire system, including the Lunar Excursion Module (LEM). Studies have concluded that running all-systems checks from the beginning is a quicker and more certain method of checkout than to add additional systems flight by flight. Also the deletion of the four flights would save $50 million in FY 1964. The changes would delay the first manned Apollo flights by some nine months. (NASA Release 63-246; *NYT*, 10/31/63, 18; *Wash. Post*, 10/31/63, 5)

October 30: Thrust-Augmented Thor space booster developed by the USAF would be added to the National Launch Vehicle program as booster for Agena and Delta vehicles, DOD and NASA announced. TAT would have three XM-33 solid-propellant rockets, each 31-in. in diameter and producing 54,000 lbs. thrust, strapped to the sides of the basic Thor vehicle. This is the same rocket developed by Thiokol Chemical Corp. as the second stage of NASA's Scout booster. The three solid rockets would be fired at the same time as the Thor engine, would burn for about 40 sec., and then drop off as the Thor engine continued powered flight. Raising Thor's total thrust to 330,000 lbs., TAT would represent a 20 to 30% increase in capability over the Delta, putting 1,000 lbs. into earth orbit compared with 800 lbs. for Delta. (DOD Release 1434-63)

- Dr. Edward C. Welsh, Executive Secretary of the National Aeronautics and Space Council, speaking before the Air Transport Association Engineering and Maintenance Conference in Washington, said: "While much has been said about the Congressional cuts in the NASA budget request, it is noteworthy that the debate was not on cutting back the program. Rather, it was on how much should the increase be. This is, I believe, the key to the public outlook on space." (Text)
- NASA Administrator James E. Webb, speaking at Montana State University, Missoula, Mont., listed six reasons for making Project Apollo the major manned space flight effort of the 1960's:

    "1) The goal of lunar exploration is feasible from an engineering standpoint . . . .

    "2) By planning ahead, and adopting a realistic schedule, we avoid the wastefulness of indecision and stop-and-go financing.

    "3) We have set ourselves a clear goal which the entire world can understand, and one in which we have a good chance of being first . . . .

    "4) We have set a goal which will focus our efforts, and at the same time enable us to build up a broad base of space power . . . .

    "5) In Project Apollo we protect ourselves against the great psychological advantage the Soviet Union would have if it alone could occupy and use the moon . . . .

    "6) Project Apollo gives us our first good chance of overcoming the lead in manned space exploration which the Soviets now hold." (Text)
- Sen. A. S. Monroney (D.-Okla.), Chairman of the Subcommittee on Aviation of the Senate Commerce Committee, speaking at the Air Transport Association Engineering and Maintenance Conference in Washington, said that the Government's plan of cost sharing for the development of the supersonic transport aircraft ($1 billion, the U.S. putting up $3/4$, the aviation industry the other $1/4$) was "a firm figure that must be considered final . . . .

"I am afraid that many of our finest businessmen still hope to consider this project in the nature of a defense contract. This effort must stand alone as a civil aviation need . . . ." The Senator did concede that if the market turned out to be for 100 aircraft instead of 200, the Government "might be able to give the companies some bail-out on this." (*NYT*, 10/31/63, 58)

October 30: In column strongly supporting the U.S. space program, Roscoe Drummond stated that the United States "should not make the mistake of allowing Nikita Khrushchev to determine the size of the U.S. space program. This is a pitfall to be avoided at all costs . . . .

"Naturally we must leave it to the concensus of the scientists and the technicians to decide whether the manned flight to the moon is the 'best way and the urgent next step to insure preeminence in outer space. The controlling and compelling objective is not a manned lunar landing on the moon for its own sake, but the achievement of mastery of the element of outer space for all that it will mean to the kind of earth we will inhabit for a long time to come." (*Wash. Post*, 10/30/63)

- Editorial in *The Houston Post:*

"Why Khrushchev's word should be accepted on this [commenting that Russia does not have a definite program for putting a man on the moon] any more than on anything else is hard to understand . . . . It . . . must be kept in mind that Soviet policy in any area is a zig-zag affair and subject to quick change. Moreover, under the Soviet system, Khrushchev does not have to worry about congressional attitudes or reaction if he should decide upon a change . . . .

". . . why should Khrushchev be permitted to set the policy on this country's space program? That is something for the Legislative and Executive branches of the United States government to do on the basis of careful study and evaluation.

"The fact is that the United States government made its decision several years ago. It stands committed to a program of trying to put a man on the moon in this decade. What the Russians do or don't do should be a matter of secondary importance at this point in the program. Much more is involved than merely a contest with Russians as to time." (Text)

- A new program was being established to encourage the development of skilled writers who would explain science to the layman. The Richard Prentice Ettinger Program for Creative Writing would have three participating institutions—The Rockefeller Institute, New York Univ., and Univ. of Pa., and would be headed by Dr. Loren C. Eiseley, on leave from his post as Chairman of the Department of the History and Philosophy of Science at the Univ. of Pa. Beginning in 1964, it would award fellowships for the purpose of enabling the fellows to produce high-quality books and articles on science free from the pressure of earning a living. It also awards a Richard Prentice Ettinger Medal, with $1,000, to writers or others who have fostered understanding of science. (*NYT*, 10/30/63, 40)

October 31: In Presidential news conference, the question was asked whether the President construed Premier Khrushchev's statement of Oct. 26 as taking Soviet Union out of the moon race. The President replied: "I didn't read that into his statement. I thought his statement was rather cautiously worded . . . .

"I think it is remarkable that some people who were so unwilling to accept our test ban treaty, where there was a very adequate area of verification of whatever the Soviet Union was doing, were perfectly ready to accept Mr. Khrushchev's very guarded and cautious remark that he was taking himself out of the space race, and use that as an excuse for us to abandon our efforts.

"The fact of the matter is that the Soviets have made an intensive effort in space, and there is every indication that they are continuing and that they have the potential to continue. I would read Mr. Khrushchev's remarks very carefully. I think that he said before anyone went to the moon there should be adequate preparation. We agree with that.

"In my opinion, the space program we have is essential to the security of the United States, because, as I have said many times before, it is not a question of going to the moon. It is a question of having the competence to master this environment. I would not make any bets on Soviet intentions." (Transscript, *Wash. Post*, 11/1/63)

- Addressing the Southwest Conference on Arms Control at the University of Oklahoma, William C. Foster, Director, U.S. Arms Control and Disarmament Agency, discussed the complex relationship between national strategy, military security, and arms control in a thermonuclear world.

  " . . . certain types of weapons which could be developed in the future might be extremely difficult to make comparably safe. I am thinking in particular of weapons of mass destruction which might be stationed in outer space. Present analysis indicates that such weapons would be both more expensive and less effective than conventional ICBM delivery systems. Moreover, once placed in space such weapons would constitute a permanent risk. There would always be a possibility that normal mechanical failure, collision with a meteorite, or interaction with unforeseen solar radiations might trigger it by accident.

  "In our agency we have long believed that it would be desirable if we could reach an understanding that such weapons would not be placed in space. Without such an understanding we believe there would be a greater risk that an arms race in space might develop simply for prestige reasons rather than because of any real military utility." (*This Changing World*, 1/1/64)

- A second Aerobee-Hi research rocket in the NASA-French joint program investigating propagation of very low frequency waves in the ionosphere was launched from Wallops Station. The 193-lb. payload went to an altitude of 115 mi. and yielded seven minutes of telemetry data before impact. The first experiment in this series was conducted on Oct. 17, 1963. (Wallops Release 63-98)

October 31: NASA launched a Nike-Apache sounding rocket from Wallops Station. Payload was a 77-lb. package sent to 84-mi. altitude to measure mechanical and thermal stresses on a scientific payload during launch. (Wallops Release 63-99)

- NASA Director of Plans and Program Evaluation Abraham Hyatt resigned, effective this date. His office would be absorbed on Nov. 1 into the office of the Assistant Administrator for Technology Utilization and Policy Planning as part of the Headquarters reorganization. (*Space Bus. Daily*, 10/17/63, 100)
- Development testing of the supersonic 500-mile-range Hound Dog at low altitudes was successfully concluded with a launch over the Eglin Test Range, Fla. (AFSC Release 313-R-114, 1/1/64, 4)
- Three ruby-colored spots appeared on the surface of the moon on the night of Oct. 29, 1963, and might have been volcanic eruptions, Dr. John Hall, Director of Lowell Observatory, Flagstaff, Ariz., announced. Hall said that James Greenacre found the spots while studying the moon through a 24-in. refracting telescope. The spots stood out against the yellow-white surface of the moon. Greenacre said that in three years of studying the moon's surface he had never seen spots similar to these, although a Russian astronomer had reported sighting a similar phenomenon in Nov. 1958 or 1959. (UPI, *San Diego Union*, 11/1/63; *Christian Science Monitor*, 11/2/63)
- Gordon Bain, FAA Deputy Administrator for Supersonic Transport Development, issued statement rejecting Oct. 1 proposal by Continental Airlines President Robert F. Six that U.S. join with U.K. and France in supersonic transport development. Regarding U.S. participation in U.K.-French Concorde development, Bain said that "at the present stage in the development program of the Concorde, there is little, if anything, that United States manufacturers could contribute, except possibly money," and monetary contribution would aggravate U.S. balance-of-payments problem. Regarding U.K.-French-U.S. joint development of mach 3 airliner, Bain said: "At no time have the British-French indicated any interest in such a project. (*Wash. Post*, 10/4/63)

During October: Centaur AC-2 vehicle completed its last two major tests prior to launch. At Lewis Research Center the new stage separation system underwent successful test. At General Dynamics/Astronautics Sycamore Canyon test site in Calif., two RL-10 engines underwent full duration, full power test (380 sec. at 30,000 lbs. thrust). (*Av. Wk.*, 10/14/63, 32)

- Tests continued at Pacific Missile Range to select and qualify one of five sounding rockets for use in more than 200 solar experiments planned for the forthcoming IQSY. Rockets being tested were Canadian Bristol Aerojet, Ltd., Black Brant III; Douglas Aircraft Co. DAC-Roc; British Aircraft Corp. Thunderbird; the Seagull; and Atlantic Research Corp. Archer. (*M&R*, 10/28/63, 16)
- NASA had decided not to begin development of the Multi-Mission Module (MMM) at this time. The module would use two RL-10

engines with tanks configured for different missions. Serving as the third stage of the Saturn I-B or the fourth stage of the Saturn V, it would give the big boosters the flexibility needed for planetary and other missions. Present decision was based on the fact that the stage is not necessary for the Apollo program. (*Av. Wk.*, 10/14/63, 23)

During October: Project Stabilization Agreement was ratified for NASA Mississippi Test Operations facilities by representatives of business, labor, and Government. Worked out by the President's Missile Sites Labor Commission, the unusual agreement established a standard for working conditions, hours, hiring practices, grievance procedures, etc., in return for the promise of no strikes or work stoppages. Agreement would run to July 1, 1966. (*Space Bus. Daily*, 10/11/63, 65)

- NASA was about to enter the prolonged-flight phase of its bioscience program, according to Dr. Homer E. Newell, NASA Director of Space Sciences. He cited negotiations now underway with GE to provide six 1,000-lb. recoverable biosatellites. These would be used for such flights as three-to-five day flights to study effects of weightlessness and radiation on plants and animals; 21-day orbital flights to study effects of weightlessness and absence of earth's rotation on biological rhythms in plants and animals; and 14-to-30 day flights by small monkeys to study effects of prolonged weightlessness on the cardiovascular system, central nervous system, and general physiology and behavior. All studies would have importance for prolonged manned space flight. (Macomber, *San Diego Union*, 10/12/63)

- USAF was reported by *Aviation Week and Space Technology* to be planning an expansion of its ballistic missile early warning system (BMEWS) to cover the southern approaches to the U.S. and possibly also to detect launches from submarines. Named Project Red Mill, the proposed system being developed by Raytheon Co. under USAF contract would use high-frequency ionospheric radar bounce techniques to detect changes in the ionospheric critical frequency made by rocket exhaust. (*Av. Wk.*, 10/14/63, 23)

- *Harvard Business Review* conducted poll of corporation executives on the space program, found that seven out of ten believe U.S.S.R. is ahead in race to send a man to the moon, six out of ten that the U.S.S.R. will beat the U.S. to the moon by a year and a half; only four out of ten would speed up the space program. (*Harvard Business Review*, Sept.–Oct. 1963)

- Dr. Henry J. E. Reid, who was Director of NACA's Langley Research Center for 34 years prior to his retirement, was named "Elder Statesman of Aviation" by the National Aeronautic Association. (Langley Release)

- AFOAR proposed a five-year plan for its support of basic research, including a 40% increase in support of geophysics and a 30% increase for the other physical sciences. (*M&R*. 10/21/63, 21)

- U.S.S.R. MARS I probe lost contact with tracking stations when 65 million miles out because of malfunction in the spacecraft attitude control system, according to Alexandre A. Mironov, Soviet Embassy scientific adviser in Washington. (*Av. Wk.*, 10/21/63, 28)

During October: H. H. Koelle, Director of Future Projects Office, MSFC, in "Trends in Earth-to-Orbit Transportation Systems," in *Aeronautics and Aerospace Engineering*, noted that the next year would be interesting in that both the U.S. and U.S.S.R. were expected to come up with big increase in payload in orbit. The U.S. Saturn I was expected to put 10 tons in orbit. "Also, I would not be a bit surprised if the U.S.S.R. launched a payload in the 40- to 50-ton class before this [next] year is over. The next step increase will be provided by the Saturn V in about 1967, when a 100-ton capability will be reached by the U.S. Many people hope that at least by that time the U.S.S.R. single-flight capabilities will have been equalled or exceeded."

Space transportation was developing fast, Koelle concluded: "While air transportation needed two generations to improve the efficiency of passenger transport by three orders of magnitude, we expect to improve the economy of Earth-to-orbit cargo transportation by four orders of magnitude in one generation." (*A&AE*, Nov. 63, 25–30)

NOVEMBER 1963

November 1: U.S.S.R. announced launching of POLET I (Flight I), a new type of maneuverable spacecraft for use in manned orbital rendezvous flight. Initial orbital data were: apogee, 592 km. (368 mi.); perigee, 339 km. (211 mi.). After what were described as "repeated" changes in altitude and inclination, the spacecraft on Nov. 2 attained "final orbit": apogee, 1,437 km. (893 mi.); perigee, 343 km. (213 mi.); period, 102.5 min.; inclination, 58°55'. (*Pravda*, 11/2/63, AFSS-Trans.; Shabad, *NYT*, 11/2/63, 1; 11/3/63, 33)

- Premier Nikita Khrushchev announced the launching of POLET I at a Moscow reception in honor of Laotian Prince Souvanna Phouma: ". . . the present spaceship is really new. While the previous ships placed into orbit made flights mainly in the direction imparted to them when they were launched from Earth, the spacecraft that was lofted today is making wide maneuvers in space, varying the orbital plane and altitude.

  "The fact that we have launched such a ship bears testimony that human ingenuity has reached a higher stage. Now man in space is no longer a prisoner of his ship. He controls it and guides its flight. The spacecraft has become ever more responsive to man's will."

  Stating that the U.S.S.R. had been given serious attention to President Kennedy's proposal for a joint lunar landing program, Premier Khrushchev added: "What could be better than to send a Russian and an American to the moon together, or better yet, a Russian man and an American woman?

  "If we could agree on further easing of tension, not just in moral but in concrete terms such as disarmament, this would give greater means, namely international means, to the development of science." (*Pravda*, 11/2/63, AFSS-Trans.; Shabad, *NYT*, 11/2/63, 1, 9)

- U.S. reaction to U.S.S.R.'s launching of POLET I and Premier Khrushchev's subsequent expression of interest in U.S.-U.S.S.R. cooperation of manned space flight was summed up by NASA Administrator James E. Webb, who pointed out "the wisdom of President Kennedy's remarks on Wednesday that there is every indication that the Soviet Union is proceeding with a vigorous space program.

  "It also points out how important it is for us to pursue our broad-based program as assurance against surprise."

  U.S. space experts likened POLET I to the U.S. Gemini spacecraft, scheduled to make its first unmanned flight in late 1963 or early 1964. (Simons, *Wash. Post*, 11/2/63)

November 1: GSFC awarded contract to Yale Univ. to design and develop a worldwide radio monitoring network for study of planet Jupiter. Four stations would comprise the global network, located at approximately every 90° longitude around the earth—one at GSFC in Greenbelt, Md., and the other three at U.S. satellite tracking stations in Hartesbeesthoek, South Africa; Carnarvon, Australia; and South Point, Hawaii. Primary duty of the stations would be to maintain a 24-hr. radio monitor of the mysterious low-frequency radio noises sporadically emitted from the planet. The data should provide information on Jupiter's magnetosphere, the interplanetary medium, and the earth's ionosphere. (GSFC Release G-25-63)

- Major reorganization of NASA Hq. became effective (see Oct. 9 for details), consolidating the four major program offices into three and delineating and elevating certain staff functions. (NASA Releases 63-225, 242)
- MSC announced a contract amendment to General Dynamics/Convair in the amount of $2,247,174 for two additional Little Joe II solid boosters for use in the Apollo test program. The two new boosters would be used in high-altitude abort test—around 60,000 ft.—testing the capability of the launch escape system to separate the command module from the booster. This amendment would bring total Little Joe II procurement costs to $8,928,637, including the four vehicles originally ordered and nearly $500,000 for design, development, and installation of two launchers at the White Sands Missile Range, N.M. (MSC Release 63-223)
- Dr. Edward C. Welsh, Executive Secretary of the National Aeronautics and Space Council, spoke before the Society of Experimental Test Pilots in Washington: ". . . it occurs to me that perhaps some of our best known astronauts took greater risks, with less thoroughly tested equipment, when they were flying aircraft in the atmosphere than when they went into space . . . . I simply want to mention that our astronauts, just as all test pilots, had done many courageous things before the Space Age descended upon us and too few were recognized as heroes for those feats." (Text)
- In address at a space exploration symposium at New York University, Dr. Eugene M. Emme, NASA Historian, reviewed the birth of the space age and its first half decade. Referring to the impact of space exploration upon society, he said:

  "Space exploration—with all of its novelty and drama and future potential for society—continues to jar the intellect, stir the emotions, and stimulate actions among peoples everywhere . . . . Our children, of course, have no doubt whatever that man will soon set foot on the surface of the moon, and then the nearby planets; it is only a question of exactly how soon. The rest of us old folks, conditioned as we are to the scientific and technological realities of the recent past, have a little difficulty comprehending either the reality or the significance concerning the mobility of mankind in space today and tomorrow . . . .

  "To the historian, the closing of the present kindergarten era of astronautics, with its tender philosophy, offers the germ for a new renaissance in the mind and spirit of mankind. It could well be a renaissance for mankind as was sparked with the new geography of Columbus and Magellan and the new astronomy of Copernicus and Galileo which helped loosen Europe from the Dark Ages; or when the new biology of Darwin for the physical organism, the challenges of Marx and the new psychology of Bergson and Freud for the conscious man assisted the great intellectual stimulus of the late nineteenth century as well as the humanitarianism and the technological boom of the twentieth century . . . ." (Text; *Airpower Historian*, 1/64, 6–10)

November 1: Arecibo Ionospheric Observatory was dedicated by ARPA and USAF in Arecibo, Puerto Rico. Largest radar-radio telescope of its kind in the world, AIO is a 1,000-ft.-diameter bowl constructed in a natural bowl formed by several mountain peaks. As a radiotelescope, it should detect radiation from galactic sources considerably more distant than any yet detected. As a radar, it would be the most sensitive instrument yet built for ionospheric research. Dr. Thomas Gold, head of the astronomy department of Cornell Univ., which built and would operate the new facility, predicted that it would help make clear "the grand architecture of the Universe." (DOD Release 1358–63; Schmeck, *NYT*, 11/2/63, 9)

- Titan II was launched 5,700 mi. down AMR in test of vibration levels prior to its employment as a booster for Gemini manned space flights. Also along on the flight was a pickaback capsule of instruments to study the exhaust plume of the missile. (UPI, *NYT*, 11/2/63, 9)

- DOD announced that the Army would begin in November a series of "graduation firings" of the Pershing ground-to-ground solid-propellant missile by tactical units. Firings would be from Fort Sill, Okla. (DOD Release 1442–63)

- Army Nike-Zeus antimissile missile was successfully fired from White Sands Missile Range, N.M. Continuation of tests were to aid in development of the more advanced Nike X system. (DOD Release 1444–63)

- Chinese Communists claimed they had shot down an American-built, Nationalist-Chinese-operated U–2 reconnaissance aircraft near Shanghai in East China. This was the second U–2 the Chinese Communists claimed to have shot down over their territory. (*NYT*, 11/2/63, 6)

- Defense Documentation Center (DDC) for Scientific and Technical Information was transferred from operational control of USAF to Defense Supply Agency (DSA), effective this date. Physical location remained at Cameron Station, Alexandria, Va. DDC was previously known as Armed Services Technical Information Agency (ASTIA). (DOD Release 1371–63)

- Columnist William S. White, writing in the Washington *Evening Star*, charged that Premier Nikita Khrushchev was being supported in his efforts to slow down the U.S. space program by "one of the strangest coalitions we have ever known" in Congress, made up of conservatives who want to save money and liberals who want to spend the money on welfare. "The conservatives . . . ought to ponder what they are about here. For apart from the almost indescribable strategic and scientific significance of this program, there is the bottom fact that it is already nearly indispensible to the American economy and may later become indispensible in the absolute sense.

"Automation, when fully launched, will create huge pools of unemployables. Politically, these must and will be cared for, under any forseeable regime, Republican or Democratic. Is it not better to spend the money for space than to speed the day when all this money and more will have to be thrown about for the most gigantic—and also permanent—leaf-raking schemes in the world's history?

"The space program is the precise opposite of economic crackpotism. It is sensible conservatism's greatest future weapon against just such crackpotism." (Wash. *Eve. Star*, 11/1/63)

November 2: The international communications conference convened in Geneva by the International Telecommunications Union had been in session a month. Representatives from 170 countries and territories had been working out positions on a number of technical problems in committees and working groups. During the fifth and last week, these findings would be submitted to the plenary sessions for approval. (Reuters, *NYT*, 11/3/63, 12F)

- European interest in a cooperative communications satellite effort was continuing. The U.K. and 12 other European countries met on the subject in Paris in May 1963 and in London in July. Various committees were formed to study the technical, economic, and political problems involved and these committees were to report to a meeting to be held in Rome on Nov. 27. The general direction of the discussions has been toward some sort of partnership between the European countries and the U.S. ComSatCorp. (Farnsworth, *NYT*, 11/3/63, 12F)

- Humorous columnist Arthur Hoppe commented on Premier Khrushchev's statement of Oct. 26 that Russia was not competing to land a man on the moon, said this marked the beginning of a new weapon in Soviet diplomacy—"Competitive unresistance."

  "Oh, it strikes at the one weakness in the American character. For while we are imbued with the will to win, we can't stand licking somebody who isn't trying. Competitive unresistance, let me warn you, will sap our will to resist.

  ". . . Of course we could do some things for other reasons. Like maybe feeding people because they're hungry. Or reaching for the moon because we believe the future of our race lies in the stars. But would Congress buy that? Nonsense. It's unrealistic.

  "So I say we must meet this new Soviet threat head-on. We must abolish our moon program, knock off foreign aid and do our utmost, fellow Americans, to make our Nation a second-rate power. But cheer up. I'm certain we can count on Congress." (Hoppe, Wash. *Eve. Star*, 11/2/63, 5)

- USAF launched Minuteman ICBM from Vandenberg AFB, Calif., the 20th of the solid-fueled missiles launched from the West Coast base. (AP, Wash. *Sun. Star*. 11/3/63)

November 2: Soviet reported POLET I had gone into fixed elliptical earth orbit after flying through several orbits and orbital planes and had increased its maximum distance from 368 to 892 mi. from earth while keeping a minimum distance of about 210 mi. (*A&A*, Jan. 1964, p. 5)

November 3: Cosmonauts Capt. Valentina V. Tereshkova and Maj. Andrian G. Nikolayev were married in a civil ceremony in Moscow, followed by an emotional four-hour reception for 300 guests. Premier Nikita Khrushchev acted as toastmaster at the reception, and relatives and friends were crowded into the background by Soviet dignitaries. A crowd of 1,000 stood outside the state wedding palace prior to the marriage and watched guests arrive. Moscow Radio announced the ceremony 90 min. after it had taken place and interrupted its program frequently for recorded bulletins from the reception. Fellow cosmonauts were in attendance and signed the marriage register as witnesses. (Shabad, *NYT*, 11/4/63, 1)

- U.S.S.R. was designing a supersonic transport aircraft from the start, not modifying a bomber design as they had done with the TU-104 and were rumored to be doing again, according to FAA Deputy Administrator Gordon Bain. (*NYT*, 11/4/63, 1)
- ComSatCorp was planning its initial offering of stock in the spring of 1964, probably on the order of $200 million. Half of this stock would by law have to be made available to the general public at no more than $100 per share, the remaining half being taken by the communications industry. Wall Street predicted that the stock would meet with unprecedented enthusiasm for a large issue by a new and speculative corporation. Meanwhile the ComSatCorp technical staff was still studying the technical and economic factors involved in establishing a global communications net "at the earliest practicable date" that Congress directed. The goal was for an operating satellite system by 1966–67, employing either medium-altitude or synchronous-orbit satellites. (*NYT*, 11/3/63, 1F)
- A roundup of world press opinion on the manned lunar landing race, following Premier Khrushchev's statement of Oct. 26 that Russia was not presently planning manned lunar flight, indicated that the majority of the newspapers thought Khrushchev had made a sensible decision which the U.S. ought to follow. (*NYT*, 11/3/63, 9E)
- U.S.S.R.'s maneuverable spacecraft POLET I, launched Nov. 1, 1963, was put through its maneuvers by radio commands from the ground, according to K. Gilzin, writing in the Soviet armed forces newspaper *Krasnaya Zvezda:* "Our new maneuverable spaceship, heeding radio commands from earth, obediently turned first to one side, then to the other, soared up and dived, changing its position in space." (*NYT*, 11/4/63)
- Editorial in the Washington *Sunday Star* commented on U.S.S.R.'s latest space feat, the launching of the maneuverable spacecraft POLET I: "Russia's newest venture in space has come at exactly the right time. Although not so intended, it pulls the rug from under those among us—such as economizing Congressmen and exces-

sively zealous scientists like Linus Pauling—who have been clamoring for an end to our country's program to place Americans on the moon by 1970.

"These people, with much naivete (to use a polite word), have attached an extravagant degree of importance to Premier Khrushchev's recent remarks vaguely suggesting that he may be withdrawing the Soviets from the lunar race .... their latest space shot plainly, and disturbingly, indicates that they are well ahead of us with the kind of rendezvous capability that is essential to efforts to place men on the moon and bring them back safely ....

"Clearly, we must run this race as swiftly as we possibly can." (*CR*, 11/4/63, A6869)

November 3: Reviewing the U.S. manned lunar landing program, John Finney of the *New York Times* concluded that the race to the moon had always been a one-sided one, that there was no evidence that the Russians were building the large rocket that would be necessary for such a venture, that they were rather proceeding on a building-block program in space that eventually would lead to a lunar landing, but not as a special, high-priority project. This, Finney said, was also the approach of the U.S. until May 1961. "Then, in the wake of the Bay of Pigs fiasco and the first Soviet manned space flight, the Kennedy Administration ordered an abrupt change in course." The original justification was that U.S. prestige was at stake. As time went on and budget resistance developed in Congress, the Administration argument shifted to one of attaining "a position of preeminence" in space. Also the potential military benefits were emphasized more and more. Finally, the President's proposal for the joint U.S.-U.S.S.R. manned lunar program seemed to undercut all preceding positions. To Mr. Finney, all this added up to the certainty that the U.S. would continue its program for a manned lunar landing "but it will be pursued with less competitive zeal and at a more leisurely pace." (Finney, *NYT*, 11/3/63, 5E)

- Former President Dwight D. Eisenhower, interviewed on NBC–TV, said he was surprised in May 1960 when the U.S.S.R. publicized the downing of the U.S. U–2 reconnaissance aircraft over Russia. Secretary of State John Foster Dulles also "believed just as I did that it was necessary to use it but he also believed that it would cause no real publicity if one of them should fall." Both of them felt that the Soviets would not be willing to concede publicly that for "three or four years we've been doing this . . . ." On the question of admitting U.S. responsibility for the flights, Mr. Eisenhower said: " . . . I just thought as long as the thing had come out, the best thing to do—and I don't believe I asked anyone's advice on this—I just said 'I am responsible and that's that.'" (*Wash. Post*, 11/4/63)

November 4: USAF launched an Abres (advanced ballistic re-entry system) vehicle on an Atlas booster from Vandenberg AFB, Calif. Like the two previous launches in the series from AMR, Abres carried scientific experiments designed to advance the state of the art in re-entry systems. No attempt was made to recover the payload. (*Space Bus. Daily*, 11/5/63, 202)

- *November 4:* U.S.S.R. and U.S. were reported by the *Washington Post* to have reached agreement on frequency allocation for communications satellites at the International Telecommunications Union-sponsored conference in Geneva. In previous years the Soviet Union had opposed any space activity by nongovernmental organizations, but in recent months had dropped its opposition to private firms, such as the U.S. ComSatCorp, operating in space. Original positions at the conference had seen the U.S. ask for reservation of 2,725 megacycles in a highly sought-after portion of the microwave spectrum, while the U.S.S.R. had favored only 1,600 mc. Details of the compromise position would emerge when both nations went before the full convention. (*Wash. Post*, 11/5/63)
- PCH-1, patrol craft hydrofoil built by Boeing, flight-tested by U.S. Navy at speeds of more than 45 knots. (*A&A*, January 1964, p. 5)
- Partial results announced by Univ. of Stockholm scientists from the U.S.-Swedish series of sounding rocket experiments made in July and Aug. 1963 showed extremely low temperatures to be associated with the presence of noctilucent clouds. Sounding rocket data from the Kronogård firings indicated that up to 35 mi. temperatures closely followed the profile found over the U.S. and other countries. The temperature increased from about $-50°C$ ($-58°F$) at 10 mi. to slightly above $0°C$ ($32°F$) at 32 mi., where absorption of the sun's ultraviolet radiation by ozone molecules causes a warming. Above this level, temperatures decreased again to a normal minimum at 50 mi. of $-120°C$ ($-184°F$). When noctilucent clouds were present at this altitude, however, the temperature dropped to what appears to be the lowest temperature ever recorded in nature— $-143°C$ ($-225°F$). This would confirm earlier findings in 1962 that noctilucent clouds are composed of cloud particles coated with ice. (NASA Release 63-248)
- Laser beam bounced off the moon by Soviets was detected by Crimean Astrophysical Observatory. (*A&A*, January 1964, 5)
- Editorial in *Aviation Week and Space Technology* by Editor Robert Hotz criticized the national space policy as "badly shredded" by President Kennedy's recent statements, called for a new statement of policy "based on solid elements of national interest" before the FY 1965 budget debate begins in Congress. (*Av. Wk.*, 11/4/63)
- Gen. Bernard A. Schriever, Commander of AFSC, spoke to the AIAA/ASD Vehicle Design and Propulsion Meeting in Dayton, Ohio, on the future of areonautical systems. He predicted a continued need for aircraft in the weapons inventory, cited particular improvement potential in cargo aircraft and V/Stol. Another potential was that "future space operations may require the use of recoverable boosters, for reasons of both efficiency and economy. The first stage of such a booster system would probably be a recoverable air-breathing vehicle."

  Two technical areas of particular promise were identified as materials and propulsion. "We have identified new materials that are stronger, lighter, stiffer, and able to withstand higher temperatures. For example, new types of metallic fibers may be feasible in forming extremely strong, lightweight composite structures. If such projected applications prove to be practical, then we may be in for a revolution in aircraft design and fabrication.

"There also appear to be high payoffs in the propulsion area. New technology should enable us to achieve better thrust-to-weight ratios and to move away from single point design.

". . . Engineers are almost always overly optimistic about the next two or three years and unduly conservative in their estimates of a 15-year period or more. There were study groups in the 1950's that denied the practicality and feasibility of ICBM's, the turbofan, and hypersonic flight. Ten years ago, so-called 'sound engineering judgment' would not have forecast the present generation of systems. The point is that 'engineering judgment' does not always give the best evaluation of long-range predictions. The crucial question to ask is, 'Does a new concept defy known scientific principles?' If it does not, then the technical problems will eventually be overcome." (AFSC Release 310-R-97)

November 4: NASA was examining the nuclear-pulse propulsion concept (Orion) which had been studied under USAF contracts for some five years. Nuclear-pulse—a series of controlled nuclear explosions—was considered a possible means of propulsion for fast manned trips to Venus, Mars, and Jupiter in the 1975–1995 time period. The fast trip would be desirable because component lifetimes would run around 10,000 hrs., somewhat over one year. It would therefore be highly desirable that the round-trip be possible in one year's time. If Orion were chosen as the propulsion system, some amendment to the nuclear test-ban treaty might be necessary to legalize its use. NASA was studying the Orion potential through a contract issued in June 1963 to the General Atomics Div. of General Dynamics. (*M&R*, 11/4/63, 34)

November 5: MSC announced a reorganization designed to strengthen Apollo and Gemini management structure and to assimilate Project Mercury personnel into these programs. Under Dr. Robert R. Gilruth as Director and James C. Elms as Deputy Director would be four assistant directors, managers of major programs, and manager of MSC Florida Operations. The four assistant directors: Assistant Director for Engineering and Development (Maxime A. Faget), Assistant Director for Flight Operations (Christopher C. Kraft, Jr.), Assistant Director for Flight Crew Operations (Donald K. Slayton), and Assistant Director for Administration (Wesley L. Hjornevik). Major program managers: Apollo Spacecraft Program Office (Dr. Joseph F. Shea) and Gemini Program Manager (Charles W. Mathews). (MSC Release)

• Various Russian sources had listed a variety of possible uses for the Polet type of maneuverable satellite, according to *Space Business Daily.* Space expert Vladimir Dobronravov had said that the spacecraft would be used for rendezvous and docking experiments. Tass mentioned the potential for reconnaissance, weather, or communications satellite. M. Litvin-Sedoi said it was a test of a system for construction of an orbiting space platform. (*Space Bus. Daily*, 11/5/63, 201)

November 5: Dr. Wernher von Braun, Director, Marshall Space Flight Center, addressing Chamber of Commerce, Charleston, West Virginia, said:

"In support of Project Apollo major manufacturing, testing, and launching facilities are now under construction. Years of construction, personnel buildup, and training for activation will be necessary before they can be put to use. If Project Apollo should be short-changed, these unfinished buildings would stand as monuments to America's folly. You simply cannot turn a complex project like Apollo off and on like a faucet." (Text)

- Dr. Jerome B. Wiesner, the President's Scientific Advisor, would resign from government this winter and return to MIT, the White House announced. (*Wash. Post*, 11/5/63)
- Sen. William Proxmire (D.-Wis.) said on the Senate floor that "we can and should dedicate America's great resources to the magnificent challenge of space" but charged NASA with waste and inefficiency. He inserted in the *Congressional Record* an article of his in which he recommended (1) NASA build into its organization more effective safeguards against waste, duplication, and other "leakage" of funds; (2) Congress acquire skilled investigators to analyze the space program in depth; and (3) "Private individuals and groups should apply their concern about excessive and unnecessary Government spending to the space program." (*CR*, 11/5/63, 19990–91)
- Rep. Melvin Price (D.-Ill.), speaking before AIAA-AFSC meeting in Dayton, Ohio, warned that "the honeymoon in research and development is over" on Capitol Hill. "The day of ill-defined objectives for research and development programs, of gross overruns in costs, of blurred and overlapping management responsibility is rapidly coming to an end." As proof of the hardening attitude in Congress, he cited the cuts in the NASA budget, in the National Science Foundation budget, and the appointment of the Select Committee to examine the Government's research programs. Improvements must be made, he said, "before the Congress and the people lose faith in those who are responsible for carrying out our vital research and development programs and implementing their results." Otherwise, he warned, corrective action would fall into "hands which are less considerate and less understanding." Rep. Price has for 10 years been chairman of the Subcommittee on Research and Development of the Joint Committee on Atomic Energy. He was recently named chairman of the new Research Subcommittee of the House Armed Services Committee and a member of the Select Committee on Government Research. (Finney, *NYT*, 11/6/63, 15)
- 1963 Nobel Prize in Physics was announced in Stockholm to have been won jointly by American Prof. Maria Goeppert Mayer, of the Univ. of Calif. at La Jolla, and American Prof. Eugene Paul Wigner, of Princeton Univ., the former sharing her half of the prize with German Prof. J. Hans D. Jensen, of the Univ. of Heidelberg. Mrs. Mayer and Prof. Jensen, working independ-

ently, devised nuclear models portraying the particles as arranged in levels or shelves which increase abruptly in number of particles at each level further out from the nucleus. Prof. Wigner was cited for devising symmetry principles explaining the interaction of the proton and neutron in accordance with the direction of their spinning motion. The prize in Chemistry was also announced, being awarded jointly to Italian Prof. Guilo Natta, of the Institute of Technology at Milan, and German Prof. Carl Ziegler, of the Max Planck Institute for Carbon Research in Mulheim. They were cited for devising a system of controlling the polymerizing of simple hydrocarbons into large molecule substances. Their discovery had important commercial results, making possible the development of many kinds of plastics, synthetic detergents, antiknock mixtures for high-octane fuels, etc. (*NYT*, 11/6/63, 1)

November 6: Premier Nikita Khrushchev told a group of visiting U.S. businessmen that U.S.S.R. was continuing its manned lunar landing program. Clarifying his remarks of Oct. 26, he said that "when we have the technical possibilities of doing this, and when we have complete confidence that whoever is sent to the moon can safely be sent back, then it is quite feasible, quite possible." He said that POLET I had made space rendezvous possible but that "no definite date" had been set for such an experiment. On the supersonic transport aircraft, he said: "I can say that we are designing and building a supersonic plane. I can't give you the details because I don't remember the technical data on that plane." (*NYT*, 11/7/63, 1; AP, *Wash. Post*, 11/7/63)

- Utmost care would be taken in the assembly and checkout of the Mariner B spacecraft intended for a landing on Mars in 1966 so that the Mars lander would be completely free of earth microbes that might contaminate the Martian environment. Workers with even minor colds would be barred from the assembly area. The final assembly would be subjected to a dry heat at 275°F for 24 hours, then sealed into its special can and placed aboard the booster for launching. The special can would not open until the spacecraft was some 350 mi. from earth, considered to be the limit of earth microbes. (*Wash. Post*, 11/7/63)

- Sen. Clinton P. Anderson (D.-N.M.), Chairman of the Senate Committee on Aeronautical and Space Sciences, told the Senate that it was unwarranted to assume from Premier Khrushchev's Oct. 26 remarks that the Soviets had pulled out of the space race. "The competition in space between the United States and the Soviet Union is still intense. It would be foolish indeed if this country were to attempt to reorient its vast space program each time the Soviet Union made some pronouncement about its goals. We should take cognizance of Russian statements but we should not let ourselves be diverted from sound objectives on the ground that competition no longer exists." (*CR*, 11/6/63, 20140)

- By the beginning of 1964 the U.S. would have 475 ICBM's as against U.S.S.R.'s 100, according to a study released by the Institute of Strategic Studies in London. U.S. was also estimated to have 1,300 strategic bombers and 10 Polaris submarines with 16 missiles each. U.S.S.R. was said to have 1,200 bombers—1,000 of which were medium bombers of only 3,500 mi. range—and no Polaris-type submarines but 90 400-mi. missiles capable of being fired from a surfaced submarine. (AP, *NYT*, 11/6/63, 3)

November 6: MSFC awarded Rocketdyne Div. of North American Aviation an $8,441,956 contract to extend the duration run of the J-2 liquid-hydrogen rocket engine from 250 sec. to 500 sec. (*Space Bus. Daily*, 11/7/63, 218)

- Dr. George E. Mueller, Associate Administrator for Manned Space Flight, speaking before the Washington chapter of AIAA, said that funds for the space program thus far in this decade had (1) created a team of trained people; (2) developed basic technology; (3) constructed facilities; and (4) been used up in consumption of consumable items. Consumables, he said, represented the cost of launches and costs of the recovery forces. This category amounted to less than eight per cent of the total. "The remaining 92 per cent of the funds are used for creating permanent capital . . . ." (Text)

- Ford Foundation announced a grant of $300,000 to take young engineering professors from the university campus into industry for one year to give them experience in "manufacturing, marketing, financing, and other considerations that influence engineering decisions in practice." The program would cover three years and would involve some 60 faculty members. Program would be directed by Clarence E. Watson, former vice president of Columbia Broadcasting System Laboratories. Another $200,000 grant was made to the Univ. of Tenn. to take outstanding scientists and engineers into parttime teaching. (*NYT*, 11/7/63, 41)

November 7: MSC successfully conducted the first off-the-pad abort test of the Apollo launch escape system at White Sands Missile Range, N.M. A boilerplate version of the Apollo command module (Boilerplate No. 6) was lifted off the pad by the 155,000-lb.-thrust launch escape motor mounted on a launch escape tower attached to the Apollo module. Thrust continued to a height of 4,100 ft. (T+8 sec.), escape tower jettisoned at 4,900 ft. (T+15.5 sec.), and parachutes deployed for the recovery. Primary purpose of the test was to determine the stability and operational characteristics of the escape configuration during a pad abort. (MSC Release 63-247; *NYT*, 11/8/63, 16)

- U.S.S.R. celebrated the 46th anniversary of the Bolshevik Revolution with a Moscow parade that featured the first public display of its antimissile missile. Marshal Sergei S. Biryuzov, Chief of the Soviet General Staff, hailed the weapon on a radio broadcast as capable of destroying "the enemy in the air." At a reception Premier Khrushchev said: "Before the Revolution Russia was a beggar. Now she has become the Soviet Union and is a king in the land. From the lowest she has become the second country in the world, and in a maximum of seven years we will be first." (Tanner, *NYT*, 11/8/63, 1)

- X-15 No. 3 was flown to 2,925 mph (mach 4.40) and 82,300-ft. altitude by Major Robert Rushworth (USAF). (X-15 Project Office)

November 7: U.S. and U.S.S.R. were said near agreement on a declaration of principles of space law. Agreement was said to have been reached in a series of talks at the U.N. Both nations were to present their positions to the 28-member Committee on the Peaceful Uses of Outer Space in hopes that the entire committee would sponsor the declaration in the General Assembly. Earlier in the year the U.S. had found four Soviet demands with regard to space law to be unacceptable and talks on the declaration had seemed deadlocked. (Hamilton, *NYT*, 11/8/63, 1)

- Perfection in manufacture of engines for Project Gemini would be sought through a Verification Instruction Program (VIP) to train and qualify manufacturing and test personnel at Aerojet's Liquid Rocket Plant. The program would include 30 days of classroom instruction followed by on-the-job training for top management and senior hourly personnel. Program would be operated by NASA Astronaut Frank Borman and Aerojet plant manager Ray C. Stiff, Jr. (*Space Bus. Daily*, 11/8/63, 228)

- President Kennedy announced that he had appointed Dr. Donald F. Hornig, Chairman of the Chemistry Department at Princeton Univ., to succeed Dr. Jerome B. Wiesner as the President's science adviser and director of the White House Office of Science and Technology. Dr. Hornig has previously been a member of the President's Science Advisory Committee. The President also announced that Robert H. Charles, a special assistant to the NASA Administrator, would succeed Joseph S. Imirie as Assistant Secretary of the Air Force for Materiel. (*Wash. Post*, 11/8/63)

- DOD announced U.S. Army had filed with Fédération Aeronautique Internationale for certification of six new world speed records for rotocraft. A Hiller OH–23G light observation helicopter, piloted by Capt. Bertram G. Leach (USA), flew record flights in two weight classes on three courses for speed over a straight course and over a closed circuit. In the E1B (1,102 to 2,204 lbs.) class, the OH–23G was timed at 123.67 mph on a 3-km. straight course, 123.58 mph on a 15-km. straight course, and 119.81 mph on a 100-km. closed circuit. In the E1C (2,204 to 3,858 lbs.) class, the OH–23G was clocked at 123.44 mph, 123.77 mph, and 121.70 mph on the same three courses. (DOD Release 1460–63)

- USAF Minuteman missile was launched from AMR on an intended 1,700-mi. flight, but went out of control and exploded several hundred feet above the launch pad. Several fires were started on Cape Canaveral by flaming fragments but without injury or property damage. (AP, *NYT*, 11/8/63, 7)

- Rep. George P. Miller (D.-Calif.) spoke on the House floor of the series of meetings with leading scientists that had been conducted by the Subcommittee on Science, Research and Development of the House Subcommittee on Science and Astronautics. "I cannot overemphasize the work of the Daddario Subcommittee in the effect it will have on the future progress of the United States, progress not only in science but also in the proper relationship of science and technology to our entire society. The members of the subcommittee are assuming a very weighty responsibility because, I believe, results of the subcommittee work will set the framework for science and technology in relation to the overall resources that we have to guide ourselves in the difficult years that lie ahead of us." (*CR*, 11/7/63, 20332)

November 7: Argentina was planning a series of science fairs for secondary school students on a national basis. The Argentine Department of Education would operate the program, with assistance from the National Commission for Scientific and Technical Investigation. (Science Service, *NYT*, 11/7/63, 67)

November 8: ITU conference on communications frequency allocation concluded in Geneva on a successful note. Joseph McConnell, president of the Reynolds Metal Co. and head of the U.S. delegation, talked from Geneva via SYNCOM II communications satellite about the results of the conference. Final agreement had set aside some 2,800 megacycles of the spectrum for use of space communications. This was somewhat more than the U.S. had requested, although not all of it in the portions of the spectrum that the U.S. had wanted. (NASA Press Conference Transcript; *Wash. Post*, 11/9/63)

- West Germany joined the list of nations participating in satellite communications with the opening of its narrowband station at Raisting. A wideband station to permit television transmission was under construction. Raisting became the seventh station in the satellite communications network. Other narrowband stations were at Nutley, N.J.; Rio de Janeiro, Brazil; Fucino, Italy. Wideband stations were at Andover, Me.; Goonhilly Downs, U.K.; and Pleumeur-Bodou, France. (NASA Release 63-250)

- Army Corps of Engineers issued invitations to bid on construction of a Vibration Test Laboratory to be built at NASA Manned Spacecraft Center for an approximate cost of $1.5 million. The laboratory would contain some 13,700 sq. ft. and have a general vibration test area 42 ft. high, a spacecraft vibration test area 115 ft. high, and provision for offices, shops, and control rooms. Bids would be opened on Jan. 9, 1964. (MSC Release 63-226)

- Sen. Philip A. Hart (D.-Mich.) introduced a bill (S.2298) to establish a Commission on the Application of Technology to Community and Manpower Needs. The Commission would be composed of members from Government, science, industry, and labor. It would study the current technology and its growth potential, looking to selection of those areas of benefit to national and community needs. It would also study and make recommendations on the impact of technological application. (*CR*, 11/8/63, 20376-79)

- In lecture on "The Practical and Impractical Uses of Space" at Catholic University, Dr. Edward Teller supported the U.S. lunar program but parried questions on the reasons. "I just don't know why . . . I wanted to go to the moon before Sputnik."

  Teller stated that he believed that water could be drawn from lunar rocks by underground nuclear explosions: "One hundred tons of water on the moon would be the equivalent of 100 tons of gold here." He said that contamination of the water would not be a problem as the U.S. had already developed "clean" bombs and that even cleaner bombs are possible. (Ayres, *Wash. Post*, 11/9/63, C14)

November 8: MSC officials stated that the reason the first Little Joe II did not terminate at the proper time and altitude was because the primer cord joining the initiator to the charge that would have exploded the casing of the Algol engine, thus terminating propulsion, was not connected. (Maloney, *Houston Post,* 11/9/63)

- Large fireball streaked across San Francisco Bay area sky, landing in the Pacific Ocean several miles off-shore, and witnessed by thousands of residents. Fireball was presumed to be a meteor. (AP, *Wash. Post,* 11/8/63, A3)
- Cleveland Clinic kept a calf alive for 20 hours after its heart had been replaced by a mechanical heart system. The controls to drive the pump of this mechanical heart were designed for the Clinic by NASA Lewis Research Center pump specialists. (Lewis Chronology, 11)

November 9: USAF's Rome Air Development Center (RADC), Griffiss AFB, N.Y., would take part in the tracking and experimentation with NASA's Echo II passive communications satellite scheduled for launching shortly. A 60-ft.-diameter steerable parabolic antenna had been erected at the Center's Passive Satellite Research Terminal near Rome, N.Y. The terminal would be able to transmit on S and X bands for 5,000-mi. tracking and communications range. USAF would investigate passive communications satellites for their possibilities as a relatively nonjammable military communications system. (*A-N-AF Journ. & Reg.,* 11/9/63, 28)

- USAF Aerospace Research Pilot School, Edwards AFB, Calif., was using F-104 and F-106 fighter aircraft to simulate portions of space vehicle re-entry and landing profiles, Capt. Edward J. Dwight (USAF) said at Howard Univ. in Washington. The F-104 is usually accelerated to mach 2 at 35,000 ft. and continued in powered climb to 75,000, coasts to a top altitude of 90,000 to 95,000 ft., and starts down. "We usually relight the engine at about 60,000 feet and recover from a spin." When the NF-104, modified to add a rocket engine, was available, this maneuver would be extended to 120,000 to 140,000 ft. The F-106 is used to simulate the landing pattern of a Dyna Soar-type space vehicle, dropping from 20,000 ft. to a high-speed landing. (*NYT,* 11/10/63, 88)
- USAF launched an unidentified satellite from Vandenberg AFB, Calif., on board a Thor-Agena booster. (*M&R,* 11/18/63, 12)
- An Eastern Air Lines DC-8 jet transport aircraft flying over Texas in clear weather suddenly dropped more than two miles—from 20,000 ft. to 6,000 ft.—in an unexplained dive so violent that it tore one of the four jet engines loose from its wing mounting and injured 17 of the 124 passengers. The aircraft made an emergency landing without incident. An investigation immediately began as to the effect of clear-air turbulence on jet aircraft. One theory was that the conventional means of dealing with turbulence—that is, slowing down to soften the impact of the turbulence—might in fact be dangerous in jet aircraft because the slower air speed would lessen the pilot's control of the aircraft and make it too easy for the aircraft to go out of control. (Hudson, *NYT,* 11/15/63, 57)

November 9: USAF launched a Titan II ICBM from an underground silo at Vandenberg AFB, Calif. Test, sixth Titan II launch from Vandenberg, was to check operation of the total weapon system. (AP, *Wash. Post*, 11/10/63)

During early November: USAF selected General Dynamics/Ft. Worth, Boeing, and North American Aviation, Inc., to study feasibility of an Advanced Manned Penetrator (AMP), revision of previous Low Altitude Manned Penetrator (LAMP). All three had proposed use of kerosene-type fuels in the engine. (*Av. Wk.*, 11/11/63, 25)

November 10: X-15 No. 2, being remodeled under USAF contract by North American Aviation, Inc., would have as one of its most important research objectives the testing of a ramjet engine modified to burn liquid hydrogen. Using the basic configuration of the Navy's Typhon missile engine, the ramjet would slow the supersonic airstream of the X-15 to subsonic speed for burning in the engine. Theoretically the ramjet flying at 4,600 mph should be twice as efficient as a jet engine. (Clark, *NYT*, 11/10/63, 88)

- Sen. Albert Gore (D.-Tenn.) wrote in the *New York Times Magazine* about space law and the Dryden-Blagonravov agreements: "This agreement, though limited in nature, is significant if for no other reason than that it indicates there are some areas in which our two countries have found it possible to cooperate in outer space development. If successfully implemented it can, perhaps, lead to other and broader agreements. The Dryden-Blagonravov agreement relates to scientific and technical matters. It was achieved despite political differences. Similarly, agreement should be possible on some legal questions if we can isolate those questions from political considerations.

    "Perhaps if we can build on these small areas of agreement an escalation of the arms race into outer space can be avoided. It is conceivable that science and technology, so often in history energized by the threat or event of war, may, through space research and development, pave the way for removal of the barriers erected by hate, fear, and suspicion." (*NYT Magazine*, 11/10/63, 23 ff)

November 11: U.S.S.R. orbited COSMOS XXI, "intended for continued space research." Orbital data: apogee, 229 km. (142 mi.); perigee, 195 km. (121 mi.); period, 88.5 min.; and inclination, 64°50'. (Tass, *Komsomolskaya Pravda*, 11/13/63, 1, AFSS-T Trans.)

- Academician Mstislav V. Keldysh, president of the Soviet Academy of Sciences, was interviewed by Soviet newsmen on the significance of POLET I, the maneuverable spacecraft orbited by U.S.S.R. on November 1, 1963. In speaking of its implications for manned space flight, Keldysh said: "Maneuverable spacecraft will permit us to execute a landing from any orbit to a given 'kosmodrom'; carry out a meeting in space of ships which are flying in different orbits; and also allow astronauts to select the most advantageous landing area.

    "The ability of a ship to maneuver will make it possible for us to create heavy orbital scientific research stations in space so that we can exchange crews, replace scientific equipment and maintain a continuous supply of all that is necessary . . . ." (*Pravda*, 11/11/63, 1, AFSS–T Trans.)

November 11: NASA postponed indefinitely the launch of the first interplanetary monitoring probe (Imp) after ground tests of the third-stage launch vehicle indicated that a change would have to be made in the separation sequence to prevent contamination of the spacecraft. (AP, *NYT*, 11/12/63, 42; Wash. *Daily News*, 11/11/63)

- U.S. tracking data on the Russian POLET I maneuverable spacecraft substantiated the Russian claim of change in apogee but indicated that any substantial change in plane had to be made before completion of the first orbit. According to *Aviation Week*, NORAD's initial orbital plane for the Russian satellite was 59.99°, later revised to 58.89°. Whether this 1.1° change was a result of a satellite maneuver or was merely a refinement of earlier data is uncertain. (*Av. Wk.*, 11/11/63, 28)

- Rocketdyne had completed test firings of the X–8 experimental engine, one of a series of experimental rocket engines being designed and tested by Rocketdyne for USAF's Ballistic Systems Div. in a search for advanced propulsion techniques. The X–8 engine had tested the feasibility of using liquid hydrogen as coolant for the thrust chamber. The tests showed liquid hydrogen to be a remarkably good coolant. An X–12 engine would now be built to test the possibility of achieving an increase of four to five times the pressure in the thrust chamber using coolant. This would not alter thrust at sea level but would give considerably increased thrust at altitude. (*Av. Wk.*, 11/11/63, 99)

- A number of unofficial reports from Moscow have identified two key figures in the Soviet space program, according to Theodore Shabad, the *New York Times* Moscow correspondent. One of the men was Valentin P. Glushko, known as a rocket propulsion expert who worked with Friedrich Zander in the first Soviet research in liquid fuel rockets in the early 1930's. The other was Sergei P. Korolov, a mechanical engineer long associated with mechanics and structural problems of flight. Both are full academicians in the Soviet Academy of Sciences. (Shabad, *NYT*, 11/12/63, 2)

- *Aviation Week* reported that USAF and USN were pressing NASA to improve the reliability of the Scout launch vehicle. The Navy was said to have had only 50% success with it in their Transit navigational satellite program. USAF was said to have had a little better performance but only because it had undertaken a complete and expensive checkout of components. Government-furnished equipment was said to be especially unreliable. (*Av. Wk.*, 11/11/63, 34)

- Chemist S. Stephen Papell, of NASA's Lewis Research Center, had applied for a patent on magnetic rocket fuel. Mr. Papell had found a means of magnetizing JP–4 and other liquid hydrocarbons by mixing small quantities of magnetic iron oxide with the liquids. Under weightlessness, an electromagnet located near the propellant pumps could pull the fuel into the pumps. Papell said much work remained to be done on the process before it could be considered operationally feasible. (*Av. Wk.*, 11/11/63, 30; Lewis Chronology, 11)

November 11: U.S. Army had contracted Lockheed California for further development of the Ping Pong short-range reconnaissance missile concept. Under this concept, a missile would be fired a short distance behind enemy lines, would take photos of enemy positions, and then a motor on the other end of the missile would fire it back over U.S. lines for recovery. (*Av. Wk.*, 11/11/63, 34)

- Editorial by Editor Robert Hotz in *Aviation Week and Space Technology* on U.S. supersonic transport aircraft: "One of the most significant points developed by the [Senate] hearings is that the FAA has vastly underestimated the development costs of the supersonic transport, and that the aircraft manufacturing industry cannot and will not assume even the fraction of these costs that the government has assigned to it." (Hotz, *Av. Wk.*, 11/11/63, 21)

- The comet Burnham 1960 II "wagged" its tail during its trip past earth and to the point in its orbit nearest the sun. Photographs taken by Daniel Malaise in April and May 1960 from the Astrophysical Institute of the Univ. of Liege, Belgium, showed that the tail of the comet moved in a four-day cycle, swinging about eight degrees to each side of the radius vector each four days. The most obvious explanation of the movement would be that it was caused by rotation of the comet's nucleus, but so far no explanation had been fully satisfactory. (Schmeck, *NYT*, 11/11/63, 29; cf. article in *The Astronomical Journal*, Oct. 1963)

- USN launched a Polaris A-3 missile under water from the nuclear submarine U.S.S. *Andrew Jackson* in a flight of more than 2,000 mi. down the AMR. (UPI, *NYT*, 11/12/63, 18)

November 12: NASA launched an Aerobee 150A sounding rocket from Wallops Station, Va., sending a 156-lb. payload to an altitude of 137 mi. Payload was a photospectrometer to measure ultraviolet light emissions from atomic oxygen at various altitudes in the upper atmosphere to learn more about the amounts and properties of the oxygen atoms. (Wallops Release 63-101; *NYT*, 11/13/63, 29)

- As ComSatCorp Board Chairman Leo D. Welch and President Dr. Joseph V. Charyk left the U.S. for a meeting in Bonn, Germany, with the European Conference of Posts and Telecommunications, there was increasing evidence that European nations would demand a substantial role in the management of a global communications satellite operation in return for joining the U.S. ComSatCorp effort rather than build a competing system of their own. (Finney, *NYT*, 11/13/63, 17)

- Soviet Cosmonaut Major Andrian Nikolayev, touring India with his bride, Cosmonaut Valentina Nikolavev, said in New Delhi that U.S.S.R. was planning a spacecraft capable of sustaining the life of a crew for three years. This was the estimated time for a manned flight to Mars or Venus and return, with some 14 months of this actual roundtrip flight time and the other two years waiting on the planet for the proper orbital relationship for the return flight. (AP, Wash. *Eve. Star*, 11/12/63)

November 12: Rep. Thomas M. Pelly (R.-Wash.) read into the *Congressional Record* an editorial from *Life* Magazine on priorities in the U.S. space program: "The United States needs a reasoned and balanced space program that is more than a series of jumpy reflexes to things Khrushchev may or may not be doing. The first priority is not the moon. It is the conquest of inner space (up to 500 miles high), the true military high ground from which our world can be photographed, weather surveyed, perhaps atom policed. Beyond that lie vast challenges in the quest for knowledge of space, where the moon is only one way station. The challenges must be met, the United States must be a space-faring nation; but at a rate and in an order set by our own capacity and deliberation." (*CR*, 11/12/63, A7016)

November 13: Senate Appropriations Committee completed action on NASA's FY 1964 appropriation by approving a $5.19 billion NASA budget. This was only $90 million more than approved by the House of Representatives. It had been thought that the Senate would raise the House figure to about $5.3 billion. The Senate Committee also retained its version of the Pelly amendment, forbidding a joint lunar program without consent of Congress. (Finney, *NYT*, 11/14/63, 21)

- NASA's Mississippi Test Operations (MTO) had a work force of nearly 900 as of this date, of which nearly 700 were contractor employees, 81 were Army Corps of Engineers personnel, and 34 were NASA employees. Active contracts at MTO now stood at some $28 million, with another $10 million out on bids. (*Marshall Star*, 11/13/63, 1)
- Dr. Jerome B. Wiesner, retiring scientific adviser to the President, was named Dean of the School of Science, MIT, succeeding Dr. George R. Harrison. (AP, *NYT*, 11/14/63, 31)
- MSC announced new system of procurement actions, designed to give the Government a better product at lower prices and to save industry time and money in competing for contracts. Under the new Two-Step Formal Advertising system, companies would submit only technical proposals in the first stage, instead of both technical and business management aspects as was formerly done. MSC would evaluate the technical proposals, notify those companies whose proposals were acceptable, and invite them to submit a formal bid. This system was thought to provide all the advantages of competition while retaining the Government's right to pick the company with the best solution to the problem. (MSC Release 63-236)
- DOD announced that Army solid-fuel Pershing missiles would be sent to Europe early in 1964 to begin replacing the Redstone missiles now in the field. Advantages of the Pershing were cited as being greater mobility and reliability and an all-weather capability. (DOD Release 1474-63; UPI, *NYT*, 11/14/63, 6)
- Dr. Vannevar Bush, honorary chairman of the corporation of MIT. long-time scientific adviser to the Government, and Director of the Office of Scientific Research and Development from 1941-46, wrote a letter to the *New York Times* on the manned lunar program. Noting that Premier Khrushchev said recently that a

manned lunar landing would not be worth the cost in money or in lives, Dr. Bush stated: "This is an aspect of the space program which the American public has by no means faced." That lives would be lost in the program was inevitable, he asserted, and would occur with the whole world watching.

"What will happen when the tragedy occurs? The public is now in doubt on the program. They will be impatient as it proceeds very slowly, with little of popular appeal to report. When a grim tragedy occurs they may throw the whole experiment out the window.

"I have just as much enthusiasm as the next man for the parts of the program which make sense: the communications and weather satellites, for example. But the whole program has been blown up to absurd dimensions. Instruments on the moon would gather scientific information, nearly as much as a man would, and the information would not be worth the cost.

"To put a man on the moon is folly, engendered by childish enthusiasm. It will backfire on those who drive it ahead." (*NYT*, 11/17/63, 8E)

November 13: Titan II launch from AMR on Nov. 1, 1963, testing the effectiveness of new damping devices installed to reduce surges of propellant and oxidizer that created vibration levels unacceptable to Project Gemini manned space flight, was termed successful by *Space Business Daily*. Vibration level had dropped to a level acceptable to NASA, although USAF wanted further tests to confirm reliability. (*Space Bus. Daily*, 11/14/63, 251)

- MSC officials said the Apollo Pad Abort Test No. 1, conducted at WSMR on Nov. 7, 1963, appeared from data to have been successful in all respects and showed the aerodynamics of the Apollo escape system "to be just what we expected." Liftoff and subsequent flight sequencing were within one second of program. (MSC *Space News Roundup*, 11/13/63, 1)
- USAF launched a Minuteman missile from AMR in a successful 5,000-mi. flight. Missile had an advanced second stage intended to extend the range of the solid-fuel ICBM. (*NYT*, 11/14/63, 13)
- USAF launched an Atlas D ICBM from Vandenberg AFB, Calif. (*M&R*, 11/18/63, 11)

November 14: X-15 No. 1 was flown to 3,286 mph (mach 4.75) and 90,800-ft. altitude by Capt. Joseph Engle (USAF), his second X-15 flight. The X-15 was put through two sharp rolling turns to test aircraft stability at high speeds and as part of the pilot's familiarization training. (St. Louis *Post-Dispatch*, 11/15/63; X-15 Project Office)

- An MSC operations team finished a study at Ling-Temco-Vought, Inc., that indicated an astronaut could effect rendezvous in space using only his piloting skill and simple spacecraft radar. Although Apollo flight would have automatic guidance and navigation equipment and continuous assistance from ground tracking stations, developers of backup systems needed to know the minimum with which the astronaut could operate. (MSC Release 63-238)

November 14: MIT scientists reported the first observation of the elements of water in space. Supported by a USAF 84-ft. parabolic radio telescope, a digital computer, and a radiation measuring device, the scientists detected hydroxyl radicals (OH) in space. Since the free hydrogen atom was detected in space in 1951, this means that the constituents of water have now been detected. MIT would now study the Milky Way on the frequencies absorbed by the hydroxyl radical and compare the resulting map of the dust and gas clouds with that already constructed from observation on the hydrogen absorption line. Information gained was expected to add greatly to knowledge of the structure and dynamics of the cosmos. (*NYT*, 11/14/63, 47)

- The Chemistry and Energy Conversion Division at the Lewis Research Center began a major study of the feasibility of using thin-film solar cells for space power applications. (Lewis Chronology, 11)
- President Kennedy was asked in his weekly news conference about the significance of the Soviet antimissile missile shown in Moscow for the first time on Nov. 7. He replied: "I don't think there is any doubt that they have an anti-missile, as do we. The problem, of course, is what you do with saturation. I don't think the Soviet Union or the United States have solved the problem of dealing, as I said before, with a whole arsenal of missiles coming at us at maximum speed with decoys. That, up to now, has been the impossible task." (*Wash. Post*, 11/15/63)
- FAA's Miami air route traffic control center began area positive control (APC) radar separation service to jet aircraft flying above 24,000 ft., bringing 22 out of 25 FAA centers into the APC network and covering 90% of continental U.S. Centers at Great Falls, Mont., and Boston will join in 1964 and the St. Louis center would be phased out. The next step in the program would be to lower the APC floor from 24,000 ft. to 18,000 ft. (FAA Release 63-97)
- Dr. Wernher von Braun, Director of NASA's Marshall Space Flight Center, said in Kansas City that there was no conflict between science and religion. "Science stands to better understand creation. Religion stands to better understand the creator. Science seeks to harness the forces of nature around us; religion and ethics seek to harness the forces of nature within us." (*Kansas City Times*, 11/15/63)
- U.K. announced it had established a space research station for satellite observation at the Sembawang naval air station in Singapore. (UPI, Wash. *Daily News*, 11/15/65)
- Dr. Fred L. Whipple, Director of the Smithsonian Astrophysical Observatory, stated his conviction that a belt of comets never seen from earth lies near the path of the planet Pluto. It was this comet belt, he though, that accounted for the disturbances in the orbit of the planet Neptune. Pluto had sometimes been credited with this effect, but Dr. Whipple said that Pluto could scarcely weigh as much as the earth, while the objects producing such an effect would have to weigh ten times as much as the earth. (Science Service, *NYT*, 11/14/63, 3)

November 14: London *Daily Herald* reported Rolls Royce, Ltd., had submitted proposals to the British government for sending space probes to the moon and Mars by 1970, by converting the Blue Streak booster to use of liquid hydrogen. (UPI, *Wash. Post*, 11/15/63)
- SAC crew launched a Titan I ICBM from Vandenberg AFB, Calif., in a routine training launch. (AP, *Wash. Post*, 11/15/63)
- Soviet scientists reported that radio soundings had indicated that the surface of the moon was "hard and extremely porous," in contrast to other theories that the moon was covered with a thick layer of dust. (UPI, Wash. *Daily News*, 11/14/63; *Izvestia*, 11/15/63, 6, AFSS–T Trans.)

November 15: NASA's Flight Research Center, FAA, and U.S. Navy announced completion of a series of test flights simulating the supersonic transport aircraft (SST) flying in the air traffic control network. A Navy A5A jet aircraft was used to simulate the SST because its thrust-to-weight ratio enabled it to simulate closely the performance of the SST up to 50,000 ft. Some 21 flights were made into Los Angeles along Federal airways, starting from level flight at 50,000 ft. and mach 1.7, then decelerating to mach 1.4 and descending, dropping below the speed of sound before reaching 30,000 ft., and flying at 340 mph at 20,000 ft., not much faster than present-day jet transports. Flight controllers reported no problems with descent and landing, although takeoff and climb did present more problems because of the high performance of the aircraft. (FRC Release 31–63)

November 16: U.S.S.R. launched COSMOS XXII into earth orbit. Initial orbital data: apogee, 394 km. (245 mi.); perigee, 205 km. (127 mi.); period, 90.3 min.; inclination, 64°56'. (*Pravda*, 11/17/63, 1, AFSS–T Trans.)
- President Kennedy visited Cape Canaveral, his third tour of the test site within two years. Flying in from Palm Beach, where he was spending the weekend, the President spent three hours watching the launching of a Polaris missile from the submerged nuclear submarine U.S.S. *Andrew Jackson*, riding in a helicopter over the Merritt Island Launch Area (MILA) where NASA is constructing facilities for Saturn V launches, visiting the Saturn I rocket on its launch site. Speaking of the Saturn I flight scheduled for early December, the President said that if successful it would give the U.S. "the largest booster in the world and show significant progress in space.

"I think that would be an important milestone for us."

The President was briefed on Projects Gemini and Apollo and the related booster programs by a group of NASA officials including NASA Administrator James E. Webb, NASA Associate Administrator Dr. Robert C. Seamans, Jr., and MSFC Director Dr. Wernher von Braun. At one point in the illustrated lecture, the President lagged behind his party to examine models of the various boosters in the space program. He picked up a model of the 72-ft. Atlas D which had boosted the Project Mercury astronauts into orbit and compared it with a model of the 281-ft. Saturn V intended to propel Project Apollo astronauts to a manned lunar landing. "This is fantastic," the President said. (*NYT*, 11/17/63, 1; *Wash. Post*, 11/17/63, 2)

November 16: U.S. and U.S.S.R. were reported to have a draft of a resolution on legal aspects of space exploration and to be circulating it for comment to other members of the 28-nation U.N. Committee on the Peaceful Uses of Outer Space prior to formal submission. It was announced on Nov. 8 that the U.S. and U.S.S.R. had reached general agreement. (Brewer, *NYT*, 11/17/63, 34)

• Soviet Marshal Nikolai I. Krylov, writing in *Izvestia*, claimed that Soviet ICBM's "in many respects exceed both quantitatively and qualitatively" the strategic rocket potential of any other nation. The new-model tests in the Pacific in the spring of 1963 had produced "fabulous super-sniper accuracy" over a range of 8,000 mi., he said. Among characteristics of Soviet ICBM's, the Marshal listed "unlimited range," relative ease of operation, combat readiness "within minutes," and capability of launching from mobile field installations having no preliminary engineering work. (*NYT*, 11/17/63, 39)

November 17: NASA launched an Aerobee 150A sounding rocket from NASA Wallops Station, Va., sending a 200-lb. payload 115 mi. into the upper atmosphere to measure ultraviolet radiation. Experiment was part of a continuing program to develop techniques for measuring ultraviolet dayglow and from the measurements determining the physical processes taking place in the upper atmosphere. Such techniques would also be useful as instrumentation on satellites studying the atmospheres of other planets. (Wallops Release 63–103)

• First international art program via communications satellite was presented over NBC–TV. Entitled "Museums Without Walls," the half-hour program interchanged views of paintings from the Louvre in Paris and the National Gallery in Washington via Relay communications satellite. The program went off without any significant technical difficulties and reception was excellent. (Gould, *NYT*, 11/18/63, 59)

November 18: Speaking in Tampa, Fla., at a ceremony celebrating the 50th anniversary of the first scheduled air service in the U.S., President Kennedy contrasted the first air service between Tampa and neighboring St. Petersburg with today's jet travel. Looking ahead, he said that "within our sight, if not yet within our grasp, is the day when men will routinely fly through space at 25 times the speed of sound." As soon as 1975, the President predicted, "metroplanes" would take off from the center of one city and land in the center of another, while supersonic jet transports would be crossing the oceans at 1,400 mph. And beyond that would be "hypersonic transports" flying at mach 6 to 8. (*Wash. Eve. Star*, 11/18/63)

• The House Select Committee on Government Research opened its hearings with nine witnesses from Government and university scientific circles. Created by the House last September "to make a complete, full and thorough investigation of the numerous research programs" conducted by or sponsored by the Federal Government, the Committee would by November 1964 have listened to some 70 scientific leaders from Government, universities, industry, and labor. (Finney, *NYT*, 11/19/63, 18)

November 18: NASA Administrator James E. Webb, testifying before the Select Committee on Government Research of the House of Representatives, said: "Something more than four billion dollars will be paid industry for work it will do under NASA contracts this year. About half of this will pay for the design of this advanced equipment and for its fabrication, using the latest methods and newest technologies in its production. . . . Just to make sure that the design is right, the fabrication is right, and the equipment sufficiently reliable, the other half of this four billion dollars will be paid to industry to test these machines and advanced systems of equipment once they have been fabricated. Thus American industry is moving into an entirely new period where the importance of very advanced design, manufacture, fabrication, and testing for utmost reliability will replace some of the practices of the past." (Text)

- Former NASA Director of Manned Space Flight, D. Brainerd Holmes, said in Hartford address that industry and government leaders must discover better ways to manage the U.S. space program. He discussed Project Apollo plans and said the space flight programs must be based on better mutual confidence between Government and industry. (Hartford [Conn.] *Courant*, 11/19/63)
- Dr. Frederick Seitz, President of the National Academy of Sciences, testified at the initial hearing held by the House Select Committee on Government Research. Dr. Seitz warned Congress against starving basic research and individual or small-team research "in order to divert funds for research which appears to be more spectacular from a popular standpoint . . . . There appear to be signs of such diversion in the present session of the Congress." As evidence of this trend, Dr. Seitz pointed out the relative size of the cuts in the NASA budget compared with the cuts of the National Science Foundation. Also, he pointed out, even the NASA cuts were heavier in space sciences than in manned space flight or applications. NASA, he thought, should play a role like that of the Office of Naval Research ten years ago, offering broad support to basic research in chemistry, physics, astronomy, and other sciences. (Loory, N.Y. *Herald Tribune*, 11/19/63)
- Secretary of Defense Robert S. McNamara, in address before Economic Club of New York, summarized current status of "the balance of strategic nuclear forces":

"The U.S. force now contains more than 500 operational long-range ballistic missiles—Atlas, Titan, Minuteman, Polaris—and is planned to increase to over 1,700 by 1966. There is no doubt in our minds and none in the minds of the Soviets that these missiles can penetrate to their targets. In addition, the U.S. has Strategic Air Command bombers on air alert and over 500 bombers on quick reaction ground alert. By comparison, the consensus is that today the Soviets could place about half as many bombers over North America on a first strike. The Soviets are estimated to have today only a fraction as many intercontinental missiles as we do. Furthermore, their submarine-launched ballistic missiles are short range, and generally are not comparable to our Polaris force. The Soviets pose a very large threat against Europe, including hundreds of intermediate and medium-range ballistic missiles. This threat is today and will continue to be covered by the clear superiority of our strategic forces . . . ." (Text)

November 18: Willis B. Foster became Director of new Manned Space Science Div. of NASA's Office of Space Science and Applications. Formerly Deputy Assistant Director for Research in DOD's Directorate of Defense Research and Engineering, Mr. Foster would direct liaison and scientific support of the manned space flight program, including the criteria and selection of scientist-astronauts and scientific R&D in support of manned space flight. (NASA Release 63-242)

- NASA had launched only six of a projected 40 major satellites and probes so far in 1963, according to *Aviation Week*. Though all six had been successful, it was a big drop from the 18 launches in 1962. NASA Associate Administrator Dr. Robert C. Seamans, Jr., attributed some of the drop in the launch rate to "squaring away difficulties which existed before NASA was created or which developed in the early days of the agency." He included Centaur and Nimbus in this category.

  "As for some of the other delays and postponements, I think they are due to our involvement in projects where we are doing things for the first time. In building spacecraft such as EGO, and the Orbiting Astronomical Observatory, we are learning how to design, assemble and test advanced spacecraft—a new generation of spacecraft that are much more complex than the ones we have been flying so successfully." (*Av. Wk.*, 11/18/63, 30, 31)

- NASA Administrator James E. Webb was showing interest in acquiring USAF Titan III boosters to launch future heavy payloads such as observatory satellites, according to *Aviation Week*. Webb was said to have discussed with DOD's Dr. Harold Brown the possibility of adding Titan III to the list of standard launch vehicles in the national space program. (*Av. Wk.*, 11/18/63, 25)

- Japan's Science and Technology Agency announced a five-year space program that included development of Japanese boosters and launching of three satellites. It would be funded at $59,138,888 and employ some 300 Japanese engineers. Programs included a 60-km. weather observation rocket; a 150-km. two-stage weather observation rocket; 400-km. three-stage rockets; 1,200-km. two-stage rockets; navigation satellites, to be ready for launching by NASA in 1967. (*M&R*, 11/18/63, 13)

- The ratio of NASA scientists and engineers to NASA-contractor scientists and engineers should be 1 to 4 by early 1964. It was 1:2 in Jan. 1960, 1:3 in Jan. 1963. (*Space Bus. Daily*, 11/18/63, 265)

- Astronomers at McMath Hulbert Observatory of the Univ. of Michigan and at the U.S. Naval Observatory reported the first signs of a new 11-yr. solar cycle. Indications of a new cycle were the appearance of sun spots well above the sun's equator and the reversed magnetic polarity of the disturbance. (Science Service, *NYT*, 11/18/63, 15)

November 18: Two USAF U-2 aircraft have been used at AMR for more than three years in a program gathering data on infrared characteristics of the exhaust plumes of ballistic missiles, according to *Aviation Week*. The program had been sponsored by Aeronautical Systems Div. of AFSC. Known as Smokey Joe, it was coordinated with ARPA's Project Tabstone, which includes aircraft-user programs Lookout (using RCAF CF-100's) and RAMP (Radiation Air Borne Measurement Program). One of the U-2's was equipped with 400-lb. radiometer, the other with spectrometer. One or both covered 80% of AMR launchings, flying at 60,000 ft. to observe the plume of the rocket as it passed on its upward flight. Data was forwarded to agencies concerned with ballistic missile defense. (*Av. Wk.*, 11/18/63, 53)

- USAF-sponsored solar hydrogen rocket engine (SOHR) was successful in series of ground tests, producing up to one pound thrust in runs of several hours. Scale-up to 10-lbs. thrust now appeared feasible. Built by Electro-Optical Systems, Inc., SOHR uses large, lightweight solar concentrators to focus solar energy into a cavity absorber containing a heat exchanger using hydrogen. The heated hydrogen provides thrust as it is accelerated through a nozzle. One SOHR generating one to two lbs. thrust could move a 6,000-lb. payload from a 300-mi. orbit to a 22,400-mi. synchronous orbit in 20 days, compared with half a day for a chemical powerplant and 45 days for an electrical one. Specific impulse of SOHR is 800 sec., compared with 400 sec. for chemicals, 1,200 for arc jets, and 5,000 sec. up for ion rockets. (*M&R*, 11/18/63, 35)

- U.S. had produced a total of 53 U-2 reconnaissance aircraft, according to *Aviation Week*. The first model, of which 48 were produced, was a single seater; most of them were powered by the Pratt & Whitney J57-P-37A (JT3) jet engine, a few—including the one flown by Francis Gary Powers—by P&W J75's. The second model was two-place, powered by a modified P&W J75 twin-spool engine. Five were produced. (*Av. Wk.*, 11/18/63, 57)

- Editorial in *Missiles and Rockets* took issue with widespread assertions that the space program was a serious drain on the Nation's scientific and technical manpower. Asserting that no other part of the economy was known to have a serious shortage of scientists or engineers and that even if the space program were stopped tomorrow the crosstraining of scientists and engineers into some other part of the economy would be long and difficult if not impossible, the editorial concluded that the big problem of the future would be a shortage of skilled manpower against a background of a rising threat of unemployment. The Nation would need an estimated 34.5 million new jobs by 1970. "New jobs develop from new technologies. And one of the fastest and most promising routes to development of new technologies lies in the missile/space industry.

    "With such a demand for new jobs ahead and with a concurrent need for skilled workers, the answer seems obvious—yet has received little attention. Training of surplus manpower to fit the requirements for skilled workers is the only sensible route—not a cutback in what will soon be the nation's largest industry.

But we have yet to see any large-scale government program for training of unemployables.

"Nor is it solely a government problem. It must be shared by the nation's industrial management and labor unions." (*M&R*, 11/18/63, 46)

November 18: French government report recognized the direct and indirect economic benefits to the French economy from space and nuclear work. Of 13,000 persons employed in rocket development, 5,000 were in electronics, 5,000 in chemical-metallurgy. Of 14,000 working at the Pierrelatte nuclear center, 92% are employees of private industry. (*M&R*, 11/18/63, 9)

- USAF was reported by *Aviation Week* to have canceled its specific operational requirement (SOR) for the CX-4 cargo aircraft (100,000-lb. payload, 4,000-mi. range) in favor of an upgraded requirement for a CX-X concept that would provide 180,000-lb. payload and 10,000 to 12,000-mi. range by taking advantage of new state of the art improvements expected to be available within the next two years, including the high-bypass-ratio turbofan engine for lower fuel consumption, boundary-layer control for increased lift and range, and lighter structural materials, including plastic and metal honeycombs, ceramic-metal combinations for use in high-temperature portions of the engines, and other plastic parts. CX-X would be able to deliver troops and equipment overseas, airdrop them, and return to U.S. without landing or refueling. Cost per ton mile was expected to be less than the long-sought five cents per ton mile; indeed, it was thought to be cheaper than sea lift when antisubmarine forces are included. (*Av. Wk.*, 11/18/63, 26)

November 18-20: In a paper presented to the national symposium of the Society of Aerospace Material and Process Engineers in Seattle, Wash., D. J. Levy, a Lockheed Missiles and Space Co. scientist, discussed a revolutionary new gold-plating technique. His own invention, the technique makes it possible to goldplate by the simple method of spraying a gold solution from an ordinary spray gun, or from an aerosol bomb. Gold is the ideal substance for thermal control because of its optical properties and resistance to tarnish. This new method, "Lockspray-Gold," has the advantage of great flexibility since there is no limit to the size of the object that can be coated. Levy also described how his gold-spray process was successfully used in the design of special lightweight furlable antennas developed by Lockheed for space vehicles. (*Space News Roundup*, 12/11/63, 2)

November 19: A NASA Hq press conference confirmed that the combustion instability that had plagued the 1.5 million-lb.-thrust F-1 rocket engine had been corrected. NASA had assembled a team of the best propulsion experts in the country to work on the problem. The nature of the corrective action was redesign of a part of the engine behind the injector, so that fuel and oxydizer were no longer subject to surging as they entered the thrust chamber. Further redesign was underway to simplify the rather complicated series of baffles in the new part. (*Wash. Eve. Star*, 11/19/63)

November 19: NASA Associate Administrator Dr. Robert C. Seamans, Jr., speaking before the Chamber of Commerce, Albuquerque, N.M., said: "All of you saw the headlines saying, 'Khrushchev Announces Withdrawal from Moon Race' or words to that effect; but I wonder how many read the subsequent stories in which Khrushchev said his remarks had been misinterpreted, and that the Soviet Union intended to explore the moon as soon as this could be done with confidence of success.

" 'Even our opponents realize that we have the leading role in space,' Khrushchev said recently. 'They haven't overtaken us yet, and we are not going to let them. Our people are covering themselves with glory. We shall give the capitalist world no peace, since it has to go.'

"We have a balanced, fast paced space program geared to our national needs and resources. I can assure you that the Soviets are also working hard on a well conceived program, even though they throw a veil of secrecy over their failures and try to confuse us about their objectives." (Text)

- FAA announced a plan for allocation of the U.S. supersonic airliners as they came off the production line. U.S. airlines would receive 44 of the first 70 aircraft, with preference going to airlines flying the Atlantic. Foreign airlines would receive the remaining 26 of the first 70, again with preference going to airlines on the Atlantic Ocean run. (*NYT*, 11/20/63, 70)

- In the Senate debate on the Independent Offices Appropriations bill, Sen. J. W. Fulbright (D.-Ark.) made a motion that the NASA appropriation be cut an additional $519 million—a flat 10% cut in each of the three NASA budget categories: research and development, construction of facilities, and administrative operations. "Simply stated," the Senator said, "the purpose of the amendment is to allow time to reevaluate the goal of trying to reach the moon in this decade and to proceed on a more deliberate and thoughtful basis." (*CR*, 11/19/63, 21270–299)

- Dr. Joseph F. Shea, Program Manager, Apollo Spacecraft, MSC, spoke before the National Rocket Club in Washington. He explained the philosophy behind the recent changes in the Apollo flight-test program, when Saturn I launch vehicles were replaced by Saturn IB's and the first manned flight thereby delayed for some nine months. Dr. Shea pointed out that the space age had advanced to the point that the space environment is understood well enough so that components can be designed, built, and ground-tested with confidence that they can perform well in space. The flight test program has become "the relatively small portion of the development iceberg which is clearly visible to the public." This had led to a desire to flight-test the entire spacecraft configuration from the outset, rather than the previous method of adding additional systems flight by flight, when "every spacecraft is a somewhat different design containing the potential of new problems.

"The discipline of a single configuration—configuration control, if you will—not only increases the carry over from test to test so necessary for establishing design confidence, but also provides repeated opportunities for early spacecraft tests of substantial scope." (Text)

November 19: Dr. Philip Abelson, Director of the Carnegie Institution's Geophysical Laboratory and editor of the AAAS magazine *Science,* criticized the mechanism for providing the President with scientific advice. In a speech in Houston, he said: "Too much power is concentrated in a few overworked people, most of whom are of limited competence. The secrecy of the operations and the power invoked must inevitably lead to corruption, either of the intellect or of the purse. . . ." As to the President's Scientific Adviser, Dr. Jerome B. Wiesner, Abelson charged he "has accumulated more visible and invisible power than any scientist in the peacetime history of this country . . ." and was a man who had failed "to engage in forward planning" of science policy. (Simons, *Wash. Post,* 11/20/63)

November 20: Senate passed (72–1) the appropriations bill for Independent Offices (H.R. 8747), which included the NASA FY 1964 appropriations, and sent the bill to a conference committee to reconcile Senate and House versions. Senate amendments included one by Sen. Proxmire cutting NASA appropriations by $90 million ($80 million from R&D, $10 million from construction of facilities), bringing the total NASA appropriation approved by the Senate down to the $5.1 billion approved by the House; and one by Sen. Anderson requiring NASA to notify Congress 30 days in advance of any transfer of R&D funds in excess of $250,000 to construction of facilities. Amendments defeated included the one by Sen. Fulbright to cut the NASA appropriation by 10% ($519 million). (*CR,* 11/20/63, 21339–382; *NYT,* 11/21/63; *Wash. Post,* 11/21/63)

- President Kennedy congratulated the U.S. delegation to the recent International Telecommunications Union conference in Geneva, said the conference "has been one of the most successful of its kind held in recent times." The U.S. and the ComSatCorp "can now take practical steps, in cooperation with other governments and foreign business entities to develop a single global communications system." Reassuring small countries on their role in the communications satellite system, the President said: "It continues to be the policy of the United States that all countries which wish to participate in the ownership, management and use of this system will have the opportunity to do so." (MacKenzie, *Wash. Post,* 11/21/63)

- President's report to Congress on U.S. participation in the U.N. during 1962 stated:

    "In two other fields the United Nations has continued to be a vital instrument to effect a disengagement in important sectors of the great power confrontation. The Organization has served as a forum for encouraging an agreement for the cessation of nuclear weapon testing and for promoting progress toward general disarmament. It has served, as well, as a mechanism for negotiating legal principles and technical cooperation in outer space. We must be no less concerned with these persistent efforts to shape the future within the framework of the United Nations Charter than we are with United Nations operations designed to respond to the alarm bells of the present." (*CR,* 11/20/63, 21321)

November 20: Lewis scientists reported on studies of radiation phenomenon they believe caused VANGUARD I's voice to change. Dr. Jim Blue, Lewis scientist, attributed the "voice change" to a structural change inside Vanguard's quartz crystal. VANGUARD I, launched March 17, 1958, first noted the earth's pear shape, (LRC Release 63-93, Lewis Chronology, 11)

- Tass reported that a leading Soviet space scientist, Prof. Gleb Chebotarev, claimed U.S.S.R. would put satellites in orbit around the moon, Mars, and Venus "in the near future.... The regions of outer space around the moon, Venus, Mars and Mercury where stable movement of artificial satellites is possible have been determined by means of electric machines at the Theoretical Astronomy Institute." (UPI, *NYT*, 11/21/63)

- DOD announced the transfer of the Naval Missile Facility at Point Arguello, Calif., and Navy tracking stations in the Pacific to the USAF and the establishment within the USAF of a central authority to act as single manager for ICBM and space tracking activities, including AMR, ICBM and space activities at Point Arguello and Vandenberg AFB, Calif., and the Air Force Satellite Control Facility, Sunnyvale, Calif. USAF also acquired on-orbit control of DOD satellites except Navy navigational satellites and military communications satellites. (DOD Release 1494-63)

- Dr. Edward C. Welsh, Executive Secretary of the National Aeronautics and Space Council, speaking before the Atomic Industrial Forum Conference in New York, said that nuclear energy was essential to the space program and offered larger potential in space applications than in terrestrial ones. "One of the major defects in space thinking today is that it is too 'short-run'. There is a tendency to think only from budget to budget rather than from decade to decade. It has taken great effort—not altogether successful—to get people to think as long-run as a manned flight to the moon. In terms of what space travel and space exploration mean to this country, the lunar objective is really a short-run target. Yet, some people think of it as the final space goal which is so far in the future that it can be further delayed without serious effects. If we are to think clearly about space, it is important that we recognize the lunar program for what it is, i.e., only one of the initial steps in a very young program. I believe that those who understand and support nuclear energy in space have a responsibility to educate others in the long-run, broad-based view of space.... I am suggesting that, if we can get people to realize now that space travel and space exploration are permanent features of the economy, the idea of nuclear energy will automatically be considered a major element in the program." (Text)

- An American U-2 reconnaissance aircraft believed to be returning from a mission over Cuba had crashed into the Gulf of Mexico 40 mi. north of Key West. An air-sea rescue search failed to recover the pilot. (UPI, *Wash. Post*, 11/21/63)

- Sen. Clinton P. Anderson (D.-N.M.), Chairman of the Senate Committee on Aeronautical and Space Sciences, spoke to the 1963 Annual Conference of the Atomic Industrial Forum in New York on the relationship between Congress and science. Sen. Anderson offered four main reasons for the new level of Congressional concern over the Federal investment in science and technology: (1) cost consciousness; (2) "the belief among some Members that Congress has lost the ability to oversee effectively the vast diffusion of R&D activities for which it appropriates funds"; (3) "concern that the procedures of Congress, in some respects, may not measure up to the demands of 'big science' "; and (4) "criticism of the space program as 'moon madness' and 'lunacy' gets lumped in with the criticism of heavy spending for research and development." Sen. Anderson agreed with the aims of those who were cost conscious, although he noted that many R&D programs that had been canceled as "failures" had in fact left a valuable technological legacy to subsequent programs. He agreed that Congress needed more advice on technical matters, but doubted that the need would be solved by Congress hiring its own technical staff; rather it needed more help from specially convened panels; more engineering advice—since the bulk of the Nation's $15 billion annual investment in science and technology is in the engineering of hardware rather than in science; improved presentations by the Executive Department, including more briefing by the Office of Science and Technology; expansion of channels of information such as Library of Congress, NAS, NSF; reports by the scientific and engineering communities to Congress on the state of science and of engineering; and more long-term planning. (*CR*, 12/3/63, 21997–22000)

November 20: First award of the recently established Burroughs International Test Pilot Award was made to Joseph T. Tymczyszym, Chief of FAA's West Coast Supersonic Transport Office. The award presented by FAA Administrator Najeeb E. Halaby at the Wings Club in New York was established by United Aircraft Corp. in honor of Richard H. Burroughs, Chance Vought test pilot who was killed when he stayed with his disabled experimental aircraft to guide it away from an inhabited area. Award would be administered by Flight Safety Foundation, Inc., of New York. (FAA Release 63–96)

- Army fired a Pershing solid-fuel missile from Fort Wingate, N.M., 260 mi. to White Sands Missile Range, N.M., in a public demonstration of the battlefield missile prior to its deployment to Europe early in 1964. This was the 15th firing of the Pershing to WSMR from various distances away. (Langguth, *NYT*, 11/21/63, 25)

- USAF accepted the first two F–4C tactical fighter aircraft into the Air Force inventory in ceremonies at MacDill AFB, Fla. The F–4C is the Air Force version of the Navy's F–4B, combines into one aircraft the capability for close air support, interdiction, and air superiority. (DOD Release 1496–63)

November 21: President Kennedy visited San Antonio and Houston. In San Antonio, he participated in the dedication of the new $6-million Aerospace Medical Health Center at Brooks AFB. "Too many Americans make the mistake of assuming that space research has no value here on earth. Nothing could be further from the truth," the President said, and then went on to enumerate some of the advances in medical science and technique that had come from space research.

Referring to the reduced NASA budget passed by the Senate the previous day, the President said: "There will be pressures for our country to do less and temptations to do something else. But this research must and will go ahead. That much we know. That much we can say with confidence and conviction . . . .

"Our effort in space is not, as some have suggested, a competitor for the national resources needed to improve our living standards. It is instead a working partner and coproducer of those resources." (Wicker, *NYT*, 11/22/63; Kiker, N.Y. *Herald Tribune*, 11/22/63)

November 21: USAF Outstanding Unit Award was presented to the Air Force School of Aerospace Medicine for having "formulated new scientific concepts and performed original research of great national and international significance." Award was made during ceremonies in which President Kennedy dedicated the school's new facilities at Brooks AFB, Tex. (*Air Force Magazine*, 1/63, 84)

- First rocket to be launched from India was achieved as the result of the coordinated efforts of France, India, and the U.S. The Nike-Apache, launched from Thumba, the site near the southern tip of India that would become an international rocket launching facility for IQSY experiments, reached an altitude of 106 mi., and was the first of four sodium-vapor experiments to determine speed and direction of upper atmosphere winds. (NASA Release 63-105; NASA Rpt. of S. Rkt. Launching, 12/12/63)

- GSFC announced that under a $2 million contract now under final negotiation, Sperry Rand Corp.'s Univac Div. would deliver eleven Model 1218 computer systems to manned space flight tracking stations for operation by July 1964. These computers would automatically summarize telemetry from the spacecraft, provide summaries for display in the mission control center so that the controllers can select and examine certain data on a real-time basis, and prepare the telemetry data for final processing in the more elaborate computers at GSFC and MSC. During the Mercury program, controllers at the tracking stations had to select data manually. (GSFC Release G-26-63)

- Text of U.S.-U.S.S.R. agreement on legal principles of space exploration was released at the U.N. in New York. The document would be brought before the U.N. Committee on the Peaceful Uses of Outer Space on Nov. 22. Even if approved, the statement of principles would not be legally binding on U.N. members. The plan was for two treaties based on the principles to be submitted to member nations early in 1964. The treaties would cover (1) rescue and return of astronauts and space vehicles, and (2) liability for death, injury, or damage caused on earth by space vehicles. (*Wash. Post*, 11/22/63; Text, *NYT*, 11/22/63, 18)

- British scientist Dr. John F. Kerridge of Univ. of London reported at the cosmic dust conference in New York sponsored by the New York Academy of Sciences that he had found material in meteorites that closely resembled the kinds of clays formed by water action on earth. This would argue that the meteorites were once part of a body large enough to have its own atmosphere and hence its own weather. This supported previous findings by Dr. Bartholomew Nagy of Fordham Univ., who also claims to have found fossil-like objects in meteorites. (Sullivan, *NYT*, 11/22/63, 32)

November 21: Dr. Jerome B. Wiesner, Scientific Adviser to the President, defended himself and his staff against charges by Dr. Philip Abelson the previous day. Testifying before the Senate Committee on Aeronautical and Space Sciences in favor of continuing NASA's graduate training program, Dr. Wiesner said of Abelson: "His statement is so inconsistent that it's hard to answer. He accused me of being a czar . . . but also said I haven't done much." Wiesner defended his staff against Abelson's charge that they were not outstanding scientists: "A Nobel Prize winner is not necessarily the best man to advise the President." (*Wash. Post*, 11/22/63)

- Testifying before the House Select Committee on Government Research, Dr. Vannevar Bush said:

  "The spectacular success of applied research during the war led to a fallacy entertained by many. It is that any problem can be solved by gathering enough scientists and giving them enough money . . . . It is folly to thus proceed. The great scientific steps forward originate in the minds of gifted scientists, not in the minds of promoters. The best way to proceed is to be sure that really inspired scientists have what they need to work with, and leave them alone.

  "If the country pours enough money into research, it will inevitably support the trivial and the mediocre. The supply of scientific manpower is not unlimited.

  "In any broad program of research the keyword in regard to any one aspect of the program is 'relevance.' It is a good word to have in mind in examining any research program. Competent directors of research know what it means. Probably 'conducive to progress toward the main object of a program' is as good a definition as any. Just finding out something new is not by itself sufficient justification for research. It needs to mean something when we find it.

  "When scientific programs are judged by popular acclaim we inevitably have over-emphasis on the spectacular. That is just what we have today. The deeply important scientific advances moving today are not easy to understand. If they were they would have been accomplished long ago. Outstanding scientific progress, which will most affect the lives and health of our children, is not grasped by many."

  Objecting to the practice of the government and the armed forces of utilizing universities in the management of secret programs, Bush said:

  "It should never be forgotten that the main task of the universities is to educate men. The country will need skilled professional men in the future as much as it will need new knowledge.

As we now go we are not meeting this challenge sufficiently. Every research program placed in a university should be so ordered that its product is not only new knowledge but skilled educated men." (Text)

November 21: The House Republican task force on space and aeronautics issued a 15-page report on the Administration's space program. Main criticism was that the military space program was being neglected. "Too little progress has been made toward the development of a strategic space capability. In fact, the tendency of the present Administration is in the opposite direction—away from any military capability in space whatsoever other than conventional unmanned missiles." Meanwhile, the report charged, the civilian space program "has grown so big so fast that waste and inefficiency have been all too inevitable." Although NASA was taking some steps to tighten controls, it "could and should do more." In international space agreements, the report charged that the U.S. had given more than it received in the agreements reached with the Soviet Union. Furthermore it contended that the Administration was usurping the powers of Congress by entering into a series of space agreements without submitting them to the Senate for its advice and consent. Finally, the report expressed concern over rising costs of Federal R&D and its competition with industry. "Competition between Government and industry in research and development appears to be increasing rather than decreasing, with industry operating at a natural and unnecessary disadvantage. This apparent imbalance should be corrected." (*NYT*, 11/22/63, 21)

- The old question of whether sun spots affect human behavior received new empirical evidence. Drs. Howard Friedman, Robert O. Becker, and Charles H. Bachman of the State Univ. of New York and Syracuse Univ., reported on a study they had made in which the daily admission rate in seven New York psychiatric hospitals was compared with the daily variation in the earth's magnetic field over a period of four years. The comparison revealed that "the greater the intensity of the earth's magnetic field due to 'sun spots' and other natural interference coming out of the cosmos, the higher the rate of psychiatric hospital admissions." The effect was explained by the fact that all creatures with central nervous systems operate on self-generated, direct-current electricity. The same is true of radio transmitters and receivers. In both cases the electric impulses must travel through the earth's magnetic field. This magnetic field is normally smooth flowing. Only when it is disturbed does radio transmission suffer. The doctors reasoned that the same thing might occur in the human nervous system. The doctors concluded: "Speculatively, the results are in keeping with the conception of the behavior of an organism being significantly influenced, through the direct-current control system, by external force fields. Attention is thus invited to a hitherto neglected dimension in the complexity of psychopathology specifically, and perhaps generally in all human behavior." (UPI, *Wash. Post*, 11/22/63)

November 21: Donald K. Slayton was sworn into his Civil Service appointment as Assistant Director for Flight Crew operations, MSC. His resignation from the USAF had become effective the previous day. (MSC Release 63-242)

November 22: President John F. Kennedy was assassinated in Dallas, Tex. Within two hours of the President's death, Vice President Lyndon B. Johnson had taken the oath as the 36th President of the U.S., aboard Air Force No. 1, the Presidential jet aircraft. (*NYT*, 11/24/63, E1)

- President Kennedy, speaking at a breakfast in Ft. Worth a few hours before his assassination, mentioned as part of the growing strength of the U.S. the development of the TFX aircraft:

    "There's been a good deal of discussion about the long and hard-fought competition to win the TFX contract; but very little discussion about what this plane will do.

    "It will be the first operational aircraft ever produced that can literally spread its wings through the air. It will thus give us a single plane capable of carrying out missions of speed as well as distance; able to fly very far in one form, or very fast in another.

    "It can take off from rugged, short airstrips, enormously increasing the Air Force's ability to participate in limited wars. The same basic plane will serve the Navy's carriers, saving the taxpayers at least one billion dollars in costs if they built separate planes for the Navy and the Air Force." (*NYT*, 11/24/63, 2)

- In the speech which President Kennedy was on his way through Dallas to deliver when he was assassinated, the President had prepared a statement on U.S. preparedness. After reviewing the improvements in military readiness, the President would have said:

    "I have spoken of strength largely in terms of the deterrence and resistance of aggression and attack. But, in today's world, freedom can be lost without a shot being fired, by ballots as well as bullets. The success of our leadership is dependent upon respect for our mission in the world as well as our missiles—on a clear recognition of the virtues of freedom as well as the evils of tyranny . . . .

    "And that is also why we have regained the initiative in the exploration of outer space—making an annual effort greater than the combined total of all space activities undertaken during the fifties—launching more than 130 vehicles into earth orbit—putting into actual operation valuable weather and communications satellites—and making it clear to all that the United States of America has no intention of finishing second in space.

    "This effort is expensive—but it pays its own way, for freedom and for America. For there is no longer any fear in the free world that a Communist lead in space will become a permanent assertion of supremacy and the basis of military superiority. There is no longer any doubt about the strength and skill of American science, American industry, American education and the American free enterprise system. In short, our national space effort represents a great gain in, and a great resource of, our national strength . . . ." (*NYT*, 11/24/63, 2)

November 22: The first live transmission of TV signals across the Pacific Ocean was accomplished via RELAY I communications satellite. The U.S. ground station in the Mohave desert transmitted the pictures, which were relayed by the satellite to Japan's new Space Communications Laboratory north of Tokyo. Viewers in Japan saw and heard taped messages from Japanese Ambassador Ryuji Takeuchi and NASA Administrator James E. Webb, as well as scenic and cultural sequences. A message of greeting that President Kennedy had taped for the occasion was deleted when word came in of his assassination a few hours before the broadcast was to take place. ABC- and NBC-TV shared in producing the program. (NASA Release 63–256; *NYT*, 11/24/63, 14)

November 23: Communications satellites have been used for some time by AT&T to transmit a small portion of the overseas telephone calls that normally would have gone by underwater cable. After each call the telephone company would call up the user and ask him about the quality of his call. None of the users were told that they had talked via satellite, since part of the experiment was to test the user's psychological reactions and these might have been affected by such knowledge. Although the communications satellites are not licensed for commercial use, AT&T would pay the U.S. government a fee for the time used for the experiments. Only the medium-altitude Telstar and Relay satellites were used. (Finney, *NYT*, 11/24/63, 16)

November 24: The space program would not change greatly under the new President, Lyndon B. Johnson, William Hines of the Washington *Sunday Star* predicted: "The reasons are twofold and simple.

"First, Mr. Johnson is fully as enthusiastic about the space program as was Mr. Kennedy, if not more so. Second, the new President has been in on the space program from its start, and through the 34 months of President Kennedy's tenure was 'Mr. Space.' "

Mr. Johnson was head of the Senate Preparedness subcommittee at the time the U.S.S.R. launched SPUTNIK I, Hines pointed out. As Senate Majority Leader he caused the inquiry into U.S. space posture and sparked the special Senate committee that later became the standing Committee on Aeronautical and Space Sciences.

"More than any other one man, President Johnson is personally responsible for the National Aeronautics and Space Act of 1958 and the sprawling agency which it created . . . .

"With the advent of the Kennedy administration, the space program immediately grew in importance, largely through Mr. Johnson's urging. The Space Act was amended to make the Vice President head of the [Space] council . . . .

"Under then Vice President Johnson, the council took on real meaning. . . .

"The Johnson-organized council staff prepared the studies which the Vice President used to persuade President Kennedy in the spring of 1961 that a manned lunar landing should be undertaken 'before the decade is out.' " (Hines, *Wash. Sun. Star*, 11/24/63)

November 24: A group of scientists met in Seattle to discuss policy for the Pacific Science Center. Formerly the U.S. exhibit at Seattle's Century 21 Exposition, the Center is located in the six-building complex covering more than six acres, it having been leased by the Government to the nonprofit Pacific Science Center Foundation. A 12-man science advisory committee headed by Paul A. Scherer, former associate director of the National Science Foundation, would advise on policy and exhibit content as part of the effort to convert the area into a permanent institution. It would become part of the 74-acre Seattle Center. (Davies, *NYT*, 11/24/63, 18)

November 25: President John F. Kennedy was buried in Arlington National Cemetery in a state funeral attended by the largest gathering of foreign dignitaries ever to visit Washington. Throughout the four days of stark events, the Nation and the world participated to a degree never before possible by means of round-the-clock TV coverage that raised that young medium to a new dimension for thoroughness, maturity, and sensitivity. RELAY I communications satellite enabled all of Europe, including the U.S.S.R., to view the funeral ceremonies. The satellite also provided transmission across the Pacific to Japan, where an estimated 95 million persons viewed the ceremonies. (*NYT*, 11/26/63)

- Stratoscope II balloon was launched from Palestine, Tex., landed near Kosciusko, Miss., some 18 hrs. later, having lifted a 3½-ton 36-in. telescope some 80,000 ft. where the telescope, viewing space from a point above 95% of the earth's interfering atmosphere, photographed the infrared light coming from Jupiter, the moon, and two giant red stars. Dr. Marvin Schwarzchild, of Princeton Univ. and head of the scientific team, said they were hopeful that the photographs would provide another piece of evidence on the life cycle of stars. NSF, which has sponsored the series of Stratoscope balloon flights along with ONR and NASA, termed the flight "very successful." (*Wash. Post*, 11/28/63, L5; *NYT*, 12/3/63, 46)

- In an article in *Missile and Rockets*, NASA Administrator James E. Webb emphasized the importance of the engineer in the space program:
    "During this period of intensive research and development, it becomes doubly important for industry, the press, and NASA to tell the workshop and test site story as well as the launch site story. The role of the space engineer must be brought out as clearly as the achievements of space scientists and the astronauts . . . . " (*M&R*, 11/25/63)

- *Missiles and Rockets* article on NASA's advanced research program said: "The greatest single obstacle to development of new and improved rocket engines—large or small—lies in certain peculiar sounds emitted by powerplants in operation.
    "Commonly called screech, screaming, squealing or buzzing, the sounds are all manifestations of combustion instability—one of least understood characteristics of a rocket engine.

"The key to the lack of knowledge and understanding of the phenomenon is in the very complexity of the combustion process within the engine." (*M&R*, 11/25/63, 63)

November 26: NASA successfully launched EXPLORER XVIII, the Interplanetary Monitoring Probe (Imp), from AMR on a Thor-Delta booster. Intended to have an apogee of 173,000 mi. and a perigee of 125 mi., EXPLORER XVIII was tracked as having an apogee of 122,800 mi. and a perigee of 120 mi., with an orbital period of about four days, and an inclination of 33.3°. Essentially a continuation of the series of energetic particle spacecraft and similar in design and in some experiments to the earlier EXPLORERS XII, XIV, and XV, its 35 lbs. of instruments would measure the major magnetic field phenomena in space, including the interplanetary magnetic field, interactions of the streaming solar plasma and the geomagnetic field, galactic and solar radiation. (NASA Releases 63–249, 261; *Wash. Post*, 11/27/63)

- Two sounding rockets were launched from Wallops Station, Va., as NASA provided launching, tracking, and data acquisition for two related Univ. of Michigan experiments investigating characteristics of the upper atmosphere. A Nike-Apache was launched to 103 mi., with a 70-lb. payload chiefly comprised of an ion mass spectrometer to measure air density and composition as a function of altitude. Shortly thereafter, a Nike-Cajun was launched to 81 mi. with a 56-lb. payload chiefly composed of three 26-in. mylar balls which were ejected at intervals and their descent and drift tracked as a measure of wind direction and velocity and atmospheric density. (NASA Release 63–106)

- First complete flight system test of the Sert (space electric rocket test) payload under simulated space conditions at Lewis Research Center. Two engines in the Sert package would be the first electric engines to operate in space. (Lewis Chronology, 11)

- RCA Victor Co. announced that it would join with Canada's Federal Transport Department in building a $5-million experimental station in eastern Canada to serve as a space relay station for voice transmissions from overseas. The station would be operated under an agreement between Canada and NASA. (UPI, *NYT*, 11/27/63, 21)

- TELSTAR II communications satellite brought live American television to the Soviet Union for the first time in the weekend of reporting on the assassination of President Kennedy and the mourning and funeral that followed. *Izvestia* devoted one whole page of its four pages to an analysis of the ironic events. Entitled "Texas and Telstar," the article pointed out that Telstar, "the technical wonder of the 20th century, came into our lives with America's mourning and brought us the cry of battle, a battle between light and darkness raging this week on the other side of the earth.

"Only the darkest Spanish Inquisition could have produced the scenes that were flashed by the American satellite . . . .

"We have seen the grief of the American nation and profoundly sympathize with it. We have seen a mad detective thriller and we reject it with contempt and anger." (Shabad, *NYT*, 11/27/63, 17)

November 26: USN launched a Polaris A–3 missile off U.S.S. *Observation Island* missile support ship, had to destroy the missile 52 sec. after launch when it went off course. (*M&R*, 12/9/63, 10)

November 27: NASA launched the first successful Atlas-Centaur space booster from AMR. The booster performed perfectly and the Centaur second stage ignited its liquid-hydrogen engines in space and went into orbit as the heaviest object (10,500 lbs.) yet orbited by the U.S. Orbital data were later announced to be: apogee, 1,050 mi.; perigee, 340 mi.; period, 108 min.; and inclination, 30°. Although another six flights remained before the Centaur could be considered operational, this first successful flight of the high-energy liquid hydrogen-liquid oxygen booster was a major landmark in a development program dogged with delays and disappointments. The previous launch attempt, on May 8, 1962, ended 55 sec. after launch in fiery explosion when the weather shield ripped off the Centaur second stage. In a press conference after the launch, NASA Deputy Associate Administrator for Space Sciences and Applications Edgar M. Cortright spoke of the launching as "the world's first successful flight of a hydrogen-oxygen rocket." Dr. Abe Silverstein, Director of NASA Lewis Research Center and the man who personally oversaw the reworking of Centaur during the past year, added that the flight was "the first test of our ability to successfully ignite hydrogen on time in space." (Hines, Wash. *Eve. Star*, 11/28/63; Benedict, *Wash. Post*, 11/28/63; Witkin, *NYT*, 11/28/63; NASA Releases 63–254–261)

- X–15 No. 3 was flown to 89,800 ft. and 3,310 mph (mach 4.94), by NASA pilot Milton O. Thompson. This was Thompson's second flight in the X–15 and was intended to familiarize him with the extreme altitude control of the rocket research aircraft. (*M&R*, 12/9/63, 10; X–15 Project Office)

- President Lyndon B. Johnson addressed a joint session of Congress, his first major speech since assuming the Presidency on Nov. 22 following the assassination of President Kennedy: "The greatest leader of our time has been struck down by the foulest deed of our time . . . .

    "The dream of conquering the vastness of space—the dream of partnership across the Atlantic and across the Pacific as well—the dream of a Peace Corps in less developed nations—the dream of education for all of our children—the dream of jobs for all who seek them and need them—the dream of care for our elderly—the dream of an all-out attack on mental illness—and above all, the dream of equal rights for all Americans, whatever their race or color—these and other American dreams have been vitalized by his drive and by his dedication . . . ." (*CR*, 11/27/63, 21734)

November 27: NASA signed Memorandum of Agreement with Military Sea Transportation Service (MSTS) providing ocean transportation services between U.S. ports for large Saturn launch vehicle stages and supporting equipment in connection with the manned lunar landing program. (NMI 739, 2/10/64)
- LRC-engineers described base-heating studies being performed on Saturn V Apollo rockets. Base-heating can be a severe problem with rocket vehicles pouring out exhaust gases at temperatures as high as 6000°F. Similar studies were performed on Saturn I. (LRC Release 63-95, Lewis Chronology, 12)
- NASA's J-2 liquid-hydrogen engine, designed to provide 200,000 lbs. of thrust for upper stages of the Saturn IB and Saturn V boosters, successfully underwent its first extended duration ground test firing, a static firing of 510 sec. The J-2 had been designed for a duration of 250 sec. and had been undergoing ground firings for that duration since Oct. 1962. Recently the requirements were changed to provide the longer burning time. (*Marshall Star*, 12/4/63, 3)
- Feasibility of operating large numbers of ion engines together was established when Lewis Research Center successfuly operated an array of nine ion engines under simulated space conditions. (LRC Chronology)
- USAF launched a Thor-Agena D booster carrying an unidentified satellite, from Vandenberg AFB, Calif. Also launched that day was a Minuteman ICBM in what was termed a routine training launch. (AP, Wash. *Eve. Star*, 11/28/63)
- The Air Force accepted the first Titan III engines from Aerojet-General Corporation in Sacramento. The engines were the first- and second-stage liquid-fuel engines for the Titan IIIA (core) of the Titan IIIC. Together they produce a combined thrust of 530,000 lbs. (*A-N-AF Journal and Register*, Jan., 1964)

November 28: President Johnson delivered a personal Thanksgiving Day message to the American people via TV and radio, less than one week after he assumed the duties of the Presidency upon the assassination of President Kennedy. "All of us have lived through seven days that none of us will ever forget . . . .

"A great leader is dead; a great nation must move on. Yesterday is not ours to recover, but tomorrow is ours to win or lose . . . .

"And to honor his memory and the future of the works he started, I have today determined that Station No. 1 of the Atlantic Missile Range and a NASA launch operations center in Florida shall hereafter be known as the John F. Kennedy Space Center.

"I have also acted today with the understanding and the support of my friend, the Governor of Florida, Farris Bryant, to change the name of Cape Canaveral. It shall be known hereafter as Cape Kennedy." (AP, *Wash. Post*, 11/29/63)
- SYNCOM II successfully performed a velocity correction maneuver, using the lateral H_2O_2 jet to correct a westward drift of .104° per day. The jet performed as expected; drift after correction was $-.0045°$ per day. (GSFC Historian)

November 29: President Johnson signed the Executive Order establishing the John F. Kennedy Space Center:

"WHEREAS President John F. Kennedy lighted the imagination of our people when he set the moon as our target and man as the means to reach it; and

"WHEREAS the installations now to be renamed are a center and a symbol of our country's peaceful assault on space; and

"WHEREAS it is in the nature of this assault that it should test the limits of our youth and grace, our strength and wit, our vigor and perseverance—qualities fitting to the memory of John F. Kennedy:

"NOW, THEREFORE, by virtue of the authority vested in me as President of the United States, I hereby designate the facilities of the Launch Operations Center of the National Aeronautics and Space Administration and the facilities of Station No. 1 of the Atlantic Missile Range, in the State of Florida, as the John F. Kennedy Space Center; and such facilities shall be hereafter known and referred to by that name." (*Marshall Star*, 12/11/63, 2)

- Soviet news agency Tass announced that U.S.S.R. would launch a test series of "new, improved types of rocket boosters for spacecraft" and designated two areas in the Pacific Ocean (Area I in the vicinity of 10° N latitude and 170° W longitude, Area II near 35° N latitude and 170° E longitude) as impact areas to be avoided by ships and aircraft from Dec. 2, 1963, through Jan. 25, 1964. (*Pravda*, 11/29/63, 1, AFSS-T Trans.)

- President Johnson signed H. J. Res. 809, joint resolution making continuing appropriations for FY 1964 and for other purposes through Jan. 31, 1964. (P.L. 88-188) (NASA Leg. Act. Rpt. II/195)

- ComSatCorp announced that John A. Johnson, NASA General Counsel since NASA was formed in 1958, would become its Director of International Arrangements. NASA in turn announced that Mr. Johnson would be succeeded as General Counsel by Walter D. Sohier, Deputy General Counsel. (Wash. *Eve. Star*, 11/29/63; ComSatCorp Release; NASA Release 63-259)

- DOD announced that the Nike-Zeus antimissile missile had scored another successful interception against an ICBM. This 10th successful Nike-Zeus interception was from Kwajalein Atoll in the Pacific, the target a Titan I ICBM launched from Vandenberg AFB, Calif. DOD refused to state the date on which the interception took place. (DOD Release 1519-63; *Wash. Post*, 11/30/63)

- Rep. Robert E. Jones (D.-Ala.) inserted in the *Congressional Record* an article by Eugene Patterson, editor of the *Atlanta Constitution*, in which Mr. Patterson noted with approval remarks by Dr. Wernher von Braun on the lunar program viewed as a " 'capital investment into the next few decades of space exploration, both military and civilian.' " Patterson concluded: "It will be an incredible American performance if the United States shrinks back now from the space age into which von Braun dragged us, kicking and protesting, and in which he is now having to struggle to keep us." (*CR*, 11/29/63, A7320)

November 29: USAF launched a Minuteman ICBM from an underground silo at Vandenberg AFB, Calif. (*Wash. Post*, 11/30/63)

November 30: GSFC announced negotiations with the Marine Engineering Dept. of Northrop-Nortronics on a contract for design and construction of a huge 230-ton test chamber to simulate the launch phase of space flight for test of unmanned spacecraft and their components. The chamber would duplicate as nearly as possible the environmental conditions and the separate or combined conditions of acceleration, noise, vibration, and vacuum. The Launch Phase Simulator was expected to cost approximately $1.9 million. (GSFC Release G-27-63)

- Japan sent a note of protest to the U.S.S.R. in reaction to the Soviet announcement of Nov. 29 designating two areas in the Pacific as test areas for a new series of Russian rocket launchings from Dec. 2, 1963, through Jan. 25, 1964. Japan reserved the right to claim damages if any were incurred by Japanese as a result of the tests. (AP, Wash. *Eve. Star*, 11/30/63)
- Two Soviet Ilyushin-18 transport aircraft completed 15,000-mile flight from Moscow to the Russian Antarctic base of Mirny. Director of the Leningrad Polar Institute, A. Treshnokov, said in New Zealand that such flights might well inaugurate regular supply flights between Moscow and the Antarctic. (Reuters, *Wash. Post*, 12/1/63, A29)
- Soviet Army newspaper *Red Star* reported an account of a rocket failure, the first public mention of a rocket launching failure in the U.S.S.R. No date or program was named, only a report of heroism of a Lt. Col. Serebreynnikov. He reportedly pulled out a starting plug by hand after mechanism had failed, and only six seconds before the rocket exploded. (AP, *Wash. Post*, 12/1/63, A35)

During November: Vice President Lyndon B. Johnson, writing in the Fall issue of *Challenge* magazine of the General Electric Co., said:
"Space is not a gambit. It is not a gimmick.

"Our national activities in space research and exploration are no longer in the category of a gamble. We are able to talk seriously about explorations and journeys, 26 million miles away to the planet Venus—or 47 million miles away to the planet Mars.

"We have come a long way. In the new Age of Space which brings us together, we are destined to go a long way further. Where the moon is a major goal today, it will be tomorrow a mere whistle stop for the space traveler. I confidently believe that the developments of the Space Age will bring the beginning of the longest and greatest boom of abundance and prosperity in the history of man." (*Challenge*, Fall 1963, 24-5)

- The Marquardt Corp. of Van Nuys, Calif., has delivered 30 production units of its Roksonde 200-I to West Germany. The sounding rockets would be used for meteorological testing at the Salto di Quirra Test Range in Sardinia. The meteorological data gathering is part of the overall European effort in support of the IQSY worldwide programs. As part of the program, the Marquardt Corp. will assist the West German government in two launching series. The first series was currently in progress, scheduled for completion by mid-December; the second would be completed in early 1964. (Marquardt Release, 11/21/63, SIN, Jan., 1964)

During November: JPL scientists Dr. Richard M. Goldstein and Roland L. Carpenter reported in *Science* magazine on recent radar experiments using an 85-ft. parabolic antenna at a tracking station in the Mohave Desert. The planet Mercury was found to be quite rugged, more so than Mars and twice as rugged as Venus. The experiments also corroborated Mercury's rotation period as 88 days. (Science Service, *NYT*, 11/6/63, 33)

- Reported in *Aviation World* (*Flugwelt*) that West Germany has a $25 million-budget to develop a series of 5 satellites to be built under NASA's established quality control standards. The satellite would consist of three parts: the satellite capsule with a recovery unit, the sun mirror, the energy converter. The first satellite recovery capsule would descend with the aid of a parachute; later models would include a winged glider version. Telemetric transmissions and subsequent evaluation would commence as soon as the satellite enters an Earth orbit. The second satellite and the remainder of the series would be equipped with engines that will permit changes in the orbital parameters. On the second satellite small solid fuel rockets would be used, but the following models could be equipped with ion engines.

  The satellite's mission would include research in the guidance and direction stabilization systems during the launching investigation of the orbital rendezvous problems, as well as a study of the radiation belts. In the case of the fifth satellite, the ion engines could be replaced with nuclear-powered engines to permit continuous orbital changes and accelerate to an escape velocity. (*Flugwelt* (*Aviation World*), Nov., 1963; SIN, Jan., 1964)

- William B. Bergen, president of the Martin Co., writing in the November issue of *Aerospace Management*, expressed his approval of DOD's new cost-plus-incentive-fee (CPIF) contracting as opposed to the more conventional cost-plus-fixed-fee (CPFF) contracting that had existed in R&D work for some years. CPFF was necessary in the early days of missile work and of space technology when "We had a critical time element and our technology was shaky." Now, he said, "there aren't enough projects . . . to fill the aerospace plants. CPIF is going to be a great catalyst in separating the men from the boys in this business. Americans recognize talent by financial reward. If you have it, you get; if you don't have, you don't get. CPIF will be a great factor in providing the measure and the reward of the industry's talents." (*CR*, 11/19/63, A7150–51)

- Navy Bureau of Weapons' astronautics group approved a study by the Johns Hopkins Applied Physics Laboratory and recommended that the Thor-Able-Star be substituted for the Scout as the launch vehicle for the Navy Transit navigational satellites. Navy had been dissatisfied with the reliability record of the NASA Scout boosters in the Transit program. (*Av. Wk.*, 11/25/63, 23)

- USAF was studying a novel method of protecting spacecraft from meteoroid hits, according to Maj. Gen. Marvin C. Demler (USAF),

Commander of the AF Research and Technology Div. USAF's Flight Dynamics Laboratory at Wright-Patterson AFB, Ohio, was studying the concept of surrounding the spacecraft with a 40-in.-thick layer of metallic dust held in place by an electrostatic field. A meteoroid hitting the vibrating dust layer would be shattered or vaporized. (*M&R*, 11/25/63, 29)

During November: Military role in the U.S. space program is "generally misrepresented and largely misunderstood," Maj. Gen. Ben I. Funk, Commander of AF Space Systems Div., told the World Affairs Council meeting in Los Angeles. The national policy of using space for peaceful purposes "serves to put the military into space, not restrict space to purely non-military activities. Whatever shape and direction our ventures into space ultimately take, military considerations cannot be separated from the political, economic, psychological, technological and sociological implications of space development." (*M&R*, 11/25/63, 30)

- NASA and the Air Force began a test of a 5-psi 100% oxygen environment in a 42-day experiment with four airmen at Brooks AFB, Tex. It would include 7 days in the chamber in a normal atmosphere to gather base-line data, 30 days in the test atmosphere, one half day for detailed lung and eye examinations outside the chamber, and then 5 more days in the test atmosphere for "follow-up" observations. Conducted jointly with NASA Manned Spacecraft Center, the experiment should lay to rest fears about the debilitating effects of pure oxygen on lungs, blood, and other organs. *A&AE*, December 1963, 91)

- A solar array characteristics test was run on the orbiting SYNCOM II synchronous-orbit commuciations satellite. The test (on Nov. 22) found a power loss of 20% from the effects of solar radiation on the solar cells during four months in orbit. The text confirmed the desirability of changing the next Syncom satellite—Syncom C—from p/n cells with .006-in. quartz cover slides to n/p cells with .012-in. quartz cover slides. The newer n/p cells had shown in ground evaluation to last up to 20 times longer than conventional p/n cells. (GSFC Historian; *Goddard News*, 12/2/63, 2)

- Reported in *Astronautics and Aerospace Engineering* that a "hitchhiker" satellite called Satar (Satellite-Aerospace Research) would be built and tested by General Dynamics. A bullet-shaped spacecraft 12 ft. long and 30 in. in diameter, Satar would be incorporated in Atlas missiles used for training missions. It would consist of a payload section capable of accommodating "several hundred pounds of experiments" and a propulsion section including a solid-propellant motor, guidance, attitude control, and related equipment. Satar would be released after Atlas burnout and coast about 15 min. before its own engine ignited to send it into orbit. First flight was scheduled for next spring. (*A&AE*, Nov. 1963, 158)

DECEMBER 1963

December 1: Air Defense Command and FAA began combined operations in Great Falls Sage (Semi-automatic Ground Environment) Direction Center at Malmstrom AFB, Mont., the event marking a milestone in military and civil air traffic control. FAA's new Great Falls Air Route Traffic Control Center makes use of ADC's operational facilities, computers, long-range radars, and related equipment to provide air traffic control services to civil and military aircraft operating in about 135,000 sq. mi. of air space over Montana and North Dakota. (FAA Release 63-102)

- RCA Board Chairman David Sarnoff, addressing American Friends of the Hebrew University in New York, predicted vast scientific and technological developments in the years ahead that "are so fundamental that they will alter the very structure of society and compel each of us to readjust some of our traditional concepts . . . .

  "Modern man—especially the would-be leader in society—has to be a culturally integrated individual, familiar with the sciences as well as the humanities. There cannot be two isolated cultures for the simple reason that there is only one society in which to live . . . .

- "I would propose that scientific and cultural education begin with the earliest school years and extend without interruption through college and into the professional schools. Specifically, courses in the social sciences and humanities should be related to the physical sciences. It seems to me highly desirable that science itself should be studied as a social phenomenon because of the great impact it has upon society . . . .

  ". . . scientific realities of tomorrow will surpass our vision of today. The great challenge before all of us is to make sure that the new knowledge and new instrumentalities shall be used constructively and not destructively. The powers of science and technology are neither good nor evil in themselves. Their capacity for good or evil lies in the use we make of them . . . ." (Text, *CR*, 12/4/63, 22087-89)

December 2: Sen. Clinton P. Anderson (D.-N.M.), Chairman of the Senate Committee on Aeronautical and Space Sciences, announced that revised testimony refuted the earlier contention that 25 per cent of U.S. scientific manpower would be working on space programs by 1970: "I challenged the accuracy of that forecast. The next day the Senate Aeronautical and Space Sciences Committee began hearings on NASA's relations with universities . . . .

"The best estimates are that 6 or 7 per cent—at most 10 per cent—might be involved. That is a long way from being an excessive drain on scientific brainpower." (UPI, *NYT*, 12/3/63, 47)

- Rep. George P. Miller (D.-Calif.), Chairman of House Committee on Science and Astronautics, reviewed Centaur development in speech on the House floor: ". . . Centaur began as a low-priority, financially austere feasibility study, in competition with high-priority defense programs. As its importance to the national space effort became more apparent, its terms of reference were changed; and, as time passed, its inherent technical difficulties came to the surface; technical difficulties, I might add, that we have learned to expect in most new programs.

  "Certainly, the original flight schedule of Centaur was overly optimistic. Hindsight also tells us that the complexity of the program was greatly underestimated . . . ."

  Referring to press articles charging $100 million was wasted in Centaur program, based on GAO investigative report, Rep. Miller pointed out:

  "That report was misinterpreted by the press, and a closer look at it will reveal that no such conclusion was drawn by the Comptroller General. The alleged $100 million waste included $76 million reportedly lost in the Advent project, the military communications satellite project . . . .

  "While there can be no doubt that certain programs incurred losses because of the unavailability of Centaur on schedule, it is incorrect and unfair to attribute the entire unrecoverable loss associated with the Advent program to Centaur, as the press apparently did. Suffice it to say that the Advent project had its own severe management and technical difficulties which led to its cancellation in June 1962."

  He called the Nov. 27 flight test of AC-2 "a significant advance in the development of a new technology upon which much of America's future space effort depends . . . ." (*CR*, 12/2/63, 21906 ff.)

- Political Committee of the U.N. General Assembly opened debate on report by the U.N. Committee on the Peaceful Uses of Outer Space.

Addressing the 111-nation Committee, U.S. Ambassador Adlai Stevenson said President Johnson instructed him to reaffirm the proposal for U.S.-U.S.S.R. cooperation in exploration of the moon that President Kennedy had offered last September. But, Stevenson continued, "If giant steps cannot be taken at once, we hope that shorter ones can. We believe there are areas of work, short of integrating the two national prorgams, from which all could benefit. We should explore the opportunities for practical cooperation, beginning with small steps and hopefully leading to larger ones."

Stressing that U.S. manned lunar program was only one part of the space program and that no more than 10 percent of the total U.S. space expenditure went directly for the lunar flight, he added: "Let me make clear that exploration of the moon is not a stunt, distinct from the outer-space program as a whole; nor is it the exclusive concern of only two nations.

"Our policy of engaging in mutually beneficial and mutually supporting cooperation in outer space—with the Soviet Union as with all nations—does not begin or end with a manned moon landing. There is plenty of work yet to come before that—and there will be even more afterward." (Teltsch, *NYT*, 12/3/63, 1, 14; *L.A. Times*, *Wash. Post*, 12/3/63, 1)

December 2: NASA selected Douglas Aircraft Co. for negotiations leading to follow-on study contract for refinement and evaluation of NASA manned orbital laboratory concept. Awarded on basis of Douglas performance of previous three-month study, the six-to-nine-month follow-on contract called for refining the NASA concept of a manned orbiting laboratory based on cylindrical six-man spacecraft. (NASA Release 63-262)

• President Johnson's active leadership in space program since early 1958 was reviewed in *Missiles and Rockets* by Hal Taylor, who predicted the new President would give the space program even greater support than did President Kennedy. Added support could mean growth of military space projects, Taylor said, possibly including assignment of manned space station development to USAF. But FY 1965 budget estimates would be virtually unchanged.

Article in *Aviation Week and Space Technology* indicated President Johnson was likely to "play a more direct role than his predecessor in the nation's military space and supersonic transport programs after the current transitional period, during which the emphasis is on showing the nation and the world that President Kennedy's commitments will be fulfilled." (*M&R*, 12/2/63, 14–15; *Av. Wk.*, 12/2/63, 26–27)

• On 21st anniversary of atomic age, Dr. J. Robert Oppenheimer was presented the Enrico Fermi award, highest honor of the Atomic Energy Commission, by President Johnson in White House ceremony. Citation honored Dr. Oppenheimer for his "contributions to theoretical physics as a teacher and originator of ideas and for leadership of the atomic energy program during critical years." Dr. Oppenheimer headed select group of scientists who in 1942 achieved first self-sustaining nuclear chain reaction at Univ. of Chicago; was director of Los Alamos Scientific Laboratory during World War II; and since 1947 was director of Institute for Advanced Study at Princeton, N.J. (Finney, *NYT*, 12/3/63, 1)

December 2: Proposal that mixed-manned NATO forces take over control of 150 Minuteman missile sites in the U.S. was included in report of committee on NATO nuclear force to Assembly of the Western European Union. The report, which strongly supported U.S. plan for a mixed-manned NATO nuclear fleet, stated that teams made up of mixed NATO forces manning the U.S. Minuteman sites would have "considerable political significance in demonstrating—to Americans as well as to Europeans—full NATO participation in the strategic deterrent based on the American Continent, which hitherto has been a purely American preserve." (Middleton, *NYT*, 12/3/63, 15)

- Battelle Memorial Institute predicted total U.S. R&D expenditures would reach $20 billion next year, $1.7 billion more than this year. Predicted breakdown: government, $13.9 billion; industry, $5.6 billion; academic and nonprofit institutions, $500 million. (*M&R*, 12/3/63, 7)
- Indian Defense Ministry spokesman announced Soviet MiG-21 aircraft to be built in India would be equipped with air-to-air guided missiles. (AP, *NYT*, 12/3/63)
- Prague radio disclosed that Czechoslovakia would buy Soviet Tu-124 jet airliners, new twin-engine craft designed for short-range and medium-range flights. (*NYT*, 12/3/63)

December 2–4: Fourth NASA Intercenter Conference on Plasma Physics held at NASA Hq., with participants from NASA Hq. and installations as well as from universities and industries. (Program)

December 3: M-2 experimental wingless glider tested at Edwards AFB, Calif., with Col. Charles Yeager (USAF) as pilot. The craft was towed to 9,000-ft. altitude, where Col. Yeager cut loose from the tow plane and glided down at rate of 4,000 ft. per min. Shortly before landing, Yeager tilted up the M-2's nose to reduce its 135-mph descent speed to 80 mph. The craft landed on its three-wheel gear and rolled to a stop in about 300 ft. This was Col. Yeager's first flight in the NASA-developed M-2; he remarked: "She handles great." (*Wash. Post*, 12/4/63; *Chic. Trib.*, 12/4/63)

- NASA scheduled the first Syncom communications satellite to be launched into orbit in which the satellite would truly remain stationary with respect to the earth in orbit—synchronous, equatorial, and circular—in the second quarter of 1964. Thrust-Augmented Delta (TAD) launch vehicle would boost the spacecraft into 22,300-mi. eccentric orbit from Cape Kennedy, and at apogee Syncom's onboard kick motor would place the satellite in circular orbit. Equatorial position would be achieved by change of orbital plane—once as third stage fires and again as kick motor fires at equatorial injection point. The spacecraft would then move to on-station position at 180° longitude over the Pacific Ocean, where it would be stopped by onboard gas jets. The satellite will provide the first continuous communications link between Asia and North America. (NASA Release 63-263)

December 3: A second disturbance on the moon has been observed by USAF lunar mappers at Lowell Observatory, USAF Aeronautical Chart and Information Center announced. Sighting was made by Observatory Director Dr. John A. Hall and four other observers Nov. 27 and lasted an hour and a half (first sighting Oct. 29 lasted 20 min.). Both sightings were near crater Aristarchus, both were ruby red, and both were seen when moon was in the same phase (sunlight returning to this surface area). Previous sighting was of three red spots, but this sighting was of single red area 12 mi. long and 1½ mi. wide, inside crater's rim where one of first spots had been seen. Observation was confirmed by astronomy graduate student Peter A. Boyce, at Ohio State Univ.'s Perkins Observatory. Dr. Hall believed spots may be gases released from lunar interior by intense heat of the returning sunlight. (AP, *San Diego Union*, 12/4/63; *Time*, 12/27/63, 54-55)

- AFSC announced recent three-day manned test of device to control the atmosphere in manned spacecraft was successfully conducted at AFSC Aerospace Medical Research Laboratories, Wright-Patterson AFB, and the evaluation proved the automatic device was feasible for aerospace applications. Developed by Northrop Space Laboratories, the "nitrogen and oxygen indicator-controller assembly" senses percentage of oxygen, as well as total air pressure, inside simulated space capsule, automatically regulating flow of gases required to maintain desired nitrogen-oxygen ratio. If controller fails, warning lights and buzzer alert crewmen to make necessary adjustments in oxygen or nitrogen manually. Volunteers for the test were S/Sgt Elvin H. Engle and A/2C Frank O. Thornton. (AFSC Release 310-23-105)
- Presidential Press Secretary Pierre Salinger announced President Johnson, in his review of the FY 1965 budget estimate, was attempting to arrive at a spending estimate "between $98 billion and $103 billion." (Kenworthy, *NYT*, 12/4/63, 1)
- Less than 24 hours after U.N. Ambassador Adlai Stevenson's reaffirmation of U.S. proposal for negotiations leading to U.S.-U.S.S.R. cooperation in lunar exploration, U.S.S.R. promised privately to Ambassador Stevenson it would conduct a "detailed scientific study" of the U.S. proposals. Soviet sources disclosed two "stages" were being planned: first, "careful scrutiny by the Soviet scientific community of possible areas of moon-flight cooperation; thereafter, if conclusions are 'positive,' a joint study by Soviet-American scientists." (Frye, Wash. *Eve. Star*, 12/4/63)
- Addressing the Kokomo (Ind.) Chamber of Commerce, NASA Adminstrator James E. Webb remarked: "We are in a continuing competition for world leadership, and leadership in space has become an important element in this struggle . . . .

"It is logical that aggressive elements in the Soviet Union may hope to enter a new period of forward thrust by some startling breakthrough in space technology. The only way we can make sure that this does not happen—the only way to safeguard our technological frontiers—is to pursue an active program of research and development over the entire range of space technology. This includes manned space flight, the development of nuclear-powered rockets, electric propulsion for spacecraft, advanced communications, satellite observations of the global weather picture, lunar and planetary exploration, and acquisition of new understanding of the basic forces operating in our solar system and the universe. The space frontier has many salients, many different points of activity. We must not lag behind in any area of space technology or our ability to deter aggression, in space and on earth, may be jeopardized . . . .

"Most of those who question the value of the lunar program fail to appreciate that it involves much more than a single expedition to the moon before 1970. Although the moon is the ultimate objective of the Apollo program, its main purpose is to develop our space competence—our space muscle and space skills —at the most rapid pace consistent with good management, crew safety, and the efficient use of funds . . . .

"Do we conclude that a vigorous effort to develop knowledge and capability in space science and technology, particularly given the long lead times involved in this development, is fruitless and extravagant until military requirements are clearly defined? Do we fly in the face of history and assert that the enormous economic potential of space research and exploration does not exist simply because we cannot pinpoint all of the precise benefits which such research will produce and which all the lessons of history tell us are there? Or do we, in the tradition which has made this country great, grasp the opportunity to cross a challenging new frontier and insure that the nation will not sacrifice the opportunity or risk the hazards which lie beyond?" (Text)

December 3: Deputy Secretary of Defense Roswell L. Gilpatric discussed management of military R&D at Chemical Engineering Achievement Awards Dinner, New York City:

"Ten years ago, expenditures for defense research and development were at a level of about $2 billion or about 5% of the U.S. military budget. Today the comparable figure is over $7 billion, or nearly 14% of the current Defense budget. Half of this total is for system development, a good deal of which represents prototype hardware. The other half represents the dollars that buy brain-power at the drawing board and in the labs—and nowadays I should add computer time . . . .

"There are indications that we are seeing the beginning of a period of increasing Congressional concern and interest that may lead to even tighter controls over R&D expenditures. This change in Congressional attitude is more than an expression of growing concern over deficit spending. It reflects an increased awareness of the national importance of R&D . . . .

"Today, fortunately, there can be no question about the sufficiency of American military power, either now or for the foreseeable future. The 100% increase of strategic alert force megatonnage in the past three years and the large and unquestioned U.S. lead in both bombers and missiles provide us with the kind of strength that permits us to go forward with our R&D effort in a more orderly way . . . ." (Text, DOD Release 1527-63)

December 3: *New York Times* editorialized: ". . . A singularly fitting international tribute to President Kennedy's memory would be the realization soon of the dream of Soviet-American cooperation in space he voiced so eloquently in what we now know was his last formal statement to all the world's peoples." (*NYT*, 12/3/63)

- Soviet Premier Khrushchev was quoted as saying U.S.S.R. had an ICBM that can fly around the globe in any direction. "You wait for it at the door but it climbs through the window," Khrushchev was quoted as saying of the missile. (UPI, *Wash. Post*, 12/4/63)
- Article from *Washington Post's Parade* magazine written by *Parade* editor Jess Gorkin was inserted in *Congressional Record* by Sen. Robert C. Byrd (D.-W.Va.). Gorkin declared:

  "The conquest of space has come to mean more to the Kremlin than a mere matter of national pride; it has become part of the Communist dogma. The Soviets are fanatically committed to outdoing the United States in this race. Having declared that nuclear war isn't necessary to achieve world domination, they hope to accomplish the same end by dominating space. Make no mistake, it is powerful fuel, this mixture of national pride, political ambition and doctrinal zeal . . . ." (*Parade*, 11/24/63, in *CR*, 12/3/63, 21995)
- H.J. Res. 787 to provide for erection of a memorial statue to the late Dr. Robert H. Goddard was ordered favorably reported to the House by the House Committee on Science and Astronautics. The legislation would require that NASA erect a bronze statue of Dr. Goddard in central Massachusetts, the memorial giving appropriate recognition to Dr. Goddard's 1914 patents, first liquid-fuel rocket flight, and other pioneering achievements. (NASA Leg. Act. Rpts. II/196 and II/198)
- U.K.'s Labor Party issued statement that if Labor came to power it would discontinue development of the TSR-2 tactical-strike-reconnaissance/strategic nuclear bomber as a strategic bomber or as a contributor to Britain's nuclear deterrent. As to the aircraft's original role as a tactical weapon, the Labor Party would reserve judgment until the aircraft's performance has been flight-tested. (*Aviation Daily*, 12/3/63)

December 3-4: 14 NASA officials participated in Florida Aerospace Industry Seminar in Orlando, with more than 1,500 Florida businessmen, industrialists, and educators attending. (NASA Release 63-260)

December 4: EXPLORER XVIII (Imp) satellite completed second of its long elliptical orbits (apogee, 122,800 mi.; perigee, 119 mi.), and all systems appeared to be functioning as planned.

Lifetime of IMP-A, launched Nov. 26 from Cape Canaveral, revised by NASA from estimated 12 months to hundreds of years. (NASA Release 63-265; A&A, February 1964, 11)

December 4: Zero gradient synchrotron—first major particle accelerator in the Midwest—dedicated at Argonne National Laboratory near Chicago. Financed by AEC, the accelerator is one of the **three largest in the world, generates 12.5-bev energy.** (Plumb, *NYT*, 12/5/63)

- NASA announced the Centaur stage (AC-2), launched from Cape Canaveral Nov. 27, was orbiting the earth with several objects varying in size, believed to be insulation panels and nose fairing of the stage. Centaur project officials believed that after attitude control was discontinued in orbit, the stage began to tumble because of venting of gaseous hydrogen, and the panels and nose fairing were separated. (NASA Release 63-265)
- NASA ordered its major manned space flight program contractors to freeze all new hirings and not replace any persons lost through normal attrition. Expected to save millions of dollars, the temporary action was taken while NASA was reassessing its budget programing for the current fiscal year, based on an expected appropriation of only $5.1 billion. (Simons, *Wash. Post*, 12/6/63)
- USN disclosed new Subroc antisubmarine missile, a submarine-launched, rocket-propelled, inertially-guided nuclear depth bomb, would be operational in 1965. Launched from standard torpedo tube, Subroc is powered by solid-fuel rocket motor which ignites underwater and propels missile up and out of the water. Rocket booster is automatically separated from depth bomb payload, which continues on ballistic trajectory until impacting water, where it sinks and detonates. (DOD Releases 1536-63 and 1537-63)
- Franco-Italian project to develop "Iris" rocket capable of delivering mail to any European city was reported by Associated Press. Powered with mixed solid-liquid fuel, the rocket is being designed to deliver 65 lbs. of mail more than 400 mi. away. Rocket would be launched vertically to predetermined altitude—around 72,000 ft.—then guided by radio to delivery point. 30 mi. from target, engine would be shut off, allowing rocket to glide; retrorockets would be fired to point its nose up, then Iris would descend backwards to earth. (AP, Wash. *Eve. Star*, 12/4/63)
- Dr. Ralph E. Lapp, in address at Queens College, Charlotte, N.C., said photographs made by U.S. military satellites were providing excellent studies of Soviet military installations. (AP, Wash. *Eve. Star*, 12/5/63)

December 4-7: 32 high school science students participated in National Youth Science Congress in Washington, sponsored by NASA and National Science Teachers Association of the National Education Association. (Wash. *Eve. Star*, 12/5/63)

December 5: X-15 No. 1 flown by Maj. Robert Rushworth (USAF) to 101,000-ft. altitude and 4,018 mph speed (mach 6.06), just 86 mph slower than record speed of 4,104 reached in June 1962 flight by Joseph Walker (NASA). Today's flight was scheduled to reach 3,900 mph, but an extra burst of power from the rocket engine propelled the aircraft close to the record speed. Purpose of flight was to test new navigation instrument, cross-ranger indicator, and to measure effects of high speed flight on photographic techniques. Photographs taken by the X-15's camera were compared with those taken by camera in U-2 aircraft, piloted by Lt. Col. Harry Andonian at the same time as the X-15 flight and along the same route—Edwards AFB to Delamar Lake, Nev. (AP, *Balt. Sun*, 12/6/63; UPI, *Chic. Trib.*, 12/6/63)

December 5: F-1 rocket engine static-fired in 10-sec. test, first in series conducted by NASA Marshall Space Flight Center. (*Marshall Star*, 12/11/63, 2)

- USAF launched two unidentified satellites using Thor-Able-Star launch vehicle from Vandenberg AFB. (AP, *Balt. Sun*, 12/6/63; *Pres. Rpt. on Space, 1963.* 12764)

- U.N. General Assembly's Political Committee approved draft resolution and declaration on international cooperation in space sciences and law.

  The resolution endorsed recommendations by U.N. Committee on the Peaceful Uses of Outer Space, which called for proposals on exchange of scientific information, consideration of agreements on legal principles "governing the activities of the states in the exploration and use of outer space," and other steps toward cooperation.

  The declaration was an endorsement of legal principles drafted jointly by U.S. and U.S.S.R. representatives last month. Provisions of the nine-point draft included statements that celestial bodies should be regarded as free territories not subject to national sovereignty; exploration of space should be for the benefit of all men of all nations; and all nations engaged in space exploration should observe international law.

  With adoption of the two drafts, the Political Committee concluded its present consideration of space exploration. (Lubasch, *NYT*, 12/6/63, 1 & 17)

- House and Senate conferees met in executive session and filed a conference report on the differences between the Senate- and House-passed versions of H.R. 8747, FY 1964 appropriations for independent offices, including $5.1-billion appropriation for NASA. Two of the NASA amendments were reported in disagreement. (House Rpt. 1004, 12563; NASA Leg. Act. Rpt. II/198; *L.A. Times*, *Wash. Post*, 12/7/63)

- NASA and Australian Dept. of Supply jointly announced manned space flight tracking and data acquisition facilities were being moved from Muchea, Australia, to a new site being built near Carnarvon, on the west coast. Move would consolidate at one site NASA's ground support facilities in Australia for manned Projects Gemini and Apollo. Site also would include tracking and data acquisition facilities for unmanned Orbiting Observatory satellites—Oao, Ogo, and Oso. (NASA Release 63-266)

- Reacting to President Johnson's Thanksgiving Day message announcing he had "acted . . . to change the name of Cape Canaveral . . . [to] Cape Kennedy," the City Council of Cape Canaveral and the Cocoa Chamber of Commerce adopted resolutions opposing changing the name of Cape Canaveral.

Florida's Governor Farris Bryant defended President Johnson's decision, in a Tallahassee press conference: ". . . the people of Florida, in the year 2063, will look back and understand what President Johnson has done and will approve."

Arthur A. Baker, chairman of the Domestic Names Committee of the U.S. Board of Geographic Names, said he regarded the name change as an accomplished fact. Edward P. Cliff, chairman of the Board, also said he regarded the change as fact. Official name changing awaited action by the Board, which was scheduled to meet next Jan. 21. (AP, *Wash. Post*, 12/6/63; UPI, N.Y. *Herald Trib.*, 12/6/63; AP, Wash. *Eve. Star*, 12/5/63)

December 5: Transmission of seismic data over long distances by telephone lines has been achieved by USAF Cambridge Research Laboratories (AFCRL) scientists, USAF OAR announced. A significant "first" in seismography, the technique permits comparison of data from relatively quiet area of seismic activity supplied by New England stations with data from active Aegean region and along West Coast of North and South America and around Japan. (OAR Release 10-63-5)

- President Lyndon B. Johnson transmitted NASA's *Eighth Semiannual Report to Congress*, reporting projects and progress for the period July 1–December 31, 1962. (*8th Semiann. Rpt.*)
- 26 persons associated with Project Mercury were presented the John J. Montgomery Award by the San Diego Chapter of the National Society of Aerospace Professionals. Recipients from NASA were Astronauts M. Scott Carpenter, L. Gordon Cooper, Jr., John H. Glenn, Jr., Virgil I. Grissom, Alan B. Shepard, Jr., Donald K. Slayton, and Walter M. Schirra; Dr. Robert R. Gilruth, MSC Director; Dr. Walter C. Williams, NASA Deputy Associate Administrator; Kenneth S. Kleinknecht, Project Mercury Manager; Christopher C. Kraft, Jr., Project Mercury Flight Operations Director; Maxime A. Faget, Assistant Director of Engineering and Development; Dr. Charles A. Berry, Medical Operations Chief; and Lt. Col. John A. Powers (USAF) Mercury Public Affairs Officer. (*San Diego Union*, 12/6/63)
- X-15 pilot Joseph A. Walker (NASA) was presented 1963 Pilot of the Year award by National Pilots Association in Miami. Award was made in recognition of Walker's contribution to aeronautical research and the general advancement of aviation. Participating in the X-15 program since its inception, Walker has flown the X-15 25 times, more than any other man; he has flown it to record speed of 4,104 mph and has attained record altitude, 354,200 ft. (FRC Release 36-63)
- High-altitude research chamber was formally opened at Atlantic Research Corp., Alexandria, Va. Described as first such private facility in the Free World, the chamber would simulate conditions at altitudes up to 50 mi. The $200,000 facility was built with aid of NASA contract. (*Wash. Post*, 12/6/63; Wash. *Eve. Star*, 12/6/63)
- U.S. is recovering more than three of every four satellites launched in its orbit-recovery program, according to Maj. Gen. Benjamin I. Funk, AFSSD Commander, writing in Bell Aerosystems

Co. magazine: "In the recovery of satellites from orbit, we achieved a 75 per cent success record last year, and this rate has since improved even more." (Troan, *Wash. Daily News*, 12/5/63)

December 5: Sen. Clinton P. Anderson (D.-N.M.), Chairman of Senate Committee on Aeronautical and Space Sciences, outlined his recommendations for improving Congressional dealings with science: (1) Congressional committees dealing with science should be strengthened; (2) these committees should make use of ad hoc groups; (3) information flow between Congress and scientists should be easier, to prevent needless repetitiousness in hearings; (4) all parts of Executive Branch should improve Congressional committee presentations; and (5) channels for gathering information through Library of Congress should be expanded and more heavily used. (*Space Bus. Daily*, 12/6/63, 354)

- Sen. Daniel B. Brewster (D.-Md.), inserting in the *Congressional Record* an article from *Fortune* magazine (Nov. 1963) on the Martin Marietta Corp., pointed out that "during the past year, the Martin Co. was one of the first to agree to an incentive type defense contract on a major project, the Air Force's standard space launch system, Titan III. This cost plus incentive type of contract has been cited by Secretary McNamara as a most effective method of reducing the enormous expense of defense in the 1960's. I believe that Martin Marietta is to be congratulated not only for its accomplishments in space technology but also for its leadership in efforts to control cost. Martin's willingness to be the first company to accept such a contract on a $275 million project is a testament to its courageous and forward-looking executives. . . ." (*CR*, 12/5/63, 22439)

- AFSC announced awarding two six-month study contracts for operations analyses and preliminary design of long-range, low-level supersonic air-to-surface missile designated the Chemical Low Altitude Missile (CLAM). Recipients of the two $300,000 contracts were North American Aviation's Space and Information Systems Division and Lockheed California Co. (AFSC Release 39-107-104)

- West German rocket firm, Weapons and Aviation Armament Corp. of Hamburg, test-fired four solid-fuel rockets to 90,000-ft. altitude and announced signing contract to deliver missiles to an unnamed foreign country. (*Wash. Post*, 12/6/63)

December 6: NASA Administrator James E. Webb, addressing the National Association of Manufacturers' 68th Annual Congress of American Industry, in New York City, said:

"Although our decision to launch a vigorous effort in space was, in part, a response to a threat to our security, it also stemmed from recognition of the changing requirements for industrial and economic growth, and the fact that to maintain leadership in the new age of science and technology, our resources must be organized on a national scale. Thus the technological advances which it produces will come not as the mere by-product of an effort launched to satisfy other needs, but as the fruit of an effort deliberately undertaken to stimulate advances in the forces that make for economic growth, and meet the demands of a nation immersed in a new flood of scientific discovery and technical progress.

"Throughout history, and even in the early years of the present century, man has been slow in the application of basic discoveries to practical use. In the past, the utilization of new technology has rarely depended so much on the worth of the discovery as on what one economist has called 'the mentality of the times.' The willingness of people to accept and appreciate change, the belated appreciation of a previously undetected need, an imaginative leadership able to foresee the potential applications and benefits of new ideas, coupled with the wisdom and determination to insure that new knowledge is put to practical use, have always been key elements in the march of human progress.

"Today, confronted with an unprecedented explosion of knowledge, and an ever shortening time lag between discovery or invention and practical use, the search for new knowledge and its application cannot proceed at a leisurely pace—certainly not if the nation and its industries are to compete successfully with rapidly expanding foreign industrial capability in the world economic competition which exists today. Languor is a luxury which an open society can ill afford in an age when the fruits of discovery, secretly obtained, may be applied against it before their existence is even known. . . ." (Text)

December 6: Senate passed joint resolution (S.J.Res. 124) designating December 17 of each year as "Wright Brothers Day," commemorating first successful flights of heavier-than-air, mechanically-propelled aircraft by Orville and Wilbur Wright on Dec. 17, 1903. (*CR*, 12/6/63, 22573)

- AEC announced second U.S. satellite wholly powered by nuclear energy was launched into orbit recently by a Thor-Able-Star booster from Vandenberg AFB, and signals from the satellite were being transmitted sucessfully with electricity from the Snap-9A isotopic power generator. Designed to provide 25 watts of direct electrical current, the Snap-9A was the same type of generator as that providing power for a satellite launched from Vandenberg earlier this year. The two Snap-9A's are designed for operating lifetime of five years. However, they are in orbits of at least 900 years so that by the time they re-enter earth's atmosphere they will be almost completely decayed. At that time they are designed to burn into minute particles which will be widely dispersed in the atmosphere, thus increasing radioactivity in the atmosphere negligibly. (AEC Release F-250)

- AT&T informed ComSatCorp that it would prefer to rely on satellites rather than additional undersea cables to meet transatlantic communications needs in the near future. In letter from James E. Dingman, AT&T Executive Vice President, made public Dec. 10 by ComStatCorp, AT&T's basic position was reiterated: need exists for both cable and satellite circuits; however, AT&T would prefer to use satellite circuits across the Atlantic until there are approximately as many satellite channels as cable circuits. (ComSatCorp Release w/ltr)

- USAF announced that Cape Kennedy would be open to the general public for three hours every Sunday beginning on December 15th, to allow for "public drive-through." Only on Armed Forces

Day in 1961 had the general public previously been allowed within the rocket and space launching complex. (UPI, *NYT*, 12/8/63, A4)

December 6: In White House ceremony, President Lyndon B. Johnson awarded 31 Presidential Medals of Freedom, then added additional awards to the late President John F. Kennedy and the late Pope John XXIII. Among those receiving this new award, created by President Kennedy, were Dr. Alan T. Waterman, Robert A. Lovett, Edwin H. Land, and Mark S. Watson. (*Wash. Post*, 12/7/63, A8)

- Lt. Gen. Leslie R. Groves (USA, Ret.), speaking at panel meeting ("Satellite Sanity—Lunar Lunacy") of National Association of Manufacturers' 68th Congress, criticized Project Apollo as a wasted effort, value of which would be only prestige. He charged the lunar landing program was too expensive and, if continued, would bankrupt the nation.

  NASA Administrator James E. Webb, attending the meeting, was invited by presiding officer Paul Bachman to the platform to answer General Groves. Mr. Webb defended values and costs of the national space program. In his remarks, Mr. Webb pointed out that the national debt ($300 billion) is only half of the annual GNP ($600 billion), so he didn't think the country was bankrupt. He defended the use of manned instead of instrumented spacecraft and remarked, "I for one don't think there is any doubt that man and his brain is superior to a machine."

  Supporting Mr. Webb's defense were panel members Dr. Robert Jastrow, Director of Goddard Institute of Space Studies, and Walter L. Lingle, Jr., NASA Deputy Associate Administrator. (Halley, *NYT*, 12/7/63)

- Addison M. Rothrock, Director of NASA Policy Planning Div., addressing Mid-Winter Conference on Science Teaching at Jekyll Island, Ga., discussed "the need for close relation between our studies in science and technology and our studies in the humanities":

  "In space we are presented with the foremost challenge in science and technology that man has known. As a nation we face other challenges, challenges presented to us in the humanities—challenges calling for a mass assemblage of knowledge and intellect equal to that supplied by the 40,000 engineers and scientists that will by next year be working on our project to land men on the moon. And we have here a problem to which we have not found the answer: how to mount a mass attack by men with the kind of training acquired in the physical sciences to work on these problems in the humanities in which detailed, repeatable, and controlled experiments cannot be made. This, I think, is a great need, if not the greatest need of our time. If the space program shows us how to assemble—and use well—efforts of this magnitude and shows us the betterment for man that can result from doing it well, we will truly have made a great and lasting contribution to our nation and to our civilization.

  ". . . we as a nation must understand more of science and its implications . . . .

"And though all children will not and should not attempt science as a calling, I think there should be an attempt as far as feasible to teach all children what science means and to teach them the importance of experimentation as a means of increasing our knowledge and understanding And for those who do not major in science this need to experiment—to learn from experience—should be heavily emphasized . . . .

"So please educate our youth well, teach them of the power that comes with knowledge of our universe and the people who inhabit it. Teach them sciences so that they can achieve triumphs in technology. Teach them the humanities so that they will use these triumphs for man's betterment. And above all help our children to grow in tolerance, help them to develop their wisdom, help them achieve self discipline. And 'Holding fast to basic concepts while discarding outmoded practices, — —let us work together, make every effort and meet every challenge to build this nation's most fundamental resource: the human mind.'" (Text)

December 6: Editorial in *Indianapolis News* stated:
"The national interest, as well as the furtherance of Indiana's economic well-being, would be served by the location in this State of the National Aeronautics and Space Administration's projected electronics research center.

". . . Indiana can offer transportation advantages, availability of trained young scientists, a favorable labor climate, coordination with established private industry, good living conditions and ample space for expansion.

"Purdue and Notre Dame provide outstanding examples of what should be a top-priority site for installation of the $50 million facility proposed by NASA . . . ." (*Ind. News*, in *CR*, 12/10/63, A7511)

* Exercise Top Rung II, consisting of a number of simulated bomber attacks on areas in North Central U.S., was conducted by NORAD/SAC. (DOD Release 1524-63)

December 7: NASA launched two sounding rockets from Wallops Station to compare measurements of pressure, temperature, density, and winds:

Nike-Cajun carried 82-lb. instrumented payload to 64.4-mi. altitude, ejecting 11 grenades which detonated at intervals from 20 to 60 mi. At 272,000-ft. altitude an inflatable sphere was ejected, ascending to 319,000-ft. apogee, and was tracked for 20 min. to 96,000 ft.

Thirty-two minutes later a Nike-Apache carried 62-lb. instrumented payload to 87.2-mi. altitude. Experiment chiefly consisted of pitot-static probe measurements. Approximately 6 min. of scientific data were telemetered to ground stations before Apache impacted in Atlantic Ocean 93 mi. downrange. (NASA Rpts. of S. Rkt. Launchings; Wallops Release 63-108)

* Special committee of the FAI, meeting in Paris, decided that Soviet cosmonaut Valentina Nikolayeva-Tereshkova was the "first absolute world champion among women in all types of competitive achievements in space," according to Tass. (Tass, *Krasnaya Zvezda*, 12/8/63, 3, ATSS-T Trans.)

December 7: Details of Soviet antimissile missile were disclosed for the first time, by Hungarian Communist Party newspaper *Nepszabadsag*. The two-stage rocket has range of between 2,500 and 3,750 miles, can operate effectively against target missiles at altitudes from 20 to 200 miles. Weapon is part of long-range radar system that automatically feeds data of approaching missiles into computers; data are passed automatically to antimissile missile stations, which take over tracking; then, when approaching missile is within 600 miles, antimissile missile is fired automatically. (UPI, *Wash. Post*, 12/8/63)

- James J. Haggerty, Jr., writing in *Army-Navy-Air Force Journal and Register*, said: "The excitement and enthusiasm which characterized the early years of space research is dimming. The barrage of criticism leveled at the Administration for its lunar landing program has taken a severe toll and brought forth a budget which will move the manned moon shot back at least two years and there seems to be a general lack of concern over this abrogation of a stated national goal . . . .

  "Space exploration may soon run into its first real barrier, which might be termed the 'megabuck barrier.' In the fantastic missions being considered for the next decade, there are myriad technical problems, the solutions of which will require a supreme effort on the part of the scientific and technological community. Yet none of these problems appears impossible of solution. The real stumbling block may be money . . . .

  "Perhaps there will be some great breakthrough—or a series of them—which will bring a dramatic reduction in the costs of space flight. But at the moment, with no such breakthrough on the horizon, the megabuck barrier appears more formidable than the speed of light." (*A-N-AF Journ. & Reg.*, 12/7/63, 14)

- Princeton-Pennsylvania Accelerator, a 3-bev proton synchrotron, dedicated at the James Forrestal Research Center of Princeton Univ. In his remarks at the dedication, AEC Chairman Dr. Glenn T. Seaborg pointed out that the accelerator "is unique among large proton accelerators in the unusually rapid pulsing of the magnet—19 cycles per second." (AEC Release S-37-63)

December 8: Rep. Edward J. Gurney (R.-Fla.), forwarded to President Johnson an appeal to preserve the name of Cape Canaveral, asking the President to clarify his executive order and make it clear that it applied only to the launch operations center and missile range station. Residents of nearby Florida communities writing to their Congressional delegation largely favored naming the John F. Kennedy Space Center as a tribute to the late President but preferred retaining 400-year-old name of Cape Canaveral. (AP, *Wash. Post*, 12/9/63; *Av. Wk.*, 12/9/63, 25)

- Dr. Caryl P. Haskins, President of the Carnegie Institution of Washington, warned in his annual report that long-range planning and general allocation of resources for scientific research were becoming "an absolute necessity.

  "We cannot, in the nature of things, simultaneously do all things well. And there is the very real danger that, without a careful and continuing general review of our over-all scientific activities in relation to their probable bearing on human welfare, we could find ourselves irrevocably overconcentrated in certain spectacular areas of science where past performance and present magnitude lend a luster that promise for the future does not match.

". . . currents of such magnitude and import as those now sweeping science on its course, if allowed to flow undirected, can bring incalculable damage . . . [in producing] massive imbalances among delicately related elements of our research structure . . . ."

He warned that "we are firmly wedded to massiveness, and our long and triumphant experience in developing vast technology leads us easily to equate success in research with sheer magnitude of effort . . . ."

Actually, the major advances "must inevitably rest, in the future as they have in the past, in the hands of a comparatively few highly original and gifted men and women—a minuscule proportion of the entire scientific population . . . ." (Finney, *NYT*, 12/9/63, 1 and 43)

December 8: Soviet newspaper *Krasnaya Zvezda* (*Red Star*) reported that Soviet cosmonauts are undergoing intensive training for "more complicated tasks." Cosmonaut Yuri Gagarin, promoted to full colonel, is directing new program of training the other veteran space flyers. (UPI, *Wash. Post*, 12/9/63)

- Tenth anniversary of the Atoms for Peace Program, which began when President Eisenhower before the United Nations had proposed making nuclear power "a great boon for the benefit of all mankind." (Finney, *NYT*, 12/8/63, A43)
- In the *Washington Post*, Howard Simons reviewed reconnaissance satellites, referring to previously released information and statements regarding a so-called "Samos" project, and inferring that "Samos" satellites were now operational. (*Wash. Post*, 12/8/63, E1–5)

December 8–18: FAA Administrator Najeeb E. Halaby visited Moscow, London, Paris, and Berlin, conferring with Aeroflot officials in Moscow, with civil aviation officials in London and Paris, and surveying U.S. air carrier operations in and out of Berlin. (FAA Release 63–104)

December 9: President Lyndon B. Johnson appointed Dr. Edward C. Welsh "acting temporary chairman" of the National Aeronautics and Space Council. As Vice President, Johnson had served as NASC Chairman. (*Space Bus. Daily*, 12/10/63, 370)

- Report by three-man NASA advisory committee concluded that "shielding [against radiation] of the crew for Project Apollo is not possible within the time and weight limitations of the project." The crew "will simply have to accept the relatively low probability of encountering a major solar flare during their relatively brief excursion to the moon." Report estimated 10,000 lbs. of polyethylene shielding would be required to protect men in 10-ft.-diameter spacecraft for a week. For flights to Mars or Venus, more than 20,000 lbs. of polyethylene would be required. "The accomplishment of manned flights to the vicinity of Mars and

Venus may have to be delayed until some means are found for reducing these shielding weights," report stated. (AP, *NYT*, 12/10/63, 24; AP, Wash. *Eve. Star*, 12/9/63)

December 9: GSFC held bidders' briefing with 350 representatives of 55 aerospace firms hearing outlined the "Unified S-Band" method of providing tracking and communications for Apollo lunar missions. Conferees were notified of firm design and production requirements for two complete systems and three partial systems. Contract would be awarded after April 1, 1964. (GSFC Release G-30-63)

- RCA announced completion of negotiations for a $23.5 million contract from Grumman Aircraft Engineering Corp. to provide a radar subsystem for Project Apollo's Lunar Excursion Module (LEM). Grumman is prime NASA contractor for building the LEM. (*Wall Street Journal*, 12/10/63)

- "It has been a long and rough road from the original upper stage studies by General Dynamics/Astronautics in 1956-57 to the orbiting Centaur of 1963," *Aviation Week and Space Technology* editor Robert Hotz wrote.

  "At Cleveland, it was a Lewis Research Center team headed by David S. Gabriel, Centaur project manager, with Edmund Jonash, Cary Nettles and Ronald Rovenger who developed and executed the rigorous ground testing program in the wind tunnels and rigs that paid off so well in the recent flight performance. And on Capitol Hill, it was Representative Joseph Karth, Democrat of Minnesota, whose investigation into Centaur's troubles shook the program out of its management difficulties and established its priority as a vital step in the lunar landing program . . . .

  "Centaur should merit a permanent page in the history of the space age as an excellent example of what can be accomplished when industry and government devote their best talents and unflagging determination to reach for a distant but desirable technical goal. It has taught lessons that must not be forgotten when new barriers loom on the frontiers of technology." (*Av. Wk.*, 12/9/63, in *CR*, 12/19/63, A7749-50)

- Christian M. Clausen, JPL's public information director since 1957, died at 49 in Altadena, Calif. (*NYT*, 12/11/63, 47)

- Dr. Glenn T. Seaborg, Chairman of AEC, spoke at Joint Meeting of the Indianapolis Engineering Societies and the Indianapolis Scientific and Engineering Foundation:

  "None of us today, I think, would presume to cover quite so wide a territory in our endeavors as Leonardo [da Vinci] did, but we are living in a time when the example of his bold imagination can be a valuable source of inspiration.

  "Much has happened to close the gaps between what Leonardo dreamed of and what we have now achieved. History has witnessed in our country two revolutions—the revolution inspired by our Founding Fathers which gave birth to this Nation, and the Industrial Revolution, inspired by man's inventive spirit, which gave us a place among the nations of the world. We are involved now in a Third Revolution—the Revolution of Science, inspired by our educated men and women—which is strengthening our position as a leader of nations.

"During the last several decades, and especially the last two decades, this Nation and the world have witnessed a remarkable expansion and extension of knowledge. This knowledge has been and is being transformed into the outlines of a new and revolutionary society. We have all been part of a new tide of exploration and discovery, not limited to the dimension of the past but creating new dimensions of its own.

"Indeed, the revolutionary accomplishments of the past decade—in space—have extended man's reach far beyond this planet. We now know that the search for new knowledge, if not restricted to subjects of foreseeable and immediate practical importance, results in an unexpected increase in our understanding of physical or biological phenomena. The increases, in turn, give rise to far-reaching practical applications which could not have been anticipated from the original basic research. Our scientific knowledge and technology are advancing at an explosive rate. The time lag between the discovery of a fundamentally new scientific principle and its application in engineering is now very short, and these rapid developments are changing the lives of all of us in many ways which we only dimly perceive . . . ." (Text)

December 9: Dr. Eugene Fubini, Deputy Director of DDR&E, was commended by *Missiles and Rockets* for speaking out in favor of "old-fashioned engineering."

"What Fubini is seeking is something that has not yet been done, the application of modern technology to conventional warfare problems on a massive scale. He has coupled this appeal for fresh thinking in R&D with a brief for simplicity and sound engineering.

" 'Old-fashioned engineering may come back in vogue,' he said, 'because one cannot have simplicity without good engineering. To make things simple takes a lot of ability and this is going to be the most important and the most difficult challenge facing us.' " (*M&R*, 12/9/63)

- Researchers at Technology Audit Corp. continued, under NASA contract extension, to study possible use of a satellite as a navigation aid for advanced aircraft such as the supersonic transport. (*M&R*, 12/9/63, 9)

December 10: Cancellation of USAF Dyna Soar (X-20) manned aero spacecraft project announced by Secretary of Defense Robert S. McNamara. Some of the funds saved would be diverted into broader, long-range exploration of the problems and potential of manned military operations in near space, chief project of which would be Manned Orbiting Laboratory (MOL). Attached to modified Gemini spacecraft (Gemini-X), MOL will be launched into orbit by Titan III launch vehicle, the two-man Gemini crew then transferring to the house-trailer-sized MOL where they will conduct experiments and observations for two to four weeks, then return to earth in the Gemini-X spacecraft. Schedule called for launching late in 1967 or early in 1968. Also affected would be Project Asset, which will be expanded to increase number of launchings, increase shapes and materials studied, and "substantially expand our knowledge of precision reentry gained through this unmanned program."

Secretary of the Air Force Eugene M. Zuckert issued statement: "We welcome the assignment of the manned orbiting laboratory project, and we will now concentrate our resources and best management effort on this job in the exciting field of space. This will assure effective Air Force participation in the manned space program."

Air Force Chief of Staff Gen. Curtis LeMay said that "the Air Force will provide the best space laboratory program in the national interest that is possible within technical and management resources available to us."

NASA Assistant Administrator for Public Affairs Julian Scheer said: "The decisions announced by Secretary McNamara today with respect to the Dyna Soar Project and Manned Orbital Lab. followed discussions with NASA and were fully coordinated with the programs of this agency. The transfer of research resources and objectives from Dyna Soar to the Asset program will accomplish objectives important to both agencies. The decisions announced by Secretary McNamara are based on the best uses of resources to maximize our national capacity in space and NASA fully supports them." (DOD Briefing Transcript; *NYT*, 12/11/63, 1, 22; *Wash. Post*, 12/11/63; *AF Policy Letter*, 1/1/64, 1; *Space Bus. Daily*, 12/12/63, 386)

December 10: Both Houses cleared for the President the Independent Offices Appropriations bill (H.R. 8747), including $5.1 billion NASA appropriation for FY 1964 ($612,000,000 less than requested). Of this amount, $3,926,000,000 was for research and development; $680,000,000 for construction of facilities; and $494,000,000 for administrative operations. House agreed to Senate amendments (unresolved in conference committee report): Amendment 82 provided that a joint venture to land a man on the moon may not be undertaken with any other nation without the consent of the Congress; Amendment 84 provided that no part of NASA's appropriation could be transferred to any other Government agency for research without approval of the Bureau of the Budget. (NASA Leg. Act. Rpt. II/202; *CR*, 12/10/63, 22788–95)

- National Science Foundation announced $393,000 grant to Univ. of Minnesota for Project Pocibo (Polar Circling Balloon Observatory), a U.S. contribution to the International Quiet Sun Year (IQSY). 20 or more balloons will be launched during IQSY from Point Barrow, Alaska, research laboratory and fly at 100,000-ft. altitude, with 45-lb. instrumented payload to observe cosmic rays and the aurora borealis. (*NYT*, 12/11/63, 22)

- Nobel Peace Prize for 1962 was presented to Prof. Linus Carl Pauling in Oslo. Gunnar Jahn, chairman of award-selection committee, emphasized Prof. Pauling's relationship with the nuclear test ban treaty: "Can anyone believe that this treaty would have been reached now if there had been no responsible scientist who tirelessly, unflinchingly, year in, year out, had impressed on the authorities and on the general public the real menace of nuclear tests?" (*NYT*, 12/11/63, 1, 20)

December 10: President Johnson briefed Congressional leaders on the defense budget, emphasizing that "the U.S. military program will continue to provide for a strategic force sufficiently large to absorb a surprise attack and survive with sufficient power to be capable of destroying the aggressor." (Kiker, *N.Y. Herald Trib.*, 12/11/63)

- Hook-up of three global satellite tracking networks into a single central console at NASA Goddard Space Flight Center was announced. Console allows combinations of operators at 51 communications points around the world in the satellite tracking and data acquisition network, the manned space flight tracking network, and the deep space instrumentation facilities. (AP, *NYT*, 12/11/63)
- Col. Charles Yeager, Commandant of Aerospace Research Pilot School at Edwards AFB, was forced to eject at 10,000 ft. after his rocket-augmented NF-104A aircraft failed to recover from a flat spin from 90,000 ft. Col. Yeager sustained first and second degree burns on face, neck, and left hand before bailing out. (*M&R*, 12/16/63, 13)

December 11: Sen. Stuart Symington (D.-Mo.) said in St. Louis that DOD's cancellation of Dyna Soar (X-20) manned space flight project probably would accelerate production of Gemini capsule by McDonnell Aircraft Corp. of St. Louis, thus benefiting employment and economy of St. Louis. (*St. Louis Post-Dispatch*, 12/11/63)

- Following Secretary McNamara's announcement of the Dyna Soar cancellation, *New York Times* editorialized: "Now the news that the expensive and difficult station-in-space project has been added to the list of tasks to be accomplished this decade makes it more imperative than ever that a new hard look be taken at what we are trying to do, the wisdom of the choices made, and the adequacy of our resources for the total effort." (*NYT*, 12/11/63, 46)
- NASA announced it would negotiate with IBM for electronic equipment and support in Saturn IB and Saturn V launch vehicle programs. Cost of work over next five years was expected to exceed $75 million. Negotiations would cover integration and checkout of the instrument units, the design and manufacture of data adapters, and development and production of digital guidance computers. (*Marshall Star*, 12/11/63, 2)
- President Johnson issued statement to Cabinet aimed at holding down Federal employment and achieving greater economy measures. He emphasized that 1965 budget totals for civilian employment "must be held below the levels in the 1964 and 1963 budgets . . . ." (Wicker, *NYT*, 12/12/63, 1, 28)
- Prof. Linus C. Pauling, in his Nobel speech at Oslo Univ., called for: U.N. supervision of nuclear stockpiles; acceptance of Communist China into the U.N. so that meaningful disarmament can take place; a control system for nuclear weapons under which a country could use them only with U.N. approval; a general agreement to halt R&D of biological and chemical weapons and prohibit their use. (*NYT*, 12/12/63, 3)

December 11: Dr. Gertrude Blanch, USAF Aerospace Research Laboratories scientist, was awarded "Exceptional Civilian Service" medal for her work in mathematics. (USAF OAR Release 12-63-2)

December 12: NASA announced postponement of Saturn SA-5 launching from mid-December to mid-January because of technical difficulties. First flight test of two-stage Saturn I, SA-5 would place in orbit the heaviest payload to date—38,000 lbs. (including S-IV upper stage) and would flight-test performance of RL-10 liquid hydrogen engines of S-IV stage. (UPI, *Wash. Post,* 12/13/63)

- Rep. Edward J. Gurney (R.-Fla.) announced renaming of Cape Canaveral to Cape Kennedy "will not be official unless it is acted on by the Florida Legislature." Gurney quoted Arthur Baker, Chairman of Board of Geographic Names, as saying name change will apply only to federal maps and documents, is not compulsory for the people of Florida. (AP, *Kansas City Times,* 12/13/63)

- Japan launched Lambda-2 sounding rocket with instrumented payload to 376-mi. altitude in experiment designed to gather data on the upper atmosphere for IQSY. Rocket was built by Tokyo Univ. Industrial Engineering Research Center and fired from Kagoshima Space Center in southern Japan. (*M&R,* 12/23/63, 10; *Space Bus. Daily,* 12/19/63, 428)

- At news conference on the phenomena of pneumatic tire hydroplaning, at NASA Hq., Walter B. Horne and Robert C. Dreher of NASA Langley Research Center described results of experiments conducted with automobiles traveling on wet roads. On highway curve where maximum safe speed is posted as 45 mph, automobile with tire pressure of 25 psi did not reach hydroplaning speed until about 52 mph; automobile with tire pressure of 16 psi attained hydroplaning speed at 40 mph and swung out of control. Mr. Horne said: "In this country, most automobiles use tire inflation pressures between 16 and 30 pounds per square inch. This is a hydroplaning speed between 42 miles an hour and 55 miles an hour.

    "This hydroplaning speed range is well within the higher legal speed limits on roads in this country. Therefore, automobiles are real susceptible to hydroplaning." He exploded the popular notion that reducing tire pressure makes driving on wet roads safer. "About the only place where it does any good to reduce the tire pressure is when you are driving on sand and you want to get higher flotation." (AP, Wash. *Eve. Star,* 12/17/63; NASA Fact Sheet; NASA News Conf. Transcript; NASA TN D-2056)

- Titan II missile launched from AFMTC in successful flight test down the Atlantic Missile Range. (UPI, *Wash. Post,* 12/13/63)

- NASA awarded Douglas Aircraft Co. two Saturn contract modifications: $48,064,658 contract for delivery of four S-IVB flight stages and support equipment and modification of S-IVB dynamic test stages; and $1,090,000 contract for modifications to S-IV stage. (*Space Bus. Daily,* 12/13/63, 397)

- AEC announced low-yield nuclear device was exploded underground at Nevada test site, 22nd underground nuclear test announced by AEC this year and 103rd explosion since U.S. resumed testing Sept. 15, 1961. (AP, *N.Y. Herald Trib.,* 12/13/63)

December 12: Secretary of Defense Robert S. McNamara initiated action to discontinue or substantially reduce activities at 33 defense installations—26 in U.S. and 7 overseas—no longer required by the armed forces. Action, when completed, would produce more than $100 million per year savings. (DOD Release 1562-63)

December 13: U.S.S.R. launched COSMOS XXIII into earth orbit (apogee, 381 mi.; perigee, 149 mi.; period, 92.9 min.; inclination to the equator, 49°), the satellite carrying scientific instrumentation for "continuing the study of outer space" in accordance with March 16, 1962, Tass announcement. Tass said onboard equipment was functioning normally. (Tass, *Krasnaya Zvezda*, 12/14/63, 1, ATSS-T Trans.)

- U.N. General Assembly approved broad program for cooperation in exploration of outer space, unanimously adopting two resolutions on outer space presented by committees: (1) draft setting forth U.N. space program for 1964, expressing belief that international cooperation can be fruitful in space exploration, and calling for use of space satellites for communications and meteorological purposes; (2) draft setting groundwork for legal guidelines covering space exploration, urging formal international agreements to provide for such issues as return of astronauts and space vehicles forced to land on foreign territory. (AP, *Wash. Post*, 12/14/63)

- Cancellation of five follow-on flights of the Ranger lunar impact spacecraft announced by NASA Associate Administrator for Space Sciences and Applications Dr. Homer E. Newell. Decision to cancel the follow-on series represented $90 million saving in the Ranger project. Dr. Newell said: "By this move, we are placing greater reliance on the remaining four Rangers, the Surveyor landers and the Lunar Orbiters for unmanned lunar exploration prior to manned landings on the Moon. At the same time, this move will contribute to necessary economies in the overall NASA program." (NASA Release 63-276)

- Launch of the SA-5 vehicle was delayed due to failure of connecting fittings in critical high-pressure pneumatic and hydraulic lines. Primary cause of the failures was attributed to the use of a specified heat treatment that did not provide optimum stress corrosion resistance for the specific stainless steel alloy used. The defective parts were replaced with parts heat-treated by a different process designed to provide more consistent properties and much improved resistance to stress corrosion failure. (MSFC, R-P-VE-N comment)

- Two sounding rockets launched at White Sands Missile Range in U.S.-Sweden coordinated study of ion and electron concentrations in D-region of the ionosphere. Arcas sounding rocket carrying Sweden's scientific experiment was fired to 60-mi. altitude, its 12-lbs. of instruments to obtain measurements of ion densities from peak altitude down to 25 mi. At apogee, parachute was deployed and instrumentation exposed.

  Nike-Apache sounding rocket, launched 62 min. later, carried U.S. experiments in 100-lb. payload of instrumentation designed to measure ion and electron densities, ultraviolet radiation in the upper atmosphere, and a camera supplied by Univ. of Leicester, England, to photograph x-ray emission; payload did not achieve desired altitude (about 110 mi.) because of malfunction in the launch vehicle. Cause of malfunction was not yet known.

U.S.-Sweden experiments were part of program sponsored by Swedish Committee for Space Research and NASA, whose Goddard Space Flight Center has responsibility for U.S. coordination. Both firings were conducted by Physical Science Laboratory of New Mexico State Univ. under U.S. Army contract. (NASA Release 63-277)

December 13: Appointment of Dr. W. Randolph Lovelace II as consultant to Dr. George E. Mueller, NASA Associate Administrator for Manned Space Flight. President of the Lovelace Foundation for Medical Education and Research, Dr. Lovelace would be consultant on medical aspects of current manned space programs. (NASA Release 63-274)

- Dr. Wernher von Braun, Director of NASA Marshall Space Flight Center, said he believed Soviet Union is building super-rocket that would dwarf any flown before. "We would be foolish to think they are standing still," he remarked. Dr. von Braun was at Michoud Operations, near New Orleans, formally accepting the first Saturn S-I booster built by private industry (Chrysler Corp.). (UPI, *Chic. Trib.*, 12/15/63; *Marshall Star*, 12/18/63, 1)

- McDonnell Aircraft Corp. of St Louis shipped the first of two Gemini mission simulators to Cape Kennedy, Fla., to provide NASA astronauts and ground crews with realistic training before actual launching of the two-man spacecraft. (AP, *Kansas City Times*, 12/14/63)

- Approximately 110 persons in NASA Manned Spacecraft Center's Computation & Data Reduction Div. moved into new offices at Clear Lake, second large group to occupy new quarters. With completion of Technical Information Div.'s move later this month, more than 250 MSC employees would have been relocated at Clear Lake. (MSC Release 63-254)

- Rep. Bob Wilson (R.-Calif.) said on House floor that U.S. aircraft manufacturers should achieve mach-3 supersonic transport "through logical evolution. If we begin today on the mach 2-plus commercial transport we will see that within no more than a year after the first one flies it will be modified—at relatively little expense—to give substantially better performance. Our aircraft industry has demonstrated this sort of design flexibility throughout its history . . . . There is a natural evolution that takes place when economic and engineering considerations become compatible. There is no reason to fear that this will not take place in the case of the supersonic aircraft. When we upset this normal progression we incur additional expense and introduce both operational problems and safety hazards . . . .

"Our aircraft industries and our airlines have competed successfully in the world marketplace for more than 50 years. There can be no logical justification for risking destruction of these industries by insisting that they accept more governmental help than they need. And that is just what we shall be doing if we persist in demanding a mach 3 commercial transport at this time." (*CR*, 12/13/63, 23391-92)

December 13: Army Corps of Engineers awarded $1,701,368 contract to C&B Construction Co., Memphis, for construction of NASA Marshall Space Flight Center's components test facility at Huntsville, Ala. (DOD Release 1569-63)

- Final test-firing of Army's Nike-Zeus antimissile missile at White Sands Missile Range was announced by DOD. The successful firing was test of propulsion and guidance components of the missile. Army would continue to fire Nike-Zeus at Kwajalein Island in the Pacific (DOD Release 1566-63)

December 14: First anniversary of MARINER II fly-by of planet Venus, providing data that clouds surrounding Venus are cold and dense and that Venus has surface temperature of 800°F. After passing the planet, MARINER II continued in perpetual solar orbit. Since launch Aug. 27, 1962, MARINER II has traveled 754 million mi. in 474 days. It is currently ahead of earth at straight line distance of 76,172,000 mi. Analysis of MARINER II tracking data provided new determinations of the Astronomical Unit, the mass of Venus and the moon, and locations of points on earth. Communications records included sending commands to the spacecraft flying by Venus (36 million mi.), and reception of signal from the spacecraft at distance of nearly 54 million mi. (NASA Release 63-275)

- USAF successfully launched Minuteman ICBM from Vandenberg AFB in flight down the Pacific Missile Range. (*M&R*, 12/23/63, 10)

- Marie Marvingt, 88-year-old sportswoman and aviatrix, died in Nancy, France. With copilot she flew English Channel in balloon in 1909. First Frenchwoman to hold pilot's license (1910), she participated in many air shows, held women's records for distance and endurance. (AP, *NYT*, 12/16/63, 35)

December 15: President Johnson announced the five recipients of 1963 National Medal of Science: Dr. Vannevar Bush, president of MIT corporation and formerly president of Carnegie Institution of Washington, director of Office of Scientific Research and Development during World War II, NACA Chairman 1939-41; Luis W. Alvarez, Associate Director of Lawrence Radiation Laboratory; John R. Pierce, Executive Director of Bell Telephone Laboratories; Dr. Cornelius B. van Niel of Hopkins Marine Station, Stanford Univ., and Dr. Norbert Wiener, Professor Emeritus of MIT. (AP, *NYT*, 12/16/63, 36)

- Interview by Howard Simons of Dr. Donald F. Hornig appeared in the *Washington Post*. Chairman of the Chemistry Dept. at Princeton Univ., Dr. Hornig was appointed Science Adviser to the President on November 7, accompanied President Kennedy and Vice President Johnson on a tour of military and space installations in November, and was recently reaffirmed as Science Adviser by President Johnson. Dr. Hornig has been one of the 18 members of the President's Science Advisory Committee since 1960.

Dr. Hornig was reported as saying that the place of basic research was distorted by the legacy of World War II when concentrated scientists were able to convert the most abstract scientific notions into impressive useful devices: "The days of pie-in-the-sky promises of basic research are over. The mood 20 years ago was one that suggested by turning loose great numbers of scientists miracles happened. Well they did happen, and I suppose it was logical that after the war the same notion would carry over.

"Only it doesn't work that way today. Miracles of the kind performed during the war are a little harder to find. Scientists are no longer concentrated nor are their efforts. They do come, of course, but over periods of years . . . .

"The meaning of Sputnik," Hornig said, "was that it demonstrated the Soviets were capable of big-time technology; capable, if you will, of playing in the big leagues." He declined to label the American attempt to send men to the moon as a "race."

"The moon is, after all, the first important object beyond the earth. . . . There are valid scientific objectives to be pursued on the moon's surface all linked to gaining new insights into cosmology and to providing clues to man's destiny. That is, where we came from. . . . (H. Simons, *Wash. Post*, 12/15/63, E3)

December 15: DOD decision to develop military communications satellite system, probably beginning operation in 1966, reported in *New York Times*. Funds were contained in proposed DOD budget for FY 1965, recently presented to the White House. Plans called for 24–30 satellites in random polar orbits with altitudes ranging from 5,000–12,000 mi., linking military forces around the world. No contracts would be awarded, however, until discussions with ComSatCorp resolve question of whether DOD should use commercial facilities for its military purposes. (Finney, *NYT*, 12/16/63, 1)

December 16: Modifications to X–15 No. 2, made when the aircraft had to be repaired after accident in Nov. 1962, were aimed at performance goal of mach 8 at 100,000-ft. altitude, 30% faster than basic design but lower than X–15's 354,000-ft. record altitude. X–15 No. 2 ultimately will make possible 4 min. 15 sec. of flight above mach 5 with more extensive experimental payload capability. (*M&R*, 12/16/63, 39)

• Activities at Kitty Hawk, N.C., commemorating 60th anniversary of powered flight (Dec. 17, 1903):

Youth Seminar on Flight, with 200 state delegates, observers, and representatives of youth attending, was moderated by Astronaut John H. Glenn, Jr. (Lt. Col., USMC). Panel members were Max Conrad, pilot of stock-model business aircraft; Capt. Harold Blackburn, TWA jet pilot; James Nields, business pilot; Jacqueline Cochran, record-holding aviatrix; and Maj. Robert Rushworth (USAF), X–15 pilot.

Full-scale replica of the Wright Flyer, built by volunteers from National Capitol Section of AIAA, was installed in museum at Wright Brothers National Memorial. Secretary of Commerce Luther H. Hodges was featured speaker at annual banquet of Kill Devil Hills Memorial Society at Nags Head, N.C.

Soaring Society of America presented to Wright Memorial Museum and Visitor's Center a plaque commemorating Wright Brother's glider flights. Presentation was made by Floyd Sweet of NASA Hq., past president of the Society. (60th Anniv. Committee Release; *Wash. Post*, 12/15/63; Soaring Society of America, Inc. Release, 12/16/63)

December 16: At ceremonies commemorating 60th anniversary of powered flight at Kitty Hawk, N.C., Walter Bonney, director of public information for Aerospace Corp. and former director of public information for NACA and NASA, recounted historic events of Dec. 17, 1903. He quoted Orville Wright on his initial, 120-ft. flight: " 'This flight lasted only 20 seconds, but it was, nevertheless, the first in the history of the world in which a machine carrying a man had raised itself by its own power into the air in full flight, had sailed forward without reduction of speed, and had finally landed at a point as high as that from which it started.'

"There were three additional flights that bleak December day, with the brothers alternating at the controls. The longest and last, with Wilbur the pilot, was of 59 seconds' duration and was for a distance of 852 feet. . . ." (Text, *CR*, 12/17/63, 23663)

- Although USAF was assigned responsibility for manned orbiting laboratory project, NASA would continue providing technical and information support, USAF officials said. Monthly NASA–USAF coordination meetings, "mutual information sessions," would continue to be held. (*Space Bus. Daily*, 12/16/63, 401)
- Nine enormous star-like objects emitting strong radio energy have been observed by astronomers in recent months, according to scientists reporting at International Symposium on Gravitational Collapse in Dallas, sponsored by Southwest Center for Advanced Studies, Univ. of Texas, and Yeshiva Univ. Identification of the new objects was accomplished by British-American-Australian astronomers using interferometry technique. Discoveries were described as the most important in astronomy in many decades. (Sullivan, *NYT*, 12/17/63, 1, 35)
- AFCRL research project to find if there were any correlation between phases of the moon and rainfall on earth involved checking 1,000 dates over 92 years of observations, checking each date against the lunar phase. Result of research, reported AFCRL's Dr. Fred Ward, indicated there was no correlation between rainfall and the lunar cycle. (*OAR Research Review*, 12/16/63, 2)
- USAF launched Titan II ICBM from silo at Vandenberg AFB, Calif., in successful flight down Pacific Missile Range. (*M&R*, 12/23/63, 10)
- USN launched Hydac sounding rocket, built by Douglas Aircraft and Lockheed, from Pt. Arguello, Calif. (*M&R*, 12/23/63, 10)
- Prof. Leonid I. Sedov, chief Soviet space scientist, quoted as saying he "believes" that by 1968 there would be 100-ton space stations and an automatic scientific station on the moon. (*Space Bus. Daily*, 12/16/63, 400)
- Using a selected cross-section of Americans, the Harris Survey indicated that the majority of Americans favored the exchange of

musicians and artists, and students with the Soviet Union, as well as favoring the sale of wheat and other food.

With regard to "Scientific Cooperation":

- *Sending Man to the Moon Jointly:*
  - Favor _____ 35%
  - Oppose _____ 54%
  - Not sure_____ 11%
- *Exchanging Scientists and Engineers:*
  - Favor _____ 34%
  - Oppose _____ 54%
  - Not sure_____ 12%

Selling equipment with potential military application to Russia was adamantly opposed. (L. Harris, *Wash. Post*, 12/16/63, 1)

December 16: Edward M. Shafer, NASA Attorney-Adviser since September 1959, was appointed NASA Assistant General Counsel. (NASA Announcement 63-275)

- 13-man Army and USAF team of parachutists jumped from USAF C-130 flying at 43,500 ft., delaying their parachute openings until below 2,500 ft., to claim new world record for mass free-fall. Previous record was 36,650 ft. set by 9-man Soviet team in 1961. The chutists were all participating in series of tests called "Operation Halo" (high altitude/low opening) conducted at El Centro, Calif. (DOD Release 1578-63)
- Four airmen ended simulated 30-day space flight mission inside a space cabin at USAF School of Aerospace Medicine, San Antonio, Tex. The airmen breathed pure oxygen at simulated altitude of 27,500 ft. in preparation for future manned space flights. (AP, Wash. *Eve. Star*, 12/17/63)

December 17: President Johnson addressed U.N. General Assembly, emphasizing continuation of efforts to achieve world peace.

". . . The great transition from colonial rule to independence has been largely accomplished. The decade of development has successfully begun. The world arms race has been slowed. The struggle for human rights has been gaining new force.

"And a start has been made in furthering mankind's common interest in outer space, in scientific exploration, in communications, in weather forecasting, in banning the stationing of nuclear weapons, and in establishing principles of law . . . .

"The greatest of human problems, and the greatest of our common tasks, is to keep the peace and save the future. All that we have built in the wealth of nations, and all that we plan to do toward a better life for all, will be in vain if our feet should slip, or our vision falter, and our hopes ended in another worldwide war. If there is one commitment more than any other that I would like to leave with you today, it is my unswerving strengthening of the peace. Peace is a journey of 1,000 miles, and it must be taken one step at a time . . . ." (Text)

- Sen. Leverett Saltonstall (R.-Mass.) speaking on Wright Brothers Day in the Senate, said: ". . . The attention to detail and the years of experimentation were fundamental to the triumph which we commemorate on Wright Brothers Day.

"The pioneering spirit, the determination, and the ability exemplified by the Wright brothers are the essence of human progress. They were neither inhibited by the predictions of failures from those who had not tried, nor were they dissuaded by the tragedies of those who had tried. As we recall that historic day, let us draw inspiration from the two men who made this aged dream a reality. In this space age, our efforts must also be a combination of bold design and prudent preparation." (*CR*, 12/17/63, 23705)

December 17: In a letter to NASA Administrator James E. Webb, Rep. Clarence D. Long (D.-Md.) criticized NASA for failure to fulfill Public Law 88–113 whereby "'a detailed study of the geographic location of, the need for, and the nature of, the proposed center'" should be made. Long was referring to the selection of a site for and the building of a $50-million electronics research center. He implied that the NASA "factfinding committee" set up to conduct this study had not given sufficient consideration to all suitable locations—the Baltimore area in particular—and, thereby, had failed to serve the national interest. (*CR*, 1/22/64, 934)

- Dr. Nicholas E. Golovin of the President's Office of Science and Technology told National Space Club in Washington that low reliability of Scout booster (50%) was because of difficulties in four-stage booster technology. A basic difficulty to be overcome in future space systems, he said, was development of long-lifetime components and subsystems. Today's missile system technology requires operational lifetime of less than a half hour; but Apollo would require 400 times that long and Mars fly-by mission, up to 20,000 times that long. The other basic problem for future space systems, he said, was achieving greater reliability with less flight testing. (*Space Bus. Daily*, 12/18/63, 417)

- Wright Brothers Day. Events commemorating the 60th anniversary of powered flight were highlighted by dedication of First Flight Airport at Kitty Hawk, N.C., the 3,000-ft. runway paralleling flight path of Wright Flyer. Airport was built by State of North Carolina, FAA, and National Park Service. Dedication speeches were made by Governor Terry Sanford and Secretary of Commerce Luther H. Hodges. Other participants in ceremony included Sen. A. S. Monroney (D.-Okla.), CAB Chairman Alan Boyd, TAC Commander Gen. Walter C. Sweeney, Jr. (USAF), and Park Service Director Conrad L. Wirth. Army, Navy, Air Force, Marine, and Coast Guard joined for first time in display of air power, with more than 70 airplanes flying over the site. (*NYT*, 12/16/63; *NYT*, 12/18/63, 30)

- Luncheon in Wright brothers' hometown, Dayton, O., commemorated 60th anniversary of powered flight and 50th anniversary of Air Force Silver Wings. General Mark Bradley, Commander of AFCL, Wright-Patterson AFB, made presentation to Maj. Gen. Benjamin D. Foulois (USA, Ret.) one of first military officers to fly an airplane. National Aviation Hall of Fame president, James Jacobs, announced four selections to Aviation Hall of Fame: Octave Chanute, General Foulois, Frank Purdy Lahm, and Samuel Pierpont Langley. Luncheon was sponsored by Dayton Area Chamber of Commerce and National Aviation Hall of Fame, Inc. (*CR*, 12/17/63, 23666–67)

December 17: At Wright Brothers Memorial Banquet in Hollywood, Calif., the Timers Aero Club of NAA awarded "Elder Statesmen of Aviation" awards to John K. Northrop, Floyd B. Odlum, and (posthumously) James Kindelberger. Also FAI medals were presented to Paul F. Bikle, glider pilot and Director of NASA's Flight Research Center, and Joseph A. Walker, NASA X-15 pilot. General Curtis E. LeMay, USAF Chief of Staff, said in address to Wright Memorial Dinner: ". . . When Orville and Wilbur Wright made their historic flights, 60 years ago, little did they know that their effort would have such an impact on the nation's strategy and on its military forces.

"We should learn a lesson from this.

"We cannot in today's world stop our search for new sources of national strength. Further, if this Nation ever finds itself falling behind in the technological race—and lacking men possessing the vision and courage of the Wright Brothers, and the great Aerospace flyers who followed them, then in my opinion our days at a great Nation will be numbered and so will the days of freedom." (*Air Force and Space Digest*, 1/64, 84; Text, DOD Release 1572–63)

• In preface to 1963 edition of *Jane's All the World's Aircraft*, editor John W. R. Taylor wrote that U.S. had achieved "a measure of equality" with U.S.S.R. in "space race."

"This may seem a strange comment to make . . . .

"But the real measure of achievement in aviation is a day-to-day service, not once-only exploits, and here the Soviet picture looks less self-assured than it did at the end of 1962 . . . .

"Nor has Russia's space program gone with its customary smoothness and unsullied success.

"The results achieved have been spectacular and highly creditable but there is good reason to believe that the 'space twin' flights of Lt. Col. Bykovsky and Miss Tereshkova failed to achieve all their objectives, which may have included orbital rendevous, and contact with the Mars probe was lost before it could add to our knowledge of that exciting planet . . . .

"By comparison, America's Mercury man-in-space program far exceeded its original—more limited—design objectives with the final 22-orbit flight of Gordon Cooper, while the Mariner 2 space probe sent back invaluable, if depressing, data concerning conditions on the planet Venus ending our hope of finding any form of life known on earth." (UPI, *Wash. Post*, 12/18/63)

• Donald Wills Douglas, founder, board chairman, and chief executive officer of Douglas Aircraft Co., was presented 1963 Wright Brothers Memorial Trophy at Aero Club of Washington's annual Wright Brothers Memorial Banquet. (NAA Release)

• Speaking at Boston College Seminar, Franklyn W. Phillips, Director of NASA's North Eastern Office, said:

"At the request of the Congress, NASA is conducting additional studies on a need for an Electronics Center, the nature of the Center, and its geographical location. The results of these three studies will be presented to Mr. Webb, who after making NASA's decision on the location of the Center will forward these reports to the House and Senate Space Committees for their consideration on or about the first of February. These Committees have 45 days in which to take action on NASA's report. If they take favorable action or for that matter if they take no action in the 45 days after submittal of the report, NASA can proceed with its plans. On the other hand, should these committees take negative action of any kind, it would force a further reconsideration of the proposal.

"The proposed Electronics Research Center will employ approximately 2,000 people of which six to seven hundred would be scientists and engineers. Estimates of the investment to be made in constructing the Center amount to 56 million dollars. When the Center is fully operative, it will have an annual operating budget of $20 to $30 million, and will be expected to disburse considerably more than that through a contract research and development program with industry and universities." (Text)

December 17: First successful flight of the C-141 Starlighter took place at Dobbins AFB, Ga. Featuring four fan-jet engines, the plane can cross any ocean nonstop at speeds of more than 500 mph. (*AFSC Operational Highlights*, 13)

- U.S. launched an unidentified probe using Blue Scout booster. (*Pres. Rpt. on Space, 1963*, 1/27/64)

December 18: In press conference, President Johnson discussed preparation of FY 1965 budget:

"I am working from a budget of $98.8 billion this year. It appears that we will expend about that amount, and maybe a little under or a little over, but substantially $99 billion will be the expenditures this year. That was the amount of Mr. Kennedy's budget. There are built-in increases of $1,790 million that are mandatory—military pay increase for 9 months, military retired pay, civilian pay, National Aeronautics and Space contracts, the Agency for International Development, Post Office rise, Federal Aviation Agency, urban renewal, and public assistance grants. . . .

"In the Roosevelt war years we spent as high as 46 percent of our gross national product for administrative budgets. During the transition period under Mr. Truman we spent 29 percent. In 1954, under Mr. Eisenhower, we spent 18.6 percent. We are hoping that we can come under that figure in our budget next year—in other words, that our budget expenditure as a percentage of our gross national product will be less for the fiscal year 1965 than it has been for any of these periods I mentioned . . . ."

To the question, "Mr. President, have you reached a hard decision on changing the space program, including eliminating the Rover project?" the President replied: "No." (Transcript, *CR*, 12/19/63, 24005-007)

- Flight of X-15, with Maj. Robert Rushworth (USAF) as pilot, canceled minutes before scheduled drop from mother ship because of mechanical troubles in the rocket aircraft. (UPI, *Wash. Post*, 12/19/63)

December 18: Editorializing about possible cuts in Rover project, *New York Times* said:

> ". . . The effort to build a nuclear-powered rocket for eventual manned exploration beyond the moon is a longer-term project [than Project Apollo], one which might eventually be accomplished by joint work of Soviet and American scientists, or perhaps even by a broader international team. Any proposal to kill this program in its entirety appears to us unsound; while to continue it at a substantially reduced rate would retain the gains of past research and continue slowly to increase our capabilities in this field. Such a course of action would mean minimum loss for the nation's near- and medium-term space goals, give us bargaining power in negotiations for a joint effort with the Soviet Union and still provide useful savings." (*NYT*, 12/18/63, 40)

- USAF launched three Atlases from Vandenberg AFB: Atlas D fired down the Pacific Missile Range in Abres (advanced ballistic reentry system) test; Atlas-Agena D booster rocket combination launched an unidentified satellite; and Atlas F fired down the PMR with mock nuclear warhead. (*NYT*, 12/19/63, 18)
- NASA Administrator James E. Webb testified before Senate Select Committee on Small Business, Subcommittee on Retailing, Distribution, and Marketing Practices' hearings on the role of effects of technology on the nation's economy. Webb said that NASA's "pull on the nation's pool of scientific and technical manpower has been substantially less than many alarmists have stated." Reiterating that NASA is a "management agency," he pointed out that 93% of NASA's space research money goes to private industry, "thus encouraging the growth of research and development under private industry." (*Space Bus. Daily*, 12/19/63, 425)
- NASA-Military Sea Transportation Service agreement signed for ocean transportation of Saturn V's three stages from California and Mississippi production and test sites to Florida launch site. Two ships of the LSD class, USNS *Point Barrow* and possibly the USNS *Taurus*, will be modified to carry S-IC stage and the smaller S-II and S-IVB stages. (NASA Release 63–278)
- NASA grant of $493,150 to Dept. of Interior's Geological Survey for lunar mapping was reported. (*Space Bus. Daily*, 12/18/63, 420)
- USAF's AFSC/SSD awarded $1,800,000 contract to Aerojet General Corp. for development and fabrication of Titan II propulsion system for Project Gemini. Award was made under existing cost-plus-fixed-fee contract. (DOD Release 1581–63)
- USAF launched Minuteman ICBM from Cape Kennedy in successful 5,000-mi. flight to target area near Ascension Island. (*M&R*, 12/23/63, 10)
- U.S.S.R. has issued new postage stamps showing routes of Soviet space flights, Radio Moscow reported. (UPI, *Miami Herald*, 12/18/63)

December 19: EXPLORER XIX was launched by NASA from PMR on a Scout booster. The 17.8-lb. polka-dot balloon satellite was placed in orbit by a new fourth stage (X–258) of the Scout booster. Apogee was lower than planned and resulted in an orbital period almost six min. less than planned. This, plus a very weak signal from the tracking beacon on the satellite, caused several days' delay in confirming satellite inflation and in defining the orbital data. Optical and radar tracking confirmed orbital data as: apogee, 1,488 mi.; perigee, 366 mi.; period, 115.8 min.; inclination, 78.62°. EXPLORER XIX, like EXPLORER IX, is an atmospheric density satellite. Because of its large surface area and very low mass such a satellite is very sensitive to fluctuations in density in the upper atmosphere. These changes, largely caused by variations in solar radiation, can be measured by fluctuations in the satellite's orbit. EXPLORER XIX was launched into polar orbit and, when tracked by NASA and the Smithsonian Astrophysical Observatory for a good portion of an 11-yr. solar cycle, should provide atmospheric density data on the earth's high latitudes comparable to those provided by EXPLORER IX for the lower latitudes. (EX. XIX Project Office, LARC, W. J. O'Sullivan; NASA Releases 63–271 and 281)

December 19: COSMOS XXIV earth satellite orbited by U.S.S.R. (apogee, 254 mi.; perigee, 131 mi.; period, 90.5 min.; inclination to the equator, 65°). Tass said the satellite carried scientific instrumentation for "continuing the study of outer space" in accordance with program announced March 16, 1962, and the onboard equipment was functioning normally. (Tass, *Komsomolskaya Pravda*, 12/20/63, 1, ATSS-T Trans.)

- President Johnson signed H.R. 8747, FY 1964 Appropriations for Independent Offices (P.L. 88–215). The President issued a special statement criticizing Congressional amendment prohibiting U.S. participation in manned lunar landing program with any other country "without the consent of the Congress," calling the restriction unnecessary and undesirable. (NASA Leg. Act. Rpt. II/211; Finney, *NYT*, 12/20/63)
- AFSC Commander Gen. Bernard A. Schriever (USAF) wired congratulations to William Bergen, President of Martin Co., and to Maj. Generals Waymond A. Davis and Benjamin Funk of AFSC for solving "pogo-stick action" in the Titan II booster for Project Gemini. The sharp surging accelerations during early stages of flight were noted early this year. To solve the problem, Martin Co. engineers redesigned the fuel distribution system to include "surge chambers." In the two test-flights of Titan II's incorporating the modification, the booster met NASA-USAF requirements for manned space flight. (Hines, Wash. *Eve. Star*, 12/20/63)
- The Black Report was submitted to President Johnson by Presidential advisers Eugene R. Black, former World Bank president, and industrialist Stanley de J. Osborne. The report said:
  "We conclude that one of the basic philosophies of the current program, namely that of tying the United States effort to the Concorde and therefore compressing the time of development and construction, is dangerous, technically and economically.
  "We feel that a superior aircraft which is available in two to three years of the first 'Concorde' deliveries will still be able to catch the bulk of the world market." (Wash. *Daily News*, 2/13/64; *CN*, 2/14/64)

December 19: James Reston, writing of President Johnson's attack on poverty, said the President was faced with having "to choose not necessarily between the moon and the slums, but between the problems of today and the problems of tomorrow.

". . . some of the President's scientific advisers have recommended that . . . [the FY 1965 budget request for Project Rover] be cut again to $100 million, which is less than half the Rover project budget for the current [fiscal] year.

"There is, of course, no assurance that money saved on this project would be approved by the Congress for an attack on poverty and illiteracy in the slums, but President Johnson's decision on this will give some indication of the set of priorities he proposes to establish for his Administration . . . ." (Reston, *NYT*, 12/20/63, 28)

- NASA and Government of Malagasy Republic announced agreement for installation of transportable tracking station at Majunga, Madagascar, for monitoring injection phase of such U.S. satellites as Nimbus, Echo, and Orbiting Geophysical Observatory. (NASA Release 63-279)
- $9,847,000 contract for electrical and mechanical equipment for three launcher umbilical towers for Saturn V launch vehicle was awarded by John F. Kennedy Space Center, NASA, to two-company team of Paul Smith Construction Co. and E. C. Ernst Co., Inc. (*Space Bus. Daily*, 12/20/63, 432)
- Despite renaming of Cape Canaveral to Cape Kennedy, the nearby town of Cape Canaveral would retain its name by choice of its 4,000 residents. Committee on Domestic Geographic Names said it would never approve a new name for a town over protests of its populace. (Diggins, *Wash. Post*, 12/20/63)
- Martin-Marietta Corp. of Baltimore announced its research efforts had demonstrated continuous electrical current generated by magnetohydrodynamics (MHD). Scientists of Martin Marietta's nuclear division said they achieved this goal—passing a stream of high-temperature ionized gas through a magnetic field to generate electricity—by solving the major problem of equipment's melting from the heat necessary in the MHD process. They used principle of magnetically induced nonequilibrium ionization of gas, in which temperature of the electrons is made higher than that of the gas itself; the electrons then conduct electricity without penalty of overheating. Experiments employed helium gas seeded with molten cesium droplets to enhance ionization. Dr. Mostafa Talaat, project director, said current generated was a few thousandths of an ampere. First practical use of MHD power generation, combined with heat derived from nuclear reactor, was estimated at 5–10 years away. (Schmeck, *NYT*, 12/20/63, 31)
- Dr. J. N. Bahcall of Cal Tech's Kellog Radiation Laboratory outlined at American Physical Society meeting in Pasadena a proposed experiment to determine temperature of sun's interior. A tank the size of an Olympic-size swimming pool, sunk at bottom of 5,000-ft. shaft drilled into the earth to screen cosmic rays, would be filled with 100,000 gals. of chlorine-37 isotopes. Neutrinos from sun bombarding the pool would change chlorine-

37 into argon-37. Measuring rate of change would help physicists determine solar temperature as well as sun's structure and composition. (Becker, *NYT*, 12/20/63, 31)

December 20: NASA selected Boeing Co. for negotiation of incentive-type contract to build Lunar Orbiter spacecraft, contract expected to exceed $60 million. One of the five bidding firms, Boeing would provide probes that are three-axis stabilized, use Eastman Kodak-developed camera system, and RCA power and communication subsystems. Five Lunar Orbiter spacecraft would be launched to take close-range photographs of lunar surface, as well as to measure such conditions as radiation and micrometeoroid density. Lunar Orbiter complements Ranger and Surveyor lunar probe projects, providing scientific data about the moon in preparation for Project Apollo manned lunar landing. (NASA Release 63-280)

- NASA Manned Spacecraft Center announced Grumman Aircraft Engineering Corp., prime contractor for development of Apollo lunar excursion module (LEM), had awarded $8,370,000 contract to Hamilton Standard Div. for design, development, and production of environmental control system for LEM. (MSC Release 63-257)

- 25th Minuteman ICBM fired by USAF from Vandenberg AFB made routine training flight down PMR. (*M&R*, 12/6/63, 10)

- Gordon M. Bain, FAA Deputy Administrator for Supersonic Transport, suggested at meeting of Society of American Travel Writers in New York that he believed the U.S. supersonic transport aircraft could be economically used on routes as short as "perhaps 1,200 miles." Maximum range of the U.S. SST would be around 4,000 mi. (Clark, *NYT*, 12/21/63)

December 21: TIROS VIII (A-53) meteorological satellite was placed in orbit by NASA, using a Thor-Delta booster at AMR. Initial orbital data: apogee, 474 mi.; perigee, 431 mi.; period, 99.3 min.; inclination, 58.49°. TIROS VIII, in addition to the usual wide-angle TV camera transmitting cloud-cover pictures once an orbit to a rather sophisticated ground station, featured the first orbiting of the automatic picture transmission system (APT), designed to provide real-time local weather information to any area in the world by means of a simple, inexpensive ($32,000) ground station. APT equipment in the satellite was a 24-lb. package including a new wide-angle (108°) Tegea-lens TV camera, with its storage and slow-scan transmission system. Preliminary results from the new system were excellent. (NASA Release 63-269; GSFC Tracking Off.)

- President Johnson ordered establishment of nine-member Committee of Defense and Economic Agencies in the Executive Branch, to study problems arising out of shifts and reductions in defense spending, or possible disarmament, and to assure they are accomplished with as little economic dislocation as possible. (NASA Leg. Act. Rpt. II/215)

- USAF launched Thor-Agena D booster rocket with unidentified satellite from Vandenberg AFB, Calif. (UPI, *Wash. Post*, 12/22/63; *Pres. Rpt. on Space, 1963*, 1/27/64)

December 22: Communications Satellite Corp. issued to industry a Request for Proposals for Engineering Design of a Commercial Communications Satellite, to include satellite design for either a medium-altitude system or a synchronous-orbit system. The RfP's were directed to development of ComSatCorp's basic comsat system, with initial global coverage to be achieved not later than 1967 for medium-altitude system or 1968 for synchronous-orbit system. Design proposals, due Feb. 10, would be evaluated and ComSatCorp would select one or more for six-month design contracts. (ComSatCorp Release)

December 23: NASA announced it would negotiate contract extension with General Electric Co. for plant- and test-support services at NASA Mississippi Test Facility. Extension was expected to exceed $25 million. NASA had selected GE in February 1962 to support Project Apollo including systems integration, checkout, and reliability—including plant and test support at MTF. (NASA Release 63–283)

- JPL awarded contract to Motorola, Inc., for fabrication and system integration of receiving subsystems operating at S-band frequencies in the Deep Space and Manned Space Flight tracking networks. Contract was follow-on to present contract awarded March 1962 under which Motorola designed and fabricated modules for S-band receiving subsystems for Deep Space Network. The additional equipment was for the NASA Unified S-Band System, in which receiving equipment for both networks are identical except for slight difference in operating frequency. Unified S-Band System allows all communications functions between ground stations and spacecraft to be accomplished at single frequency; provides more efficient power transfer between ground station's directive antennas; and reduces by hundreds of times the galactic noise received on UHF frequencies. (NASA Release 63–284)
- General Dynamics was proposing combination of fluorine-and-liquid-oxygen Atlas and Centaur upper stage as a cheaper substitute for Titan III booster for USAF Manned Orbiting Laboratory (MOL), *Missiles and Rockets* reported. (*M&R*, 12/23/63, 9)
- Manned Spacecraft Center announced building contractors had been invited by Corps of Engineers to submit contract proposals for construction of Spacecraft Control Technology Laboratory at MSC. The laboratory will be a $1,500,000 addition to the Spacecraft Research Office and Laboratory and will contain facilities for noise and vibration testing. (MSC Release 63–261)
- Aerojet General Corp. reported it had developed new solid fuels that can propel a missile a mile in one second after ignition. Based on combination of mechanical and chemical methods, the propellants have fastest burning rate ever obtained in nonexplosive solid fuels, Aerojet said. (*M&R*, 12/23/63, 9)
- Dr. Glen Wilson, staff member of Senate Committee on Aeronautical and Space Sciences, said in *Missiles and Rockets* interview: "Contrary to some published reports, President Johnson is not predisposed to a military emphasis on space. He will judge the re-

quirements of DOD and NASA strictly on merit, if his past performance is any measure." Dr. Wilson recalled Johnson was Chairman of Senate Armed Services Committee's Preparedness Investigating Subcommittee, later was Chairman of Senate Committee on Aeronautical and Space Sciences from its inception until he became Vice President. Dr. Wilson stressed that "it must be noted that Johnson was one of the few men on the Hill to recognize the supreme importance of the SPUTNIK I launch, and he instigated an extraordinary session of Senate hearings a few weeks later." (*M&R*, 12/23/63, 16)

December 23: Deputy Assistant Secretary of Defense for Arms Control Arthur Barber advised defense industry to seek nondefense markets as insurance against declining defense budgets. Barber estimated there was a better than even chance of some East-West arms control agreement with mutual inspection provisions during 1964. (*M&R*, 12/23/63, 16)

- Article in *Steel, the Metal-Making Weekly*, included remarks by Louis B. C. Fong, Director, NASA's Technology Utilization Program, and Dr. Earl P. Stevenson, member of NASA's Industrial Application Advisory Board:

    "Spin-off is in the form of ideas and innovations, not finished products ready to roll off production lines and change American buying habits overnight," explained Mr. Fong. "A recent study by Denver Research Institute, Denver, uncovered examples (of spin-off) and added that these examples do not include all, nor can they be considered statistically representative of, the transfers which have taken place.

    "NASA centers have come up with about 850 innovations thought to have industrial potential. The majority have been in the mechanical and electrical fields, 34 percent each, followed by materials, 22 percent, energy sources, nine percent and life sciences, one percent.

    "Time (the interval between the inception of an innovation and its commercial application) is an inherent problem. It takes four to seven years to make an idea into something practical. It will be tough to speed up that process that occurs between conception and use," Mr. Fong stated.

    Dr. Stevenson commented that:

    "Small firms, as well as large aerospace contractors, can benefit from spin-off. Techniques . . . being developed in the fabrication of rockets and space vehicles are not . . . of the assembly line and mass production [varieties], but rather those of the job shop." (*Steel, the Metal-Making Weekly*, 12/23/63)

- General Electric Co. received $2,000,000 increment to existing USAF cost-plus-fixed-fee contract for work on Titan II launch vehicle program. (DOD Release 1593–63)

December 24: NASA and AEC announced revision in Project Rover, canceling development of Rift (Reactor-In-Flight-Test) stage and deferring flight objectives of Nerva (Nuclear Engine for Rocket Vehicle Application). Rover's Kiwi ground reactor program was unchanged. Work towards development of flight systems was deferred until technology learned from Kiwi and Nerva is "satisfactorily established." Revision was expected to save as much as $180 million of planned and programed funds in FY 1964 and FY 1965. NASA and AEC now will concentrate on ground reactor and experimental engine research, development, and testing "with particular emphasis on analyzing and understanding power levels, temperatures, operating life, and problems of frequent and reliable restarts." (NASA Release 63-285)

December 24: Perth, Australia, was ablaze with Christmas lights far into the night. Perth was celebrating its new tradition as "City of Lights" begun Feb. 20, 1962, when its lights were a beacon for Astronaut John H. Glenn's orbital Mercury space flight. (Lukas, *NYT*, 12/28/63, 6)

- U.S.S.R. fired missile into Pacific target area south of Hawaii and east of Johnston Island, part of testing series beginning early this month and extending through Jan. 25. (AP, *Wash. Post*, 12/27/63)

December 25: FAA Administrator Najeeb E. Halaby, just returned from discussions in Moscow on technical aspects of a Moscow-New York air agreement, reported that U.S.S.R. plans to develop a supersonic transport appeared "very similar" to the U.S. SST program in scope and progress. What he learned of Soviet SST plans gave him neither cause for panic nor "any sense of confidence." Halaby stated: "They could have spoofed me but I have the impression their progress is about the same as ours."

Halaby said that the Soviets planned to develop a new engine and were persuaded that aluminum would not be a suitable airframe metal. Soviet target date for commercial flight, Halaby said, is in the 1970-73 time-period, paralleling U.S. estimates. The SST appeals to the Soviets because of long distance runs from Moscow to domestic points as well as New York, New Delhi, and Djakarta. (MacKenzie, *Wash. Post*, 12/26/63, 1)

- Secretary of Defense Robert S. McNamara was considering USAF proposals for initiating research on a new manned bomber, *New York Times* reported, apparently favoring a low-altitude manned penetrator aircraft. (Raymond, *NYT*, 12/26/63, 1)

December 26: NASA announced it would negotiate extension to existing contract with Control Data Corp. for procurement of computers and associated equipment for Apollo spacecraft Pre-Flight Acceptance Checkout Equipment (PACE) stations. Cost of the equipment for the nine stations was expected to exceed $25 million. General Electric Co. would receive the equipment as government-furnished equipment, would deliver first two PACE systems to North American Aviation, Inc., for checkout of Apollo command and service modules. Four systems would be delivered to John F. Kennedy Space Center, two to Grumman Aircraft Engineering Corp., and one to NASA Manned Spacecraft Center. (NASA Release 63-286)

- Spectroscopic studies of Mars indicate Martian atmosphere is too thin (10 millibars pressure) for wing- or parachute-landings by spacecraft, according to Dr. Gerard P. Kuiper, Director of Lunar and Planetary Laboratory of the Univ. of Arizona. Until recently, pressure on Mars was generally considered to be 85 millibars, based largely on observations by Dr. Audouin Dollfus of the Paris Observatory. Earlier this year, Dr. Kuiper had estimated surface pressure to be 17 millibars, but more precise analysis led him to 10-millibar conclusion. NASA scientists planned further observations for March 1965, when Mars is closest to earth, to gain decisive evidence. (Sullivan, *NYT*, 12/26/63, 1)

December 26: NASA Administrator James E. Webb submitted to Sen. Richard B. Russell, Chairman of Committee on Armed Services, NASA's recommendations on S. 960, bill to establish joint committee to investigate defense and space contracts, and S. 961, bill to amend U.S. Code on procurement to require NASA to establish central file on contract communications. Regarding S. 960, Mr. Webb said "it is not apparent to this agency that more effective machinery is needed for the investigation of the NASA procurement program by the Congress than exists today . . . ." Regarding S. 961, Mr. Webb said "it is felt that current practices provide ample opportunity for the committees of the Congress to review and investigate the handling of any particular procurement action of NASA without resorting to the assembling of a mass of miscellaneous communications in a central file." (NASA Leg. Act. Rpt. II/217; Letter, 12/26/63)

- Unnamed NASA official said NASA had reviewed Pratt and Whitney's prototype of fuel cell for Apollo spacecraft and had accepted the fuel cell. Recent criticism had previously been reported that the fuel cells fell short of NASA's requirements. (Simons, *Wash. Post*, 12/27/63)
- U.S.S.R. has not yet named delegates for working groups to work out details implementing U.S.-U.S.S.R. space agreements reached this year by Dr. Hugh L. Dryden and Academician A. A. Blagonravov, Howard Simons reported in *Washington Post*. U.S. and U.S.S.R. had agreed on coordinated communications satellite experiments, weather and magnetic mapping cooperation. Simons also pointed out Soviet lack of definite response to proposal for negotiations leading to U.S.-U.S.S.R. cooperation in lunar exploration, presented by President Kennedy Sept. 20 and reiterated this month by President Johnson. (Simons, *Wash. Post*, 12/26/63)
- AFSC/SSD awarded $4,396,000 increment to existing contract to Aerojet General Corp. for work on Titan III liquid rocket engines for space programs. (DOD Release 1599-63)

December 27: NRL astrophysicists reported on "galactic X-ray astronomy" at American Astronomical Society meeting held at Georgetown University. By launching rocketborne "telescopes," NRL scientists have detected two x-ray sources previously unreported. One, a mysterious x-ray source near the constellation Scorpius, is so strong that its x-ray output is one billion times as bright as the earth's sun. This may be the hitherto elusive neutron star theorized by S. Bowyer, E. T. Byram, T. A. Chubb, and H. Friedman. The other x-ray source, located near Scorpius in the Crab Nebula, is the remnant of a super stellar explosion that occurred in 1054 A.D., according to NRL reports.

Detectors in a rocket launched on April 29, 1963, spotted the mysterious x-ray emitter near Scorpius. "One possible hypothesis," Friedman said, "is that the x-rays come from an invisible cloud of extremely hot gas, of the order of a million degrees . . . ." Friedman concluded that the development of more sensitive detectors for use on rockets and orbiting satellite observatories would help make less mysterious the strange phenomena now being observed in space. (Simons, *Wash. Post*, 12/28/63, B2)

December 27: At AAS meeting in Washington, S. Karrer and C. C. Kiess reported new evidence suggesting planet Mars has toxic clouds of oxygen and nitrogen compounds, probably nitrogen dioxide gas. By spectroscopic techniques splitting the light from the planet's telescopic image, they found the pattern yielded by nitrogen dioxide gas. Kiess said that phenomena of the Martian appearance through earth telescopes can be explained by changes in the nature and color of nitrogen and oxygen compounds under changing Martian temperatures. (Barbour, *Wash. Post*, 12/28/63, B2)

- At AAS meeting in Washington, Dr. G. E. Moreton described photographs of H-Alpha rays showing that solar flares produce shock waves which depress the sun's surface. The photographs were taken last Sept. 20 at Lockheed Solar Observatory, Burbank, Calif. (Wash. *Eve. Star*, 12/28/63)

- Application of space research in consumer products reported by Dept. of Agriculture researchers Mary Jane Ellis and Philip B. Dwoskin. Aiding the housewife are much advanced "programmed cooking"—ovens with automatic temperature holds—and processed foods—for example, 51 varieties of processed potato products. (Slevin, HTNS, *Wash. Post*, 12/27/63, A20)

- At American Association for the Advancement of Science meeting in Cleveland, Prof. Kenechi Maeda of Kyoto Univ. described proposal for Japanese space booster submitted by Japanese engineers. Headed by Prof. H. Itokawa of Tokyo's Institute of Industrial Science, the engineers are designing a multistage, solid-fueled booster capable of putting about 100 lbs. into low earth orbit. Development of the rocket, known by Greek letter "Mu," is not yet approved by Japanese Government. (Hines, Wash. *Eve. Star*, 12/27/63)

- At AAAS meeting in Cleveland, Dr. Allan Brown, biology prof. at Univ. of Pennsylvania, urged that U.S. initiate a program to land a scientific probe on Mars in 1969, because the Martian "surface could be sampled during spring and summer when there would be the best chance of detecting plantlife organisms." He recommended adapting Saturn IB booster to carry a several-hundred-pound instrumented payload. Dr. Brown estimated significant scientific probe of Mars could be achieved in 1969 at cost of $2.9 billion over six years. He suggested international cooperation on the project "should be more easily achieved than in any other area of space research or exploration in which both U.S. and Soviet programs are involved." (*M&R*, 1/6/64, 14)

December 28: NASA Manned Spacecraft Center announced award of $272,522 contract to S.I.P., Inc., to build and install space environmental chamber and associated equipment at MSC for use in thermochemical tests. (MSC Release 63-263).

December 29: Soviet newspaper *Pravda* announced Soviet radio-astronomy station had bounced radio signals off planet Jupiter when the planet was 370 million mi. from earth last September and October. The radio signals took one hour and six minutes to travel the 740-million-mi. round trip from the earth to Jupiter and back. Academician Vladimir A. Kotelnikov, director of the Institute of Radio Engineering and Electronics, was quoted as saying the experiment was designed to study reflective properties of Jupiter's surface and radiowave propagation over long distances. (Shabad, *NYT*, 12/30/63)

- In letter to NASA Administrator James E. Webb, New York Congressional delegation urged NASA to select site in New York state for its proposed electronics research center. NASA had received more than 30 bids from areas in at least 19 states for location of the center. (AP, *NYT*, 12/30/63; *M&R*, 12/23/63, 9)

- FCC Chairman E. William Henry said in year-end statement: "Continued experiments with 'Telstar', 'Relay' and 'Syncom' hold forth the promise that space satellites, in the next few years, can provide reliable and economic communications." (UPI, *Wash. Post*, 12/30/63)

December 30: Astronomers at Smithsonian Astrophysical Observatory announced it was virtually impossible for EXPLORER XIX balloon satellite to be sighted by the naked eye, because even in the most favorable conditions the satellite appears only of the fourth magnitude (on borderline of visibility). So far, all sightings reported to the Observatory were from the Southern Hemisphere and were assisted by powerful binoculars or telescopes. (AP, *NYT*, 12/31/63, 4)

- Department of Defense announced that the last of the Titan II ICBM's had become operational, giving the U.S. 534 ICBM's ready to fire. Titan II's can be fired in less than one minute, and reportedly can carry 24-megaton warheads 9,000 miles or more. (DOD Release 1606-63; UPI, *Wash. Post*, 12/31/63, A1)

- At AAAS meeting in Cleveland, biophysicist Dr. Harold C. Urey expressed disappointment with cutback in scientific exploration of the moon preparatory to manned lunar landings. Criticizing advocates of exploring the moon by instruments instead of men, he repeated his view that only personal observations by trained men will provide answers to scientific questions about the nature of the moon. In this connection, he deplored criteria for selecting astronauts, arguing that the emphasis should be on trained geologists rather than jet pilots. Later at press conference, Dr. Urey stated that there is no scientific reason to believe the moon is devoid of water and that "most probably" there are pools of water below the rocky surface.

Dr. Urey discussed eruptions on moon near the crater Aristarchus, observed by Lowell Observatory and others. He said sightings may have been carbon in a form not found on earth.

Action of water on calcium carbide beneath moon's surface may have released acetylene gas, which was then broken down by the sunlight into a molecular form of carbon in which the atoms are paired. Such substance could exist in moon's near-vacuum atmosphere, but on earth the atoms would join and form graphite. (Sullivan, *NYT*, 12/31/63, 5; *Wash. Post*, 12/31/63; *NYT*-CTS, *Chic. Trib.*, 12/31/63)

December 30: Plans for orbiting 18 unhatched chicken eggs in Biosatellite were described at AAAS meeting by Dr. G. N. Hoover, Chief of Life Sciences, North American Aviation, Inc. One of several Biosatellite experiments planned for life specimens, the experiment was expected to yield data on effects of weightlessness on incubating eggs. (Troan, *Wash. Daily News*, 12/30/63)

- C. Leo DeOrsey, attorney for the original seven NASA astronauts, announced the astronauts had sold their interest in the Cape Colony Inn in Cocoa Beach, Fla. (AP, *Wash. Post*, 12/31/63)
- Dr. Richard Sharp of Lockheed Missiles & Space Co. told American Geophysical Union meeting in Boulder, Colo., that a polar-orbiting satellite had discovered last May that most of the Aurora Borealis is invisible to the naked eye. The seemingly thin sheets of shimmering light actually are several hundred miles thick. The satellite studies also confirmed that auroral light is produced by electrons. (Sci. Serv., *NYT*, 12/31/63, 20)

December 31: Brig. Gen. Samuel C. Phillips (USAF), Vice Commander of Air Force Ballistic Systems Div., was appointed Deputy Director of Project Apollo in NASA Office of Manned Space Flight, effective Jan. 15. General Phillips would assume major responsibility for management and administration of Apollo, while Dr. George E. Mueller, Associate Administrator for Manned Space Flight and Director of the Office of Manned Space Flight, would continue having major responsibility for technical direction of Project Apollo. (NASA Release 63–287; DOD Release 1605–63)

- USAF awarded to United Technology Center a $15,390,000 increment to existing contract for design, development, fabrication, delivery, and flight testing large segmented solid-propellant motors. USAF also obligated to Thiokol Chemical Corp. a $1,380,000 increment to existing contract for R&D on solid-propellant motors. (DOD Release 1609–63)

During December: In its "Statement of Purpose," the Subcommitee on Science, Research and Development of the House Committee on Science and Astronautics outlined scope of Federal Government's support of R&D and reviewed responsibilities of the Congress in answering questions of basic policy and evaluating the Federal R&D effort. Report pointed out that Government provides "more than two-thirds of the total research and development money being spent in the United States. Of this total, 75 percent is spent in industry, 8 percent in colleges and universities, and 15 percent within the Government itself, according to studies of the National Science Foundation." ("Government and Science No. 1: A Statement of Purpose," Rpt. of the House Subcomm. on Science, R&D, 1963)

During December: Marshall Space Flight Center engineer Paul J. deFries spent two weeks at U.S. scientific stations in Antarctica gathering data on their experiences. Some factors might relate to the problem of supplying astronauts during long stays on the moon. (*Marshall Star*, 1/1/64, 1)
- AEC Chairman Dr. Glenn T. Seaborg received the 1963 Franklin Medal, highest honor of The Franklin Institute, citing him "for outstanding contributions to our country as a distinguished scientist, as a thoughtful and imaginative educator, and now as the leading representative of science in our nation's service." (*A&A*, 1/64, 7)
- NASA Manned Spacecraft Center issued $49,000 contract amendment to North American Aviation's Space and Information Systems Div. for extension of study on Apollo spacecraft modifications for manned space sciences laboratory. Amendment called for studies and concepts of linking two or more modified Apollo spacecraft. (*Space Bus. Daily*, 12/4/63, 338)
- Ten senior executives, representing NASA's major Apollo program contractors, visited the Manned Spacecraft Center in Houston for a briefing on the current status of Apollo. (*Space News Roundup*, 1/8/64, 2)
- Staff members of the Committee on Science and Astronautics, U.S. House of Representatives, visited the Manned Spacecraft Center and were given thorough briefings by MSC key personnel. (*Space News Roundup*, 1/8/64, 2)
- 1,000th radiosonde transmitter released by Atmospheric Measuring Group at NASA's Mississippi Test Operations. Milestone came just over one year after the group was established at MTO. The radiosonde transmitters are sent aloft several times daily attached to six-ft. balloons to study temperature, humidity, and wind direction and velocity in the atmosphere. (*Marshall Star*, 12/11/63, 1)
- Summary article entitled "Ames Hypervelocity Free-flight Research" by Alvin Seiff in *Astronautics and Aerospace Engineering* reviewed shock-tube wind tunnel and light-gas gun experiments at Ames Research Center. Model speeds up to 41,000 fps and at air densities from 0.0002–0.16 times that at sea level have been achieved. (*A&AE*, 12/63, 16–23)
- Preparations continued for NASA-CNES launching of U.S. instrumented payload with French Centaure rocket from Colomb-Bechar, Algeria, next February. (*M&R*, 12/2/63, 8–9)
- A gimbaled integral nozzle (GIN) was sucessfully tested on the Lockheed "Char" motor by experts at the AF Rocket Propulsion Lab, designers of the exhaust system. Partially submerged in the aft section of the rocket case, the flanges and actuating components were shielded from the exhaust gases by the case itself. The concept would lead to a gimbaling system requiring less operating power than present operational swivel-type nozzles. A heavy, high-temperature-resistant silicone grease was used to lubricate the ball-and-socket gimbaling system. (*M&R*, 2/17/64, 21)
- U.K. completed first static-firings of Blue Streak rocket planned as first stage of three-stage booster for ELDO. Tests were conducted at Woomera, Australia. (*M&R*, 12/23/63, 9)

During December: NASA's trend toward more incentive-type contracts was outlined in Commerce Clearing House report, pointing out that in early 1962 NASA had no incentive-type contracts, by end of June 1963 NASA had awarded 10 large incentive contracts with others in negotiation, and by June 1964 the number of incentive contracts was expected to be increased by a large margin. Incentive-type contracts were expected to lower costs and improve work performance, in place of cost-plus-fixed-fee type. (Goodman, *Wash. Post*, 12/18/63)

- Dr. Willard F. Libby received patent for space vehicle construction designed to resist "flash heating" of re-entry into the atmosphere, the design relying largely on reflection by highly polished glass fibers. (NYTNS, Louisville *Courier-Journal*, 12/13/63)
- Trap III (Terminal Radiation Airborne Program) system for measuring radiation involved in high-speed re-entry experiments was delivered to USAF Aeronautical Systems Div. for installation in modified C-121 aircraft. (*M&R*, 12/23/63, 13)
- Choice of 5-psi, 100% oxygen environment for Apollo spacecraft was documented in *Journal of Aerospace Medicine* by Edward L. Michel, George B. Smith, Jr., and Richard S. Johnston of NASA Manned Spacecraft Center. They said it ranks higher than 7-psi, 50/50 oxygen-nitrogen environment in weight, leakage rates, system simplicity, reliability, and decompression protection, lower in fire dysbarism and atelectasis protection. (*M&R*, 1/27/64, 27)
- U.S.S.R. has failed in at least 13 attempts to send probes to Mars, Venus, and the moon since October 1960, *Astronautics and Aerospace Engineering* reported, and these failures were chiefly caused by inability of fourth stage to send the payload out of parking orbit. *A&AE* concluded, "it seems clear that the Russians have need for an improved upper stage to launch deep-space vehicles, and that Polyot I fills the bill." (*A&AE*, 12/63, 90)

During 1963: Man's conquest of space moved from spectacular hazard toward routine accomplishment, from the exhilarating peaks and slumps of first-generation hardware and experiments to the steadily improving reliability and capability of second- and third-generation launch vehicles and satellites. NASA launched 13 major space shots, with mission results of 11 successes, 1 partial success and 1 unsuccessful, compared with the 1962 record of 19 successes, 4 partial successes, and 3 unsuccessful. Project Mercury ended with L. Gordon Cooper's 22-orbit flight; Project Apollo began its flight tests with two unmanned capsule tests. Communications satellites demonstrated near-synchronous orbit and Saturn and Centaur boosters moved toward operational status with successful test flights. U.S.S.R. conducted a 129-orbit tandem flight and orbited an unmanned prototype of a new manned maneuverable space capsule. NASA Administrator James E. Webb cited as NASA's greatest specific accomplishment of 1963 the fact that his key staff people in Headquarters and the Centers had "learned to supervise the efforts at hand," that he now has "the men matched with the job and the job matched with the men." (NASA HHR-8; HHR-10; *Space Bus. Daily*, 1/3/64, 10)

During 1963: During the year the BMD delivered two Minuteman wings to the Strategic Air Command, making an average of one ICBM delivered per day in 1963. (AFSC Release 312-R-114, 1/1/64, 1)
- On a date not released, DOD launched a pickaback satellite attached to a principal satellite payload boosted by Thor-Agena D. (*Pres. Rpt. on Space, 1963*, 1/27/64)
- On a date not released, DOD launched a pickaback satellite attached to a principal satellite payload boosted by an Atlas-Agena B. (*Pres. Rpt. on Space, 1963*, 1/27/64)
- The highlight in delivery of missile systems to the operational inventory by the Ballistic Missile Division (BMD), Norton AFB, Calif., was the turnover of six squadrons of Titan II missiles to the using command in 1963. The first squadron was delivered on June 13, 1963, at Davis-Monthan AFB, Ariz., and the last two squadrons were turned over on December 30, 1963, at Little Rock AFB, Ark. The turnover completed the Titan II program and gave the nation an inventory of 54 of the powerful weapons. (AFSC Release, #312-R-114, 1/1/64, 1)
- The X-19, first of a new tri-Service family of vertical take-off and landing (Vtol) aircraft, flew for the first time. (AFSC Release, #312-R-114, 1/1/64, 2)
- More than 150,000 visitors visited NASA Marshall Space Flight Center, more than 50,000 visits recorded by persons on business. Biggest single attraction was Space Orientation Center, drawing about 83,000 visits during the year. (*Marshall Star*, 1/22/64, 1)
- In coordinating the military and civilian scientific and industrial efforts of the United States toward the development of aerospace weapon systems, AFSC had managerial responsibility for some 40 per cent of the Air Force budget, or approximately 10 cents of each Federal dollar. (*CR*, 2/18/64, 2910)

Through 1963: Since March 1959, more than 460 space probes have been conducted from Air Proving Ground Center's (APGC) Eglin AFB, Fla., aerospace launch facility on Santa Rosa Island. The probes included at least 87 in the "Project Firefly" program. The project involved intense investigation of the earth's atmosphere and near space. (AFSC Release, 312-R-114, Atch #2, 1/1/64, 1)

APPENDIX A

SATELLITES, SPACE PROBES, AND MANNED SPACE FLIGHTS

A CHRONICLE FOR 1963

The following tabulation was compiled from open sources by Dr. Frank W. Anderson, Jr., Assistant NASA Historian. Sources included the United Nations Public Registry, the *Satellite Situation Report* issued by the Space Operations Control Center at Goddard Space Flight Center, public information releases of the Department of Defense, NASA, and other agencies, and the *Report to the Congress from the President of the United States: United States Aeronautics and Space Activities, 1963.* Russian data are from the U.N. Public Registry, the *Satellite Situation Report*, translations of Tass News Agency statements in the Soviet press, and international news services' reports.

Launch dates are local time at the launch site. Orbital data are as near initial data as possible, being actual initial orbit, or those first reported in the *Satellite Situation Report*, or those found in the *President's Report on Space, 1963.* The new satellite identification system (adopted for 1963 by COSPAR and used by the U.S. in official listings) shows one Soviet satellite—1963-1—that was reported and tracked by U.S. sources but not acknowledged by the U.S.S.R. either in the press or in reports to the U.N. Public Registry.

It is of interest that 1963 was the first year in annual editions of this appendix that the total number of U.S. launches fell below that of the previous year (44 compared to 57 in 1962). On the other hand, the total number of payloads put into space continued to increase (66 to 62 in 1962) because of the large step-up in DOD's use of multiple satellites on a single launch vehicle (21 secondary satellites compared with 5 in 1962).

As we have cautioned in previous years, the "Remarks" column of these appendixes is never complete because of the inescapable lag behind the flight of the analysis and interpretation of the scientific results.

Launch Date	Name	International Designation	Vehicle	Payload Data	Apogee (st. mi.)	Perigee (st. mi.)	Period (minutes)	Inclination	Remarks
Jan. 4	U.S.S.R. Spacecraft (U.S.S.R.).	1963-1	Not available.	Not available.		Not available.			Re-entered 1/5/63.
Jan. 7	DOD Spacecraft (United States).	1963-2	Thor-Agena D.	Total weight: Not available. Objective: Develop space flight techniques and technology. Payload: Not available.	244	130	90.5	82°	Re-entered 1/24/63.
Jan. 16	DOD Spacecraft (United States).	1963-3	Thor-Agena D.	Total weight: Not available. Objective: Develop space flight techniques and technology. Payload: Not available.	334	286	94.6	81.89°	Still in orbit.
Feb. 1	DOD Probe (United States).		Blue Scout, Jr.			Not applicable.			Orbit not intended.
Feb. 14	SYNCOM I (United States).	1963-4	Thor-Delta.	Total weight: 150 lbs, including fuel for satellite-stage motor; 86 lbs. after fuel was exhausted. Objective: Place communications satellite in near-synchronous orbit; test communications capability of satellite and new transportable ground stations. Payload: 28" (dia.) x 15¼" cylinder, plus apogee motor on one end and antennas on the other; 2 transponders, command systems; attitude and orbit control systems; 3,840 solar cells.	22,762	21,303	1,424.8	33.3°	Near-synchronous orbit was achieved, confirmed 3/4/63 by visual means after satellite transmission failed toward end of launch phase. Still orbiting; never transmitted from orbit.
Feb. 19	DOD Spacecraft (United States).	1963-5	Blue Scout.	Total weight: Not available. Objective: Develop space flight techniques and technology. Payload: Not available.	493	313	97.8	100.47°	Still in orbit.
Mar. 13	DOD Probe (United States).		Blue Scout, Jr.			Not applicable.			Orbit not intended.
Mar. 21	COSMOS XIII (U.S.S.R.).	1963-6	Not available.	Total weight: Not available. Objective: Continuation of Cosmos scientific satellite series. Payload: Not available.	209	127	89.77	64.58°	Re-entered 3/29/63.
Apr. 1	DOD Spacecraft (United States).	1963-7	Thor-Agena D.	Total weight: Not available. Objective: Develop space flight techniques and technology. Payload: Not available.	254	129	90.6	75.37°	Re-entered 4/26/63.

Date	Name	Designation	Launch vehicle	Payload/Objective					Remarks
							Still being computed		
Apr. 2	LUNIK IV (U.S.S.R.)	1963-8	Not available.	Total weight: 3,135 lbs. Objective: Not available. Payload: Not available.					LUNIK IV launched into parking orbit around earth, then fired on escape trajectory to moon. Jodrell Bank Observatory reported indications of complicated maneuvers with the probe as it neared the moon on 4/5/63, suggesting attempt to land an instrument capsule on moon; passed 5,281 mi. from moon on 4/6/63. U.S.S.R. said "broad experimental data" were obtained.
Apr. 2	EXPLORER XVII (United States).	1963-9	Thor-Delta.	Total weight: 405 lbs. Objective: Measure atmospheric density, composition, pressure, and temperature at altitudes from 155 to 550 mi. Payload: 35" sphere, containing 2 neutral mass spectrometers, 4 vacuum gauges, 2 electrostatic probes; aspect control system; telemetry system; 2 radios; 150 lbs. of silver-zinc batteries.	571	158.9	96.4	57.61°	Obtained first measurements of the neutral-helium belt, surrounding the earth; confirmed day and night temperature fluctuations of electrons. Stopped transmitting 7/10/63. Still in orbit.
Apr. 13	COSMOS XIV (U.S.S.R.).	1963-10	Not available.	Total weight: Not available. Objective: Continuation of Cosmos scientific satellite series. Payload: Not available.	298	160	91.7	48.9°	Re-entered 8/29/63.
Apr. 22	COSMOS XV (U.S.S.R.).	1963-11	Not available.	Total weight: Not available. Objective: Continuation of Cosmos scientific satellite series. Payload: Not available.	153	100	98.0	64.95°	Re-entered 4/27/63.
Apr. 28	COSMOS XVI (U.S.S.R.).	1963-12	Not available.	Total weight: Not available. Objective: Continuation of Cosmos scientific satellite series. Payload: Not available.	249	129	90.4	65.01°	Re-entered 5/8/63.
May 7	TELSTAR II (United States)	1963-13	Thor-Delta.	Total weight: 175 lbs. Objective: Continue research on low-to-medium-altitude active communications satellites. Payload: 34½" sphere, containing communications receivers and transmitters; tracking beacons; electron detector; 10 semiconductors for sensing radiation damage; 3,600 solar cells; batteries.	6,717	604	226.3	42.7°	Second AT&T-owned satellite launched by NASA; efforts to minimize radiation damage suffered by TELSTAR I—putting TELSTAR II in more elliptical orbit, evacuating gas from around transistors—were successful; satellite went silent 7/16-8/12/63, reasons unknown. Still transmitting.

Launch Date	Name	International Designation	Vehicle	Payload Data	Apogee (st. mi.)	Perigee (st. mi.)	Period (minutes)	Inclination	Remarks
May 10	DOD Spacecraft (United States)	1963-14A	Atlas-Agena B.	Total weight: Not available. Objective: Develop space flight techniques and technology. Payload: Not available.	2,290	2,249	166.6	87.44°	Still in orbit; 1 of 4 satellite payloads orbited with the same booster.
	and TRS I	1963-14B		Total weight: 1½ lbs. Objective: Measure radiation damage to solar cells. Payload: 6½" tetrahedron, containing solar power and test cells, telemetry system.	2,297	2,241	166.5	87.43°	Still in orbit.
	and TRS I	1963-14C		Total weight: 1½ lbs. Objective: Measure radiation damage to solar cells. Payload: 6½" tetrahedron, containing solar power and test cells, telemetry system.	2,282	2,238	166.5	87.34°	Still in orbit.
	and WEST FORD.			Total weight: 50 lbs. Objective: Deploy in space around the earth a belt of metal fibers for passive communications experiments. Payload: Canister containing 400 million copper filaments.	2,275	2,275		87.44°	Still in orbit since ejection on 5/10/63. Communications experiments completely successful, effect of belt on radio astronomy found to be negligible; belt decaying as planned.
May 15	FAITH 7 Mercury-Atlas IX (MA-9) (United States)	1963-15	Atlas D.	Total weight: 3,000 lbs. in orbit; 4,000 lbs. at launch; 2,640 lbs. at recovery. Objective: Orbit and recover manned spacecraft after extended orbital flight; study effects of extended orbital flight on astronaut and spacecraft; verify that man can function as a primary system in space flight. Payload: Bell-shaped 9½' x 6' (diameter at base) capsule, containing, in addition to astronaut, still and TV cameras; life-support equipment; attitude control system; radio and telemetry equipment; drag balloon; radiation experiments; HF antenna experiments.	165.9	100.2	88.7	32.55°	Fourth U.S. manned orbital flight exceeded combined flight time of other flights, with 22 orbits lasting 34 hrs. 20 min. Astronaut L. Gordon Cooper made manual re-entry 5/16, landing 7,000 yds. from U.S.S. Kearsarge, 80 mi. SE of Midway Island. Flight confirmed engineering and scientific feasibility of long-duration manned space flight and demonstrated man's value as an active system in space flight.
	and FLASHING LIGHT.			Total weight: 10 lbs. Objective: Orbit near FAITH 7 so astronaut could experiment with observations of nearby object in space. Payload: Balloon and flashing light.	165.9	100.2	88.7	32.55°	Put into independent orbit by astronaut on his 3rd orbit; he could not spot them on orbit 4, but did observe them from time to time on orbits 5 and 6.

Date	Name	ID	Vehicle	Description					Orbit
May 17	DOD Probe (United States)	------	------	Not available.			Not applicable.		Orbit not intended.
May 18	DOD Spacecraft (United States)	1963-16	Blue Scout, Jr.	Total weight: Not available. Objective: Develop space flight techniques and technology. Payload: Not available.			Not available.		Re-entered 5/27/63.
May 22	COSMOS XVII (U.S.S.R.)	1963-17	Thor-Agena D.	Total weight: Not available. Objective: Continuation of Cosmos scientific satellite series. Payload: Not available.	488	161	94.82	49.02°	Still in orbit.
May 24	COSMOS XVIII (U.S.S.R.)	1963-18	Not available.	Total weight: Not available. Objective: Continuation of Cosmos scientific satellite series. Payload: Not available.	187	130	89.44	65.01°	Re-entered 6/2/63.
Jun. 12	DOD Spacecraft (United States)	1963-19	Not available.	Total weight: Not available. Objective: Develop space flight techniques and technology. Payload: Not available.	262	120	90.7	81.98°	Re-entered 7/12/63.
Jun. 14	VOSTOK V (U.S.S.R.)	1963-20	Thor-Agena D.	Total weight: 10,406 lbs. Objective: Continue study of influence of space flight on human organisms; conduct biomedical experiments in connection with prolonged space flight; and improve spacecraft equipment and pilotage. Payload: Metal spacecraft containing, in addition to cosmonaut, 2 shortwave radio transmitters, 2 receivers; 2 ultrashortwave receivers and transmitters; 1 standard broadcast receiver; 1 tape recorder; electro-acoustical equipment; retrorockets; parachute; capsule landing gear.	146	112	88.27	64.56°	Launched from Baikonur, VOSTOK V completed a record 81 orbits (119 hrs. 6 min.), the last 48 of them in tandem flight with VOSTOK VI, to which it came as close as 3 mi.; Cosmonaut L/Col. Valery F. Bykovsky landed by parachute 6/19 some 540 km. NW of Karaganda, Kazakhstan.
Jun. 15	LOFTI IIA (United States)	1963-21B	Not available.	Total weight: Not available. Objective: Conduct low-frequency communications experiments. Payload: Not available.	550	109	96.2	69.87°	Re-entered 7/18/63.
	and SOLARAD IV	1963-21C		Total weight: Not available. Objective: Measure solar radiation. Payload: Not available.	546	109	95.1	69.87°	Re-entered 8/1/63.
	and DOD Spacecraft	1963-21D		Total weight: Not available. Objective: Develop space flight techniques and technology. Payload: Not available.	549	109	96.2	69.87°	Re-entered 7/30/63.

Launch Date	Name	International Designation	Vehicle	Payload Data	Apogee (st. mi.)	Perigee (st. mi.)	Period (minutes)	Inclination	Remarks
	DOD Spacecraft and	1963-21E		Total weight: Not available. Objective: Develop space flight techniques and technology. Payload: Not available.	541	109	95.0	69.87°	Re-entered 7/27/63.
	DOD Spacecraft and	1963-21F		Total weight: Not available. Objective: Develop space flight techniques and technology. Payload: Not available.	533	114	94.9	69.86°	Re-entered 7/5/63.
June 15	DOD Spacecraft (United States).	1963-22	Blue Scout.	Total weight: Not available. Objective: Develop space flight techniques and technology. Payload: Not available.	475	450	99.8	90.01°	Still transmitting.
June 16	VOSTOK VI (U.S.S.R.).	1963-23	Not available.	Total weight: 10,370 lbs. Objective: Continue study of influence of space flight on human organisms, including a comparison of effect on male and female; conduct biomedical experiments and improve spacecraft equipment and pilotage. Payload: Metal spacecraft containing, in addition to cosmonaut, 2 shortwave radio transmitters, 2 receivers; 2 ultrashortwave receivers and transmitters; 1 standard broadcast receiver; 1 tape recorder; electro-acoustical equipment; retrorockets; parachute; capsule landing gear.	141	112	88.3	65°	Launched from Baikonur, VOSTOK VI completed 48 orbits (70 hrs. 50 min.) as part of 129-orbit tandem flight with VOSTOK V, launched 6/14. Cosmonaut Jr. Lt. Valentina V. Tereshkova, first woman to fly in space, landed by parachute 6/19 some 620 km. NE of Karaganda, Kazakhstan.
June 19	TIROS VII (United States).	1963-24	Thor-Delta.	Total weight: 297 lbs. Objective: Continue development of hardware and techniques for operational weather satellite; obtain global weather data during tropical storm season; obtain infrared heat-balance data and electron temperature measurements. Payload: 22″ x 42″ cylindrical 18-sided polygon, containing 2 wide-angle-lens TV cameras; 2 infrared detectors; electron temperature probe; 4 transmitters and 2 tracking beacons; 9,260 solar cells; 63 nickel-cadmium batteries.	405	385	97.4	58.22°	Orbited with all equipment functioning; sighted Hurricane Flora before any other weather sensor enabling the warning that saved lives and property in the Caribbean. Still transmitting.

Date	Name	ID	Vehicle	Payload/Objective				Remarks
June 26	DOD Spacecraft (United States)	1963-25A	Thor-Agena D.	Total weight: Not available. Objective: Develop space flight techniques and technology. Payload: Not available.	247	124	90.5	81.58°	Re-entered 7/26/63.
	and RADIATION MONITOR.	1963-25B		Total weight: 176 lbs. Objective: Measure magnetically trapped electrons and protons of all significant energy levels. Payload: 3' octagon, containing proton detector, plasma probe, electron detector, electrostatic analyzers, Geiger counter, magnetometers; command receiver; tape recorders; telemetry equipment; solar cells.	2,561	201	132.6	82.13°	Orbited with satellite 25A on 6/26, separated on 7/1, fired kick motor to attain higher apogee; instrument performance was outstanding, providing high degree of resolution; showed relation between solar flares and low-energy particles in solar plasma; found a gradual transition between inner and outer Van Allen belts. Still in orbit.
June 28	GEOPHYSICS RESEARCH (United States).	1963-26	Scout.	Total weight: Not available. Objective: Measure gas composition in space. Payload: Mass spectrometer; retarding potential analyzer; telemetry equipment.	815	255	102.1	49.72°	Returned telemetry data for 13 orbits; still in orbit.
June 29	DOD Spacecraft (United States).	1963-27	Thor-Agena B.	Total weight: Not available. Objective: Develop space flight techniques and technology. Payload: Not available.	319	308	94.8	82.32°	Still in orbit.
July 12	DOD Spacecraft (United States).	1963-28	Atlas-Agena D.	Total weight: Not available. Objective: Develop space flight techniques and technology. Payload: Not available.	116	114	88.3	95.87°	Re-entered 7/18/63.
July 19	DOD Spacecraft (United States).	1963-29	Thor-Agena D.	Total weight: Not available. Objective: Develop space flight techniques and technology. Payload: Not available.	210	120	89.9	82.85°	Re-entered 8/13/63.
July 19	DOD Spacecraft (United States).	1963-30A	Atlas-Agena B.	Total weight: Not available. Objective: Develop space flight techniques and technology. Payload: Not available.	2,316	2,274	167.9	88.40°	Still in orbit.
	and TRS 1	1963-30B		Total weight: 1¼ lbs. Objective: Measure radiation damage to solar cells. Payload: 6½" tetrahedron, containing solar power and test cells, telemetry system.	2,319	2,276	167.9	88.40°	Still transmitting.
	and DOD Spacecraft.	1963-30D		Total weight: Not available. Objective: Develop space flight techniques and technology. Payload: Not available.	2,326	2,270	168	88.41°	Still in orbit.

Launch Date	Name	International Designation	Vehicle	Payload Data	Apogee (st. mi.)	Perigee (st. mi.)	Period (minutes)	Inclination	Remarks
July 26	SYNCOM II (United States)	1963-31	Thor-Delta	Total weight: 147 lbs. with apogee-motor fuel; 86 lbs. without fuel. Objective: Place active communications satellite in near-synchronous orbit and gain experience in its operation; test new attitude and period control system. Payload: 28" (diameter) x 15¼" cylinder, plus apogee motor on one end and antennas on the other; 2 transponders, command systems; attitude and orbit control systems; 3,840 solar cells.	Initial, 22,548 Final, 22,260	Initial, 22,300 Final, 22,242	1,436	33.09°	Launch put satellite in good initial position; series of maneuvers culminated 8/15 in definite synchronous, figure-8 orbit over Brazil and South Atlantic Ocean. Wide range of communications experiments were conducted with SYNCOM II, including voice, teletype, data relay transmissions; even TV was relayed, a capability not designed into its transponders. SYNCOM II would have enough gas fuel for minor orbital corrections to remain on station about 2 yrs.
July 30	DOD Probe (United States)		Blue Scout, Jr.	Total weight: 50 lbs. Objective: Measure extraterrestrial radio noise and total radio energy input. Payload: 4 radio receivers tuned on 2.2, 4, 1, and 3 mc; 2 30' dipoles; FM telemetry link.		Not applicable.			Rose to altitude of 8,000 mi.; returned 3½ hrs. of data; landed 9,000 mi. from launch area.
July 31	DOD Spacecraft (United States)	1963-32	Thor-Agena D.	Total weight: Not available. Objective: Develop space flight techniques and technology. Payload: Not available.	283	94	90.6	74.92°	Re-entered 8/13/63.
Aug. 6	COSMOS XIX (U.S.S.R.)	1963-33	Not available.	Total weight: Not available. Objective: Continuation of Cosmos scientific satellite series. Payload: Not available.	311	163	92.1	48.99°	Still in orbit.
Aug. 25	DOD Spacecraft (United States)	1963-34	Thor-Agena D.	Total weight: Not available. Objective: Develop space flight techniques and technology. Payload: Not available.	261	108	90.5	75°	Re-entered 9/12/63.
Aug. 29	DOD Spacecraft (United States)	1963-35A	Thor-Agena D.	Total weight: Not available. Objective: Develop space flight techniques and technology. Payload: Not available.	202	183	90.8	81.90°	Re-entered 11/7/63.
	and DOD Spacecraft.	1963-35B		Total weight: Not available. Objective: Develop space flight techniques and technology. Payload: Not available.	262	195	92	81.90°	Re-entered 9/28-29/63.

ASTRONAUTICS AND AERONAUTICS, 1963

Sep. 6	DOD Spacecraft (United States).	1963-36	Atlas-Agena D.	Total weight: Not available. Objective: Develop space techniques and technology. Payload: Not available.		Not available.		Re-entered 9/13/63.
Sep. 24	DOD Spacecraft (United States).	1963-37	Thor-Agena D.	Total weight: Not available. Objective: Develop space flight techniques and technology. Payload: Not available.	257	109	90.5	74.94°	Re-entered 10/12/63.
Sep. 28	DOD Spacecraft and DOD Spacecraft.	1963-38B	Thor-Able-Star.	Total weight: Not available. Objective: Develop space flight techniques and technology. Payload: Not available.	714	676	107.4	89.90°	Still in orbit.
		1963-38C		Total weight: Not available. Objective: Develop space flight techniques and technology. Payload: Not available.	705	667	107.4	89.89°	Still transmitting 1/31/64.
Oct. 17	DOD Spacecraft (United States).	1963-39A	Atlas-Agena D.	Total weight: Not available. Objective: Develop space flight techniques and technology. Payload: Not available.	68,905	68,301	105 hrs.	38.30°	Still in orbit.
	TRS II and	1963-39B		Total weight: 3 lbs. Objective: Measure intensity of charged particles in Van Allen belts. Payload: 9" tetrahedron, containing proton and electron detectors; telemetry system; solar cells.	68,388	129	39 hrs.	38.69°	Still in orbit.
	DOD Spacecraft.	1963-39C	Not available.	Total weight: Not available. Objective: Develop space flight techniques and technology. Payload: Not available.		Not available.			Still in orbit.
Oct. 18	COSMOS XX (U.S.S.R.).	1963-40		Total weight: Not available. Objective: Continuation of Cosmos scientific satellite series. Payload: Not available.	193	128	89.5	65°	Re-entered 10/30-31/63.
Oct. 25	DOD Spacecraft (United States) and DOD Spacecraft.	1963-41A	Atlas-Agena D.	Total weight: Not available. Objective: Develop space flight techniques and technology. Payload: Not available.		Not available.			Re-entered 10/29/63.
		1963-41B		Total weight: Not available. Objective: Develop space flight techniques and technology. Payload: Not available.		Not available.			Re-entered 10/28-29/63.

Launch Date	Name	International Designation	Vehicle	Payload Data	Apogee (st. ml.)	Perigee (st. ml.)	Period (minutes)	Inclination	Remarks
Oct. 29	DOD Spacecraft (United States) and DOD Spacecraft.	1963-42A	Thor-Agena D.	Total weight: Not available. Objective: Develop space flight techniques and technology. Payload: Not available.	218	172	90.9	89.90°	Re-entered 1/21/64.
		1963-42B		Total weight: Not available. Objective: Develop space flight techniques and technology. Payload: Not available.	349	193	93.4	89.94°	Still in orbit.
Nov. 1	POLET I (U.S.S.R.).	1963-43	Not available.	Total weight: Not available. Objective: Orbit maneuverable spacecraft, gain experience in operating it through a series of orbital changes. Payload: Not available.	Initial, 306 Final, 893	Initial, 211 Final, 213	102.5	58.55°	POLET I was described as unmanned prototype of manned maneuverable spacecraft, one also capable of a variety of unmanned satellite mission configurations. After "repeated" changes in altitude and inclination, spacecraft apparently exhausted its fuel on 11/2/63. Still in orbit. Re-entered 11/14/63.
Nov. 11	COSMOS XXI (U.S.S.R.).	1963-44	Not available.	Total weight: Not available. Objective: Continuation of Cosmos scientific satellite series. Payload: Not available.	142	121	88.5	64.60°	Re-entered 11/22/63.
Nov. 16	COSMOS XXII (U.S.S.R.).	1963-45	Not available.	Total weight: Not available. Objective: Continuation of Cosmos scientific satellite series. Payload: Not available.	245	127	90.3	64.56°	Re-entered 11/22/63.
Nov. 26	EXPLORER XVIII (United States).	1963-46	Thor-Delta.	Total weight: 138 lbs. Objective: Measure interplanetary magnetic fields, solar wind, and cosmic radiation in area between earth and the moon. Payload: 28" x 12" octagon, with 6' telescoping boom rising out of the center to support a rubidium-vapor magnetometer; 2 7' booms extending from the sides of the octagon and mounting flux-gate magnetometers; 4 paddles mounting solar cells; 13 silver-cadmium batteries; transmitter; telemetry; 4 energetic particle sensors; 3 solar wind sensors.	122,793	119	5,666	33°	EXPLORER XVIII, an interplanetary monitoring probe (IMP), was successfully launched into highly elliptical orbit, although apogee was some 50,000 ml. less than hoped for; returned data of importance to Project Apollo. Still transmitting, 1/31/64.
Nov. 27	CENTAUR II (United States).	1963-47	Atlas-Centaur.	Total weight: No payload, only 2nd stage and ballast. Objective: Test-fly Atlas-Centaur vehicle, principally testing structural	1,095	304	107.7	30.40°	Second attempt, first success in flight testing Centaur; first ignition of liquid hydrogen in space and on time. Centaur stage was

			integrity, stage separation, in-space ignition of Centaur's liquid-hydrogen engines, and accuracy of Centaur's guidance system. Payload: None.				put into orbit, its 10,500 lbs. being heaviest weight orbited by U.S. to that date.
Nov. 27	DOD Spacecraft (United States)	1963–48	Thor-Agena D.	Objective: Develop space flight techniques and technology. Payload: Not available.	240	109	90.2	69.99°	Re-entered 12/15/63.
Dec. 5	DOD Spacecraft (United States) and DOD Spacecraft.	1963–49B 1963–49C	Thor-Able-Star.	Total weight: Not available. Objective: Develop space flight techniques and technology. Payload: Not available. Total weight: Not available. Objective: Develop space flight techniques and technology. Payload: Not available.	692 690	667 667	107.2 107.2	89.98° 89.99°	Still transmitting, 1/31/64. Still transmitting, 1/31/64.
Dec. 13	COSMOS XXIII (U.S.S.R.)	1963–50	Not available.	Total weight: Not available. Objective: Continuation of Cosmos scientific satellite series. Not available.	348	144	92.4	48.99°	Still in orbit.
Dec. 17	DOD Probe (United States)		Blue Scout.			Not applicable.			Orbit not intended.
Dec. 18	DOD Spacecraft (United States)	1963–51	Atlas-Agena D.	Total weight: Not available. Objective: Develop space flight techniques and technology. Payload: Not available.		Not available.			Re-entered 12/20/63.
Dec. 19	COSMOS XXIV (U.S.S.R.)	1963–52	Not available.	Total weight: Not available. Objective: Continuation of Cosmos scientific satellite series. Payload: Not available.	254	181	90.5	65°	Re-entered 12/28/63.
Dec. 19	EXPLORER XIX (United States)	1963–53	Scout.	Total weight: 17.8 lbs. for inflated balloon satellite; 77 lbs. for canister prior to satellite ejection. Objective: Orbit air density satellite and by tracking it obtain measurements of fluctuations in atmospheric density above the earth's high latitudes and extending through most of the solar cycle; perform engineering test of new X-258 4th stage of Scout booster. Payload: 12' (when inflated) sphere of 4 alternating layers of ¼-mil aluminum foil and mylar; painted with white polka dots for temperature control; tracking beacon, solar cells, and nickel-cadmium batteries attached to skin of balloon satellite.	1,468	366	116.8	78.62°	Orbit achieved, but signal from tracking beacon was so weak that optical tracking was needed to define orbit and confirm inflation of balloon satellite; EXPLORER XIX would provide for the high latitudes the kind of atmospheric density data that EXPLORER IX did for the lower latitudes.

Launch Date	Name	International Designation	Vehicle	Payload Data	Apogee (st. mi.)	Perigee (st. mi.)	Period (minutes)	Inclination	Remarks
Dec. 21	TIROS VIII (United States).	1963-54	Thor-Delta.	Total weight: 265 lbs. Objective: Continue development of hardware and techniques for operational weather satellite; test new APT (automatic picture transmission) subsystem and its related ground equipment. Payload: 22″ x 42″ cylindrical 18-sided polygon, containing 1 104° wide-angle-Tegea-lens TV camera, picture storage, and transmission system; 1 108° wide-angle-Tegea-lens camera and its slow-scan transmitting system; 9,260 solar cells; 63 nickel-cadmium batteries.	474	431	99.3	58.49°	APT would transmit on call pictures of local cloud conditions to inexpensive ground stations on a real-time basis. System performed as planned. Some 44 ground stations were involved in initial experiments. Still transmitting, 1/31/64.
Dec. 21	DOD Spacecraft (United States) and DOD Spacecraft.	1963-55A	Thor-Agena D.	Total weight: Not available. Objective: Develop space flight techniques and technology. Payload: Not available.	189	107	89.3	64.87°	Re-entered 1/9/64.
		1963-55B		Total weight: Not available. Objective: Develop space flight techniques and technology. Payload: Not available.	245	196	91.7	64.52°	Still in orbit.
		1963-55C		Total weight: Not available. Objective: Develop space flight techniques and technology. Payload: Not available.	Not available.	Not available.	Not available.		
During 1963	DOD Spacecraft (United States).		Thor-Agena D.	Total weight: Not available. Objective: Develop space flight techniques and technology. Payload: Not available.					Minor pickaback satellite carried by a payload already described.
During 1963	DOD Spacecraft (United States).		Atlas-Agena B.	Total weight: Not available. Objective: Develop space flight techniques and technology. Payload: Not available.					Minor pickaback satellite carried by payload already described.

APPENDIX B

CHRONOLOGY OF MAJOR NASA LAUNCHINGS
JANUARY 1, 1963, THROUGH DECEMBER 31, 1963

This chronology of major NASA launchings during 1963 is intended to provide an accurate and ready historical reference, one compiling and verifying information previously scattered over several sources. It includes launchings of all vehicles larger than sounding rockets launched either by NASA or under "NASA direction" (e.g., NASA launched and tracked RFD–1 for AEC).

An attempt has been made to classify the performance of both the launch vehicle and the payload and to summarize total results in terms of primary mission. Three categories have been used for vehicle performance and mission results—successful (S), partially successful (P), and unsuccessful (U). A fourth category, unknown (Unk), has been provided for payloads where launch vehicle malfunctions did not give the payload a chance to exercise its main experiments. These divisions are necessarily arbitrary, since many of the results cannot be neatly categorized. Also they ignore the fact that a great deal was learned from shots that may have been classified as unsuccessful.

A few unique items require separate treatment. Their dates have been kept in sequence, but their history has been relegated to footnotes.

Dates of launchings are referenced to local time at the launch site.

Sources used were all open ones, verified where in doubt from the project offices in NASA Hq. For further information on each item, see Appendix A of this volume and the entries in the main chronology as referenced in the index. Prepared February 1964 by Dr. Frank W Anderson, Jr., Assistant NASA Historian (ATPH).

Date	Name	NASA Code	General Mission	Launch Vehicle (Site)	Performance Vehicle	Performance Payload	Mission Results	Remarks
1963 Feb. 14	SYNCOM I	A-25	Communications earth satellite.	Thor-Delta (AMR).	S	P	P	First attempt for near-synchronous 24-hr. orbit was successful. Satellite transmitted data during launch, went silent in orbit, was lost until orbit was confirmed photographically, Mar. 4, 1963.
Mar. 28	Saturn	SA-4	Launch vehicle development test.	Saturn I (AMR).	S	S	S	First programed in-flight cut-off of 1 of 8 engines in cluster; last of 4 1st-stage flight tests.
Apr. 2	EXPLORER XVII	S-6	Scientific earth satellite.	Thor-Delta (AMR).	S	S	S	Atmospheric structure satellite. Measured density, composition, neutral particle temperature, and electron temperature of earth's atmosphere.
May 7	TELSTAR II	A-41	Communications earth satellite.	Thor-Delta (AMR).	S	S	S	Active-repeater communications satellite; owned by AT&T, launched by NASA.
May 15	FAITH 7	MA-9	Orbital manned Mercury flight.	Mercury-Atlas (AMR).	S	S	S	Cooper flight, 22 orbits. Final flight in Project Mercury.
May 22[1] June 19	TIROS VII	A-52	Meteorological earth satellite.	Thor-Delta (AMR).	S	S	S	2 wide-angle TV cameras, infrared sensors, and electron temperature probe performed as planned.
July 20	Re-entry		Re-entry flight demonstration.	Scout (WS).	U	Unk	U	Scout veered off course seconds after launch, had to be destroyed; had heat-shield payload.
July 26	SYNCOM II	A-26	Communications earth satellite.	Thor-Delta (AMR).	S	S	S	Positioned correctly in launch; series of maneuvers culminated Aug. 15 with figure-8 synchronous orbit; transmitted voice, teletype, etc., between North America, Africa, and South America.
Aug. 28	Little Joe II	LJ-II-1	Suborbital Apollo capsule test.	Little Joe II (WSMR).	P	S	S	Booster qualification test with dummy payload, met 5 of 6 test objectives; nonoperation of destruct system caused impact 47,000 ft. downrange instead of planned 32,000 ft.
Nov. 7[2] Nov. 26	EXPLORER XVIII	Imp A	Scientific earth satellite.	Thor-Delta (AMR).	S	S	S	Highly elliptical orbit achieved with new 3rd stage, although some 50,000 mi. less than planned for apogee; mapped radiation most of the way to the moon for Apollo.
Nov. 27	Centaur	AC-2	Launch vehicle development test.	Atlas-Centaur (AMR).	S	S	S	First successful launch of large liquid-hydrogen booster; first ignition of hydrogen on time in space; Centaur stage orbited with remaining liquid hydrogen on board.
Dec. 19	EXPLORER XIX		Scientific earth satellite.	Scout (PMR).	S	b	S	Polar orbit to measure solar-cycle effect on atmospheric density at high latitudes.
Dec. 21	TIROS VIII	A-53	Meteorological earth satellite.	Thor-Delta (AMR).	S	S	S	First Tiros to include the automatic picture transmission (APT) system for worldwide instantaneous readout of local weather data.

[1] May 22, 1963. BFD-1 re-entry test; primarily an AEC test flight to confirm that space nuclear reactors would safely disintegrate on re-entry. AEC bought Scout booster, furnished payload, and most impact-area data acquisition. NASA launched the Scout, tracked it. BFD-1 traveled suborbital trajectory from Wallops, re-entered 800 mi. down range; telemetry lost at critical point in re-entry, making flight only partial success.

[2] Nov. 7, 1963. Boilerplate No. 6 command module of the Apollo spacecraft was launched to 5,000-ft. altitude by the Apollo launch escape motor in a pad abort test at WSMR. Test confirmed stability and operational characteristics of escape system.

INDEX

A5A, 433
A-52, 247
AAAS. See American Association for the Advancement of Science.
AACB. See Aeronautics and Astronautics Coordinating Board.
Aachen, Germany, 185
ABC. See American Broadcasting Corp.
Abelson, Dr. Philip H., 148, 149, 164, 331, 440
ABL. See Allegheny Ballistics Laboratory.
ABMA. See Army Ballistic Missile Agency.
Abres (advanced ballistic re-entry systems), Project, 78
    launch, 173, 418, 486
Acceleration conditions, 453
Accelerator, 278, 321, 463, 470
Accident, aircraft, 426
AC Spark Plug Div. (General Motors), 242, 392
ADC. See Air Defense Command.
Adcole digital sun sensor, 244
Adenaur, Chancellor Konrad, 105
Adenosine triphosphate (ATP) molecule, 315
Adey, Dr. W. R., 227
Advanced ballistic re-entry systems (Abres). See Abres, Project.
"Advanced hypersonic manned aircraft." See Aerospace plane.
Advanced Manned Penetrator (AMP), 427
Advanced Orbiting Solar Observatory (Aoso), 73, 264, 397
Advanced range instrumentation ship. See Aris.
Advanced research and technology program, 381
Advanced Research Projects Agency (ARPA), 19, 173, 184, 275, 299, 311, 402, 415, 437
Advent, Project, 457
Advisory Committee on Biotechnology and Human Research, 221
Advisory Group for Aeronautical Research (NATO). See AGARD.
Adzhubey, A., 218
AEC. See Atomic Energy Commission.
AEDC. See Arnold Engineering Development Center.
Aegean region, 465
Aero Club (Washington, D.C.), 484
Aero-Space Luncheon Club, 6
Aerobee, 56
    launch, 95, 156, 247, 388
        Hi, 409
        Jr., 131
        150, 107, 195, 244, 250, 258, 279, 333, 366, 384, 387
        150A, 30, 123, 131, 139, 184, 188, 268, 278, 281, 336, 357, 391, 429, 434
        300A, 146, 280
"Aerodynamic Characteristics of Airfoils at High Speeds", 106
Aerodynamics, 280, 369
Aeroflot, 20, 471
Aerojet-General Corp., 68, 145, 281, 313, 322, 424, 451, 490
    contracts, 340, 486, 493
Aeromedical Research Laboratory (USAF), 393
Aeromedicine, 317

516 INDEX

Aeronautical Systems Div. (ASD) (USAF), 28, 360, 498
Aeronautics, 110, 282, 342, 419, 420
Aeronautics and Aerospace Engineering, 412
Aeronautics and Astronautics Coordinating Board (AACB), 315, 391
Aeronomy, 262
Aeronutronics Div. (Ford Co.), 232, 239
Aerospace industry, 374, 379
Aerospace (magazine), 362
Aerospace Corp., 99, 295, 362, 481
Aerospace Industries Association (AIA), 218, 373
Aerospace Management, 454
Aerospace Medical Association, 175, 222
Aerospace Medical Health Center (Brooks AFB, Tex.), 442
Aerospace Medical Research Laboratories (AFSC), 393, 460
Aerospace medicine, 41, 46, 70, 272, 442
Aerospace plane (ASP), 31, 168, 251
Aerospace Research Pilot School (USAF), 114, 207, 475
Aerospace Test Wing, 6555th, 263
Aerospace Test Wing, 6595th, 360
Aerospace weapon systems, 499
Aerothermodynamic/Elastic Structural Systems Environmental Tests (Asset). See Asset, Project.
Aetna Life Insurance Co., 187
AFA. See Air Force Association.
AF Aero-Propulsion Laboratory, 263
AF Avionics Laboratory, 263
AFCRL Space Physics Laboratory. See Air Force Cambridge Research Laboratories.
AF Flight Dynamics Laboratory, 263, 455
AFLC. See Air Force Logistics Command.
AFL-CIO (American Federation of Labor-Congress of Industrial Organizations), 32
AF Materials Laboratory, 263
AFMDC. See Air Force Missile Development Center.
AFMTC. See Air Force Missile Test Center.
AFOAR. See Air Force Office of Aerospace Research.
Africa, 295, 296, 298, 342, 345, 353
AF Rocket Propulsion Laboratory (Edwards AFB, Calif.), 497
AFSC. See Air Force Systems Command.
AFSSD. See Air Force Space Systems Div.
AFSWC. See Air Force Special Weapons Center.
AF Weapons Laboratory (Kirtland AFB, N.M.), 176
AGARD (NATO) (Advisory Group for Aeronautical Research), 348
Agena (See also Thor-Agena), 38, 74, 272, 311, 342, 397, 407
    B, 334
    NASA-DOD agreement, 334
Agency for International Development (AID), 485
Agriculture, Dept. of, 494
AIA. See Aerospace Industries Association.
AIAA. See American Institute of Aeronautics and Astronautics.
AIAA/ASD Vehicle Design and Propulsion Meeting, 419, 421
AID. See Agency for International Development.
AIO. See Arecibo Ionospheric Observatory.
Air Defense Command (ADC), 456
Air density, 295, 299, 342, 449, 497
Air Force and Space Digest, 329
Air Force Association (AFA), 107, 338, 340
Air Force Cambridge Research Laboratories (AFCRL), 56, 84, 244, 465
    instrument, 147, 377, 390
        research
            ionosphere, 103, 279
            radiation, 130, 138
            weather, 25, 201, 481
Air Force Caribbean Air Command Tropic Survival School, 229
Air Force Flight Test Center (AFFTC) (Edwards AFB, Calif.), 338
Air Force Historical Foundation, 73, 279

Air Force Logistics Command (AFLC), 383, 483
Air Force Missile Development Center (AFMDC), 68
Air Force Missile Test Center (AFMTC), 21, 254, 476
Air Force No. 1 (Presidential jet aircraft), 446
Air Force Office of Aerospace Research (AFOAR), 9, 65, 84, 91, 279, 335, 390, 411, 465
Air Force Satellite Control Facility (Sunnyvale, Calif.), 441
Air Force School of Aerospace Medicine. See USAF School of Aerospace Medicine.
Air Force Silver Wings, 483
Air Force Space Systems Division (AFSSD), 25, 35, 258, 350, 362, 455, 465, 486, 493
Air Force Special Weapons Center (AFSWC), 335
Air Force Systems Command (AFSC)
    aeronautical systems, 419
    air defense, 42
    citation, 301
    contract data management, 384, 499
    contract, 466, 486, 493
    development, 179
    facilities, 68, 331, 338
    organization, 38, 176, 263
    procurement (management), 189
    R&D programs, 4, 112, 168, 287, 315, 326, 340, 342, 437, 487
    report, 342, 360, 460
    Vandenberg AFB transfer, 307
    X-20, 19
    X-21A, 183
Air France, 229
Air power display, 483
Air pressure, 269
Air Products and Chemicals, Inc., 400
Air Proving Ground Center (APGC) (Eglin AFB, Fla.), 499
Air Space Transportation (Astra), 260
Air traffic control
    centers, 256, 432, 456
    operations, 377, 456
    requirements, 94
Air Training Command (ATC), 207
Air Transport Association Engineering and Maintenance Conference, 407
Air travel, 20, 153, 434
Aircraft, 28, 116, 306, 378, 439, 492
    commercial, 333, 367, 410, 439
    configuration, 151, 185, 281, 338
    development, 283, 306, 338, 394, 397
    research goal, 6, 122, 283, 419
    supersonic, 278, 434, 439
    test, 48, 56
    transport, 378, 426, 433, 434
    use of, 335, 395
    Wright brothers, 62, 343, 467
Aircraft carrier, 401
Aircraft, supersonic commercial air transport (SCAT) (See also Aircraft, supersonic transport), 367, 380, 478
Aircraft, supersonic transport (SST), 141, 248, 434, 489, 458
    cost, 13, 306, 311, 378, 384, 429
    development, 71, 78, 209, 253, 282, 287, 312, 320, 394, 407, 474, 478, 487, 492
    Government-industry program, 251, 306, 336, 397, 407
    plans, 6, 230, 251, 252, 253, 314, 397, 406
    research, 16, 34, 69, 90, 94, 114, 119, 136, 281, 328, 344
Aircraft supersonic transport, foreign
    France-U.K. (Concorde), 140, 141, 194, 198, 229, 234, 253, 312, 320, 367, 406, 410
    U.S.S.R. 379, 417, 422, 492
Aircraft, tactical fighter, experimental (TFX) (F-111), 16, 283, 332, 400, 446
Aircraft, vertical take-off and landing (Vtol), 282
Airglow, 14, 257
Airlift system, 321, 398

Akher Saa, 295
Alabama Research Institute, Univ. of, 83
Alabama Society of Professional Engineers, 66
Alabama, Univ. of, 83, 240
Alamogordo, N.M., 106
Albania, 391
Albrook AFB, C. Z., 229
Albuquerque, N.M., 65, 79, 367, 439
Aldrin, Maj. Edwin E., Jr. (USAF), 392
Alexander, George, 68
Alexandria, Va., 323, 415, 465
Alexandrov, Yevgeny, 11
Alford, William J., 278
Alfred P. Sloan Fellowship, 136
Algae, 78
Algeria, 212
Algol engine, 426
Alhazen (moon crater), 12
Allegheny Ballistics Laboratory (ABL), 165, 184
Allen, William M., 178, 306
Allis-Chalmers Manufacturing Co., 337
All-Union Symposium of Radioastronomers, 88
All-Woman Transcontinental Air Race, 273
ALOUETTE I, 359
    birthday, 357
    results, 92, 182, 263, 357
Alouette (program), 32, 263, 334
Alpha radiation measurements, 107
Alphonsus (moon crater), 235
Altadena, Calif., 472
Altimeter, 390
Altitude, measurements, 390
Alvarez, Luis W., 479
AMA. See American Medical Association.
Amateur Rocketeers of America (ARA), 155, 362
"America in Space", 136
American Airlines, 207, 397
American Association for Contamination Control, 178
American Association for the Advancement of Science (AAAS), 148, 259, 440, 494
American Astronautical Society, 14, 16, 17, 136
American Astronomical Society, 232, 493, 494
American Bar Association, 312
American Bridge Div. (U.S. Steel Corp.), 268
American Broadcasting Company (ABC), 106, 132, 375, 447
American Citizenship Award, 214
American Council on Education, 369, 372
American Editorial Cartoonists, 192
American Expeditionary Force, 257
American Geophysical Union, 144, 146, 147, 151, 170, 496
American Helicopter Society, 179
American Institute of Aeronautics and Astronautics (AIAA), 38, 74, 232
    meetings, 55, 98, 124, 136, 155, 156, 157, 159, 161, 245, 248, 366, 373, 395, 419
    National Capital Section, 343, 423, 480
American Institute of Biological Sciences, 136, 232, 326, 327
American Institute of Chemical Engineers, 184
American Legion, 285
American Management Association, Inc., 352
American Mathematical Society, 27
American Medical Association (AMA), 176
American Meteorological Society (AMS), 20, 24, 25, 147
American Newspaper Publishers Association Research Institute, 235
American Nobel Memorial Foundation, 152
American Ordnance Association, 188, 199
American Physical Society, 23, 26, 155, 157, 256, 488
American Psychological Association, 331

American Red Cross, 245
American Revolution, 167, 472
American Rocket Society (ARS) (See also American Institute of Aeronautics and Astronautics), 28, 38
American Science and Engineering Corp., 388
American Society of International Law (ASIL), 40
American Society of Lubrication Engineers, 251
American Society of Mechanical Engineers (ASME), 55, 204
American Society of Newspaper Editors, 149, 150, 151
American Society of Quality Control (ASQC), 96
American Telephone and Telegraph Co. (AT&T), 10, 104, 134, 184, 248, 274, 283, 299, 353, 447, 467
American University, 188, 236
Amery, Julian, 112
"Ames Hypervelocity Free-Flight Research", 497
Ames, Milton B., Jr., 133, 257
Ames Research Center (ARC), 221, 257, 264, 280, 365, 381, 404
    contract, 77, 175, 184, 203
    research programs, 1, 70, 271, 283, 336, 387, 497
    supersonic transport, 16, 94, 344
Amherst, Mass., 326
AMP. See Advanced Manned Penetrator.
Ampex Corp., 237
AMR. See Atlantic Missile Range.
AMS. See American Meteorological Society.
Amsterdam, Netherlands, 30
Anaheim, Calif., 379
"An Approach to American Strategy", 289
Anders, Capt. William A. (USAF), 392
Anderson, Sen. Clinton P., 215, 244, 272, 372
    appropriations, 293, 300, 373, 440
    Congress-science, 441, 466
    lunar program, 131, 209, 291, 304
    space program, 90, 143, 190, 319, 395, 422, 457
Anderson, Adm. George W. (USN), 137
Andover, Me., 3, 184, 186, 189, 241, 267, 353, 425
Andrews, Rep. George W., 384
Andreyev, Maj. Yvgeniy, 218
Andromeda Nebula, 40
Anemia, 176
Animal, 244, 355, 392, 411
ANNA I (geodetic satellite), 63, 308
Annapolis, Md., 117
Anniversary, 32, 33, 85, 96, 118
Antarctic, 347, 367, 394, 453
Antarctica, 332, 348, 396
Antelope Valley College, 191
Antenna, 27, 35
    furlable, 438
    parabolic, 426, 454
    tracking, 325, 402
"Anti-xi-zero" antiparticle, 310
Antigua, B.W.I., 37
Antimissile defenses, 287
Antiradiation, 96
Antisatellite defense systems, 394
Antisubmarine, 14
AOSO. See Advanced Orbiting Solar Observatory.
AP. See Associated Press.
APC (Area positive control). See Air traffic control.
APGC. See Air Proving Ground Center.
APL. See Applied Physics Laboratory.
"Apollo and Its Critics", 165

Apollo, Project, 78, 149, 180, 193, 210, 233, 254, 276, 290, 316, 353, 393, 433, 461, 483
    astronaut, 4, 169, 180, 298, 388, 392, 431
    criticism, 165, 179, 190, 191, 215, 237, 468
    facilities, 29, 51, 421, 464
    funds, 14, 112, 383, 384
    lunar landing program, 17, 226, 293, 326, 339, 378, 487, 489
    management, 54, 125, 135, 380, 420, 496
    plans for, 276, 327, 435
    progress, 84, 375, 382, 386, 497
    reliability, 156, 174, 388
    support of, 142, 192, 214, 237, 238, 245, 407
Apollo (spacecraft), 25, 72, 392, 403, 406, 471, 493, 498
    equipment, 25, 33, 51, 189, 219, 350, 359, 397, 423, 431, 490, 492
    launch vehicle, 273, 411, 414, 451
    modules
        command 1, 51, 62, 107, 178, 242, 325, 334, 382, 392, 406, 423
        lunar excursion (LEM), 1, 25, 90, 180, 219, 406, 472, 489
    simulators, 221, 343, 387, 497
    test, 62, 126, 132, 178, 257, 397, 431, 439, 498, 514
"Apollostok", 358
Applications Notes, 203
Applications Reports, 203
Applied Physics Laboratory (APL) (John Hopkins Univ.), 308, 454
APT. See Automatic picture transmission.
ARA. See Amateur Rocketeers of America.
Aral Sea, 259
ARC. See Ames Research Center.
Arcas (sounding rocket), 209, 301, 322, 477
Archer (sounding rocket), 410
Arc of the meridian, 37
Arc jets, 437
Arctic Circle, 301
Arctic Ocean Basin, 231
Area positive control (APC), 432
"Area rule", 297
Arecibo Ionospheric Observatory (AIO), 415
Arecibo, Puerto Rico, 415
"A Report Covering Evaluation of Areas Considered for Proposed NASA Electronics Research Center", 311
Argentina, 32, 212, 425
Argentine Atomic Energy Commission, 213
Argentine Department of Education, 425
Argo (Javelin) (sounding rocket)
    D-4, 263, 357, 359
    D-8, 50
Argon-37, 489
Argonne, Ill., 179
Argonne National Laboratory, 179, 463
Argus, Project, 383
Ariel (program), 112
ARIEL I, 92, 182
Aris (Advanced Range Instrumentation Ship), 140, 176
Aristarchus (moon crater), 235, 460, 495
Arizona, 6
Arizona, Univ. of, 492
Arlington National Cemetery, 448
Armaments, 391, 393
Armed Forces Day, 467
Armed Forces Dinner, 202
Armed Services Technical Information Agency (ASTIA), 415
Armour Research Foundation, 100
Arms control agreement, 491

Armstrong, Neil A., 28, 236
Army Air Force Scientific Advisory Group, 176
Army Ballistic Missile Agency (ABMA), 34
Army Corps of Engineers, 221, 319, 345, 425, 479, 490
Army Engineer Research and Development Laboratory, 271
Army Materiel Command, 328, 401
Army-Navy-Air Force Journal and Register, 358, 470
Army Transportation Corps, 115
Arnold Engineering Development Center (AEDC), 95
Arnold, General H. H. ("Hap"), 176
ARPA. See Advanced Research Projects Agency.
ARS. See American Rocket Society.
Arsentyev, V., 119
Arthur S. Flemming Award, 56
Artificial cloud particle measurements, 301
Artificial Satellites Conference, 308
Artists' Cooperation Program (NASA), 255
Ascension Island, 173, 230, 486
ASD. See Aeronautical Systems Division.
Asia, 50, 353, 460
ASIL. See American Society of International Law.
ASME. See American Society of Mechanical Engineers.
ASP. See Aerospace plane.
ASQC. See American Society of Quality Control.
Assassination, 446
Asset, Project (Aerothermodynamic/Elastic Sturctural Systems Environmental Tests), 4, 345, 474
Associated Press (AP), 87, 132, 463
Association of Symbolic Logic, 27
ASTIA. See Armed Services Technical Information Agency.
ASTRA. See Air Space Transportation.
Astrobee 1500 (sounding rocket), 132
Astronaut, 5, 56, 80, 173, 177, 187, 286, 340, 359, 379, 431, 443, 477, 496
    achievement, 61, 373, 414
    award, 246, 283, 385
    life stories, sale of, 48, 268, 269, 343
    performance, 36, 41, 245–246, 296, 388
    recruitment, 197, 243, 246, 259, 273, 322, 361, 392
    training, 4, 28, 136, 186, 221, 268, 271, 298, 340, 478
    women, 34, 256, 273, 277
Astronautics, 136
Astronautics and Aerospace Engineering, 74, 115, 129, 171, 455, 498
Astronautics Award (AIAA), 246
Astronautics Engineer Award, 104
Astronautics Space Research Medal (BIS), 330
Astronomers, 56, 116, 481
Astronomical Unit (AU), 2, 479
Astronomy, 73, 274, 481, 493
Astrophysical Institute (Univ. of Liege, Belgium), 429
Astrophysical Journal, The, 368
Astrophysics, 317
Astrophysics Institute (Kazakhstan Academy of Sciences), 193
Astroscience Center (Armour Research Foundation), 100
"Astrovoice", 54
ASWA. See Aviation/Space Writers Association.
AT&T. See American Telephone and Telegraph Co.
Athena, Project, 42
Atkinson, Dr. James D., 289
Atlanta Constitution, 452
Atlantic City, N.J., 251, 273

Atlantic Missile Range (AMR) (See also Cape Kennedy, Cape Canaveral, and
    John F. Kennedy Space Center)
    awards, 76, 205, 365
    conferences and reports, 197, 200, 202
    facilities, 169, 221
    launches
        manned space flight, 195
        missiles, 5, 9, 25, 44, 46, 50, 57, 59, 62, 70, 78, 134, 147,
            163, 173, 188, 201, 202, 213, 219, 230, 231, 257, 263, 266, 275, 278,
            283, 298, 318, 320, 327, 377, 406, 415, 424, 429, 431, 476, 486
        probes, 449, 462
        satellites, 54, 119, 184, 247, 283, 389, 449, 459, 462, 489
        vehicles, launch, 110, 290, 418, 431, 450, 463
    launching programs, 4
    management, 21, 390, 441
    name change, 451, 452
    operation of, 307
    prelaunch activities
        Gemini, 371
        Mercury, 44, 100, 186, 192, 194
        Saturn, 18, 19, 143, 271, 320, 348
    research, 437
    strikes, 147
    tours, 230, 433
    tracking, 37
Atlantic Monthly, 291, 329
Atlantic Ocean, 111, 134, 149, 278, 297, 312, 336, 344, 345, 348, 357, 360, 406,
    439, 469
Atlantic Research Corp., 203, 322, 410, 465
Atlas
    booster, 100, 147, 153, 296, 324, 342, 418, 433, 455
    missile, 28, 53, 72, 78, 89, 105, 163, 242, 266, 272, 285, 290, 323,
        325, 327, 335, 337, 380, 406
        D, 43, 379, 431, 486, 504
        E, 161, 229
        F, 105, 173, 373, 486
    operational, 435
    R&D program, 68
    USAF responsibility, 334
Atlas-Agena, 252, 326, 334
    launch, 334, 342, 389, 486
Atlas-Agena B, 12, 504, 507, 512
    launch, 189, 239, 279, 499
Atlas-Agena D, 133, 272, 335, 391, 400, 486, 509, 511
Atlas-Centaur. See Centaur.
Atlas, Dr. David, 25
Atmosphere, 262, 284, 290, 396
    measurement, 77, 449, 498
    research, 15, 282, 342, 434, 449, 476, 477–478, 499
Atmospheric Measuring Group (NASA Mississippi Test Operations), 497
Atom, 26, 74, 149, 270, 370
Atomic age, 458
Atomic Energy Commission (AEC), 78, 109, 116, 127, 134, 212, 230, 236, 351, 463,
    467, 470, 472, 491
    awards, 130, 390, 458, 497
    budget, 15, 20, 315, 337
    nuclear agreement, 203, 206, 220
    organization, 213, 237, 253, 258
    reactor, 3, 20, 208, 343
    reports, 41, 360, 383
    research, 138, 234, 310, 328, 402, 476
Atomic Industrial Forum Conference, 441
Atomic weapons ban, orbital, 353
Atoms for Peace Program, 471
ATP. See Adenosine triphosphate molecule.

AU. See Astronomical Unit.
Aubiniere, Gen., 362
AURORA 7 (see also MA-7), 55, 238
Aurora borealis, 342, 474, 496
Aurora, N.Y., 225
Australia, 7, 95, 175, 241, 279, 295, 311, 400, 481
    radio wave measurement, 139, 188, 241
    tracking stations, 32, 295, 325
Australian Commonwealth Scientific and Industrial Research Organization, 188
Australian Weather Office, 318
Austria, 197
Auter, Henry F., 243
Automatic data processing equipment. See Computer.
Automatic picture transmission (APT), 77, 489
Automation, 378
Automobiles, 390, 476
Avco Corp., 54, 132, 253
    Research and Advanced Development Div., 361
Aviation, 342
Aviation Education Workshop, 254
Aviation Hall of Fame, 483
Aviation/Space Writers Association (ASWA), 147, 206, 209, 373
Aviation Week. See *Aviation Week and Space Technology.*
Aviation Week and Space Technology, 12, 42, 68, 98, 155, 165, 204, 226, 237, 252, 268, 306, 318, 403, 411, 419, 428, 429, 436, 437, 438, 458, 472
Aviation World (Flugwelt), 454
Awards
    civic, 301
    government, 56
        foreign, 30, 130, 153, 386, 390, 474
        U.S., 38, 56, 76, 90, 156, 273, 278, 281, 287, 338, 365, 368, 385, 442, 443, 476
    institutions
        general, 136, 379, 390
    societies
        astronautics, 262, 340, 367, 465
        aviation, 107, 265, 279, 324, 330, 359, 483, 484
        engineering, 128, 279
        foreign, 330
        industry, 47
        scientific, 18, 24, 25, 64, 385
        space, 22, 76, 104
Award for Outstanding Achievement in Bioclimatology, 25
Award for Outstanding Contribution to the Advance of Applied Meteorology, 25
Award for Outstanding Services to Meteorology by a Corporation, 25
Azikiwe, Governor General Dr. Nnamdi (Nigeria), 296
B-17 (Flying Fortress), 306
B-29 (Superfortress), 252
B-47 (Stratojet), 27, 400
B-52 (Stratofortress), 27, 107, 189, 256, 321, 387
B-58 (Hustler), 209, 302, 335, 389
B-70. See RS-70.
Baar, James, 396
BAC-111 (British aircraft), 397
Bachman, Dr. Charles H., 445
Bachman, Paul, 468
Back-up interceptor control (BUIC)   See BUIC.
Bacteria, 294, 326, 331, 355
Bade, William G., 27
Bahcall, Dr. J. N., 488
Baikonur, U.S.S.R., 241, 244
Bain, Gordon M., 287, 320, 380, 406, 410, 417, 489
Baker, Arthur A., 465, 476
Bakersfield, Calif., 273
Baldwin, Hanson W., 285
Balewa, Prime Minister Sir Abubaker Tafawa, 321

Balfe, Capt. Paul J. (USAF), 186
Ball Bros., 244
Ballistic Missile Div. (BMD) (AFSC) (Norton AFB, Calif.), 499
Ballard, Frank, 353
Balloon, 65, 106, 140, 142, 150, 216, 219, 474
    flight, 76, 314, 317, 323, 331
        history of, 30
        landing, 78, 328
        mylar, 244, 295
Balloon-launched sounding rocket. See Rockoon.
Balloon-parachute. See Ballute.
Balloon Research Center, 331
Ballute (balloon-parachute), 127
Baltimore, Md., 117, 483
Banana River, Fla., 57
Bangor, Wash., 83
Barbados, B.W.I., 29
Barber, Arthur, 491
Bardeen, John, 150
Barges, 115
"Barnard's Star B," 145
Barnes, J. H., 182
Barnicki, Roger, 275
Barre, Dr. Robert L., 250
Barrett, Maj. John O. (USAF), 389
Barron, Dr. Charles I., 163
Barron's National Business & Financial Weekly, 193
Bartholomae, Mrs. Sara, 86
Bartlett, Sen. E. L., 289, 310
Baseball, 106
Base-heating studies, 451
Basic Construction Co., 303
Basic Research Resumes for 1961 and 1962, 91
Bassett, Capt. Charles A., II (USAF), 392
Battelle Memorial Institute, 459
Batten, Harry A., 48, 269
Batteries, 406
Bauer, Siegfried J., 227, 367
Baumholder, Germany, 88
Bay of Pigs, 418
Bayonne, N.J., 140
Beacon-Arrow (sounding rocket), 23
Beale, Ambassador Sir Howard, 325
Bean, Lt. Alan L. (USN), 392
Beatty, Nev., 322
Becker, John V., 22
Becker, Dr. Robert O., 445
Belem, Brazil, 177
Belgium, 133, 197, 332
Belgium National Space Research Committee, 367
Bell, Jack, 260
Bell Aerosystems Co., 17, 170, 240, 242, 275, 357, 465
Bellcomm, Inc., 101, 135
Bell Telephone Hour, 186
Bell Telephone Laboratories (BTL), 1, 2, 23, 71, 150, 172, 184, 222, 304, 479
Belokonev, A., 218
Beloussov, Vladimir V., 328
Beloyarsk, U.S.S.R., 52
Bendix Eclipse-Pioneer Div., 305
Bendix Systems Div., 153
Benedict, Dr. Manson, 237
Bennett, Sen. Wallace F., 159, 163, 273
Bennett, William O., 169
Berea, Ohio, 240
Berenice (French rocket), 273

Bergen, William B., 454, 487
Berger, Rainer, 16
Bergson, Henri, 415
Berkeley, Calif., 27, 131, 141, 252, 328
Berkner, Dr. Lloyd V., 24, 238
Berlin, Germany, 471
Berlin's Institute for Elements of Space Travel, 7
Berlin, Univ. of (Germany), 7
Berlin Wall, Germany, 394
Bermuda, 32, 168, 192, 194, 208, 302, 348
Bernardo, James V., 342
Berry, Dr. Charles A., 296, 465
Bethesda, Md., 240
Beverly Hills, Calif., 252
Bevins, Postmaster General John, 107
Beynon, Dr. W. G., 314
Big Dipper, 304
Big Lift, Exercise (DOD), 398
Bikle, Paul F., 22, 235, 265, 282, 484
Bilaniok, Dr. O. M., 213
Billingham, Dr. John, 70
Billings, Mont., 403
Bioastronautics, 275, 354
"Bioastronautics Review", 354
Biological effects, 354-355
Biologists, 56
Biomedical, 272, 393
Bios. See Biosatellite.
Biosatellite (Bios), 77, 320, 395, 411, 496
Bioscience, 49, 244, 254, 326-327, 411
Birdling's Flat, N.Z., 209
Biryusov, Marshal Sergei S., 65, 111, 141, 423
BIS. See British Interplanetary Society.
Bisplinghoff, Dr. Raymond L., 19, 109, 161, 163, 177, 182, 198, 226, 281, 283, 381
Bisson, Edmond E., 251
Black Brant (Canadian sounding rocket), 279, 410
Blackburn, Capt. Harold (TWA), 480
Black, Eugene R., 311, 487
Black Knight (British rocket), 295, 311
Black Mesa, Utah, 351
"Black Report", 487
Blacksburg, Va., 127, 308
Blagonravov, Dr. Anatoli A., 20, 89, 100, 198, 212, 219, 227, 303, 314, 318, 336, 350, 376, 427, 493
Blanch, Dr. Gertrude, 476
Bleymaier, B/G Joseph S. (USAF), 311
Bleymaier, Col. Joseph S. (USAF), 57
Block, Louis, 377
Bloemfontein, South Africa, 72, 79, 196
Blue, Dr. James, 441
Blue Scout, 502, 505, 506, 508, 511
    launch, 59, 129, 163, 242, 485
    Jr., 38, 91, 201, 290
    malfunction, 129
Blue Streak (British rocket), 433, 497
BMD. See Ballistic Missile Div.
BMEWS (Ballistic Missile Early Warning System). See Red Mill, Project.
BOAC (British Overseas Airways Corp.), 229
Board of Geographic Names, 476, 488
BOB. See Bureau of Budget.
Bochum Institute for Satellite and Space Research (Germany), 265
Bode, Dr. Hendrick W., 123
Boeing 377 (Stratocruiser), 271
Boeing 707 (Stratoliner), 306

Boeing Airplane Co., 94, 220, 222, 275, 296, 334, 419
    aircraft, 48, 119, 202, 206, 344, 387
        contracts, 16, 68, 107, 114, 178, 197, 230, 234, 262, 271, 283, 292, 306, 427, 489
        space chamber, 275, 280
        strikes, 23, 28, 143, 148, 176
Bologna, Joseph M., 86
Bolling AFB, D.C., 176, 263
Bolshevik Revolution, 423
Bomb detection device, 387
Bomb, nuclear, 210, 294, 425
Bomber attacks, simulated, 469
Bonn, Germany, 270, 429
Bonney, Walter, 481
Bono, Philip, 246
Boone, Adm. Walter F. (USN, Ret), 48, 70, 382
Booster, 271, 315, 357, 419, 483
Booz, Allen Applied Research, Inc., 335
Borman, Maj. Frank (USAF), 28, 424
Bossi, Enea, 7
Boston College, 148, 484
Boston Globe, 47
Boston, Mass., 16, 20, 47, 62, 97, 98, 106, 131, 178, 211, 240, 254, 311, 400, 432
Boston Univ., 39
Boulder, Colo., 35, 496
Boundary layer, 281, 438
Bourdeau, R. E., 367
Bow, Rep. Frank T., 82
Bowyer, S., 493
Boyce, Peter A., 460
Boyd, Alan S., 390, 483
Boyden Observatory (South Africa), 72, 80
Bozeman, Mont., 405
Brackett, Ernest W., 49, 51, 99, 178
Bramlet, James B., 157
Bradley, Gen. Mark (USAF), 483
Branch, B/G Irvin L. (USAF), 236
Brandenburg Gate, Berlin, 394
Brasilia, Brazil, 164
Brattain, Walter Houser, 150
Brazil, 32, 177, 284, 287, 312, 425
Brazilian Commission for Space Activities, 177
Breen, Whilden P., Jr., 80, 143
Brewster, Sen. Daniel B., 466
Bridgeport, Conn., 96
Briefing Conference on National Patent Policies and Practices, 207
Briggs, Dr. Lyman J., 106
Brigham Young Univ., 273
Brisbane, Australia, 193
Brissenden, Roy F., 23
Bristol, U.K., 161
Britain. See United Kingdom.
Britain, Battle of, 205
British Aeronautical Board, 12
British Aircraft Corp., 140, 231, 397, 410
British Astronomical Association, 116
British Aviation Corp., 229
British-French supersonic airliner (Concorde), 194, 253
British Interplanetary Society (BIS), 205, 206, 330
British Overseas Airways Corp. (BOAC). See BOAC.
British West Indies (B.W.I.), 37
Brookhaven National Laboratory, 141, 213, 258, 310
Brooks AFB, Tex., 55, 139, 393, 443, 455
Brown, Dr. Allen, 494
Brown Deer, Wis., 59
Brown, Dr. Harold, 240, 242, 269, 394, 436

Brown, Harrison, 335
Brown, Gov. Pat, 73
Brown, Walter L., 23
Brussels, Belgium, 270, 378
Bryant, Farris, 451, 465
BTL. See Bell Telephone Laboratories.
Buchanan, Roger I., 365
Buchheim, Dr. Robert W., 263
Buckley, Edmond C., 163
Budapest, Hungary, 83, 185
Budget, Bureau of (BOB), 66, 145, 294, 360, 384, 474
Buettner, Prof. Konrad J. K., 25, 58
Buffalo, N.Y., 184
BUIC (Back-Up Interceptor Control), 42
Bulletin of the Atomic Scientists, 172
Bulova Watch Co., 169
"Bungling in Space", 380
Burbank, Calif., 494
Bureau of Aeronautics (USN), 377
Bureau of Mines, 212
Bureau of Naval Weapons (USN), 322, 454
Burlington, Mass., 153
Burnham 1960 II comet, 429
Burroughs International Test Pilot Award, 442
Burroughs, Richard H., 442
Busemann, Dr. Adolf, 344, 456
Bush, Dr. Vannevar, 430, 444, 479
"Business Man of the Year", 10
Butylbenzene, 390
B.W.I. See British West Indies.
Byram, E. T., 493
Byrd, Sen. Robert C., 224, 462
Bykovsky, L/C Valery F., 241, 244, 251, 254, 259, 372, 484
C-47 (Dakota), 331
C-97 (Stratocruiser), 348
C-121 (Constellation), 498
C-130 (Hercules), 367, 482
C-133 (Cargomaster), 178, 397
C-135 (Stratolifter), 398
C-140A (Jet Star), 34. 252
C-141A (Starlifter), 321, 485
C&B Construction Co., 479
CAB. See Civil Aeronautics Board.
Cadle, Don D., 65
Cahill, Dr. L., 380
Cairo, U.A.R., 179, 295, 319
Cajun (sounding rocket), 326
Caldwell, N.J., 282
Calf, 426
California, 107–109, 110, 111, 229, 272, 279, 486
California Institute of Technology (Cal Tech), 17, 44, 95, 229, 238, 321, 335, 385
    Kellog Radiation Laboratory, 488
California, Univ. of (Berkeley), 27, 76, 214, 222, 237, 246, 268, 275, 334
California, Univ. of (La Jolla), 421
California, Univ. of (Los Angeles) (UCLA), 227, 328
Callaghan, Richard, 382
Cal Tech. See California Institute of Technology.
Cal Tech Guggenheim Laboratory, 185
Cal Tech Radio Observatory, 95
Cambridge, Mass., 272, 388
Cambridge Univ. (U.K.), 155
Camera, 63, 101, 244, 270, 280, 281, 306, 317, 326, 388, 478, 489
Cameron Station, Alexandria, Va., 415
Campbell, Joseph, 116
Campbell, William Wallace, 56

Canada, 32, 115, 206, 212, 213, 257, 279, 357, 372
Canadian Bristol Aerojet, Ltd., 410
Canadian Defence Research Board, 32
Canadian space program 279, 321
Canada's Dept. of Transport, 321, 449
Canal Zone, 229
Canard control surfaces, 344
Canary Islands, 44, 298
Canberra. See RB-57F.
Canberra, Australia, 82, 295, 325
Canberra (British bomber), 371, 389, 400
Cannon, 29
Cannon, Rep. Clarence, 374, 376, 384
Cannon, Sen. Howard W., 46, 243
Canoga Park, Calif., 6
Canterbury Plains, N.Z., 209
Canterbury, Univ. of (N.Z.), 209
Cape Canaveral, Fla. (See also Cape Kennedy and Atlantic Missile Range), 147, 176, 268, 319, 345, 402, 424, 464
    construction, 57, 401
    name change, 451, 464–465, 470, 476, 488
    tours, 230, 433
Cape Colony Inn, 496
Cape Kennedy (See also Cape Canaveral and Atlantic Missile Range), 451, 464–465, 467, 476, 488
Cape Town, South Africa, 18, 367
Capitol Hill, 421, 472
Capitol Radio Engineering Institute, 128
Capsules, ballistic, 361
Carey, Pvt. 1/C R., 271
Carey, William D., 360, 361
Carl-Gustaf Rossby Award, 24
Carnarvon, Australia, 82, 326, 414, 464
Carnegie Foundation for Advancement of Teaching, 266
Carnegie Institution of Washington, 331, 440, 470, 479
Carnegie Museum, 260
Carpenter, LCdr. M. Scott (USN), 26, 28, 238, 257, 280, 296, 339, 465
    appearances, 3, 55, 155
Carpenter, Roland L., 454
Carson Sink, Nev., 298
Carswell AFB, Tex., 302
Carter, Dr. Launor F., 263
Case, Sen. Clifford P., 58, 97, 131, 143
Caspian Sea, 259
Cassidy, William B., 156
Catalogue of Radio Sources, Third Cambridge (U.K.), 117
Catalytic Construction Co., 56
Cathode experiment, 280
Catholic Univ., 192, 425
Catlin, Ephron, 47
CBS. See Columbia Broadcasting System.
CDA. See Command and Data Acquisition Station.
Celestial bodies, 464
Cell, solar, 41
Centaur, 324, 336, 401, 510, 514
    AC-2, 410, 450, 457, 463, 490
    development, 7, 100, 320, 457, 472, 498
    GAO report, 370, 374, 376, 380, 457
    program, 38, 436
    tests, 257, 296, 315
Centaure (French rocket), 497
Central Inertial Guidance Test Facility, 68
Central Methodist College (Mo.), 334
Central Pacific, 214
Centre National d'Etudes Spatiales (CNES). See National Center for Space Studies (France).

Centrifuge, 68, 106
Century 21 Exposition (Seattle, Wash.), 448
Century of Progress, 30
Ceramics, advanced, 305
Cernan, Lt. Eugene A. (USN), 392
Certificate of Outstanding Achievement (NASA), 90
Cervenka, A. J., 73
Cesium, 488
Cessna 180 (aircraft), 231
CF-100 (RCAF), 437
CH-3B, (helicopter), 230
Chadwick, Sir James, 149
Chaffee, Lt. Roger B. (USN), 392
Challenge magazine (GE), 453
Chamber of Commerce
    Albuquerque, N.M., 439
    Billings, Mont., 403
    Charleston, W. Va., 421
    Charlotte, N.C., 17
    Cocoa, Fla., 464
    Dayton, Ohio, 483, 484
    Houston, Tex., 289
    Kokomo, Ind., 460
    Los Angeles, Calif., 296, 367
    New York, N.Y., 98
    U.S., 24, 323
    Waco, Tex., 64
Chance Vought Astronautics Div. (Ling-Temco-Vought, Inc.), 25
Chance Vought Corp., 197, 442
Chanute, Octave, 483
Chapel Hill, N.C., 18
Chapman, Dr. Dean R., 365
"Char" motor, 497
Charles Franklin Brooks Award, 24
Charles, Robert H., 239
Charleston, W. Va., 424
Charlotte, N.C., 17, 463
Charyk, Dr. Joseph V., 72, 91, 116, 213, 429
Chebotarev, Prof. Gleb, 9, 441
Chemical Engineering Achievement Awards, 461
Chemical Industry Medal, 356
Chemical Low Altitude Missile (CLAM). See Missile, air-to-surface.
Chemical trails (see also Sodium-vapor experiments), 222
Chemistry, 435
Cherokee (rocket), 326
Chicago and Midwest Space Month, 136
Chicago, Ill., 100, 105, 136, 149, 177, 230, 235, 463
    meetings, 7, 31, 39, 180-181, 235, 249, 253, 312
Chicago International Trade Fair, 249
Chicago, Univ. of, 177, 270, 459
Chico, Calif., 244
China, Communist, 194, 298, 353, 379, 415, 475
Chinese
    Communist, 415
    Nationalist, 415
Christchurch, N.Z., 112
Christopher Columbus International Prize for Communication, 386
Christ und Welt, 140
Christmas, 492
Chrysler Corp., 173, 387, 478
Chubb, T. A., 493
Chula Vista, Calif., 27
Churchill, Manitoba, 270
Churchill Research Range (USAF) (Ft. Churchill, Canada), 278
Churchill, Sir Winston, 134, 205

"City of Lights" (Perth, Australia). 492
City of Los Angeles Commendation Award, 301
Civil Aeronautics Board (CAB), 19, 254, 255, 333, 390, 483
Civil Aeronautics Research Institute. See FAA Civil Aeronautics Research Institute.
Civilian Industrial Technology Program, 92
Civilian pay, 485
Civil Service Commission. See U.S. Civil Service Commission.
CLAM (Chemical Low Altitude Missile). See Missile, air-to-surface.
Claremont, Calif., Air Museum, 191
Clark, Frank L. 365
Clark, Sen. Joseph S., 303, 378
Clausen, Christian M., 472
Clauser, Dr. Milton, 275
Clay, Gen. Lucius D. (USA, Ret), 279
Clayton, Dr. Donald D., 155
Clear, Alaska, 343
Clear Lake, Tex., 282, 290, 333, 337, 478
Clemence, Gerald M., 327
Clement, Gov. Frank, 95
Clement, Stuart H., Jr., 379
Cleveland Abbe Award, 24
Cleveland Clinic, 426
Cleveland, Ohio, 256, 354, 375, 472, 494, 495
Cliff, Edward P., 465
Clifford, George, 379
Chlorine-37 isotopes, 489
Cloud seeding, 401
Clouds, cumulonimbus, 293
Clouds, gas, 88, 114
Clouds, hurricane, 293
Clouds, noctilucent, 301, 388, 419
CNES. See National Center for Space Studies (France).
Coastal Sentry, 84, 224
Cochran, Jacqueline, 74, 480
Cocoa Beach, Fla., 98, 252, 496
Cocoa Chamber of Commerce, 464
COIN (Counter-Insurgency). See Aircraft, counter-insurgency.
Cold war, 347, 351, 393
College of Osteopathic Medicine & Surgery, 232
Collier Trophy, 385
Collin, Everett E., 290
Collins, Capt. Michael (USAF), 392
Colomb-Bechar, Algeria, 497
Colorado, 65
Colorado, Univ. of, 222, 456
Columbia Broadcasting System (CBS), 132, 389, 423
Columbian Women, 18
Columbia Univ., 25, 141, 173, 204, 235, 237, 345
    School of Journalism, 272, 278
Columbus, Christopher, 185, 194, 200, 201, 217, 360, 370, 415
Columbus Day, 386
Combustion instability, 231, 438, 448
Comcor, Inc., 38
Comet, 114, 332, 432
    Burnham 1960 II, 429
    experiments, 382, 383
    man-made, 139
Command and data acquisition station (CDA), 341
Commerce Clearing House report, 498
Commerce, Dept. of, 9, 19, 92, 212, 222, 333, 385, 480, 483
Commission of Exploration and Utilization of Outer Space (Soviet Academy of Sciences), 314
Commission on Science and Technology, 60, 87, 212
Committee of Defense and Economic Agencies (Executive Branch), 489

INDEX 531

Committee for Economic and Cultural Development of Chicago, 84
Committee for Economic Development, 191
Committee on Domestic Geographic Names (Board of Geographic Names), 488
Committee on Space Research (COSPAR), 1, 16, 226, 227, 314, 330
Committee, Select House, on Research, 298, 337
Commoner, Dr. Barry, 171, 234
Communications
    carriers, 262
    deep space, 278
    exchange, 296, 314, 347, 482
    experiments, 190, 219, 287, 383
    frequency allocation, 419, 425
    global, 35, 304, 317, 396, 417, 429, 440, 467, 490
    international
        companies, 269
        systems, 312, 327, 347, 390
    laser, optical, 44, 101
    NASA facilities, 371
    problems, 355
    satellites, 265, 267, 269, 274, 379, 416, 459–460, 461, 477, 495
        system, 112, 319
        transmissions via, 267, 326, 339
    tests, 222, 341
    "Unified S-Band" method, 472, 490
Communications Satellite Corp. (ComSatCorp)
    communications
        national, 229
        international, 213, 216–217, 247–248, 416, 440, 467
    contract, 271, 299, 317, 490
    operation of, 358, 403, 419
    organization, 429, 452
    stock issuance, 285, 301, 417
Communist
    Bloc, 353
    rulers, 405
    system, 405, 462
    world, 381
Communist China. See China, Communist.
Complex 12 (at AMR), 334
Complex 39 (at MILA), 51, 371, 468
Computer
    biomedical, 105
    complex, 272
    defense, 42
    digital, 432
    electronic, use of, 335
    library, 172
    model 1218, 443
    Saturn IB and V, 475
    services award, 317
    technology, center for, 272
    use of, in Federal Gov't., 401
    X–15, 38
ComSatCorp. See Communications Satellite Corp.
CONAD. See Continental Air Defense Command.
Concord College (W. Va.), 224
Concorde (see also Aircraft, supersonic transport), 140–141, 229, 253, 312, 320, 379, 390, 398, 410, 487
Condon, John E., 336
Conference on Communications Frequency Allocation (ITU) (see also World Conference on Space Communications), 416, 419, 425
Conference for Engineers & Architects, 190
Conference of European Posts and Telecommunications, 429
Conference of National Organizations, 251
Conference on Higher Education, 253
Conference on Outer Space (Third Midwest Assembly), 345

Conference on Peaceful Uses of Space, 131, 180–183
Conference on Plasma Physics (NASA), 459
Conference on Science Teaching, Mid-Winter, 468
Conference on Space Science & Space Law, 247, 250
Congressional Office of Science and Technology (COST), 289, 310
Congressional Record, 187, 353, 377, 384, 466
    aircraft, 183, 306
    science and technology, 212, 253, 360
    space program
        astronaut, 256
        facilities, 273, 311
        manned lunar landing, 291
        objective, 48, 288, 430, 452
        research, 56, 169, 319
        U.S.S.R., 303, 462
        waste, 185, 378, 421
Congress of American Industry (68th Annual). See National Association of Manufacturers, 68th Congress.
Connecticut General Life Insurance Co., 197
Conner, John T., 91
Conrad, Lt. Charles, Jr. (USN), 28
Conrad, Max, 480
Constan, Dr. George N., 256
Constitution, 167
Construction, Office of (NASA Hq.), 323
Consultant, 268
Contamination of spacecraft, 173
Continental Air Defense Command (CONAD), 377
Continental Airlines, 367, 410
Contracts (see also under agencies, such as NASA, USAF), 51, 383, 454, 493
Control Data Corp., 492
Control system, 281, 377, 411
Cook Electric Co., 162
Cook, Gen. Orval R. (USAF, Ret), 13, 251
Coolidge AFB, Antigua, B.W.I., 37
Cooling, 336
Cooper, Dr. John Cobb, 40
Cooper, L/C K. B., 184
Cooper, Maj. Leroy Gordon (USAF), 28, 198, 200, 202, 218, 226, 238, 303–304, 354
    appearances, 202, 209, 257
    awards, 205, 208, 219, 340, 465
    flight preparations, 1, 101, 131, 192
    flight results, 386, 498
    simulated flights, 186, 189, 192
    space flight, 196, 198, 200, 227, 245, 255, 344, 371, 484
Coopers Island, Bermuda, 168
Copenhagen, Denmark, 340
Copernicus, 167, 415
Cornell Aeronautical Laboratory, Inc., 136
Cornell Univ., 136, 415
Corona, 257, 258, 280, 310, 339, 366
Coronagraph, 280
Corporate Leadership Dinner, 342
Corp., Space Communications (CSC) (see also Communications Satellite Corp.), 27, 91
    and U.S., 59, 90, 99, 163, 169
    officers, 72, 91, 161
    organization, 17, 39, 139
    stock, 57, 114, 155
Corps of Engineers (USA), 98, 138
Cortright, Edgar M., 52, 56, 90, 182, 355, 450
Cosmic dust, 304
Cosmic noise, 335
Cosmic radiation, 355
Cosmic ray, 74, 79, 328, 331, 332, 474, 488
Cosmologist, 117

Cosmology, 480
Cosmonaut
    death of, 213, 218
    flight, 347
    life of, 355
    marriage of, 417
    medical, 20, 45-46, 355
    wings, 83
    woman as, 11, 374
Cosmonautics Day, 139
Cosmos (Universe), 432
Cosmos (program), 119, 227
COSMOS XIII, 101, 502
COSMOS XIV, 140, 503
COSMOS XV, 154, 503
COSMOS XVI, 165, 503
COSMOS XVII, 209, 505
COSMOS XVIII, 212, 505
COSMOS XIX, 298, 375, 508
COSMOS XX, 392, 509
COSMOS XXI, 427, 510
COSMOS XXII, 433, 510
COSMOS XXIII, 477, 511
COSMOS XXIV, 487, 511
COSPAR. See Committee on Space Research.
COST. See Congressional Office of Science and Technology.
Cost-plus-fixed-fee (CPFF) contracting (DOD), 454
Cost-plus-incentive-fee (CPIF) contracting (DOD), 454
Coughlin, William J., 361
CPFF. See Cost-plus-fixed-fee contracting (DOD).
CPIF. See Cost-plus-incentive-fee contracting (DOD).
Crab Nebula, 493
Crafton, Lt. Robert W. (USN), 222
Crash, airplane, 271, 441
"Crater hopper", 260
Crayons, use of, 287
Creative writing in science, 408
Cree (payload), 87
Creve Coeur Club, 65
Crimea, U.S.S.R., 274
Crimean Astrophysical Observatory (U.S.S.R.), 419
Crocker, E. A. 170
Crossfield, A. Scott, 384
Cross-ranger indicator, 464
Cruz, E. S., 179
Cryogenic environment, 276
Cryomagnet, 95
CSC. See Corp., Communications Satellite.
Cuba, 347, 374, 391, 393, 441
Cumberland Plateau Seismological Observatory, 58
Cunningham, R. Walter, 392
"Current Developments in Space Law", 40
Curtis, Sen. Carl T., 187
Curtiss-Wright Corp., 256
Cutler, C. C., 222
Cuxhaven, Germany, 177
CVA 69. See Aircraft carrier.
CX-4 (cargo aircraft), 438
CX-X (cargo aircraft), 438
Czechoslovakian Academy of Sciences, 30
Czechoslovakian Socialist Republic, 30
D-558-II (Skyrocket), 191
DAC-Roc (sounding rocket), 410
Da Rosa, Aldo Vieira, 177
Da Vinci, Leonardo, 472
Daddario, Rep. Emilio Q., 84, 322, 332, 424

Daily Herald (London), 433
D'Aiutolo, Charles T., 53
Dallas, Tex., 25, 155, 156, 157, 179–180, 206, 209, 277, 446, 481
Damping device, 431
Danville, Ill., 114, 151
Dark Ages, 415
Darwin, Australia, 82
Darwin, Charles R., 415
Data processing equipment (see also Computers), 277
David, Charles E., 365
David C. Schilling Trophy, 340
Davis, M/G Leighton I. (USAF), 105, 205, 254
Davis-Monathan AFB, Ariz., 499
Davis, Noah S., 278
Davis, M/G Waymond A. (USAF), 487
Day, Postmaster General J. Edward, 255
Daytona Beach Operations, (NASA), 252
Dayton, Ohio, 419, 421, 483
Dazzle, Project, 295, 311
DC-8, 257, 279, 306, 339, 426
D.C. Board of Zoning Adjustment, 17
DDC. See Defense Documentation Center.
De Broglie, Prince Louis, 74
De Fries, Paul J., 497
De Gaulle, President Charles, 12, 234
De Vaucouleurs, Dr. Gerald, 234
Deacon-Judi (sounding rocket), 168
Dearborn, Mich., 30
Deauville, France, 341
Debus, Dr. Kurt, 115, 175
Decker, James L., 54
Decker, Martin M., 231, 265
Declaration of Independence, 167
Decomposition unit, 278
Decontamination, spacecraft, 339
Deep Freeze, Operation (USN), 367
Deep Space Instrumentation Facility (DSIF), 32, 179
Deep Space Network, 27
Defender, Project, 184
Defense, Department of (see also USA, USAF, USN), 212, 216, 220, 269, 318, 319, 356, 398, 399, 401, 475, 491
   appropriations, 168, 350
   awards, 205
   bomber, 492
   budget, 215, 255, 400, 461–2
   communications satellite system, 319, 327, 403
   contracts, 59, 162, 239, 255, 466
   defense installations, 477
   flight tests, 203, 231
   funds, 14, 15, 68, 83–4, 206, 249, 262, 315
   international relations, 400
   launch, 202, 257, 351, 389, 391, 400, 499
   management, 116, 234, 272–3, 441, 453, 461–2
   NASA–DOD, 21, 56, 59, 87, 98, 121–2, 199, 207, 277, 300, 315, 319, 334, 391, 400
   nuclear propulsion projects, 367
   organization, 22, 216, 260
   reports, 18, 41, 414, 424, 431, 452, 478, 495
   research, 337
   satellites, 360, 441
   Skybolt, 1
   solid rocket propellant, 203
   space projects, 72, 90, 167–168, 193, 199, 200, 210, 299, 316, 338, 377, 399
   submarines, 111
   Titan III, 262
   transport, supersonic, 7, 19
   weapons, 394

Defense Documentation Center (DDC), 415
Defense Supply Agency (DSA), 415
Delamar Lake, Nev., 256
Delaware, Univ. of, 235
Delta (see also Thor-Delta), 76, 235, 352, 395, 407
Delta Dart. See F-106.
Delta Day, 76
Delta, Thrust-Augmented (TAD), 395, 459
Dembling, Paul, 382
Demler, M/G Marvin C. (USAF), 454
Demon. See F-3B.
Denmark, 197, 340
Dennison, James T., 278, 354
Dennison Manufacturing Co., 278
Density, atmospheric, 20, 61, 110, 469, 487
Denver, Colo., 136, 232, 256, 270, 291, 491
Denver Research Institute, 327, 491
DeOrsey, C. Leo, 48, 187, 379, 496
Dept. of Commerce Gold Medal for Exceptional Service, 56
Dept. of Space Science (Rice Inst.), 2
Design Engineering Conference, 204
Des Moines, Iowa, 232, 340, 398
Dessler, Dr. Alexander J., 2
Destroyer. See WB-66.
Detergents, synthetic, 422
Detroit, Mich., 342, 398
Deutsch, Dr. Stanley, 246
Die Technik, 9
Dietz, Maj. Harold W. (USAF), 42
Diluzio, F. C., 367
Dingman, James E., 467
Diode, 399
Dipole, 22, 341
Disarmament, 346, 353, 393, 440, 475, 489
    test ban, 53, 58, 91, 103, 298, 347
Disconnector, 278
Discoverer (satellite), 63
Disney, Walt, 98
Distance estimation, 37
Distinguished Civilian Service Award (DOD), 368
Distinguished Public Service Medal (DOD), 234
Distinguished Service Medal (NASA), 205
Distinguished Service Medal (USAF), 148
Dixon, Thomas F., 34, 47, 77
Djakarta, Indonesia, 492
Dobbins AFB, Ga., 485
Dobronravov, Vladimir, 420
Dobrynin, Ambassador Anatoli, 389
Doctor, Norman J., 56
DOD. See Defense, Dept. of.
Dolgov, Petr Ivanovich, 218
Dollfus, Dr. Audouin, 493
Domestic Names Committee, 465
Donlan, Charles J., 365
Donnelly, Paul C., 367, 371
Doolittle, Dr. James, 373
Doss, Joseph H., 296
Dossin, Dr. Francois V., 332
Douglas Aircraft Co., 18, 191, 240, 271, 280, 386, 484
    contract, 76, 230, 242, 251, 292, 458, 476
    Missile and Space Systems Div., 98, 246, 370
    research, 48, 246
    rocket, 9, 410, 481
    Saturn IV, 143, 297
    studies, 98, 257, 266, 370, 383
Douglas, Donald Wills, 484

Douglas Space Center (Calif.), 404
Downey, Calif., 126, 382
Downing, Rep. Thomas N., 127
Draper, Dr. Charles Stark, 22, 355, 359
Draper Medal, 156
Dreher, Robert C., 476
Drewry, Col. I. O. (USA), 62
Drummond, Roscoe, 408
Dryden, Dr. Hugh L., 61–62, 67, 106, 129, 158–159, 284, 363, 395
    awards, 18, 156, 243, 249, 379
    awards from, 365
    engineering and science, 55, 178, 288
    international program, 81–82, 89, 100, 249, 285, 303, 314, 318, 325, 336, 350, 375, 376, 427, 493
    space cooperation, 198–199, 212
    space program, 59–60, 122–123, 165, 316
Dryden-Blagonravov agreements, 427, 493
DSA. See Defense Supply Agency.
DSIF. See Deep Space Instrumentation Facility.
Dual compensation law, 298
Dual employment law, 298
Dubin, Maurice, 211
Dubna, U.S.S.R., 321
DuBridge, Dr. Lee A., 91, 238, 321
Dulles International Airport, Washington, D.C., 387
Dulles, John Foster, 418
Dunning, Dr. John R., 204
Dunsmuir House, 110, 113
Dust, cosmic, 144, 306, 432, 443
Dust, lunar, 306, 387, 433
Dust, metallic, 455
Dust, meteoritic, 388
Dwight, Capt. Edward J. (USAF), 426
Dwoskin, Philip B., 494
DX priority, 89
Dyna Soar. See X-20.
Dyson, Dr. Freeman J., 260
Early Manned Planetary-Interplanetary Roundtrip Expedition. See EMPIRE, Project.
"Early U.S. Satellite Proposals," 377
Earth
    albedo (whiteness), 342
    atmosphere, 222, 331, 366, 367, 388, 400, 414, 498
    horizon scanning, 403
    magnetic
        field, 282, 342, 356, 445
        pole, 270, 333
    magnetosphere, 356, 380
    rock layer, 328
    rotation, effect of, 411
    shape, 37, 364
    water resources, 328, 331
Earthquake, 402
East Berlin Radio-TV, 394
East China, 415
Eastern Air Lines, 20, 426
Eastern Hemisphere, 396
Eastern Pyrenees Dept. (France), 8
East Germany, 9, 13, 252, 394
Eastman-Kodak, 489
East-West arms control agreement, 491
Eccentric Geophysical Observatory. See Ego.
Eccentric Orbiting Geophysical Observatory. See Eogo.
E. C. Ernst Co., Inc., 488
ECHO I, 231, 297, 304, 488
ECHO II, 100, 158, 304, 314, 406, 426, 488

Eclipse, 252
Eclipse, solar, 18–19, 257, 273, 279, 332, 339
École Polytechnique (Paris, France), 330
Economic Club of New York, 435
Economist, need for, 403
Economy in Government, 475, 489
ECS. See Environmental control system.
Ecuador, 33
Eden, Prime Minister Sir Anthony, 270
Edmondson, Sen. J. Howard, 240
Edmonton, Alberta, 279, 342
Education
    astronaut, 136
    engineers, 59, 423
    inducements to, 267, 343
    money for, 26, 172, 266, 318, 372, 423
    new institutions, 3, 9, 95
    need for, 24, 110, 118–119, 148, 249–250, 343, 369–370, 378, 437, 444, 450, 456
    programs, 391
    science teachers, 26, 423, 468–469
Edward R. Sharp Medical Library, 240
Edwards AFB, Calif., 114, 146, 148, 186, 203, 207, 237, 296, 322, 336, 337, 338, 371, 426, 459, 475
Edwards-Wendover-Holloman-White Sands complex, 296
Eggs, 496
Eglin AFB, Fla., 222, 499
Eglin Gulf Test Range, Fla., 16, 19, 56, 87, 410
Ego (Eccentric Geophysical Observatory), 36, 248, 436
Egypt. See United Arab Republic.
E. H. Rietzke Achievement Awards, 128
Einstein, Albert, 360
Eisele, Capt. Donn F. (USAF), 392
Eisely, Dr. Loren C., 408
Eisenhower, President Dwight D., 116, 200, 240, 253, 270, 418, 471
    space program
        budget, 112, 124, 132, 485
        manned lunar landing, 239
        need for, 215–216, 374, 403
        pace, 124, 166, 201
El Centro, Calif. 35, 49, 192, 397, 482
Elderly, care for, 450
"Elder Statesman of Aviation," award, 411, 484
ELDO. See European Launcher Development Organization.
Electric current, 488
Electricity, thermal, 342
Electric propulsion, 461
Electrojet, 11, 336
Electromagnet, 428
"Electromagnetic Effects on World Communications", 383
Electromagnetism, 149
Electron, 23, 26, 41, 86, 95, 111, 156, 270, 308, 356
    concentration, 477–478
    density, 263, 273, 279, 351, 357, 360, 391
    temperature, 263, 273, 279, 325, 351, 356, 357, 360
Electronic, 272, 312, 363, 391
Electronic Components Conference, 186
Electronic data processing. See Computer.
Electronic News, 42, 237, 254
Electronic recovery control center, Sunnyvale, Calif., 296
Electronics Research Center (NASA)
    congressional action, 240, 254, 283, 293, 302, 310, 384, 484, 495
    location of, 47, 97, 106, 127, 178, 394, 469, 483, 484–485, 495
    plans for, 16, 62, 98, 121, 129, 178, 272, 311–312
Electro-Optical Systems, Inc., 437
Electrostatic field, 455

Ellington AFB, Tex., 64
Elliott, Rep. Carl, 337
Ellis, Mary Jane, 494
Ellsworth AFB, S.D., 399
Elms, James C., 1, 190, 420
Ely, Nev., 322
Emme, Dr. Eugene M., 414
EMPIRE, Project (Early Manned Planetary-Interplanetary Roundtrip Expedition), 232
Employment, 391, 450, 475
Encephalogram, 161
Endeavour magazine, 37
Energy collecting device, 333
Engine
    development, 267, 492
    electric, 449
    ion, 451
    liquid-fuel, 450, 451
    ramjet, 427
    solid-propellant rocket, 66, 426
    testing, 491–2
Engineering Societies of New England, 62
Engineers
    and scientists, 177, 190, 214
    challenges, 313
    education, 59
    in space effort, 204, 448
    shortage of, 46, 48
    use of, 50
Engineers Joint Council, 36, 179
England. See United Kingdom.
Engle, S/Sgt. Elvin H. (USAF), 460
Engle, Capt. Joseph H. (USAF), 235, 378, 431
English Channel, 146, 479
Enrico Fermi Award, 130, 458
Enrico Fermi Institute for Nuclear Studies, 177, 270
Enthoven, Dr. Alain C., 239
Environmental conditions, simulation of, 453, 455
Environmental control system (ECS), 107
Environmental sciences, 287
Eogo (Eccentric Orbiting Geophysical Observatory), 123, 268
Equator, 297
Eratosthenes (moon crater), 12
Erickson, Wayne D., 365
Erkin, Feridon Cemal, 25
Escape systems, 264, 383
ESRO. See European Space Research Organization.
Essen, West Germany, 140
Estep, Dr. Samuel D., 269
Estes, L/G Howell M., Jr. (USAF), 334
Esther, Hurricane, 293
Europe, 194, 196, 206, 211, 213, 299, 325, 342, 430, 436, 442, 448
European Launcher Development Organization (ELDO), 2, 112, 133, 206, 211, 271, 497
European Postal and Telecommunications Administrations, 197, 429
European Space Flight Symposium, 205, 211, 214
European Space Research Organization (ESRO), 206, 271
European Space Vehicle Congress, 212
Eurospace, 133
Eurovision, 241
Evening Moscow, 138
Evans, Albert J., 401
Ewing, Dr. W. Maurice, 214
E. W. Muller, Contractor, Inc., 136
Exceptional Bravery Medal (NASA), 186

Exceptional Civilian Service Medal (USAF), 476
Executives' Club, 39
Exercise Big Lift (DOD). See Big Lift, Exercise.
Exhaust plume, 415, 437
Exhibit
    Chicago, 136
    Mercury spacecraft, 95
    NASA program, 315
    World's Fair, 99
    X-15, 95
Exotech, Inc., 323
Expenditures, 286, 316, 325, 337
EXPLORER I, 32, 34, 364
EXPLORER VIII, 367
EXPLORER IX, 487
EXPLORER XII, 95, 317, 449
EXPLORER XIV, 7, 252, 449
    failure, 380
    performance, 12, 30, 295
    results, 95, 182, 295, 356, 380
EXPLORER XV, 83, 182, 449
EXPLORER XVI, 53, 341, 400
EXPLORER XVII, 119, 134, 146, 182, 503, 514
EXPLORER XVIII (IMP-A), 449, 462, 510, 514
EXPLORER XIX, 487, 495, 511, 514
Explosion, 318
F-1 (engine)
    combustion instability, 263, 438
    development, 6, 42, 84, 97-98, 221
    test, 464
F-3B (Demon), 296
F-4B (Phantom II), 216, 442
F-4C (USAF), 216, 442
F-104 (Starfighter), 190, 252, 426
F-106 (Delta Dart), 426
F-111 (TFX). See Aircraft, Tactical fighter, experimental.
FAA. See Federal Aviation Agency.
FAA Civil Aeronautics Research Institute, 228
Faget, Maxime A., 420, 465
FAI. See Fédération Aeronautique Internationale.
FAI Paul Tissandier Diploma, 266
Fairbanks, Alaska, 248, 341
Fairchild Stratos Corp., 43
Fairhall, Minister of Supply Allen (Australia), 295
FAITH 7 (see also MA-9), 147, 154, 192, 196, 205, 218, 238, 386, 504, 514
Fallon, Nev., 402
Farnborough, U.K., 112
Farrell, Richard, 275
"Father of the Common Market", 270
Faulkner, William, 199
Fayette, Mo., 334
FBM (Fleet Ballistic Missile Submarine), 262
FCC. See Federal Communications Commission.
Federal Aviation Act of 1958, 314
Federal Aviation Agency (FAA), 122, 256, 271, 279, 299, 442, 483, 485
    accident investigations, 254
    air carrier operations, 59, 471
    aircraft, commercial, 333, 387
    air traffic control, 377, 432, 456
    Supersonic Transport Authority, 251, 271
    transport, supersonic
        cooperation, international, 410
        U.S., 6, 19, 69, 90, 94, 119, 136, 141, 198, 209, 248, 251, 253, 287, 307, 312, 314, 320, 336, 380, 390, 398, 406, 429, 432, 439, 489
        U.S.S.R., 417, 492

Federal Bar Association, 229
Federal budget. See United States, budget.
Federal Communications Commission (FCC), 114, 151, 285, 312, 372, 495
   ComsatCorp, 39, 57, 114, 155, 247–248, 285, 301
Federal Government. See United States, Government.
Federal Personnel Associations of N.Y. and N.J., 213
Federal Radiation Council, 220
Federal Register, 229
Federal Women's Award, 131, 188
Fédération Aeronautique Internationale (FAI), 30, 32, 48, 341, 372, 424, 469, 484
Federation of American Scientists, 152
Fedorenko, Nikolai T., 142, 231
Feldman, George, 91, 139, 229, 247
Ferdinand and Isabella, 217
Ferguson, L/G James (USAF), 35, 77
Fermi, Enrico, 149
Fiberglass cloth, 287
Field Enterprises Educational Corp., 48, 268, 269, 343
"Financial Protection Against Risks of Major Harm in Government Programs", 141
Finger, Dr. Harold B., 89, 137, 169, 183, 233
Finland, 8, 197
Finney, John W., 1, 19, 327, 332, 374, 406, 418
Fireball, 426
Fire-fighting equipment, 271
Firefly III, Project (USAF), 16, 499
First Flight Airport (Kitty Hawk, N.C.), 483
"First Man-Rocket Belt, The," 104
Fission, 284
Flag, 361
Flagstaff, Ariz., 192, 396, 410
Flanders, Sen. Ralph, 333
Flare, solar, 92, 93, 263, 404, 471, 494
Fleet Ballistic Missile (FBM) submarine, 111, 263
Flies, fruit, 355
Flight dynamics, 327
Flight, hypersonic, 420
Flight, powered, 480
Flight Research Center (FRC), 282, 322, 381
    awards, 360, 364, 484
    contracts, 17, 34, 38, 190
    flight test, 218, 252, 282, 331, 395, 433
    models, 170
    sonic boom, 59
    X–15 pilots, 235
Flight Safety Foundation, Inc., 442
Flood, Rep. Daniel J., 168
Florida, 451, 452, 470, 476
Florida Aerospace Industry Seminar, 462
Florida East Coast Railway, 337
Florida Legislature, 476
"Flow separation spike", 326
Flox (Liquid Fluorine and Liquid Oxygen), 342, 379, 490
Flugwelt (Aviation World), 454
Fluorescent lighting, 385
Fluorine/hydrogen propellant, 277
Fluorine, liquid. See flox.
Fluorine pump, 397
Fluorine/liquid oxygen. See flox.
Fluorine-oxygen/hydrogen propellant, 277
"Flying carpet" (escape system), 383
Flying Fortress. See B–17.
Flying saucer, 58
Fong, Louis B. C., 102, 183, 352, 491

Food and Agriculture Organization, 396
Food supply, 273
Ford Co. See Aeronutronics Div.
Ford Foundation, 50, 488
Ford, Rep. Gerald R., 315
Fordham Univ., 444
Forecast, Project (USAF), 216
Fort Belvoir, Va., 271
Fort Churchill, Canada, 71, 88, 111, 206, 208, 278, 333
Fort Collins, Colo., 35
Fort Dix, N.J., 349, 353
Fort Sill, Okla., 415
Fort Wingate, N.M., 399, 442
Fort Worth, Tex., 446
Fortune magazine, 466
Foster, William C., 409
Foster, Willis B., 436
Foulois, M/G Benjamin D. (USA, Ret.), 73, 257, 383
Fowler, Dr. William A., 155
Fragment, 3, 59, 138, 175
France
    aircraft, 229, 230, 234, 367
        supersonic transport "Concorde" (Fr.-U.K.), 320, 379, 390, 398, 406, 410
    communications, 186, 189, 197, 425
    launch, 350, 392
    metric system, 37
    nuclear, 176, 298, 391, 438
    sounding rocket, 19, 171, 212, 273, 391, 410
    space
        budget, 362
        cooperation, 7, 89, 133, 443, 463
        program, 8, 335, 438
    tracking, 298
Frankfurt, W. Germany, 164
Franklin Institute (Philadelphia), 390, 497
Franklin Medal, 390, 497
Frazitta, Frank M., 470
FREEDOM 7, 238
Freeman, Capt. Theodore C. (USAF), 392
Freeth, Denzil (Parliamentary Secretary for Science, U.K.), 275
Free World, 334, 337, 465
Freeze, NASA personnel, 463
Freitag, Capt. Robert F. (USN), 70, 181
French Academy of Sciences, 336
French Association Pour l'Étude et la Recherche Astronautique Gold Medal, 330
French Astronautics Society, 205
French National Academy, 36
French National Center for Space Studies (CNES). See National Center for Space Studies (France).
French Revolution, 36
Frequency, 34, 419, 425
Freud, Sigmund, 415
Frick, Charles W., 54, 125
Friction drag, 281
Friedman, Dr. Howard, 445, 493
FRIENDSHIP 7 (see also MA-6), 16, 61, 67, 238, 361
Frutkin, Arnold W., 115, 182, 211, 306, 382
Fubini, Dr. Eugene G., 260, 265, 319, 473
Fucino, Italy, 3, 5, 425
Fuel, 422, 427, 428, 463, 490
Fuel cell, 337, 386, 405, 493
Fulbright, Sen. J. William, 179, 368, 391, 439, 440
Fulton, Rep. James G., 84, 287
Fulton, Rep. Richard, 35

Fundamental Methods Associates, 340
Funk, M/G Benjamin I. (USAF), 25, 287, 301, 455, 465, 487
Fuqua, Rep. Don, 35
Fylingdales, U.K., 343
Gabriel, David S., 472
Gagarin, L/C Yuri, 12, 355, 374, 471
    appearances, 138, 367, 390, 394
Galileo, 167, 182, 191, 415
Gallium, liquid, 129
Gamma rays, 397
GAO. See General Accounting Office.
Garbarini, Dr. Robert F., 240, 256
Gardner, John H., 273
Gardner, Richard N., 147
Gardner, Trevor, 359
Garrett-AiResearch Los Angeles, 107
Garrett Corp., 252
Garrett, J. C., 252
Gauge, pressure, 390
Gault, Dr. Donald E., 144
Gautrand, John A., 61
Gavin, Joseph G., Jr., 100
GD/A. See General Dynamics/Astronautics.
GE. See General Electric Co.
Geiger counter, 273, 387
Geissler, Dr. Ernst D., 365
Gelatin, 287
Gell-Mann, Dr. Murray, 142
Geologist, 56
Gemini Program Planning Board (NASA-USAF), 21, 300, 332
Gemini, Project, 78, 316, 353, 433
    and X-20, 94, 114
    astronauts, 4, 177, 298, 333, 340, 392
    biomedical support, 393
    facilities, 28, 464
    management, 21, 47, 87, 102, 102, 190, 199, 277, 301, 319, 332, 424
    plans for, 45, 127, 174, 180, 208, 252, 332, 473
    problems, 40, 309, 405
    progress, 85, 164, 174, 238
    purpose, 174, 181, 207, 226, 254
Gemini (spacecraft)
    booster, 415, 487
    engines, 424
    equipment, 162, 273, 328
    escape system, 127, 156, 264
    flight tests, 370–371, 413
    landing, 297, 366
    malfunction detection system, 368
    modified version, 473
    parachute, recovery, 35, 49, 127, 192, 302, 366
    problems, 406
    production, 119, 475
    recovery system, 302
    rendezvous-docking, 334, 397
    simulators, mission, 478
    spacesuit, 393
    tracking, 325
    vibration, 431
    vehicles, launch, 320, 334, 368, 400, 431, 486, 488
    waste disposal system, 273
Geneen, Harold S., 247
General Accounting Office (GAO), 116, 370, 374, 376, 380
General Atomics Div. (General Dynamics), 420

General Dynamics/Astronautics (GD/A), 67, 455
    Atlas, 12, 100, 380, 490
    Centaur, 315, 370, 401, 410, 472, 490
    contract, 251, 268
    flox oxidizer, 379
    life-support system, 229, 230
    Nova, 232
    RL-10 engines, 410
    TFX, 178, 283
General Dynamics/Convair
    Apollo, 62, 273, 414
    contract, 62, 414
    Little Joe II, 273, 414
General Dynamics/General Atomics Div., 420
General Dynamics/Ft. Worth, 292, 427
General Electric Co. (GE), 135, 287
    contract, 132, 142, 164, 194, 345, 411, 490, 491, 492
    Missile and Space Div., 320
    research, 36, 267
    Space Technology Center, 268
General Electric Forum, 363
Gen. H. H. Arnold, 140, 176
General Henry H. "Hap" Arnold Aviation Gold Medal Award, 324
General Motors Co. See AC Spark Plug Div. (General Motors).
General Precision, Inc., 169, 221
General Services Administration (GSA), 277, 369
General Telephone and Electronics Corp., 155
General Thomas D. White USAF Space Trophy, 226
Generator, nuclear (see also Snap), 345, 357, 467
Geneva, Switzerland, 43, 50, 58, 78, 89, 90, 98, 131, 164, 195, 200, 314, 346, 372, 379, 393, 416, 419, 425, 440
Genisco, Inc., 263
Genoa, Italy, 386
Geologist, 495
Geomagnetic, 314, 449
Geophysical Laboratory (Washington, D.C.), 237
Geophysics, 258, 267, 308, 411
Geophysics Corp. of America, 19, 104, 241
Geophysics Research Board of NAS-NRC, 307
Georgetown Univ., 354, 493, 494
Georgetown Univ. Forum, 267
George Washington Univ., 18
Georgia Tech (Georgia Institute of Technology), 52
Georgia, Univ. of, 313
German American Day, 215
German Ministry of Defense, 322
German Society for Rocket Technology and Space Flight, 205
Germany, 379, 398
Geronimo, 345
Gerrity, L/G Thomas P. G. (USAF), 168, 188, 338
Getting, Ivan A., 99, 295
Gifford, Dr. Franklin A., Jr., 56
Gill, Dr. Jocelyn R., 257, 280
Gillespie, R. W., 14
Gilpatric, Roswell L., 167, 242, 319, 322, 461, 462
Gilruth, Dr. Robert R., 47, 202, 258, 291, 343, 373, 420
    awards, 22, 465
Gilzin, K., 417
Gimbaled integral nozzle (GIN), 497
GISS. See Goddard Institute of Space Studies.
GIN. See Gimbaled integral nozzle.
Glenn, L/C John H., Jr. (USMC), 12, 26, 28, 86, 238, 303, 492
    appearances, 61, 67, 136, 153, 154, 224, 342, 361, 480
    awards, 104, 330, 465
    politics, 266, 278, 379
Glennan, Dr. T. Keith, 395

Glider, 282, 459, 481
Global weather. See Weather, global.
Glushko, Valentin P., 428
Goddard Day, 41
Goddard Historical Essay Competition, 377
Goddard Historical Prize Essay, 377
Goddard Institute of Space Studies (GISS) (NASA), 45, 149, 329, 468
Goddard Memorial Scholarship Award, 188
Goddard, Dr. Robert H., 41, 462
Goddard, Mrs. Robert H., 104
Goddard Scientific Satellite Symposium, 92, 92, 94–95
Goddard Space Flight Center (GSFC), 32, 33, 54, 185, 211, 352, 368, 404
    awards, 76, 188, 365
    communications, 44, 98
    contracts, 258, 345, 397, 414, 443, 443, 453, 472
    experiments, 70, 186, 195, 279, 333, 351, 357, 384, 388
    organization, 381
    programs, 38, 61, 70, 71, 73, 119, 139, 156, 184, 279, 301, 478
    radiation, 50, 95
    reports, 2, 14, 210, 220, 221, 243
    research, 227, 241, 267, 312–313, 317
    results, 11, 117, 146, 308, 332, 386
    satellite monitoring, 134, 146. 247, 304, 312–313, 339, 353, 380, 381, 402, 475
    studies, 5, 38
Goett, Dr. Harry J., 111
Goland, Martin, 46
Gold, 399, 438
Gold Medal, 18
Gold Medal for Exceptional Service (Dept. Commerce), 222
Gold-plating, 438
Gold Star Medals (U.S.S.R.), 251
Gold, Dr. Thomas, 144, 415
Goldstein, Dr. Richard M., 454
Goldstone Tracking Station, 27, 63, 69, 179
Goldwater, Sen. Barry, 72, 279, 310
Golobachev, V., 372
Golomb, Dr. Soloman W., 17
Golovin, Dr. Nicholas E., 483
Goodyear Aerospace Corp. (formerly Goodyear Aircraft Corp), 406
Goodyear Aircraft Corp. (see also Goodyear Aerospace Corp.; named changed), 12
Goonhilly Downs, U.K., 3, 5, 184, 235, 241, 425
Gordon, John A., 186
Gordon, LCdr. Richard F., Jr. (USN), 392
Gore, Sen. Albert, 99, 427
GORID (Ground Optical Recorder for Intercept Determination) telescope, 399
Gorki, U.S.S.R., 88
Gorkin, Jess, 462
Gorman, H. H., 328
Gorshkov, Adm. Sergei, 44
Governors' Conference (Miami Beach), 282
Grachev, A., 218
Graduate Research Center of the Southwest, 238
Graham, Beardsley, 91
Graham, Gary A., 296
Graham, Philip L., 27
Grand Central Station, N.Y., 196
Grand Kremlin Palace (Moscow), 297
Grand Turk Island AFB, B.W.I., 37
Grants
    Government, 83, 286, 337, 485, 486
    non-government, 50, 396
Graveline, Dr. D. E., 228
Gravelines, France, 140
Gray, Edward Z., 387

Gray, Horace M., 86
Great Britain. See United Kingdom.
Greater Boston Chamber of Commerce. 487
Great Falls Air Route Traffic Control Center (FAA), 456
Great Falls, Mont., 432
Great Falls SAGE (Semi-Automatic Ground Environment) Direction Center, 456
Great Slave Lake (Canada), 280
Greece, 197
Green Bank, W. Va., 131
Green River, Utah, 42
Greenacre, James, 410
Greenbelt, Md., 247, 414
Greenhouse effect, 151
Greenstein, Dr. Jesse L., 160
Gregory, Donald T., 323
Griffiss AFB, N.Y., 426
Grillo, Stephen J., 169
Grissom, Maj. Virgil I. (USAF), 28, 238, 465
Gromyko, Andrei A., 283, 297, 349, 353, 365, 368
Groton, Conn., 165
Ground Optical Recorder for Intercept Determination (GORID). See GORID.
Group Achievement Award (NASA), 76, 205
Groves, L/G Leslie R. (USA, Ret), 183, 468
Gruening, Sen. Ernest, 256
Grumman Aircraft Engineering Corp., 90, 219, 257, 472, 489, 492
GSA. See General Services Administration.
Gubser, Rep. Charles S., 315
Guderley, Dr. Karl G., 368
Guggenheim, Harry F., 74
Guggenheim International Astronautics Award, 262, 367
Guidance and control data, 327, 350, 355
Guinea, 101
Guinea, Gulf of, 345
Gulfport, Miss., 175
Gulliver, 49
Gurney, Rep. Edward J., 35, 470, 476
Gyro, 244
H-Alpha rays, 494
Haeussermann, Dr. Walter, 365
Haggerty, James J., Jr., 358, 470
Hahnemann Medical College (Philadelphia), 234
Hailey, Ida., 282
Hailsham, Lord, 79, 283
Halaby, Najeeb E., 279, 299, 442, 471, 492
  transport, supersonic, 13, 71, 78, 114, 119, 141, 198, 209, 248, 251, 253, 287, 307, 312, 390, 406
Hall, Dr. Albert C., 269, 399
Hall-current plasma accelerator, 278
Hall, G. F., 106
Hall, Dr. John A., 410, 460
Hall, R. Cargill, 104, 377
Hall of Science building, 375
Halleck, Rep. Charles A., 112
Ham (chimpanzee), 128
Hamburg, Germany, 466
Hamilton Standard Div. (United Aircraft Corp.). See United Aircraft Corp.
Hammock, D. M., 193
Hampton, Va., 366
"Handbook on Space Radiation Effects to Solar Cell Power Systems", 323
Haney, Paul, 291
Hardtack, Operation, 383
Harriman, W. Averell, 283
Harrington, Dr. John V., 170
Harris, Mrs. Grace M., 265
Harris, Rep. Oren, 247

Harris, Sam, 91
Harris Survey, 481
Harrison, Dr. George R., 430
Harry E. Salzberg Memorial Medal, 379
Hart, Sen. Philip A., 425
Hartesbeesthoek, South Africa, 414
Hartford, Conn., 197, 435
Harvard Business Club, 187
Harvard Business Review, 332, 411
Harvard Business School, 332
Harvard College Observatory, 39, 72
Harvard Univ., 53, 228
Harwell Nuclear Research Institute (U.K.), 153
Haskins, Dr. Caryl P., 470
Hasson, Dennis F., 214
Hattiesburg, Miss., 254
Hauty, Dr. George T., 228
Havana, Cuba, 11
Havighurst, Prof. Robert J., 278
Hawaii, 107, 492
Hawk (missile), 230
Hawker Aircraft, Ltd., 48
Hawkins, G. S., 39
Hawkins, Willis M., Jr., 324
Hawkinson, John, 62
Haworth, Dr. Leland J., 103, 253, 258
Hawthorne, Calif., 146
Hayden, Stone and Co, Inc., 379
Hayes, L/G Charles H. (USMC), 361
Haywood, Dr. Oliver G., Jr., 278
Heart, mechanical system, 426
Heat exchanger, 437
Heating, 281, 336, 451
Heat resistance, 333
Heat shield, 280
Health, Education, and Welfare, Dept. of (HEW), 212, 337
Hearst, William R., Jr., 218
Hebrew Univ., 456
Heidelberg, Univ. of (Germany), 421
Helicopter
    CH-3B, 230
    future, 6
    OH-23G, 424
    S-61, 222
    speed records, 424
    use in airplane crash survival, 271
    XH-51A, 179
Helios (see also Advanced Orbiting Solar Observatory), 73
Helios, Project (ONR), 30
Helium, 146, 262, 325, 367, 488
Helium 2, 213
Helsinki, Finland, 241
Helvey, Dr. William M., 176
Henderson, Arthur, 199
Henry, E. William, 301, 312, 495
Henry, Thomas R., 6
Heppe, R. R., 179
Herbede, West Germany, 140
Hercules. See C-130.
Hercules Powder Co., 165
Hermann Oberth Medal, 212
Herold, Curt P., 278
"Hero of the Soviet Union", 218, 251
Herzfeld, Charles M., 56
Hess, Dr. H. H., 238

Hess, Robt. V., 278
Hess, Dr. Wilmot N., 227, 308
H. eutropha microbe, 326
HEW. See Health, Education, and Welfare, Dept. of.
Heyden, Rev. Francis J., 64
Heyl, Dr. Paul R., 106
Hickam AFB, Hawaii, 287, 307
Hicks and Greist, Inc., 314
High-altitude research chamber, 465
"High Range", 322
High Voltage Engineering Corp., 153
Hilburn, Earl D., 256, 382, 385
Hines, William, 6, 132, 135, 372, 376, 447
Hirsch, Dr. Jules, 58
Historian, 403
Hitler, Adolf, 205
Hjornevik, Wesley L., 420
HL-10 (lifting body), 292
HMS *Ark Royal*, 48
Hochstim, Adolph R., 331
Hodges, Luther H., 24, 127, 290, 480, 483
Holifield, Rep. Chet., 131
Holloman AFB, N.M., 69, 142, 150, 237, 296, 326, 393
Holloman, Dr. J. Herbert, 24, 385
Hollywood, Calif., 484
Holmes, D. Brainerd, 48, 149, 190, 225, 239, 281, 299
    award, 334
    Gemini, Project, 174
    manned lunar landing
        cost, 111
        importance, 149, 165, 166
        progress, 71, 85
        program, 121, 186, 435
        purpose, 45, 65, 124, 198
    organization, 62, 375
Holy Loch, Scotland, 111
Home, Foreign Secretary Lord, 297
Honest John (missile), 230
Honest John-Nike (rocket), 328
Honeycomb materials, 438
Honolulu, Hawaii, 220
Honorary American Fellowship (IAS), 22
Hoover, Dr. G. N., 496
Hopkins Marine Station (Stanford Univ.), 479
Hopkins, Philip S., 271
Hoppe, Arthur, 416
Horizon-scanning techniques, 403
Hornbeck, John A., 101
Horne, Walter B., 476
Hornig, Dr. Donald F., 424, 480
Hospitals, psychiatric, 445
"Hot spot in the sky", 117
Hotz, Robert B., 165, 204, 226, 419, 429, 472
Houbolt, Dr. John C., 365
Hough, A3/C Larry J. (USAF), 186
Houghton, Prof. Henry, 279
Houghton, Dr. Karl H., 456
Hound Dog (missile), 410
Houston *Chronicle*, 9
Houston Post, 271, 408
Houston, Tex., 94, 289, 347, 351, 356, 364, 371, 375, 383, 442, 497
Hovercraft, 275
Howard Univ. (Washington, D.C.), 426
Howe, Dr. John P., 279

Howlett, Dr. Duncan, 358
"How to Fly to Mars", 9
Hoy, W., 389
Hoyle, Dr. Fred., 155
H. ruhladi microbe, 326
Hsinhua (press agency). 353
Hughes Aircraft Co., 101, 299, 325, 362
    contracts, 141, 170
    missiles, 16
Human behavior, 445
Human body, 397
Human rights, 482
Humanity, 468, 471
Humphrey, Sen. Hubert H., 344, 382
Hungary, 83, 333
Hungarian Communist Party, 470
Huntington Beach, Calif., 404
Hunstville, Ala., 19, 66, 83, 201, 479
Hurricane, 293, 344, 401
Hurricane Arlene, 302, 344
Hurricane Beulah, 401
Hurricane Debra, 348
Hurricane Flora, 360, 367
Hyatt, Abraham, 410
Hydac (sounding rocket), 481
Hydraulic press intensifier, 297
Hydrocarbon, 422, 428
Hydrofoil, patrol craft (PCH-1). See PCH-1.
Hydrogen
    atmosphere, upper, 367
    atom, 498
    atomic, 14, 15, 107
    gaseous, 463
    ignition, 143
    J-2, 451
    liquid
        and oxygen, liquid, 450
        radiant heating, 247, 336
        supply, 400
        use of, 428, 433, 437, 451
        zero gravity, 247
    peroxide jets, 304
    rocket-engine chamber cooling, 336
Hydroplaning, pneumatic tire, 476
Hydroskimmer (SKMR-1). See SKMR-1.
Hydroxyl airglow measurements, 15
Hydroxyl radicals (OH), 432
Hygiene, 273
Hyman, William A., 172
Hyperon, 26
"Hypersonic organosynthesis", 331
Hypnosis Quarterly, 170
Hypsometer, 390
IAA. See International Academy of Astronautics.
IAF. See International Astronautical Federation.
IAM. See International Association of Machinists.
IAS. See Institute of Aerospace Sciences.
IATA. See International Air Transportation Association.
IBGA. See International Bureau of General Aviation.
IBM. See International Business Machine Corp.
ICAO. See International Civil Aviation Organization.
ICBM (intercontinental ballistic missile). See Missile, intercontinental ballistic.
Ice, 344, 419
Iceland, 197

INDEX 549

IDA. See Institute for Defense Analysis.
"If Columbus Had Never Sailed", 217
IFR. See Instrument flight rules.
Ignition, 143
Igor I. Sikorsky International Trophy, 222
IGY. See International Geophysical Year.
Ikeda, Prime Minister Hayato, 224
Ikeye (comet), 114
Ikeye, Kaoru, 114
Il-18 (Ilyushin 18) (transport aircraft), 453
Ile du Levant launch center (France), 273
Illinois, 156
Illinois Institute of Technology, 100, 172
Illinois Institute of Technology Research Institute, 249
Illinois, Univ. of, 86, 114, 372
Ilyushin 18 (Il-18). See Il-18.
Ilyushin and Tupolev teams, 379
IMCC. See Integrated Mission Control Center.
Imirie, Joseph S., 424
Imp (Inflatable Micrometeorite Paraglider). See Paraglider.
Imp. See Interplanetary Monitoring Platform.
IMP-A. See EXPLORER XVIII.
India, 230, 429
    cooperation, space, 11, 68, 336
    defense, 142
    launch, 443
    tracking, 33
Indian Defense Ministry, 459
Indiana, 469
Indianapolis Engineering Societies, 472, 473
Indianapolis, Ind., 155
Indianapolis News, 469
Indianapolis Scientific and Engineering Foundation, 472
Indies, 194
Industrial Management Assistance Survey, 189
Industrial Research, 212, 260
Industrial Revolution, 472
Industry, 44, 108, 115, 239, 491
    and government, 356, 472, 478, 479
    requirements, 466, 467
Inflatable micrometeorite paraglider (Imp). See Paraglider.
Information
    exchange of, 57, 58, 91, 127, 328, 390, 464
    release of, 22, 108, 304
    reporting, 8, 15
    retrieval, 154
    sale of, 48
Infrared, 150, 342, 377, 437, 448
Ingalls Iron Works, 268
INJUN I, 15
INJUN III, 95, 299
Innovations, 491
"Inspection System Provisions for Suppliers of Space Materials, Parts, Components, and Services" (NASA), 362
Institute for Advanced Study (Princeton Univ.), 40, 130, 260, 459
Institute for Defense Analysis (IDA), 331
Institute for Elements of Space Travel (Berlin), 7
Institute of the Aerospace Sciences (IAS) (see also American Institute of Aeronautics and Astronautics), 19, 22, 23, 25, 63
Institute of Industrial Science (Tokyo), 494
Institute of Radio Engineering and Electronics (U.S.S.R.), 495
Institute of Radio Engineers (IRE), 31
Institute of Strategic Studies (London), 422
Institute of Technology, Milan, Italy, 422

Instrument
 altitude predictor, 321
 coronagraph, 280
 gauge, pressure, 390
 geiger counter, 273
 ionospheric measuring, 263
 magnetometer, 377
 radiometer, 377
 spectrometer, 280, 377
 spectrophotometer, 279
Instrument flight rules (IFR), 377
Instrumentation, 177, 281, 379
Integrated Mission Control Center (IMCC) (MSC), 272
Intensifier, hydraulic press, 297
Inter-American Bar Association, 172
Inter-Church World Peace Seminar, 375
Intercontinental ballistic missile. See Missile, intercontinental ballistic.
Intercontinental television, 267
Interferometry, 481
Intergovernmental Oceanographic Commission, 396
Interior, Dept. of, 212, 277, 486
International Academy of Astronautics (IAA), 74, 220, 262, 359
International Aeronautical Film Festival, First, 32
International Air Pageant, Inc., 32
International Air Show, 234
International Air Transportation Association (IATA), 40
International Art Show, 434
International Association of Machinists (IAM), 9, 23, 28, 29, 143, 147, 148, 176, 197, 375
International Astronautical Congress, XIVth, 354
International Astronautical Federation (IAF), 354, 355, 367
International Astronomical Union, 396
International Atomic Energy Agency, 351, 396
International Bank for Reconstruction and Development, 311
International Benjamin Franklin Society, 18
International Brotherhood of Electrical Workers, 14
International Bureau of General Aviation (IBGA), 31
International Business Machine Corp. (IBM), 272, 369, 475
International Civil Aviation Organization (ICAO), 31
International Code, 1
International communications, 312, 326, 390
International Conference on Atmospheric and Space Electricity, 185
International Congress of Aeronautical and Space Medicine, VIth, 375
International Congress on Electronics, Nuclear Energy, Radio, Television, and Cinema, 257
"International Cooperation in Space Research", 115
Internation cooperation, 45, 152, 344–45, 353, 396, 452, 471
 airliner, commercial, 367, 410, 438, 492
 accelerators, 321
 communications satellite network, 107, 321, 416, 425, 440
 electrojet program, 336
 IGY, 314
 nuclear test-ban treaty, 298, 348, 350, 389
 rocket, 20, 463, 497
 supersonic transport, 410
 tracking, 54
International cooperation, space (see also International space programs)
 cooperative programs, 2, 20, 82, 139, 212, 269, 270, 272, 274, 275, 276, 279, 285, 314, 317, 319, 329, 348, 350, 364, 365, 369, 393, 405, 445, 478, 494
 proposals, 162, 358, 464
International Council of Scientific Unions, 98
International Electronic Circuit Packaging Symposium, Fourth, 311
International Geophysical Year (IGY), 63, 68, 80, 131, 307, 314
International Geophysics Bulletin, 35

INDEX 551

International Hydrological Decade, 328
International Latex Corp., 403
International law, 347
International Northwest Aviation Council, 342
International Scientific Radio Union, 168, 341
International sounding rocket facilities, 329, 447
International space programs (see also International cooperation, space)
    U.S.-Australia, 82, 139, 295, 295, 464
    U.S.-Canada, 32, 61, 71, 115, 279, 321, 357, 410, 449
    U.S.-Denmark, 340
    U.S.-France, 89, 335, 391, 410, 487, 497
    U.S.-France-India, 443
    U.S.-Germany, West, 454
    U.S.-India, 11
    U.S.-Italy, 148, 295
    U.S.-Japan, 267, 351, 357, 436
    U.S.-Nigeria, 296
    U.S.-Norway, 340
    U.S.-Scandinavia, 340
    U.S.-Sweden, 301, 340, 419, 478
    U.S.-U.A.R., 319
    U.S.-U.K., 112, 295, 311
    U.S.-U.K.-U.S.S.R., 283, 353, 388, 410
    U.S.-U.S.S.R., 89, 100, 158, 276, 303, 306, 313, 314, 317, 319, 332, 336, 343, 344, 347, 348, 350, 351, 353, 355, 358, 365, 366, 368, 369, 373, 375, 376, 381, 386, 389, 390, 393, 394, 413, 419, 424, 427, 434, 443, 458, 460, 462, 464, 482, 486, 493, 494
International Space Science Symposium, Fourth, 226
International Symposium on Gravitational Collapse, 481
International Symposium on Quantum Electronics, Third, 54
International Symposium on the International Geophysical Year, 307
International Technical-Scientific Meeting on Space, 257
International Telecommunication Union (ITU), 329, 416, 419, 425
International Telephone and Telegraph Corp. (IT&T), 57, 87, 247, 262
International Union of Geodesy and Geophysics, 252, 329
International Union of Theoretical and Applied Mechanics, 344
International Year of the Quiet Sun (IQSY) 98, 443
    plans for
        Germany, 322, 453
        U.S. 8, 35, 80, 474
        U.S.-U.S.S.R., 332
    research during, 227, 328
    rocket test, 410, 476
Inter-Parliamentary Union, 78
Interplanetary exploration, 153, 366
Interplanetary Monitoring Platform (Imp) (see also EXPLORER XVIII), 36, 395
    launch, 428, 449
    performance, 462
Interplanetary Monitoring Probe. See Interplanetary Monitoring Platform (Imp).
Interplanetary travel, 316
Interstellar Communication, 260
Intervision, 241
Inventions and inventors (see also Patent), 135, 174, 297, 385, 438
Inventions and Contributions Board (NASA), 281
Ion
    chamber, 107, 273
    concentration, 477, 478
    density, 87, 263, 357
    engine, 280, 451
    rocket, 437
    spectrometer, 449
    temperature, 263, 357
Ionized gas, 488

Ionosphere, 188, 279, 332, 357, 414
    communications, 241
    cosmic noise, 335
    D & E layers, measurement of, 87, 279, 477, 478
    flights in, 16, 70, 87, 111, 156, 203
    measurement, 222, 273, 290, 351, 357, 359
    noise, galactic, 335
    pollution of, 19
    radio wave propagation, 391, 409
    research, 415
    results, 92
    storms, effect on, 185
Iowa Bankers Association, 398
Iowa City Daily Press Association, 340
Iowa, State Univ. of., 15
Iowa Summer Study, 180
IQSY. See International Year of the Quiet Sun.
Iran, 33
IRE. See Institute of Radio Engineers.
Ireland, 197
Iris (rocket, Franco-Italian), 463
Iron oxide, magnetic, 428
Isakov, Adm. of the Soviet Fleet Ivan S., (Ret), 140
"Is Mercury Program Headed for Disaster?", 396
Isotopes, radioactive, 351
Israel, 105, 105
Italian National Research Council, 295
Italy
    communications, 197, 425
    Jupiter missile, 168
    San Marco Project, 295
    sodium experiments, 212
    space cooperation, 133
    space program, 68, 203, 463
IT&T. See International Telephone and Telegraph Corp.
Itokawa, Prof. H., 494
ITU. See International Telecommunication Union.
Iven C. Kincheloe Award, 359
Izvestia, 20, 32, 76, 123, 140, 193, 218, 401, 434, 449
J-2 (rocket engine), 53, 423, 451
J57-P-37A (JT3) (jet engine), 437
J-60 (jet engine), 371
J75 (jet engine), 437
Jacksonville, Ill., 224
Jackson, John E., 227
Jackson, Sen. Henry M., 107, 114, 271
Jacobs, James, 483
Jaffe, Leonard, 70, 120, 163, 265
Jahn, Gunnar, 474
James Forrestal Research Center (Princeton Univ.), 470
James, Jack Norval, 104, 365
Jane's All the World's Aircraft, 484
Japan, 447, 453
    booster, 436, 465
    defense agreement, 164
    launch, 203, 476
    probe, radio-frequency resonance, 351
    Relay, use of, 227
    rocket, 212, 307, 436
    satellite, 224, 436
    TV transmission, 447
Japanese Space Development Council, 224
Japan's Radio Research Laboratory, 351
Japan's Science and Technology Agency, 436
Japan's Space Communications Laboratory, 447

Jastrow, Dr. Robert, 149, 214, 291, 329, 468
Jato (jet-assisted takeoff), 185
Javelin (rocket). See Argo D-4.
Jefferson-Jackson Day Dinner, 203
Jefferson, President Thomas, 403
Jekyll Island, Ga., 468
Jensen, Rep. Ben F., 185
Jensen, Prof. J. Hans D., 421
Jet-assisted takeoff. See Jato.
Jet Propulsion Laboratory (JPL), 2, 5, 185, 381, 472
    awards, 104, 365
    contract, 361, 490
    Mariner 3, 310
    radar experiments, 454
    radarastronomy, 63
    radioastronomy, 6
    Ranger, Project, 87, 264, 310, 399
Jet, rocket (see also Rocket), 312
Jet Star. See C-140A.
Jicamarca, Peru, 9
Jodrell Bank Experimental Station (U.K.), 130, 185, 303, 305
Jodrell Bank Observatory (U.K.), 95, 274
Johannesburg, South Africa, 196
John Crerar Library, 172
John F. Kennedy Space Center. See Kennedy Space Center.
John Fritz Medal, 279
John J. Montgomery Award, 465
Johns Hopkins Univ., 184, 279, 308, 454
Johnson, Caldwell C., 54
Johnson, Charles B., 365
Johnson, Dr. Howard W., 136
Johnson, President Lyndon B. (see also Johnson, Vice Pres. Lyndon B.), 451, 464
    appointments, 471, 479
    appropriations, 452
    awards, 458, 468, 479
    budget, 475, 485
    Cape Canaveral, 451, 470
    Cape Kennedy, 451, 464
    economy, 475, 489
    expenditures, 489
    Federal employment, 475
    Kennedy, John F., 450, 451
    Kennedy Space Center, 451, 452
    lunar flight program, 458, 493
    military in space, 491
    peace, 482
    President, 446
    space programs, 447, 453, 458
    supersonic transport, 458, 487
Johnson, Vice President Lyndon B. (see also Johnson, President Lyndon B.), 71, 249, 340
    lunar landing, 359, 369
    President, 446
    space cooperation, 349, 404
    space impact, 156, 171, 192
    space objectives, 33, 162, 234, 362
    space pace, 132, 224, 282
    space policy, 364
    space progress, 29, 58, 106, 205, 277, 290
    space program, 400, 404
    space understanding, 104, 203
Johnston Island, 492
Johnston, Richard S., 498
Joint Meeting of Service Clubs, 405

Joint Systems Analysis Group, 136
Jonash, Edmund, 472
Jones, Rep. Robert E., 452
Journal of Aerospace Medicine, 498
Journal-American, N.Y., 213, 218
Journal of Engineering Education, 59
Journal of Geophysical Research, 39
Journal of the American Medical Association, 254
Journeyman (sounding rocket). See Argo D-8.
JP-4 (fuel), 428
JPL. See Jet Propulsion Laboratory.
Junin, Argentina, 48
Junior Chamber of Commerce, Washington, D.C., 56
Jupiter (missile), 25, 168
Jupiter (planet), 6, 16, 52, 86, 197, 414, 420, 448, 495
Jupiter C (booster), 34
Justice, Dept. of, 28
Kac, Dr. Mark, 279
Kachur, I., 218
Kagoshima, Japan, 203, 476
Kagoshima Space Center (Japan), 476
Kaiser, Edward P., 91
Kaiser Engineers, 110
Kakunin, Alexander, 171
Kalitinsky, Andrew, 232
Kamanin, Gen. Nicolai Petrovich, 355
Kamchatka (Siberia), 244
Kansas, 86
Kansas City Aero Club, 265
Kansas City, Kan., 256
Kansas City, Mo., 323, 432
Kansas State Univ., 86
Kantrowitz, Dr. Arthur R., 142
Kaplan, Dr. Carl, 279
Kaplan, Dr. Joseph, 63, 278
Kaplan, Dr. Lewis B., 151
Kappel, Frederick R., 10, 47, 98
Kapustin Yar, U.S.S.R., 259
Kardashev, N.S., 40
Karrer, S., 494
Karth, Rep. Joseph E., 324, 396, 472
Katz, Dr. Ludwig, 130
Kavanau, Dr. Lawrence L., 48, 72, 206, 269
Kazakhstan Academy of Sciences, 193
Kazakhstan, U.S.S.R., 241, 244
Keck, Capt. Louis K. (USMC), 222
Keeling, B/G Gerald F. (USAF), 38, 189
Kefauver, Sen. Estes, 45
Kehlet, Alan B., 214
Keldysh, Prof. Mstislav V., 30, 40, 42, 303, 305, 427
Kelley, Dr. Albert J., 42, 51, 62, 127, 272, 311
Kelley, Dr. Marvin J., 123
Kellog Radiation Laboratory (Cal Tech), 488
Kelly, Gen. Joseph W. (USAF), 321
Kennedy, David M., 91
Kennedy, Sen. Edward M., 54
Kennedy, President John F., 27, 288, 393, 447, 450, 451, 452, 470
    appointment, 103, 287, 311, 324, 424, 479
    assassination, 446
    award, 16, 205, 385, 468
    budget, 14, 66, 252, 253, 262, 316, 334, 442, 485
    burial, 448, 449
    communications, 39, 199, 229, 247, 265, 296, 321, 440
    defense, 116
    education, 109, 148

Kennedy, President John F.—Continued
    industry, 241, 318
    international cooperation, 349, 353
    jet transport, 321, 434
    lunar flight, manned, joint U.S.-U.S.S.R., 319, 346–349, 350–351, 353, 358, 365, 368–369, 374, 376, 381, 413, 418, 458, 462, 493
    missile, U.S.S.R., 432
    missile program, 294
    nuclear test-ban treaty, 284, 348
    patent policy, 384
    peace, 236
    program, space, 10, 67, 123–124, 157, 191, 197, 198, 201, 208, 214, 217, 230, 243, 276, 303, 310. 346, 347, 348, 349, 369, 374, 403, 409, 419, 442, 447, 458, 462
    reports, 29, 57, 82, 361
    science in government, 396, 402
    strike, 23, 28, 176
    supersonic transport, 209, 458
    technology, 43, 48, 105
    TFX, 446
    tour, 433, 442
    United Nations, 347, 440
Kennedy Space Center, NASA (see also Launch Operations Center), 486, 492
    contracts, 488
    establishment, 451, 452, 470
Kerner, Gov. Otto, 156
Kernochan, John M., 141
Kerr, Sen. Robert S., 1, 77
Kerridge, Dr. John F., 443
Kershner, Richard B., 308
Kettering, Charles, 176
Key West, Fla., 441
Khabakov, Aleksandr, 123
Khrushchev, Premier Nikita, 415
    cooperation, 349
    cosmonauts, 244, 351
    manned lunar landing, 303, 349, 386, 401, 402, 404, 409, 416, 417, 418, 422, 430
    meetings, 66, 252
    missile, 462
    nuclear test-ban treaty, 236, 264
    spacecraft, 413, 422
    space race, 422, 439
    supersonic transport, 422
Kidd, Charles V., 50
Kidnapped, 98
Kiess, C. C., 494
Kill Devil Hills, N.C., 343, 480
Kill Devil Hills Memorial Society, 480
Killian, Dr. James R., 128
Killion, George L., 91
Kimberly Mountains, Australia, 318
Kincheloe Award. See Iven C. Kincheloe Award.
Kincheloe, Mrs. Dorothy, 359
Kindelberger, James, 384
Kinzel, Dr. Augustus B., 62
Kirk, Richard L., 169
Kirtland AFB, N.M., 176
Kittenger, Capt. Joseph, Jr. (USAF), 106, 151
Kitty Hawk, 16
Kitty Hawk, N.C., 481, 483
Kiwi (nuclear reactor), 89, 491
Kleinknecht, Kenneth S., 61, 205, 465
Knaff, Dr. George M. (Col. USAF, Ret), 273
Koelle, Dr. Heinz H., 212, 412
Kokomo, Ind., 460
Kollsman Instrument Corp., 334, 392

Komsomolskaya Pravda, 40, 101
Kondrashov, Stanislav, 358
Konecci, Dr. Eugene B., 135, 163, 175, 187, 233, 246, 354
Koppenhaver, James T., 336
Korolov, Sergei P., 428
Kosciusko, Miss., 448
"Kosmodrom", 427
Kotelnikov, Vladimir A., 495
Kovach, Dr. Robert L., 44
Kovda, Prof. V., 50
Kozyrev, Nikolai, 235
Kraft, Christopher C., 200, 205, 309, 420, 465
Kraft, Joseph, 191
Kramer, S. B., 133
Krasnaya Zvezda (see also *Red Star*), 3, 44, 58, 63, 83, 111, 417, 471
Krieger, F. J., 171, 259
Krisberg, Dr. Nathan L., 279
Kronogard Rocket Range (Sweden), 295, 301, 419
Krylov, Marshal Nikolai I., 141, 434
Krypton, 37
Kubesch, Maj. Sidney (USAF), 389
Kubica, Andrew J., 278
Kuchel, Sen. Thomas, 313
Kuiper, Dr. Gerard P., 279, 492
Kurzweg, Dr. Hermann H., 22
Kusch, Dr. Polycarp, 237
Kwajalein Atoll, 220, 242, 272, 324, 327, 452, 479
Kyoto Univ. (Japan), 494
L-19 (USA aircraft), 395
Labor, Dept. of, 338
Laborite, 199
Labor Union Local 872, 393
Lagos Harbor, Nigeria, 55, 284, 285, 296, 302, 321, 339, 345, 349
LaGuardia Airport, 153
Lahm, B/G Frank Purdy (USA, Ret), 268, 483
Laird, Rep. Melvin R., 298
La Jolla, Calif., 421
Lake Erie, 357
Lakehurst, N.J., 284, 296, 304, 339, 345
Lambda-2 (sounding rocket), 476
Lambert Field, Mo., 216
Laminar flow, 146, 338
Lamont Geological Observatory, 214
LAMP. See Low Altitude Manned Penetrator.
Lancaster, Calif., 191
Land, Edwin H., 468
Landing gear, 281
Lang, David W., 289
Langley Research Center (LARS), 283, 307, 308, 344, 381, 387, 411
    aerodynamics, 94, 170
    award, 205, 235, 338, 365, 366
    contract, 16, 136, 158, 164, 230, 255, 299, 303
    experiment, 195
    hydroplaning, 476
    invention, 242, 297
    proposal, 326, 403
    research, 51, 71, 234, 297
    Space Radiation Effects Laboratory, 56, 163, 251, 303
    space station, manned, 251
    supersonic transport, 253
Langley, Samuel Pierpont, 483
Langmuir, Dr. David B., 279
Langmuir probe, 241, 273, 357
Lapp, Dr. Ralph E., 463

INDEX   557

LARC. See Langley Research Center.
Larsen, Finn J., 324
Laser (Light Amplification by Stimulated Emission of Radiation)
    application of, 330
    development, 36, 48, 92, 116, 157, 292
    patent. 362
    use of, 44, 54, 88, 101, 388, 400, 419
Las Vegas, Nev., 58
Latin America, 50, 353
Launch complex, 307, 335
Launch Operations Center (LOC) (see also Kennedy Space Center), 16, 21, 209, 268, 328
    awards, 205
    construction, 29, 319
    name change, 451, 452, 470
    organization, 65, 175, 252, 381, 398
    programs, 38
Launch Phase Simulator, 453
Launch tower, 268
Launch vehicle, 297
Launcher umbilical tower, 488
"Launching the Moon Rocket", 115
Lausanne, Switzerland, 78
Lawrence B. Sperry Award, 22
Lawrence, David, 361
Lawrence Radiation Laboratory, 141, 213, 479
Leach, Capt. Bertram G. (USA), 424
Lear-Siegler, Inc., 101, 252, 379
Le Bourget Airport (Paris), 230, 234
Lechtman, Max D., 326
"Lectures in Aerospace Medicine", 46
Lederberg, Dr. Joshua, 49, 214
Lee, Dr. William A., 61, 387
Legion of Merit (USAF), 70, 273
Legislative Drafting Research Fund, 141
Lehner, Francis E., 44
Leicester, Univ. of (U.K.), 478
Le Journal du Dimanche, 355
LEM. See Lunar excursion module.
LeMay, Gen. Curtis E. (USAF), 39, 64, 200, 209, 216, 219, 257, 283, 338, 474, 484
Lemke, George, 33
Lemnitzer, Gen. Lyman (USA), 3
Leningrad Cosmonaut Club (U.S.S.R.), 296
Leningrad Polar Institute (U.S.S.R.), 453
Lenin Prize, 154
Lenz, Hans, 242
Le Soir, 378
Levy, D. J., 438
Lewis and Clark Expedition, 86
Lewis Research Center (LRC), 99, 381, 426
    Centaur
        development, 450, 472
        test, 257, 410
    contract, 110, 143, 269, 340, 342, 379, 401
    experiment, 129, 180, 211, 231, 247, 280, 397
    invention, 297
    Plum Brook Station, 20, 95
    program, 15, 38, 95, 208, 240, 325, 336, 370, 382
    studies, 432, 441, 451
    test, 449, 451
Lexington, Mass., 51
Libby, Dr. Willard F., 214, 498
LIBERTY BELL 7, 238
Library of Congress, 442, 466

Lick Observatory, 56, 73,
Liege, Univ. of (Belgium), 429
Life
    extraterrestrial, 4, 16, 17, 25, 40, 66, 260
    protective measures, 362
    raft, 281
Life magazine, 200, 256, 343, 430
Life science (see also Bioscience)
    atmospheric pressures, 393
    bacteria, use of, 326
    bones, 46, 228
    deceleration effect, 393
    equipment, 162, 393
    experiment, 155, 176, 195, 196, 225, 227, 241, 254, 275, 411
    heart, 50, 228, 245
    impact velocity effect, 393
    isolation, 80
    nutrition, 393
    optics, 260
    oxygen environment test, 455
    program, 135, 225, 268
    radiation, 96, 204
Life support system, 275, 326, 332
Light-gas gun experiment, 497
Light, intensity measurement, 278
Lighting, ball, 211
Lilienthal, David E., 183
Lilienthal Medal, 265
Lilly, William E., 61
Lincoln Laboratory. See Massachusetts Institute of Technology.
Lindberg, Judge William J., 28
Lindbergh, Charles A., 67, 196, 204, 205
Linde Co., 184, 400
Lindsay, Dr. John, 93
Lindstrom, Robert E., 157
Lingle, Walter L., Jr., 81, 175, 323, 381, 468
Ling-Temco-Vought, Inc., 36, 179, 431
    Chance Vought Astronautics Div., 25
Linne (moon crater), 12
Linsley, John, 74, 79
Lippmann, Walter, 351, 368
Lipschutz, Dr. Michael J., 151
Liquid fluorine and liquid oxygen. See Flox.
Litchfield, Edward H., 9, 260
Litschgi, A. Byrne, 91
Little, Col. James W. (USAF), 207
Little Joe II, 514
    contract, 62, 414
    flight test, 257, 273, 426
    launch, 325
Little John (missile), 230
Little Rock AFB, Ark., 499
Little, Stafford, 183
Littlewood, William, 207
Litvin-Sedoi, M., 420
Load, 281
Lobogo magazine, 333
LOC. See Launch Operations Center.
Locke, John W., 335
Lockheed Aircraft Corp., 12, 179, 324, 481, 497
    contract, 16, 232, 276
    strike, 9, 29
Lockheed-California Co., 16, 18, 119, 190, 234, 266, 344, 429
Lockheed Electronics Co., 189
Lockheed-Georgia Co., 34

Lockheed Missile and Space Co., 95, 175, 186, 204, 438, 496
    contract, 114, 320, 397
    patent, 133
Lockheed Propulsion Co., 33, 145, 357
Lockheed Solar Observatory, 494
"Lockspray-Gold", 438
LOFTI IIA, 242, 1505
London *Times*, 63
London, U.K., 389, 422, 432, 471
    communications, 86, 235, 241, 243, 270, 389
    meetings, 31, 91, 175, 245, 248, 307, 314, 367, 455
London, Univ. of (U.K.), 37, 443
Long, Rep. Clarence D., 483
Long Construction Co., 256, 324
Long, Robert W., 256, 323
Long, Sen. Edward V., 229
Long, Sen. Russell, 174
Lookout, Program (Canada), 437
Lorenz, Prof. Edward N., 24
Los Alamos Scientific Laboratory, 244, 459
Los Angeles, Calif., 14, 86, 175, 301, 433
    meetings 31, 91, 175, 245, 248, 307, 314, 367, 455
Los Angeles Times, 73
Louisiana, 318, 335
Louisiana Purchase, 403
Louisville, Ky., 202
Louis W. Hill Space Transportation Award, 22
Louvre (Paris, France), 434
Lovelace Foundation for Medical Education and Research, 478
Lovelace, Dr. W. Randolph, II, 478
Lovell, LCdr. James A., Jr., (USN), 28
Lovell, Sir Bernard, 128, 153, 185, 274, 276, 303, 305, 317
Lovett, Robert A., 468
Low, George M., 56, 61, 366, 380, 382
Low Altitude Manned Penetrator (LAMP), 427
Lowell Observatory, 192, 234, 396, 410, 460, 495
Lowery, Harold R., 365
Lowman, Paul D., Jr., 220
Lowry, R. H., 275
LRC. See Lewis Research Center.
Lubricant, 216, 497
Lubrication, 129, 216
Luce, Clare Booth, 256
Ludlum, Dr. David M., 24
Lufthansa, 379
Lunar (see also Moon and Lunar program)
    base, 220, 275
    contract, 262, 345
    crater, 12, 116, 387
    dust, 387
    environment, 387
    exploration, 83, 148, 194, 291, 316, 392, 407, 477
    gases, 460
    gravity, 275, 387
    landing site 101
    maps, 25, 245, 486
    materials, research, 306, 396
    orbiter, 275, 326, 356, 477
    photography, 396
    probe, 165, 326, 339
    rendezvous, 355, 365
    scooter, 260
    simulation chamber, 306
    spacecraft, 165, 339
    surface, 83, 85, 170, 275, 306, 387, 402, 489

Lunar Excursion Module (LEM), 406
    cost, 392
    development, 380, 384, 387
    environmental control system, 489
    escape system, 25
    guidance and navigation system, 392
    landing engine, 219
    radar subsystem, 472
Lunar Landing Research Facility (LARC), 387
Lunar Orbiter (spacecraft), 274, 326, 356, 477, 489
Lunar program (U.S.) (see also Moon and Lunar)
    advantage, 362
    comment, 185–186, 198, 216, 237, 239, 285, 290–291, 416, 417
    criticism, 148, 191, 214, 215, 237, 240, 250, 290–291, 331, 397, 452, 468, 470
    funds, 191, 215, 231, 300, 376, 385, 392, 468, 475
    manned flight and landing, 67, 71–72, 304, 317, 408, 409, 430–431, 433–434, 447, 458, 472, 495
        restrictions, 487
    objective, 66, 275, 337, 340, 361, 461
    pace, 165, 250–251, 290, 304, 383, 417, 418
    priority, 190, 191
    support of, 214, 442
Lunar program (joint U.S.-U.S.S.R.)
    Apollo flight, 347–348
    comment, 275–276, 303, 342–343, 344–345, 351, 356, 358, 359, 365, 366, 369
    cooperations, 275, 303, 306, 317–318, 319, 336, 344–345, 347, 349–350, 351–352, 359, 365–366, 368–369, 376, 381, 390, 458, 460, 482, 493
    criticism, 347, 350, 351, 358, 386
    funds for, 384, 386, 430
    proposal, 347, 348, 350, 351, 353, 365–366, 368, 381, 413, 418, 460
    restrictions, 487
Lunar Surface Materials Conference, 211
"Lunescape climber", 260–261
LUNIK III, 130
LUNIK IV, 121, 124, 132, 135, 141, 1503
    flight, 119–120, 123, 127, 129
    performance, 130
Luster, Project (NASA), 387
Lyman-Alpha rays, 279
Lynn, Vernon L., 170
Lyon, Curtis C., 186
Lyster Award, 222
M-2 (lifting body), 70, 291, 331, 336, 459
M-82 (galaxy), 368
MA-6 (see also FRIENDSHIP 7), 238
MA-7 (see also AURORA 7), 238
MA-8 (see also SIGMA 7), 26, 238, 245, 286
MA-9 (see also FAITH 7), 257, 504, 514
    award, 340
    comments on, 204, 208, 209, 224, 227–228, 246, 354–355
    flight, 195–196, 202, 245
    plans, 43, 131–132, 200
    preparations, 1–2, 27, 29, 31, 43, 51, 84, 100, 136, 146–147, 153, 154, 186, 192, 194
    problems, 218, 371
    results, 238, 245
    simulation, 189, 192
    sketches of, 255
MA-10, 202, 208, 238
McAndrew, William R., 131–132
MacArthur, General of the Army Douglas (USA), 39
McClellan, Sen. John L., 60, 135, 212
McConnell, Joseph, 425
McCormack, James, 123
McCormack, Rep. John W., 360–361
McCracken, Curtis W., 211

McDermott, Dr. Walsh, 43
MacDill AFB, Fla., 442
McDivitt, Capt. James A. (USAF), 28
MacDonald, Dr. Gordon J. F., 214, 275
MacDonald, Capt. Robert (USAF), 209
McDonald, Rt. Rev. William J., 192
McDonnell Aircraft Corp., 328, 475, 478
    award, 365
    contract, 120
    strike, 14
McGill Univ. (Canada), 29
McGlathery, Dave M., 240
McGraw-Hill, 187
Mach 2 airliner, 270, 398
Mach 3 airliner (see also Aircraft, supersonic transport), 271, 398
McIntyre, Sen. Thomas J., 333
McKay, John, 161, 191
MacKay Trophy, 209, 302
McKee, Col. Daniel D. (USAF), 331
McKee, Gen. William F. (USAF), 221, 243
Mackenzie, R/A Hugh Stirling (RN), 23
McKie, Dr. Douglas, 37
McLaughlin, Dr. Merlyn, 232
McLellan, Charles H., 365
McMahon, Sen. Brian, 183
McMath Hulbert Observatory (Univ. of Mich.), 463
McMillan, Dr. Brockway, 21, 48, 127, 300, 302
Macmillan, Prime Minister Harold, 136, 199, 236
McMinnville, Tenn., 58
McMurdo Sound, Antarctica, 367
MacMurray College (Ill.), 224
McNamara, Robert S., 21, 31, 32, 84, 94, 114, 151, 168, 199, 252, 401, 403, 435, 466, 473, 475, 477, 492
MACS. See Medium Altitude Communications Satellite.
MacTaggart, J. A., 12
Macy, John W., Jr., 99
Madey, Jesse M., 365
Maeda, Prof. Kenechi, 494
Magellan, Ferdinand, 370, 415
"Magna Carta of Space—The Legal Lodestar, The", 172
Magna Corp., 326
Magnetic, 391, 399, 493
    disturbances, 310
    field, 95, 257, 295, 356, 380, 449
    polarity, 270, 436
Magnetohydrodynamics (MHD), 488
Magnetometer, 377
Magnetosphere, 356, 380, 414
Magnuson, Sen. Warren G., 90, 107, 114, 247, 271
Mail, delivery by rocket, 463
Maine, 1, 186, 189, 425
Maine, Univ. of, 396
Mainz, Germany, 21
Majunga, Madagascar, 488
Malagasy Republic, 488
Malaise, Daniel, 429
Malfunction-detection system, 320
Malina, Dr. Frank J., 220, 359
Malinovsky, Marshal Rodion, 66
Mallick, Donald L., 23
Malmstrom AFB, Mont., 170, 266, 399, 456
Malone, Lee B., 365
Maltby, Per, 88
Malvern, U.K., 63
Management Services, Inc., 390

Manchester, U.K., 116
Manes, T/Sgt. Charles L. (USAF), 186
Manhattan, Kan., 86
Manhattan Project, 106, 210
Man High Project (USAF), 30
Manitowoc, Wis., 3
"Manned aircraft gap", 338
Manned Flight Mission Control Center (MFMCC), 29
Manned Orbital Laboratory (MOL) (USAF), 251, 316, 458, 473, 481, 490
Manned Orbital Research and Development System (MORDS), 315
Manned orbital space station, 251, 346, 391
Manned Spacecraft Center (MSC), 16, 47, 78, 258, 366, 371, 439, 443, 492
    Apollo, 497
    astronaut, 28, 175, 221, 238, 245, 246, 273, 296, 322, 340, 392
    award, 205, 273, 281, 364, 365
    contract, 1, 25, 51, 62, 162, 169, 170, 184, 213, 219, 239, 258, 289, 292, 334, 343, 361, 366, 387, 414, 495, 497
        proposal, 333
    facilities, 30, 64, 255, 272, 282, 331, 333, 337, 375, 425, 478, 490
    Florida Operations, 420
    Gemini, 94, 292
    Mercury, 200, 309, 372, 382
    organization, 1, 70, 190, 281, 290, 380, 381, 398, 420, 446
    personnel, 125, 315, 323, 375
    report, 425, 430, 489
    research, 35
    spacecraft plans, 74, 193, 254
    studies, 9, 101, 184, 431, 498
    test, 107, 170, 178, 192, 264, 302, 423, 455
    visit, 45, 48, 94, 497
Manned space flight
    astronaut-scientist, 180, 436
    bioscience program, 411
    centrifuge facility, 271
    cooperation, 139, 366
    equipment, 140, 239
    funds, 316
    human factors, 180, 228
    importance of, 4, 67, 148, 155, 162, 164, 165, 340, 398, 409
    life support systems, 326
    lunar landing, 14, 23, 100, 101, 119, 148, 153, 190, 274, 303, 319, 323, 337, 365, 374, 384, 388, 392, 402, 408, 447, 451
    Mars landing, 14, 21, 228, 429
    medical, 45, 228
    military use, 319, 332
    plans, 49, 206, 276
    program, 71, 207, 283, 307, 309, 335, 375, 376, 381, 398, 436, 461
    radiation, 41, 96
    re-entry, 112, 228
    reliability requirements, 309
    research and development, 110
    simulator, 44
    tandem, 304, 372
    U.S.-U.S.S.R. cooperation, 319, 336, 343, 344, 346, 347, 348, 349, 350, 351, 353, 355, 358, 360, 365, 366, 368, 369, 376, 381, 386, 390, 393
    U.S.S.R., 305, 362
    Venus capture, 14, 228, 429
Manned Space Flight Meeting, Second, 157
Manpower, scientific, 110
Mansfield, Sen. Mike, 281
Man-spacecraft combination, 354
March on Washington for Jobs and Freedom, 325
Marconi wireless, 149
Marcotte, Paul G., 12
Marietta, Ga., 321

Marine Corps Museum, 361
Mariner (program), 7, 8, 58, 334
    B, 5, 422
    cost, 96
    Mars flyby, 137
    R, 5
MARINER II, 16, 96
    performance, 2, 8, 293, 341
    results, 47, 69, 144, 151, 310, 484
    Venus flyby, 479
Mariner 3, 310
Mariner, Ruth, 315
Marinin, Yuri, 36
Marion Power Shovel Co., 51
Mark I (see also Space station), 46
Mark II (see also Space station), 46
Marks, Leonard, 91
Marquardt Corp., 453
Mars, 150
    atmospheric conditions, 331, 492, 494
    contamination of, 339, 422
    exploration, 232
    flight, criticism of, 331
    flyby mission, 483, 484
    landing mission, 339, 361, 366, 494
    life on, 4, 49, 74, 138, 161, 228, 246, 294, 494
    manned flight to, 21, 52, 56, 184, 189, 193, 220, 232, 237, 239, 292, 317, 336, 354, 370, 420, 429, 453, 471
    study of, 136, 199, 246, 317, 320
    surface of, 63, 80, 179, 193, 228, 454, 494
    unmanned flight to, 8, 248, 259, 310, 422, 433, 441, 498
MARS I
    flight, 9, 35, 40, 63, 76, 96, 411, 484
    results, 58, 102, 199, 227, 248
Mars excursion module (MEM), 239, 366
Marshak, Dr. Robert E., 172
Marshall, C. H., 192
Marshall, Mrs. George C., 116
Marshall Space Flight Center (MSFC), 45, 110, 116, 143, 157, 188, 240, 313, 320, 384, 400, 412
    award, 5, 212, 214
    contract, 258, 266, 271, 292, 379, 423
    facilities, 115, 260, 421, 479
    meeting, 16
    organization, 115, 243, 256, 327, 381, 398
    procurement, 430
    programs, 115, 370, 433
    research, 5, 71
    studies, 184, 232, 245, 497
    test, 71, 91, 132, 464
    visit, 123, 259, 499
Martin Co., 255, 269, 368, 401, 487
Martin-Marietta Corp., 232, 466
    contract, 68, 162, 164
    research, 330, 488
Martynov, Prof. A., 74
Marvingt, Marie, 479
Marx, Karl, 415
Maryland, 368
Maryland, Univ. of., 80, 143, 164, 234
Marymount College (N.Y.), 34
Maschal, W. E., 386
Mason, Dr. Edward A., 65
Mason-Rust Co., 387
Massachusetts, 54, 65, 230, 381, 462
Massachusetts Associated Industries, 400

Massachusetts Institute of Technology (MIT), 3, 51, 53, 79, 105, 136, 170, 237, 323, 328, 359, 388, 392, 421, 430, 479
    Center for Space Research, 3
    Instrumentation Laboratory, 350, 359
    Lincoln Laboratory, 170, 180, 193, 341
    School of Science, 430
Massachusetts Turnpike Authority, 106
Massachusetts, Univ. of, 166
Massena, N.Y., 213
Mass free-fall record, 482
Materials, space, 338, 342, 390, 438
    development, 104, 155, 305, 419
    importance of, 226
    re-entry, 280
Mathematical, 287, 368
Mathematical Association of America, 27
Mathematicians, 27
Mathews, Charles W., 420
Mathews, T. A., 88
MATS. See Military Air Transport Service.
Matter, chemical structure of, 397
Maus, Hans, 328
Max Planck Institute for Carbon Research (Mulheim, Germany), 422
Maxwell AFB, Ala., 279
Maxwell, James Clerk, 149
Mayer, Prof. (Mrs.) Maria Goeppert, 421
Mayer, Max, 242
Maynard, Owen E., 54
Mayo-Wells, Wilfred J., 104
Maytag, L. B., Jr., 270
Medal for Exceptional Scientific Achievement (NASA), 365
Medal for Outstanding Leadership (NASA), 38, 205, 240, 365
Medal of Honor (N.Y.C.), 208
Medal of Merit (U.S.), 106
Medicine, aerospace, 41, 45, 273, 442
Mediterranean, 111
Medium Altitude Communications Satellite (MACS) (military), 403
"Megabuck barrier", 470
Meisinger Award, 24
Melbourne, Australia, 164
Melekess, U.S.S.R., 220
MEM. See Mars excursion module.
Memphis, Tenn., 479
Mendel Medal, 222
Mental illness, 450
Mentzer, W. C., 194
Merck, Sharp and Dohme Research Laboratories, 356
Mercury-Atlas, 238
Mercury Control Center, 30
Mercury (planet), 2, 42, 179, 228, 441, 454
Mercury, Project, 174, 194, 254, 281, 353, 366, 433, 492
    achievement, 55, 61, 72, 155, 195, 196, 245, 309, 341, 354, 484
    astronaut, 257, 296, 359, 385, 389
    award, 260, 465
    capsule, 58, 139, 218, 366
    completion, 277, 284, 498
    computer, 116
    history, 205, 252, 371
    industry workmanship, 371, 372, 374, 376, 378, 380, 382
    launch, 196, 245-46
    personnel, 393
    program, 1, 15, 29, 31, 77, 78, 200, 208, 226, 316, 396
    progress, 238
    results, 202, 245, 303, 332, 340, 364
    simulation, 186, 189, 192
    tracking, 84, 443

Mercury, Project, Summary Report Conference, 364, 371, 372
Mercury-Redstone suborbital flights, 238, 365
Merritt Island Launch Area (MILA) (NASA), 21, 221, 337
    contract, 209, 268, 371
    construction, 29, 319, 433
Meteor, 106, 314, 426
Meteorite, 138, 151, 331, 335
    composition of, 443
    dust, 388
    impact, 39, 387
Meteoroid
    detection, 43, 54, 120, 195
    satellite impact, 153, 399
    spacecraft protection, 340, 454
Meteoroid detection satellite. See Interplanetary Monitoring Platform.
Meteorological Rocket Network, 20, 227
Meteorology, 314, 317, 477
Metric system, 37
"Metroplanes", 434
Metropolitan Board of Trade (Washington), 53
Meudon, France, 396
Mexico, 372, 391
Mexico, Gulf of, 441
Meyer, Andre J., Jr., 77
Meyer, Dr. Peter. 270
MFMCC. See Manned Flight Mission Control Center.
MHD. See Magnetohydrodynamics.
Miami, Fla., 323, 432, 465
Miami Beach, Fla., 270, 282
Michel, Edward L., 498
Michel, Dr. F. Curtis (Capt., USAFR), 267
Michels, Charles, 255
Michigan Industry-University Research Conference, 398
Michigan State Univ., 246
Michigan. Univ. of, 235, 269, 436, 449
Michoud Operations (NASA), 318, 335, 387, 478
Micrometeorite penetration, 240
Micrometeoroid, 241, 489
Microwave spectrum, 419
Midas (satellite, USAF), 242, 399
Midway Island, 195
Midwest Regional Conference of Society of American Military Engineers, 385
Mid-Winter Conference on Science Teaching, 468
Miezkuc, Bernard, 327
MiG-21, 8, 459
Mikhailov, A. A., 56
Mikhaylov, G., 218
MILA. See Merritt Island Launch Area (NASA).
Milan, Italy, 7, 422
Military Air Transport Service (MATS), 321, 398
Military Factory 333 (Cairo, U.A.R.), 179
Military Orbital Development Systems. See MODS.
Military pay increase, 485
Military research and development, 461
Military retired pay, 485
Military Sea Transportation Service-NASA, 451, 486
Military Strategy, 232
Milky Way, 6, 25, 117, 160, 260, 432
Miller, Mrs. Betty, 279
Miller, Rep. George P., 43, 56, 108, 245, 272, 277, 322, 323, 377, 424
    Centaur, 457
    education, scientific, 236, 274
    facilities, 127
    science, 424
Miller, Sen. Jack, 288
Miller, Prof. Maynard M., 246

Millikan, W. W., 266
Millsaps, Dr. Knox, 9
Miner, Marcia S., 104, 188
Miniaturization progress, 363
Ministry of Aviation (U.K.), 63
Minitrack, 32, 346
Minneapolis, Minn., 30, 161
Minnesota Museum of Natural History, 74
Minnesota, Univ. of, 30, 50, 177, 303, 474
Minow, Newton N., 21, 154, 247
Minuteman, 138, 170, 266, 298, 395, 399, 459, 499
    operational, 435
        R&D, 5, 25, 62, 188, 202, 213, 230, 257, 258, 263, 266, 271, 276, 283, 286, 298, 326, 416, 424, 431, 451, 453, 479, 486, 489
Miracle Mile Association, 91
Mirage IV (fighter aircraft), 54
Mirny, 453
Mironov, Alexandre A., 411
Missile, 164, 274, 318, 367, 399, 483, 492
Missile and Space Systems Div. (Douglas Aircraft Co.), 98
Missile, air-to-air, 8, 16, 459
Missile, air-to-surface (CLAM), 466
Missile, antimissile, 266
    development, 310, 380
    program, 18, 294, 311
    firing, 415, 479
    U.S.S.R., 423, 432, 470
Missile, antisatellite, 138
Missile, antitank, 8
Missile, ballistic, 437
Missile, intercontinental ballistic (ICBM), 42, 43, 103, 138, 170, 310, 359, 420, 422, 431, 435, 452
    U.S.S.R., 434, 462
Missile warning system. See Red Mill Project.
Missiles and Rockets, 5, 12, 49, 59, 89, 90, 184, 193, 216, 237, 252, 298, 307, 317, 319, 324, 350, 355, 361, 387, 395, 396, 437, 448, 458, 473, 490
Mississippi Test Operations (MTO) (NASA), 115, 142, 175, 243, 260, 400, 411, 430, 486, 490, 497
Missoula, Mont., 407
Missouri, Univ of, 214
"Mr. Space", 447
MIT. See Massachusetts Institute of Technology.
MIT Center for Space Research, 3
MIT Corporation, 479
MIT Instrumentation Laboratory, 350
MIT Laboratory for Nuclear Science, 79
MIT Lincoln Laboratory, 170, 180, 193, 341
MMM. See Multi-Mission Module.
MODS (Military Orbital Development Systems), 43, 168
Moeckel, Wolfgang E., 232
Moffett, Alan T., 88
Mohave Desert, 447
MOL. See Manned Orbital Laboratory.
Molecular bond, 385
Molecule, 332
Mona Lisa, 7
Monkey, 411
Monnet, Jean, 270
Monroney, Sen. A. S., 407, 483
Montana, 403, 456
Montana State Univ., 407
Montgolfier Award, 265
Montgomery, J. B., 338
Montreux, Switzerland, 185
Moody, Alton B., 126, 197

Moon (see also Lunar)
    atmosphere, 495
    contamination of, 175
    craters, 12, 387
    dust, 144, 211, 306, 387
    effects of, 481
    exploration of, 148, 150, 275, 316, 331, 439, 495
    gravity, 275, 380
    impacts on, 387
    infrared light, 448
    landing, manned, 45, 67, 101, 194, 198, 203, 210, 215, 217, 220, 225, 245, 275, 276, 300, 304, 329, 337, 354, 356, 364, 374, 376, 378, 385, 401, 413, 414, 418, 422, 458, 469, 471, 474, 477, 480, 496
        U.S. program
            comment, 148, 185, 190, 214, 215, 231, 240, 250, 290, 351, 383, 397, 416, 430
            objective, 267, 340, 361, 452, 453, 461
            priority, 190, 231
    laser impact, 419
    life on, 4, 165
    maps, 192, 245
    military use of, 3, 140
    origin, 104
    photos, 18, 129, 326
    shape, 19
    soft landing, 83, 305
    space station, 481
    surface of, 12, 44, 83, 85, 88, 91, 170, 220, 228, 245, 275, 326, 410, 433, 460, 479, 480
    unmanned flight to, 119, 130, 310, 392, 433, 440, 498
    volcanic processes, 235
    water on, 131, 425, 495
MORDS. See Manned Orbital Research and Development System.
Moreton, Dr. G. E., 494
Morrow, Dr. W. E., Jr., 22, 341
Morse, Sen. Wayne, 92
Moscow-New York air agreement, 492
Moscow Mining Institute, 11
Moscow, U.S.S.R., 164, 202, 205, 206, 220, 236, 251, 252, 254, 283, 324, 376, 394, 423, 428, 432, 453, 471
    air routes, 11
    communications, 129, 327
    meeting, 417, 492
Moss, Sen. Frank E., 353
Moss, John E., 211
Motion generator, 263
Motorola, Inc., 490
Motz, Prof. Lloyd, 25
Mt. Everest, 246
Mt. Palomar Observatory, 144, 160, 229, 368
Mt. Wilson Observatory, 95, 116, 160, 229
Moyer, Xopher W., 365
Mrazek, Dr. William A., 365
MSC. See Manned Spacecraft Center.
MSFC. See Marshall Space Flight Center.
MTO. See Mississippi Test Operations.
"Mu" (Japanese rocket), 494
Muchea, Australia, 464
Mud Lake, Nev., 177
Mueller, Dr. George E., 281, 381, 382, 406, 478
    Apollo, 496
    lunar program, 348, 392
    space program, 423
Mulheim, Germany, 422
Multer, Rep. A. J., 207

Multi-Mission Module (MMM), 410
Multivator, 49
Mundelein College (Ill.), 230
Municipal aerospace park, 366
MUR (Mock-Up Reactor). See Plum Brook Station.
Murphy, Dr. Thomas P., 278
Murray, Dr. Bruce, 144
Mururoa Atoll, 176
Muscle Shoals, Ala., 201
Museum, 61
"Museums Without Walls", 434
Mylar, 304
Myoelectric, 376
"Myopia in Space", 260
NAA. See National Aeronautic Association.
NAA. See North American Aviation, Inc.
NACA. See National Advisory Committee for Aeronautics.
Nags Head, N.C., 480
Nagy, Dr. Bartholomew, 444
Nairobi, Kenya, 164
NAMAC. See National Amateur Missile Analysis Center.
NANA. See North American Newspaper Alliance.
Nance, Hershel M., 365
Nancy, France, 479
NAS. See National Academy of Sciences.
NASA. See National Aeronautics and Space Administration.
NASA-DOD Contractor Performance Rating System, 49
NASA-DOD Gemini Program Planning Board, 102
NASA Fifth Anniversary Banquet, 364
NASA Fifth Anniversary Honor Awards Ceremony, 364
NASA Industrial Applications Advisory Committee, 62
NASA-Industry Program Plans Conference, 51, 52, 54
NASA Intercenter Conference on Plasma Physics, 459
NASA Management Advisory Committee on Manned Space Flight, 123
NASA-Military Sea Transportation Service, 451, 486
NASA Research Advisory Committee on Missile and Space Vehicle Structures, 361
NASA Space Vehicle Review Board, 16
NASA-USAF Gemini Program Planning Board, 86, 102, 332
NASC. See National Aeronautics and Space Council.
Nashville, Tenn., 47, 201
Nassau, B.W.I., 23, 136
Nasser, President Gamal A., 274
Natal, Brazil, 177
National Academy of Engineering, 179
National Academy of Neurology, 161
National Academy of Sciences (NAS), 3, 80, 179, 226, 238, 252, 310, 372, 395, 402
    award, 156
    code, space, 1
    fellowship, 332
    hearing, 388
    meeting, 73, 161, 162, 168, 307, 396, 397
    Space Science Board, 275, 339
    studies, 50, 435, 442
National Advisory Committee for Aeronautics (NACA), 73, 240, 253, 373, 411, 479, 481
    history, 197, 394
    research, 19, 207, 333
National Aero Clubs, 31
National Aeronautic Association (NAA), 32, 184, 231, 265, 302, 411, 484
National Aeronautics and Space Act, 108, 174, 447
National Aeronautics and Space Administration (NASA), 294, 382, 462, 493
    accomplishments, 39, 352, 373
    aeronautics, 19, 94, 207, 283
    agreements, 11, 21, 89, 102, 127, 177, 212, 314, 391, 411, 451, 485
    anniversary, 364, 373

National Aeronautics and Space Administration—Continued
   Apollo, 471, 472, 497
   appointments, 40, 62, 70, 278, 281, 336, 398
   appropriations, 16, 81, 82, 83, 90, 98, 112, 172, 210, 215, 222, 273, 283, 286, 293, 300, 301, 302, 316, 322, 323, 325, 334, 351, 373, 374, 376, 384, 391, 392, 407, 430, 440, 464, 474
   astronauts, 259, 269, 278, 340, 361, 389, 392, 478
   awards, 17, 56, 90, 104, 136, 186, 191, 245, 278, 334, 338, 365, 379, 465
   bioscience program, 411
   budget, 231, 254, 255, 258, 262, 269, 283, 323, 383, 395, 421, 463
   comment re, 300, 305
   conference, 54, 84, 131, 136, 283, 368, 399, 438
   consultant, 268, 478
   contracts, 49, 59, 73, 99, 153, 178, 258, 289, 305, 435, 485, 490, 492, 498
      boosters, 68, 76, 197, 271, 305, 475, 476
         engine, 221, 320
      facilities, 51, 56, 136, 175, 213
      fuel, 277, 400
      ground systems, 27, 30, 105, 272, 345, 379, 397
      management, 87, 142, 210
      spacecraft
         manned, 90, 120, 343, 392
         unmanned, 43, 52, 141, 235, 264, 304, 320, 489
      space equipment, 229, 272, 361, 392
      studies, 132, 143, 219, 230, 252, 262, 263, 281, 345
   criticism, 288, 370, 380, 421
   DOD–NASA
      applications, 87, 300
      Atlas-Agena, 252, 334
      contracting, 49
      Gemini, 12, 102, 199, 277, 319
      space station, 277, 315, 391, 400
      Titan III, 436
      X–20 (Dyna-Soar), 319, 474
   educational program, 216, 255, 290, 402
   equipment, electronic data processing, 277, 369
   exhibit program, 315, 357
   facilities
      communication system, 371
      construction, 98, 175, 337, 387
      liquid hydrogen, 400
      location, 47, 106, 118, 127, 297, 302, 394, 469, 483, 484, 486
      tracking, 402
   funds, 222, 316
   grant 143, 236, 272, 278, 362, 396, 486
   international programs, 152, 211, 295, 306
      Australia, 464
      Canada, 321
      France, 391, 409, 487, 497
      Japan, 351, 357
      West Germany, 454
   laser, use of, 400
   launches, 182, 313, 318, 325, 352, 399, 449, 497, 498
      manned, 364
      satellite, 54, 184, 357, 436, 459, 489
      sounding rocket, 14, 30, 50, 58, 60, 63, 70, 71, 87, 88, 95, 107, 110, 134, 168, 184, 186, 188, 212, 241, 247, 263, 268, 295, 336, 351, 366, 388, 391, 409, 410, 429, 434, 469
      test, 110, 132, 280, 476
   lunar orbiter, 275, 489
   management, 272, 289, 390, 424, 486, 498
   meeting, 356, 462
   organization, 34, 61, 65, 81, 102, 175, 283, 287, 291, 324, 380, 381, 401, 410, 414, 424, 452, 455, 482, 496
   patent, 59, 86, 92, 214, 237, 428

National Aeronautics and Space Administration—Continued
   personnel
      legislation, 298
      recruitment, 163, 259, 463
      requirements, 99
      risk, 140, 444
      training course, 204
   procurement, 175, 430, 493
   program, interplanetary, 5, 8, 90, 428, 477
   program, solar, 8
   programs, study
      lunar base, 178, 262
      lunar program, 290
      Luster, Project, 387
      space environment simulation, 385
   report, 41, 68, 287, 296, 323, 341, 353, 356, 376, 390, 391, 392, 404, 406, 433, 439, 463, 465, 476, 490, 493
   research, electronic, 312
   research philosophy, 19
   research program, 92, 109, 337, 436, 447, 448, 486
   rocket, nuclear, 153
   satellite program, 45, 395
   Satellite Situation Report, 2, 14, 210, 243, 267
   scientists and engineers, 210, 436
   space cooperation, 198
   spacecraft, 173, 339, 355
   sponsorship, 76, 404, 463
   studies, 3, 271, 278, 315, 379, 448
   support of, 197, 290
   training program, graduate, 372, 444
   universities, 3, 210, 274, 293, 372, 456
   USAF-NASA, 78, 129, 131, 145, 168, 428, 455, 491
      ECHO II, 426
      Gemini, Project, 47, 332, 487
      MOL, 481
   USN-NASA, 216, 428, 454
National Aeronautics and Space Council (NASC), 13, 318, 414, 471
   budget, 20, 249, 407
   chairman, 106, 277, 291, 359, 362, 414, 471
   manned lunar landing, 267, 363
   meeting, 277, 290
   space competition, 259, 337, 359, 402
   space program, 373, 441
National Aerospace Education Council, 270
National Airlines, 270
National Air Museum, 16, 266, 277
National Amateur Missile Analysis Center (NAMAC), 155
National Association of Business Economists, 354
National Association of Counties, 291
National Association of Manufacturers' 68th Congress of American Industry, 466, 468
National Aviation Facilities Experimental Center (FAA), 94
National Aviation Hall of Fame, 484
National Broadcasting Co. (NBC), 98, 131, 186, 342, 418, 434, 447
National Bureau of Standards, 9, 35, 106, 212, 234, 331
National Business Publications, 47
National Capital Section of AIAA, 343
National Center for Atmospheric Research, 323, 328
National Center for Space Studies (CNES) (France), 19, 89, 350, 362, 391
National Commission for Scientific and Technical Investigation, 425
National Conference on Citizenship, 342
National Conference on Peaceful Uses of Space, Third, 84
National Defense Education Act, 236
National Editorial Association, 306

National Education Association (NEA), 118, 463
National Engineering Co., 48
National Engineers Week, 48
National Federation of Business and Professional Women's Clubs, 277
National Flight Forum Symposium, 197
National Frank G. Brewer Trophy, 231
National Gallery of Art (Washington, D.C.), 434
National Geographic Society, 226, 257, 280, 317
National Industrial Conference Board, 316
National Information Center, 154
National Institutes of Health (NIH), 50, 212, 272
National Inventors Council, 385
National Launch Vehicle program, 407
National Medal of Science, 185, 479
National Meteorological Center, 118
National Observer, The, 49, 155
National Oceanographic Data Center (NODC) (Washington, D.C.), 345
National Office of Space Research (French Army), 335
National Orbital Space Station (NOSS), 400
National Park Service, 483
National Pilots Association, 465
National Radio Astronomy Observatory, 131
National Research Council, 50, 80, 307
National Research Data Processing and Information Retrieval Center, 239
National Rocket Club, 13, 104, 188, 246, 274, 319, 343, 377, 439
National Rocket Club Award, 104
National Science Foundation (NSF), 6, 103, 109, 253, 307, 322, 337, 448
    data systems, 234
    funds, 15, 172, 377, 421, 435
    grant, 236, 474
    personnel, 258
    project, 76, 332, 442, 448, 496
National Science Teachers Association, 171, 368, 404, 463
National Science Youth Congress, 404
National Scientific Balloon Flight Station, 76
National Security Industrial Association, 92, 141
National Security Industrial Forum, 8
National Security Political, Military, and Economic Strategy in the Decade Ahead, 289
National Severe Storms Project, 201
National Society of Aerospace Professionals, San Diego Chapter, 465
National Space Research Committee, of the Royal Society of New Zealand, 209
National Space Club, 483
"National survival industry", 379
National Transporation Institute, 31
National Weather Council, 145
National Weather Satellite Center, 53, 247, 330, 341, 344
National Youth Science Congress, 463
National Zoological Park (Washington, D.C.), 128
Nationalist China, 194
Nation's Engineering Research Needs, 1965–85, The, 36
Nativitas (meteorite), 151
NATO (North Atlantic Treaty Organization), 25, 185, 252, 348, 459
Natta, Prof. Guilo, 422
Natural History, 74
Nature, 56, 235
Naugle, John E., 182
Naval Gun Factory, 154
Naval Missile Facility (Pt. Arguello, Calif.), 441
Naval Ordnance Test Station, China Lake, Calif., 264
Naval Research Laboratory (NRL), 6, 86, 493
Naval Weather Service, 293
Navigation, 310
Nazi, 205
NBC. See National Broadcasting Co.

NBS. See National Bureau of Standards.
NEA. See National Education Association.
Nebular spectra experiment, 384
Neiss, M/G O. K. (U.S. Surgeon General), 273
Nelson P. Jackson Aerospace Award, 104
NEO. See North Eastern Office (NASA)
Neon gas, 292
Neporozhni, Pyotr, 52
Nepszabadsag, 470
Neptune (planet), 6, 432
Nerva (Nuclear Engine for Rocket Vehicle Application), 115, 491
Netherlands, 33, 133, 197, 389
Nettles, Cary, 472
Neuman, Temple, 366
Neutrino, 154
Neutron, 149, 276, 285, 455
Nevada, 134
Nevada Test Site, 393, 476
New Delhi, India, 164, 246, 429, 492
New England, 465
New Hampshire, Univ. of, 11, 380
New Haven, Conn., 125
New Jersey, 1, 55, 97, 143, 368, 425
New Jersey Society of Professional Engineers, 178
New Mexico, 65, 74
New Mexico State Univ., 478
New Orleans, La., 478
New South Wales, Australia, 175
New York, 3, 106, 311, 495
New York Academy of Sciences, 443
New York *Herald Tribune*, 290, 386
New York International Airport, 153
New York *Journal-American*, 218
New York, N.Y., 164, 208
    air route, 20, 153, 196, 204, 492
    communications, 86, 106, 137, 186, 241, 389
    meeting, 19, 21, 26, 45, 58, 98, 124, 152, 194, 204, 379, 390, 395, 441, 442, 443, 456, 461, 466, 489
New York Society of Security Analysts, 379
New York, State Univ. of., 445
New York Times, 19, 48, 128, 129, 175, 187, 214, 216, 244, 258, 275, 285, 288, 296, 299, 303, 327, 332, 346, 374, 375, 378, 380, 386, 394, 402, 405, 418, 428, 430, 462, 475, 486, 492
New York Times Magazine, 153, 427
New York Univ., 408, 414
New York World's Fair, 375
New Zealand, 209, 453
Newark Airport, N.J., 153
Newell, Dr. Homer E., 76, 158, 381
    budget, 111, 111
    moon, 85, 291
    space, peaceful uses of, 219
    space programs, scientific, 83, 152, 159, 200, 307, 329, 334, 411, 477
Newfoundland, 247
Newman, James B., 365
Newmann, Dr. William F., 228
Newport News, Va., 56, 63, 136, 242, 255, 288, 309
Newton, Sir Isaac, 11
Ney, Dr. Edward P., 303
NF–101 (Voodoo), 19
NF–104 (Starfighter), 426, 475
Niagara Falls, Ontario, 12
Nice, France, 32
Nicholson, Dr. Seth B., 264
Nicks, Oran W., 52

Nicolet, Prof. Marcel 262, 367
Nields, James, 480
Nigeria, 33, 296, 321
NIH. See National Institutes of Health.
Nike (missile), 87
Nike-Ajax (missile), 88
Nike-Apache (sounding rocket), 11
    launch, 60, 70, 87, 110, 134, 186, 203, 206, 209, 212, 241, 273, 278, 336, 382, 383, 443, 449, 477
Nike-Asp (sounding rocket)
    launch, 58
Nike-Cajun (sounding rocket)
    launch, 14, 15, 60, 71, 88, 241, 295, 301, 449
Nike-Hercules (missile), 230
Nike X (missile), 18, 161, 302, 380, 415
Nike-Zeus (missile), 14, 57, 62, 138, 147, 161, 192, 220, 230, 242, 263, 266, 272, 285, 302, 310, 324, 327, 415, 452, 479
Nikolayev, Maj. Andrian G., 305
    appearance, 88
    condition of, 20, 45, 355
    marriage, 376, 417
Nikolayeva-Tereshkova, Valentina V., 429, 469
Nimbus, Project, 77, 125, 131, 371, 488
Nimbus (satellite), 248, 306, 345, 436, 488
Nina, 226
Nirenberg, Dr. Marshall, 65
"Nitrogen and oxygen indicator-controller assembly", 460
Nitrogen dioxide gas, 494
Nobel Prize, 185, 444
    Chemistry, 422
    Peace, 385, 397, 474, 475
    Physics, 421, 422
NODC. See National Oceanographic Data Center.
Noise, 268, 281, 414, 453, 490
Nomenclature, 35
NORAD. See North American Defense Command.
Nordberg, Dr. William, 117
Nored, Donald L., 143
Norfolk, Va., 90
North America, 42, 460, 465
North American Aviation, Inc. (NAA), 53, 77, 384, 496
    contract, 114, 119, 221, 251, 342, 379, 387, 404, 427
    program, Apollo, 382, 492
    Space and Information Systems Div., 92, 126, 169, 260, 366, 378, 497
North American Defense Command (NORAD), 210, 428, 469
North American Newspaper Alliance (NANA), 379
North Atlantic, 217
North Atlantic Treaty Organization. See NATO.
North Bergen, N.J., 215
North Carolina, 106, 483
North Carolina Citizens Association, 111
North Carolina Press Association, 18
North Dakota, 456
North Eastern Office (NEO) (NASA), 484
North Pole, 231
Northrop Corp., 87, 146, 178, 183, 203, 266, 320
Northrop, John K., 484
Northrop-Nortronics, 453
Northrop Space Laboratories, 460
Northrop Ventura, 178
North, Warren J., 156
Northwestern Univ., 243, 327
Norton AFB, Calif., 176, 263, 499
Norway, 133, 197, 340
Norwood, Joseph, Jr., 245

NOSS. See National Orbital Space Station.
Notre Dame, Univ. of, 469
Nova (booster), 67, 157, 188, 266, 365
Nowitsky, A. M., 175, 204
Noyes, W. Albert, 26
Nozzle, gimbaled integral (GIN). See Gimbaled integral nozzle.
NRDS. See Nuclear Reactor Development Station.
NRL. See Naval Research Laboratory.
NSF. See National Science Foundation.
Nuclear, 328, 422, 441, 459
    detonation detection systems, 286, 299, 389
    detonations, high-altitude, 89, 285, 286, 308
    economic impact, 438, 441
    forces, U.S., 435
        fleet, 459
    generator, 345, 357
    industries, 318
    peaceful uses, 143, 471
    propulsion, 103, 115, 137, 266, 370, 401, 420
    reactor, 52, 89, 137, 140, 168, 343, 367
    rocket, 153, 486
    test-ban treaty, 58, 264, 283, 284, 298, 322, 326, 337, 338, 347, 348, 350, 365, 367, 368, 383, 393, 394, 409, 420, 440, 474
    test, 312, 385, 402, 487
    test stand, rocket, 269
    underground test, 353, 393, 402
    war, 379, 397, 462
    warhead, 78, 105, 285, 344, 486
    weapon, 333, 351, 365, 368, 375, 381, 388, 389, 462, 463, 474, 475, 482
Nuclear powered ballistic missile submarine. See SSBN.
Nuclear Reactor Development Station (NRDS), 32
Nutley, N. J., 3, 5, 15, 87, 339, 425
Oakland, Calif., 107, 110, 113, 193
Oakland Corp., 260
Oak Ridge, Tenn., 56
Oao. See Orbiting Astronomical Observatory.
OAR Research Review, 306, 377
Oberth, Prof. Hermann, 7
Observatories, 274, 402, 494
Ocean, 396
Oceanographic data, 345
Odlum, Floyd B., 484
Office of Naval Research (ONR), 76, 435, 448
Office of Scientific Research and Development (OSRD), 430, 479
Ogo. See Orbiting Geophysical Observatory.
(OH) hydroxyl radicals, 432
OH–23G (helicopter), 424
O'Hara, Rep. Barrett, 253
O'Hare International Airport, 105
Ohio, 65
Ohio State Univ. 190
O'Keefe, Dr. John A., 104, 386
Oklahoma City, Okla., 118, 203
Oklahoma, Univ. of, 247, 249, 250, 409
Olin Mathieson Chemical Corp., 311
Olivine, 131
Olympic Games, 227
Olympics of the Air, 32
Omaha, Neb., 385
O'Neill, Eugene S., 2
ONR. See Office of Naval Research.
"Open skies" policy, 334
"Operation Halo" (High Altitude/Low Opening), 482
Ophiuchus (constellation), 145
Oppenheimer, Dr. J. Robert, 130, 458
Optical communications, 44

Optical systems, 396
Optical tracking, 44
Optics, 355
Orbital scientific research stations, 427
Orbiting Astronomical Observatory (Oao), 12, 248, 326, 436, 464
Orbiting Geophysical Observatory (Ogo), 12, 326, 464, 488
Orbiting Solar Observatory (Oso), 36, 298, 341, 395, 464
Order of Lenin, 130, 251
Order of Railroad Telegraphers, 337
Ordnance, 259
Organic molecular compounds, 331
"Origins of Space Telemetry", 104
Orion, Project (ARPA), 103, 367, 420
Orlando, Fla., 330, 462
Orleans, France, 54
Osborne, Stanley de J., 311, 487
Oslo, Norway, 474
Oslo, Univ. of (Norway), 475
Osmium-186, 155
Osmium-187, 155
Oso. See Orbiting Solar Observatory.
OSO I, 85, 93, 182, 298
Oso-B 250, 258
OSRD. See Office of Scientific Research and Development.
Ostrander, M/G Don R. (USAF), 27, 279
Otis AFB, Mass., 230
Outstanding Unit Award (USAF), 287, 443
Owens Valley, Calif., 95
Oxygen
    atomic, 429
    environment tests, 455, 482, 498
    in stratosphere, 323
    liquid (lox), 68, 143
    supply, 326, 367, 498
Oyster Point, Newport News, Va., 164, 303
Ozone, 14, 143, 342, 419
P-1127 (aircraft), 48
Pace, Dr. Nello, 268
PACE. See Pre-Flight Acceptance Checkout Equipment.
Pacific, Central, 202
Pacific Launch Operations Office (PLOO), 381
Pacific Missile Range (PMR), 57, 105, 163, 188, 193, 296, 479, 481, 486
    launch, 57, 239, 266, 291, 299, 323, 326, 327, 334, 344, 351, 427, 486, 487, 489
    test, sounding rocket, 410
    transfer of, 307
Pacific Ocean, 57, 176, 193, 195, 220, 224, 242, 244, 272, 279, 285, 296, 312, 314, 327, 344, 426, 434, 447, 452, 460, 479, 492
Pacific Science Center, 448
Pacific Science Center Foundation, 448
Packard, Robert F., 249
Pakistan, 212
Palestine, Tex., 65, 76, 80, 314, 323, 328, 331, 448
Palewski, Gaston, 8, 171
Palm Beach, Fla., 433
Palomar Observatory, 95, 116, 264
Pan American World Airways, 229, 253
Panama, 172
Paoli, Pa., 352
Papago Indian Reservation, Ariz., 6
Papell, S. Stephen, 428
Parachute, 77, 87, 170, 178, 477
    jumping, 218, 267, 482
    landing, 268, 340, 492
    recovery system, 302, 332
    test, 35, 49, 74, 192, 397
Parade magazine (*Washington Post*), 179, 359, 462

Paraglider, 49, 214
Parin, Vasily V., 130, 375
Paris, France, 434, 471
    air route, 196, 204, 230
    communications, 86, 106, 137
    meeting, 54, 355, 362, 367, 416, 469
Paris Observatory (France), 493
Parks Air College (Univ. of St. Louis), 281
Parks, Robert J., 47, 165, 365
Pasadena, Calif., 6, 488
Paso Robles, Calif., 302
Pastore, Sen. John O., 59
Patent (see also Inventions), 36, 77, 214, 260, 296
    legislation, 174
    policies, 86, 92, 93, 135, 207, 384–385
    regulations, 45, 59
Paterson, N.J., 114, 151
Patrick AFB, Fla., 78, 263
Patten, Rep. Richard J., 35
Patterson, Eugene, 452
Pauling, Dr. Linus C., 385, 397, 418, 474, 475
Paul Smith Construction Co., 488
Paul Tissandier Diploma (FAI), 266
Pawtucket, R.I., 289
PCH–1 (hydrofoil), 419
Peace Corps, 450
Peace, world, 40, 236, 482
Pearson, Drew, 216
Pecos, Tex., 314
Peewee, Project (USAF), 371
Peking, China, 353
Pelly, Rep. Thomas M., 369, 373, 430
Penetrometer, 84
Peninsula Engineers Club, 63
Pennington, Jack E., 23
Pennsylvania, 368
Pennsylvania, Univ. of, 375, 408, 470, 494
Pensacola, Fla., 74
Pentagon, 14, 219
Peoria, Ill., 65
Perkins, Prof. Courtland D., 348
Perkins Observatory (Ohio State Univ.), 460
Pershing (missile), 430
Perth, Australia, 492
Peru, 33
Petersen, Norman V., 119
Petrone, Rocco A., 98
Petrosyants, Andronik M., 143, 206
Petterssen, Dr. Sverre, 162
PFRT (Preliminary Flight Rating Tests), 97
Phantom II. See F–4B.
Philadelphia, Pa., 56, 171, 226, 234, 390
Philco Corp., 29, 164, 193, 366
Philippine Islands, 302
Phillips, Franklyn W., 178, 481
Phillips, B/G Samuel C. (USAF), 176, 496
Phi-meson, 141
Phoenix (missile), 16
Photoelectric measurement, 206
Photography, 56, 162, 206, 396, 429
Photometer, 14, 15, 107
Photon, 95
Photospectrometer, 429
Photovoltaic Specialists Conference, 323
Phouma, Prince Souvanna, 413
Physical Review Letters, 74, 79, 86, 141, 317

Physical sciences, 390
Physics, 11, 435, 458
Physicists, role of, 288
Piccard, Auguste, 30
Piccard, Donald L., 140, 265
Piccard, Jean Felix, 30
Pic du Midi, France, 234
Pickering, Dr. William H., 16, 38
Pierce, Dr. John R., 10, 172, 429
Pierrelatte Nuclear Center (France), 438
Piland, Robert O., 22, 54, 125
Pilot-Astronaut citation (U.S.S.R.), 251
Pilot license, 479
Pilot of the Year Award, 465
Pilot, space, 114, 471
Piloting techniques, 281
Pi-meson, 26
Ping Pong (short-range reconnaissance missile concept), 429
Pinta, 226
Pioneer (rocket), 295
Pioneer (space probe), 1, 8, 141, 235
PIONEER V, 2
Pittendrigh, Dr. Colin S., 161, 237
Pittman, Dr. Frank K., 169
Pittsburgh, 57
Pittsburgh, Pa., 122, 260
Pittsburgh, Univ. of, 9, 225, 260
Pitzer, Kenneth S., 2, 268
Planetary research materials, 396
Plants, 411
Plasma gun, coaxial, 255
Plasma propulsion devices, 211
Plasma shield, 51, 56
Plasma streams, solar, 317, 317
Plastic film, 406
Plastic laser, 116
Plastics, 422
Platform, jet-propelled, 260
Plato (moon crater), 12
Pleasant Pond, Me., 280, 332
Pleumeur-Bodou, France, 189, 270, 425
PLOO.   See Pacific Launch Operations Office.
Plowshare, Project, 103
Plum Brook Station (NASA Lewis Research Center), 20, 95, 269, 275, 296, 299, 303
Plum Brook Research Reactor (NASA), 153, 276, 299
Pluto (planet), 52, 432
Pluto, Project (USAF), 367
PMF-PAC.   See Polaris Missile Facility—Pacific.
PMR.   See Pacific Missile Range.
Pocibo, Project (Polar Circling Balloon Observatory) (NSF), 474
Pogo (Polar Orbiting Geophysical Observatory), 36, 248
"Pogo-Stick" action, 487
Point Arguello, Calif., 307, 441
    launch, 38, 50, 59, 91, 129, 163, 189, 201, 239, 242, 272, 279, 335, 342, 481
Point Barrow, Alaska, 474
Pt. Mugu, Calif., 296, 307, 341
Polar Circling Balloon Observatory.   See Pocibo, Project.
Polar Ionosphere Beacon Satellite (NASA) (S-66), 54, 177
Polar Orbiting Geophysical Observatory.   See Pogo.
Polar orbiting satellites, 306, 487, 496
Polaris, 78, 124, 138, 263, 423, 433
    launch
        A-2, 367
        A-3, 10, 46, 50, 59, 134, 137, 201, 230, 231, 245, 251, 266, 278, 318, 327, 354, 402, 429, 450
    operational, 435
    use of, 12, 23, 25, 136

Polaris Missile Facility—Pacific (PMF-PAC), 83
POLET I, 510
    launch, 413, 417–418
    performance, 417, 418, 428
    use, 420, 422, 427, 498
Polhemus, Edward C., 277
Polyethylene shielding material, 471
Polymers, 385
Polyot I. See POLET I.
Ponnamperuma, Dr. Cyril, 315
Pontecorvo, Bruno, 153
Pope John XXIII, 468
Pope Paul VI, 251, 264, 354, 375
Popovich, L/C Pavel R., 304
    appearances, 88, 101, 138, 369
    condition of, 20, 48, 355
Porfirov, Boris, 101
Port Canaveral, Fla., 140
Port of New York Authority, 153
Porter, Dr. Richard W., 226
Positron, 26, 270
Possony, Dr. Stefan T., 103
Post-Echo passive communications satellite, 406
Potter, Andrew E., Jr., 139
Power, Gen. Thomas S. (USAF), 161
Power system, 346, 355, 405
Powers, Francis Gary, 437
Powers, L/C John A. (USAF), 29, 136, 284, 291, 315, 465
Prague, Czechoslovakia, 303, 459
Pratt and Whitney Aircraft Co., 277, 320, 336, 437, 493
Pravda, 11, 12, 35, 88, 259, 300, 305, 368, 372, 495
Pre-Flight Acceptance Checkout Equipment (PACE), 492
Preliminary Flight Rating Tests. See PFRT.
"Prelude to Independence", 224
Presentations, 16, 116
Presidential Boeing 707, 202, 206
Presidential Medal of Freedom, 468
President's Award for Distinguished Civilian Service, 239
President's Missile Sites Labor Commission, 411
President's Science Advisory Committee (PSAC), 8, 57, 73, 78, 109, 213, 236, 424, 479
President's Scientific Advisor, 93, 152, 234, 390, 421, 430, 440, 444, 479
Press and Union League Club (San Francisco), 111
Press conferences, 15, 69
"Press Conference USA", 365
Pressly, Eleanor C., 188
Pressure, air, 20, 60, 269, 301, 306, 469
Pressure suit, 360
Preston, G. Merritt, 205
Price, Rep. Melvin, 421
Price, Nathan, 260
Prince Motors Ltd., of Tokyo, 307
Princeton, N.J., 260, 459
Princeton Univ., 40, 53, 76, 130, 161, 164, 183, 214, 237, 238, 421, 424, 470, 479
Printing press, 21
Probes, 110
    deep space, 274
    Langmuir, 273, 357
    lunar, 339
    Lunar-orbiter, 326
    Mars, 339, 494
    space, 271, 308, 335, 365, 366, 499
    three-axis stabilized, 489
    traversing, 369
Processed foods from space research, 494

Proctor, Charles, 275
Procurement policy, 430, 493
Project Stabilization Agreement, 411
Propellant, 277, 333, 490, 496
Propulsion, 232, 279, 363, 428
Propulsion, space
    atomic propulsion, 367
    chemical, 129, 154, 266
    control data, 327
    nuclear reactor, 367
    nuclear system, 266
    technology, 419–20
Propulsion system, 281, 338, 361
Proton, 26, 356
Proxmire, Sen. William F., 48, 421, 440
Pruss, Judge Thaddeus, 59
Psychopathology, 445
Public Health Service, 105
Public Relations Society of America, 136
Public Service Award (NASA), 365
Public Works Administration, 164
Pucinski, Rep. Roman C., 239, 314
Puerto Rico, 37
Pugwash Conference, 11th, 353
Pulaski, Tenn., 78
Pulkovo Observatory (U.S.S.R.), 56, 197, 235
Purdue Univ., 136, 469
"Pygmy" satellite, 389
Pyrennees, French, 234
Quality control standards, 376
"Quality Program Provisions for Space Systems Contractors" (NASA), 362
Queens College (N.C.), 463
Quesada, E. R., 253, 253
Quilon, India, 68
Rabinowitz, Dr. Samuel, 173
Radar, 3, 54, 63, 162, 330, 406, 411, 454, 472
Radartelescope, 9, 42, 170, 387
RADC. See Rome Air Development Center.
RADC Passive Satellite Research Terminal, 426
Radhakrishnan, President Sarvepalli, 230
Radiant heating, 336
Radiation, 320, 403, 441
    alpha, 107
    detectors, 256, 357
    galactic, 335, 415, 449
    measurement, 489, 498
    shielding, 471
Radiation Air Borne Measurement Program (RAMP), 437
Radiation belts, 45, 227, 263, 308, 312, 346, 364, 389, 391
    and ionosphere, 92
    Saturn, 86
    study of, 50, 58
Radiation, corpuscular, 299
Radiation, cosmic, 355
Radiation damage, 189, 279, 346, 357
Radiation effects, 244, 355, 411
RADIATION MONITOR (satellite), 256, 263, 281, 507
Radiation, nuclear, 23, 137, 140, 220
Radiation particles, 281
Radiation, solar
    effects of, 305, 455
    measurement, 87, 242, 279, 487
    study of, 258, 388, 449

Radiation, space, 303, 343
    artificial, 41
    damage, 306, 346
    effect on satellites, 12, 96
    Mars, 138
    measurement, 111, 432
    simulated, 56
Radiation, ultraviolet, 258, 287, 296, 434
Radiative material, 388
Radio Corp. of America (RCA), 98, 116, 170, 216, 247, 257, 287, 299, 456, 472, 489
    Astro-Electronics Div., 25, 52, 219
    Data Systems Div., 305
    Electron Tube and Semiconductor Div., 345
    Services Co., 1
    Victor Co., 449
Radio, development of, 162
Radio energy, 481
Radio frequencies, 379, 414
Radio monitoring network, 414
Radio Moscow, 5, 196, 369, 417, 486
Radio Research Laboratory (Japan), 351, 357
Radio-Television Industry, 104
Radio tracking, 54
Radio waves, 265, 304, 341
Radioactive material, 387
Radioastronomy, 86, 88, 151, 154, 190, 372, 495
Radiometer, 377, 437
Radiosonde transmitter, 290, 497
Radiotelescope, 6, 40, 86, 114, 415
Radnofsky, Matthew I., 281
Rainfall, effect of moon phases on, 481
Raisting, West Germany, 425
Raleigh, N.C., 111
Ramey AFB, Puerto Rico, 37
Ramiah, K. Raghu, 142
Ramo, Dr. Simon, 138, 237, 329
RAMP. See Radiation Air Borne Measurement Program.
Ramsey, Dr. Norman F., 213
RAND Corp., 171, 259, 263, 377
Randolf AFB, Tex., 222
Ranger (program), 5, 83, 87, 204, 275, 326, 334, 489
    decontamination procedure, 339
    follow-on flight, 477
    sterilization, 339
RANGER V, 264, 364
Ranger 6, 165, 310, 399
Raven Industries, 265
Rayle, Warren, 211
Raymond, Dr. Arthur E., 123
Raytheon Co., 51, 292, 299, 375, 392, 411
RB–57F (Canberra), 371
RCA. See Radio Corp. of America.
RCA Astro-Electronics. See Radio Corp. of America.
RCA Data Systems Div. See Radio Corp. of America.
RCA Electron Tube and Semiconductor Div. See Radio Corp. of America.
RCA Service Co. See Radio Corp. of America.
RCA Victor Co. See Radio Corp. of America.
Reactor-in-flight-test. See Rift.
Reactor, nuclear, 20, 343, 455, 491
Reader's Digest, 288
Ream, Harold E., 23
Rebound, Project (NASA), 67
Rechtin, Dr. Eberhardt, 2, 6, 70, 278
Record, aircraft, 101
Record, space, 96
Recorder, magnetic tape, 177

Recovery sites, 296
Recovery systems, 302
Red Mill, Project (USAF), 343, 411
Red Star (see also *Krasnaya Zvezda*), 453, 471
Redstone Arsenal, Ala., 201
Redstone (missile), 318, 430
Reedy, R/A James R. (USN), 367
Re-entry, 112, 366, 384
    communications, 55
    material, 280, 403, 498
    research, 311
    simulation, 426
    test, 4, 23, 168, 295, 345, 418, 447
    vehicle, 287, 336
Rees, Dr. Eberhard, 328
Reeves, Maj. J. L. (USAF), 40
"Reexamining the Soviet Scientific Challenge", 172
Reformation, 166
Reichelderfer, Dr. Francis W., 287, 290, 360
Reid, Dr. Henry J. E., 411
Reiffel, Dr. Leonard, 100
Reinhart, Dr. Bruce L., 65
Reinhardt, G. Frederick, 386
Relative ionospheric opacity meter. See Riometer.
Relativity, 26
RELAY I, 60
    performance, 1, 3, 5, 6, 86, 91, 98, 103, 264, 287, 300, 339; 342
    transcaribbean, 15, 161
    transatlantic, 7, 86, 106, 134, 137, 143, 194, 196, 229, 235, 251, 434, 448
    transpacific, 447, 448
Relay (program), 227, 495
Reliability, 42, 52, 99, 280, 361
"Reliability Program Provisions for Space System Contractors" (NASA), 361
Religion and science, 432
Renaissance, 166
Rendezvous
    and docking, 334, 355, 420
    choice of, 36, 210
    earth-orbit, 36, 484
    lunar-orbit, 366
    problems, 355
    technique, 27, 72, 355, 362, 431
Republic Aviation Corp., 38, 176, 253, 264, 397, 398
Republican Conference (House), 315
Republican Party, 2, 203
Rescue, 390, 443
Research Advisory Committee on Biotechnology and Human Research (NASA), 163.
Research and development
    Armed Services, 326
    DOD, 14
    engineering, 473
    Federal, 302, 360, 434, 445, 496
    industry. 128. 327
    information distribution, 308
    management, 36, 151, 421, 444, 461
    planning, long-range, 342, 470
"Research and Development and the Federal Budget", 360
Research and Technology Div. (RTD) (AFSC), 176, 263
Research, basic. 28, 31, 110, 370, 480
Research Institute for Physics of Jet Propulsion (Stuttgart, Germany), 7
Research Institute of America, 363
Research Triangle Institute, Durham, N.C., 281
Reston, James, 128, 488
Reusable Orbital Module—Booster and Utility Shuttle. See ROMBUS.

Review of Space Research, 3
Revolution of Science, 472
Revolution, scientific, 167, 472
Reynolds Electrical and Engineering Co., 32, 393
Reynolds Metal Co., 425
RFD-1 (re-entry flight demonstration-1), 208, 514
Rhode Island State Dept. of Public Instruction, 290
Rice, Theron J., 323
Rice Institute, 2, 268, 289
Richard Prentice Ettinger Medal, 408
Richard Prentice Ettinger Program for Creative Writing, 408
Rift (Reactor-in-Flight-Test), 491
Rio de Janeiro, Brazil, 15, 88, 143, 235, 339, 425
Riometer (relative ionospheric opacity meter), 335
Ritland, M/G O. J. (USAF), 43, 48
RL-10 (rocket engine), 5, 297, 315, 320, 336, 410
Roadman, B/G Charles (USAF-MC), 61, 180
Roanoke, Va., 285
Robert H. Goddard Historical Essay Award, 104
Robert H. Goddard Memorial Dinner, 104
Robert H. Goddard Memorial Trophy, 104
Robert H. Goddard Scholarship, 104
Roberts, Dr. James E., 46
Roberts, Dr. Leonard, 23
Roberts, Walter Orr, 154
Robinson, Marvin W., 326
Rochen, Herbert D., 53
Rochester, N.Y., 200
Rochester, Univ. of, 26, 172, 213, 228
Rockefeller Institute, The, 408
Rockefeller, Nelson, 3
Rocket, 271, 273, 362, 387, 405
    booster, 271
    engine, 267
        liquid-propellant, 185, 320, 336, 448, 453, 462
        solid-propellant, 231, 271, 280, 322
    ion, 437
    launch from aircraft, 335
    liquid hydrogen-oxygen, 450
    meteorological, 20, 307, 436
    nuclear, 461, 486
    sled, 326
    solar-powered, 36
    solid-propellant, 395, 407, 466
    sounding rocket, 72, 222, 263, 273
    test stand, nuclear, 269
Rocket Center, W. Va., 165
Rocket engine, solar hydrogen (SOHR), 437
Rocketdyne Div. (NAA), 379, 387
    F-1 (engine), 6, 42, 97
    J-2 (engine), 53, 423
    X-8 (engine), 428
    X-12 (engine), 428
Rockoon (balloon-launched sounding rocket), 30
Rogallo, Francis M., 278
Rogallo, Gertrude S., 278
Rogers, Rep. Paul G., 367
Rohr Corp., 27
Roksonde (sounding rocket), 453
Rolls Royce, Ltd., 433
Roman, Dr. Nancy C., 34, 188
ROMBUS (Reusable Orbital Module-Booster and Utility Shuttle), 246
Rome Air Development Center (RADC), 426
Rome, Italy, 16, 86, 89, 100, 257, 375, 389
Rome, Univ. of (Italy), 171

Roosevelt, President Franklin D., 205
Roosevelt war years, 485
Rose, Dr. Frank A., 83
Rose Knot, 84
Rosen, Milton, 61, 70
Rosetta Stone of the solar system, 150
Rosman, N.C., 248, 402
Rotary Club, 156
Rothrock, Addison M., 225, 340, 468
Rotocraft (see also Helicopter), 424
Roudebush, Rep. Richard L., 293
Roush, Rep. J. Edward, 405
Rovenger, Ronald, 472
Rover, Project (NASA), 89, 485, 486, 488, 491
Rowe, Dr. Lynn B. (Capt., USAF), 186
Roy, Maurice M., 330
Royal Aircraft Establishment, 112
Royal New Zealand Air Force, 209
Royal Radar Establishment, 63
Royal Society of New Zealand, 209
Royal Society (U.K.), 63
RS-70 (Valkyrie), 27, 68, 84, 380
RTD. See Research and Technology Div. (AFSC).
Rubel, John H., 49, 133, 234, 260
Ruina, Dr. J. P., 275
Rumsfeld, Rep. Donald, 35, 298
Ruppe, Dr. H. O., 232
Rural Electric Cooperatives Association, 58
Rushworth, Maj. Robert (USAF), 137, 189, 194, 246, 256, 277, 283, 387, 423, 480, 485
Rusk, Dean, 32, 33, 297, 368
Russell, Sen. Richard B., 332, 493
Ryan Aeronautical Co., 220
Rye, U.K., 140
S-IV (Saturn I stage) (see also Saturn I)
    delivery, 348, 476
    engine, 320, 336, 475
    test, 297, 304
    transportation, 18, 143, 324, 451, 486
S-6 (atmospheric structure satellite) (see also EXPLORER XVII), 110
S-17 (satellite). See Oso-B.
S-53 (satellite). See UK-3.
S-61 (helicopter), 222
S-66 (Polar Ionosphere Beacon Satellite). See Polar Ionosphere Beacon Satellite.
SA-4 (see also Saturn I), 110
SA-5 (see also Saturn I)
    launch, 476, 477
    stage, 297, 304, 320, 348
    test, 71, 91, 304
SAC. See Strategic Air Command.
Sacramento, Calif., 281, 297, 451
Saenger, Prof. Eugen, 7, 140
Sagan, Dr. Carl, 228, 315
Sage (Semi-Automatic Ground Environment), 42
Sahara, 350
St. Clair, Wade, 216
St. Louis Citizen's Committee for Nuclear Information, 73
St. Louis, Mo., 25, 62, 234, 345, 366, 432, 475, 478
St. Louis, Univ. of., 281, 345
St. Petersburg, Fla., 434
Saint (program), 399
Sakurai, Dr. Jun John, 142
Salinger, Pierre, 460
Salisbury, Dr. John W., 130, 306-307
Sallebert, Jacques, 137

Salto di Quirra Test Range (Sardinia), 453
Salton Sea, Calif., 302
Saltonstall, Sen. Leverett, 41, 482–3
Salzberg Lecturer, 379
SAM. See USAF School of Aerospace Medicine.
Samos (satellite), 471
Sampler, 328
Sampling Aerospace Nuclear Debris. See Sand, Project.
San Angelo, Tex., 323
San Antonio, Tex., 41, 46, 442, 482
San Diego, Calif., 465
San Francisco Bay, 426
San Francisco, Calif., 56, 107, 111, 321
San Jose, Calif., 104
San Jose dos Campos, Brazil, 177
San Marco floating launch facility, 68, 148
San Marco, Project (Italy), 295
Sand, Project (Sampling Aerospace Nuclear Debris), 328
Sanders, John C., 36
Sandford, John W., 260
Sandia Corp., 23, 328
Sandusky, Ohio, 95, 153
Sanford, Terry, 483
Santa Maria, 226
Santa Monica, Calif., 143
Santa Rosa Island, Fla., 499
São Paulo, Brazil, 95, 140
Sardinia, Italy, 203, 453
Sarnoff, David, 10, 216, 247–248, 262, 342, 456
Saskatchewan, 282
Satar (Satellite-Aerospace Research), 455
Satellite, 310, 377
    artificial, 307, 308
    defense, 394
    recovery, 287, 465
    systems, 222
    uses of, 74, 97, 290, 473
Satellite-Aerospace Research. See Satar.
Satellite, applications, 120, 197, 334–335, 473
Satellite, atmospheric density, 486–7
Satellite, balloon (see also EXPLORER XIX), 486–487
Satellite, communications, 365, 406, 467, 490
    cooperation, international, 107, 416, 493
    design, 362
    frequency allocation, 419
    future of, 10, 47, 91, 133, 139, 171, 227, 247, 265, 267, 312, 446
    military use of, 426
    performance and durability, 287
    program, 120, 171, 214, 282, 304, 340
    R&D, 60, 67, 90, 110, 163, 164, 247, 365, 420, 431
    results, 45, 70, 229, 235, 241, 251, 264, 267, 302, 325, 426, 434, 447, 499
    systems, 112, 197, 216–217, 247, 269, 271, 299, 417, 440
    testing, 274, 321
Satellite communications network, 358, 425, 490
Satellite, geodetic, 308
Satellite, inflatable, 297, 299, 304
Satellite, ionosphere, 115
Satellite, Medium Altitude Communications (MACS). See Medium Altitude Communications Satellite.
Satellite, meteoroid detection, 43
Satellite, meteorological, 111, 219, 242, 247, 306
Satellite, military, 463
Satellite, navigational, 97, 126, 224, 280, 319–20, 365, 436
Satellite, nuclear detection, 299
Satellite, nuclear powered, 467

Satellite, observatory, 436
Satellite, polar orbiting, 306
Satellite, radiation monitor, 256, 263, 281
Satellite, reconnaissance, 471
"Satellite Sanity-Lunar Lunacy", 468
Satellite Situation Report, 2, 14, 210, 243, 267
Satellite stabilization, 280
Satellite Test Center (Sunnyvale, Calif.), 237, 307
Satellite, Tracking and Data Acquisition Network (STADAN) (NASA), 402, 475
Satellite, unidentified
    U.S., 263, 282
        launch vehicle
            Atlas-Agena, 189, 272, 279, 334, 400, 486, 499
            Blue Scout, 129, 485
            Scout, 258, 267
            Thor-Able-Star, 357, 464
            Thor-Agena, 5, 14, 118, 163, 202, 239, 242, 256, 279, 292, 323, 326, 349, 351, 426, 451, 489, 499
            Thor-Agena (TAT), 98, 406
        USN, 311, 357
    U.S.S.R., 31
Satellite, weather, 156, 344, 365
    data exchange, 314
    future of, 53, 112, 219, 365, 431, 446
    requirement, 24, 420
    results, 45, 365
    use of, 100, 107, 282, 319, 367
"Satmos" satellite (France), 335
Saturday Evening Post, 201, 243
Saturday Review, 10
Saturn (planet), 52, 86, 114
Saturn (program), 72, 182, 320, 324, 365, 400, 451
Saturn, Advanced (C-5). See Saturn V.
Saturn I, 514
    booster, 143, 406, 451, 478, 498
    capability, 110, 406, 433
    facilities, 335
    flight, 433, 475
    launch vehicle, 439
    stage, 320
    test, 71, 132
    use of, 43, 46
Saturn IB
    booster, 406, 411, 486
    engine, 451
    equipment, 305
    facilities, 335
    launch vehicle, 439
Saturn V, 411, 451
    booster, 271
    development, 405
    engines, 6, 451
    equipment, 305, 379, 475
    facilities, 51, 98, 269, 319, 335, 371, 433, 488
    stage, 68, 131, 335, 411, 486
    use of, 22, 266, 411
Saturn C-1. See Saturn I.
Saturn C-1B. See Saturn IB.
Saudi Arabia, 344
Saunders, Joseph F., 56
Scandinavia, 340
Scandinavian Committee for Satellite Telecommunication, 340
Scanner, Project (NASA), 403
SCAT. See Aircraft, supersonic commercial air transport.
Scheer, Julian, 40, 102, 315, 370, 382, 474

Scherer, Paul A., 448
Schieff, Prof. Leonard, 279
Schirra, LCdr. Walter M., Jr. (USN), 28, 140, 226, 238
    appearances, 32, 33
    award, 465
    condition, 245, 296
    flight, space, 26, 344
Schjeldahl, G. C. Co., 304
Schlesinger, Arthur, 366
School of Aerospace Medicine (SAM). See USAF School of Aerospace Medicine.
School of Aviation Medicine (USN) (Pensacola, Fla.), 74
Schrader, James H., 365
Schriever, Gen. Bernard A. (USAF), 216, 487
    aeronautical systems, future, 338, 419
    award, 324
    instrumentation, 176
    manned space flight, 96
    space, importance of, 31, 168, 187, 202
    support of NASA, 64
Schurmann, Carl W. A., 389
Schwartz, Dr. Sidney, 155
Schwarzchild, Dr. Martin, 76, 78, 160, 238, 448
Schweickart, Russell L., 392
Science, 172, 408, 411, 425, 432, 466
Science, 138, 148, 149, 164, 259, 331, 440, 454
Science and technology
    and study of humanities, 468
    future of, 456, 470
    Government support of, 85, 86, 388, 390, 392, 396, 441
    national policy and goals, 332, 424
    need for, 110, 250
    plans and programs, 152, 332, 382
    requirements, 466, 470
Science and Technology Agency (Japan), 436
Science and technology institutions
    Astroscience Center, 100
    New York, 3
    Pittsburgh, Univ. of, 9
    Tennessee Space Research Institute, 95
Science Digest, 260
Science, Government, and Information, 8
Science Information Exchange, 382
Scientific Advisory Group (USAF OAR), 278
Scientific Research Society, 352
Scientist
    and Congress, 463
    and engineers, 177, 214, 465
    as astronaut, 4, 246
    challenge, 313
    emigration of, 50, 63, 79
    importance of, 210, 250, 402, 444, 468
    medical, 55
    quality, 162
    shortage of, 26, 46, 437
    use of, 50, 105, 159
    woman, 34
Scientists Institute for Public Information, 58
Scorpius (constellation), 493
Scott, Capt. David R. (USAF), 392
Scout, 89, 168, 182, 197, 297, 324, 335, 407, 428, 454
    failure, 280, 352
    launch, 208, 258, 280, 299, 357, 486, 507, 511, 514
    problems, 483
Screech, rocket engine, 448
Screvane, Paul R., 99

Seaborg, Dr. Glenn T., 143, 166, 169, 202, 205, 220, 230, 237, 253, 351, 390, 470, 472, 497
Seagull, 410
Seal, 397
Seamans, Dr. Robert C., Jr., 21, 48, 120, 158, 202, 433
    award, 186, 365
    budget, 67, 82
    contracting, 134
    facilities, 118
    launch program, 436
    launch vehicle, 76, 334
    personnel, 99, 256, 323, 381
    program, lunar, 392
    program, space, 8, 102
    programs, space, 8, 15, 245, 347, 369, 439
Seattle *Post-Intelligencer*, 114
Seattle, Wash., 59, 94, 324, 448
Security, 334, 375, 381, 404
Sedov, Prof. Leonid I., 378, 481
See, Elliott M., Jr., 28
Seeds, 355
Seiff, Alvin, 497
Seismic data transmission, 353, 465
Seismography, 465
Seitz, Dr. Frederick, 50, 238, 279, 372, 388, 435
Self-hypnosis, 170
Seliger, Berthold, 177
Sellers, Robert C., and Associates, 311
Sembawang Naval Air Station, Singapore, 432
Semi-Automatic Ground Environment. See Sage.
Semiannual Report to Congress, Eighth (NASA), 82
Semiannual Report to Congress, Seventh (NASA), 465
Senegal, West Africa, 19
Sensor, 244, 310, 341
Serebreynnikov, Lt. Col., 453
Sergeant (missile), 230
Serpukhov, U.S.S.R., 220
Sert (Space Electric Rocket Test), 449
Sevareid, Eric, 217
Shabad, Theodore, 494
Shafer, Edward M., 482
Shanghai, China, 415
Sharnov, V. V., 228
Sharp, Dr. Edward R., 240
Sharp, Dr. Richard, 496
Shawnee, Okla., 257
Shcheglov, Pyoter, 114
Shea, Dr. Joseph F., 61, 66, 156, 181, 251, 380, 420, 439
Shepard, Cdr. Alan B., Jr. (USN), 28, 238, 465
Shepherd, Dr. L. R., 206
Shewmake, Glenn A., 281
Sheilding, 471
Shiner, A. James, 233
Shklovskiy, I. S., 40
Shoal, Project (AEC), 402
Shockley, William Bradford, 150
Shock waves, 494
Short take-off and landing (Stol)   See Stol.
Shotput (sounding rocket), 295
Shriver, Rep. Garner E., 306
Shternberg State Astronomical Institute (U.S.S.R.), 40
Sidwell Friends School Forum (Washington, D.C.), 45
Siepert, Albert F., 65
SIGMA 7 (see also MA-8), 140, 238
Sigma Xi, 125

Signals Research Development Establishment, 112
Silicate powders, 306
Silo launch, 283
Silver iodide, use of, 293, 401
Silver Quill Award, 47
Silverstein, Dr. Abe, 7, 99, 100, 450
Simons, Howard, 6, 286, 347, 471, 479, 493
Simpson, Dr. George L., Jr., 40, 92, 102, 126, 161, 210, 278, 382
Simulator, rendezvous-docking, 334
Simulator, space orbital, 171
Simulator, spaceship, 338
Singapore, 432
Singer, Dr. S. Fred, 53, 107, 146, 330
Sinton, Dr. W. M., 234
Sioux Falls, S.D., 140
S.I.P., Inc., 495
Siri, William F., 246
Sirons, Janis A., 362
Six, Robert F., 367, 410
Skifter, Dr. Hector R., 123
SKMR-1 (Hydroskimmer), 357
Skybolt (missile), 1, 32, 41, 136, 240
Skyhook, Project (USAF), 30
Skylark (rocket), 77
Skywatch, Project (Weather Bureau), 226
Slayton, Maj. Donald K. (USAF), 28, 389, 420, 446, 465
Sled, rocket, 326
Sloan, James E., 61
Sloanaker, Russell M., 86
Slobodrian, Dr. Rodolfo J., 213
Sloop, John L., 129, 154
Smith, C. R., 397
Smith, Ernie, 107
Smith, Dr. George B., Jr., 45, 70, 498
Smith, Madison B., 213
Smith, Sen. Margaret Chase, 18, 159, 173, 188, 191, 231
Smith, Sheldon, 280
Smith, Wendell, 117
Smithsonian Astrophysical Observatory, 32, 315, 399, 432, 487, 495
Smithsonian Institution, 16, 61, 67, 191, 364
Smokey Joe, Project (USAF), 437
SMS. See Synchronous Meteorological Satellite.
Smull, Dr. Thomas L. K., 313
Snap (Space Nuclear Auxiliary Power), 169, 343, 357, 455, 467
Snap-8, 53, 340
Snap-9A, 360, 437
Snap-50, 53
Snow, Sir Charles (C. P.), 250, 403
Soaring Society of America, 481
Social science and scientists, 402
Society of Aerospace Material and Process Engineers, 226, 438
Society of American Military Engineers, 122, 385
Society of American Travel Writers, 489
Society of Chemical Industry, American Section, 356
Society of Experimental Test Pilots, 359
Society of Professional Engineers, 66
Sociologist, need for, 402
Sodium airglow, 15
Sodium-vapor experiment, 206, 212, 222, 443
Sohier, Walter D., 452
Sohigian, M. D., 170
Sohio Research Center, 375
SOHR. See Rocket engine, solar hydrogen.
SOLARAD IV, 242, 505

Solar cell, 323, 345, 346, 432
    radiation damage, 189, 279
    radiation effects, 455
Solar concentrator, 437
Solar corona. See Corona.
Solar cycle, 436
Solar eclipse, 18, 273, 279, 332, 339
Solar energy, 437
Solar flare. See Flare, solar.
Solar hydrogen rocket engine (SOHR). (See Rocket engine, solar hydrogen.)
Solar plasma stream, 317, 449
Solar radiation. See Radiation, solar.
SOLAR RADIATION I, 87
Solar simulation, 180
Solar system, history of, 149, 355
Solar wind. See Wind, solar.
Solid propellant, 34
    development, 13, 145
    rocket motor, 5, 271, 280, 357, 496
Solid Propellant Rocket Conference, 34
"Some International Aspects of Communications Satellite Systems", 269
Sonic boom, 59, 140, 271, 314, 390
SOR. See Specific operational requirement.
Sounding rocket
    experiment, 419
    launch, 365
        Japan, 203, 476
        NASA, 14, 30, 50, 58, 60, 63, 70, 71, 87, 88, 95, 107, 110, 111, 134, 168, 184, 186, 188, 212, 241, 247, 263, 268, 336, 351, 366, 388, 391, 410, 429, 434, 469
        NASA-Italy, 295
        New Zealand, 209
        Sweden, 301
        USAF, 57
South America, 360, 465
South Pacific, 44
South Point, Hawaii, 414
Southern California, 32
Southern Governors' Conference, 318
Southern Hemisphere, 495
Southern Interstate Nuclear Board, 318
Southern Mississippi, Univ. of, 254
Southern States Work Conference, 225
Southwest Center for Advanced Studies, 481
Southwest Community Hospital, 240
Southwest Conference on Arms Control, 409
Southwest Research Institute, 46
Soviet Academy of Medical Sciences, 130
Soviet Academy of Sciences, 2, 30, 42, 212, 274, 303, 305, 314, 427, 428
Soviet Air Force, 355
Soviet Armed Forces Day, 65, 66
Soviet Embassy, 411
Soviet Ministry of Communications, 245
Soviet Navy, 44
"Soviet Space Technology", 116
Sowers, Maj. Robert S. (USAF), 209
Space (medium), 334
    exploration of, 45, 61, 117, 232, 264, 288, 305, 313, 328, 329, 342, 346, 355, 359, 368, 379, 413, 446, 452, 464, 470, 477, 482, 486
    impedance, 342
    nature of, 314
    pollution of, 19
    study of, 476
    water in, 432
Space Age, 162, 371, 395, 414, 453, 472, 483
Space and Information Systems Div. (Raytheon Co.), 51

Space Business Daily, 84, 368, 383, 399, 400, 420, 431
Space capability, 17, 40, 308, 352, 383, 445, 461, 462
Space capsule, 287, 460
Space, challenge of, 153, 156, 160, 224, 329, 354, 358, 363, 421, 460, 462, 468
Space chamber, 1, 110, 275, 280
Space, economic impact, 438
    fallout, 169
    investment, 62, 329, 352, 398, 415, 455
    leverage, 111, 356,
Space Electric Rocket Test (Sert). See Sert.
Space environmental conditions, 1, 333, 385, 495
Space failures, 30, 31, 130, 135, 175, 248
Space flight, 263, 268, 350, 453
Space Flight Testing Conference (AIAA), 98
Space, impact of, 87, 110, 113, 156, 204, 274, 414
Space Industry Assistance Symposium, Second, 289
Space, interplanetary, 293
Space law
    agreement on, 78, 424, 434, 443
    and U.N., 142, 145, 177, 464, 477, 482
    nature of, 427
    need for, 40
    questions, 52
    ruling, 59
Space medicine. See Life science; Bioscience.
Space, military use of, 279
    manned flight, 39, 114, 319, 332
    need for, 2, 3, 64, 168, 221, 243, 254, 307, 338, 363, 418, 430, 445, 455, 461
    objectives, 25, 35, 63, 64, 71, 79, 254, 287
    patrol, 77
    satellite, 137
    space station, 43, 46, 346
Space news, 272, 278
Space Nuclear Auxiliary Power. See Snap.
Space observatories, 365
Space Orientation Center, MSFC, 115, 499
Space, peaceful uses of, 40, 172, 205, 219, 349, 404
Space platform, 305, 355, 421
Space, politics in, 355, 358, 374, 389, 400, 416, 417, 418, 427, 455
Space, possibilities, 147, 224, 307, 329, 461
Space program, national
    accomplishment, 44, 47, 58, 91, 226, 249, 282, 289, 291, 319, 321, 345, 352, 362, 373, 446, 453, 473, 484, 498
    civilian, 302, 307
    cost of, 131, 142, 157, 215, 300, 329, 335, 359, 360, 363, 373, 374, 383, 384, 422, 423, 430
    criticism, 191, 195, 240, 275, 279, 288, 302, 360, 380, 396, 403, 421, 442, 445, 470
    education, 290, 372
    industry in, 289, 372, 373, 435, 445, 466
    manpower, 237, 457
    military, 39, 72, 168, 221, 232, 302, 338, 363, 374, 455, 474
    need, 18, 46, 125, 128, 191, 206, 235, 236, 249, 329, 377, 409, 466
    objectives, 8, 13, 14, 65, 66, 78, 79, 85, 142, 149, 159, 234, 257, 274, 285, 334, 340, 345, 348, 404, 408, 430, 443, 450
    pace of, 17, 29, 44, 112, 121, 132, 282, 291, 324, 383, 416
    policy, 249, 363, 369, 404, 408, 419
    prospects for, 18, 52, 224, 232, 234, 316, 329, 379
    requirements, 46, 267, 395, 419, 441
    support of, 262, 269, 368, 373, 391, 398, 407, 411, 415, 458, 485
    women in, 34, 225, 273
Space race, 58, 201, 279, 288, 304, 411, 422, 427, 439, 462, 480
    meaning, 187, 200, 258
    moon landing, 6, 274, 302, 402, 404, 405, 409, 416, 418
    need for, 48, 200, 215, 258, 259
    rate of progress, 29, 141, 352, 396, 484

"Space Race with U.S.S.R.", 141, 276
Space Radiation Effects Laboratory (NASA), 56, 136, 164, 288, 308, 309
Space Rendezvous, Rescue and Recovery Symposium, 336, 337
Space research, 162, 494
Space results
    artificial radiation, 130
    flares, solar, 92, 93
    medical, 20
    plasma, solar, 69
    proton, 95
    radiation belts, 58, 86
    temperature, 47
    Venus, 69
    wind, 11
Space science, 67, 159, 302, 381, 464
Space, Science, and Urban Life Conference, 110, 113
Space Science Board (NAS–NRC), 275, 339
Space science building (Univ. of Calif.), 222
Space Science Summer Group, 234
Space Science Summer Study, 3
Space scooter, 260
Space station
    agreement, NASA–DOD, 391
    auxiliary power, 9
    designs, 22, 43, 46, 77, 96, 158, 170, 266, 346
    development, 458
    escape system, 383
    patent, 133
    plans, 82, 239
    possibilities, 142, 481
    program, 300
    research, 74, 268
Space suit, 12, 196, 393, 403
Space systems, 483
Space technology, 110, 302, 355, 466
Space Technology Laboratories (STL), 184, 193, 219, 235, 237, 266, 275, 277, 281
Space travel
    equipment testing, 268, 337
    future, 316, 412
    obstacles to, 192
    women in, 244, 254, 256
Space vehicle orientation, 278
Spacecraft, 444, 460
    contamination, 173
    development of, 355, 361
    environmental test, 385
    logistics, study of, 292
    protection, 340, 346, 454
    U.S.S.R., 413, 417, 418
Spacecraft Control Technology Laboratory (MSC), 490
Spacecraft Research Office and Laboratory (MSC), 490
Spaceflight, 116
Spaceship, 338, 413
Spain, 33, 194, 197
Spanish Inquisition, 449
Sparkman, Sen. John, 323
Sparoair (research rocket), 296
Specific operational requirement (SOR), 438
Spectrometer, 279, 377, 437, 449
Spectrophotometer, photoelectric, 278
Spectrophotometric instrument, 279
Spectroscopic techniques, 494
Speed brakes, 321

Sperry Gyroscope Co., 256, 376, 392
Sperry Rand Corp., 317, 443
Sperry Utah Co., 353
Sphere, inflatable, 469
Spin-off from space research, 491
Spinrad, Bernard I., 179
Spirit of St. Louis, 16, 67
Spitzer, Dr. Lyman, Jr., 214
Sporadic-E disturbances, 241
Sproul Observatory, 145
SPUTNIK I, 112, 197, 313, 369, 371, 377, 404, 425, 447, 480, 491
SPUTNIK IV, 59
SRI. See Stanford Research Institute.
SSBN (nuclear powered ballistic missile submarine), 263
SSGS. See Standardized Space Guidance System.
SST. See Aircraft, Supersonic Transport.
Stability and control, 244, 321
Stabilization, attitude, 311
Stack, John, 253, 398
Staebler, Rep. Neil, 35
Stafford, Capt. Thomas E. (USAF), 28
Stagg Field (Chicago), 149
Stamp, postage, on space program, 353, 486
STANDAN. See Satellite Tracking and Data Acquisition Network.
Standard Oil Co. of New Jersey, 72
Standard Oil of Ohio, 84
Standardized Space Guidance System (SSGS), 350
Stanford, Neal, 162
Stanford Research Institute (SRI), 123, 143, 218
Stanford Univ., 49, 214, 334, 479
Star, 25, 88, 116, 151, 160, 278, 281, 317, 448, 481, 493
STARAD, 130
Starfighter. See F-104.
Starfish Prime, 383
Starfish, Project, 106
Stargazer, Project (USAF), 130, 142, 150, 154
Starlifter. See C-141A.
Stasser, Bruce, 71
State Astronomical Institute (U.S.S.R.), 74, 119
State, Dept. of, 3, 31, 91, 205, 249, 397
State of the Union, 10
"Statement of Purpose", 496
States-General (France), 37
Statler, W. H., 179
Stead AFB, Nev., 298
Steam-diffusion chamber, 315
Stedman Hall of Science, Central Methodist College (Mo.), 334
Steel, 491
Steele, Lindell E., 65
Stellar photography, 56
Stellenwerf, Col. William A. (USAF), 292
Sterilization, space vehicle, 161, 165, 339
Sternglass, Ernest J., 26
Stevenson, Adlai E., 231, 458, 460
Stevenson, Dr. Earl P., 491
Stiff, Ray C., Jr., 424
STL. See Space Technology Laboratories.
Stockholm, Sweden, 421
Stockholm, Univ. of (Sweden), 301, 419
Stoddard, Dr. David H., 181
Stoddard, Laurence, 18
Stol (Short Take-Off and Landing), 6
Stoller, Morton J., 38, 240
Stone, John W., 157
Stoney, William E., Jr., 282

Storm, 25, 344, 367, 401
Storm Radar Data Processor (Stradap), 25
Stormfury, Project (Weather Bureau), 293, 401
Storms, Harrison, 378
Stradap. See Storm Radar Data Processor.
Strand, Kaj Aa., 327
Strategic Air Command (SAC), 307, 435, 469
    award, 302
    Atlas, launch, 28, 161, 290, 323, 325
    Minuteman, 399, 499
    launch, 188, 266, 286, 326
    Titan I, launch, 30, 314
Strato-Lab, Project, 30
Stratocruiser. See C-97.
Stratolifter. See C-135.
Stratoscope II (balloon), 76, 78, 82, 448
Stratosphere, 323
Stratton, Julius A., 3
Stress, 410
Strikes
    Cape Canaveral, 147
    McDonnell Aircraft Corp., 14
    Merritt Island, 337
    Mississippi Test Operations, labor agreement, 411
    NASA Wallops Station, 325
    Nevada Test Site, 393
Structures, 281, 287, 355, 361, 498
Struve, Dr. Otto, 131
Stuttgart, Germany, 7, 205, 211, 212, 214
Submarine, 44, 402, 411, 422
Submarine cable, 37
Submarine, Fleet Ballistic Missile. See FBM.
Submarine, Nuclear Powered Ballistic Missile. See SSBN.
Subroc (submarine rocket) (anti-submarine missile), 463
Sud-Aviation, 140, 229
Suez, U.A.R., 137
Suitland, Md., 247, 341
Sun
    corona, 279, 339, 366
    influence, 390, 445
    photos, 388
    polar regions, 257
    probe, 164
    radiation, 397, 406, 419
    spots, 436
    structure, 314, 488, 494
    temperature, 488
Sundlun, Bruce G., 91
Sunflower turboalternator unit, 339
Sunnyvale, Calif., 441
Supernovae, 155
Supersonic transport. See Aircraft, supersonic transport (SST).
Supersonic Transport Advisory Group (FAA), 253, 271
Surge acceleration, 487
Surgeon, flight, 56
Surgeon General, 455
Survey of Business Plans for New Plants and Equipment, 187
Surveyor (program), 5, 7, 83, 275, 326, 477, 489
Survival, 271, 298, 332
Swanson, Andrew G., 136
Swarthmore College (Pa.), 145
Sweden, 133, 197, 295, 301, 340
    launch, 388, 477
Swedish Committee for Space Research, 478
Sweeney, Gen. Walter C., Jr. (USAF), 483

Sweet, Floyd, 481
Swift Current, Saskatchewan, 282
Swindal, Col. James B. (USAF), 202
Switzerland, 133, 185
Sycamore Canyon, Calif., 410
Sylvania Electric Products, Inc., 219
Sylvanus Albert Reed Award, 22
Symington, Sen. Stuart, 165, 302, 475
Symposium on Applications of the Theory of Functions in Continuum Mechanics, 344
Symposium on "open space and peace", 334
Symposium on the Physics of Solar Flares, 404
Symposium on Space Rendezvous, Rescue and Recovery, 336, 337
Symposium on the Exploration of Mars, 232
Synchrocyclotron, 56, 303
Synchronization system, 271
Synchronous Meteorological Satellite (SMS), 38
Synchrotron, 463, 470
Syncom (program), 60, 171, 362, 459
SYNCOM I, 502, 514
    launch, 54
    photos of, 72, 79
    status, 57, 60, 341, 495
SYNCOM II, 508, 514
    control of, 284, 297, 312
    launch, 284
    performance, 284, 287, 304, 312, 333, 339, 340, 451, 455
    transmission, 296, 302, 321, 342, 345, 349, 353, 425
Syncom C, 455
Synon, Capt. George D. (USCG), 220
Syracuse, N.Y., 187
Syracuse Univ., 379, 445
Systems Program Management Survey, 189
Syvertson, Clarence A., 336
Tabstone, Project (ARPA), 437
TAC. See Tactical Air Command.
Tactical Air Command (TAC), 483
Tactical fighter, experimental (TFX). See Aircraft, tactical fighter, experimental.
Tactical-strike-reconnaissance/strategic nuclear bomber. See TSR-2.
TAD. See Delta, Thrust-Augmented.
Taft-Hartley Act, 23
Tahiti, 176
Takeuchi, Ambassador Ryuji, 447
Talaat, Dr. Mostafa, 488
Tallahassee, Fla., 465
Tampa, Fla., 434
Tape, Dr. Gerald F., 253
Tarrytown-on-the-Hudson, N.Y., 34
Tart, James J., 128
Tass
    spaceflight, 369, 441, 469
        COSMOS, 101, 140, 154, 209, 212, 477, 478
        LUNIK IV, 127, 128, 130
        MARS I, 40, 58, 63, 96, 192
        POLET, 420
        VOSTOK, 244
    space science, 114, 153
    test, 271, 452
TAT. See Thor, Thrust-Augmented.
Tawes, J. Millard, 273
Taylor, Prof. George W., 375
Taylor, Hal, 458
Taylor, John W. R., 484
Taylor, Gen. Maxwell D. (USA), 84
Tbilisi, U.S.S.R., 344
Teachers' workshop, 328

Teague, Rep. Olin E., 350
Technology, 425, 472, 473
Technology Audit Corp., 473
Technology Utilization and Policy Planning (NASA), 278
Tektite, 39, 104, 365
Telecommunication systems, 355
Telecomputing Services, 387
Telephone, 162, 447
Telemetry, 268, 274, 281, 297, 380, 409, 443
    facilities, 308, 355
Telescope, 274, 395, 399
    balloon-borne, 76, 78, 106, 448
    monocular, 395
    radar-radio, 415, 432
    use of, 274, 305, 388, 405, 410, 493, 494
Telespazio, 5
Television, 21, 448
    channels, 114, 151, 154, 372
    color, 98, 131
    development, 162
    laser, 92
    manned space flight, 44, 131
    moon, 129
    satellite, via, 354, 447
Television Shares Management Corp., 62
Teller, Dr. Edward, 3, 334, 392, 425
Telstar (program), 10, 22, 495
TELSTAR I, 60, 98, 184
    development, 104
    performance, 1, 2
    radiation, 23, 41
    results, 13, 47
TELSTAR II, 274, 283, 503, 514
    launch, 184
    plans for, 98
    transmission, 186, 189, 241, 267, 270, 304, 325, 354, 389, 447, 449
Temperature measurement
    atmosphere, 20, 61, 71, 206, 208, 469
    electron, 325
    lunar surface material, 306
Tennessee, 95
Tennessee Space Research Institute, 95
Tennessee, Univ. of, 95, 423
Tennessee Valley Authority (TVA), 201
Tenney Engineering, Inc., 171
Tepper, Dr. Morris, 71, 125
Tereselic, Richard, 297
Tereshkova, Valentina V.
    appearances, 252, 254, 303, 374, 390, 394
    award, 251
    flight, 241, 244, 246, 256, 259, 268, 372, 484
    marriage, 376, 417
    (See Nikolayeva, Valentina V.)
Terhune, Robert W., 157
"Test Ban, The", 103
Test chamber, 453
Test stand, nuclear rocket, 271
Teterboro Airport ,N.J., 153
Tetrahedral Research Satellite (TRS), 189, 279, 346, 391, 504, 507, 509
Texas, 381, 426
"Texas and Telstar", 449
Texas A&M College, 47
Texas Associated Press Managing Editors Association, 29
Texas Mid-Continent Oil and Gas Association, 351
Texas, Univ. of, 234, 481
TFX. See Tactical fighter, experimental.

"The Bug", 176
"The Christian Revolution", 389
The Economist (London), 224
"The Mastery of Space", 341
"The Practical and Impractical Uses of Space", 425
Thermal electricity, 342
Thermal Radiation Airborne Program. See Trap III.
Thermochemical test, 333, 495
Thermoelectric conversion technique, 455
"Thermonuclear converter", 57
Thermonuclear research, 325
Thermonuclear world, future in, 409
"The X-15 Story", 341
Thiokol Chemical Corp., 145, 175, 322, 407, 496
Thiry, Gen. Jean (France), 176
This Week, 89
Thom, Herbert C. S., 25
Thom, Dr. Karlheinz, 245
Thomas, Rep. Albert, 349, 353, 384
Thomas, David D., 239
Thomas D. White National Defense Award, 107
Thomas, N. E., 242
Thomas, Richard N., 222
Thompson, Dr. Floyd L., 90, 163, 205, 235, 251, 308, 366
Thompson, Milton O., 235, 331, 404, 450
Thompson Ramo Wooldridge, 138, 219, 237, 329, 339
Thompson, Robert F., 254
Thor (missile), 9, 103, 168, 176, 258, 324, 334, 345
Thor-Able (booster), 258
Thor-Able-Star (booster), 258, 357, 360, 454, 464, 467, 509, 510, 511
Thor-Agena (booster), 502, 505, 507, 508, 509, 510, 511, 512
    launch
        B, 98, 163, 258, 263, 342, 349, 426, 489
        D, 5, 14, 118, 202, 239, 242, 257, 279, 292, 323, 326, 351, 406, 451, 489, 499
    use of by NASA, 334
Thor-Delta (see also Delta), 182, 385, 502, 503, 506, 508, 512, 514
    launch, 54, 119, 184, 247, 258, 284, 449, 489
Thor, Thrust-Augmented (TAT) (booster) (see also Thor-Agena), 395, 407
    launch, 72, 124, 406
Thornton, A/2C Frank O. (USAF), 460
Thrust vector control data, 327
Thule, Greenland, 343
Thumba, India, 11, 443
Thunderbird, 410
Thurmond, Sen. Strom, 138
Tibet, 227
Tidewater Science Congress, 90
Tikhov, G. A., 74
Timerbaev, Roland M., 177
Timers Aero Club (NAA), 484
Times Herald (Newport News, Va.), 242
Tiros (program), 125, 170, 226, 341, 395
    contract, 52, 219
    research, 219
    results, 72, 229, 360
    value of, 367
TIROS I, 118, 150
TIROS II, 147
TIROS V
    launch, 297
    performance, 72, 118, 147, 179, 297, 344
TIROS VI, 344
    launch, 297
    performance, 72, 118, 242, 297, 302, 391
    results, 344

TIROS VII, 506, 514
    launch, 297
    performance, 247, 297, 302, 348
    use of, 371
TIROS VIII, 489, 512, 514
Tischler, A. O., 262
Tishler, Dr. Max, 356
Titan I (missile), 435
    launch, 30, 57, 175, 220, 276, 314, 324, 344, 433, 452
Titan II (missile and booster), 9, 40, 164, 286, 491, 495, 499
    booster, 48
    Gemini launch vehicle, 291, 320, 368, 386, 401, 431, 487
    instrumentation pod, 377
    launch, 44, 57, 59, 147, 164, 188, 193, 213, 219, 250, 320, 350, 415, 427, 476, 481
    performance, 377
    problems, 174, 219, 309
    propulsion system, 486
Titan III (booster), 46, 57, 68, 162, 262, 311
    booster, 66, 436, 490
    contract, 262
    cost, DOD, 466
    launch, 473
    rocket engine, liquid, 281, 493
    rocket motor, solid-propellant, 280
Titov, L/C Gherman S., 138, 300
    award, 136
    condition of, 355
Titov, Tamara, 350
Titov, Tanya, 350
TNT manufacturing facilities, 303
Tobago, 367
"Today", 7
Tokaty, Dr. G. A., 116
Tokyo, Japan, 164, 194, 227, 267, 341, 389, 447
Tokyo's Institute of Industrial Science, 494
Tokyo Univ. Industrial Engineering Research Center (Japan), 476
Toll, Thomas A., 278
Tolobko, Col. Gen. Vladimir Fedorovich, 63
Tonopah, Nev., 23
Tonopah Test Range, Nev., 328
Toong, Dr. Tau-Yi, 50
Topeka, Kan., 86
Topographic data, 326
Top Rung II, Exercise, 469
Torrey, Ray, 280
Tousey, Dr. Richard, 156
"Toward the Orbital Launch Facility", 115
Town Meeting of the World, 270, 389
Toynbee, Arnold, 229
Tracking, 487
    Australian station, 325, 464
    Ball Bros. system, 244
    deep space, 27, 43, 82, 326
    Egyptian station, 319
    facilities, 308, 335, 426, 475
    French network, 298
    international network, 32
    Malagasy Republic station, 488
    optical, 44, 54, 63
    "Unified S-Band" method, 472, 490
    U.S. station, 414, 441, 443
    U.S.S.R. cooperation, 390
"Trackless Void: The U.S. Space Program Had Better Come Down To Earth", 193
Traffic control system, 377

Transistor, 150, 162, 357, 385
Transit (program), 87, 115, 129, 428, 454
TRANSIT II, 87
TRANSIT V-B, 357
Transmitter, 35
Transonic flow theory, 368
Transonic wind tunnel, 297
Transpolar flight, 367
Transportation, 378, 387, 451, 486
Transport, jet, 321
Trans-World Airlines (TWA), 480
Trap III (Thermal Radiation Airborne Program), 498
Travelers Research Center, Inc., 287
Treaty, 297
Tregaron, 17
Trendex survey, 314
"Trends in Earth-to-Orbit Transportation Systems", 412
Treshnokov, A., 453
Trimpi, Robert L., 365
Trinidad, 367
Trippe, Juan T., 299
TRS. See Tetrahedral Research Satellite.
Truman, President Harry S., 106, 198, 485
Truszynski, Gerald M., 309
Tsarapkin, Semyon K., 90
TSR-2 (tactical-strike-reconnaissance/strategic nuclear bomber) (U.K.), 462
Tu-104 (aircraft), 417
Tu-124 (aircraft), 459
Tufts Univ., 179
Tullahoma, Tenn., 95, 341
Tungsten disenlenide, 216
Tupolev and Ilyushin teams, 379
Turboalternator (see also Sunflower), 339
Turbofan engine, 420, 438
Turbulence, 222, 426
Turkey, 25, 168, 197
Tucson, Ariz., 145
TV channel 3, 7, 372
TVA. See Tennessee Valley Authority.
TWA. See Trans-World Airlines.
Two-Step Formal Advertising System, 430
Tymczyszym, Joseph J., 442
Typhon (missile), 427
Typhoon, 344
Typhoon Bess, 302
Typhoon Carman, 302
Tyuratam, U.S.S.R., 259
U-2 (aircraft), 74, 194, 371, 415, 418, 437, 441
U.A.R. See United Arab Republic.
UCLA. See Calif., Univ. of (Los Angeles).
Udall, Stewart L., 192
UHF (ultra-high frequency), 490
U.K. See United Kingdom.
UK-3 (S-53), 231
Ulam, Dr. Stanislaw, 367
Ultra-high frequency. See UHF.
Ultrasonics, 32
Ultraviolet, 87, 280, 366
    light emission, 429
    radiation, 287, 296, 339, 406, 419, 434, 477
U.N. See United Nations.
UNCAST (United Nations Conference on Applications of Science and Technology), 43, 50
UNESCO (United Nations Educational, Scientific and Cultural Organization), 50, 329

"Unified S-Band", 472
Union Carbide Corp., 62
Union of Soviet Socialist Republics. See U.S.S.R.
Unitarian minister, 358
United Aircraft Corp., 222, 239, 442
    Hamilton Standard Div., 387, 403, 489
    Pratt & Whitney Div., 320
United Air Lines, 194
United Arab Republic (U.A.R.), 105, 137, 156, 274, 295, 319
United Auto Workers Union, 374
United Kingdom (U.K.), 149
    aircraft
        commercial jet, 397
        Concorde, 312, 320, 367, 379, 390, 398, 406, 410
        man-powered, 378
    astronomer, 481
    communications network, 425
    communications, satellite, 107, 112, 197, 235, 416
    cooperation, defense, 8, 23
    cooperation, space, 5, 7, 133, 295, 348
    cosmic dust, 443
    flight, 389
    House of Commons, 199, 205, 275
    Labor Party, 462
    launching, 487
    missile program, 311
    nuclear test-ban treaty, 284, 297, 350, 388
    Parliament, 37, 283
    Polaris, 136
    satellite, 197, 231
    satellite observation station, 433
    scientist, 275
    space program, 91, 112, 433
    Thor, 168
    threat to, 205
    tracking, 33
United Nations (U.N.), 43, 321, 348, 471, 475
    and space, 40, 63, 366, 440
    charter, 347, 440
    Committee on the Peaceful Uses of Outer Space
        Scientific and Technological Subcommittee, 68, 98, 142, 145, 159, 172, 177, 195, 200, 219, 221, 326, 329, 424, 434, 443, 457, 464
    General Assembly, 162, 342, 351, 368, 369, 381, 389, 390, 391, 447, 482
        Disarmament Committee, 388
        Political Committee, 388, 389, 457, 464
United Nations Conference on Applications of Science and Technology. See UNCAST.
United Nations Educational, Scientific and Cultural Organization See UNESCO.
United Press International (UPI), 194
United States (U.S.) (see also appropriate agencies)
    award, 341
    budget, 360, 373, 390
    defense
        budget, 475, 485
        economy, 485, 486
        effort, 288, 397, 461, 475
        installations, 477
    foreign policy, 394
    Government, 111, 356
        criticism of, 439
        research and development, 325, 496
        science and technology, 390, 393, 396
    military, 475
    nuclear test-ban treaty, 284, 337, 338, 350, 389
    peace, 381, 394

United States—Continued
    preparedness, 446
    security, national, 190, 215, 216, 221, 237, 334, 356, 375, 404, 409, 466
    space capability, 352
    weapons ban, nuclear, 388
United States Information Agency (USIA), 321
United Technology Center, 5, 66, 280, 496
United Technology Corp., 68
Univac Div. (Sperry Rand Corp.), 317, 443
Universe, 155, 314
Universities, 272, 273, 274, 278, 456
    and space effort, 372
    importance of, 444
UPI. See United Press International.
Upper atmosphere, 282, 301
Upper Mantle Project, 328
Upper Volta, 138
Uranus (planet), 6
Urban renewal, 195, 485
Urey, Dr. Harold C., 26, 162, 214, 237, 495
U.S. See United States.
USA. See U.S. Army.
U.S. Aeronautics and Space Activities, 29
USAF. See U.S. Air Force.
USAF Aeronautical Chart & Information Center, 25, 460
USAF Aerospace Research Laboratories, 368, 476
USAF Aerospace Research Pilot School, 426
USAF Ballistic Systems Div., 428, 496
USAF OAR. See USAF Office of Aerospace Research.
USAF Office of Aerospace Research (USAF OAR), 278
USAF School of Aerospace Medicine (Brooks AFB, Tex.), 40, 41, 42, 43, 44, 45, 46, 47, 393, 443, 482
USAF School of Aviation Medicine (Brooks AFB, Tex.), 41, 46, 55, 361
U.S. Air Corps Training Center, Randolph Field, Tex., 268
U.S. Air Force (USAF) (see also Defense, Department of), 232, 482, 483, 484
    AF-Army-Navy program, 282
    aircraft, 321, 438, 442, 492
    astronaut, 136, 243, 388
    award, 205, 286, 443
    biomedical support, 393
    budget, 499
    contract, 68, 107, 251, 253, 259, 322, 357, 404, 406, 427, 486, 490, 496
    experiment, 16, 56, 332
    facilities, space, 57, 307, 341, 441, 467
    flight, 389, 427
    launch, 59, 154, 163, 239
        missiles, operational, 28, 30, 53, 72, 89, 105, 188, 263, 266, 272, 285, 286, 290, 298, 314, 323, 325, 327, 337, 431, 451, 489
        probes, 38, 91, 290, 418
        R&D, 9, 25, 44, 57, 59, 62, 72, 77, 105, 148, 161, 163, 164, 173, 176, 181, 193, 201, 202, 213, 220, 230, 239, 251, 257, 258, 276, 282, 320, 344, 345, 350, 357, 373, 416, 424, 427, 431, 453, 479, 481, 485, 486
        satellite, 5, 14, 98, 118, 129, 163, 189, 242, 258, 267, 272, 279, 282, 292, 323, 326, 335, 346, 349, 406, 426, 451, 464, 486, 489
        sounding rocket, 56, 222
    management, 415, 441
    NASA–USAF, 64, 78, 90, 129, 131, 428, 454, 481
        ECHO II, 426
        Gemini, Project, 47, 252, 292, 368, 401
    nuclear test-ban treaty, 388
    pilot, space, 114
    report, 68, 335, 340
    research, 31, 78, 130, 154, 216, 437
    rocket engine, 13, 145, 437
    satellite communications 133, 180, 185, 187, 193, 212, 406
    space booster, 129, 168

INDEX 601

U.S. Air Force—Continued
    space effort, 39, 167, 199
    space station, manned orbital, 346
    spacecraft protection, 453
    telescope, radar, 415, 431
    test, 216, 222, 337
    Thor, 9, 407
    Titan III, 451, 466
    X-20, 107, 311
    X-21A, 203
U.S. Air Force Academy, 136, 230
U.S. Arms Control and Disarmament Agency, 409
U.S. Army (USA), 29, 88, 95, 133, 179, 220, 430, 482, 483
    contract, 380, 429, 478
    launch, missile, 57, 62, 70, 147, 161, 242, 263, 266, 272, 285, 318, 324, 351, 399, 415, 442
    rotocraft, 424
    test, 220, 399, 483
U.S. Army Satellite Communications Agency, 285
U.S. Board of Geographic Names, 465
USCG. See U.S. Coast Guard.
U.S. Civil Service Commission, 100
U.S. Coast Guard (USCG), 220, 483
U.S. Congress, 419, 445, 475
    action, prior, 253, 384, 401, 407, 417
    criticism, 381, 403, 416, 487
    Joint Committee on Atomic Energy
        Subcommittee on Research & Development, 421
    Joint Committee on Economics, 36
    joint session, 205, 321, 464
    opinion, 402
    report to, 222, 272, 394, 440
    science & technology, 441, 466
U.S. Congress, House of Representatives, 259, 277, 311, 315, 386, 404, 405, 464, 478
    bills introduced, 239, 314
    bills passed, 255, 277, 325, 337, 474
    Committee on Appropriations, 168, 376, 377, 384, 430
        Subcommittee on DOD Appropriations, 168
        Subcommittee on Independent Offices, 316, 373, 374, 384, 439, 440, 474
    Committee on Armed Services, 31, 63, 64, 68, 83, 286, 326, 421
        Subcommittee on Military Appropriations, 242
    Committee on Government Operations, 337, 434, 435
        Subcommittee on Foreign Operations & Government Information, 210, 304
    Committee on Interstate & Foreign Commerce, 247
    Committee on the Judiciary, 208
    Committee on Post Office and Civil Service, 298, 401
    Committee on Rules, 286, 298, 325
    Committee on Science and Astronautics, 35, 43, 48, 168, 245, 269, 272, 274, 283, 286, 315, 322, 323, 370, 373, 377, 395, 445, 457, 462, 485, 497
        bills approved, 293, 325
        bills introduced, 288
        hearings, 22, 26, 67, 70, 71, 77, 81, 82, 84, 97, 99, 100, 101, 110, 118, 119, 120, 121, 122, 125, 126, 127, 129, 133, 134, 137, 154, 169, 174, 178, 180, 193, 223, 254, 298
        Subcommittee on Applications and Tracking and Data Acquisition, 99, 120, 125, 126, 133, 134, 169, 178, 269
        Subcommittee on Manned Space Flight, 97, 101, 110, 119, 121, 154, 174, 180
        Subcommittee on NASA Legislative Oversight, 382
        Subcommittee on Science, Research and Development, 322, 332, 388, 400, 424, 496
        Subcommittee on Space Sciences and Advanced Research and Technology, 99, 100, 109, 118, 121, 122, 129, 133, 137, 240, 390, 392
    Select Committee on Government Research, 444

U.S. Congress, Senate, 77, 266, 311, 337, 379, 464
    bills introduced, 41, 60, 135, 174, 229, 310, 425
    bills passed, 87, 257, 302, 325, 350, 351, 440, 467, 474
        Commission on the Application of Technology to Community and Manpower Needs, 425
        Committee on Aeronautical and Space Sciences, 1, 90, 99, 158, 159, 161, 163, 167, 169, 173, 191, 209, 231, 236, 238, 244, 283, 291, 293, 300, 319, 322, 325, 367, 372, 395, 422, 441, 444, 447, 466, 485, 490
        Committee on Appropriations, 388, 392, 430, 474
            Subcommittee on DOD Appropriations, 319
        Committee on Armed Services, 135, 322, 493
            Subcommittee on Preparedness, 240, 242, 147, 491
        Committee on Commerce, 90, 91, 247, 262
            Subcommittee on Aviation, 390, 394, 397, 406, 407
            Subcommittee on Communications, 59
        Committee on Foreign Relations, 326, 368, 391
        Committee on Government Operations, 369, 382
            Subcommittee on Investigations, 178, 283
        Committee on the Judiciary, 135, 174
            Antitrust Subcommittee, 45
        Committee on Rules and Administration, 310
        Committee on Small Business, 86
            Subcommittee on Monopoly, 92, 93
            Subcommittee on Retailing, Distribution & Marketing Practices, 486
    confirmation, 161, 253
    debate, 138, 281, 289, 302, 378, 421, 439, 482
    Republican Policy Committee, 190, 215, 288
    resolutions, 198, 467
    Select Committee on Technological Developments, 294, 421
U.S. Geological Survey, 138, 192, 245
USIA. See United States Information Agency.
U.S. Marine Corps (USMC), 483
USMC. See U.S. Marine Corps.
USN. See U.S. Navy.
U.S. Naval Academy, 197
U.S. Naval-Air Force Parachute Facility, 397
U.S. Naval Observatory, 327, 436
U.S. Navy (USN), 10, 367, 483
    aircraft, 179, 216, 433
    aircraft carrier, 401
    cooperative program
        Army-Navy-Air Force, 282
        NASA–USN, 192, 216, 428
    launch
        missile, operational, 318
        missile, R&D, 46, 50, 59, 134, 137, 201, 230, 231, 245, 251, 296, 327, 354, 367, 429, 450
        satellite, 357
        sounding rocket, 481
    Pacific Missile Range transfer, 307
    reports, 401, 463
    satellite, navigational, Transit, 87, 115, 129, 454
    satellite stabilization, 280
    test, 419
    tracking stations (Pacific), transfer of, 441
U.S. News & World Report, 142, 253, 298, 305
USNS *Kingsport*, 60, 80, 284, 296, 302, 321, 339, 345, 349
USNS *Point Barrow*, 486
USNS *Taurus*, 486
U.S. Post Office, 485
U.S.S. *Andrew Jackson*, 367, 402, 429, 433
U.S.S. *Daniel Webster*, 165
U.S.S. *Kearsarge*, 195, 198
U.S.S. *Nathan Hale*, 10
U.S.S. *Observation Island*, 137, 245, 251, 354, 450

U.S. Space Recovery Center, 296
U.S.S.R. (Union of Soviet Socialist Republics), 111, 321, 348, 359, 361, 362, 374, 418, 423, 448, 459, 486
    aircraft, supersonic transport, 379, 417, 422, 492
    armament, 5, 44, 63, 65, 66, 393, 435, 459
    award, 341
    bioastronautics, 355
    booster, space, 9, 157, 181, 452
    communications, 327, 419
    cooperation, defense, 8, 142
    cooperation, space, 89, 100, 196, 198, 205, 212, 227, 303, 306, 317, 319, 348, 358 361, 366, 369, 374, 381, 390, 393, 404, 445, 462, 482, 486, 493
    cosmonaut, 172, 218, 296, 303, 347, 367, 374, 376, 390, 394, 417, 469, 471
    electronics equipment, 237
    IQSY, 332
    laser, 419
    launch
        COSMOS, 140, 154, 165, 209, 298, 392, 427, 433, 477, 487
        LUNIK, 119
        POLET, 413
        test, 202, 214, 492
        VOSTOK, 241, 244
    lunar
        exploration timetable, 333
        flight, manned, 374, 378, 389, 401, 402, 404, 405, 408, 417, 418
        joint U.S.–U.S.S.R., 319, 336, 343, 344, 347, 348, 350, 351, 353, 354, 355, 358, 359, 365, 366, 368, 369, 376, 381, 386, 390, 393, 418, 422, 458, 460
    military installations, 464
    missile, antimissile, 423, 432, 470
    missile program, 294, 310, 400, 422, 434, 435, 462
    nuclear, 143
    nuclear test, 231, 264, 284, 294, 402
    nuclear test-ban treaty, 283, 284, 298, 350, 389
    nuclear test, effect of, 23, 288
    observations by, 192, 215
    parachutist, 267
    radioastronomy, 495
    record, air, 101
    rendezvous program, 324, 355, 358, 413
    research, 200, 203, 480
    resources, 36
    rocket, 405, 478
    satellite, 171, 306, 375, 416, 417, 420, 427, 498
    science, 172, 389
    security, 334, 375, 376
    space
        craft, 428, 429
        effort, 191, 215, 231, 243, 245, 254, 259, 408, 409, 412, 461
        failures, 31, 130, 135, 248, 498
        flight, manned, 304, 362, 372, 407, 486, 498
        law, 145, 219, 424, 434, 443, 464
        plans, 83, 107, 138, 141, 171, 232, 246, 254, 276, 337, 439, 441, 481
    space politics, 172, 389, 424, 453
    space program, 116, 119, 124, 164, 197, 221, 227, 234, 246, 254, 286, 288, 303, 304, 305, 317, 319, 375, 411, 413, 428, 462, 484
        results, 58, 74, 88, 244, 256, 259, 282, 362, 409, 433, 484
        tracking observatory, 274
    test, 192, 202, 214, 453
    weapons ban, atomic orbital, 353, 381, 388
U.S. *Standard Atmosphere, 1962*, 62
U.S. Steel Corp. See American Bridge Div.
U.S.S. *Thresher*, 137, 140
U.S. Weather Bureau, 201, 226, 242, 290, 293, 307, 330, 360, 367
    facilities, 77, 341
    funds, R&D, 15, 371
    reports, 68, 401

U Thant, U.N. Secretary Gen., 231, 328, 379
V-1 (missile), 177
V-2 (missile), 154, 176, 177
VAB. See Vertical Assembly Building.
Vacuum chamber, 333
Valley Forge Space Technology Center (GE), 90
Van Allen radiation belt, 45, 227, 263, 308, 312, 346, 364, 389, 391
Van Allen, Dr. James A., 22, 30, 73, 94, 130, 214, 364
Van de Camp, Dr. Peter, 145
Vandenberg AFB, Calif., 307, 441
    launch
        booster, 72, 357
        missile
            R&D
                Titan II, 9, 57, 164, 193, 213, 250, 427, 481
            operational, 325, 327, 337
                Atlas, 28, 53, 72, 89, 105, 242, 266, 285, 290, 323
                Atlas D, 239, 379, 431, 486
                Atlas E, 161, 229
                Atlas F, 105, 373, 486
                Minuteman, 188, 258, 266, 286, 416, 451, 453, 479
                Titan I, 30, 175, 276, 314, 324, 327, 344, 433, 452
        satellite, 464, 467, 486, 489
            Atlas-Agena D, 486
            Blue Scout, 129
            Scout, 258
            TAT, 98, 406
            Thor-Able-Star, 357
            Thor-Agena B, 258
            Thor-Agena D, 5, 14, 256, 326, 349, 426, 451, 489
        test, 389
Vanderbilt Univ., 47
VANGUARD I, 32, 96, 364, 441
Van Niel, Dr. Cornelius B., 479
Van Nuys, Calif., 28, 453
VARC. See Virginia Associated Research Center.
Variable thrust, 5
Vatican, 264
Vehicle, crawler-transport, 51, 338
"Vehicle Design for Earth Orbit to Mars Orbit & Return", 232
Vela (program), 299, 389
"Velocities that are Possible in the Universe", 192
Venus, 42, 150, 293, 365, 453, 479
    knowledge of, 96, 144, 228
    landing, 361
    life on, 4, 151, 481
    manned flight to, 52, 317, 420, 429, 471
    surface, 9, 179, 454, 479
    temperature, 69, 75, 144
    unmanned flight to, 8, 259, 441, 498
Verification Instruction Program (VIP), 424
Vernov, S. V., 227
Veronique (rocket) (France), 392
Vershinin, Chief Marshal of Aviation Konstantin, 5
Vertical Assembly Building (VAB), 51, 319
Vertical take-off and landing (Vtol) aircraft, 282
Very low frequency. See VLF.
Veterans of Foreign Wars, 324
Vibration, 415, 431, 453, 490
Vibration Test Laboratory (MSC), 425
Victor, Walter K., 227
Vienna, Austria, 348, 351
Villanova Univ., 222
Vilter, H. A., 9
VIP. See Verification Instruction Program.
Virginia Associated Research Center (VARC), 288, 309
Virginia Engineering & Technical Societies, 251

Virginia Polytechnic Institute (VPI), 127, 288, 307
Virginia, Univ. of, 164, 288
Viscount (aircraft), 397
Vishniac, Dr. Wolf, 49
Visibility, 395
Visscher, E. H., 133
Visual performance in space, 36
Vitro Laboratories, 76
VLF (very low frequency), 388, 391, 409, 495
Voice of America, 15, 21, 321, 365
Voinova, T., 267
Volcano Ranch, N.M., 79
"Volga" (balloon), 218
Von Beckh, Dr. Harold J., 50
Von Braun, Dr. Wernher, 7, 313, 432, 433, 478
    award, 6, 214
    space benefits, 45
    space future, 188, 452
    space program, 384, 392, 395, 421
    space progress, 33, 97
    U.S.S.R. space program, 6, 157, 478
Von Brentano, Dr. Heinrich, 270
Von Kármán, Dr. Theodore, 107, 176, 185, 220
Von Kármán Memorial Citation, 249
Von Tiesenhausen, Georg F., 22, 115
Voodoo. See NF-101.
Voronezh, U.S.S.R., 52
"Vostapollo", 358
Vostok (program), 259, 324, 333, 362
VOSTOK II, 300, 350
VOSTOK III & IV, 9, 119, 304
VOSTOK V, 241, 243, 244, 245, 372, 505
VOSTOK VI, 241, 244, 372, 506
Voyager (program), 5, 90, 132
VPI. See Virginia Polytechnic Institute.
V/Stol, 198, 328, 338, 419
Vtol (vertical take-off and landing), 282, 315
Vyenikov, V., 88
W. A. Benjamin, Inc., 260
Waco, Tex., 64
Wagner, Robert, 208
Walker, Dr. Eric, 179
Walker, Joseph A.
    award, 281, 393, 465, 484
    flight, 14, 146, 177, 218, 254, 269, 278, 299, 320, 360
Wallops Station, Va. (NASA), 50, 89, 117, 215, 247, 341, 359, 364
    launch, 280, 352
        reactor, 168, 208
        RFD-1 192, 208
        satellite, 258, 267
        sounding rocket
            Aerobee, 95, 156, 247, 409
            Aerobee 150A, 30, 123, 139, 184, 188, 268, 278, 280, 336, 351, 357, 391, 429, 434
            Aerobee 300A, 146, 280
            Argo D-4 (Javelin), 357, 359
            Astrobee, 132
            Hopi-Dart, 71
            Nike-Apache, 60, 70, 87, 110, 134, 186, 208, 212, 272, 336, 382, 383, 410, 449, 469
            Nike-Asp, 58
            Nike-Cajun, 14, 15, 60, 71, 88, 295, 449, 469
            Shotput, 148, 295
    organization, 381
    strikes, 325
Wall Street, 417
Wall Street Journal, 47, 162, 378

Walton, Capt. John T. (USAF), 209
Ward, Barbara, 224
Ward, Dr. Fred, 481
Warhead, 406, 486, 495
Warner, Dr. Dwain W., 74
Warren, Chief Justice Earl, 52
Warren, Dr. Charles R., 138
Warren, L/C Robert E. (USA), 60, 80
Warsaw, Poland, 32, 100, 226, 294
Washington, D.C., 1, 128, 186, 188, 194, 201, 202, 206, 219, 226, 243, 278, 289, 321, 358, 361, 368, 402, 434, 448
    agreement, 136
    air route, 20
    communications, 327
    meeting, 6, 8, 13, 18, 24, 28, 45, 51, 53, 56, 64, 92, 99, 104, 136, 144, 147, 149, 152, 155, 161, 164, 168, 175, 179, 186, 189, 192, 199, 207, 213, 229, 239, 245, 249, 319, 324, 337, 338, 342, 343, 364, 369, 373, 377, 395, 396, 399, 404, 407, 414, 426, 439, 463, 494
Washington (D.C.) Academy of Sciences, 64
Washington *Daily News*, 369, 379, 383
Washington *Evening Star*, 6, 135, 163, 195, 216, 217, 245, 271, 305, 372, 376, 415, 417
Washington Post, 1, 66, 114, 179, 185, 216, 351, 359, 374, 376, 419, 471, 479, 493
Washington *Sunday Star*, 447
Washington Univ. (St. Louis), 171, 234
Washington, Univ. of, 59, 305
Wasielewski, Eugene, 352
Waste disposal system, 272
Water, 130, 425, 432
Waterman, Dr. Alan T., 468
Watson, Clarence E., 423
Watson, Mark S., 468
Watson, Thomas J., 369
Waves, electromagnetic, 89
WB-66 (Destroyer) (aircraft), 146
Weapons
    ban, nuclear orbital, 353, 368, 381, 388, 389, 391, 482
        multi-megaton, 365
    biological, 475
    chemical, 475
    ground-to-space, 394
    military, 390
Weapons and Aviation Armament Corp. of Hamburg, 466
Weather
    action of, 150
    data exchange with U.S.S.R., 319
    forecasting, 147, 201, 481
    global, 390, 461
    international exchange, 347, 493
    systems, 290
Weather Bureau. See U.S. Weather Bureau.
Weatherwise, 24
Weaver, Rep. James D., 35
Weaver, Rep. Phil, 404
Webb, James E., 131, 188, 222, 448, 493, 498
    Apollo, Project, 386, 407, 433
    astronaut-scientist, 180
    award from, 16, 38, 334, 365
    award to, 208
    budget, 15, 67, 81, 215, 316, 374, 376
    contract, 73, 313
    Government research, 435
    industry and Government, 107, 373, 466
    lunar flight, joint U.S.-U.S.S.R., 365, 402
    lunar program, national, 392, 398
    Mercury Project, 372
    NASA-DOD, Gemini, 199, 252

Webb, James E.—Continued
    NASA Electronics Research Center, 106, 272, 484, 495
    patent, 92
    regional development, 88, 107, 277
    science and Government, 85, 86, 113, 298
    science education, 118
    space achievements, 3, 426
    space cooperation, 16, 303, 305, 413, 460
    space, military role in, 31
    space program
        nature of, 144, 149
        pace, 112, 164, 225, 235, 238, 245, 291
        support of, 158, 197, 240, 262, 269, 300, 345, 383, 395, 398, 404, 405, 461, 466, 468
    supersonic transport, 284, 394
    technology effects, 486
    Titan III, 436
    tracking, 32
    TV transmissions, 447
    university activities, 372
    Voice of America, 365
    World's Fair exhibit, 99
Weidner, Hermann, 328
Weightlessness, 41, 49, 50, 355, 386, 411
Weinberg, Sidney J., 91
Weiss, Howard M., 96
Welch, B. E., 228
Welch, Leo D., 72, 91, 169, 213, 216, 247, 285, 301, 429
Welch, Nat, 318
Welch, W. Va., 216
"Welding Tips", 203
Wells College (N.Y.), 225
Welsh, Dr. Edward C., 471
    space race, 124, 141, 259
        lunar landing, 337, 362, 402
    U.S. space program, 13, 159, 267, 373, 414, 441
        budget, 249, 407
Wendover AFB, Utah, 237, 296
"We're Running the Wrong Race with Russia", 288
West Berlin's Technical University (see also Berlin's Institute for Elements of Space Travel), 6, 7
West Coast, 374, 416
West Ford, Project (USAF), 231, 341, 504
    interference with astronomy, 185, 187, 199, 219
    launch, 189
    life, orbital, 73
    plans, 180
    results, 192, 212
    secrecy, 146
    test, 22
West German Bundestag, 270
West German Cabinet, 271
West Germany, 105, 133, 177, 197, 214, 252, 270, 388, 425, 454, 466
West Point Corps of Cadets, 39
Western Electric Corp., 380
Western Electronic Show and Convention, 321
Western Europe, 184
Western European Union, 459
Western Hemisphere Center, 396
Western Operations Office (WOO) (NASA), 258
Western Test Station, Ida., 343
Westinghouse Astronuclear Laboratory, 345
Westinghouse Defense Center, 309
Westinghouse Electric Corp., 203, 213, 216
Westinghouse Research Laboratories, 26
Westlake, Maj. Edward (USAF), 275

Westland Co., 275
Wexler, Dr. Harry, 24
Wheat, surplus to U.S.S.R., 393
Whipple, Elden C., Jr., 185, 367
Whipple, Dr. Fred L., 239, 399, 432
Whirlpool Corp., 273, 361
White, Edward H., II, 28
White House, 8, 19, 72, 134, 205, 321, 348, 366, 385, 421, 458, 468, 475
White House Office of Science and Technology, 21, 388, 400, 424, 442, 444, 483
White, M/G M. S. (USAF), 222
White, Dr. M. Samuel, 299
White, Maj. Robert M. (USAF), 66, 148, 236
    award, 148, 222, 226
White, Robert M., 287, 360
White Sands Missile Range (WSMR), 62, 273, 296, 415, 442
    flight test, 195, 250, 325
    launch
        missile, 57, 62, 107, 147, 161, 192, 230, 244, 258, 263, 266, 280, 285, 302, 318, 351, 366, 388, 477
        test, 257, 399, 414, 423, 431, 479
White Sands, N.M., 42, 107
White, L/C Stanley C. (USAF), 45, 139
White, William, 93
White, William C., 106, 151
White, William S., 195, 415
Whitehead, J. D., 241
Whitford, Dr. Albert E., 73
Whittaker three-axis gyro, 244
"Why Land on the Moon?", 329
Wichita, Kan., 88
Wichita, Univ. of, 88
Wiener, Dr. Norbert, 479
Wiesner, Dr. Jerome B., 71, 73, 234, 388, 421, 424, 430, 440, 444
    education, 108, 152
    Federal role in science and technology, 390
    Mars program, 21
    NASA-DOD, 93, 400
Wigner, Prof. Eugene Paul, 421
William and Mary, College of (Va.), 235, 288
William F. Patterson Memorial Award, 338
William L. Clayton Lecture, 179
Williams, Capt. Clifton C. (USMC), 392
Williams, Thomas, 59
Williams, Walter C., 1, 22, 51, 190, 245, 382, 398, 465
Williamsburg, Va., 224
Williamson, Capt. Gerald R. (USAF), 389
Wilson, Rep. Charles H., 240
Wilson, Dr. Glen, 490
Wilson, Dr. Raymond H., Jr., 191
Wilson, Rep. Bob, 2, 478
Wind, 117
    measurement of, 11, 20, 60, 71, 212, 301, 469
    solar, 80, 310, 406
    upper atmosphere, 206, 208, 222, 443
Wind tunnel, 170, 297, 497
Wing, delta, 344, 345
Wing, variable-sweep, 278, 344, 379
Wings Club (N.Y.), 194, 442
Winter Conference on Military Electronics, 31
Wire grid, use of, 406
Wirth, Conrad L., 483
Witkin, Richard, 394
WMO. See World Meteorological Organization.
Wolf Trap, 49
Women, 34, 279

"Women in Aviation Week", 273
Women's National Democratic Club, 395
woo. See Western Operations Office (NASA).
Woodcock, Leonard, 91
Wood, Lysle A., 147
Woomera Rocket Range, Australia, 295, 311
Woomera, Australia, 77, 112, 497
World Affairs Council, 455
World arms race, 482
World Bank, 487
"World-Circling Spaceships—Satellite Studies in the U.S. During the 1940's", 104
World Conference on Space Communications (see also Conference on Communications Frequency Allocation (ITU)), 372, 379, 393, 416
World Conference on World Peace Through Law, First, 265
World Federation of Scientific Workers Center, 453
World Festival of Aeronautic and Space Films, 341
World Food Congress, 229
World Health Organization, 396
World Magnetic Survey (1965), 314
World Meteorological Organization (WMO), 131, 162, 164, 329, 396
World's Fair, 99
World War I, 257, 396
World War II, 303, 396, 459, 479, 480
World weather watch, 164, 396
World Woman's Congress, 252
Wright brothers, 343, 481, 483, 484
"Wright Brothers Day", 467, 482, 483
Wright Brothers Memorial Banquet, 484
Wright Brothers Memorial Trophy, 484
Wright Brothers National Memorial, 480
Wright Flyer, 480, 483
Wright, Harold W., 286
Wright Memorial Museum, 343, 481
Wright, Orville, 436, 481, 484
Wright-Patterson AFB, Ohio, 28, 222, 263, 360, 368, 393, 455, 460, 483
Wright, Wilbur, 73, 225, 268, 467, 481, 484
WSMR. See White Sands Missile Range.
WWVB and WWVL transmitting stations (NBS). See National Bureau of Standards.
Wyatt, D. D., 181, 214, 382
Wydler, Rep. John W., 35, 293, 311
X-8 (engine), 428
X-12 (engine), 428
X-15 (rocket research aircraft), 122, 364, 384, 404
    accident, 186
    air pressure measurement, 269
    altitude density measurement, 278
    award, 148, 222, 226
    climb angle measurement, 320
    computer, 38
    flight
        canceled, 189, 387, 485
        successful
            No. 1, 137, 161, 196, 254, 269, 277, 378, 404, 431
            No. 3, 14, 146, 177, 194, 218, 246, 256, 278, 299, 320, 360, 423, 450
    Follow-on program, 161
    heat transfer, 146, 281
    "High Range" proposal, 322
    inertial guidance system, 387
    modification, 404, 427, 480
    pilot, 235, 281, 283, 386, 480, 484
    research, 137, 189, 237, 281, 328
    rocket engine energy measurement, 320
    stability test, 277
    test, planned, 252

X-19 (Vtol), 282, 315, 499
X-20 (Dyna Soar), 19, 31, 90, 94, 107, 114, 255, 404, 426
    cancellation, 473, 475
    pressure suit, 360
    recovery, 237
X-21A (aircraft), 146, 183, 203
X-22 V/Stol, 240
X-22A (Vtol), 242, 315
X-258 (rocket stage), 486
XC-142A (Vtol), 315
XH-51A (helicopter), 179
Xenon gas, 293
X-ray, 87, 92, 388, 397, 478, 493
Yagoda, Herman, 138
Yak-18P (aircraft), 101
Yale Univ., 125, 414
Yardley, John F., 365
Yeager, Col. Charles E. (USAF), 66, 186, 459, 475
Yeshiva Univ. (N.Y.), 481
Yomiuri, 224
Yorty, Samuel, 301
Yost, Ed, 140
Young, John D., 65, 382
Young, LCdr. John W. (USN), 28
Young, R. S., 228, 294
Young, Robert B., 313, 328
Young, Sen. Stephen M., 301, 379
Youth Science Congress, 368
Youth Science Fair, 65
Youth Seminar on Flight, 480
Zander, Friedrich, 428
Za Rubezhom (Soviet magazine), 358
Zenoff, Judge David, 134
Zero gravity, 336
Zhukovsky Academy (U.S.S.R.), 116
Ziegler, Prof. Carl, 422
Zimmerman, Charles H., 6, 122, 328
Zodiacal light, 257
Zonshayn, Prof. S., 192
Zuckert, Eugene M., 148, 200, 216, 226
    NASA–DOD, 63, 78, 131
    nuclear test-ban treaty, 338
    space, 232, 233, 474
    TFX, 283

www.ingramcontent.com/pod-product-compliance
Lightning Source LLC
Chambersburg PA
CBHW081713170526
45167CB00009B/3563